ENVIRONMENTAL CHEMISTRY

ENVIRONMENTAL CHEMISTRY FOURTH EDITION

- **COLIN BAIRD**
  University of Western Ontario

- **MICHAEL CANN**
  University of Scranton

W. H. FREEMAN AND COMPANY

New York

Publisher: Clancy Marshall
Senior Acquisitions Editor: Jessica Fiorillo
Marketing Manager: Anthony Palmiotto
Media and Supplements Editor: Samantha Calamari
Assistant Editor: Kathryn Treadway
Photo Editor: Ted Szczepanski
Cover and Text Designer: Blake Logan
Project Editor: Vivien Weiss
Illustrations: Macmillan Publishing Solutions
Senior Illustration Coordinator: Bill Page
Production Coordinator: Paul W. Rohloff
Composition: Macmillan Publishing Solutions
Printing and Binding: RR Donnelley

Library of Congress Preassigned Control Number: 2007939344
ISBN-13: 978-1-4292-0146-9
ISBN-10: 1-4292-0146-0

Third printing

W. H. Freeman and Company
41 Madison Avenue
New York, NY 10010
Houndmills, Basingstoke RG21 6XS, England
www.whfreeman.com

BRIEF CONTENTS

CONTENTS

PART V   METALS, SOILS, SEDIMENTS, AND WASTE DISPOSAL   661

PREFACE

To the Student

There are many definitions of environmental chemistry. To some, it is solely the chemistry of Earth's natural processes in air, water, and soil. More commonly, as in this book, it is concerned principally with the chemical aspects of problems that humankind has created in the natural environment. Part of this infringement on the natural chemistry of our planet has resulted from the activities of our everyday lives. In addition, chemists, through the products they create and the processes used to create them, have also had a significant impact on the chemistry of the environment.

Chemistry has played a major role in the advancement of society and in making our lives longer, healthier, more comfortable, and more enjoyable. The effects of human-made chemicals are ubiquitous and in many instances quite positive. Without chemistry there would be no pharmaceutical drugs, no computers, no automobiles, no TVs, no DVDs, no lights, no synthetic fibers. However, along with all the positive advances that result from chemistry, copious amounts of toxic and corrosive chemicals have also been produced and dispersed into the environment. Historically, chemists as a group have not always paid enough attention to the environmental consequences of their activities.

But it is not just the chemical industry, or even industry as a whole, that has emitted troublesome substances into the air, water, and soil. The fantastic increase in population and affluence since the Industrial Revolution has overloaded our atmosphere with carbon dioxide and toxic air pollutants, our waters with sewage, and our soil with garbage. We are exceeding the planet's natural capacity to cope with waste, and, in many cases, we do not know the consequences of these actions. As a character in Margaret Atwood's recent novel *Oryx and Crake*, stated, "The whole world is now one vast uncontrolled experiment."

During your journey through the chapters in this text, you will see that scientists do have a good handle on many environmental chemistry problems and have suggested ways—although sometimes very expensive ones—to keep us from inheriting the whirlwind of uncontrolled experiments on the planet. Chemists have also become more aware of the contributions of their own profession and industry in creating pollution and have created the concept of *green chemistry* to help minimize their environmental footprint in the future.

To illustrate these efforts, case studies of their initiatives have been included in the book. However, as a prelude to these studies, in the Introduction we discuss some of the history of environmental regulations—especially in the United States—as well as the principles and an illustrative application of the green chemistry movement that has developed.

Although the science underlying environmental problems is often maddeningly complex, its central aspects can usually be understood and appreciated with only introductory chemistry as background preparation. However, students who have not had some introduction to organic chemistry are encouraged to work through the Appendix on Background Organic Chemistry, particularly before tackling Chapters 10 to 12. Furthermore, the listing of general chemistry concepts that will be used in each chapter should assist in identifying topics from earlier courses that are worth reviewing.

To the Instructor

Environmental Chemistry, Fourth Edition, has been revised and updated in line with comments and suggestions from various users and reviewers of the third edition. In particular, where possible, we have fulfilled the request that larger chapters be split so that they can be covered in a week or two. In several places, some of the more advanced material has been placed in boxes that have been reordered to appear at the end of a series of related chapters.

Some instructors prefer to cover chapters in a different order than we have used, so the list of concepts that opens each chapter, describing material covered elsewhere in the book, should help facilitate restructuring.

As in previous editions, the background required to solve both in-text and end-of-chapter problems is either developed in the book or would have been covered previously in a general chemistry course—as listed at the beginning of each chapter. Where appropriate, hints are given to start students on the solution. The *Solutions Manual* to the text includes worked solutions to all problems (except for Review Questions, which are designed to direct students back to descriptive material within each chapter).

New to This Edition

•   All chapters now start with an outline of the concepts and methods from introductory chemistry and from previous chapters of the text that will be used in the chapter. This will give students a better idea of what background material to review, and it will give instructors a better idea of what is assumed in the material.

•   Detailed applications of several of the newsworthy topics in the text have been formulated as Case Studies and are available on the website for this text (www.whfreeman.com/envchem4).

• Much of the detail concerning CFCs (Chapter 2) has been cut, as it is now largely irrelevant since CFCs have been banned. However, the background and politics of the controversy about the remaining ozone-depleting substance have been documented in the new web-based Case Study *Strawberry Fields—The Banning of Methyl Bromide*.

• Material on converters and traps for diesel engines and lawn mowers, as well as the problems of two-stroke engines, together with the initiatives to produce low-sulfur gasoline and diesel fuel, have been added to Chapter 3.

• The relatively advanced material on the distribution of particle sizes in tropospheric air has been rewritten, with a new and more informative illustration, and placed in Box 3-2.

• A new green chemistry case concerning the use of ionic liquids has been added to Chapter 3.

• The information concerning the aqueous-phase oxidation of atmospheric SO_2 has been moved to a more relevant position in Chapter 3.

• Material concerning particulates in air as a health risk has been updated and reformulated as the web-based Case Study *The Effect of Urban Air Particulates on Human Mortality*, associated with Chapter 4.

• Material on the acid rain problem in China has been added and the concept of critical load introduced (Chapter 4).

• The Indoor Air Pollution section on benzene has been expanded to introduce the term *air toxics* and to discuss methylated benzenes as well as the parent compound (Chapter 4).

• The term *albedo* is now introduced early in Chapter 6, and sunlight reflection is further discussed.

• Material introducing the greenhouse effect and the discussion of Earth's energy balance, previously split between these two areas, has logically been combined and expanded by introducing a simple physical model for greenhouse warming in Chapter 6. A more sophisticated model of the greenhouse effect has been added as Box 6-1.

• A phase diagram for CO_2, showing the supercritical region, has been included to illustrate the green chemistry case in Box 6-2.

• The section on methane has been rewritten and updated, with new graphs provided, especially with regard to its current concentration plateau (Chapter 6).

• Updates on climate change based on the IPCC Fourth Assessment Report are included in Chapter 6.

- Chapter 7 now begins with a more extensive discussion of energy usage and CO_2 emissions by country, with an analysis of the factors that affect them globally and with particular reference to the United States and China.

- A separate Box 7-1, on fractional distillation of petroleum, has been added.

- Material on carbon sequestration—particularly on the chemistry involved—has been expanded and updated, given the growing importance of this technique (Chapter 7).

- Material in Chapter 8 concerning renewable energy sources—wind, hydroelectricity, biofuels, and geothermals—has been added, along with the web-based Case Study *Mercury Pollution and the James Bay Hydroelectric Project (Canada)*.

- A new green chemistry case concerning the production of biodiesel has been added to Chapter 8.

- The chapter Radioactivity, Radon, and Nuclear Energy (now Chapter 9) has been repositioned into Part II, Energy and Climate Change.

- In the discussions about radioactivity (Chapter 9), definitions of the *bequerel* and the *curie* are now provided. The controversial *hormesis* theory is also introduced, and the debate about the existence of a threshold for health damage from radioactivity has been expanded. The material on the reprocessing of nuclear waste has been developed into a fuller discussion of the chemistry involved.

- The history of DDT, and an expanded discussion about banning it, have been incorporated into the web-based Case Study *To Ban or Not to Ban DDT? Its History and Future*, associated with Chapter 10.

- The term *lethal concentration* has been introduced in Chapter 10.

- The Chapter 10 material on organophosphate insecticides has been reorganized so that a description of their molecular structure now precedes explanation of their uses, etc. Pesticide toxicities are now discussed in terms of World Health Organization (WHO) categories.

- An extended treatment of the controversial insecticide endosulfan has been developed in the new Box 10-1.

- At the suggestion of several reviewers, a short discussion of the degradation of pesticides is now incorporated in Chapter 10.

- Chapter 11 now includes additional help for students in solving problems that involve the production of dioxins from chlorophenols.

- The section concerning PBDEs (Chapter 12) now incorporates a description of the chemical mechanism by which brominated fire retardants operate, and it has been extensively updated with new information concerning the uses and regulatory status of PBDEs.

- Following the suggestion of several reviewers, a new Environmental Instrumentation Analysis box on the widespread use of GC/MS for pesticide analysis has been added at the end of Part III.

- The new Box 13-1 reviews the assignment of oxidation numbers and the balancing of redox equations.

- The sections on sulfur compounds and acid mine drainage have been revised and repositioned (Chapter 13) so that the pE concept need not be covered as background to them.

- At reviewer request, the species diagram for the CO_2–carbonate system and its derivation have been added (Chapter 13).

- A new web-based Case Study, *Mercury Emissions from Power Plants*, a topic of considerable current interest, has been added (Chapter 15).

- The information concerning the arsenic-contaminated drinking water in Asia has been updated, and more information has been added concerning the ways As can be removed from water (Chapter 15).

Scientific American Feature Articles

We are proud to feature several *Scientific American* articles in this edition. The topics discussed in these articles are highly relevant to topics covered in the book and allow students access to the most current thinking about contemporary environmental issues by active researchers.

Supplements

The *Solutions Manual* (1-4292-1005-2) includes worked solutions to virtually all problems (except for Review Questions, which are designed to direct students back to the appropriate material within each chapter).

Students and instructors interested in pursuing specific topics in more detail should consult the Further Readings section at the end of each chapter, as well as the Websites of Interest that are given for each chapter on the website, www.whfreeman.com/envchem4.

To All Readers of the Text

The authors are happy to receive comments and suggestions about the content of this book from instructors and students via e-mail: Colin Baird at cbaird@uwo.ca and Michael Cann at cannm1@scranton.edu.

Acknowledgments

The authors wish to express their gratitude and appreciation to a number of people who in various ways have contributed to this edition of the book:

To Professor Thomas Chasteen of Sam Houston State University for masterfully writing the Environmental Instrumental Analysis boxes, which shed light on the perspective of scientists working in this discipline and add much to the book.

To Professor Brian D. Wagner of the University of Prince Edward Island, for supplying some of the Additional Problems included in the book.

To the students and instructors who have used previous editions of the text and who, via their reviews and e-mails, have pointed out subsections and problems that needed clarifying or expanding.

To W. H. Freeman and Company Senior Acquisitions Editor for the third and fourth editions, Jessica Fiorillo; Project Editor Vivien Weiss; and Assistant Editor Kathryn Treadway—for their encouragement, ideas, insightful suggestions, patience, and organizational abilities. To Margaret Comaskey for her careful copyediting and suggestions again in this edition, to Ted Szczepanski for finding the photographs, to Nancy Walker for obtaining permissions for figures and photographs, to Blake Logan for design, and to Paul Rohloff for coordinating production.

Colin Baird wishes to express his thanks . . .

To Ron Martin and Martin Stillman, his colleagues at the University of Western Ontario who used the first two editions and have made valuable suggestions for improvement, and to his colleagues at Western and elsewhere who supplied information or answered queries on various subjects: Myra Gordon, Duncan Hunter, Roland Haines, Edgar Warnhoff, Marguerite Kane, Currie Palmer, Rob Lipson, Dave Shoesmith, Felix Lee, Peter Guthrie, Geoff Rayner-Canham, and Chris Willis.

To his secretaries through the years—Sandy McCaw, Clara Fernandez, Darlene McDonald, Diana Timmermans, Elizabeth Moreau, Shannon Woodhouse, Wendy Smith, and Judy Purves—for their brave attempts to decipher his writing and for dealing with the always-urgent problems that authors seem to have.

To his daughter Jenny—and others of her generation, and those following them—for whom this subject matter really matters.

Mike Cann wishes to express his thanks . . .

To his students (especially Marc Connelly and Tom Umile) and fellow faculty at the University of Scranton, who have made valuable suggestions and contributions to his understanding of green chemistry and environmental chemistry.

To Joe Breen, who was one of the pioneers of green chemistry and one of the founders of the Green Chemistry Institute.

To Paul Anastas (Center for Green Chemistry and Green Engineering at Yale) and Tracy Williamson (U.S. Environmental Protection Agency), whose boundless energy and enthusiasm for green chemistry are contagious.

To Debra Jennings, who for over 30 years as the chemistry department secretary has, to his amazement, managed to decode his handwriting and simply put up with him, always in good spirit.

To his loving wife, Cynthia, who has graciously and enthusiastically endured countless discussions of green chemistry and environmental chemistry.

To his children, Holly and Geoffrey, and his grandchildren, McKenna, Alexia, Alan Joshua, Samantha, and Arik, who, along with future generations, will reap the rewards of sustainable chemistry.

Both authors wish to express thanks to the reviewers of the fourth edition of the text for their helpful comments and suggestions:

Ann Marie Anderssohn, *University of Portland*
D. Neal Boehnke, *Jacksonville University*
Nathan W. Bower, *Colorado College*
Michael Brabec, *Eastern Michigan University*
Patrick J. Castle, *U.S. Air Force Academy*
Jihong Cole-Dai, *South Dakota State University*
Arlene R. Courtney, *Western Oregon University*
James Donaldson, *University of Toronto-Scarborough*
Jennifer DuBois, *University of Notre Dame*
Robert Haines, *University of Prince Edward Island*
Yelda Hangun-Balkir, *California University of Pennsylvania*
Michael Ketterer, *Northern Arizona University*
John J. Manock, *University of North Carolina-Wilmington*
Steven Mylon, *Lafayette College*
Myrna Simpson, *University of Toronto*
Chuck Smithhart, *Delta State University*
Barbara Stallman, *Lourdes College*
Steven Sylvester, *Washington State University, Vancouver*
Brian Wagner, *University of Prince Edward Island*
Feiyue Wang, *University of Manitoba*
Z. Diane Xie, *University of Utah*
Chunlong (Carl) Zhang, *University of Houston-Clear Lake*

ENVIRONMENTAL CHEMISTRY

Introduction

In this book you will study the chemistry of the air, water, and soil, and the effects of anthropogenic activities on the chemistry of the Earth. In addition, you will learn about green chemistry, which aims to design technologies that lessen the ecological footprint of our activities.

Environmental chemistry deals with the reactions, fates, movements, and sources of chemicals in the air, water, and soil. In the absence of humans, the discussion would be limited to naturally occurring chemicals. Today, with the burgeoning population of the Earth, coupled with continually advancing technology, human activities have an ever-increasing influence on the chemistry of the environment. To the earliest humans, and even until less than a century ago, humans must have thought of the Earth as so vast that human activity could scarcely have any more than local effects on the soil, water, and air. Today we realize that our activities can have not only local and regional but also global consequences.

There are now many indications that we have exceeded the carrying capacity of the Earth, i.e., the ability of the planet to convert our wastes back into resources (often called *nature's interest*) as quickly as we consume its natural resources and produce waste. Some say that we are living beyond the "interest" that nature provides us and dipping into nature's capital. In short, many of our activities are not sustainable.

As we write these introductory remarks, we are reminded of the environmental consequences of human activities that impact the areas where we live and beyond. Colin spends his summers on a small island just off the North Atlantic coast in Nova Scotia, while Mike spends a few weeks each winter on the west coast of southern Florida a few kilometers from the Gulf of Mexico. Although these locations are a great distance apart, if predictions are correct, both may be permanently submerged by the end of this century as a result of rising sea levels brought about by enhanced global warming (see Chapters 6 and 7). The public footbridge that links Colin's island to the mainland is treated with creosote, and the local residents no longer harvest mussels from the beds below for fear they may be contaminated with PAHs (Chapter 12). Colin's well on this island was tested for arsenic, a common pollutant in that

> *If mankind is to survive, we shall require a substantially new manner of thinking.*
>
> Albert Einstein

1

area of abandoned gold mines (Chapter 15). To the north, the once robust cod fishing industry of Newfoundland has collapsed due to overfishing. Mike lives in northeastern Pennsylvania on a lake where the wood in his dock is preserved with the heavy metals arsenic, chromium, and copper (Chapter 15). Within a short distance are two landfills (Chapter 16), which take in an excess of 8000 tonnes of garbage per day (from municipalities as far as 150 kilometers away), two *Superfund* sites (Chapter 16), and a nuclear power plant that generates plutonium and other radioactive wastes for which there is no working disposal plan in the United States (Chapter 9). Colin's home in London, Ontario, is within an hour's drive of Lake Erie, famous for nearly having "died" of phosphate pollution (Chapter 14), and nuclear power plants on Lake Huron. Nearby farmers grow corn to supply to a new factory that produces ethanol for use as an alternative fuel (Chapter 8); in Ottawa, a Canadian company has built the first demonstration plant to convert the cellulose from agricultural residue to ethanol (Chapter 8).

On sunny days we apply extra sunscreen because of the thinning of the ozone layer (Chapters 1 and 2). Three of the best salmon rivers in North America in Nova Scotia must be stocked each season because the salmon no longer migrate up the acidified waters. Many of the lakes and streams of the beautiful Adirondack region of upstate New York are a deceptively beautiful crystal clear, only because they are virtually devoid of plant and animal life, again because of acidified waters (Chapters 3 and 4).

Environmental issues like these probably have parallels with those that exist where you live; learning more about them may convince you that environmental chemistry is not just a topic of academic interest, but one that touches your life every day in very practical ways. Many of these environmental threats are a consequence of anthropogenic activities over the last 50 to 100 years.

In 1983 the United Nations charged a special commission with developing a plan for long-term sustainable development. In 1987 the report titled "Our Common Future" was issued. In this report (more commonly known as the *The Brundtland Report*), the following definition of **sustainable development** is found:

> Sustainable development is development that meets the needs of the present without compromising the ability of future generations to meet their own needs.

Although there are many definitions of sustainable development (or sustainability), this is the most widely used. The three intersecting areas of sustainability are focused on society, the economy, and the environment. Together they are known as the *triple bottom line*. In all three areas, consumption (particularly of natural resources) and the concomitant production of waste are central issues.

The concept of an "ecological footprint" is an attempt to measure the amount of biologically productive space that is needed to support a particular human lifestyle. Currently there are about 4.5 acres of biologically productive space for each person on the Earth. This land provides us with the resources that we need to support our lifestyles and to receive the waste that we generate and convert it back into resources. If the entire population of 6.5 billion people lived like Colin and Mike (rather typical North Americans), the total ecological footprint would require five planet Earths. Obviously, everyone on the planet can't live in as large and inefficient a house, drive as many kilometers in such an inefficient vehicle, consume as much food (particularly meat) and energy, create as much waste, etc., as those living in the most developed regions. As countries such as China and India, the two most populous countries in the world (with a combined total of over 2.3 billion people) and two of the fastest growing economies in the world, continue to develop, they look to the lifestyles of the 1 billion people on the planet who live in already developed countries. Factor in the expected increase in population to 9 billion by 2050, and clearly this is not sustainable development.

The people of the world (in particular, those in developed countries) must strive to lead a lifestyle that is sustainable. This does not necessarily mean a lower standard of living for those in the developed world, but it does mean finding ways (more efficient technologies along with conservation) to reduce our consumption of natural resouces and the concomitant production of waste. A widespread movement toward the development and implementation of sustainable technologies or green technologies currently seeks to reduce energy and resource consumption, to use and develop renewable resources, and to reduce the production of waste. In chemistry, these developments are known as *green chemistry*, which is described later in this Introduction and which we will see as a theme throughout this text.

A Brief History of Environmental Regulation

In the United States, many environmental disasters came to a head in the 1960s and 1970s. In 1962, the deleterious effects of the insecticide DDT were brought to the forefront by Rachel Carson in her seminal book, *Silent Spring*. In 1969, the Cayahoga River, which runs through Cleveland, Ohio, was so polluted with industrial waste that it caught fire. The Love Canal neighborhood in Niagara Falls, New York, was built on the site of a chemical dump; in the mid-1970s, during an especially rainy season, toxic waste began to ooze into the basements of area homes, and drums of waste surfaced. The U.S. government purchased the land and cordoned off the entire Love Canal neighborhood. These distressing events were brought into the homes of Americans on the nightly news and, along with other environmental disasters, became rallying points for environmental reform.

This era saw the creation of the U.S. Environmental Protection Agency (EPA) in 1970, the celebration of the first Earth Day, also in 1970, and a mushrooming number of environmental laws. Before 1960, there were approximately 20 environmental laws in the United States; now there are over 120. Most of the earliest of these were focused on conservation, or setting aside land from development. The focus of environmental laws changed dramatically, starting in the 1960s. Some of the most familiar U.S. environmental legislation includes the *Clean Air Act* (1970) and the *Clean Water Act* (originally known as the *Federal Water Pollution Control Act Amendments of 1972*). One of the major provisions of these acts was to set up pollution control programs. In effect, these programs attempted to control the release of toxic and other harmful chemicals into the environment. The Comprehensive Environmental Response, Compensation and Liability Act (also known as the *Superfund Act*) set up a procedure and provided funds for cleaning up toxic waste sites. These acts thus focused on dealing with pollutants after they were produced and are known as "end-of-the-pipe solutions" and "command and control laws."

The risk due to a hazardous substance is a function of the exposure to and the hazard level of the substance:

$$\text{risk} = f\,(\text{exposure} \times \text{hazard})$$

The end-of-the-pipe laws attempt to control risk by preventing our exposure to these substances. However, exposure controls inevitably fail, which points out the weakness of these laws. The Pollution Prevention Act of 1990 is the only U.S. environmental act that focuses on the paradigm of *prevention* of pollution at the source: If hazardous substances are not used or produced, then their risk is eliminated. There is also no need to worry about controlling exposure, controlling dispersion into the environment, or cleaning up hazardous chemicals.

Green Chemistry

The U.S. Pollution Prevention Act of 1990 set the stage for **green chemistry.** Green chemistry became a formal focus of the U.S. EPA in 1991 and became part of a new direction set by the EPA, by which the agency worked with and encouraged companies to voluntarily find ways to reduce the environmental consequences of their activities. Paul Anastas and John Warner defined green chemistry as *the design of chemical products and processes that reduce or eliminate the use and generation of hazardous substances*. Moreover, green chemistry seeks to

- reduce waste (especially toxic waste),

- reduce the consumption of resources and ideally use renewable resources, and

- reduce energy consumption.

Anastas and Warner also formulated the *12 principles of green chemistry*. These principles provide guidelines for chemists in assessing the environmental impact of their work.

The 12 Principles of Green Chemistry

1. It is better to **prevent waste** than to treat or clean up waste after it is formed.

2. Synthetic methods should be designed to **maximize the incorporation of all materials** used in the process into the final product.

3. Wherever practicable, synthetic methodologies should be designed to use and generate **substances that possess little or no toxicity** to human health and the environment.

4. Chemical products should be designed to **preserve efficacy of function while reducing toxicity.**

5. The use of **auxiliary substances** (solvents, separation agents, etc.) **should be made unnecessary** whenever possible and innocuous when used.

6. **Energy requirements** should be recognized for their **environmental and economic impacts and should be minimized.** Synthetic methods should be conducted at ambient temperature and pressure.

7. A raw material **feedstock should be renewable** rather than depleting whenever technically and economically practical.

8. Unnecessary **derivatization** (blocking group, protection/deprotection, temporary modification of physical/chemical processes) should be **avoided** whenever possible.

9. **Catalytic reagents** (as selective as possible) are superior to stoichiometric reagents.

10. **Chemical products** should be designed so that at the end of their function they **do not persist in the environment** and ultimately break down into innocuous degradation products.

11. Analytical methodologies need to be further developed to allow for **real-time in-process** monitoring and control prior to the formation of hazardous substances.

12. Substances and the form of a substance used in a chemical process should be chosen so as to **minimize the potential for chemical accidents,** including releases, explosions, and fires.

In many of the chapters that follow, real-world examples of green chemistry are discussed. During these discussions, you should keep in mind the 12 principles of green chemistry and decide which of them are met by the particular example. Although we won't consider all of the principles at this point, a brief discussion of some of them is beneficial.

- Principle 1 is the heart of green chemistry and places the emphasis on the prevention of pollution at the source rather than cleaning up waste after it has formed.

- Principles 2–5, 7–10, and 12 focus on the materials that are used in the production of chemicals and the products that are formed.

  ○ In a chemical synthesis, unwanted by-products are often formed in addition to the desired product(s); these compounds are usually discarded as waste. Principle 2 encourages chemists to look for synthetic routes that maximize the production of the desired product(s) and at the same time minimize the production of unwanted by-products (see the synthesis of ibuprofen discussed later).

  ○ Principles 3 and 4 stress that the toxicity of materials and products should be kept to a minimum. As we will see in later discussions of green chemistry, Principle 4 is often met when new pesticides are designed with reduced toxicity to nontarget organisms.

  ○ During the course of a synthesis, chemists employ not only compounds that are actually involved in the reaction (reactants) but also auxiliary substances such as solvents (to dissolve the reactants and to purify the products) and agents that are used to separate and dry the products. These materials are usually used in much larger quantities than the reactants, and they contribute a great deal to the waste produced during a chemical synthesis. When chemists are designing a synthesis, Principle 5 reminds them to consider ways to minimize the use of these auxiliary substances.

  ○ Many organic chemicals are produced from petroleum, which is a non-renewable resource. Principle 7 urges chemists to consider ways to produce chemicals from renewable resources such as plant material (biomass).

  ○ As we will see in Chapter 10, DDT is an effective pesticide. However, a major environmental problem is its stability in the natural environment. DDT degrades slowly. Although it has been banned in most developed countries since the 1970s (in the United States since 1972), it can still be found in the environment, particularly in the fatty tissues of animals. Principle 10 stresses the need to consider the lifetimes of chemicals in the environment and the need to focus on materials (such as pesticides) that degrade rapidly in the environment to harmless substances.

- Many chemical reactions require heating or cooling and/or a pressure higher or lower than atmospheric pressure. Performing reactions at other than ambient temperature and pressure requires energy; Principle 6 reminds chemists of these considerations when designing a synthesis.

Presidential Green Chemistry Challenge Awards

To recognize outstanding examples of green chemistry, the **Presidential Green Chemistry Challenge Awards** were established in 1996 by the U.S. EPA. Generally five awards are given each year at a ceremony held at the National Academy of Sciences in Washington, D.C. The awards are given in the following three categories.

1. The use of alternative synthetic pathways for green chemistry, such as:

   • catalysis/biocatalysis,

   • natural processes, such as photochemistry and biomimetic synthesis, and

   • alternative feedstocks that are more innocuous and renewable (e.g., biomass).

2. The use of alternative reaction conditions for green chemistry, such as:

   • solvents that have a reduced impact on human health and the environment and

   • increased selectivity and reduced wastes and emissions.

3. The design of safer chemicals that are, for example:

   • less toxic than current alternatives and

   • inherently safer with regard to accident potential.

Real-World Examples of Green Chemistry

To introduce you to the important and exciting world of green chemistry, we provide you with real-world cases of green chemistry throughout this book. These examples are winners of Presidential Green Chemistry Challenge Awards. As you explore these examples, it will become apparent that green chemistry is very important in lowering the ecological footprint of chemical products and processes in the air, water, and soil.

We begin our journey into this important topic by briefly exploring how green chemistry can be applied to the synthesis of *ibuprofen*, an important everyday drug. We will see how the redesign of a chemical synthesis can eliminate a great deal of waste/pollution and reduce the amount of resources required.

Before discussing the synthesis of ibuprofen, we must first take a brief look at the concept of **atom economy.** This concept was developed by Barry Trost of Stanford University and won a Presidential Green Chemistry Challenge Award in 1998. Atom economy focuses our attention on Green Chemistry Principle 2 by asking the question: *How many of the atoms of the reactants* *are incorporated into the final desired product and how many are wasted?* As we will see in our discussion of the synthesis of ibuprofen, when chemists synthesize a compound, not all the atoms of the reactants are utilized in the desired product. Many of these atoms may end up in unwanted products (by-products), which are in many instances considered waste. These waste by-products may be toxic and can cause considerable environmental damage if not disposed of properly. In the past, waste products from chemical and other processes have not been disposed of properly, and environmental disasters such as the Love Canal have resulted.

Before we take on the synthesis of ibuprofen, let us look at a simple illustration of the concept of atom economy using the production of the desired compound, 1-bromobutane (compound 4) from 1-butanol (compound 1).

$$H_3C-CH_2-CH_2-CH_2-OH + Na-Br + H_2SO_4 \longrightarrow$$
$$1 \qquad\qquad 2 \qquad 3$$

$$H_3C-CH_2-CH_2-CH_2-Br + NaHSO_4 + H_2O$$
$$4 \qquad\qquad 5 \qquad\quad 6$$

If we inspect this reaction, we find that not only is the desired product formed, but so are the unwanted by-products sodium hydrogen sulfate and water (compounds 5 and 6). On the left side of this reaction, all the atoms of the reactants that are utilized in the desired product are printed in green and the remainder of the atoms (which become part of our waste by-products) in black. If we add up all of the green atoms on the left side of the reaction, we get 4 C, 9 H, and 1 Br (reflecting the molecular formula of the desired product, 1-bromobutane). The molar mass of these atoms collectively is 137 g/mol, the molar mass of 1-bromobutane. Adding up *all* the atoms of the reactants gives 4 C, 12 H, 5 O, 1 Br, 1 Na, and 1 S, and the total molar mass of all these atoms is 275 g/mol. If we take the molar mass of the atoms that are utilized, divide by the molar mass of all the atoms, and multiply by 100, we obtain the % **atom economy,** here 50%. Thus we see that half of the molar mass of all the atoms of the reactants is wasted and only half is actually incorporated into the desired product.

% **atom economy** = (molar mass of atoms utilized/molar mass of
all reactants) × 100

= (137/275) × 100 = 50%

This is one method of accessing the efficiency of a reaction. Armed with this information, a chemist may want to explore other methods of producing 1-bromobutane that have a greater % atom economy. We will now see how the concept of atom economy can be applied to the preparation of ibuprofen.

Ibuprofen is a common analgesic and anti-inflammatory drug found in such brand name products as *Advil*, *Motrin*, and *Medipren*. The first commercial synthesis of ibuprofen was by the Boots Company PLC of Nottingham, England. This synthesis, which has been used since the 1960s, is shown in Figure In-1. Although a detailed discussion of the chemistry of this synthesis

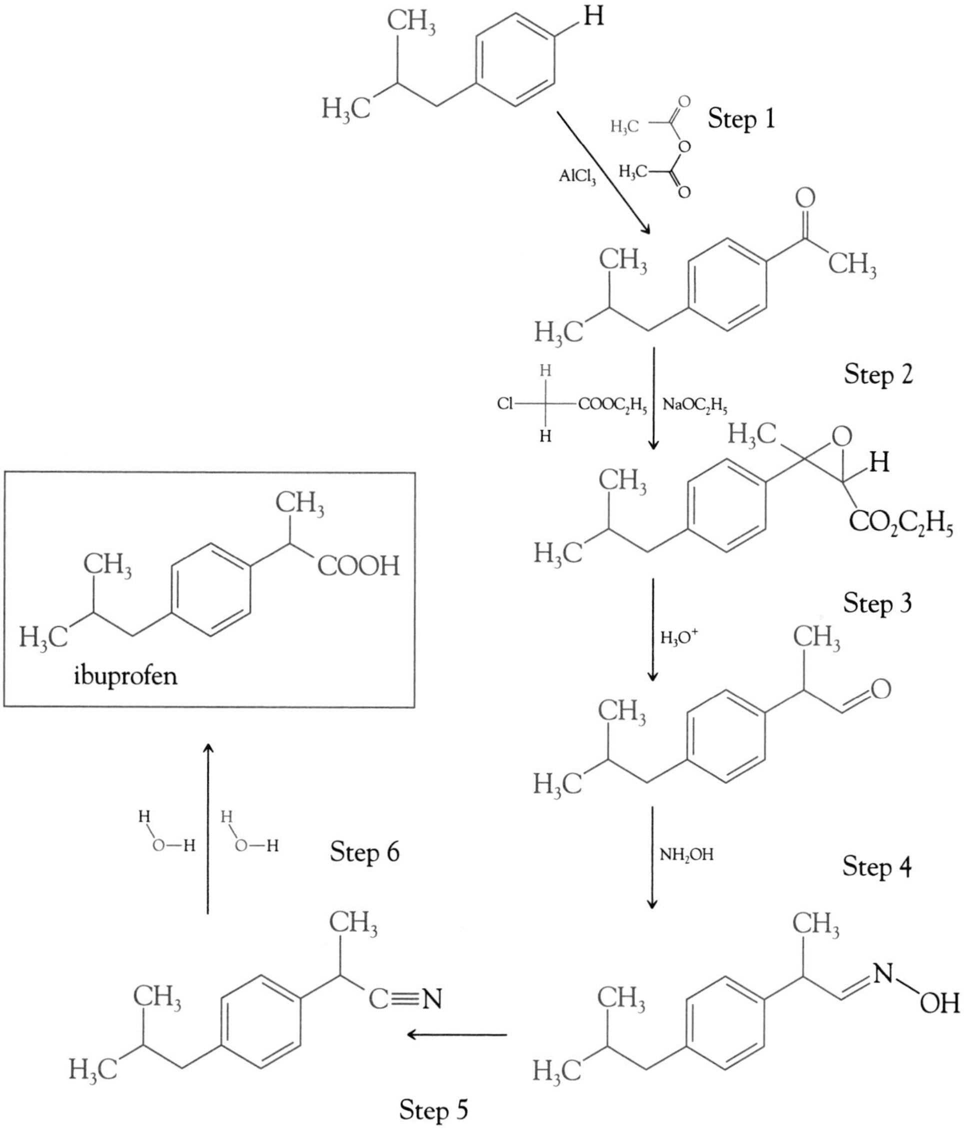

FIGURE In-1 The Boots Company synthesis of ibuprofen. [Source: M. C. Cann and M. E. Connelly, *Real-World Cases in Green Chemistry* (Washington, D.C.: American Chemical Society, 2000).]

is beyond the scope of this book, we can calculate its atom economy and obtain some idea of the waste produced. In Figure In-1, the atoms printed in green are those that are incorporated into the final desired product, ibuprofen, while those in black end up in waste by-products. By inspecting the structures of each of the reactants, we determine that the total of all the atoms in the reactants is 20 C, 42 H, 1 N, 10 O, 1 Cl, and 1 Na. The molar mass of all these atoms totals 514.5 g/mol. We also determine that the number of atoms of the reactants that are utilized in the ibuprofen (the atoms printed in green) is 13 C, 18 H, and 2 O (the molecular formula of ibuprofen). These atoms have a molar mass of 206.0 g/mol (the molar mass of the ibuprofen). The ratio of the molar mass of the utilized atoms to the molar mass of all the reactant atoms, multiplied by 100, gives an atom economy of 40%:

$$\textbf{\% atom economy} = \text{(molar mass of atoms utilized/molar mass of all reactants)} \times 100$$

$$= (206.0/514.5) \times 100 = 40\%$$

Only 40% of the molar mass of all the atoms of the reactants in this synthesis ends up in the ibuprofen; 60% is wasted. Because more than 30 million pounds of ibuprofen are produced each year, if we produced all the ibuprofen by this synthesis, there would be over 35 million pounds of unwanted waste produced just from the poor atom economy of this synthesis.

A new synthesis (Figure In-2) of ibuprofen was developed by the BHC Company (a joint venture of the Boots Company PLC and Hoechst Celanese Corporation), which won a Presidential Green Chemistry Challenge Award in 1997. This synthesis has only three steps as opposed to the six-step Boots synthesis and is less wasteful in many ways. One of the most obvious improvements is the increased atom economy. The molar mass of all the atoms of the reactants in this synthesis is 266.0 g/mol (13 C, 22 H, 4 O; note that the HF, Raney nickel, and the Pd in this synthesis are used only in catalytic amounts and thus do not contribute to the atom economy), while the utilized atoms (printed in green) again weigh 206.0 g/mol. This yields a % atom economy of 77%.

$$\textbf{\% atom economy} = \text{(molar mass of atoms utilized/molar mass of all reactants)} \times 100$$

$$= (206.0/266.0) \times 100 = 77\%$$

A by-product from the acetic anhydride (reactant 2) used in step 1 is acetic acid. The acetic acid is isolated and utilized, which increases the atom economy of this synthesis to more than 99%. Additional environmental advantages of the BHC synthesis include the elimination of auxiliary materials (Principle 5), such as solvents and the aluminum chloride promoter

FIGURE In-2 The BHC Company synthesis of ibuprofen. [Source: M. C. Cann and M. E. Connelly, *Real-World Cases in Green Chemistry* (Washington, D.C.: American Chemical Society, 2000).]

(replaced with the catalyst HF, Principle 9), and higher yields. Thus the green chemistry of the BHC Company synthesis lowers the environmental impact for the synthesis of ibuprofen by lowering the consumption of reactants and auxiliary substances while simultaneously reducing the waste.

Other improved syntheses that are winners of Presidential Green Chemistry Challenge Awards include the pesticide *Roundup*, the antiviral agent *Cytovene*, and the active ingredient in the antidepressant *Zoloft*.

Green chemistry provides a paradigm for reducing both the consumption of resources and the production of waste, thus moving toward sustainability. One of the primary considerations in the manufacture of chemicals must be the environmental impact of the chemical and the process by which it is produced. Sustainable chemistry must become part of the psyche of not only chemists and scientists, but also business leaders and policymakers. With this in mind, real-world examples of green chemistry have been incorporated throughout this text to expose you (our future scientists, business leaders, and policymakers) to sustainable chemistry.

Further Readings

1. P. T. Anastas and J. C. Warner, *Green Chemistry Theory and Practice* (New York: Oxford University Press, 1998).

2. M. C. Cann and M. E. Connelly, *Real-World Cases in Green Chemistry* (Washington, D.C.: American Chemical Society, 2000).

3. M. C. Cann and T. P. Umile, *Real-World Cases in Green Chemistry,* vol. 2 (Washington, D.C.: American Chemical Society, 2008).

4. M. C. Cann, "Bringing State of the Art, Applied, Novel, Green Chemistry to the Classroom, by Employing the Presidential Green Chemistry Challenge Awards," *Journal of Chemical Education* 76 (1999): 1639–1641.

5. M. C. Cann, "Greening the Chemistry Curriculum at the University of Scranton," *Green Chemistry* 3 (2001): G23–G25.

6. M. A. Ryan and M. Tinnesand, eds., *Introduction to Green Chemistry* (Washington, D.C.: American Chemical Society, 2002).

7. M. Kirchhoff and M. A Ryan, eds., *Greener Approaches to Undergraduate Chemistry Experiments* (Washington, D.C.: American Chemical Society, 2002).

8. World Commission on Environment and Development, *Our Common Future* [The Bruntland Report] (New York: Oxford University Press, 1987).

9. M. Wackernagel and W. Rees, *Our Ecological Footprint: Reducing Human Impact on the Earth* (Gabriola Island, BC: New Society Publishers, 1996).

Websites of Interest

1. EPA "Green Chemistry": http://www.epa.gov/greenchemistry/index.html

2. The Green Chemistry Institute of the American Chemical Society website: http://www.chemistry.org/portal/a/c/s/1/acsdisplay.html?DOC=greenchemistryinstitute\index.html

3. University of Scranton "Green Chemistry": http://academic.scranton.edu/faculty/CANNM1/greenchemistry.html

4. American Chemical Society Green Chemistry Educational Activities: http://www.chemistry.org/portal/a/c/s/1/acsdisplay.html?DOC=education\greenchem\index.html

5. *Our Common Future* (Report of the World Commission on Environment and Development): http://ringofpeace.org/environment/brundtland.html

6. Ecological Footprint: http://www.myfootprint.org/

LITTLE GREEN MOLECULES

By Terrence J. Collins and Chip Walter

POLLUTION CONTROL: Catalysts called TAMLs
work with hydrogen peroxide to break down
chlorophenols, which contaminate the wastewater
from many industrial sources.

Terrence J. Collins and Chip Walter, "Little Green
Molecules," *Scientific American*, March 2006,
82–90.

Chemists have invented a new class of catalysts that can destroy some of the worst pollutants before they get into the environment

The fish that live in the Anacostia River, which flows through the heart of Washington, D.C., are not enjoying its waters very much. The Anacostia is contaminated with the molecular remnants of dyes, plastics, asphalt and pesticides. Recent tests have shown that up to 68 percent of the river's brown bullhead catfish suffer from liver cancer. Wildlife officials recommend that anyone who catches the river's fish toss them back uneaten, and swimming has been banned.

ics and even birth-control hormones [*see illustration on opposite page*]. The amounts are often infinitesimal, measured in parts per billion or trillion (a part per billion is roughly equivalent to one grain of salt dissolved in a swimming pool), but scientists suspect that even tiny quantities of some pollutants can disrupt the developmental biochemistry that determines human behavior, intelligence, immunity and reproduction.

Fortunately, help is on the way. Over the past decade researchers in

American Chemical Society, the first principle of this community is: "It is better to prevent waste than to treat or clean up waste after it has been created." As part of this effort, however, researchers have also made discoveries that promise cost-effective methods for purging many persistent pollutants from wastewater.

In one example of this work, investigators at Carnegie Mellon University's Institute for Green Oxidation Chemistry (one of us, Collins, is the institute's director) have developed a group of designer catalyst molecules called TAML—tetra-amido macrocyclic ligand—activators that work with hydrogen peroxide and

Green chemistry can lessen some of the environmental damage caused by traditional chemistry.

The Anacostia is just one of dozens of severely polluted rivers in the U.S. The textile industry alone discharges 53 billion gallons of wastewater—loaded with reactive dyes and other hazardous chemicals—into America's rivers and streams every year. New classes of pollutants are turning up in the nation's drinking water: traces of drugs, pesticides, cosmet-

the emerging field of green chemistry have begun to design the hazards out of chemical products and processes. These scientists have formulated safer substitutes for harmful paints and plastics and devised new manufacturing techniques that reduce the introduction of pollutants into the environment. As outlined by the Green Chemistry Institute of the

other oxidants to break down a wide variety of stubborn pollutants. TAMLs accomplish this task by mimicking the enzymes in our bodies that have evolved over time to combat toxic compounds. In laboratory and real-world trials, TAMLs have proved they can destroy dangerous pesticides, dyes and other contaminants, greatly decrease the smells and color from the wastewater discharged by paper mills, and kill bacterial spores similar to those of the deadly anthrax strain. If broadly adopted, TAMLs could save millions of dollars in cleanup costs. Moreover, this research demonstrates that green chemistry can lessen some of

Overview/*Catalysts for Cleaning*

- Many pollutants released into waterways, such as dyes and pesticides, have become so omnipresent that they pose a serious threat to human health.
- Chemists have recently created enzymelike catalysts called tetra-amido macrocyclic ligand activators (TAMLs, for short) that can destroy stubborn pollutants by accelerating cleansing reactions with hydrogen peroxide.
- When applied to the wastewater from pulp mills, TAMLs have reduced staining and hazardous chemicals. The catalysts may also someday be used to disinfect drinking water and clean up contamination from bioterror attacks.

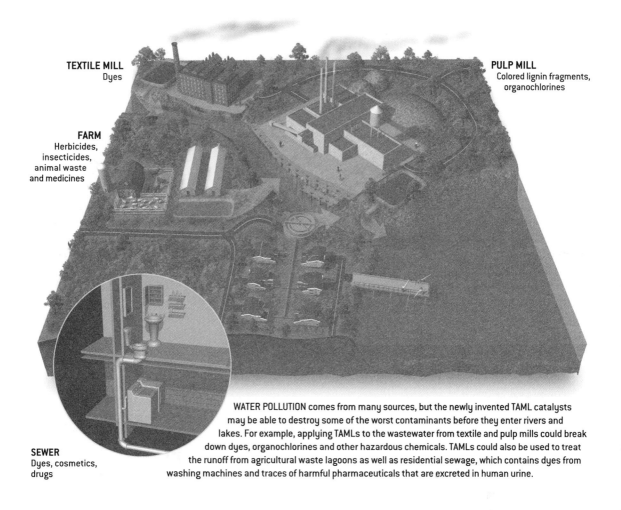

TEXTILE MILL
Dyes

FARM
Herbicides,
insecticides,
animal waste
and medicines

PULP MILL
Colored lignin fragments,
organochlorines

SEWER
Dyes, cosmetics,
drugs

WATER POLLUTION comes from many sources, but the newly invented TAML catalysts may be able to destroy some of the worst contaminants before they enter rivers and lakes. For example, applying TAMLs to the wastewater from textile and pulp mills could break down dyes, organochlorines and other hazardous chemicals. TAMLs could also be used to treat the runoff from agricultural waste lagoons as well as residential sewage, which contains dyes from washing machines and traces of harmful pharmaceuticals that are excreted in human urine.

the environmental damage caused by traditional chemistry.

The Need to Be Green

A FUNDAMENTAL CAUSE of our escalating environmental plight is that people perform chemistry in different ways than Mother Nature does. For eons, biochemical processes have evolved by drawing primarily on elements that are abundant and close at hand—such as carbon, hydrogen, oxygen, nitrogen, sulfur, calcium and iron—to create everything from paramecia to redwoods, clown fish to humans. Our industries, in contrast, gather elements from nearly every corner of the planet and distribute them in ways natural processes never could. Lead, for example, used to be found mostly in deposits so isolated and remote that nature never folded it into living organisms. But now lead is everywhere, primarily because our paints, cars and computers have spread it around. When it finds its way into children, even at minuscule doses, it is severely toxic. The same can be said for arsenic, cadmium, mercury, uranium and plutonium. These elements are persistent pollutants—they do not degrade in animal bodies or in the surrounding environment—so there is a pressing need to find safer alternatives.

Some of the new synthetic molecules in medicines, plastics and pesticides are so different from the products of natural chemistry that it is as though they dropped in from an alien world. Many of these molecules do not degrade easily, and even some biodegradable compounds have become omnipresent because we use them so copiously. Recent research indicates that some of these substances can interfere with the normal expression of genes involved in the development of the male reproductive system. Scientists have known for several years that prenatal exposure to phthalates, compounds used in plastics and beauty products, can alter the reproductive tract of newborn male rodents; in 2005 Shanna H. Swan of the University of Rochester School of Medicine and Dentistry reported similar

facturing processes with more environment-friendly alternatives [*see box on page 21*]. The work of Collins's team at Carnegie Mellon traces its origins back to the 1980s, when public health concerns about chlorine were intensifying. Chlorine was then, and still is, often used for large-scale cleaning and disinfection in manufacturing, as well as for the treatment of drinking water. Although chlorine treatment is inexpensive and effective, it can create some ugly pollutants. The bleaching of wood pulp with elemental chlorine in paper mills had been a major source of cancer-causing dioxins until the Environmental Protection Agency banned the process in

receptor system that regulates the production of critical proteins.

Rather than relying on chlorine, we wondered if we could put nature's own cleansing agents—hydrogen peroxide and oxygen—to the work of purifying water and reducing industrial waste. These cleansers can safely and powerfully obliterate many pollutants, but in nature the process usually requires an enzyme—a biochemical catalyst that vastly increases the rate of the reaction. Whether natural or man-made, catalysts act as old-fashioned matchmakers, except that rather than bringing two people together they unite specific molecules, enabling and accelerating the chemistry among them. Some natural catalysts can boost chemical reaction rates a billionfold. If not

Whether natural or man-made, catalysts act as
old-fashioned matchmakers.

effects in male infants. Another study headed by Swan found that men with low sperm counts living in a rural farming area of Missouri had elevated levels of herbicides (such as alachlor and atrazine) in their urine. Starting from our factories, farms and sewers, persistent pollutants can journey intact by air, water and up the food chain, often right back to us.

To confront this challenge, green chemists at universities and companies are investigating the feasibility of replacing some of the most toxic products and manu-

2001. (Most mills now bleach wood pulp with chlorine dioxide, which reduces the production of dioxins but does not eliminate it.) By-products created by the chlorination of drinking water have also been linked to certain cancers. Chlorine in its common natural form—chloride ions or salts dissolved in water—is not toxic, but when elemental chlorine reacts with other molecules it can generate compounds that can warp the biochemistry of living animals. Dioxins, for instance, disrupt cellular development by interfering with a

for an enzyme called ptyalin, found in our saliva, it would take several weeks for our bodies to break down pasta into its constituent sugars. Without enzymes, biochemistry would move at a numbingly slow pace, and life as we know it would not exist.

In nature, enzymes called peroxidases catalyze reactions involving hydrogen peroxide, the familiar household chemical used to bleach hair and remove carpet stains. In forests, fungi on rotting trees use peroxidases to marshal hydrogen peroxide to break down the lignin polymers in the wood, splitting the large molecules into smaller ones that the fungi can eat. Another family of enzymes, the cytochrome p450s, catalyzes reactions involving oxygen (*also called oxidation reactions*). Cytochrome p450s in our livers, for example, use oxygen to efficiently destroy many toxic molecules we inhale or ingest.

For decades, chemists have been struggling to build small synthetic molecules that could emulate these enormous enzymes. If scientists could create designer molecules with such strong catalytic abilities, they could replace the chlorine- and metal-based oxidation technologies that produce so many pollutants. In the early 1980s, however, no one was having much luck developing test-tube versions of the enzymes. Over billions of years of evolution, nature had choreographed some wonderfully elegant and extremely complex catalytic dances, making our efforts in the laboratory look clunky. Yet we knew that we could not achieve our goal of reducing pollution unless we found a way to mimic this molecular dance.

Catalytic Converters

CREATING SYNTHETIC enzymes also meant assembling molecules that would be robust enough to resist the destructive reactions they were catalyzing. Any chemistry involving oxygen can be destructive because the bonds it makes with other elements (especially hydrogen) are so strong. And because each molecule of hydrogen peroxide (H_2O_2) is halfway between water (H_2O) and molecular oxygen (O_2), this compound is also strongly oxidizing. In water, hydrogen peroxide often produces a kind of liquid fire that demolishes the organic (carbon-containing) molecules around it. A lesson from the enzymes was that a working catalyst would probably need to have an iron atom placed inside a molecular matrix of organic groups. So we had to toughen the molecular architecture of such groups to ensure they could survive the liquid fire that would result from the activation of hydrogen peroxide.

Borrowing further from nature's design, we eventually solved this problem by creating a catalyst in which four nitrogen atoms are placed in a square with a single iron atom anchored in the middle [*see box on next page*]. The nitrogen atoms are connected to the much larger iron atom by covalent bonds, meaning that they share pairs of electrons; in this kind of structure, the smaller atoms and attached groups surrounding the central metal atom are called ligands. Next we linked the ligands to form a big outer ring called a macrocycle. Over time we learned how to make the ligands and linking systems tough enough to endure the violent reactions that the TAMLs trigger. In effect, the ligands we invented became a kind of firewall that resisted the liquid fire. The longer it resisted, the more useful the catalyst. Of course, we did not want to create an indestructible catalyst, which could end up in effluent streams and perhaps produce a pollution problem of its own. All our existing Fe-TAML catalysts (TAMLs with iron as the central metal atom) decompose on timescales ranging from minutes to hours.

Building the ligand firewalls was not easy. It required developing a painstaking four-step design process in which we first imagined and then synthesized ligand constructions that we

THE AUTHORS

TERRENCE J. COLLINS and CHIP WALTER have worked together to educate the public about the challenges and possibilities of green chemistry. Collins is Thomas Lord Professor of Chemistry at Carnegie Mellon University, where he directs the Institute for Green Oxidation Chemistry. He is also an honorary professor at the University of Auckland in New Zealand. Walter is a science journalist and author of *Space Age* and *I'm Working on That* (with William Shatner). He teaches science writing at Carnegie Mellon and is a vice president of communications at the University of Pittsburgh Medical Center.

A MOLECULAR CLEANING MACHINE

Chemists designed TAMLs to emulate the natural enzymes that catalyze reactions involving hydrogen peroxide. TAMLs, though, are hundreds of times smaller than enzymes, so they are easier and cheaper to manufacture.

At the center of each TAML is an iron atom bonded to four nitrogen atoms; at the edge are carbon rings linked to form a big outer ring called a macrocycle. This linking system acts as a firewall, enabling the molecule to endure the violent reactions it triggers. In its solid state the TAML also has one water molecule (H_2O) attached to the iron atom. (The attached groups are called ligands.)

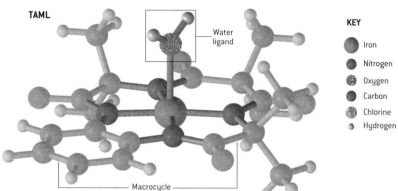

TAML

Water ligand

Macrocycle

KEY

- Iron
- Nitrogen
- Oxygen
- Carbon
- Chlorine
- Hydrogen

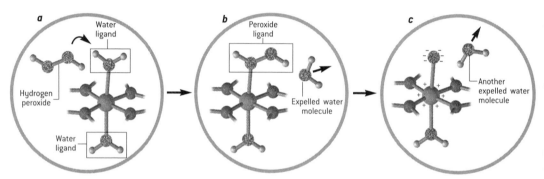

a

Water ligand

Hydrogen peroxide

Water ligand

b

Peroxide ligand

Expelled water molecule

c

Another expelled water molecule

When a TAML dissolves in water, another molecule of H_2O connects to the catalyst (*a*). If hydrogen peroxide (H_2O_2) is also in the solution, it can replace one of the water ligands, which are loosely attached and easily expelled (*b*). The peroxide ligand then discards both its hydrogen atoms and one oxygen atom in the form of a water molecule, leaving one oxygen atom attached to the iron (*c*). The oxygen pulls electrons farther away from the iron atom, turning the TAML into a reactive intermediate.

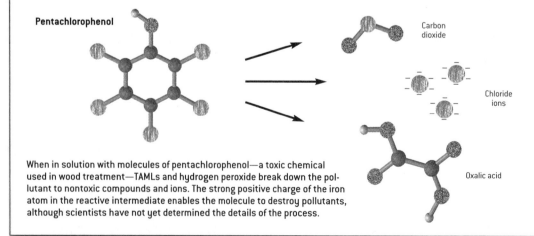

Pentachlorophenol

Carbon dioxide

Chloride ions

Oxalic acid

When in solution with molecules of pentachlorophenol—a toxic chemical used in wood treatment—TAMLs and hydrogen peroxide break down the pollutant to nontoxic compounds and ions. The strong positive charge of the iron atom in the reactive intermediate enables the molecule to destroy pollutants, although scientists have not yet determined the details of the process.

MELISSA THOMAS; SOURCE: *Institute for Green Oxidation Chemistry*

CHEMISTRY GOES GREEN

The invention of TAML catalysts is just one of the many achievements of green chemistry, which strives to develop products and processes that reduce or eliminate the use and generation of hazardous substances. Some other accomplishments are listed below.

PROJECT	PARTICIPANTS	STATUS
Using plant sugars to create polylactic acids (PLAs), a family of biodegradable polymers that could replace many traditional petroleum-derived plastics	Patrick Gruber, Randy L. Howard, Jeffrey J. Kolstad, Chris M. Ryan and Richard C. Bopp, NatureWorks LLC (a subsidiary of Cargill)	NatureWorks has built a factory in Nebraska to manufacture PLA pellets, which are used to make water bottles, packaging materials and other products
Discovering synthesis reactions that allow manufacturers to substitute water for many common organic solvents, some of which can cause cancer	Chao-Jun Li, McGill University	Pharmaceutical and commodity chemical companies are investigating the process
Developing metathesis chemistry, a method of organic synthesis that can produce drugs, plastics and other chemicals more efficiently and with less waste	Robert H. Grubbs, California Institute of Technology; Richard R. Schrock, Massachusetts Institute of Technology; Yves Chauvin, French Petroleum Institute	Widely applied in the chemical, biotechnology and food industries, this research was awarded the 2005 Nobel Prize in Chemistry
Replacing toxic petroleum-based solvents with supercritical carbon dioxide, a high-temperature, high-pressure fluid that has the properties of both a liquid and a gas	Martyn Poliakoff, Michael George and Steve Howdle, University of Nottingham, England	Thomas Swan & Co., a British manufacturer of specialty chemicals, has built a plant that uses supercritical fluids
Inventing a new method for producing sertraline, the key ingredient in the antidepressant Zoloft	James Spavins, Geraldine Taber, Juan Colberg and David Pfisterer, Pfizer	The process has reduced pollution, energy and water use while improving worker safety and product yield

hoped would keep the firewall in place. Second, we subjected the catalyst to oxidative stress until the firewall disintegrated. Third, we looked for the precise location where the breakdown began. (We found that ligand degradations always start at the most vulnerable site.) And in the final step, once we had pinpointed the weakest link, we replaced it with groups of atoms we believed would hold up longer.

Then we started the whole design cycle again.

After 15 years, we finally created our first working TAML. We knew we had succeeded one morning when Colin Horwitz, a research professor at our institute, showed off the results of a bleaching experiment that featured our most advanced design at the time. We looked at the results, and there it was: every time Horwitz squirted dark dye into a solution containing the TAML catalyst and hydrogen peroxide, the solution quickly turned colorless. We now knew that our firewalls were finally holding up long enough to allow the TAMLs to do their job. The molecules were acting like enzymes, and yet they were much, much smaller: the molecular weight of a TAML is about 500 daltons (a dalton is equal to one twelfth the mass of carbon 12, the most abundant isotope of carbon), whereas the weight of horseradish peroxidase, a relatively small enzyme, is about 40,000 daltons. The diminutive TAML activators are easier and cheaper to make, and much more versatile in their reactivity, than their natural counterparts.

Since then, we have evolved more than 20 different TAML activators by reapplying the same four-step design process that enabled us to create the first working model. Each TAML has its own reaction rate and lifetime, allowing us to tailor the catalysts to match the tasks we want them to perform. Most of the catalysts incorporate elements such as carbon, hydrogen, oxygen, nitrogen and iron, all chosen for their low toxicity. We call some of the molecules "hunter TAMLs" because they are designed to seek out and lock onto specific pollutants or pathogens, in much the same way that a magnetized mine seeks out the metal hull of a ship. Other TAMLs act as blowtorches that aggressively burn most of the oxidizable chemicals with which they come into contact. Still others are less aggressive and more selective, so that they will, for example, attack only certain parts of molecules or attack only the more easily oxidized molecules in a group. We expect to adapt TAMLs to advance green chemistry for decades to come. Although more toxicity testing must be done, the results so far indicate that TAMLs break down pollutants to their nontoxic constituents, leaving no

detectable contamination behind. We now have more than 90 international patents on TAML activators, with more in the pipeline, and we also have several commercial licenses.

Interestingly, we still do not know all the details of how the TAMLs work, but recent studies have provided deep insights into the key reactions. In their solid state, Fe-TAMLs generally have one water molecule attached as a ligand to the iron atom, oriented perpendicularly from the four nitrogen ligands; when put in solution, another water molecule connects to the opposite side of the iron atom. These water ligands are very loosely attached— if hydrogen peroxide is also in the solution, a molecule of it easily replaces one of the water molecules. The peroxide ligand swiftly reconstitutes itself, expelling both its hydrogen atoms and one oxygen atom (which escape as H_2O, a water molecule) and leaving one oxygen atom attached to the iron at the center of the Fe-TAML, which is now called the reactive intermediate (RI).

Oxygen is much more electronegative than iron, which means that its nucleus pulls most of the electrons in the complex bond toward itself and away from the iron nucleus. This effect increases the positive charge of the iron at the center of the TAML, making the RI reactive enough to extract electrons from oxidizable molecules in the solution. We have not yet deter-

mined how the RI breaks the chemical bonds of its targets, but current investigations may soon reveal the answer. We do know, however, that we can adjust the strength of the TAMLs by changing the atoms at the head and tail of the molecule; putting highly electronegative elements at those locations draws even more negative charge away from the iron and makes the RI more aggressive.

Industrial Strength

BUILDING TAMLS in the laboratory is one thing; scaling them up for commercial use is another. So far the lab tests and field trials have been promising. Tests funded by the National Science Foundation, for example, demonstrated that Fe-TAMLs plus peroxides could clean up the contamination from a bioterror attack. We found that when we combined one TAML with tertiary butyl hydroperoxide—a variation of hydrogen peroxide that replaces one of the hydrogen atoms with a carbon atom and three methyl (CH_3) groups—the resulting solution could deactivate 99.99999 percent of the spores of *Bacillus atrophaeus*, a bacterial species very similar to anthrax, in 15 minutes. In another important potential application, we hope to use Fe-TAMLs and hydrogen peroxide to someday create an inexpensive disinfectant to tackle the infectious waterborne microbes that account for so much death and disease worldwide.

In three field trials, we explored how well TAMLs can alleviate the pollution created when paper is manufactured. Every year the paper and wood pulp industry produces more than 100 million metric tons of bleached pulp, which is turned into white paper. Besides generating dioxins, chlorophenols and other hazardous organochlorines, many pulp mills discharge a coffee-colored effluent that stains streams and rivers and blocks light from penetrating the water. The reduction of light interferes with photosynthesis, which in turn affects organisms that depend on plants for food. The sources of the staining are large colored fragments of lignin, the poly-

pulp mills in the U.S. and one in New Zealand. In New Zealand we combined Fe-TAMLs and peroxide with 50,000 liters of effluent water. In the U.S. we directly injected Fe-TAMLs into a pulp-treatment tower or an exit pipe over the course of several days to bleach the wastewater. Overall, the Fe-TAMLs reduced the staining of the water by up to 78 percent and eliminated 29 percent of the organochlorines.

The development of other TAML applications also looks exciting. Eric Geiger of Urethane Soy Systems, a company based in Volga, S.D., has found that Fe-TAMLs do an excellent job processing soybean oil into useful polymers that display physical properties equal to, if not better than, those of current polyurethane products. TAMLs may even find

drugs and agricultural chemicals to pass intact into drinking water.

Despite the success of these trials, we have not resolved all the questions about TAML activators. More testing on industrial scales remains to be done, and it is important to ensure that TAMLs do not create some form of pollution we have not yet observed. Too often chemical technologies have seemed completely benign when first commercialized, and the devastating negative consequences did not become clear until decades later. We want to do everything in our power to avoid such surprises with TAMLs.

Cost is also an issue. Although TAMLs promise to be competitive in most applications, large corporations are deeply invested in the chemical processes they currently use. Shifting to new systems and techniques, even if they work,

Building TAMLs in the laboratory is one thing
scaling them up for commercial use is another.

mer that binds the cellulose fibers in wood. Bleaching with chlorine dioxide removes the lignin from the cellulose; the smaller lignin fragments are digested by bacteria and other organisms in treatment pools, but the larger pieces are too big to be eaten, so they end up in rivers and lakes.

We have tested the effectiveness of Fe-TAMLs at decolorizing these fragments at two

their way into washing machines: in another series of tests, we found that a tiny quantity of catalyst in certain household laundry products eliminated the need to separate white and colored clothing. TAMLs can prevent staining by attacking dyes after they detach from one fabric but before they attach to another. We are also working on a new family of TAMLs that can break the very stable molecular bonds that allow

usually requires significant investments. One great advantage of TAML technology, though, is that it does not require major retooling. What is more, TAMLs may ultimately save companies money by offering a cost-effective way to meet increasingly stringent environmental laws in the U.S., Europe and elsewhere.

The advances of green chemistry to date represent only a few interim steps on the road to dealing with the many environmental challenges of the 21st century. The deeper question is, Are we going to practice acute care or preventive medicine? Right now most chemists are still trained to create elegantly structured compounds that solve the specific problem for which they have been engineered, without regard to their broader impact. We are in effect performing global-scale experiments on our ecosystems and ourselves, and when these experiments fail the cost can be catastrophic. New green chemical techniques offer an alternative. The Industrial Revolution has unfolded, for the most part, without design or forethought. Perhaps now we can take some creative steps to reverse that trend and help make a world, and a future, that we can live with. ▪

MORE TO EXPLORE

Toward Sustainable Chemistry. Terrence J. Collins in Science, Vol. 291, No. 5501, pages 48–49; January 5, 2001.
Rapid Total Destruction of Chlorophenols by Activated Hydrogen Peroxide. Sayam Sen Gupta, Matthew Stadler, Christopher A. Noser, Anindya Ghosh, Bradley Steinhoff, Dieter Lenoir, Colin P. Horwitz, Karl-Werner Schramm and Terrence J. Collins in Science, Vol. 296, pages 326–328; April 12, 2002.
More information can be found online at www.cmu.edu/greenchemistry and www.chemistry.org/portal/a/c/s/1/acsdisplay.html?DOC=greenchemistryinstitute\index.html

PART I

ATMOSPHERIC CHEMISTRY AND AIR POLLUTION

Contents of Part I

Environmental Instrumental Analysis I

- Instrumental Determination of NO_X via Chemiluminescence

CHAPTER 1

STRATOSPHERIC CHEMISTRY:
The Ozone Layer

In this chapter, the following introductory chemistry topics are used:

- Moles; concentration units including mole fraction
- Ideal gas law; partial pressures
- Thermochemistry: ΔH, ΔH_f; Hess' law
- Kinetics: rate laws; reaction mechanisms, activation energy, catalysis

Introduction

The **ozone layer** is a region of the atmosphere that is called "Earth's natural sunscreen" because it filters out harmful ultraviolet (UV) rays from sunlight before they can reach the surface of our planet and cause damage to humans and other life forms. Any substantial reduction in the amount of this ozone would threaten life as we know it. Consequently, the appearance in the mid-1980s of a large "hole" in the ozone layer over Antarctica represented a major environmental crisis. Although steps have been taken to prevent its expansion, the hole will continue to appear each spring over the South Pole; indeed, one of the

A young girl applies sunscreen to protect her skin against UV rays from the Sun. [Source: Lowell George/CORBIS.]

largest, deepest holes in history occurred in 2006. Thus it is important that we understand the natural chemistry of the ozone layer, the subject of this chapter. The specific processes at work in the ozone hole and the history of the evolution of the hole are elaborated upon in Chapter 2. We begin by considering how the concentrations of atmospheric gases are reported and the region of the atmosphere where the ozone is concentrated.

Regions of the Atmosphere

The main components (ignoring the normally ever-present but variable water vapor) of an unpolluted version of the Earth's atmosphere are **diatomic nitrogen,** N_2 (about 78% of the molecules); **diatomic oxygen,** O_2 (about 21%); *argon,* Ar (about 1%); and **carbon dioxide,** CO_2 (presently about 0.04%). (The names of chemicals important to a chapter are printed in bold, along with their formulas, when they are introduced. The names of chemicals less important in the present context are printed in italics.) This mixture of chemicals seems unreactive in the lower atmosphere, even at temperatures or sunlight intensities well beyond those naturally encountered at the Earth's surface.

The lack of noticeable reactivity in the atmosphere is deceptive. In fact, many environmentally important chemical processes occur in air, whether clean or polluted. In the next two chapters, these reactions will be explored in detail when we discuss reactions that occur in the **troposphere,** the region of the atmosphere that extends from ground level to about 15 kilometers altitude and contains 85% of the atmosphere's mass. In this chapter we will consider processes in the **stratosphere,** the portion of the atmosphere from approximately 15 to 50 kilometers (i.e., 9–30 miles) altitude that lies just above the troposphere. The chemical reactions to be considered are vitally important to the continuing health of the ozone layer, which is found in the bottom half of the stratosphere. The ozone concentrations and the average temperatures at altitudes up to 50 kilometers in the Earth's atmosphere are shown in Figure 1-1.

The stratosphere is defined as the region that lies between the altitudes where the temperature trends display reversals: The bottom of the stratosphere occurs where the temperature first stops decreasing with height and begins to increase, and the top of the stratosphere is the altitude where the temperature stops increasing with height and begins to decrease. The exact altitude at which the troposphere ends and the stratosphere begins varies with season and with latitude.

Environmental Concentration Units for Atmospheric Gases

Two types of concentration scales are commonly used for gases present in air. For *absolute* concentrations, the most common scale is the number of **molecules per cubic centimeter** of air. The variation in the concentration of ozone on the molecules per cubic centimeter scale with altitude is illustrated in Figure 1-1a. Absolute concentrations are also sometimes expressed in

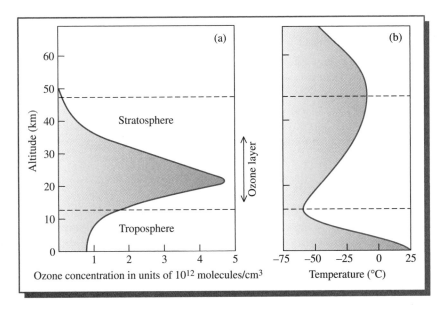

FIGURE 1-1 Variation with altitude of (a) ozone concentration (for mid-latitude regions) and (b) air temperature for various regions of the lower atmosphere.

terms of the **partial pressure** of the gas, which is stated in units of atmospheres or kilopascals or bars. According to the **ideal gas law** ($PV = nRT$), partial pressure is directly proportional to the molar concentration n/V, and hence to the molecular concentration per unit volume, when different gases or components of a mixture are compared at the same Kelvin temperature T.

Relative concentrations are usually based on the chemists' familiar **mole fraction** scale (called *mixing ratios* by physicists), which is also the molecule fraction scale. Because the concentrations for many constituents are so small, atmospheric and environmental scientists often re-express the mole or molecule fraction as a *parts per* _____ value. Thus a concentration of 100 molecules of a gas such as carbon dioxide dispersed in one million (10^6) molecules of air would be expressed as 100 parts per million, i.e., 100 ppm, rather than as a molecule or mole fraction of 0.0001. Similarly, ppb and ppt stand for parts per billion (one in 10^9) and parts per trillion (one in 10^{12}).

It is important to emphasize that for *gases*, these relative concentration units express the number of *molecules* of a pollutant (i.e., the "solute" in chemists' language) that are present in one million or billion or trillion *molecules* of air. Since, according to the ideal gas law, the volume of a gas is proportional to the number of molecules it contains, the "parts per" scales also represent the *volume* a pollutant gas would occupy, compared to that of the stated *volume of air*, if the pollutant were to be isolated and compressed until its pressure equaled that of the air. In order to emphasize that the concentration scale is based upon molecules or volumes rather than upon mass, a v (for volume) is sometimes shown as part of the unit, e.g., 100 ppm$_v$ or 100 ppmv.

The Physics and Chemistry of the Ozone Layer

To understand the importance of atmospheric ozone, we must consider the various types of light energy that emanate from the Sun and consider how UV light in particular is selectively filtered from sunlight by gases in the air. This leads us to consider the effects on human health of UV light, and quantitatively how energy from light can break apart molecules. With that background, we can then investigate the natural processes by which ozone is formed and destroyed in air.

Absorption of Light by Molecules

The chemistry of ozone depletion, and of many other processes in the stratosphere, is driven by energy associated with light from the Sun. For this reason, we begin by investigating the relationship between light absorption by molecules and the resulting activation, or energizing, of the molecules that enable them to react chemically.

An object that we perceive as black in color absorbs light at all wavelengths of the visible spectrum, which runs from about 400 nm (violet light) to about 750 nm (red light); note that one nanometer (nm) equals 10^{-9} meter. Substances differ enormously in their propensity to absorb light of a given wavelength because of differences in the energy levels of their electrons. Diatomic molecular oxygen, O_2, does not absorb visible light very readily, but it does absorb some types of **ultraviolet** (UV) light, which is that having wavelengths between about 50 and 400 nm. The most environmentally relevant portion of the electromagnetic spectrum is illustrated in Figure 1-2. Notice that the UV region begins at the violet edge of the visible region, hence the name *ultraviolet*. The division of the UV region into components will be discussed later in this chapter. At the other end of the spectrum, beyond the red portion of the visible region, lies **infrared** light, which will become important to us when we discuss the greenhouse effect in Chapter 6.

An **absorption spectrum** such as that illustrated in Figure 1-3 is a graphical representation that shows the relative fraction of light that is absorbed by a given type of molecule as a function of wavelength. Here, the efficient light-absorbing behavior of O_2 molecules for the UV region between 70 and 250 nm is shown; some minuscule amount of absorption continues beyond 250 nm, but in an ever-decreasing fashion (not shown). Notice that the fraction of light absorbed by O_2 (given on a logarithmic scale in Figure 1-3) varies quite dramatically with wavelength. This sort of selective absorption behavior is observed for all atoms and molecules, although the specific regions of strong absorption and of zero absorption vary widely, depending upon the structure of the species and the energy levels of their electrons.

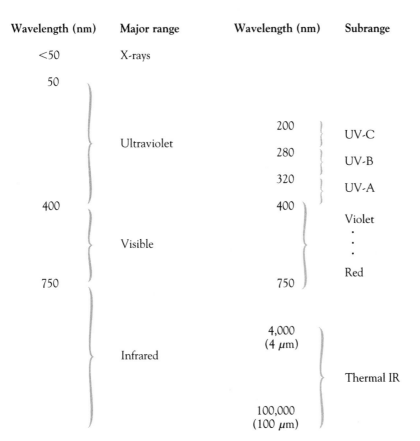

Wavelength (nm)	Major range	Wavelength (nm)	Subrange
<50	X-rays		
50			
	Ultraviolet	200	UV-C
		280	UV-B
		320	UV-A
400		400	Violet
	Visible		.
			.
			.
750		750	Red
	Infrared	4,000 (4 μm)	
			Thermal IR
		100,000 (100 μm)	

FIGURE 1-2 The electromagnetic spectrum. The ranges of greatest environmental interest in this book are shown.

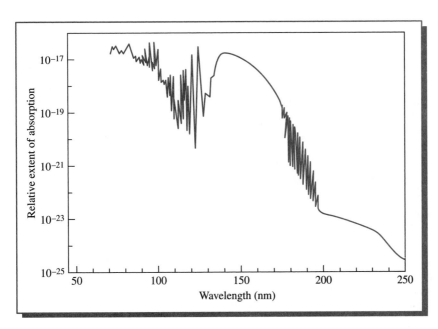

FIGURE 1-3 Absorption spectrum of O_2. [Source: T. E. Graedel and P. J. Crutzen, *Atmospheric Change: An Earth System Perspective* (New York: W. H. Freeman, 1993).]

Filtering of Sunlight's UV Component by Atmospheric O_2 and O_3

As a result of these absorption characteristics, the O_2 gas that lies *above* the stratosphere filters from sunlight most of the UV light from 120 to 220 nm; the remainder of the light in this range is filtered by the O_2 in the stratosphere. Ultraviolet light that has wavelengths shorter than 120 nm is filtered in and above the stratosphere by O_2 and other constituents of air such as N_2. Thus no UV light having wavelengths shorter than 220 nm reaches the Earth's surface. This screening protects our skin and eyes, and in fact protects all biological life, from extensive damage by this part of the Sun's output.

Diatomic oxygen also filters some, but not all, of sunlight's UV in the 220–240-nm range. Instead, ultraviolet light in the whole 220–320-nm range is filtered from sunlight mainly by ozone molecules, O_3, that are spread through the middle and lower stratosphere. The absorption spectrum of ozone in this wavelength region is shown in Figure 1-4. Since its molecular constitution, and thus its set of energy levels, is different from that of diatomic oxygen, its light absorption characteristics also are quite different.

Ozone, aided to some extent by O_2 at the shorter wavelengths, filters out all of the Sun's ultraviolet light in the 220–290-nm range, which overlaps the 200–280-nm region known as **UV-C** (see Figure 1-2). However, ozone can absorb only a fraction of the Sun's UV light in the 290–320-nm range, since, as you can infer from Figure 1-4b, its inherent ability to absorb light of such wavelengths is quite limited. The remaining amount of the sunlight of such wavelengths, 10–30% depending upon latitude, penetrates the atmosphere to the Earth's surface. Thus ozone is not *completely* effective in shielding us from light in the **UV-B** region, defined as that which lies from 280 to 320 nm (although different authors vary slightly on the limits for this parameter). Since the absorption by ozone

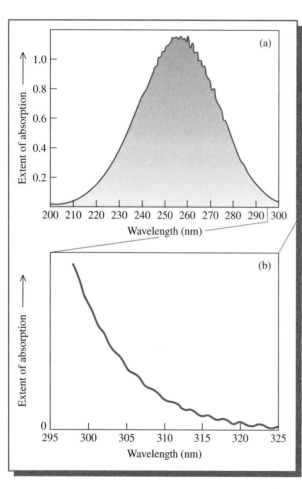

FIGURE 1-4 Absorption spectrum of O_3: (a) from 200 to 300 nm and (b) from 295 to 325 nm. Note that different scales are used for the extent of absorption in the two cases. [Sources: (a) Redrawn from M. J. McEwan and L. F. Phillips, *Chemistry of the Atmosphere* (London: Edward Arnold, 1975). (b) Redrawn from J. B. Kerr and C. T. McElroy, *Science* 262: 1032–1034. Copyright 1993 by the AAAS.]

falls off in an almost exponential manner with wavelength in this region (see Figure 1-4b), the fraction of solar UV-B that reaches the troposphere increases with increasing wavelength.

Because neither ozone nor any other constituent of the clean atmosphere absorbs significantly in the **UV-A** range, i.e., 320–400 nm, most of this, the least biologically harmful type of ultraviolet light, does penetrate to reach the Earth's surface. (Nitrogen dioxide gas does absorb UV-A light but is present in such small concentration in clean air that its net absorption of sunlight is quite small.)

The net effect of diatomic oxygen and ozone in screening the troposphere from the UV component of sunlight is illustrated in Figure 1-5. The curve at the left corresponds to the intensity of light received outside the Earth's atmosphere, whereas the curve at the right corresponds to the light that is transmitted to the troposphere (and thus to the surface). The vertical separation at each wavelength between the curves corresponds to the amount of sunlight that is absorbed in the stratosphere and outer regions of the atmosphere.

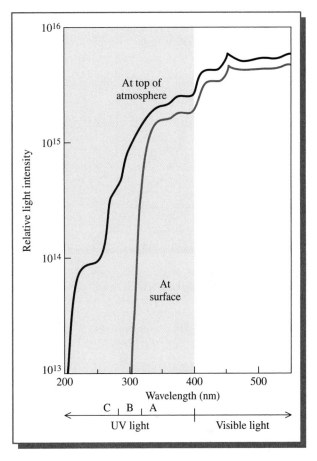

FIGURE 1-5 The intensity of sunlight in the UV and in part of the visible regions measured outside the atmosphere and in the troposphere. [Source: W. L. Chameides and D. D. Davis, *Chemical and Engineering News* (4 October 1982): 38–52. Copyright 1982 by the American Chemical Society. Reprinted with permission.]

Biological Consequences of Ozone Depletion

A reduction in stratospheric ozone concentration allows more UV-B light to penetrate to the Earth's surface. A 1% decrease in overhead ozone is predicted to result in a 2% increase in UV-B intensity at ground level. This increase in UV-B is the principal environmental concern about ozone depletion, since it leads to detrimental consequences to many life forms, including humans. Exposure to UV-B causes human skin to sunburn and suntan; overexposure can lead to skin cancer, the most prevalent form of cancer. Increasing amounts of UV-B may also adversely affect the human immune system and the growth of some plants and animals.

Most biological effects of sunlight arise because UV-B can be absorbed by DNA molecules, which then may undergo damaging reactions. By comparing the variation in wavelength of UV-B light of differing intensity arriving at

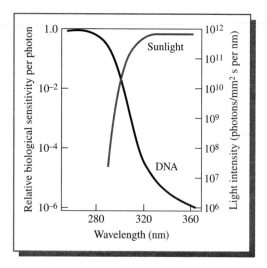

FIGURE 1-6 The absorption spectrum for DNA and the intensity of sunlight at ground level versus wavelength. The degree of absorption of light energy by DNA reflects its biological sensitivity to a given wavelength. [Source: Adapted from R. B. Setlow, *Proceedings of the National Academy of Science USA* 71 (1974): 3363–3366.]

the Earth's surface with the absorption characteristics of DNA as shown in Figure 1-6, it can be concluded that the major detrimental effects of sunlight absorption will occur at about 300 nm. Indeed, in light-skinned people, the skin shows maximum UV absorption from sunlight at about 300 nm.

Most skin cancers in humans are due to overexposure to UV-B in sunlight, so any decrease in ozone is expected eventually to yield an increase in the incidence of this disease. Fortunately, the great majority of skin cancer cases are not the often-fatal (25% mortality rate) **malignant melanoma,** but rather one of the slowly spreading types that can be treated and that collectively affect about one in four Americans at some point in their lives. The plot in Figure 1-7, which is based on health data from seven countries at different latitudes that therefore receive different amounts of ground-level UV, shows that the rise in the incidence of nonmelanoma skin cancer with exposure to UV is exponential; the reason is that the logarithm of the incidence is linearly related to the UV intensity. For example, the skin cancer rate in Europe is only about half that in the United States.

The incidence of the malignant melanoma form of skin cancer, which affects about 1 in 100 Americans, is thought to be related to short periods of very high UV exposure, particularly early in life. Especially susceptible are fair-skinned, fair-haired, freckled people who burn easily and who have moles with irregular shapes or colors. The incidence of malignant melanoma is also related to latitude. White males living in sunny climates such as Florida or Texas are twice as likely to die from this disease as those in the more northerly states, although part of this greater incidence is probably due to different patterns of personal behavior, such as choice of clothing, as well as to increased UV-B content in the sunlight. Curiously, indoor workers—who have intermittent exposure to the Sun—are more susceptible than are tanned, outdoor workers! The lag period between first exposure and melanoma is 15–25 years. If malignant melanoma is not treated early, it can spread via the bloodstream to body organs such as the brain and the liver.

The phrase *full spectrum* is sometimes used to denote sunscreens that block UV-A as well as UV-B light. The use of sunscreens that block UV-B, but not UV-A, may actually lead to an increase in melanoma skin cancer, since sunscreen usage allows people to expose their skin to sunlight for prolonged periods without burning. The substances used in sunscreen lotions (e.g., particles of inorganic compounds such as *zinc oxide* or *titanium oxide*) either reflect or scatter sunlight or absorb its UV component (e.g., water-insoluble organic compounds such as *octinoxate*—octyl methoxycinnamate—for UV-B absorption and *oxybenzone* for UV-A) before it can reach and be absorbed by the skin. Sunscreens were one of the first consumer products to use nanoparticles, i.e., tiny particles only a few dozen or a few hundred nanometers (10^{-9} m) in size. Since such particles are so tiny and do not absorb or reflect visible light, the sunscreens appear transparent.

Potential sunscreen compounds are eliminated if they undergo an irreversible chemical reaction when they absorb sunlight, because this would quickly reduce the effectiveness of the application and because the reaction products could be toxic to the skin. Also, the commonly used sunscreen component *PABA* (*p*-aminobenzoic acid) is no longer generally used because of evidence that it can itself cause cancer.

The **SPF** (Sun Protection Factor) of a sunscreen measures the multiplying factor by which a person can stay exposed to the Sun without burning. Thus an SPF of 15 means that he or she can stay in the Sun fifteen times longer than without the sunscreen. To receive that protection, however, the sunscreen must be reapplied at least every few hours.

Because of the long time lag (30–40 years) between exposure to UV and the subsequent manifestation of nonmalignant skin cancers, it is unlikely that effects from ozone depletion are observable as yet. The rise in skin cancer that has occurred in many areas of the world—and that is still occurring, especially among young adults—is probably due instead to greater amounts of time spent by people outdoors in the Sun over the past few decades. For example, the incidence of skin cancer among residents of Queensland, Australia, most of whom are light-skinned, rose to about 75% of the population as lifestyle changes increased their exposure to sunlight years before ozone depletion began. As a consequence of its experience with skin cancer, Australia has led the world in public health awareness of the need for protection from ultraviolet exposure.

In addition to skin cancer, UV exposure has been linked to several other human conditions. The front of the eye is the one part of the human anatomy where ultraviolet light can penetrate the human body. However, the cornea and lens filter out about 99% of UV from light before it reaches the retina. Over time, the UV-B absorbed by the cornea and lens produces highly reactive molecules called free radicals that attack the structural molecules and can produce cataracts. Indeed, there is some evidence that increased UV-B levels give rise to an increased incidence of eye cataracts, particularly among the nonelderly (see Figure 1-8). UV exposure has also been linked to an increase in the rate of macular degeneration, the gradual death of cells in the

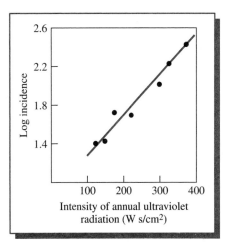

FIGURE 1-7 Incidence (logarithmic scale) for nonmelanoma skin cancer per 100,000 males versus annual UV light intensity, using data from various countries. [Source: Redrawn from D. Gordon and H. Silverstone, in R. Andrade et al., *Cancer of the Skin* (Philadelphia: W. B. Saunders, 1976), pp. 405–434.]

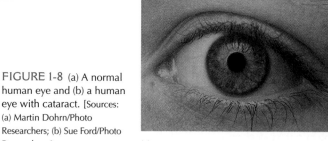 

(a)   (b)

FIGURE 1-8 (a) A normal human eye and (b) a human eye with cataract. [Sources: (a) Martin Dohrn/Photo Researchers; (b) Sue Ford/Photo Researchers.]

central part of the retina. Increased UV-B exposure also leads to a suppression of the human immune system, probably with a resulting increase in the incidence of infectious diseases, although this has not yet been extensively researched.

However, sunlight does have some positive effects on human health. Vitamin D, which is synthesized from precursor chemicals by the absorption of UV by the skin, is an anticancer agent. Recent research has established that sunlight during the winter is too weak a source of vitamin D synthesis for people living in mid- to high latitudes and that supplementary sources of the vitamin may be advisable. Insufficient vitamin D can reduce the rate of bone regeneration—since the vitamin is required for calcium utilization by the body—and thereby lead to increased fragility among middle-aged and elderly adults. Some controversial research indicates that moderate exposure to the Sun can reduce the incidence of multiple sclerosis and several types of cancer.

Humans are not the only organisms affected by ultraviolet light. It is speculated that increases in UV-B exposure can interfere with the efficiency of photosynthesis, and plants may respond by producing less leaf, seed, and fruit. All organisms that live in the first five meters or so below the surface in bodies of clear water would also experience increased UV-B exposure arising from ozone depletion and may be at risk. It is feared that production of the microscopic plants called phytoplankton near the surface of seawater may be at significant risk from increased UV-B; this would affect the marine food chain for which phytoplankton forms the base. Experiments indicate that there is a complex interrelationship between plant production and UV-B intensity, since the latter also affects the survival of insects that feed off the plants.

Variation in Light's Energy with Wavelength

As Albert Einstein realized, light can not only be considered a wave phenomenon but also to have particle-like properties in that it is absorbed (or emitted) by matter only in finite packets, now called **photons.** The quantity of energy, E, associated with each photon is related to the frequency, ν, and the wavelength, λ, of the light by the formulas

$$E = h\nu \quad \text{or} \quad E = hc/\lambda \quad \text{since} \quad \lambda\nu = c$$

Here h is Planck's constant (6.626218×10^{-34} J s) and c is the speed of light (2.997925×10^8 m s^{-1}). From the equation, it follows that *the shorter the wavelength of the light, the greater the energy it transfers to matter when absorbed.* Ultraviolet light is high in energy content, visible light is of intermediate energy, and infrared light is low in energy. Furthermore, UV-C is higher in energy than UV-B, which in turn is more energetic than is UV-A.

For convenience, the product hc in the equation above can be evaluated on a molar basis to yield a simple formula relating the energy absorbed by 1 mole of matter when each molecule in it absorbs one photon of a particular wavelength of light. If the wavelength is expressed in nanometers, the value of hc is 119,627 kJ mol^{-1} nm, so the equation becomes

$$E = 119{,}627/\lambda$$

where E is in kJ mol^{-1} if λ is in nm.

The photon energies for light in the UV and visible regions are of the same order of magnitude as the enthalpy (heat) changes, $\Delta H°$, of chemical reactions, including those in which atoms dissociate from molecules. For example, it is known that the dissociation of molecular oxygen into its monatomic form requires an enthalpy change of 498.4 kJ mol^{-1}:

$$O_2 \longrightarrow 2\,O \qquad \Delta H° = 498.4 \text{ kJ mol}^{-1}$$

In general, we can calculate enthalpy changes for any reaction by recalling from introductory chemistry that for any reaction, $\Delta H°$ equals the sum of the enthalpies of formation, $\Delta H_f°$, of the products minus those of the reactants:

$$\Delta H° = \Sigma\Delta H_f° \text{ (products)} - \Sigma\Delta H_f° \text{ (reactants)}$$

In the case of the reaction above,

$$\Delta H° = 2\,\Delta H_f° \text{ (O, g)} - \Delta H_f° \text{ (O}_2\text{, g)}$$

From data tables, we find that $\Delta H_f°$ (O, g) $= +249.2$ kJ/mol, and we know that $\Delta H_f°$ (O$_2$, g) $= 0$ since O$_2$ gas is the stablest form of the element. By substitution,

$$\Delta H° = 2 \times 249.2 - 0 = 498.4$$

To a good approximation, for a dissociation reaction, $\Delta H°$ is equal to the energy required to drive the reaction. Since all the energy has to be supplied by one photon per molecule (see below), the corresponding wavelength for the light is

$$\lambda = 119{,}627 \text{ kJ mol}^{-1} \text{ nm}/498.4 \text{ kJ mol}^{-1} = 240 \text{ nm}$$

Thus any O_2 molecule that absorbs a photon from light of wavelength 240 nm or shorter has sufficient excess energy to dissociate.

$$O_2 + \text{UV photon } (\lambda < 240 \text{ nm}) \longrightarrow 2\,O$$

If energy in the form of light initiates a reaction, it is called a **photochemical reaction.** The oxygen molecule in the above reaction is variously said to be *photochemically dissociated* or *photochemically decomposed* or to have undergone *photolysis.*

Atoms and molecules that absorb light (in the ultraviolet or visible region) immediately undergo a change in the organization of their electrons. They are said to exist temporarily in an electronically **excited state;** to denote this, their formulas are followed by a superscript asterisk (*). However, atoms and molecules generally do not remain in the excited state, and therefore do not retain the excess energy provided by the photon, for very long. Within a tiny fraction of a second, they must either use the energy to react photochemically or return to their **ground state**—the lowest energy (most stable) arrangement of the electrons. They quickly return to the ground state either by emitting a photon themselves or by converting the excess energy into heat that becomes shared among several neighboring free atoms or molecules as a result of collisions (i.e., molecules must "use it or lose it").

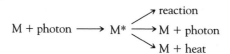

Consequently, molecules normally cannot accumulate energy from several photons until they receive sufficient energy to react; all the excess energy required to drive a reaction usually must come from a single photon. Therefore, light of 240 nm or less in wavelength can result in the dissociation of O_2 molecules, but light of longer wavelength does not contain enough energy to promote the reaction at all, even though certain wavelengths of such light can be absorbed by the molecule (see Figure 1-3). In the case of an O_2 molecule, the energy from a photon of wavelength greater than 240 nm can, if absorbed temporarily, raise the molecules to an excited state, but the energy is rapidly converted to an increase in the energy of its own motion and that of the molecules that surround it.

$$O_2 + \text{photon} (\lambda > 240 \text{ nm}) \longrightarrow O_2{}^* \longrightarrow O_2 + \text{heat}$$

$$O_2 + \text{photon} (\lambda < 240 \text{ nm}) \longrightarrow O_2{}^* \longrightarrow 2\,O \quad \text{or} \quad O_2 + \text{heat}$$

PROBLEM 1-1

What is the energy, in kilojoules per mole, associated with photons having the following wavelengths? What is the significance of each of these wavelengths? [Hint: See Figure 1-2.]

(a) 280 nm   (b) 400 nm   (c) 750 nm   (d) 4000 nm

PROBLEM 1-2

The $\Delta H°$ for the decomposition of ozone into O_2 and atomic oxygen is $+105 \text{ kJ mol}^{-1}$:

$$O_3 \longrightarrow O_2 + O$$

What is the longest wavelength of light that could dissociate ozone in this manner? By reference to Figure 1-2, decide the region of sunlight (UV, visible, or infrared) in which this wavelength falls.

PROBLEM 1-3

Using the enthalpy of formation information given below, calculate the maximum wavelength that can dissociate NO_2 to NO and atomic oxygen. Recalculate the wavelength if the reaction is to result in the complete dissociation into free atoms (i.e., $N + 2\,O$). Is light of these wavelengths available in sunlight?

$\Delta H_f°$ values (kJ mol^{-1}): NO_2: $+33.2$; NO: $+90.2$; N: $+472.7$; O: $+249.2$

Of course, in order that a sufficiently energetic photon supply the energy to drive a reaction, it must be absorbed by the molecule. As you can infer from the examples of the absorption spectra of O_2 and O_3 (Figures 1-3 and 1-4), there are many wavelength regions in which molecules simply do not absorb significant amounts of light. Thus, for example, because ozone molecules do not absorb visible light near 400 nm, shining light of this wavelength on them does not cause them to decompose, even though 400-nm photons carry sufficient energy to dissociate them to atomic and molecular oxygen (see Problem 1-2). Furthermore, as discussed above, the fact that molecules of a substance absorb photons of a certain wavelength and such photons are sufficiently energetic to drive a reaction does not mean that the reaction necessarily will occur; the photon energy can be diverted by a molecule into other processes undergone by the excited state. Thus the availability of light with sufficient photon energy is a necessary, but not a sufficient, condition for reaction to occur with any given molecule.

Creation of Ozone in the Stratosphere

In this section, the formation of ozone in the stratosphere and its destruction by noncatalytic processes are analyzed. As we shall see, the formation reaction generates sufficient heat to determine the temperature in this region of the atmosphere. *Above* the stratosphere, the air is very thin and the concentration of molecules is so low that most oxygen exists in atomic form, having been dissociated from O_2 molecules by UV-C photons from sunlight. The eventual collision of oxygen atoms with each other leads to the re-formation of O_2 molecules, which subsequently dissociate photochemically again when more sunlight is absorbed.

In the stratosphere itself, the intensity of the UV-C light is much lower because much of it is filtered by the diatomic oxygen that lies above. In addition, since the air is denser than it is higher up, the molecular oxygen concentration is much higher in the stratosphere. For this combination of reasons, most stratospheric oxygen exists as O_2 rather than as atomic oxygen. Because the concentration of O_2 molecules is relatively large and the concentration of atomic oxygen is so small, the most likely fate of the stratospheric oxygen atoms that are created by the photochemical decomposition of O_2 is *not* their mutual collision to re-form O_2 molecules. Rather, they are more likely at such altitudes to collide and react with undissociated, intact diatomic oxygen molecules, an event that results in the production of ozone:

$$O + O_2 \longrightarrow O_3 + heat$$

Indeed, this reaction is the source of all the ozone in the stratosphere. During daylight hours, ozone is constantly being formed by this process, the rate of which depends upon the amount of UV light and consequently the concentration of oxygen atoms and molecules at a given altitude.

At the bottom of the stratosphere, the abundance of O_2 is much greater than that at the top because air density increases progressively as one approaches the surface. However, relatively little of the oxygen at this level is dissociated and thus little ozone is formed because almost all the high-energy UV has been filtered from sunlight before it descends to this altitude. For this reason, the ozone layer does not extend much below the stratosphere. Indeed, the ozone present in the lower stratosphere is largely formed at higher altitudes and over equatorial regions and is transported there. In contrast, at the top of the stratosphere, the UV-C intensity is greater, but the air is thin and therefore relatively little ozone is produced since the oxygen atoms collide and react with each other rather than with the small number of intact O_2 molecules. Consequently, the production of ozone reaches a maximum where the product of UV-C intensity and O_2 concentration is maximum. The maximum density of ozone occurs lower—at about 25 km over tropical areas, 21 km over mid-latitudes, and 18 km over subarctic regions—since much of it transported downward after its production. Collectively, most of the ozone is located in the region between 15 and 35 km, i.e., the lower and middle stratosphere, known informally as the **ozone layer** (see Figure 1-1a).

A third molecule, which we will designate as M, such as N_2 or H_2O or even another O_2 molecule, is required to carry away the heat energy generated in the collision between atomic oxygen and O_2 that produces ozone. Thus the reaction above is written more realistically as

$$O + O_2 + M \longrightarrow O_3 + M + \text{heat}$$

The release of heat by this reaction results in the temperature of the stratosphere as a whole being *higher* than the air that lies below or above it, as indicated in Figure 1-1b .

Notice from Figure 1-1b that within the stratosphere, the air at a given altitude is cooler than that which lies above it. The general name for this phenomenon is a **temperature inversion.** Because cool air is denser than hot air (ideal gas law), it does not rise spontaneously, due to the force of gravity; consequently, vertical mixing of air in the stratosphere is a very slow process compared to mixing in the troposphere. The air in this region therefore is *strati*fied—hence the name *strato*sphere.

In contrast to the stratosphere, there is extensive vertical mixing of air within the troposphere. The Sun heats the ground, and hence the air in contact with it, much more than it does the air a few kilometers higher. It is for this reason that the air temperature falls with increasing altitude in the troposphere; the rate of decline of temperature with height is called the *lapse rate*. The less dense, hotter air rises from the surface and gives rise to extensive vertical exchange of air within the troposphere.

PROBLEM 1-4

Given that the total concentration of molecules in air decreases with increasing altitude, would you expect the *relative* concentration of ozone, on the ppb scale, to peak at a higher or a lower altitude or the same altitude compared to the peak for the absolute concentration of the gas?

Destruction of Stratospheric Ozone

The results for Problem 1-2 show that photons of light in the visible range and even in portions of the infrared range of sunlight possess sufficient energy to split an oxygen atom from a molecule of O_3. However, such photons are not efficiently absorbed by ozone molecules; consequently, their dissociation by such light is not important, except in the lower stratosphere where little UV penetrates. As we have seen previously, ozone does efficiently absorb UV light with wavelengths shorter than 320 nm, and the excited state thereby produced does undergo a dissociation reaction. Thus absorption of a UV-C or UV-B photon by an ozone molecule in the stratosphere results in the decomposition of that molecule. This photochemical reaction accounts for much of the ozone destruction in the middle and upper stratosphere:

$$O_3 + \text{UV photon } (\lambda < 320 \text{ nm}) \longrightarrow O_2{}^* + O^*$$

The oxygen atoms produced in the reaction of ozone with UV light have an electron configuration that differs from the configuration that has the lowest energy, and they therefore exist in an electronically excited state; the oxygen molecules also are produced in an excited state.

PROBLEM 1-5

By reference to the information in Problem 1-2, calculate the longest wavelength of light that decomposes ozone to O^* and $O_2{}^*$, given the following thermochemical data:

$$O \longrightarrow O^* \quad \Delta H° = 190 \text{ kJ mol}^{-1}$$
$$O_2 \longrightarrow O_2{}^* \quad \Delta H° = 95 \text{ kJ mol}^{-1}$$

[Hints: Express the overall reaction of O_3 decomposition as a sum of simpler reactions for which $\Delta H°$ values are available, and combine their $\Delta H°$ values according to Hess' law, which states that $\Delta H°$ for an overall reaction is the sum of the $\Delta H°$ values for the simpler reactions that are added together.]

Most oxygen atoms produced in the stratosphere by photochemical decomposition of ozone or of O_2 subsequently react with intact O_2 molecules to re-form ozone. However, some of the oxygen atoms react instead with intact ozone molecules and in the process destroy them, since they are converted to O_2:

$$O_3 + O \longrightarrow 2 O_2$$

In effect, the unbonded oxygen atom extracts one oxygen atom from the ozone molecule. This reaction is inherently inefficient since, although it is an exothermic reaction, its activation energy is 17 kJ mol^{-1}, a sizable one for atmospheric reactions to overcome. Consequently, few collisions between O_3 and O occur with sufficient energy to result in reaction.

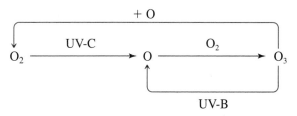

FIGURE 1-9 The Chapman mechanism.

To summarize the processes, ozone in the stratosphere is constantly being formed, decomposed, and re-formed during daylight hours by a series of reactions that proceed simultaneously, though at very different rates depending upon altitude. Ozone is produced in the stratosphere because there is adequate UV-C from sunlight to dissociate some O_2 molecules and thereby produce oxygen atoms, most of which collide with other O_2 molecules and form ozone. The ozone gas filters UV-B and UV-C from sunlight but is destroyed temporarily by this process or by reaction with oxygen atoms. The average lifetime of an ozone molecule at an altitude of 30 kilometers is about half an hour; in the lower stratosphere, it lasts for months.

Ozone is not formed below the stratosphere due to a lack of the UV-C required to produce the O atoms necessary to form O_3, because this type of sunlight has been absorbed by O_2 and O_3 in the stratosphere. Above the stratosphere, oxygen atoms predominate and usually collide with other O atoms to eventually re-form O_2 molecules.

The ozone production and destruction processes discussed above constitute the so-called **Chapman mechanism** (or cycle), shown in Figure 1-9. Recall that the series of simple reaction steps that document how an overall chemical process, such as ozone production and destruction, occurs at the molecular level is called a **reaction mechanism.**

Even in the ozone-layer portion of the stratosphere, O_3 is not the gas of greatest abundance or even the dominant oxygen-containing species; its relative concentration never exceeds 10 ppm. Thus the term *ozone layer* is something of a misnomer. Nevertheless, this tiny concentration of ozone is sufficient to filter all the remaining UV-C and much of the UV-B from sunlight before it reaches the lower atmosphere. Perhaps the alternative name *ozone screen* is more appropriate than ozone layer.

As in the case of stratospheric ozone, it is not uncommon to find that the concentration of a substance, natural or synthetic, in some compartment of the environment or in an organism does not change much with time. This does not necessarily mean that there are no inputs or outputs of the substance. More often, the concentration does not vary much with time because the input rate and the rate at which the substance decays or is eliminated from some compartment in the environment have become equal: we say that the substance has achieved a **steady state.** Equilibrium is a special case of the steady state; it arises when the decay process is the exact opposite of the input. Box 1-1 explores the mathematical implications of the steady state in common situations involving reactive substances.

BOX 1-1 | The Steady-State Analysis of Atmospheric Reactions

The Steady-State Approximation

If we know the nature of the creation and destruction reaction steps for a reactive substance, we can often algebraically derive a useful equation for its steady-state concentration.

As a simple example, consider the formation and destruction of oxygen atoms *above* the stratosphere. As mentioned previously, the atoms are formed by the photochemical dissociation of molecules of diatomic oxygen:

$$O_2 \longrightarrow 2\,O \qquad (i)$$

The atoms re-form diatomic oxygen when two of them collide simultaneously with a third molecule, M, which can carry away most of the energy released by the newly formed O_2 molecule:

$$O + O + M \longrightarrow O_2 + M \qquad (ii)$$

Recall from introductory chemistry that the rates of the individual steps in reaction mechanisms can be calculated from the concentrations of the reactants and from the **rate constant**, k, for the step. Thus the rate of reaction (i) equals $k_i[O_2]$. The rate constant k_i here incorporates the intensity of the light impinging upon the molecular oxygen. Since two O atoms are formed for each O_2 molecule that dissociates,

rate of formation of O atoms $= 2\,k_i[O_2]$

The rate of destruction of oxygen atoms by reaction (ii) is

rate of destruction of O atoms $= 2\,k_{ii}[O]^2[M]$

where we square the oxygen atom concentration because two of them are involved as reactants in the step.

The net rate of change of O atom concentration with time equals the rate of its formation minus the rate of its destruction:

rate of change of $[O] = 2\,k_i[O_2] - 2\,k_{ii}[O]^2[M]$

When atomic oxygen is at a steady state, this net rate must be zero, and thus the right-hand side of the equation above must also be zero. As a consequence, it follows that

$$k_{ii}[O]^2[M] = k_i[O_2]$$

By rearrangement of this equation, we obtain a relationship between the steady-state concentrations of O and of O_2:

$$[O]_{ss}^{\,2}/[O_2]_{ss} = k_i/(k_{ii}[M])$$

We see now why the ratio of oxygen atoms to diatomic molecules increases as we go higher and higher above the stratosphere: it is because the air pressure drops, and therefore so does [M], so the O_2 re-formation rate decreases.

Steady-State Analysis of the Chapman Mechanism

After this introduction, we now are ready to apply the steady-state analysis to the Chapman mechanism. The four reactions of concern are shown again below. Notice that the recombination of O atoms, i.e., reaction (ii) above, is not included because its rate in the mid- and low stratosphere is not competitive with other reactions, since the oxygen atom concentration is small there.

$$O_2 \longrightarrow 2\,O \qquad (1)$$

$$O + O_2 + M \longrightarrow O_3 + M \qquad (2)$$

$$O_3 \longrightarrow O_2 + O \qquad (3)$$

$$O_3 + O \longrightarrow 2\,O_2 \qquad (4)$$

Noting that O is produced or consumed in all four reactions, we obtain four terms in its overall rate expression and assume it is in a steady state:

$$\text{rate of change of } [O] = 2\,\text{rate}_1 - \text{rate}_2$$
$$+ \text{rate}_3 - \text{rate}_4$$
$$= 0 \qquad (A)$$

Other useful information about concentrations can be obtained by considering the steady-state expression for the ozone concentration:

$$\text{rate of change of } [O_3] = \text{rate}_2 - \text{rate}_3$$
$$- \text{rate}_4$$
$$= 0 \qquad (B)$$

If we add together the expressions for the rates of change in [O] and in $[O_3]$, i.e., equations (A) and (B) above, we find that the rates for reactions 2 and 3 cancel, and we obtain

$$2\,\text{rate}_1 - 2\,\text{rate}_4 = 0$$

Using the expressions for these two rates in terms of reactant concentrations, we find

$$2\,k_1[O_2] - 2\,k_4[O_3][O] = 0$$

or

$$[O_3][O] = k_1[O_2]/k_4 \qquad (C)$$

Another useful expression can be obtained by subtracting equation (B) from (A). We obtain

$$2\,\text{rate}_1 - 2\,\text{rate}_2 + 2\,\text{rate}_3 = 0$$

which by rearrangement and cancellation becomes

$$\text{rate}_3 = \text{rate}_2 - \text{rate}_1$$

It is known from experiment that rate_2 (and rate_3) is much larger than rate_1, so the latter can be neglected here, giving simply

$$\text{rate}_3 = \text{rate}_2$$

Using the expressions for these two reaction rates in terms of the concentrations of their reactants,

$$k_3[O_3] = k_2[O][O_2][M]$$

Rearranging this equation, we can solve for the ratio of ozone to atomic oxygen:

$$[O_3]/[O] = k_2[O_2][M]/k_3 \qquad (D)$$

Equations (C) and (D) give us two equations in the two unknowns, [O] and $[O_3]$. Multiplying their left sides together and equating the result to the product of their right sides eliminates [O] and leaves us with an equation for the ozone concentration:

$$[O_3]^2 = [O_2]^2\,[M]\,k_1 k_2/k_3 k_4$$

or, taking the square root of both sides, we obtain an expression for the steady-state concentration of ozone in terms of the diatomic oxygen concentration:

$$[O_3]_{ss}/[O_2]_{ss} = [M]^{0.5}\,(k_1 k_2/k_3 k_4)^{0.5} \qquad (E)$$

Thus the steady-state ratio of ozone to diatomic oxygen depends on the square root of the air density through [M]. The ratio is also proportional to the square root of the product of the rate constants for the reactions, 1 and 2, in which atomic oxygen and then ozone are produced, and inversely proportional to the square root of the product of the ozone destruction reaction rate constants. Substitution of

(continued on p. 46)

BOX 1-1	The Steady-State Analysis of Atmospheric Reactions *(continued)*

numerical values for the rate constants k and for [M] into equation (E) predicts the correct order of magnitude for the ozone/diatomic oxygen ratio, i.e., about 10^{-4} in the mid-stratosphere. Ozone is never the main oxygen-containing species in the atmosphere, not even in "the ozone layer."

Equation (E) predicts that the concentration of ozone relative to that of diatomic oxygen should fall slowly as we climb in the atmosphere, given that it is proportional to the square root of the air density, through the [M] dependence. This occurs because the formation reaction of ozone, through step 2, will slow down as [M] declines. This decline with increasing altitude is observed in the upper stratosphere and above. Below about 35 km, however, the more important change in the terms of equation (E) involves k_1; consequently, the $[O_3]/[O_2]$ ratio is not simply proportional to $[M]^{0.5}$.

The rate constant k_1 incorporates the intensity of sunlight capable of dissociating diatomic oxygen into its atoms. Since the UV-C sunlight required ($\lambda < 242$ nm) is successively filtered by absorption as the light beam descends toward the Earth's surface, the value of k_1 declines especially rapidly in the low stratosphere and below. Thus the concentration of ozone predicted by applying the steady-state analysis to the Chapman mechanism successfully predicts that the ozone concentration will peak in the stratosphere. However, as discussed above, the actual peak of ozone concentration ($\sim$25 km, above the Equator) occurs rather lower in the stratosphere than the altitude of maximum production ($\sim$40 km) because horizontal air movement transports ozone downward.

Substitution of equation (E) into (C) allows us to deduce an expression for the steady-state concentration of free oxygen atoms:

$$[O]_{ss} = (k_1 k_3/k_2 k_4)^{0.5}/[M]^{0.5}$$

Thus the concentration of atomic oxygen is predicted to increase with altitude as [M] declines—as in our previous analysis for the upper atmosphere—and as k_1 and k_3 increase, since UV light intensity increases with increasing altitude. Indeed, atomic oxygen dominates over ozone at high altitudes, whereas below about 50 km, ozone is always dominant.

The production of ozone through reaction (2) is critically dependent upon the supply of free oxygen atoms in reaction (1). The rate of oxygen atom production, in turn, is highly dependent upon the intensity of UV-C sunlight. As we have noted, this intensity falls sharply as we descend through the stratosphere. The UV-C light intensity also depends strongly upon latitude, being strongest over the Equator and declining continuously toward the poles. Thus ozone production is greatest over the Equator.

The qualitative behavior of the predicted variation of ozone concentration with altitude predicted by equation (E) is correct, but the predicted amounts of ozone exceed the observed—by about a factor of 2 near the peak concentration. Scientists eventually found that they had underestimated the rate of the ozone destruction reaction (4) by about a factor of 4, since there are catalysts in the stratosphere that greatly speed up the overall reaction. These reactions are discussed in the next section.

PROBLEM 1

Consider the following 3-step mechanism for the production and destruction of excited oxygen atoms, O^*, in the atmosphere:

$$O_2 \xrightarrow{\text{light}} O + O^*$$

$$O^* + M \longrightarrow O + M$$

$$O^* + H_2O \longrightarrow 2\,OH$$

Develop an expression for the steady-state concentration of O^* in terms of the concentrations of the other chemicals involved.

PROBLEM 2

Perform a steady-state analysis for $d[Cl]/dt$ and for $d[ClO]/dt$ in the following mechanism:

$$Cl_2 \longrightarrow 2\,Cl \qquad (1)$$

$$Cl + O_3 \longrightarrow ClO + O_2 \qquad (2)$$

$$2\,ClO \longrightarrow 2\,Cl + O_2 \qquad (3)$$

$$ClO + NO_2 \longrightarrow ClONO_2 \qquad (4)$$

Obtain expressions for the steady-state concentrations of Cl and ClO and hence for the rate of destruction of ozone.

Catalytic Processes of Ozone Destruction

In the early 1960s it was realized that there are mechanisms for ozone destruction in the stratosphere in addition to the processes described in the Chapman mechanism. These additional processes all involve catalysts present in air. In the material that follows, we investigate two general reaction mechanisms by which stratospheric ozone is catalytically destroyed, paying particular attention to the role of chlorine and bromine.

There exist a number of atomic and molecular species, designated in general as X, that react efficiently with ozone by abstracting (removing) an oxygen atom from it:

$$X + O_3 \longrightarrow XO + O_2$$

In those regions of the stratosphere where the atomic oxygen concentration is appreciable, the XO molecules react subsequently with oxygen atoms to produce O_2 and to re-form X:

$$XO + O \longrightarrow X + O_2$$

The **overall reaction** corresponding to this reaction mechanism is obtained by algebraically summing the successive steps that occur in air over and over again an equal number of times. In the case of the additional steps of the mechanism, the reactants in the two steps are added together and become the reactants of the overall reaction, and similarly for the products:

$$X + O_3 + XO + O \longrightarrow XO + O_2 + X + O_2$$

Molecules that are common to both sides of the reaction equation, in this case X and XO, are then canceled, and common terms collected, yielding the balanced overall reaction:

$$O_3 + O \longrightarrow 2\,O_2 \quad \text{overall reaction}$$

Thus the species X are **catalysts** for ozone destruction in the stratosphere since they speed up a reaction (here, between O_3 and O), but they are eventually re-formed intact and are able to begin the cycle again—with, in this case, the destruction of further ozone molecules.

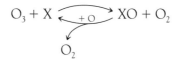

As previously discussed (Chapman cycle), the above overall reaction can occur as a simple collision between an ozone molecule and an oxygen atom even in the absence of a catalyst, but almost all such direct collisions are ineffective in producing a reaction. The X catalysts greatly increase the efficiency of this reaction, i.e., they effectively increase the value of its rate constant and thereby decrease the steady-state concentration of ozone. All the environmental concerns about ozone depletion arise from the fact that we are inadvertently increasing the stratospheric concentrations of several X catalysts by the release at ground levels of certain gases, especially those containing chlorine and bromine. Such an increase in the catalyst concentration leads to a reduction in the concentration of ozone in the stratosphere by the mechanism shown above and by one discussed later.

Most ozone destruction by the catalytic mechanism (i.e., the combination of sequential steps) described above, hereafter designated **Mechanism I,** occurs in the middle and upper stratosphere, where the ozone concentration is low to start with. Chemically, all the catalysts X are **free radicals,** which are atoms or molecules containing an odd number of electrons. As a consequence of the odd number, one electron is not paired with one of opposite spin character (as occurs for all the electrons in almost all stable molecules). Free radicals are usually very reactive, since there is a driving force for their unpaired electron to pair with one of the opposite spin, even if it is located in a different molecule. The determination of the appropriate bonding structure for simple free radicals is described in Chapter 5.

An analysis of which free-radical reactions are and are not feasible in air is given in Box 1-2.

BOX 1-2 | The Rates of Free-Radical Reactions

The rate of a given chemical reaction is affected by a number of parameters, most notably the magnitude of the activation energy required before the reaction can occur. Thus reactions with appreciable activation energies are inherently very slow processes and can often be ignored compared to alternative, faster processes for the chemicals involved. In gas-phase reactions involving simple free radicals as reactants, the activation energy exceeds that imposed by their endothermicity by only a small amount. Thus we can assume, conversely, that all exothermic free-radical reactions will have only a small activation energy (Figure 1a). Therefore, exothermic free-radical reactions usually are fast (providing, of course, the reactants exist in reasonable concentrations in the atmosphere). An example of an exothermic radical reaction with a small energy barrier is

$$Cl + O_3 \longrightarrow ClO + O_2$$

The activation energy here is only 2 kJ/mol.

Reactions involving the combining of two free radicals generally are exothermic, since a new bond is formed, so they too proceed quickly with little activation energy, provided that the radical concentrations are high enough that the reactants do in fact collide with each other at a fast rate.

In contrast, endothermic reactions in the atmosphere will be much slower since the activation barrier must of necessity be much larger (see Figure 1b). At atmospheric temperatures, few if any collisions between the molecules would have energy sufficient to overcome this

(continued on p. 50)

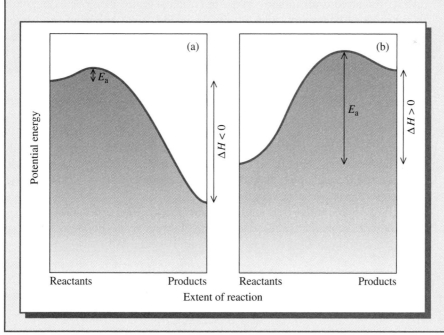

FIGURE 1 Potential energy profiles for typical atmospheric free-radical reactions, showing (a) exothermic and (b) endothermic patterns.

| BOX 1-2 | The Rates of Free-Radical Reactions *(continued)* |

large barrier and allow reaction to occur. An example is the endothermic reaction:

$$OH + HF \longrightarrow H_2O + F$$

Its activation energy must be at least equal to its $\Delta H° = +69$ kJ/mol, and consequently the reaction would be so very slow at stratospheric temperatures that we can ignore it completely.

PROBLEM 1

Draw an energy profile diagram, i.e., one similar to Figure 1b, for the abstraction from water of a hydrogen atom by ground-state atomic oxygen, given that the reaction is endothermic by about 69 kJ/mol. On the same diagram, show the energy profile for the reaction of O* with H_2O to give the same products, given that O* lies above ground-state atomic oxygen (O) by 190 kJ/mol. From these curves, predict why abstraction by O* occurs quickly but that by O is extremely slow in the atmosphere.

Catalytic Destruction of Ozone by Nitric Oxide

The catalytic destruction of ozone occurs even in a "clean" atmosphere (one unpolluted by artificial contaminants) since small amounts of the X catalysts have always been present in the stratosphere. One important "natural" version of X—i.e., one of the species responsible for catalytic ozone destruction in a nonpolluted stratosphere—is the free-radical molecule **nitric oxide,** NO. It is produced when molecules of **nitrous oxide,** N_2O, rise from the troposphere to the stratosphere, where they may eventually collide with an excited oxygen atom produced by photochemical decomposition of ozone. Most of these collisions will yield $N_2 + O_2$ as products, but a few of them result in the production of nitric oxide:

$$N_2O + O^* \longrightarrow 2\,NO$$

We can ignore the possibility that NO produced in the troposphere will migrate to the stratosphere; as explained in Chapter 3, the gas is efficiently oxidized to nitric acid, which is then readily washed out of the tropospheric air, before this process can occur.

The NO molecules that are the products of the above reaction catalytically destroy ozone by extracting an oxygen atom from ozone and forming **nitrogen dioxide,** NO_2; i.e., they act as X in Mechanism I:

$$NO + O_3 \longrightarrow NO_2 + O_2$$
$$NO_2 + O \longrightarrow NO + O_2$$
$$\text{overall}\quad O_3 + O \longrightarrow 2\,O_2$$

PROBLEM 1-6

Not all XO molecules such as NO_2 survive long enough to react with oxygen atoms; some are photochemically decomposed to X and atomic oxygen, which then reacts with O_2 to re-form ozone. Write out the three steps (including one for ozone destruction) for this process and add them together to deduce the net reaction. Does this sequence destroy ozone overall, or is it a *null cycle*, which is defined as one that involves a sequence of steps with no chemical change overall?

Another important X catalyst in the stratosphere is the **hydroxyl free radical,** OH. It originates from the reaction of excited oxygen atoms O* with water or *methane*, CH_4, molecules:

$$O^* + CH_4 \longrightarrow OH + CH_3$$

The methane originates with emissions from the Earth's surface, a small fraction of which survive sufficiently long to migrate up to the stratosphere.

PROBLEM 1-7

Write out the two-step mechanism by which the hydroxyl free radical catalytically destroys ozone by Mechanism I. By adding the steps together, deduce the overall reaction.

PROBLEM 1-8

By analogy with its reaction with methane, write a balanced equation for the reaction by which O* produces OH from water vapor.

Catalytic Destruction of Ozone Without Atomic Oxygen: Mechanism II

A factor that minimizes the catalyzed gas-phase destruction of ozone by Mechanism I is the requirement for atomic oxygen to complete the cycle by reacting with XO in order to permit the regeneration of the X catalyst in a usable form.

$$XO + O \longrightarrow X + O_2$$

As discussed above, the concentration of oxygen atoms is very low in the lower stratosphere (15–25 km altitude), so the gas-phase destruction of ozone by reactions that require atomic oxygen is sluggish there.

There is another general catalytic sequence, henceforth designated **Mechanism II,** that depletes ozone in the lower stratosphere, particularly when the concentrations of the catalysts X are relatively high. First, two ozone molecules are destroyed by the same catalysts discussed previously and by the same initial reaction:

$$X + O_3 \longrightarrow XO + O_2$$
$$X' + O_3 \longrightarrow X'O + O_2$$

We have used X' to symbolize the catalyst in the second equation to indicate that it need not be chemically identical to the X in the first equation. Either X or X' must be a chlorine atom.

In the steps that follow the first, the two molecules XO and $X'O$ that have added an oxygen atom react with each other. As a consequence, the catalysts X and X' are ultimately regenerated, usually after the combined but unstable molecule $XOOX'$ has formed and been decomposed by either heat or light:

$$XO + X'O \longrightarrow [XOOX'] \longrightarrow X + X' + O_2$$

(By convention in chemistry, a species shown in square brackets is one with a transient existence.) When we sum these steps, the overall reaction is seen to be

$$2\,O_3 \longrightarrow 3\,O_2$$

We shall see several examples of catalytic Mechanism II in operation in the ozone holes (Chapter 2) and in the mid-latitude lower stratosphere. Indeed, most ozone loss in the lower stratosphere occurs according to this net reaction. Mechanisms I and II are summarized in Figure 1-10.

Finally, we note that while the rate of production of ozone from oxygen depends only upon the concentrations of O_2 and O_3 and of UV light at a given altitude, what determines the rate of ozone destruction is somewhat more complex. The rate of ozone decomposition by UV-B or by catalysts depends upon ozone's concentration multiplied by either the sunlight intensity or the catalyst concentration, respectively. In general, the concentration of ozone will rise until the net rate of destruction just meets the rate of production, and then it will remain constant at this steady-state level as long as the intensity of sunlight remains the same. If, however, the rate of destruction is temporarily increased by the introduction of additional molecules of a

FIGURE 1-10 Summary of catalytic ozone destruction by Mechanisms I and II.

<u>Mechanism I</u>

$$X + O_3 \rightarrow XO + O_2$$
$$XO + O \rightarrow X + O_2$$
$$\overline{O_3 + O \rightarrow 2\,O_2} \quad \text{overall}$$

<u>Mechanism II</u>

$$X + O_3 \rightarrow XO + O_2$$
$$X' + O_3 \rightarrow X'O + O_2$$
$$\overline{XO + X'O \rightarrow \rightarrow X + X' + O_2}$$
$$2\,O_3 \rightarrow 3\,O_2 \quad \text{overall}$$

catalyst, the steady-state concentration of ozone must then decrease to a new, lower value at which the rates of formation and destruction are again equal. However, it should be clear from the discussion above that due to its constant re-formation reactions, atmospheric ozone cannot be permanently and totally destroyed, no matter how great the level of catalyst. It should also be realized that any decrease in the concentration of ozone at higher altitudes allows more UV penetration to lower altitudes, which produces more ozone there; thus there is some "self-healing" of total ozone loss.

Atomic Chlorine and Bromine as X Catalysts

The decomposition of synthetic chlorine-containing gases in the stratosphere over the last few decades has generated a substantial amount of **atomic chlorine,** Cl, in this region. As the stratospheric chlorine concentration increases, so does the potential for ozone destruction, since the free radical Cl is an efficient X catalyst.

However, synthetic gases are not the only suppliers of chlorine to the ozone layer. There always has been some chlorine in the stratosphere as a result of the slow upward migration of the **methyl chloride** gas, CH_3Cl (also called *chloromethane*), produced at the Earth's surface—mainly in the oceans as a result of the interaction of chloride ion with decaying vegetation. Recently another large source of methyl chloride, from tropical plants, has been discovered; this may be the missing source of the compound for which scientists have been searching.

Only a portion of the methyl chloride molecules are destroyed in the troposphere. When intact molecules of it reach the stratosphere, they are photochemically decomposed by UV-C or attacked by OH radicals. In either case, atomic chlorine, Cl, is eventually produced:

$$CH_3Cl \xrightarrow{\text{UV-C}} Cl + CH_3$$

or

$$OH + CH_3Cl \longrightarrow Cl + \text{other products}$$

Chlorine atoms are efficient X catalysts for ozone destruction by Mechanism I:

$$
\begin{aligned}
Cl + O_3 &\longrightarrow ClO + O_2 \\
\underline{ClO + O} &\underline{\longrightarrow Cl + O_2} \\
\text{overall} \quad O_3 + O &\longrightarrow 2\,O_2
\end{aligned}
$$

Each chlorine atom can catalytically destroy many tens of thousands of ozone molecules in this manner. At any given time, however, the great majority of stratospheric chlorine normally exists not as Cl nor as the free radical **chlorine monoxide,** ClO, but in a form that is not a free radical and that is inactive as a catalyst for ozone destruction. The two main **catalytically inactive** (or *reservoir*) molecules containing chlorine in the stratosphere are **hydrogen chloride** gas, HCl, and **chlorine nitrate** gas, $ClONO_2$.

The chlorine nitrate is formed by the combination of chlorine monoxide and nitrogen dioxide; after a few days or hours, a given $ClONO_2$ molecule is photochemically decomposed back to its components, and thus the catalytically active ClO is re-formed.

$$ClO + NO_2 \underset{\text{sunlight}}{\rightleftharpoons} ClONO_2$$

However, under normal circumstances, more chlorine exists at steady state as $ClONO_2$ than as ClO. (Processes similar to the reaction above occur for several other constituents of the stratosphere; as we shall see at the end of Chapter 5, the reactions are easily systematized, thereby greatly reducing the number of processes that have to be learned.)

The other catalytically inactive form of chlorine, HCl, is formed when atomic chlorine abstracts a hydrogen atom from a molecule of stratospheric methane:

$$Cl + CH_4 \longrightarrow HCl + CH_3$$

This reaction is slightly endothermic, so its activation energy is nonzero, and it therefore proceeds at a slow but significant rate (see Box 1-2). The *methyl free radical*, CH_3, does not operate like the X catalysts since it combines with an oxygen molecule and is degraded eventually to carbon dioxide by reactions discussed in Chapter 5. Eventually, each HCl molecule is reconverted to the active form, i.e., chlorine atoms, by reaction with the hydroxyl radical:

$$OH + HCl \longrightarrow H_2O + Cl$$

Again, usually much more chlorine exists as HCl rather than as atomic chlorine at any given time under normal steady-state conditions.

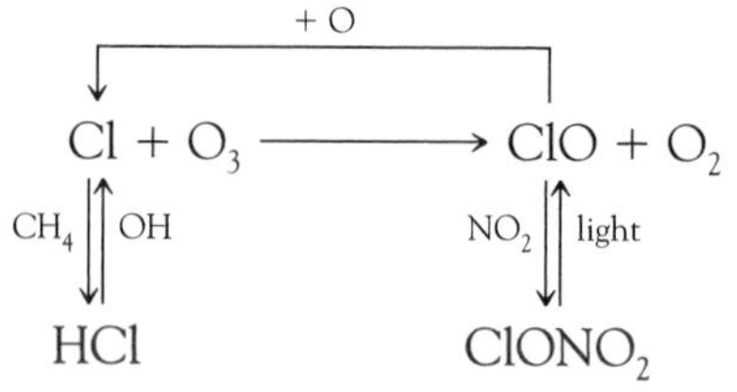

$$
\begin{array}{ccc}
& \xrightarrow{\quad + O \quad} & \\
\downarrow & & | \\
Cl + O_3 & \longrightarrow & ClO + O_2 \\
\;_{CH_4}\big\Vert^{OH} & & \;_{NO_2}\big\Vert^{light} \\
HCl & & ClONO_2
\end{array}
$$

When the first predictions concerning stratospheric ozone depletion were made in the 1970s, it was not realized that about 99% of stratospheric chlorine usually is tied up in the inactive forms. When the existence of inactive chlorine was discovered in the early 1980s, the predicted amounts of stratospheric ozone loss in the future were lowered appreciably. As we shall see, however, there are conditions under which inactive chlorine can become temporarily activated and can massively destroy ozone, a discovery that was not made until the late 1980s.

Although there has always been some chlorine in the stratosphere due to the natural release of CH_3Cl from the surface, in recent decades it has been completely overshadowed by much larger amounts of chlorine produced from synthetic chlorine-containing gaseous compounds that are released into air during their production or use. Most of these substances are *chlorofluorocarbons* (CFCs); their nature, usage, and replacements will be discussed in Chapter 2.

As with methyl chloride, large quantities of **methyl bromide,** CH_3Br, are also produced naturally and some of it eventually reaches the stratosphere, where it is decomposed photochemically to yield atomic bromine. Like chlorine, bromine atoms can catalytically destroy ozone by Mechanism I:

$$Br + O_3 \longrightarrow BrO + O_2$$
$$BrO + O \longrightarrow Br + O_2$$

In contrast to chlorine, almost all the bromine in the stratosphere remains in the active free-radical forms Br and BrO, since the inactive forms, **hydrogen bromide,** HBr, and **bromine nitrate,** $BrONO_2$, are efficiently decomposed photochemically by sunlight. In addition, the formation of HBr from attack of atomic bromine on methane is a slower reaction than is the analogous process involving atomic chlorine because it is much more endothermic and therefore has a higher activation energy:

$$Br + CH_4 \longrightarrow HBr + CH_3$$

A lower percentage of stratospheric bromine exists in inactive form than does chlorine because of the slower speed of this reaction and because of the efficiency of the photochemical decomposition reactions. For that reason, stratospheric bromine is more efficient at destroying ozone than is chlorine (by a factor of 40 to 50), but there is much less of it in the stratosphere, so overall it is less important.

When molecules such as HCl and HBr eventually diffuse from the stratosphere back into the upper troposphere, they dissolve in water droplets and are subsequently carried to lower altitudes and then transported to the ground by rain. Thus, although the lifetime of chlorine and bromine in the stratosphere is long, it is not infinite and the catalysts are eventually removed. However, the average chlorine atom destroys about 10,000 molecules of ozone before it is removed!

Review Questions

The questions below, and the comparable ones in succeeding chapters, are designed to test your knowledge mainly of some of the *factual* material presented in the chapter.

The problems within the chapter, and the more elaborate ones given below as Additional Problems, are designed to test your problem-solving abilities.

1. Which three gases constitute most of Earth's atmosphere?

2. What range of altitudes constitutes the troposphere? The stratosphere?

3. What is the wavelength range for visible light? Does ultraviolet light have shorter or longer wavelengths than visible light?

4. Which atmospheric gas is primarily responsible for filtering sunlight in the 120–220-nm region? Which, if any, gas absorbs most of the Sun's rays in the 220–320-nm region? Which absorbs primarily in the 320–400-nm region?

5. What is the name given to the finite packets of light absorbed by matter?

6. What are the equations relating photon energy E to light's frequency ν and wavelength λ?

7. What is meant by the expression *photochemically dissociated* as applied to stratospheric O_2?

8. Write the equation for the chemical reaction by which ozone is formed in the stratosphere. What are the sources for the different forms of oxygen used here as reactants?

9. Write the two reactions that, in addition to the catalyzed reactions, contribute most significantly to ozone destruction in the stratosphere.

10. What is meant by the phrase *excited state* as applied to an atom or molecule? Symbolically, how is an excited state signified?

11. Explain why the phrase *ozone layer* is a misnomer.

12. Define the term *free radical*, and give two examples relevant to stratospheric chemistry.

13. What are the two steps, and the overall reaction, by which a species X, such as ClO, catalytically destroys ozone in the middle and upper stratosphere via Mechanism I?

14. What is meant by the term *steady state* as applied to the concentration of ozone in the stratosphere?

15. Explain why, atom for atom, stratospheric bromine destroys more ozone than does chlorine.

16. Explain why ozone destruction via the reaction of O_3 with atomic oxygen does not occur to a significant extent in the lower stratosphere.

Additional Problems

1. A possible additional mechanism that could exist for the creation of ozone in the high stratosphere begins with the creation of (vibrationally) excited O_2 and ground-state atomic oxygen from the absorption by ozone of photons with wavelengths less than 243 nm. The $O_2{}^*$ reacts with a ground-state O_2 molecule to produce ozone and another atom of oxygen. What is the net reaction from these two steps? What do you predict is the

fate of the two oxygen atoms, and what would be the overall reaction once this fate is included?

2. In the nonpolluted atmosphere, an important mechanism for ozone destruction in the lower stratosphere is:

$$OH + O_3 \longrightarrow HOO + O_2$$
$$HOO + O_3 \longrightarrow OH + 2\,O_2$$

Does this pair of steps correspond to Mechanism I? If not, what is the overall reaction?

3. A proposed mechanism for ozone destruction in the late spring over northern latitudes in the lower stratosphere begins with the photochemical decomposition of $ClONO_2$ to Cl and NO_3, followed by photochemical decomposition of the latter to NO and O_2. Deduce a catalytic ozone destruction cycle, requiring no atomic oxygen, that incorporates these reactions. What is the overall reaction?

4. Deduce possible reaction step(s), none of which involve photolysis, for Mechanism II that follow(s) the $X + O_3 \longrightarrow XO + O_2$ step, such that the sum of all the mechanism's steps does not destroy or create any ozone.

5. As will be discussed in Chapter 2, atomic chlorine is produced under ozone-hole conditions by the dissociation of diatomic chlorine, Cl_2. Given that diatomic chlorine gas is the stablest form of the element and that the ΔH_f° value for atomic chlorine is $+121.7$ kJ mol^{-1}, calculate the maximum wavelength of light that can dissociate diatomic chlorine into the monatomic form. Does such a wavelength correspond to light in the visible or the UV-A or the UV-B region?

6. Under conditions of low oxygen atom concentration, the radical HOO can react reversibly with NO_2 to produce a molecule of $HOONO_2$:

$$HOO + NO_2 \longrightarrow HOONO_2$$

(a) Deduce why the addition of nitrogen oxides to the lower stratosphere could lead to an *increase* in the steady-state ozone concentration as a consequence of this reaction.

(b) Deduce how the addition of nitrogen oxides to the middle and upper stratosphere could *decrease* the ozone concentration there as a consequence of other reactions.

(c) Given the information stated in parts (a) and (b), in what regions of the stratosphere should supersonic transport airplanes fly if they emit substantial amounts of nitrogen oxides in their exhaust?

7. At an altitude of about 35 kilometers, the average concentrations of O^* and of CH_4 are approximately 100 and 1×10^{11} molecules cm^{-3}, respectively; the rate constant k for the reaction between them is approximately 3×10^{-10} cm^3 molecules^{-1} s^{-1}. Calculate the rate of destruction of methane in molecules per second per cubic centimeter and in grams per year per cubic centimeter under these conditions. [*Hint: Recall that the rate law for a simple process is its rate constant k times the product of the concentrations of its reactant concentrations.*]

8. The rate constants for the reactions of atomic chlorine and of hydroxyl radical with ozone are given by 3×10^{-11} $e^{-250/T}$ and 2×10^{-12} $e^{-940/T}$, where T is the Kelvin temperature. Calculate the ratio of the rates of ozone destruction by these catalysts at 20 km, given that at this altitude the average concentration of OH is about 100 times that of Cl and that the temperature is about $-50°C$. Calculate the rate constant for ozone destruction by chlorine under conditions in the Antarctic ozone hole, when the temperature is about $-80°C$ and the concentration of atomic chlorine increases by a factor of 100 to about 4×10^5 molecules per cubic centimeter and that of O_3 is 2×10^{12} molecules/cm^3.

9. The Arrhenius equation (see your introductory chemistry textbook) relates reaction rates to temperature via the activation energy. Calculate the ratio of the rates at $-30°C$ (a typical stratospheric temperature) for two reactions having the same Arrhenius A factor and initial concentrations, one of which is endothermic and has an activation energy of 30 kJ mol^{-1} and the other which is exothermic with an activation energy of 3 kJ mol^{-1}. In energy units, $R = 8.3$ J K^{-1} mol^{-1}.

Further Readings

1. S. A. Montzka et al., "Present and future trends in the atmospheric burden of ozone-depleting halogens," *Nature* 398 (1999): 690–693.

2. R. McKenzie, B. Connor, and G. Bodeker, "Increased summertime UV Radiation in New Zealand in Response to Ozone Loss," *Science* 285 (1999): 1709–1711.

3. T. K. Tromp et al., "Potential Environmental Impact of a Hydrogen Economy on the Stratosphere," *Science* 300 (2003): 1740.

4. (a) S. Madronich and F. R. de Gruijl, "Skin cancer and UV radiation," *Nature* 366 (1993): 23. (b) J.-S. Taylor, "DNA, sunlight, and skin cancer," *Journal of Chemical Education* 67 (1990): 835–841.

5. C. Biever, "Bring me sunshine," *New Scientist* (9 August 2003): 30–33.

Websites of Interest

Log on to www.whfreeman.com/envchem4/ and click on Chapter 1.

CHAPTER 2

THE OZONE HOLES

In this chapter, the following introductory chemistry topics are used:

■ Kinetics: Mechanisms; catalysis; reaction order

Background from Chapter 1 used in this chapter:

■ Photochemical decomposition
■ Mechanism II
■ Free radicals

Introduction

In Chapter 1, the gas-phase chemistry of the unpolluted stratosphere was explored. Since the late 1970s, however, the normal functioning of the stratosphere's ozone screen—and the protection it provides us—has been periodically upset by anthropogenic chlorine-containing chemicals in the atmosphere. Most famously, these substances now cause an ozone hole to open each spring season above the South Pole. Ozone levels in the stratosphere over the North Pole as well, and to some extent even that over our heads, have also been depleted. In this chapter, the extent of these stratospheric ozone losses are documented, and the special chemical processes that produce such destruction are described. We also document how knowledge of this chemistry led to action by humankind to prevent even more drastic loss of ozone, which should eventually heal the stratosphere.

We begin by describing how the amount of overhead ozone is reported and the history of how the ozone hole over the Antarctic was first discovered.

Dobson Units for Overhead Ozone

Ozone, O_3, is a gas that is present in small concentrations throughout the atmosphere. The total amount of atmospheric ozone that lies over a given point on Earth is measured in terms of **Dobson units** (DU). One Dobson unit is equivalent to a 0.01-mm (0.001-cm) thickness of pure ozone at the density it would possess if it were brought to ground-level (1 atm) pressure and 0°C temperature.

On average, this total overhead ozone at temperate latitudes amounts to about 350 DU; thus if all the ozone were to be brought down to ground level, the layer of pure ozone would be only 3.5 mm thick. Because of stratospheric winds, ozone is transported from tropical regions, where most of it is produced, toward polar regions. Thus, ironically, the closer to the Equator you live, the smaller the total amount of ozone that protects you from ultraviolet light. Ozone concentrations in the tropics usually average 250 DU, whereas those in subpolar regions average 450 DU, except, of course, when holes appear in the ozone layer over such areas. There is natural seasonal variation of ozone concentration, with the highest levels in the early spring and the lowest in the fall.

The Annual Ozone Hole Above Antarctica

The Antarctic ozone hole was discovered by Dr. Joe C. Farman and his colleagues in the British Antarctic Survey. They had been recording ozone levels over this region since 1957. Their data indicated that the total amounts of ozone each October had been gradually falling each year, especially during the mid-September to mid-October period, with precipitous declines beginning in the late 1970s. This is illustrated in Figure 2-1b, where the average minimum daily amount of overhead ozone is plotted against the year. The period from September to November corresponds to the spring season at the South Pole and follows a period of very cold 24-hour nights common to polar winters. By the mid-1980s, the springtime loss in ozone at some altitudes over Antarctica was complete, resulting in a loss of more than 50% of the total overhead amount. It is therefore appropriate to speak of a "hole" in the ozone layer that now appears each spring over the Antarctic and lasts for several months. The average geographic area covered by the ozone hole has increased substantially since it began (see Figure 2-1a) and now is comparable in size to that of the North American continent.

The seasonal evolution and decline of the Antarctic ozone hole in a recent year (2006) is illustrated in Figure 2-2. For reasons that will be explained later in the chapter, substantial ozone depletion does not start to occur until late August (Figures 2-2a, b) and begins to decline in November, as the stratospheric temperature rises (Figure 2-2c).

Initially it was not clear whether the hole was due to a natural phenomenon involving meteorological forces or to a chemical mechanism involving air pollutants. In the latter case, the suspect chemical was chlorine, produced mainly from gases that were released into the air in large quantities as a consequence of their use, for example, in air conditioners. Scientists had predicted that the chlorine would destroy ozone, but only to a small extent and only after several decades had elapsed. The discovery of the Antarctic ozone

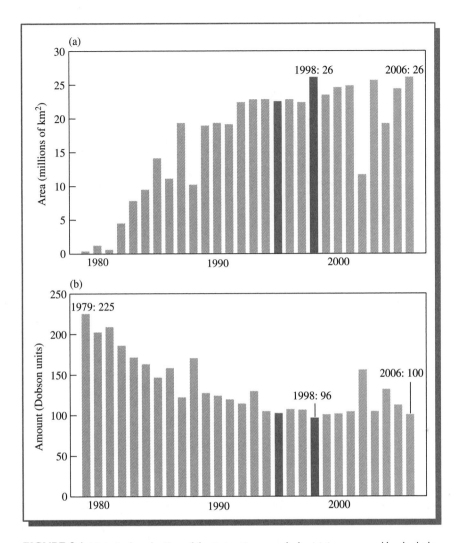

FIGURE 2-1 Historical evaluation of the Antarctic ozone hole. (a) Area covered by the hole (average for September 7 to October13), and (b) minimum overhead ozone (average for September 21 to October 16). Extreme ozone depletion occurred in 1998 and 2006, as indicated. No data were acquired during the 1995 season. [Source: NASA, at http://ozonewatch.gsfc.nasa.gov/]

hole came as a complete surprise to everyone. Subsequent research, however, confirmed that the hole indeed does occur as a result of chlorine pollution. The complicated chemical processes that cause ozone depletion are now understood and are discussed in this chapter. Based upon this knowledge, we can predict that the hole will continue to reappear each spring until about the middle of this century and that a corresponding hole may appear above the Arctic region.

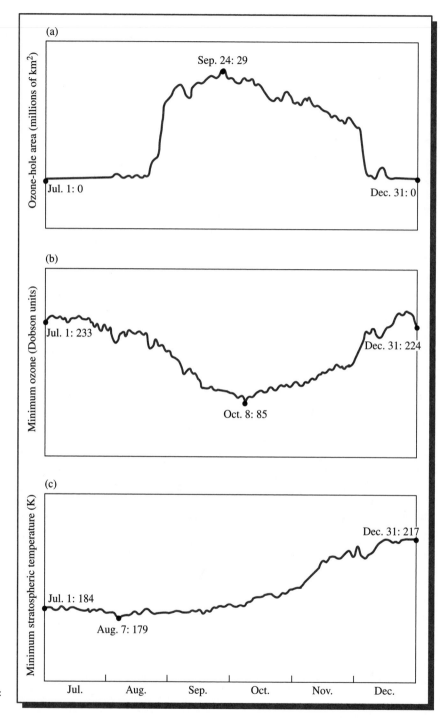

FIGURE 2-2 Evolution of the 2006 Antarctic ozone hole. (a) Area covered by the hole in millions of square kilometers; (b) minimum daily amount of overhead ozone in Dobson units; and (c) minimum daily temperature in the lower stratosphere in degrees Kelvin. [Source: NASA, at http: ozonewatch.gsfc.nasa.gov/]

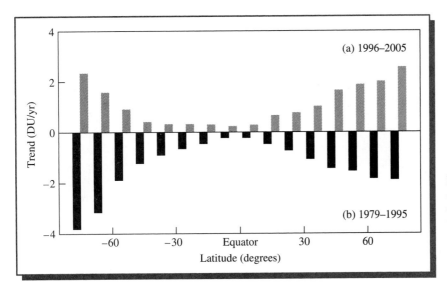

FIGURE 2-3 Changes in average overhead ozone at different latitudes. (a) Increases 1996–2005; (b) decreases 1979–1995. [Source: E. C. Weatherland and S. B. Anderson, *Nature* 441 (2006): 39.]

As a consequence of these discoveries, governments worldwide moved quickly to legislate a phase-out in production of the responsible chemicals. Thus the situation was not made much worse by the development of even more severe ozone depletion over populated areas, with the corresponding threat to the health of humans and other organisms that this increase would bring.

Ozone Depletion in Temperate Areas

Ozone was being depleted not just in the air above Antarctica but to some extent worldwide. The average overhead ozone loss at mid-latitudes amounted to about 3% in the 1980s. As indicated by the lengths of the vertical black bars in Figure 2-3b, the losses during the 1980s and early 1990s were greater the higher the latitude in both the Northern and Southern Hemispheres. However, this trend in ozone loss was reversed in the period from 1996 to 2005, the gains in the Northern Hemisphere in this period approximately canceling the earlier losses (Figure 2-3a). Although some of the recovery could be due to controls on emissions, much of it probably occurred because of natural trends in that period toward higher ozone levels due to the solar cycle and a lack of volcanic activity as well as to relatively warm Arctic winters.

The Ozone Hole and Other Sites of Ozone Depletion

As discussed previously, scientists discovered in 1985 that stratospheric ozone over Antarctica is reduced by about 50% for several months each year, due mainly to the action of chlorine. An episode of this sort, during which there is said to be a hole in the ozone layer, occurs from September to early November, corresponding to spring at the South Pole. The hole has been appearing since about 1979, as was shown in Figure 2-1, which illustrates the variation in the minimum September–October ozone concentrations above the Antarctic as a function of year. Extensive research in the late 1980s led to an understanding of the chemistry of this phenomenon. In this section, we discuss the peculiar process by which chlorine in the stratosphere becomes activated to destroy ozone and look at the detailed mechanism by which destruction occurs. We then consider the various measures of ozone-hole size, which allow us to investigate whether the hole above the Antarctic has been declining over time, whether a hole exists above the North Pole, and the effects of the holes on the amount of UV light to which we are exposed at ground level.

The Activation of Catalytically Inactive Chlorine

The ozone hole occurs as a result of special polar winter weather conditions in the lower stratosphere, where ozone concentrations usually are highest, that temporarily convert all the chlorine that is stored in the catalytically inactive forms HCl and $ClONO_2$ into the active forms Cl and ClO, all of which were discussed in Chapter 1. Consequently, the high concentration of active chlorine causes a large, though temporary, annual depletion of ozone.

The conversion of inactive to active chlorine occurs at the surface of particles formed by a solution of water, **sulfuric acid** (H_2SO_4), and **nitric acid** (HNO_3), the latter formed by combination of **hydroxyl radical** (OH) with **nitrogen dioxide** (NO_2) gas. The same conversion reactions could potentially occur in the gas phase but are so slow there as to be of negligible importance; they become rapid only when they occur on the surfaces of cold particles.

In most parts of the world, even in winter, the stratosphere is cloudless. Condensation of water vapor into liquid droplets or solid crystals that would constitute clouds doesn't normally occur in the stratosphere since the concentration of water in that region is exceedingly small, although there are always small liquid droplets consisting largely of sulfuric acid present, as well as some solid sulfate particles. However, the temperature in the lower stratosphere drops so low ($-80°C$) over the South Pole in the sunless winter months that condensation does occur. The usual stratospheric warming mechanism—the release of heat by the O_2 + O reaction—is absent because of the lack of production of atomic oxygen from O_2 and O_3 when there is total darkness. In turn, because the polar stratosphere becomes so cold during the total darkness at mid-winter, the air pressure drops since it is proportional to the Kelvin temperature, according to the ideal gas law, $PV = nRT$. This pressure phenomenon, in combination with the Earth's rotation, produces a **vortex,** a whirling mass of air in which wind speeds can exceed 300 km (180 miles) per hour. Since matter cannot penetrate the vortex, the air inside it is isolated and remains very cold for many months. At the South Pole, the vortex is sustained well into the springtime (October). (The vortex around the North Pole usually breaks down in February or early March, before much sunlight has returned to the area, but recently there have been exceptions to this generalization, as discussed later.)

The particles produced by condensation of the gases within the vortex form **polar stratospheric clouds,** or PSCs. As the temperature drops, the first crystals to form are small ones containing water and sulfuric and nitric acids. When the air temperature drops a few degrees more, below −80°C, a larger type of crystal—consisting mainly of frozen water ice and perhaps also nitric acid—also forms.

Chemical reactions that lead ultimately to ozone destruction occur in a thin aqueous layer present at the surface of the PSC ice crystals. Upon contact, gaseous **chorine nitrate,** $ClONO_2$, reacts at the surface with water molecules to produce **hypochlorous acid,** HOCl:

$$ClONO_2(g) + H_2O(aq) \longrightarrow HOCl(aq) + HNO_3(aq)$$

Also in the aqueous layer, gaseous **hydrogen chloride,** HCl, dissolves and forms ions:

$$HCl(g) \xrightarrow[\text{layer}]{\text{aqueous}} H^+(aq) + Cl^-(aq)$$

Reaction of the two forms of dissolved chlorine produces **molecular chlorine,** Cl_2, which escapes to the surrounding air:

$$Cl^-(aq) + HOCl(aq) \longrightarrow Cl_2(g) + OH^-(aq)$$

This process is illustrated schematically in Figure 2-4. Overall, when the steps are added together, the process corresponds to the net reaction

$$HCl(g) + ClONO_2(g) \longrightarrow Cl_2(g) + HNO_3(aq)$$

since the ions H^+ and OH^- re-form water. Similar reactions probably also occur on the surface of solid particles.

During the dark winter months, molecular chlorine accumulates within the vortex in the lower stratosphere and eventually becomes the predominant chlorine-containing gas. Once a little sunlight reappears in the very early Antarctic spring, or the

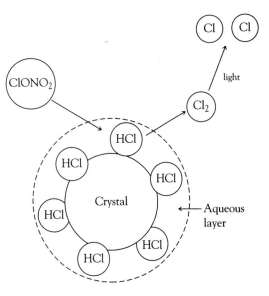

FIGURE 2-4 A scheme illustrating the production of molecular chlorine from inactive forms of chlorine in the winter and spring in the stratosphere in polar regions.

air mass moves to the edge of the vortex where there is some sunlight, the chlorine *molecules* are decomposed by the light into *atomic* chlorine:

$$Cl_2 + sunlight \longrightarrow 2\,Cl$$

Similarly, any gaseous HOCl molecules released from the surface of the crystals undergo photochemical decomposition to produce hydroxyl radicals and atomic chlorine:

$$HOCl + sunlight \longrightarrow OH + Cl$$

Massive catalytic destruction of ozone by atomic chlorine then ensues.

Since stratospheric temperatures above the Antarctic remain below −80°C even in the early spring (Figure 2-2c), the crystals persist for months. Any of the Cl that is converted back to HCl by the reaction with methane is subsequently reconverted to Cl_2 on the crystals and then back to Cl by sunlight. Inactivation of **chlorine monoxide,** ClO, by conversion to $ClONO_2$ does not occur, since all the NO_2 necessary for this reaction is temporarily bound as nitric acid in the crystals. The larger crystals move downward under the influence of gravity into the upper troposphere, thereby removing NO_2 from the lower stratosphere over the South Pole and further preventing the deactivation of chlorine. This *denitrification* of the lower stratosphere extends the life of the Antarctic ozone hole and increases the ozone depletion.

Only when the PSCs and the vortex have vanished does chlorine return predominantly to the inactive forms. The liberation of HNO_3 from the remaining crystals into the gas phase results in its conversion to NO_2 by the action of sunlight:

$$HNO_3 + UV \longrightarrow NO_2 + OH$$

More importantly, air containing normal amounts of NO_2 mixes with polar air once the vortex breaks down in late spring. The nitrogen dioxide quickly combines with chlorine monoxide to form the catalytically inactive chlorine nitrate. Consequently, the catalytic destruction cycles largely cease operation and the ozone concentration builds back up toward its normal level a few weeks after the PSCs have disappeared and the vortex has ceased, as illustrated in Figure 2-2. Thus the ozone hole closes for another year, though the ozone levels nowadays never quite return to their natural levels, even in the fall. However, before the ozone levels build back up in the spring, some of the ozone-poor air mass can move away from the Antarctic and mix with surrounding air, temporarily lowering the stratospheric ozone concentrations in adjoining geographic regions, such as Australia, New Zealand, and the southern portions of South America.

Reactions That Create the Ozone Hole

In the lower stratosphere—the region where the PSCs form and chlorine is activated—the concentration of free oxygen atoms is small; few atoms are produced there on account of the scarcity of the UV-C light that is required to dissociate O_2. Furthermore, any atomic oxygen produced in this way immediately collides with the abundant O_2 molecules to form ozone, O_3. Thus, ozone destruction mechanisms based upon the $O_3 + O \longrightarrow 2 O_2$ reaction, even when catalyzed, are not important here.

Rather, most of the ozone destruction in the ozone hole occurs via the process called Mechanism II in Chapter 1, with both X and X′ being atomic chlorine and with the overall reaction being $2 O_3 \longrightarrow 3 O_2$. Thus the sequence starts with the reaction of chlorine with ozone:

$$\textbf{\textit{Step 1:}} \quad Cl + O_3 \longrightarrow ClO + O_2$$

In Figure 2-5 the experimental ClO and O_3 concentrations are plotted as a function of latitude for part of the Southern Hemisphere during the spring of 1987. As anticipated, the two species display opposing trends, i.e., they anti-correlate very closely. At sufficient distances away from the South Pole (which is at 90°S), the concentration of ozone is relatively high and that of ClO is low, since chlorine is mainly tied up in inactive forms. However, as one travels closer to the pole and enters the vortex region, the concentration of ClO suddenly becomes high and simultaneously that of O_3 falls off sharply (Figure 2-5): Most of the chlorine has been activated and most of the ozone has consequently been destroyed. The latitude at which the concentrations both change sharply marks the beginning of the ozone hole, which continues through to the region above the South Pole. The anticorrelation of ozone and ClO concentrations shown in Figure 2-5 was considered by researchers to

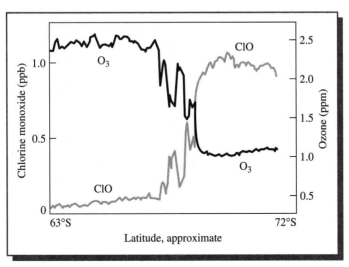

FIGURE 2-5 Stratospheric ozone and chlorine monoxide concentrations versus latitude near the South Pole on September 16, 1987. [Source: Reprinted with permission from P. S. Zurer, *Chemical and Engineering News* (30 May 1988): 16. Copyright 1988 by the American Chemical Society.]

be the "smoking gun," proving that anthropogenic chlorine compounds such as CFCs emitted into the atmosphere were indeed the cause of ozone-hole formation.

In the next reaction in the sequence, two ClO free radicals, produced in two separate step 1 events, combine temporarily to form a nonradical dimer, **dichloroperoxide,** ClOOCl (or Cl_2O_2):

$$\textbf{\textit{Step 2a:}} \quad 2\ ClO \longrightarrow Cl—O—O—Cl$$

The rate of this reaction becomes important to ozone loss under these conditions because the chlorine monoxide concentration rises steeply due to the activation of the chlorine. Once the intensity of sunlight has risen to an appreciable amount in the Antarctic spring, the dichloroperoxide molecule ClOOCl absorbs UV light and splits off one chlorine atom. The resulting ClOO free radical is unstable, so it subsequently decomposes (in about a day), releasing the other chlorine atom:

$$\textbf{\textit{Step 2b:}} \quad ClOOCl + UV\ light \longrightarrow ClOO + Cl$$

$$\textbf{\textit{Step 2c:}} \quad ClOO \longrightarrow O_2 + Cl$$

Adding steps 2a, 2b, and 2c, we see that the net result is the conversion of two ClO molecules to atomic chlorine via the intermediacy of the dimer ClOOCl, which corresponds to the second stage of Mechanism II:

$$2\ ClO \longrightarrow [ClOOCl] \xrightarrow{light} 2\ Cl + O_2$$

By these processes ClO returns to the ozone-destroying form of chlorine, Cl. If we add the above reaction to two times step 1 (the factor of 2 being required to produce the two intermediate ClO species needed in reaction 2a so that none remains in the overall equation), we obtain the overall reaction

$$2\ O_3 \longrightarrow 3\ O_2$$

Thus a complete catalytic ozone destruction cycle exists in the lower stratosphere under these special weather conditions, i.e., when a vortex is present. The cycle also requires very cold temperatures, since under warmer conditions ClOOCl is unstable and reverts back to two ClO molecules before it can undergo photolysis, thereby short-circuiting any ozone destruction. Before appreciable sunlight becomes available in the early spring, most of the chlorine exists as ClO and Cl_2O_2 since step 2b requires fairly intense light levels; such an atmosphere is said to be primed for ozone destruction.

About three-quarters of the ozone destruction in the Antarctic ozone hole occurs by the mechanism set forth above, in which chlorine is the only catalyst. This ozone destruction cycle contributes greatly to the creation of the ozone hole. Each chlorine destroys about 50 ozone molecules per day during the spring. The slow step in the mechanism is step 2a, which is the combination of 2 ClO molecules. Since the rate law for step 2a is second order in ClO concentration (i.e., its rate is proportional to the square of the ClO concentration), it proceeds at a substantial rate, and the destruction of ozone is significant, only when the ClO concentration is high. The abrupt appearance of the ozone hole is consistent with the quadratic rather than linear dependence of ozone destruction upon chlorine concentration by the Cl_2O_2 mechanism. Let us hope that there are not many more environmental problems whose effects will display such nonlinear behavior and similarly surprise us!

PROBLEM 2-1

A minor route for ozone destruction in the ozone hole involves Mechanism II with bromine as X′ and chlorine as X (or vice versa). The ClO and BrO free-radical molecules produced in these processes then collide with each other and rearrange their atoms to eventually yield O_2 and atomic chlorine and bromine. Write out the mechanism for this process, and add up the steps to determine the overall reaction.

PROBLEM 2-2

Suppose that the concentration of chlorine continues to rise in the stratosphere but that the relative increase in bromine does not rise proportionately. Will the dominant mechanism involving dichloroperoxide or the "chlorine plus bromine" mechanism of Problem 2-1 become relatively more important or less important as the destroyer of ozone in the Antarctic spring?

PROBLEM 2-3

Why is the mechanism involving dichloroperoxide of negligible importance in the destruction of ozone, compared to the mechanism that proceeds by ClO + O, in the upper levels of the stratosphere?

In the lower stratosphere above Antarctica, an ozone destruction rate of about 2% per day occurs each September due to the combined effects of the various catalytic reaction sequences. As a result, by early October almost all the ozone is wiped out between altitudes of 15 and 20 km, just the region in which its concentration normally is highest over the South Pole. This result

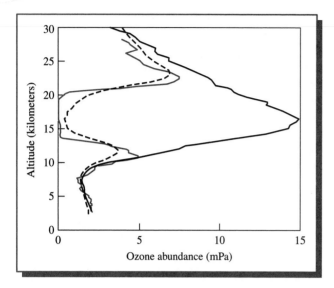

FIGURE 2-6 The typical vertical distribution of ozone over Antarctica in mid-spring (October) in 1962–1971 (black curve, before the ozone hole started), in the 1991–2001 period (dashed curve), and in 2001 (green curve). Ozone partial pressure is in millipascals. [Source: WMO/UNEP Scientific Assessment of Ozone Depletion 2006, Figure Q11-3.]

is illustrated in Figure 2-6, which shows the measured ozone partial pressure as a function of altitude over the Antarctic in mid-spring in the years preceding the ozone-hole formation (black curve) and in 2001 (green curve). Notice that the depletion from 13 to 19 km was more complete in 2001 than on average in the preceding years (dashed curve).

In summary, the special vortex weather conditions in the lower stratosphere above the Antarctic in winter cause denitrification and lead to the conversion of inactive chlorine into Cl_2 and $HOCl$. These two compounds produce atomic chlorine when sunlight appears. The chlorine atoms efficiently destroy ozone via Mechanism II. Once the vortex disappears in the late spring, the ice particles on which the activation of chlorine compounds occurs disappear, the chlorine returns to inactive forms, and the hole heals.

The Size of the Antarctic Ozone Hole

Because (as explained later) the stratospheric concentration of chlorine continued to increase until the end of the twentieth century, the extent of Antarctic ozone depletion increased from the early 1980s at least until the late 1990s. There are several relevant measures of the extent of ozone depletion.

• One measure is the *surface area* covered by low overhead ozone; Figure 2-1a shows the area within the 220-DU contour line for the mid-September to mid-October period as a function of year. This area grew rapidly and approximately linearly during the 1980s; the size of the hole in maximum depletion years (1998, 2006) is somewhat larger than in the 1980s, though smaller holes have appeared in some recent years.

• Similarly, the sharp decrease in the *minimum amount* of overhead ozone in the spring that occurred from 1978 to the late 1980s has been replaced by a slower decline, which now may have largely ceased (see Figure 2-1b).

• The average *length of time* during which ozone depletion occurs has also increased in recent years. Some reduction in ozone levels is now usually seen both in mid-winter (at least in the outer portions of the continent where

Ozone destruction step

$$O_3 + Cl \longrightarrow O_2 + ClO$$

Atomic chlorine reconstitution

Mid-stratosphere

$$ClO + O \longrightarrow Cl + O_2$$

Ozone hole/low stratosphere

$$2\ ClO \longrightarrow ClOOCl$$

$$ClOOCl + UV \longrightarrow ClOO + Cl$$

$$ClOO \longrightarrow Cl + O_2$$

Inactivation of chlorine

$$Cl + CH_4 \longrightarrow HCl + CH_3$$

$$ClO + NO_2 \longrightarrow ClONO_2$$

Activation of chlorine on particle surfaces

$$HCl(g) \xrightarrow{\text{H}_2\text{O}} H^+(aq) + Cl^-(aq)$$

$$H_2O(aq) + ClONO_2(g) \longrightarrow HOCl(aq) + HNO_3(aq)$$

$$Cl^-(aq) + HOCl(aq) \longrightarrow Cl_2(g) + OH^-(aq)$$

$$Cl_2(g) + \text{sunlight} \longrightarrow 2\ Cl(g)$$

$$H^+(aq) + OH^-(aq) \longrightarrow H_2O(aq)$$

FIGURE 2-7 A summary of the main ozone destruction reaction cycles operating in the Antarctic ozone hole.

there is some sunshine at that time) and in the summer as well as the spring, and, indeed, there is now some persistence of the depletion from one year to the next.

• The *vertical region* over which almost total ozone depletion occurs, 12–22 km, has not increased since the mid-1990s.

A review of the possible signs of recovery of the ozone layer published in 2006 pointed out that natural variations, such as the solar cycle and polar temperatures, could mask any trends in stratospheric ozone-hole recovery of the magnitude expected to date and indeed for the next few decades.

The various reactions that lead to catalytic ozone destruction by atomic chlorine by various mechanisms are summarized in Figure 2-7.

Stratospheric Ozone Destruction over the Arctic Region

Given the similarity in climate, it may seem surprising that an ozone hole above the Arctic did not start to form at the same time as one occurred in the Antarctic. Episodes of partial springtime ozone depletion over the Arctic region have occurred several times since the mid-1990s. The phenomenon is less severe than in Antarctica because the stratospheric temperature over the Arctic does not fall as low for as long and air circulation to surrounding areas is not as limited. The flow of tropospheric air over mid-latitude mountain ranges (Himalayas, Rockies) in the Northern Hemisphere creates waves of air that can mix with polar air, warming the Arctic stratosphere. Because the air is generally not as cold, polar stratospheric clouds form less frequently over the Arctic than over the Antarctic and do not last as long. In the past, only small crystals were formed; these are not large enough to fall out of the stratosphere and thereby denitrify it. However, during the extended polar night, the chlorine nitrate and hydrogen chloride do react on the surface of the small particles to produce molecular chlorine, which then dissociates to atomic chlorine and by reaction with an ozone molecule becomes chlorine monoxide, as illustrated in Figure 2-8. Notice that, although HCl is converted completely in the PSCs, the $ClONO_2$, which is present in excess, is not completely eliminated in the stratosphere above the North Pole. Once the PSCs disappear as air temperatures rise, chlorine nitrate initially dominates since it forms rapidly from ClO and nitrogen dioxide. The reaction of atomic chlorine with methane is a slower process, and consequently the HCl concentration is slower to rise.

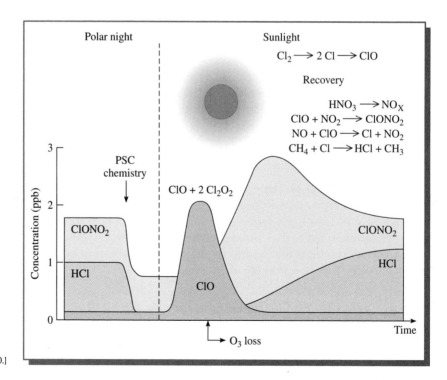

FIGURE 2-8 The evolution of stratospheric chlorine chemistry with time above the Arctic in winter and spring. [Source: Redrawn from C. R. Webster et al., *Science* 261 (1993): 1130.]

Before the mid-1990s, the vortex containing the cold air mass above the Arctic broke up by late winter; therefore NO_2-containing air mixed with vortex air before much sunlight returned to the polar region in the spring. Since the stratospheric air temperature usually rose above $-80°C$ by early March, the nitric acid in the particles was converted back to gaseous nitrogen dioxide before the intense spring sunlight could drive the Cl_2O_2 mechanism. Due to increases in NO_2 from both these sources, the activated chlorine was mostly transformed back to $ClONO_2$ before it could destroy much ozone (Figure 2-8). Thus the total extent of ozone destruction over the Arctic area was much less than that over the Antarctic in the past.

Unfortunately, there have been ominous signs in recent decades that springtime conditions above the Arctic have been changing for the worse, with the result that ozone depletion there accelerated in the lower stratosphere in some years. The Arctic vortex in the winter and spring of 1995–1996 was exceptionally cold and persistent, resulting in significant chlorine-catalyzed losses of ozone as late as mid-April. Large, nitric acid–containing particles were formed, and persisted long enough to fall out of the stratosphere, thereby denitrifying certain regions. In addition, the often-irregular shape of the Arctic vortex means that there are frequent occasions when an "arm" of it passes over a sunlit area in late winter (before the bulk of the vortex is illuminated); temporary ozone depletion occurs within such arms. For example, a portion of the vortex passed across Great Britain during March 1996, producing record lows of 195 DU in northern Scotland.

However, the extent of winter–spring ozone loss over Arctic regions has been very inconsistent, with almost no depletion in some recent winters but significant depletion in others, as indicated in Figure 2-9. The amount of ozone loss correlated linearly with the area associated with polar stratospheric clouds (Figure 2-9). Both the maximum extent of ozone depletion and the maximum vortex area appear to be increasing with time, although these extremes are achieved only every few years when the vortex of cold air above the Arctic remains stable into the late winter and early spring. The greatest ozone depletion over the Arctic observed so far, about 135 DU, occurred in the very cold winter of 2004–2005; that for 2005–2006 was considerably less since the temperatures were not as cold.

For reasons that will be explained in Chapter 6, both the depletion of ozone and the increase in carbon dioxide levels cool the stratosphere, which will lead to even more depletion if cooling occurs in the springtime and thereby extends the period in which PSCs remain. Some scientists predict that recovery from ozone depletion will be slower in the Arctic than in the Antarctic because of the cooling effects of CO_2 and O_3. Scientists do not yet know whether or not the abrupt cooling in the winter of 2004–2005 that produced record ozone depletion was due largely to the effects of increased CO_2.

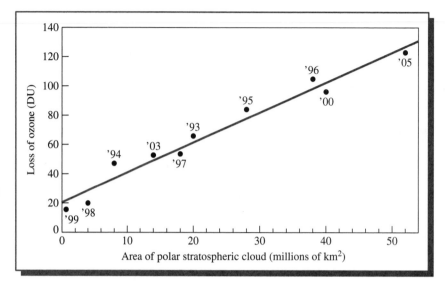

FIGURE 2-9 Loss of overhead ozone over the Artic versus the size of the polar stratospheric cloud in recent years. [Source: Redrawn from M. Rex et al., "Arctic Winter 2005: Implications for Stratospheric Ozone Loss and Climate Change," *Geophysical Research Letters* 31 (2006): L04116.]

Because the magnitude of ozone depletion above the Arctic in some recent winters was about the same as that observed over the South Pole in the early 1980s, some atmospheric scientists have stated that an Arctic ozone hole now forms in some years. Since depletion of overhead ozone is never 100% complete, the definition of what conditions constitute a hole is somewhat arbitrary.

The chemistry underlying mid-latitude losses in stratospheric ozone is discussed in Box 2-1. A systematic view of the various atmospheric chemical reactions discussed in this chapter is given in Chapter 5, after the corresponding reactions in the troposphere have been discussed.

Increases in UV at Ground Level

Experimentally, the amount of UV-B from sunlight (see Chapter 1) reaching ground level increases by a factor of 3 to 6 in the Antarctic during the early part of the spring because of the appearance of the ozone hole. Biologically, the most dangerous UV doses under hole conditions occur in the late spring (November and December), when the Sun is higher in the sky than in earlier months and low overhead ozone values still prevail. Abnormally high UV levels have also been detected in southern Argentina when ozone-depleted stratospheric air from the Antarctic traveled over the area.

Increases in ground-level UV-B intensity have also been measured in the spring months in mid-latitude regions in North America, Europe, and

BOX 2-1	The Chemistry Behind Mid-Latitude Decreases in Stratospheric Ozone

As noted earlier, there was a worldwide decrease of several percent in the steady-state ozone concentration in the stratosphere over nonpolar areas during the 1980s and an additional short-term major decrease from 1992 to 1994. The extent of depletion closely mirrored the total ozone concentration for any given month; the greatest depletion occurred in the March–April period and the least in the early fall.

Scientists have had a harder time tracking down the source of the mid-latitude ozone depletion than that over polar regions. As in Antarctica, almost all the ozone loss in non-polar regions occurs in the lower stratosphere. Some scientists have speculated that reactions leading to ozone destruction could occur not only on ice crystals but also on the surfaces of other particles present in the lower stratosphere. They suggested that the reactions could occur on cold liquid droplets consisting mainly of sulfuric acid that occur naturally in the lower stratosphere at all latitudes. The liquid droplets would have to be cold enough for them to take up significant amounts of gaseous HCl, or no net reaction would take place. There always exists a small background amount of the acid, due to the oxidation of the naturally occurring gas *carbonyl sulfide*, COS, some of which survives long enough to reach the stratosphere. However, the dominant though erratic source of the H_2SO_4 at these altitudes is direct injection into the stratosphere of sulfur dioxide gas emitted from volcanoes, followed by its oxidation to the acid. Indeed, the steep decline in ozone in 1992–1993 followed the June 1991 massive eruption of Mt. Pinatubo in the Phillipines,

and measurable ozone depletion was noted for several years after the eruption of El Chichon in Mexico in 1982. There were dips significantly below the trend for the ozone levels—both of these periods temporarily increased the concentration of sulfuric acid droplets in the lower stratosphere.

The other relevant reaction that takes place on the surface of the sulfuric acid droplets results in some denitrification of stratospheric air. In the gas-phase steps of the sequence, ozone itself converts some nitrogen dioxide, NO_2, to *nitrogen trioxide*, NO_3, which then combines with other NO_2 molecules to form *dinitrogen pentoxide*, N_2O_5:

$$NO_2 + O_3 \longrightarrow NO_3 + O_2$$
$$NO_2 + NO_3 \longrightarrow N_2O_5$$

These gas-phase processes normally are reversible and do not remove much NO_2 from the air, but in the presence of high levels of liquid droplets, a conversion of N_2O_5 to nitric acid occurs instead:

$$N_2O_5 + H_2O(\text{droplets}) \xrightarrow[\text{droplets}]{H_2SO_4} 2\,HNO_3$$

By this mechanism, much of the NO_2 that normally would be available to tie up chlorine monoxide as the nitrate $ClONO_2$ becomes unavailable for this purpose; hence a greater proportion of the chlorine atoms occur in the catalytically active form and destroy ozone. It should be realized, however, that even in the absence of particles, some NO_2 is converted to nitric acid as a result of its reaction with the hydroxyl radical. This nitric acid eventually

(continued on p. 76)

BOX 2-1	The Chemistry Behind Mid-Latitude Decreases in Stratospheric Ozone *(continued)*

undergoes photochemical decomposition in daylight hours to reverse this reaction and to produce species that are catalytically active in ozone destruction.

In the mid-latitude lower stratosphere, the most important catalytic ozone destruction reactions involving halogens employ Mechanism II, with X being atomic chlorine or bromine and X′ being the hydroxyl radical:

$$Cl + O_3 \longrightarrow ClO + O_2$$

$$OH + O_3 \longrightarrow HOO + O_2$$

$$ClO + HOO \longrightarrow HOCl + O_2$$

$$HOCl \xrightarrow{\text{sunlight}} OH + Cl$$

and similarly for the case where bromine replaces chlorine. The reaction sequence involving collision of ClO with BrO discussed for the Antarctic ozone hole is also operative here.

PROBLEM 1

Deduce the overall reaction equation for the reaction sequence shown at left.

This mechanism explains why, in the current high-chlorine lower stratosphere, large volcanic eruptions can deplete mid-latitude stratospheric ozone for a few years, but it does not account for the overall trend of decreasing ozone in the 1980s. Some of the decrease is probably due to the mechanism operating on the background concentration of sulfuric acid particles in the lower stratosphere; its magnitude would have increased continuously in this time period since the chlorine levels were continuously increasing. Chlorine and bromine increases combined resulted in about a 4% decline in mid-latitude ozone levels in the 1979–1995 period. However, much of the gradual decline over mid-latitudes is believed to be due to other factors, such as springtime dilution of ozone-depleted polar air and its transport out of the polar regions, changes in the solar cycle, and both natural and anthropogenic changes in the pattern of atmospheric transport and temperatures.

New Zealand. Calculations indicate that the extent of UV increases since the 1980s over mid- and high-latitude regions amounts to 6–14%. The most definitive experimental evidence comes from New Zealand, where long-term summertime increases in UV-B, but as expected not in UV-A, amounted to 12% by 1998–1999. The situation over mid-latitudes is complicated by the facts that some UV-B is absorbed by the ground-level ozone produced by pollution reactions (as explained in Chapter 3), thereby masking any changes in UV-B due to small amounts of stratospheric ozone depletion, and that records of UV received at the Earth's surface were started only in the 1990s.

The Chemicals That Cause Ozone Destruction

The increase in levels of stratospheric chlorine and bromine that occurred in the last half of the twentieth century was due primarily to the release into the atmosphere of organic compounds containing chlorine and bromine that are **anthropogenic,** that is to say, they are man-made. These anthropogenic contributions to stratospheric halogen levels completely overshadowed the natural input. In this section, we investigate

- why the levels of chlorine and bromine increased due to the release into the air of compounds having certain characteristics,

- how international agreements were put in place to control such substances,

- the strategy underlying the formulations of compounds to replace the original halogen compounds, and the practical difficulties and controversy about phasing out methyl bromide, and

- how two practical replacements developed by green chemistry for the now-banned chemicals can be employed.

The chlorine- and bromine-containing compounds that give rise to increased levels of the halogens in the stratosphere are those that do not have a **sink**—i.e., a natural removal process such as dissolution in rain or oxidation by atmospheric gases—in the troposphere. After a few years of traveling in the troposphere, they begin to diffuse into the stratosphere, where eventually they undergo photochemical decomposition by UV-C from sunlight and thereupon release their halogen atoms.

The variation in the total concentration of stratospheric chlorine and bromine atoms, expressed as the equivalent of chlorine in terms of ozone destruction power, measured over the course of the last quarter-century and projected to the middle of the twenty-first century, is illustrated by the topmost curve in Figure 2-10. The peak chlorine-equivalent concentration of about 3.8 ppb that occurred in the late 1990s was almost four times as great as the "natural" level due to methyl chloride and methyl bromide releases from the sea. The Antarctic ozone hole appeared first when the chlorine concentration reached about 2 ppb (dotted horizontal line).

CFC Decomposition Increases Stratospheric Chlorine

As is clear from inspection of Figure 2-10, the recent increase in stratospheric chlorine is due primarily to the use and release of **chlorofluorocarbons**—compounds containing only chlorine, fluorine, and carbon, which are commonly called **CFCs.** In the 1980s, about 1 million tonnes (i.e., metric tons, 1000 kg each) of CFCs were released annually to the atmosphere. These compounds are nontoxic, nonflammable, nonreactive, and have useful

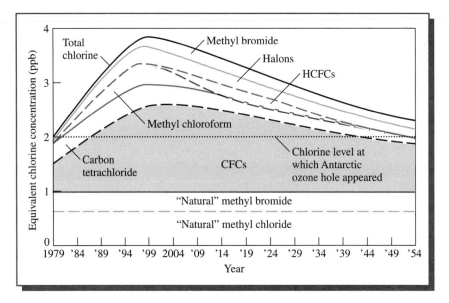

FIGURE 2-10 Actual and projected concentration of stratospheric chlorine versus time, showing the contributions of various gases. Note that ozone-depleting effects of bromine atoms in halons and methyl bromide have been converted to their chlorine equivalents. [Source: DuPont.]

condensation properties (making them suitable for use as coolants, for example); because of these favorable characteristics, they found a multitude of uses. Large volumes of several CFCs were manufactured commercially and employed worldwide throughout the mid-to-late 1900s. Most of the amounts produced eventually leaked from the devices in which they were originally placed and entered the atmosphere as gases.

CFCs have no tropospheric sink, so all their molecules eventually rise to the stratosphere. In contrast to intuitive expectation, this vertical transport in the atmosphere is *not* affected by the fact that the mass of such molecules is greater than the average molecular mass of nitrogen and oxygen in air, because the differential force of gravity is much less than that due to the constant collisions of other molecules, which randomize the directions of even heavy molecules.

The CFC molecules eventually migrate to the middle and upper parts of the stratosphere where there is sufficient unfiltered UV-C from sunlight to photochemically decompose them, thereby releasing chlorine atoms. CFCs do not absorb sunlight with wavelengths greater than 290 nm, and they generally require light of 220 nm or less for photolysis. The CFCs must rise to the mid-stratosphere before decomposing, since UV-C does not penetrate to lower altitudes. Because vertical motion in the stratosphere is slow, their atmospheric lifetimes are long. It is because of their long stratospheric lifetimes that the chlorine concentration in Figure 2-10 falls so slowly with time.

PROBLEM 2-4

Reactions of the type

$$OH + CF_2Cl_2 \longrightarrow HOF + CFCl_2$$

are conceivable tropospheric sinks for CFCs. Can you deduce why they don't occur, given that C—F bonds are much stronger than O—F bonds?

Other Chlorine-Containing, Ozone-Depleting Substances

Another widely used carbon–chlorine compound that lacks a tropospheric sink—although some of it ends up dissolving in ocean waters—is **carbon tetrachloride,** CCl_4, which also is photochemically decomposed in the stratosphere. Like CFCs, then, it is classified as an **ozone-depleting substance** (ODS). Commercially, carbon tetrachloride was used as a solvent and was an intermediate in the manufacture of several CFCs, during the production of which some was lost to the atmosphere. Its use as a dry-cleaning solvent was discontinued in most developed countries some decades ago, but until recently it has continued to be used in many other countries. Because of its relatively long atmospheric lifetime (26 years), it will continue to make a significant contribution to stratospheric chlorine for several more decades (Figure 2-10).

Methyl chloroform, CH_3—CCl_3, or *1,1,1-trichloroethane,* was produced in large quantities and used in metal cleaning in such a way that much of it was released into the atmosphere. Although about half of it is removed from the troposphere by reaction with the hydroxyl radical, the remainder survives long enough to migrate to the stratosphere. Because its average lifetime is only five years and its production has been largely phased out, its concentration in the atmosphere has declined rapidly since the 1990s. According to Figure 2-10, the contribution of methyl chloroform to stratospheric chlorine was substantial in the 1990s but by 2010 will become negligible.

 ## Green Chemistry: The Replacement of CFC and Hydrocarbon Blowing Agents with Carbon Dioxide in Producing Foam Polystyrene

Polystyrene is a common polymer that is used to make many everyday items. This polymer varies in appearance from a rigid solid plastic to foam polystyrene. Rigid plastic polystyrene is used in disposable silverware; audiocassette, CD, and DVD cases; and appliance casings. Foam polystyrene is utilized as insulation in coolers and houses, foam cups, meat and poultry trays, egg cartons; in some countries it is still used in fast-food containers. Globally, about 10 million tonnes of polystyrene are produced on an annual basis, with approximately half used to produce the foam form.

In order to produce foam polystyrene, the melted polymer is combined with a gas under pressure. This mixture is then extruded into an environment of lower pressure where the gas expands, leaving a foam which is about 95% gas and 5% polymer.

In the past, CFCs were employed as blowing agents for rigid plastic foams, and foam polystyrene is no exception. When these foams are crushed or they degrade, the CFCs are released into the atmosphere where they can migrate to the stratosphere and act to destroy ozone. Low-molecular-weight hydrocarbons, such as pentane, have also been used as blowing agents; although these compounds do not deplete the ozone layer, they do contribute to ground-level smog when they are emitted into the atmosphere, as we will see in Chapter 3. Low-molecular-weight hydrocarbons are also very flammable and reduce worker safety.

The search for a replacement for CFC and hydrocarbon blowing agents led the Dow Chemical Company of Midland, Michigan, to develop a process employing 100% carbon dioxide as a blowing agent for polystyrene foam sheets. For this discovery, Dow was the recipient of a Presidential Green Chemistry Challenge Award in 1996. **Carbon dioxide,** CO_2, is not flammable nor does it deplete the ozone layer. Nonetheless, we will see in Chapter 6 that it is a greenhouse gas and thus contributes to the environmental problem of global warming, so one might wonder whether we are trading one environmental problem for another. However, waste carbon dioxide from other processes (natural gas production and the preparation of ammonia) that would otherwise be emitted into the atmosphere can be captured and used as a blowing agent. In addition, we will see in Chapter 6 that CFCs not only dramatically affect the ozone layer but also are greenhouse gases significantly more potent than carbon dioxide.

Dow Chemical found an added advantage with the polystyrene foam sheets made with carbon dioxide in that they remained flexible for a much longer period of time than those made with CFCs. This results in less breakage during use and a longer shelf life. In addition, foam sheets made with CFCs had to be degassed of the CFCs prior to recycling them, while carbon dioxide rapidly escapes from the polystyrene, leaving a sheet composed of 95% air and 5% polystyrene within a few days.

CFC Replacements

Compounds such as CFCs and CCl_4 have no tropospheric sinks because they do not undergo any of the normal removal processes: They are not soluble in water and thus are not rained out from air; they are not attacked by the hydroxyl radical or any other atmospheric gases and so do not decompose; and they are not photochemically decomposed by either visible or UV-A light.

The compounds being implemented as the direct replacements for CFCs all contain hydrogen atoms bonded to carbon. Consequently, a majority (though not necessarily 100%) of the molecules will be removed from the troposphere by a sequence of reactions which begins with hydrogen abstraction by OH:

$$OH + H-\underset{|}{\overset{|}{C}}- \longrightarrow H_2O + \text{C-centered free radical} \longrightarrow$$
$$CO_2 \text{ and other products eventually}$$

Reactions of this type are discussed in more detail in Chapters 3 and 5. Because methyl chloride, methyl bromide, and methyl chloroform each contain hydrogen atoms, a fraction of such molecules are removed in the troposphere before they have a chance to rise to the stratosphere.

The *temporary* replacements for CFCs employed in the 1990s and the early years of the twenty-first century contain hydrogen, chlorine, fluorine, and carbon; they are called **HCFCs, hydrochlorofluorocarbons.** One important example is CHF_2Cl, the gas called *HCFC-22* (or just *CFC-22*). It is employed in modern domestic air conditioners and in some refrigerators and freezers, and it has found some use in blowing foams such as those used in food containers. Since it contains a hydrogen atom and thus is mainly removed from air before it can rise to the stratosphere, its long-term ozone-reducing potential is small—only 5% of that of the CFC that it replaced. This advantage is offset, however, by the fact that HCFC-22 decomposes to release chlorine more quickly than does the CFC, so its *short-term* potential for ozone destruction is greater than that implied by this percentage. But because most HCFC-22 is destroyed within a few decades after its release, it is responsible for almost no *long-term* ozone destruction. However, most concerns about stratospheric ozone destruction are centered on the next few decades, before substantial reduction of stratospheric chlorine occurs from the phase-out of CFCs. Notice the contributions of HCFCs to the curve in Figure 2-10. They should be significant only until about 2030.

Reliance exclusively on HCFCs as CFC replacements would have eventually led to a renewed buildup of stratospheric chlorine, because the volume of HCFC consumption would presumably rise with increasing world population and affluence. Products that are entirely free of chlorine, and that therefore pose no hazard to stratospheric ozone, will be the ultimate replacements for CFCs and HCFCs.

Hydrofluorocarbons, HFCs, substances that contain hydrogen, fluorine, and carbon, are the main long-term replacements for CFCs and HCFCs. The compound CH_2F-CF_3, called *HFC-134a*, has an atmospheric lifetime of several decades before finally succumbing to OH attack. HFC-134a is now used as the working fluid in new refrigerators and air conditioners for the North American market, including those in automobiles. All HFCs eventually react to form **hydrogen fluoride,** HF. Unfortunately, one atmospheric degradation pathway for HFC-134a, and for several HCFCs as well, produces **trifluoroacetic acid,** TFA (CF_3-COOH), as an intermediate, which is then removed from the air by rainfall. Some scientists worry that TFA represents an environmental hazard to wetlands since it will accumulate in aquatic plants and could inhibit their growth. However, some of the TFA in the environment arises from the degradation under heating of polymers such as Teflon, not from CFC replacements. Polyfluorocarboxylic acids, of which the acid form of TFA is an example, have been used in certain commercial products but are now being phased out, as discussed in Chapter 12.

Another environmental concern with HFCs involves their accumulation in air after their inadvertent release during use. While present in the troposphere, before they are destroyed, HFCs contribute to global warming by enhancing the greenhouse effect, a topic discussed in detail in Chapter 6. Outside North America, industry usually uses cyclopentane or isobutane, rather than an HFC, as a refrigerant. Such hydrocarbons have a much shorter lifetime in air than HFCs. Some environmentalists hope that developing countries will follow the hydrocarbon rather than the HFC route when they start to manufacture goods requiring coolants. Fully fluorinated compounds are unsuitable replacements for CFCs because they have no tropospheric or stratospheric sinks, and if released into the air, they would contribute to global warming for very long periods of time.

Halons

Halon chemicals are bromine-containing, hydrogen-free substances such as CF_3Br and CF_2BrCl. Because they have no tropospheric sinks, they eventually rise to the stratosphere. There they are photochemically decomposed, with the release of atomic bromine (and chlorine, if present), which, as we have already discussed, is an efficient X catalyst for ozone destruction. Thus halons also are ozone-depleting substances. Bromine from halons will continue to account for a significant fraction of the ozone-destroying potential of stratospheric halogen catalysts for decades to come (Figure 2-10).

Halons are used in fire extinguishers. They operate to quell fires by releasing atomic bromine, which combines with the free radicals in the combustion to form inert products and less reactive free radicals. The halons release their bromine atoms even at moderately high temperatures, since their C—Br bonds are relatively weak. Since they are nontoxic and leave no residues upon evaporation, halons are very useful for fighting fires, particularly in inhabited, enclosed spaces, such as military aircraft, and those housing electronic equipment, such as computer centers. The substitution of other chemicals for halons in the testing of the extinguishers drastically reduces halon emissions to the atmosphere, since only a minority of the releases are from the fighting of actual fires. Fine sprays of water can be substituted for halons in fighting many fires.

Fluorine atoms are liberated in the stratosphere as a result of the decomposition of halons, as well as CFCs, HCFCs, and HFCs. In principle, the fluorine atoms could catalytically destroy ozone (see Problem 2-6). However, the reaction of atomic fluorine with methane and other hydrogen-containing molecules in the stratosphere is rapid and produces HF, a very stable molecule.

Because the H—F bond is much stronger than the O—H bond, the reactivation of fluorine by the attack of the hydroxyl radical on hydrogen fluoride molecules is very endothermic. Consequently, its activation energy is high and the reaction is extremely slow at atmospheric temperatures (see Box 1-2). Thus fluorine is quickly and permanently deactivated before it can destroy any significant amount of ozone.

PROBLEM 2-5

The free radical CF_3O is produced during the decomposition of HCF-134a. Show the sequence of reactions by which it could destroy ozone acting as an X catalyst in a manner reminiscent of OH. (Note that it is too short-lived to actually destroy much ozone.)

PROBLEM 2-6

(a) Write the set of reactions by which atomic fluorine could operate as an X catalyst by Mechanisms I and II in the destruction of ozone. (b) An alternative to the second step of Mechanism I in the case of X = F is the reaction between FO and ozone to give atomic fluorine and two molecules of oxygen. Write out this mechanism, and deduce its overall reaction.

International Agreements That Restrict ODSs

In contrast to almost all other environmental problems, such as global warming (Chapter 7), international agreement on remedies to stratospheric ozone depletion was obtained and successfully implemented in a fairly short period of time. The use of CFCs in most aerosol products was banned in the late 1970s in North America and some Scandinavian countries. This decision was made on the basis of predictions, made by Sherwood Rowland and Mario Molina, chemists at the University of California, Irvine, concerning the effect of chlorine on the thickness of the ozone layer. There was no experimental indication of any depletion at the time of their prediction. Rowland and Molina, together with the German chemist Paul Crutzen, were jointly awarded the Nobel Prize in Chemistry in 1995 to honor their work in researching the science underlying ozone depletion.

The growing awareness of the seriousness of chlorine buildup in the atmosphere led to international agreements to phase out CFC production in the world. The breakthrough came at a conference in Montreal, Canada, in 1987 that gave rise to the **Montreal Protocol;** this agreement has been strengthened at several follow-up conferences. As a result of this international agreement, all ozone-depleting chemicals are now destined for phaseout in all nations. All legal CFC *production* in developed countries ended in 1995. Developing countries have been allowed until 2010 to reach the same goal. Figure 2-11 shows how the tropospheric concentrations of the two most

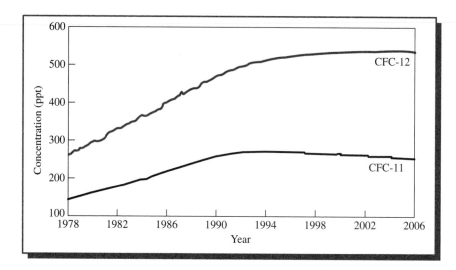

FIGURE 2-11 Tropospheric concentrations of CFC-11 and CFC-12. [Source: NOAA, at www.esrl.noaa.gov/gmd/aggi/]

widely used CFCs have changed in recent decades. The level of CFC-11 ($CFCl_3$), the average atmospheric lifetime of which is about 50 years, peaked about 1994, seven years after its production and emission started a precipitous decline, and has dropped slowly since then; the level of CFC-12 (CF_2Cl_2), which has a lifetime of more than 100 years, did not peak until about 2002.

The production of carbon tetrachloride and methyl chloroform has been phased out. Developed countries have agreed to end production of HCFCs by 2030, and developing countries by 2040, with no increases allowed after 2015.

Halon production was halted in developed countries in 1994 by the terms of the Montreal Protocol. However, use of existing stocks continues, as do releases from fire-fighting equipment. In addition, in the 1990s, China and Korea—which, as developing countries, have until 2010 to terminate production—increased their production of these chemicals. For these reasons, the atmospheric concentration of this chemical continued to rise.

The other bromine-containing ODS is the pesticide gas **methyl bromide,** CH_3Br. Scientifically we do not yet have a good handle on atmospheric methyl bromide. In particular:

• Significant new sources of natural emission of the gas to the atmosphere continue to be discovered. Consequently, even the approximate ratio of synthetic to natural emissions is uncertain, as is the lifetime of about one year.

• The tropospheric concentration of the gas has changed much more since 1999 than had been anticipated by production levels and controls. Its concentration is still declining, though now at a slower rate.

Methyl bromide was added to the Montreal Protocol during the 1992 revision of the international treaty. It was agreed that developed countries would phase out methyl bromide production and importation completely in 2005. Its consumption in all developing countries combined, which amounted to less than half the U.S. usage, was to have been frozen at 1995–1998 levels in 2002, was to have been reduced by 20% in 2005, and is to be completely eliminated by 2015. However, its phase-out has been strongly resisted by some U.S. farmers, and planned reductions have been deferred. The pros and cons of implementing the Montreal Protocol controls on this controversial chemical are discussed in the online Case Study *Strawberry Fields—The Banning of Methyl Bromide* on the website associated with this chapter.

As a direct result of the implementation of the gradual phase-out of ozone-depleting substances, the tropospheric concentration of chlorine peaked in 1993–1994, and had declined by about 5% by 2000. Much of the initial drop was due to the phase-out of methyl chloroform, which has a short atmospheric lifetime. The concentrations of CFCs are slow to drop because they were used in many applications such as foams and cooling devices that only slowly emit them to the atmosphere. The stratospheric chlorine-equivalent level was predicted to have peaked, at less than 4 ppb, at the turn of the century, with a gradual decline predicted thereafter (see Figure 2-10). Observations in 2000 indicated that the actual chlorine content in the stratosphere had peaked, but the bromine abundance was still increasing. The slowness in the decline of the stratospheric chlorine level is due to

- the long time it takes molecules to rise to the middle or upper stratosphere and to then absorb a photon and dissociate to atomic chlorine,

- the slowness of the removal of chlorine and bromine from the stratosphere, and

- the continued input of some chlorine and bromine into the atmosphere.

Because ozone is formed (and destroyed) in rapid natural processes, its level responds very quickly to a change in stratospheric chlorine concentration. Thus the Antarctic ozone hole probably will not continue to appear after the middle of the twenty-first century, that is, once the chlorine-equivalent concentration is reduced back to the 2-ppb level it had in the years before the hole began to form (Figure 2-10). Without the Montreal Protocol agreements, catastrophic increases in chlorine, to many times the present level, would have occurred, particularly since CFC usage and atmospheric release in developing countries would have increased dramatically. A further doubling of stratospheric chlorine levels would probably have led to the formation of a substantial ozone hole each spring over the Arctic region. And with significant ozone depletion over temperate areas would have come a catastrophic increase in skin cancers.

PROBLEM 2-7

Given that their C—H bonds are not quite as strong as those in CH_4, can you rationalize why ethane, C_2H_6, or propane, C_3H_8, is a better choice than methane to inactivate atomic chlorine in the stratosphere?

PROBLEM 2-8

No controls on the release of CH_3Cl, CH_2Cl_2, or $CHCl_3$ have been proposed. What does that imply about their atmospheric lifetimes, compared to those for CFCs, CCl_4, and methyl chloroform?

 Green Chemistry: Harpin Technology—Eliciting Nature's Own Defenses Against Diseases

Earlier in the chapter, we learned that methyl bromide is used as a pesticide (more specifically, as a soil fumigant), some of which finds its way into the stratosphere, where it becomes involved in the destruction of the ozone layer. An interesting development, which offers an alternative to methyl bromide, is known as *harpin technology*. This technology was developed by EDEN Bioscience Corporation in Bothell, Washington, for which it was awarded a Presidential Green Chemistry Challenge Award in 2001.

Harpin is a naturally occurring bacterial protein that was isolated from the bacteria *Erwinia amylovora* at Cornell University. When applied to the stems and leaves of plants, harpin elicits the plant's natural defense mechanisms to diseases caused by bacteria, viruses, nematodes, and fungi. Hypersensitive response (HR), which is induced by harpin, is an initial defense by plants to invading pathogens that results in cell death at the point of infection. The dead cells surrounding the infection act as a physical barrier to the spread of the pathogen. In addition, the dead cells may release compounds that are lethal to the pathogen.

Pests often build up immunity to pesticides. However, since harpin does not directly affect the pest, it is unlikely that immunity will occur with it. In addition to using traditional pesticides to control infestations in plants, more recently a second approach to this problem has been to develop genetically altered plants. The DNA in such plants has been altered to provide the plant with a means to ward off various pests. Although this approach is often quite successful, it is not without its critics, especially in Europe, where genetically altered plants face serious restrictions. In contrast, harpin has no effect on the plant's DNA: It simply activates defenses that are innate to the plant.

Traditional pesticides are generally made by chemists employing lengthy chemical syntheses, which invariably create large quantities of waste, which is often toxic. In addition, the compounds (chemical feedstocks) from which the pesticides are produced are derived from petroleum. Approximately 2.7% of all petroleum is used to produce chemical feedstocks, and thus the production of these compounds is in part responsible for the depletion of this nonrenewable resource. In contrast, harpin is made from a genetically altered benign laboratory strain of the *Escherichia coli* bacteria through a fermentation process. After the fermentation is complete, the bacteria are destroyed and the harpin protein is extracted. Most of the wastes are biodegradable. Thus the production of harpin produces only nontoxic biodegradable wastes and does not require petroleum.

Harpin has very low toxicity. In addition, it is applied at 0.002–0.06 kg/acre, which represents an approximately 70% reduction in quantity when compared to conventional pesticides. Harpin is rapidly decomposed by UV light and microorganisms, which is in part responsible for its lack of contamination and buildup in soil, water, and organisms as well as the fact that it leaves no residue in foods.

An added benefit of harpin is that it also acts as a plant growth stimulant. Harpin is thought to aid in photosynthesis and nutrient uptake, resulting in increased biomass, early flowering, and enhanced fruit yields. Harpin is sold as a 3% solution in a product called *Messenger*.

Review Questions

1. What is a *Dobson unit*? How is it used in relation to atmospheric ozone levels?

2. If the overhead ozone concentration at a point above the Earth's surface is 250 DU, what is the equivalent thickness in millimeters of pure ozone at 1.0 atm pressure?

3. Describe the process by which chlorine becomes activated in the Antarctic ozone-hole phenomenon.

4. What are the steps in Mechanism II by which atomic chlorine destroys ozone in the spring over Antarctica?

5. Explain why full-scale ozone holes have not yet been observed over the Arctic.

6. What are two effects on human health that scientists believe will result from ozone depletion?

7. Define what is meant by a tropospheric *sink*.

8. Explain what HCFCs are, and state what sort of reaction provides a tropospheric sink for them. Is their destruction in the troposphere 100% complete? Why are HCFCs not considered to be suitable long-term replacements for CFCs?

9. What types of chemicals are proposed as long-term replacements for CFCs?

10. Chemically, what are *halons*? What was their main use?

11. What gases are being phased out according to the Montreal Protocol agreements?

Green Chemistry Questions

See the discussion of focus areas and the principles of green chemistry in the Introduction before attempting these questions.

1. The development of carbon dioxide as a blowing agent for foam polystyrene won a Presidential Green Chemistry Challenge Award.

(a) Into which of the three focus areas for these awards does this award best fit?

(b) List two of the twelve principles of green chemistry that are addressed by the green chemistry of the carbon dioxide process.

2. What environmental advantages does the use of carbon dioxide as a blowing agent have over the use of CFCs and hydrocarbons?

3. Does the carbon dioxide that is used as a blowing agent contribute to global warming?

4. The development of harpin won a Presidential Green Chemistry Challenge Award.

(a) Into which of the three focus areas for these awards does this award best fit?

(b) List four of the twelve principles of green chemistry that are addressed by the green chemistry in the use of harpin.

5. Why is there little concern that pests will develop immunity to harpin?

6. Why is harpin not expected to accumulate in the environment?

Additional Problems

1. (a) Some authors use milliatmosphere centimeter (matm cm) rather than the equivalent Dobson unit for the amount of overhead ozone; 1 matm cm = 1 DU. Prove that the number of moles of overhead ozone over a unit area on the Earth's surface is proportional to the height of the layer, as specified in the definition of Dobson unit, and that 1 DU is equal to 1 matm cm.

(b) Calculate the total mass of ozone that is present in the atmosphere if the average overhead amount is 350 Dobson units, and given that the radius of the Earth is about 5000 km. [*Hints: The volume of a sphere, which you can approximate the Earth to be, is $4\pi r^3/3$. You may assume that ozone behaves as an ideal gas.*]

2. The chemical formula for any CFC, HCFC, or HFC can be obtained by adding 90 to its code number. The three numerals in the result represent the number of C, H, and F atoms, respectively. The number of Cl atoms can then be determined using the condition that the number of H, F, and Cl atoms must add up to $2n + 2$, where n is the number of C atoms. From the information, deduce the formulas for compounds with the following codes:

(a) 12   (b) 113   (c) 123   (d) 124

3. Using the information discussed in Additional Problem 2, deduce the code numbers for each of the following compounds:

(a) CH_3CCl_3 (b) CCl_4 (c) CH_3CFCl_2

4. Using the information in Additional Problem 2, show that 134 is the appropriate code number for CH_2FCF_3. Why is an a or b designation also required to uniquely characterize this compound?

5. The chlorine dimer mechanism is not implicated in significant ozone destruction in the lower stratosphere at mid-latitudes even when the particle concentration is enhanced by volcanoes. Deduce two reasons why this mechanism is not important under these conditions.

6. As discussed in the text, the seriousness of the Antarctic ozone hole in any given year can be analyzed in terms of (a) the minimum ozone concentration reached, (b) the average October (or mid-September to mid-October) ozone concentration, (c) the geographic area that the hole covers, or (d) the number of days or weeks that very low ozone values are recorded. By consulting several of the websites listed for this chapter at www.whfreeman.com/envchem/, construct up-to-date graphs of the variation in each of these parameters. Is the annual hole showing signs yet of reduction in seriousness according to any of these parameters?

7. If bromine, rather than chlorine, had been used to make CFC-like compounds, would stratospheric ozone depletion have been worse or not as severe over mid-latitude regions? [*Hint: Recall that more bromine normally exists in catalytically active form than does chlorine.*]

8. When Mechanism II for ozone destruction operates with X = Cl and X′ = Br, the radicals ClO and BrO react together to reform atomic chlorine and bromine (see Problem 2-1). A fraction of the latter process proceeds by the intermediate formation of BrCl, which undergoes photolysis in daylight. At night, however, all the bromine eventually ends up as BrCl, which does not decompose and restart the mechanism until dawn. Deduce why all the bromine exists as BrCl at night, even though only a fraction of the ClO with BrO collisions yields this product.

9. Consider the following set of compounds: $CFCl_3$, $CHFCl_2$, CF_3Cl, and CHF_3. Assuming that equal numbers of moles of each were released into the air at ground level, rank these four compounds in terms of their potential to catalytically destroy ozone in the stratosphere. Explain your ranking.

10. What would be the advantages of using hydrocarbons rather than CFCs or HCFCs as aerosol propellant to replace CFCs? What is their major disadvantage? What type of agent should be added to aerosol cans containing hydrocarbon propellants to overcome this disadvantage and make them safer?

Further Readings

1. M. J. Molina and F. S. Rowland, "Stratospheric Sink for Chlorofluoromethanes: Chlorine Atom-Catalyzed Destruction of Ozone," *Nature* 249 (1974): 810–812. [The original article concerning CFC destruction of ozone.]

2. S. Solomon, "Stratospheric Ozone Depletion: A Review of Concepts and History," *Journal of Geophysics* 37 (1999): 275–316.

3. (a) E. C. Weatherhead and S. B. Andersen, "The Search for Signs of Recovery of the Ozone Layer," *Nature* 441 (2006): 39–45. (b) S. Solomon, "The Hole Truth," *Nature* 427 (2004): 289–291.

4. (a) A. E. Waibel et al., "Arctic Ozone Loss Due to Denitrification," *Science* 283 (1999): 2064–2068. (b) G. Walker, "The Hole Story," *New Scientist* (25 March 2000): 24–28.

5. M. Rex et al., "Arctic Winter 2005: Implications for Stratospheric Ozone Loss and Climate Change," *Geophysical Research Letters* 31 (2006): L04116.

Websites of Interest

Log on to www.whfreeman.com/envchem4/ and click on Chapter 2. Several of them show trends and up-to-date satellite data (often as color contour diagrams, some with animation) concerning stratospheric ozone depletion in polar regions.

CHAPTER 3

THE CHEMISTRY OF GROUND-LEVEL AIR POLLUTION

In this chapter, the following introductory chemistry topics are used:

- Ideal gas law
- Equilibrium concept, including redox reactions and their balancing
- Acid–base theory, including pH and weak acid calculations

Background from Chapter 1 used in this chapter:

- Excited states
- Photon energies
- Gas-phase catalysis
- Sink concept

this most excellent canopy,
*   the air,*
look you, this excellent
*   o'erhanging*
firmament, this magestic roof
*   fretted*
with golden fire, why, it
*   appears*
no other thing to me than a
*   foul*
and pestilent congregation of
*   vapours*

William Shakespeare, Hamlet, Act II, Scene 2

Introduction

One of the most important features of the Earth's atmosphere is that it is an *oxidizing* environment, a phenomenon due to the large concentration of **diatomic oxygen,** O_2, that it contains. Almost all the gases that are released into the air, whether "natural" substances or "pollutants," are eventually completely oxidized in the atmosphere and the end-products subsequently deposited on the Earth's surface. The oxidation reactions are vital to the cleansing of the air. The best-known example of air pollution is the smog that occurs in many cities throughout the world. The reactants that produce the most common type of smog are mainly emissions from cars and electric power plants, although in rural areas some of the ingredients are supplied by emissions from forests. The operation of motor vehicles produces more air pollution than does any other single human activity. Ironically, diatomic oxygen is also involved in the generation of smog.

In this chapter, the chemistry underlying the pollution of tropospheric air is examined. As background, we begin the chapter by discussing the concentration units by which gases in the lower atmosphere are reported as well as the constitution and chemical reactivity of clean air. The effects of polluted air upon the environment and upon human health are discussed in Chapter 4.

Concentration Units for Atmospheric Pollutants

There is no consensus regarding the appropriate units by which to express concentrations of substances in air. In Chapter 1, ratios involving numbers of molecules—the "parts per" system—were emphasized as a measure. Other measures are often also encountered and will be used in this chapter:

Molecules of a gas per cubic centimeter of air, molecules/cm^3

Micrograms of a substance per cubic meter of air, μg/m^3

Moles of a gas per liter of air, moles/L

Given the lack of a consensus on a single appropriate scale, it is important to be able to convert gas concentrations from one set of units to another. This form of manipulation is discussed in Box 3-1. Note as well that gas partial pressures stated in atmospheres are synonymous with concentrations on the "parts per" scales, so, for example, a partial pressure of 0.002 atm in air is equivalent to 2000 ppm, since $2000 \times 10^{-6} = 0.002$.

The Chemical Fate of Trace Gases in Clean Air

In addition to the well-known stable constituents of the atmosphere (N_2, O_2, Ar, CO_2, H_2O), the troposphere contains a number of gases that are present in only trace concentrations, since they have efficient sinks and thus do not accumulate.

From biological and volcanic sources, the atmosphere regularly receives inputs of the partially oxidized gases **carbon monoxide,** CO, and **sulfur dioxide,** SO_2, and of several gases that are simple compounds of hydrogen, some of whose atoms are in a highly reduced form (e.g., H_2S, NH_3); the most important of these "natural" substances are listed in Table 3-1 on page 95.

Although most of these natural gases are gradually oxidized in air, none of them react *directly* with diatomic oxygen molecules. Rather, their reactions begin when they are attacked by the **hydroxyl free radical,** OH, even though the concentration of this species in air is exceedingly small, a few million molecules per cubic centimeter on average (see Problem 3-1). In clean tropospheric air, as in the stratosphere, the hydroxyl radical is produced when a small fraction of the excited oxygen atoms resulting from the photochemical decomposition of trace amounts of atmospheric **ozone,** O_3,

BOX 3-1	The Interconversion of Gas Concentrations

The number of moles of a substance is proportional to the number of the molecules of it (Avogadro's number, 6.02×10^{23}, is the proportionality constant), and the partial pressure of a gas is proportional to the number of moles of it. Thus, a concentration, for example, of 2 ppm for any pollutant gas present in air means

2 molecules of the pollutant in 1 million molecules of air

2 moles of the pollutant in 1 million moles of air

2×10^{-6} atm partial pressure of pollutants per 1 atm total air pressure

2 L of pollutant in 1 million liters of air (when the partial pressures and temperatures of pollutant and air have been adjusted to be equal)

Let us convert a concentration of 2 ppm to its value in molecules (of pollutant) per cubic centimeter (cm^3) of air for conditions of 1 atm total air pressure and 25°C. Since the value of the numerator, 2 molecules, in the new concentration scale is the same as in the original, all we need to do is establish the volume, in cubic centimeters, that 1 million molecules of air occupy. This volume is easy to evaluate using the ideal gas law ($PV = nRT$), since we know that

$$P = 1.0 \text{ atm}$$
$$T = 25 + 273 = 298 \text{ K}$$
$$n = (10^6 \text{ molecules})/$$
$$(6.02 \times 10^{23} \text{ molecules/mol})$$
$$= 1.66 \times 10^{-18} \text{ mol}$$

and the gas constant $R = 0.082$ L atm/mol K.

Now $PV = nRT$, so

$$V = nRT/P$$
$$= 1.66 \times 10^{-18} \text{ mol}$$
$$\times 0.082 \text{ L atm/mol K}$$
$$\times 298 \text{ K}/1 \text{ atm}$$
$$= 4.06 \times 10^{-17} \text{ L}$$

Since 1 L = 1000 cm^3, then V = 4.06×10^{-14} cm^3. Since 2 molecules of pollutant occupy 4.06×10^{-14} cm^3, it follows that the concentration in the new units is 2.0 molecules/4.06×10^{-14} cm^3, or 4.9×10^{13} molecules/ cm^3.

In general, the most straightforward strategy to use to change the value of a concentration a/b from one scale to its value p/q on another is to independently convert the units of the numerator a to the units of the numerator p (both of which involve only the pollutant) and then convert the denominator b to its new value q (both of which involve the total air sample).

To convert a value in molecules/cm^3 or ppm to mol/L, we must change the molecules of pollutant to the number of moles of pollutant; for a pollutant concentration, again of 2 ppm, we can write

moles of pollutant

$$= (2 \text{ molecules} \times 1 \text{ mol})/$$
$$(6.02 \times 10^{23} \text{ molecules})$$
$$= 3.3 \times 10^{-24} \text{ mol}$$

Thus the molarity is $(3.3 \times 10^{-24} \text{ mol})/(4.06 \times 10^{-17} \text{ L})$, or 8.2×10^{-8} M.

An alternative way to approach these conversions is to use the definition that 2 ppm

(continued on p. 94)

BOX 3-1 | The Interconversion of Gas Concentrations (continued)

means 2 L of pollutant per 1 million liters of air and to find the number of moles and molecules of pollutant contained in a volume of 2 L at the stated pressure and temperature.

A unit often used to express concentrations in polluted air is micrograms per cubic meter, i.e., $\mu g/m^3$. If the pollutant is a pure substance, we can interconvert such values into the molarity and the "parts per" scales, provided that the pollutant's molar mass is known.

Consider as an example the conversion of 320 $\mu g/m^3$ to the ppb scale if the pollutant is SO_2, the total air pressure is 1.0 atm, and the temperature is 27°C. Initially the concentration is

$$\frac{320 \ \mu g \text{ of } SO_2}{1 \text{ m}^3 \text{ of air}}$$

First we convert the numerator from grams of SO_2 to moles, since from there we can obtain the number of molecules of SO_2:

$$320 \times 10^{-6} \text{g } SO_2 \times \frac{1 \text{ mol } SO_2}{64.1 \text{ g } SO_2}$$
$$\times \frac{6.02 \times 10^{23} \text{ molecules of } SO_2}{1 \text{ mol } SO_2}$$
$$= 3.01 \times 10^{18} \text{ molecules of } SO_2$$

Then, using the ideal gas law, we can change the volume of air to moles and then molecules, using $1 \text{ L} = 1 \text{ dm}^3 = (0.1 \text{ m})^3$:

$$n = PV/RT = 1.0 \text{ atm} \times 1.0 \text{ m}^3$$
$$\times \frac{\dfrac{1 \text{ L}}{(0.1 \text{ m})^3} \Big/ 0.082 \text{ L atm}}{\text{mole K} \times 300 \text{ K}}$$
$$= 40.7 \text{ mol}$$

Now 40.7 mol $\times$ 6.02 $\times 10^{23}$ molecules/mol = 2.45 $\times 10^{25}$ molecules, or 2.45 $\times 10^{16}$ billion molecules of air.

Thus the SO_2 concentration is

$$\frac{3.01 \times 10^{18} \text{ molecules of } SO_2}{2.45 \times 10^{16} \text{ billion molecules of air}}$$
$$= 123 \text{ ppb}$$

Note that the conversion of moles to molecules was not strictly necessary, as Avogadro's number cancels from numerator and denominator. As stated previously, ppb refers to the ratio of the number of moles as well as to the ratio of the number of molecules.

It is vital in all interconversions to distinguish between quantities associated with the pollutant and those of air.

PROBLEM 1

Convert a concentration of 32 ppb for any pollutant to its value on

(a) the ppm scale,

(b) the molecules per cm^3 scale, and

(c) the molarity scale.

Assume 25°C and a total pressure of 1.0 atm.

PROBLEM 2

Convert a concentration of 6.0 $\times 10^{14}$ molecules/cm^3 to the ppm scale and to the moles per liter (molarity) scale. Assume 25°C and 1.0 atm total air pressure.

PROBLEM 3

Convert a concentration of 40 ppb of ozone, O_3, into

(a) the number of molecules per cm^3, and

(b) micrograms per m^3.

Assume the air mass temperature is 27°C and its total pressure is 0.95 atm.

PROBLEM 4

The average outdoor concentration of carbon monoxide, CO, is about 1000 $\mu g/m^3$. What is this concentration expressed on the ppm scale? On the molecules per cm^3 scale? Assume that the outdoor temperature is 17°C and that the total air pressure is 1.04 atm.

react with gaseous water to abstract one hydrogen atom from each H_2O molecule:

$$O_3 \xrightarrow{\text{UV-B}} O_2 + O^*$$

$$O^* + H_2O \longrightarrow 2\,OH$$

The average tropospheric lifetime of a given hydroxyl radical is only about one second, since it reacts quickly with one or another of many atmospheric gases. Because the lifetime of hydroxyl radicals is short and sunlight is

TABLE 3-1	Some Important Gases Emitted into the Atmosphere from Natural Sources		
Formula	Name	Main Natural Source	Atmospheric Lifetime
NH_3	Ammonia	Anaerobic biological decay	Days
H_2S	Hydrogen sulfide	Anaerobic biological decay	Days
HCl	Hydrogen chloride	Anaerobic biological decay, volcanoes	
SO_2	Sulfur dioxide	Volcanoes	Days
NO	Nitric oxide	Lightning	Days
CO	Carbon monoxide	Fires; CH_4 oxidation	Months
CH_4	Methane	Anaerobic biological decay	Years
CH_3Cl	Methyl chloride	Oceans	Years
CH_3Br	Methyl bromide	Oceans	Years
CH_3I	Methyl iodide	Oceans	

required to form more of them, the OH concentration drops quickly at nightfall. Recall from Problem 1 of Box 1-2 that because the corresponding reaction involving unexcited atomic oxygen atoms is endothermic, its activation energy is high and consequently it occurs far too slowly to be a significant source of atmospheric OH. Although OH participates in many atmospheric reactions, it has been found recently that its concentration is directly proportional to the O* concentration at any given time.

PROBLEM 3-1

In one study, the concentration of OH in air at the time was found to be 8.7×10^6 molecules per cubic centimeter. Calculate its molar concentration and its concentration in parts per trillion, assuming that the total air pressure is 1.0 atm and the temperature is 15°C.

The hydroxyl free radical is reactive toward a wide variety of other molecules, including the hydrides of carbon, nitrogen, and sulfur listed in Table 3-1, and many molecules containing multiple bonds (double or triple bonds), including CO and SO_2. Although suspected for decades of playing a pivotal role in air chemistry, the presence of OH in the troposphere was confirmed only recently since its concentration is so very small. The great importance of the hydroxyl radical to tropospheric chemistry arises because it, not O_2, initiates the oxidation of *all* the gases in Table 3-1 other than HCl. Without OH and its related reactive species HOO, most of these gases would not be efficiently removed from the troposphere, nor would most pollutant gases such as the unburned hydrocarbons emitted from vehicles. Indeed, OH has been called the "tropospheric vacuum cleaner" or "detergent." The reactions that it initiates correspond to a flameless, ambient-temperature "burning" of the reduced gases of the lower atmosphere. If these gases were to accumulate, the atmospheric composition would be quite different, as would the forms of life that are viable on Earth. Interestingly, hydroxyl is unreactive to molecular oxygen—in contrast to the behavior of O_2 with many other free radicals—and to molecular nitrogen, so it survives long enough to react with so many other species.

An example of the reactions initiated by hydroxyl radical is the net oxidation of **methane** gas, CH_4, into the completely oxidized product **carbon dioxide**, CO_2:

$$CH_4 + 2\,O_2 \xrightarrow{\text{OH catalyst}} CO_2 + 2\,H_2O$$

As we shall see in Chapter 5, this overall reaction occurs by a sequence of reactions, the first of which involves the reaction of hydroxyl radical with methane, and the next-to-last of which involves the reaction of OH with carbon monoxide. Indeed, this pair of reactions accounts for the fate of most hydroxyl radicals in a clean atmosphere. However, since a new hydroxyl radical is also produced eventually in the multistep reaction sequence, it is acting as a catalyst. Since the OH is originally produced from O_3, the case can be made that it is really ozone that causes the oxidation of most atmospheric gases. Diatomic molecular oxygen reacts with some of the free-radical species produced by OH reactions, so it does appear in the overall equation as the substance that oxidizes the reactants.

The *hydrogen halides* (HF, HCl, HBr) and fully oxidized gases such as carbon dioxide are relatively unreactive (from the oxidation–reduction point of view) in the troposphere because no further oxidation occurs with them; they eventually are deposited on the Earth's surface, often as a result of dissolving in raindrops.

Urban Ozone: The Photochemical Smog Process
The Origin and Occurrence of Smog

Many urban centers in the world undergo episodes of air pollution during which relatively high levels of ground-level ozone—an undesirable constituent of air if present in appreciable concentrations at low altitudes in the air that we breathe—are produced as a result of the light-induced chemical reaction of pollutants. This phenomenon is called **photochemical smog** and is sometimes characterized as "an ozone layer in the wrong place," to contrast it with the beneficial stratospheric ozone discussed in Chapter 1. The word *smog* is a combination of *smoke* and *fog*. The process of smog formation involves hundreds of different reactions, involving dozens of chemicals, occurring simultaneously. Indeed, urban atmospheres have been referred to as giant chemical reactors. The most important reactions that occur in such air masses will be discussed in detail in Chapter 5. In the material below, we investigate the nature and origin of the pollutants—especially nitrogen oxides—that combine to produce photochemical smog.

The chief original reactants in an episode of photochemical smog are molecules of **nitric oxide,** NO, and of unburned and partially oxidized hydrocarbons that are emitted into the air as pollutants from internal combustion engines; nitric oxide is also released from electric power plants. The concentrations of these chemicals are orders of magnitude greater than are found in clean air. Gaseous hydrocarbons and partially oxidized hydrocarbons are also present in urban air as a result of the evaporation of solvents, liquid fuels, and other organic compounds. Collectively, the substances, including hydrocarbons and their derivatives, that readily vaporize into the air are called **volatile organic compounds,** or VOCs. (Formally, VOCs are defined as organic compounds having boiling points that lie between 50°C and 260°C.) For example, vapor is released into the air when a gasoline tank is filled unless the

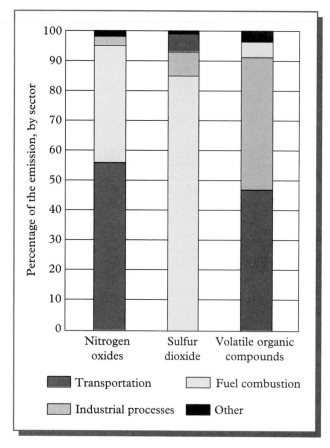

FIGURE 3-1 North American emissions of primary gaseous air pollutants from various sectors. [Source: U.S. EPA 1999 National Air Quality Trends Report.]

hose's nozzle is specially designed to minimize this loss. Evaporated, unburned gasoline is also emitted from the tailpipe of a vehicle before its catalytic converter has been warmed sufficiently to operate. Two-cycle engines such as those in outboard motor boats are particularly notorious for emitting significant proportions of their gasoline unburned into the air. Personal watercraft manufactured in the 1990s, before pollution controls came into effect, emitted more smog-producing emissions in a day's operation than an automobile of the same era driven for several years! Regulations proposed recently in California would require new lawn mowers to be outfitted with a catalytic converter, though this issue is controversial since some mower manufacturers claim that a hot converter could pose a fire hazard to the engine.

Another vital ingredient in photochemical smog is sunshine, which increases the concentration of free radicals that participate in the chemical processes of smog formation. Although the reactants—NO and VOCs—are relatively innocuous, the final products of the smog reaction—ozone, **nitric acid,** HNO_3, and partially oxidized (and in some cases nitrated) organic compounds—are much more toxic.

$$VOCs + NO + O_2 + sunlight \longrightarrow \longrightarrow mixture\ of\ O_3, HNO_3, organics$$

Substances such as NO, hydrocarbons, and other VOCs that are emitted directly into air are called **primary pollutants;** the substances into which they are transformed, such as O_3 and HNO_3, are called **secondary pollutants.** A summary of the relative importance of various economic sectors in emissions of the primary pollutants sulfur dioxide, nitrogen oxides, and VOCs in the United States and Canada is given in Figure 3-1.

Other than those that absorb sunlight and subsequently decompose, most atmospheric molecules that are transformed in air begin by reacting with the hydroxyl free radical, OH, which consequently is the key reactive species in the troposphere. The most reactive VOCs in urban air are hydrocarbons that contain a carbon–carbon double bond, C=C, and *aldehydes*, since their reactions with OH—and also with sunlight in the latter case—are very fast. Other hydrocarbons such as methane are also present in air, but due to the higher activation energy required, their reaction with OH is sluggish. However, their reaction can become important in late stages of photochemical smog episodes.

Nitrogen Oxide Production During Fuel Combustion

Nitrogen oxide gases are produced by two different reactions whenever a fuel is burned in air with a hot flame. Some nitric oxide is produced from the oxidation of nitrogen atoms contained in the fuel itself; it is called **fuel NO.** About 30–60% of a fuel's nitrogen is converted to NO during combustion. However, most fuels do not contain much nitrogen, so this process accounts for only a small fraction of NO emissions.

Nitric oxide, produced by the oxidation at high combustion temperatures of atmospheric nitrogen, is called **thermal NO.** At high flame temperatures, some of the nitrogen and oxygen gases in the air passing through the flame combine to form NO:

$$N_2 + O_2 \xrightarrow{\text{hot flame}} 2\,NO$$

The higher the flame temperature, the more NO is produced. Since this reaction is very endothermic, its equilibrium constant is very small at normal temperatures but increases rapidly as the temperature rises. One might expect that the relatively high concentrations of NO that are produced under combustion conditions would revert back to molecular nitrogen and oxygen as the exhaust gases cool, since the equilibrium constant for the above reaction is much smaller at lower temperatures. However, the activation energy for the reverse reaction is also quite high, so the process cannot occur to an appreciable extent except at high temperatures. Thus the relatively high concentrations of nitric oxide produced during combustion are maintained in the cooled exhaust gases; equilibrium cannot be quickly re-established, and the nitrogen is "frozen" as NO.

Because the reaction between N_2 and O_2 has a high activation energy, it is negligibly slow except at very high temperatures, such as occur in the modern combustion engines of vehicles—particularly when they are traveling at high speeds—and in power plants. Very little NO is produced by the burning of wood and other natural materials since the flame temperatures involved in such combustion processes are relatively low.

Two distinct mechanisms are involved in the initiation of the reaction of molecular nitrogen and oxygen to produce thermal nitric oxide; in one it is atomic oxygen that attacks intact N_2 molecules, whereas in the other it is free radicals, such as CH, that are derived from the decomposition of the fuel. The initial reaction steps of the first mechanism are

$$O_2 \rightleftharpoons 2\,O$$

$$O + N_2 \longrightarrow NO + N$$

The rate of the second, slower step is proportional to $[O]\,[N_2]$. However, since, from the equilibrium in the first step, $[O]$ is proportional to the square root of $[O_2]$, it follows that the rate of NO formation will be proportional to $[N_2]\,[O_2]^{1/2}$.

  The nitric oxide released into air is gradually oxidized to **nitrogen dioxide,** NO_2, over a period of minutes to hours, the rate depending upon the concentration of the pollutant gases present. Collectively, NO and NO_2 in air is referred to as NO_X, pronounced "nox." The yellow-brown color in the atmosphere of a smog-ridden city is due in part to the nitrogen dioxide present, since this gas absorbs visible light, especially near 400 nm (see its spectrum in Figure 3-2), removing sunlight's purple component while allowing most yellow light to be transmitted. The small levels of NO_X in clean air result in part from the operation of the above reaction in the very energetic environment of lightning flashes and in part from the release of NO_X and of **ammonia,** NH_3, from biological sources. Recently it has been discovered that NO_X is emitted from coniferous trees when sunlight shines on them and when the ambient concentrations of these gases are low.

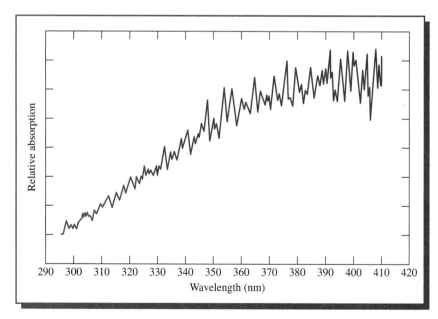

FIGURE 3-2 Absorption spectrum of NO_2 in the solar ultraviolet region.
[Source: Reprinted from A. M. Bass et al., *J. Res. U.S. Natl. Bur. Stand.* A80 (1976): 143.]

Ground-Level Ozone in Smog

Photochemical smog is a widespread phenomenon in the modern world. If we are to prevent or limit its formation, we must understand the main chemical reactions that occur in it. Although the detailed reaction mechanism in smog is quite complicated (discussed further in Chapter 5), its most important aspects are discussed below.

In order for a city to generate photochemical smog, several conditions must be fulfilled. First, there must be substantial vehicular traffic to emit sufficient NO, reactive hydrocarbons, and other VOCs into the air. Second, there must be warmth and ample sunlight in order for the crucial reactions, some of them photochemical, to proceed at a rapid rate. Finally, there must be relatively little movement of the air mass so that the reactants are not quickly diluted. For reasons of geography (e.g., the presence of mountains) and dense population, cities such as Los Angeles, Denver, Mexico City, Tokyo, Athens, Sao Paulo, and Rome all fit the bill splendidly and consequently are subject to frequent smog episodes. Indeed, the photochemical smog phenomenon was first observed in Los Angeles in the 1940s and has generally been associated with that city ever since, although pollution controls have partially alleviated the smog problem in recent decades.

As in the stratosphere, ground-level ozone is produced by the reaction of oxygen atoms with diatomic oxygen. The main source of the oxygen atoms in the troposphere, however, is the photochemical dissociation by sunlight of nitrogen dioxide molecules:

$$NO_2 \xrightarrow{\text{UV-A}} NO + O$$

$$O + O_2 \longrightarrow O_3$$

According to the results of Problem 1-3, light having wavelength less than 394 nm is capable of producing photochemical decomposition of NO_2. The absorption spectrum of nitrogen dioxide gas in the UV-A region, shown in Figure 3-2, indicates that the gas does indeed absorb in this region, and sunlight having wavelengths of about 394 nm or shorter is found to efficiently induce decomposition.

As discussed above, it is predominantly NO rather than NO_2 that is emitted from vehicles and power plants into the air. In episodes of photochemical air pollution, NO is oxidized to the dioxide gradually over a period of several hours in complex reaction sequences that involve free radicals as catalysts (see Figure 3-3). Indeed, one can see (Figure 3-4) the concentration of NO first rising from emissions from early-morning vehicle traffic and then falling during the morning as it is converted to NO_2 in urban atmospheres on smog days.

The concentration of ozone does not rise significantly in a city generating smog until late in the morning (see Figure 3-4), when the NO concentration

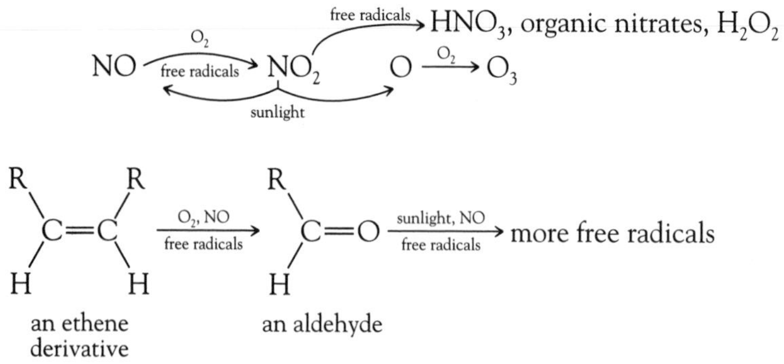

$$NO \xrightarrow[\text{free radicals}]{O_2} NO_2 \xrightarrow[\text{sunlight}]{} \begin{array}{l} \xrightarrow{\text{free radicals}} HNO_3, \text{ organic nitrates, } H_2O_2 \\ O \xrightarrow{O_2} O_3 \end{array}$$

an ethene derivative $\xrightarrow[\text{free radicals}]{O_2, NO}$ an aldehyde $\xrightarrow[\text{free radicals}]{\text{sunlight, NO}}$ more free radicals

FIGURE 3-3 Summary of photochemical smog reactions discussed in detail in Chapter 5.

has been greatly reduced. This happens because residual nitric oxide reacts with and destroys ozone formed in the morning to re-create nitrogen dioxide and oxygen, a reaction that also occurs in the stratosphere:

$$NO + O_3 \longrightarrow NO_2 + O_2$$

The sum of the last three reactions above constitutes a *null cycle*, whereby there is no net buildup of ozone or oxidation of NO to NO_2 by this mechanism.

In fact, the oxidation of NO to NO_2 does occur rapidly, before the pollution has diffused away, partly because of weather conditions and partly because of the high concentration of catalytic free radicals that are generated during a smog episode. More free radicals are produced than are consumed in

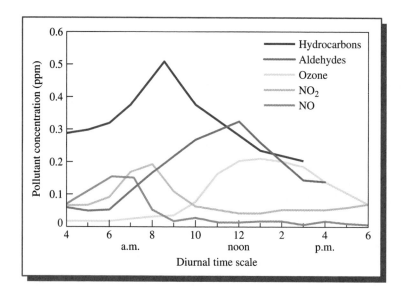

FIGURE 3-4 Time-of-day (diurnal) variation in the concentration of gases during days of marked eye irritation in Los Angeles in the 1960s. [Source: Redrawn from D. J. Speeding, *Air Pollution* (Oxford: Oxford University Press, 1974).]

smog because of the reactions of VOCs, especially those containing highly reactive bonds such as C=C and C=O. For example, ethene and its derivatives react in a complex sequence of reactions with NO, free radicals, and atmospheric oxygen to produce nitrogen dioxide and aldehydes, the concentrations of which rise rapidly in the morning (see Figure 3-4). The aldehydes absorb sunlight with $\lambda < 350$ nm, i.e., UV-B and some UV-A light; some of them photochemically decompose in sunlight to produce additional free radicals, thereby increasing their concentration (see Figure 3-3). Once produced in significant amounts by decomposition of nitrogen dioxide—and not quickly destroyed since the level of NO has abated—some of the ozone also reacts with VOCs to yield more hydroxyl radicals, further accelerating the smog reaction process.

Although our analysis above has identified ozone as the main product of smog, the situation is actually more complicated, as a detailed study in Chapter 5 indicates. Some of the nitrogen dioxide reacts with hydroxyl radical to generate nitric acid, HNO_3, and some reacts with organic free radicals to produce organic nitrates.

Governmental Goals for Reducing Ozone Concentrations

Many countries individually, as well as the World Health Organization (WHO), have established goals for maximum allowable ozone concentrations in air of about 100 ppb or less, averaged over a one-hour period. For example, the standard in Canada is 82 ppb, and that of WHO is 75–100 ppb. The United States has adopted a standard in which the ozone level over an eight-hour period is what is regulated, rather than the one-hour average; the average eight-hour limit was set at 80 ppb in 1997 for the United States, compared to the WHO eight-hour standard of 50–60 ppb. Generally speaking, the longer the period over which the concentration is averaged in a regulation, the lower the stated limit, since it is presumed that exposure to a higher level is acceptable only if it occurs for a short time.

The ozone level in clean air amounts to only about 30 ppb. By way of contrast, the levels of ozone in Los Angeles air used to reach 680 ppb, but peak levels have now declined to 300 ppb. Many major cities in North America, Europe, and Japan exceed ozone levels of 120 ppb typically for 5 to 10 days each summer.

The electrical power blackout that occurred in August 2003 in eastern North America yielded some interesting information concerning the contribution of power plants to air pollution in that region. Measurements over Pennsylvania taken 24 hours after the blackout began found that SO_2 levels were down 90%, and ozone levels down about 50%, compared to a similarly hot, sunny day a year earlier, and that visibility increased by about 40 km because haze from particulates had decreased by 70%.

Photochemical Smog Around the World

The air in Mexico City is so polluted by ozone, particulate matter, and other components of smog, and by airborne fecal matter, that it is estimated to be responsible for thousands of premature deaths annually; indeed, in the center of the city residents can purchase pure oxygen from booths to help them breathe more easily! In 1990, Mexico City exceeded the WHO air guidelines on 310 days, though peak levels have steadily declined since then. In contrast to temperate areas where photochemical smog attacks occur almost exclusively in the summer—when the air is sufficiently warm to sustain the chemical reactions—Mexico City suffers its worst pollution in the winter months, when temperature inversions prevent pollutants from escaping. Some of the smog in Mexico City originates from *butenes* that are a minor component of the liquefied gas that is used for cooking and heating in homes, some of which apparently leaks into the air.

Athens and Rome, as well as Mexico City, attempt to limit vehicular traffic during smog episodes. One strategy used by Athens and Rome is to allow only half the vehicles to be driven on alternate days, the allocation being based upon the license plate numbering (odd or even numbers).

Due to long-range transport of primary and secondary pollutants in air currents, many areas which themselves generate few emissions are subject to regular episodes of high ground-level ozone and other smog oxidants. Indeed, some rural areas, and even small cities, that lie in the path of such polluted air masses experience higher levels of ozone than do nearby larger urban areas. This occurs because in the larger cities, some of the ozone transported from elsewhere is eliminated by reaction with nitric oxide released locally by cars into the air, as illustrated previously in the reaction of NO with O_3. Ozone concentrations of 90 ppb are common in polluted rural areas.

When hot summertime weather conditions produce large amounts of ozone in urban areas but do not allow much vertical mixing of air masses as they travel to rural sites, elevated ozone levels are often observed in eastern North America and western Europe in zones that extend for 1000 km (600 miles) or more. Thus, ozone control is a *regional* rather than a local air quality problem, in contrast to what was usually assumed in the past. Indeed, on occasion, polluted air from North America moves across the Atlantic to Europe, northern Africa, and the Middle East; that from Europe can move into Asia and the Arctic; and that from Asia can reach the west coast of North America. Some analysts believe that by 2100, even the background level of ozone throughout the Northern Hemisphere will probably exceed current ozone standards.

A plot of ozone concentration contours for summer afternoon smog conditions in North America is shown in Figure 3-5a. At each point along any solid line, the concentration of ozone has the same value; hence contours connect regions having equal levels of ozone. The highest levels (100 ppb)

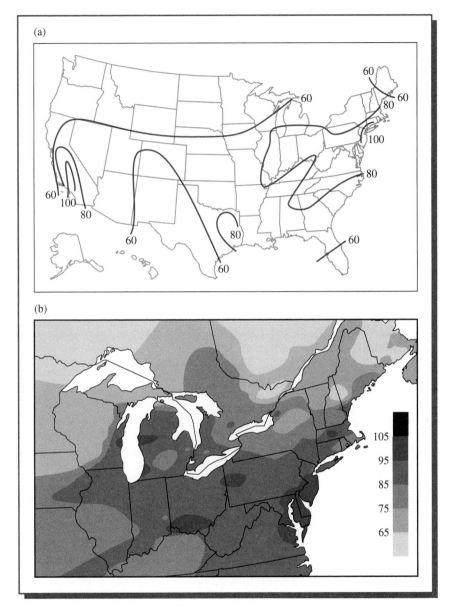

FIGURE 3-5a Ninetieth percentile contours of summer afternoon ozone concentrations (ppb) measured in surface air over the United States. Ninetieth percentile means that concentrations are higher than this 10% of the time. [Adapted from A. M. Fiore, D. J. Jacob, J. A. Logan, and J. H. Yin, "Long-Term Trends in Ground-Level Ozone over the Contiguous United States, 1980–1995," *J. Geo-phys. Res.* 103 (1998): 1471–1480.]

FIGURE 3-5b Maximum surface ozone levels, in ppb, for 1996–1998 in eastern North America. [Source: Environment Canada, "Interim Plan 2001 on Particulate Matter and Ozone," Government of Canada Publication (Ottawa: 2001).]

occur in the Los Angeles and New York–Boston areas, but note the 80-ppb contour over a wide area south of the Great Lakes and into the Southeast, as well as one surrounding Houston. Ozone levels are particularly high over Houston—reaching 250 ppb on occasion—because of emissions of highly reactive VOCs containing C=C bonds from the region encompassing the petrochemical industry. Indeed, as of the late 1990s, Houston had overtaken Los Angeles for the number of days per year in which ozone standards were exceeded.

Considerable ozone is transported from its origin in the U.S. Midwest to surrounding states and Canadian provinces, especially around the Great Lakes (see Figure 3-5b). An example of a rural area subject to high ozone concentrations is the farmland in southwest Ontario, which often receives ozone-laden air from industrial regions in the United States that lie across Lake Erie.

Elevated levels of ozone also affect materials: It hardens rubber, reducing the useful lifespan of consumer products such as automobile tires, and it bleaches color from some materials such as fabrics.

The photochemical production of ozone also occurs during dry seasons in rural tropical areas where the burning of biomass for the clearing of forests or brush is widespread. Although most of the carbon is transformed immediately to CO_2, some methane and other hydrocarbons are released, as is some NO_X. Ozone is produced when these hydrocarbons react with the nitrogen oxides under the influence of sunlight.

Limiting VOC and NO Emissions to Reduce Ground-Level Ozone

In order to improve the air quality in urban environments that are subject to photochemical smog, the quantity of reactants, principally NO_X and hydrocarbons containing $C=C$ bonds plus other reactive VOCs, emitted into the air must be reduced. The control strategies in place in the United States have resulted in some ozone level reduction over the past few decades, notwithstanding the huge increase in total vehicle-miles driven—up to 100% more in the last 25 years.

For economic and technical reasons, the most common control strategy has been to reduce hydrocarbon emissions. However, except in downtown Los Angeles, the percentage reduction in ozone and other oxidants that is achieved usually has been much less than the percentage reduction in hydrocarbons. This happens because usually there is initially an overabundance of hydrocarbons relative to the amount of nitrogen oxides, and cutting back hydrocarbon emissions simply reduces the excess without slowing down the reactions significantly. In other words, it is usually the nitrogen oxides, rather than reactive hydrocarbons, that determine the overall rate of the reaction. This is especially true for rural areas that lie downwind of polluted urban centers.

Due to the large number of reactions that occur in polluted air, the functional dependence of smog production upon reactant concentration is complicated, and the net consequence of making moderate decreases in primary pollutants is difficult to deduce without computer simulation. Computer

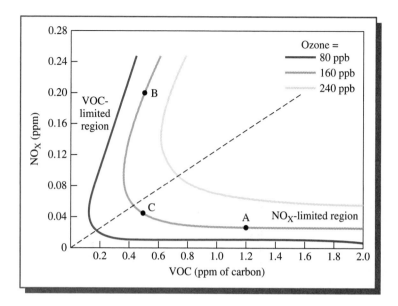

FIGURE 3-6 The relationship between NO_X and VOC concentrations in air and the resulting levels of ozone produced by their reaction. Points A, B, and C denote conditions discussed in the text. [Source: Redrawn from National Research Council, *Rethinking the Ozone Problem in Urban and Regional Air Pollution* (Washington, DC: National Academy Press, 1991).]

modeling indicates that NO_X reduction, rather than VOC reduction, would be much more effective in reducing ozone in almost all of the eastern United States. An example of the predictions that arise from the modeling studies is shown in Figure 3-6. The relationships between the NO_X and the VOC concentrations that produce contours for three different values for the concentration of ozone are shown. Point A represents a typical set of conditions in which the ozone production is NO_X-*limited*. For example, reducing the concentration of VOCs from 1.2 ppm to 0.8 ppm has virtually no effect on the ozone concentration, which remains at about 160 ppb since the curve in this region is almost linear and runs parallel to the horizontal axis. However, a reduction of the NO_X level from about 0.03 ppm at point A to a little less than half this amount, which corresponds to dropping down to the curve directly below it in the figure, cuts the predicted ozone level in half, from 160 ppb to 80 ppb. Chemically, NO_X-limited conditions occur when, due to the high concentration of VOC reactants, an abundance of peroxy free radicals HOO and ROO are produced, which quickly oxidize NO emissions to NO_2:

$$HOO + NO \longrightarrow OH + NO_2$$

The nitrogen dioxide then photochemically decomposes to produce the free oxygen atoms that react with O_2 to produce ozone, as previously discussed (see Figure 3-3).

In the portion of the VOC-*limited* region that lies to the left of the diagonal dashed line of Figure 3-6, there is a large excess of NO_X; under such conditions, the OH radical tends to react with NO_2, so less of it is available to initiate the reaction of more VOCs:

$$OH + NO_2 \longrightarrow HNO_3$$

Consequently, lowering the NO_X concentration actually produces *more* ozone, not less, since more OH is available to react with the VOCs, although production of other smog reaction products such as nitric acid is thereby reduced. Thus, for example, when the VOC concentration is about 0.5 ppm, lowering the NO_X concentration from 0.21 ppm—corresponding to point B on Figure 3-6—even by two-thirds of this amount is predicted to increase the ozone level slightly beyond 160 ppb; further reductions do not begin to decrease ozone until NO_X reaches about 0.05 ppm. In situations where VOCs are relatively plentiful, that is, to the right side of the dashed line in Figure 3-6, reducing NO_X also reduces ozone. Thus, when the VOC level is 0.5 ppm, the ozone concentration falls back to 160 ppb when the NO_X is reduced to 0.04 ppm (point C) and declines more with further decreases of NO_X.

PROBLEM 3-2

Using Figure 3-6 and assuming a NO_X concentration of 0.20 ppm, estimate the effect on ozone levels of reducing the VOC concentration from 0.5 to 0.4 ppm. Do your results support the characterization of that zone of the graph as VOC-limited?

PROBLEM 3-3

Using Figure 3-6, again with an initial VOC concentration of 0.50 ppm, estimate the effect on ozone levels of lowering the NO_X concentration from 0.20 to 0.08 ppm. Explain your results in terms of the chemistry discussed above.

Some urban areas such as Atlanta, Georgia, and others located in the southern United States include or border upon heavily wooded areas whose trees emit enough reactive hydrocarbons to sustain smog and ozone production, even when the concentration of **anthropogenic** hydrocarbons, i.e., those that result from human activities, is low. Deciduous trees and shrubs emit the gas *isoprene*, whereas conifers emit *pinene* and *limonene*; all three hydrocarbons contain C=C bonds. The blue hazes that are observed over forested areas such as the Great Smoky Mountains in North Carolina and the Blue Mountains in Australia result from the reaction of such natural hydrocarbons in sunlight to produce carboxylic acids that condense to form suspended particles of the size that scatter sunlight and thereby produce a haze. Some of the ozone molecules present above the forests react with the C=C bond in the natural hydrocarbons to first produce aldehydes, which are then further oxidized in air to the corresponding carboxylic acids. Eventually, the acids in the aerosol are attacked by hydroxyl radicals, which initiate their decomposition, if the haze is not rained out of the air beforehand.

In urban atmospheres, the concentration of these natural reactive hydrocarbons normally is much less than that of the anthropogenic hydrocarbons, and it is not until the latter are reduced substantially that the influence of these natural substances becomes noticeable. In areas affected by the presence of vegetation, then, only the reduction of emissions of nitrogen oxides will reduce photochemical smog production substantially. As an air mass moves from an urban area to a rural one downwind, it often changes from being *VOC-limited* to being *NO_X-limited*, since there are few sources of nitrogen oxides, but often substantial sources of reactive VOCs, outside cities and since the reactions that consume nitrogen oxides occur more quickly than do those that consume VOCs.

Although hydrocarbons with C=C bonds and aldehydes are the most reactive types in photochemical smog processes, other VOCs play a significant role after the first few hours of a smog episode have passed and the concentration of free radicals has risen. For this reason, control of emissions of *all* VOCs is required in areas with serious photochemical smog problems. Gasoline, which is a complex mixture of hydrocarbons, is now formulated in order to reduce its evaporation, since gasoline vapor has been found to contribute significantly to atmospheric concentrations of hydrocarbons. The control of VOCs in air is discussed in more detail in Chapter 16. New regulations in California (with Los Angeles especially in mind) limit the use of hydrocarbon-containing products such as barbecue-grill starter fluid, household aerosol sprays, and oil-based paints that consist partially of a hydrocarbon solvent that evaporates into the air as the paint dries. The air quality in this region has improved because of current emission controls, but the increase in vehicle-miles driven and the hydrocarbon emissions from nontransportation sources such as solvents have thus far prevented a more complete solution. Research has also indicated that any substantial increase in the emissions of methane to the atmosphere could prolong and intensify the periods of high ozone in the United States, even though CH_4 is usually considered to be a rather unreactive VOC.

Technological Control of Emissions
Catalytic Converters

Over the last decades, automobile manufacturers have employed several strategies to decrease VOC and NO_X emissions from their vehicles and thereby meet governmental standards. One early technique that had some success for NO_X control was to lower the temperature of the combustion flame and thereby decrease the rate of creation of thermal nitric. The temperature lowering was achieved by recirculating a fraction of the engine emissions back through the flame, which presumably lowered the flame temperature and hence the production of thermal NO by lowering the concentration of oxygen in combustion.

In recent decades, more complete control of NO_X emissions from gasoline-powered cars and trucks has been attempted using **catalytic converters** placed just ahead of the mufflers in the vehicle's exhaust system. The original **two-way converters** controlled only carbon-containing gases, including carbon monoxide, CO, by completing their combustion to carbon dioxide. However, by use of a surface impregnated with a platinum–rhodium catalyst, the modern **three-way converter** changes nitrogen oxides back to elemental nitrogen and oxygen using unburned hydrocarbons and the combustion intermediates CO and H_2 as reducing agents:

$$2 \, NO \longrightarrow N_2 + O_2 \quad \text{overall}$$

via, for example,

$$2 \, NO + 2 \, H_2 \longrightarrow N_2 + 2 \, H_2O$$

PROBLEM 3-4

Write and balance reactions in which NO is converted to N_2 by (a) CO and (b) C_6H_{14}. *[Hint: The other reaction product is CO_2, plus H_2O in the latter case.]*

The carbon-containing gases in the exhaust are catalytically oxidized almost completely to CO_2 and water by the oxygen that is present:

$$2 \, CO + O_2 \longrightarrow 2 \, CO_2$$

$$C_nH_m \, (n + m/4) \, O_2 \longrightarrow n \, CO_2 + m/2 \, H_2O$$

$$CH_2O + O_2 \longrightarrow CO_2 + H_2O$$

The catalyst is dispersed as very tiny crystallites, initially less than 10 nm in size. An oxygen sensor in the vehicle's exhaust system is monitored by a computer chip that controls the intake air/fuel ratio of the engine to the stoichiometric amount required by the fuel in order to ensure a high level of conversion of the pollutants. The whole process is illustrated in Figure 3-7a. If the air/fuel mix is not very close to the stoichiometric ratio, the warmed catalyst will not be effective for reduction (if there is too much air), causing nitrogen oxides to be emitted into the air, or for oxidation (if there is too little air), causing CO and hydrocarbons to be emitted, as illustrated in Figure 3-7b.

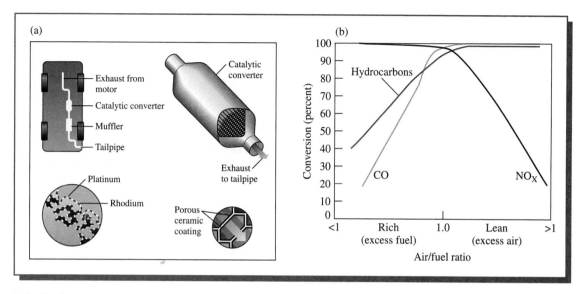

FIGURE 3-7 (a) Modern catalytic converter for automobiles, with its position in the exhaust system indicated. [Source: L. A. Bloomfield, "Catalytic Converter," *Scientific American* (February 2000): 108.] (b) Efficiency in conversion of catalytic converter versus air/fuel ratio. [Source: B. Harrison, "Emission Control," *Education in Chemistry* 37 (2000): 127.]

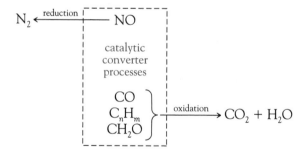

Some progress has been reported recently in the use of less valuable metals, such as copper and chromium, instead of the expensive platinum-group metals as catalysts in catalytic converters. Although the metals are recycled from old converters, a portion is inevitably lost in the process. Recently, scientists have become concerned about the environmental problem of widely broadcasting the tiny particles of platinum, palladium, and rhodium that are lost from the converters themselves during their operation.

The catalyst that reduces nitric oxide to nitrogen also reduces sulfur dioxide, SO_2, to **hydrogen sulfide,** H_2S. The emitted gases include reduced sulfur compounds such as H_2S, which often give vehicle emissions their characteristic odor of rotten eggs. In addition, the small amounts of sulfur-containing molecules in gasoline—and diesel fuel—can partially deactivate catalytic converters if sulfate particles produced during the gasoline's combustion become attached to and thereby cover the active sites of the catalyst metal. The maximum annual average sulfur levels in gasoline, amounting to several hundred parts per million in the past, have recently been reduced to 30 ppm in both the United States and Canada, and to 50 ppm in the European Union. The sulfur is removed during refining, usually by **hydrodesulfurization,** itself a catalytic process, which reacts organic sulfur-containing molecules in the gasoline with hydrogen gas, H_2, to produce hydrogen sulfide, H_2S, which is then removed. Alternatively, the sulfur-containing molecules may be removed from the fuel by absorbing them during the refining process.

In the first few minutes after a vehicle's engine has been started up, the catalysts are cold, so the converters cannot operate effectively and there are bursts of emissions from the tailpipe. Indeed, approximately 80% of all the emissions from converter-equipped cars are produced in the first few minutes after starting. Once an engine has warmed up and the catalysts have been heated to about 300°C by engine exhaust, three-way catalysts convert 80–90% of the hydrocarbons, CO, and NO_X to innocuous substances before the exhaust gases are released into the atmosphere. However, fuel-rich mixtures are fed to the engine in the first minute or so after the vehicle engine is started, and also when high acceleration occurs, so carbon monoxide and unburned hydrocarbons are emitted directly into the air under these oxygen-starved conditions.

Research and development is underway to develop catalytic converters that would convert start-up emissions so that these are not released into the air. Various approaches being investigated include

- devising a converter that will operate at lower temperatures or that can be preheated so it begins to operate immediately,

- storing pollutants until the engine and converter are heated, and

- recirculating engine exhaust through the engine until the reactions are more complete.

Older cars (with no converters or just two-way converters) still on the road continue to pollute the atmosphere with nitrogen oxides even during their normal operation.

The maximum amounts of emissions that can legally be released from light-duty motor vehicles such as cars have gradually been decreased in order to improve air quality. Some governments have recently instituted mandatory inspections of exhaust systems to ensure that they continue to operate properly.

Vehicles whose catalytic converters have been damaged or tampered with produce most of the emissions: Typically, 50% of the hydrocarbons and carbon monoxide are released from 10% of the cars on the road. For example,

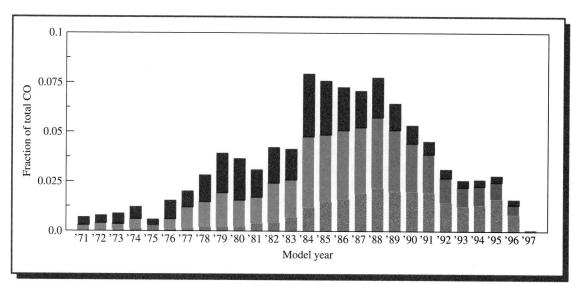

FIGURE 3-8 Fleet-weighted CO contribution of vehicles by model year. [Source: G. A. Bishop et al., "Drive-By Motor Vehicle Emissions: Immediate Feedback in Reducing Air Pollution," *Environ. Sci. Technol.* 34 (2000): 1110.]

a recent study about soot emissions from cars in the Netherlands found that 5% of vehicles accounted for 43% of pollution. A mid-1990s survey of carbon monoxide emitted by passing traffic in Denver, Colorado, produced the results shown in Figure 3-8. The great majority of Denver vehicles of ages up to about 12 years were rated "good" in terms of their control of tailpipe emissions. Most of the carbon monoxide came from cars 6–12 years old because they were so numerous and because of the presence in that fleet of cars with "poor" or "fair" emission levels.

The control over the past half-century of carbon monoxide levels in urban areas of developed countries has been one of the real success stories in environmental management. In the United States, for example, the average CO concentration declined by two-thirds in the 1983–2002 period alone. Most of the reduction has resulted from the use of catalytic converters on vehicles, from the ever tightening regulations on vehicular emissions, and from the natural continuing increase in the fraction of vehicles built after standards had been tightened. The introduction of **oxygenated** substances, which are hydrocarbons in which some of the atoms have been replaced by oxygen, into American gasoline has also reduced CO emissions from vehicles (as will be discussed further in Chapter 8). Average new-vehicle emissions before any emission controls were introduced in the United States were about 38 g CO/km, compared to the present standard of 1.5 g/km. Generally speaking, carbon monoxide emissions from vehicles are greatest from cold engines (and cold catalytic converters) and when the engine has an increased load due to rapidly accelerating or climbing a hill, since under these conditions, a rich fuel mix supplies insufficient oxygen to completely oxidize the gasoline.

There are some cities in developed countries that are still vulnerable to CO concentration excesses in winter due to meteorological and topological conditions. The temperature inversions in Fairbanks, Alaska, for example, produce several days each winter with high carbon monoxide levels. People such as traffic police who work outdoors in areas of high vehicular traffic can be exposed to elevated CO levels for long periods.

In the past, emission standards were applied only to passenger vehicles. However, starting in 2004, new U.S. regulations required for the first time that gasoline-powered sport-utility vehicles (SUVs)—which now account for about half of new-vehicle sales—and light trucks also meet emission standards.

Motorcycles and the three-wheel *tuk-tuk* taxis common in Asia are gasoline-powered, and their emissions, especially of CO and unburned hydrocarbons, make substantial contributions to air pollution. In a recent study in Switzerland, scooters and motorcycles (some fitted with catalytic converters) were found to emit more CO and hydrocarbons in urban driving, and significantly more NO_X on the highway per kilometer traveled, than did cars.

Most trucks and buses are powered by diesel engines, as are many cars in Europe and elsewhere outside North America. The catalytic converters used on vehicles with diesel engines in the past were much less effective than those on gasoline-powered vehicles. They typically removed only about half the gaseous hydrocarbon emissions, compared to 80–90% achieved for gasoline engine emissions. This difference is due to the less active catalyst formulations that had to be used with diesels because of the high sulfur content of diesel fuel; more active catalysts would have oxidized the sulfur dioxide gas to sulfate particles, which would cover the catalyst surface and render it ineffective. Much more effective catalytic converters are now being installed on diesel equipment for use with low-sulfur fuel. Also, the loss of engine exhaust gases and particles to the air—including that inside the vehicle—through crankcase emissions is being eliminated by rerouting the emissions back into the engine.

Catalytic converters used for diesel engines did not convert NO_X since diesel engines are operated "fuel lean," i.e., with excess oxygen present, and hence the required chemically reducing conditions did not exist. However, NO_X emissions from diesels are inherently lower than from gasoline vehicles since their operating temperature is significantly lower and consequently less thermal NO is formed. Lowering the engine operating temperature even further, by recirculating some of the exhaust gases through the engine, cuts back the amount of NO_X produced even more, as was also done with automobiles, as mentioned above. New emission regulations in North America that take effect in the mid- and late-2000s demand substantial reductions in NO_X emissions from diesel-powered vehicles. The U.S. Environmental Protection Agency (EPA) has recently proposed that train locomotives and ships powered by diesel should also be forced to substantially reduce their emission of nitrogen oxides. In one scheme for NO_X removal from diesel exhaust, the gas is temporarily stored on an adsorber present in the converter. Periodically, diesel fuel is injected into the stored NO_X, creating reducing conditions and promoting the reaction described above for gasoline vehicles, namely, the catalytic reduction of NO_X to N_2.

In addition to CO, hydrocarbon, NO_X, and SO_2 gases, diesel engine exhaust also includes significant quantities of solid and liquid particles. The liquid consists of unburned fuel and lubricating oil, plus some sulfuric acid produced from sulfur in the fuel. The catalytic converters formerly used on diesel engine vehicles were designed to oxidize the carbon-containing gases and liquids but without oxidizing the SO_2 further to sulfuric acid and sulfates.

Diesel fuel intended for new on-road vehicles in the United States has recently had its maximum allowed sulfur level dropped from 500 ppm to 15 ppm; and the level has dropped to 50 ppm in the European Union. Lowering the sulfur level in diesel fuel will also reduce particle emissions to some extent. Ironically, some of the soot carbon reacts with nitrogen dioxide in the exhaust gas to oxidize it:

$$C(s) + NO_2 \longrightarrow CO + NO$$

However, a filter—sometimes known as a **particle trap**—is required to achieve a suitable reduction in diesel exhaust particles. The traps can physically remove up to 90% of small particle emissions, thus preventing them from escaping into the air from the exhaust system of light-duty diesel vehicles. In order to prevent a buildup of solids, which would restrict engine exhaust flow or melt the trap, the system is designed so that the soot will ignite and burn away once a temperature of at least 500°C is attained. Alternatively, tiny amounts of a metal catalyst compound containing iron or copper are added to lower the temperature at which ignition will occur and assure more continuous regeneration of the filter. Eventually the filter requires cleaning since nonorganic components of the exhaust build up over time.

Scientists and engineers are currently developing a new type of internal combustion engine that combines the best aspects of gasoline and diesel technologies. In the *homogeneous charge compression ignition* (HCCI) engine, the fuel and the air are well mixed before ignition, thereby preventing the formation of soot particles that occurs in diesel engines. High compression of the air/fuel mix allows combustion to begin at many locations in the cylinder and produces a much greater efficiency than in gasoline engines, thereby reducing CO_2 emissions. Since the engine is run with an even larger excess of air than in a diesel, and no "hot spots" occur where combustion temperatures are high, much smaller amounts of NO_X than normal are produced.

Eventually, vehicles using internal combustion may well be replaced by emission-free ones powered by fuel cells. Such a prospect is discussed in Chapter 8 when we consider various alternative fuels.

Control of Nitric Oxide Emissions from Power Plants

In the United States, approximately equal amounts of NO_X are emitted currently from vehicles and from electric power plants; taken together they constitute the majority of the anthropogenic sources of these gases. In this section, we investigate the processes that are in use to decrease the quantity of nitrogen oxides emitted into the air by power plants.

• To reduce their NO_X production, some power plants use special burners designed to lower the temperature of the flame. Alternatively, the recirculation of a small fraction of the exhaust gases through the combustion zone has the same effect, as previously discussed for vehicles.

• Nitric oxide formation in power plants can also be greatly reduced by having the combustion of the fuel occur in stages. In the first, high-temperature stage, no excess oxygen is allowed to be present, thus limiting its ability to react with N_2. In the second stage, additional oxygen is supplied to complete the fuel's combustion but under lower temperature conditions, so that again little NO is produced.

• Other power plants, especially those in Japan and Europe, have been fitted with large-scale versions of catalytic converters that change NO_X back to N_2 before the release of stack gases into air. The reduction of NO_X to N_2 in these **selective catalytic reduction** systems is accomplished to 80–95% completion by adding ammonia, NH_3, to the cooled gas stream. This highly reduced compound of nitrogen combines with the partially oxidized compound NO to produce N_2 gas in the presence of oxygen:

$$4\,NH_3 + 4\,NO + O_2 \longrightarrow 4\,N_2 + 6\,H_2O$$

However, tight control is needed to regulate the addition of ammonia in order to prevent its inadvertent oxidation to NO_X. The same reaction can be accomplished without expensive catalysts, though with much less efficient nitric oxide removal, by using the uncooled gases at about 900°C. The catalytic process occurs at 250°C to 500°C, depending upon the catalyst used.

• The wet scrubbing of exhaust gases by an aqueous solution can also be used to prevent NO_X from being emitted into outside air. Since NO itself is rather insoluble in water and in typical aqueous solutions, half or more of the NO_X must be in the form of the much more soluble NO_2 for such techniques to be effective. Solutions of *sodium hydroxide*, NaOH, react with equimolar amounts of NO and NO_2 to produce an aqueous solution of *sodium nitrite*, $NaNO_2$:

$$NO + NO_2 + 2\,NaOH \longrightarrow 2\,NaNO_2 + H_2O$$

PROBLEM 3-5

Deduce the balanced reaction in which ammonia reacts with nitrogen dioxide to produce molecular nitrogen and water. Using the balanced equation, calculate the mass of ammonia that is required to react with 1000 L of air at 27°C, 1 atm pressure, containing 10 ppm of NO_2. [*Hint: Recall from introductory chemistry that $PV = nRT$ and that a balanced equation indicates the number of moles of the substances that react with each other.*]

PROBLEM 3-6

In a related technology, reduced nitrogen in the form of the compound urea, $CO(NH_2)_2$, is injected directly into the combustion flame to combine there, rather than later in the presence of a catalyst, with NO to produce N_2. Deduce the balanced equation that converts urea and nitric oxide into N_2, CO_2, and water.

Future Reductions in Smog-Producing Emissions

Although direct emissions of five (CO, VOCs, SO_2, particulate matter, and lead) of the six major air pollutants in the United States fell significantly between 1970 and 2000, emissions of NO_X grew by 20%, with half that increase occurring during the 1990s. Since energy consumption grew 45% and vehicle distance traveled grew 143% in that period, restrictions on nitrogen oxide emissions have had some success in controlling some of the growth in this pollutant, but not enough to prevent an overall increase. NO emissions from electric power plants, their other major source, have recently been falling somewhat.

As a consequence of the increase in overall NO emissions, ground-level ozone concentrations increased in the southern (especially around Houston) and northcentral regions of the United States in the 1990s. The latter effect can be seen in Figure 3-5a, where high ozone levels center around New York–Boston and a few midwestern sites, and somewhat lower levels cover most of the East Coast and the Midwest, extending into southern Ontario. High ozone levels have been a problem in southern California for many decades.

To help reduce the incidence of summertime smog in southcentral Canada and the northeastern United States, the two nations have signed an annex to their Air Quality Accord. The United States committed to reduce NO_X emissions originating from northern and northeastern states by 35% by 2007 and also to reduce VOC emissions during summer months, when most smog forms in this region. Canada agreed to reduce its NO_X emissions from power plants in southern Ontario by 50% by the same date. The drastic lowering of the allowed sulfur levels in gasoline should assist in the reduction of NO_X from vehicles. Emission standards for SUVs, trucks, and buses are also being tightened to bring them more in line with those for regular automobiles.

The *Gothenburg Protocol*, which controls the release of many pollutants in Europe, is expected to reduce NO_X emissions there by more than 40% by 2010, compared to 1990 levels; a 30% reduction had already been achieved by 2000. Great Britain—which saw its emissions decline by the late 1990s by almost 40% compared to their peak in the late 1980s—will have to reduce its emissions by another third from 1998 levels by 2010 in order to meet these regulations. European VOC emissions are due to drop by 40% according to the protocol.

The trends in anthropogenic NO_X emissions into the air over North America, Europe (including Russia and the Near and Middle East), and Asia (east, southeast, and south) in the past few decades are shown in Figure 3-9. As discussed above, European contributions have fallen, North American emissions displayed growth in some periods but declined in others, remaining about constant overall, and those from Asia have risen from relatively small to largest of all. Consistent with these trends, a comparison of measurements from space of tropospheric NO_2 levels in 2002 compared to those in 1996 show significant reductions over much of Europe, declines over the Ohio Valley (where emissions had dropped 40% by 2006, compared to 1999 levels), but increases in the northeastern states and in southern Ontario, and a large increase over industrial areas of China.

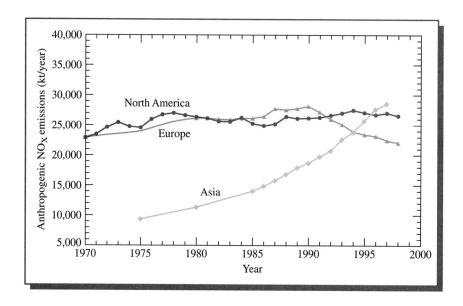

FIGURE 3-9
Anthropogenic NO_X atmospheric emissions.
[Source: H. Akimoto, "Global Air Quality and Pollution," *Science* 302 (2003): 1716.]

Green Chemistry: The Replacement of Organic Solvents with Supercritical and Liquid Carbon Dioxide; Development of Surfactants for This Compound

In addition to their role in paints, organic solvents are used in many different products and processes in both commercial and household applications. It is estimated that over 15 billion kilograms of organic solvents are used worldwide each year in such areas as the electronics, cleaning, automotive, chemical, mining, food, and paper industries. These liquids include not only hydrocarbon solvents but also halogenated solvents. Both of these types of solvents contribute not only to air pollution as VOCs but also to water pollution (Chapter 14). Some halogenated solvents contribute to the depletion of the ozone layer, as seen in Chapter 2. Discovering solvents with less environmental impact and even designing processes that use no solvents at all are the subjects of many green chemistry initiatives.

Carbon dioxide, CO_2, is one solvent that is receiving considerable attention as a replacement for traditional organic solvents. Although carbon dioxide is a gas at room temperature and pressure, it can be liquefied easily by the application of pressure. In addition to liquid carbon dioxide, there is considerable interest in supercritical carbon dioxide (a discussion of supercritical fluids can be found in Chapter 16) as a solvent in the electronics industry, as discussed in the green chemistry section of Chapter 6. The decaffeination of coffee and tea with carbon dioxide is a well-known application of this solvent.

Liquid carbon dioxide is attractive as a solvent due to its low viscosity and polarity and its wetting ability. Because of its low polarity, carbon dioxide is able to dissolve many small organic molecules. However, larger molecules including oils, polymers, waxes, greases, and proteins are generally insoluble in it. To increase the solubility of compounds in water, surfactants such as soaps and detergents have been developed which allow this very polar solvent to dissolve less polar materials such as oils and grease. In an analogous fashion, surfactants for carbon dioxide have been developed that increase the range of materials that will dissolve in it.

Joseph DeSimone, of the University of North Carolina and North Carolina State University, earned a Presidential Green Chemistry Challenge Award in 1997 for his preparation and development of polymeric surfactants for carbon dioxide. DeSimone is currently the director of the National Science Foundation Science and Technology Center for Environmentally Responsible Solvents and Processes. This center focuses on discovering ways to replace conventional organic solvents and water with carbon dioxide in a multitude of processes. An example of a surfactant developed by DeSimone is the block copolymer shown in Figure 3-10a. This molecule has nonpolar regions, which are CO_2-philic, and polar regions, which are CO_2-phobic. When dissolved in

(a)

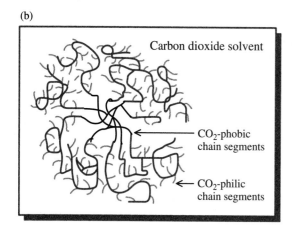

CO$_2$-phobic
chain segment

CO$_2$-philic
chain segment

(b)

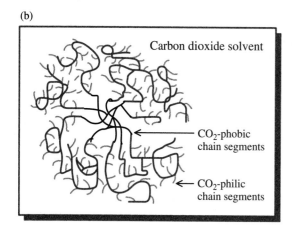

Carbon dioxide solvent

CO$_2$-phobic
chain segments

CO$_2$-philic
chain segments

FIGURE 3-10 (a) A copolymer surfactant for carbon dioxide. (b) A micelle in liquid carbon dioxide. [Source: M. C. Cann and M. E. Connelly, *Real-World Cases in Green Chemistry* (Washington, DC: American Chemical Society, 2000).]

carbon dioxide, the CO$_2$-philic regions orient themselves to interact with the surrounding carbon dioxide solvent, while the CO$_2$-phobic regions aggregate with one another. The overall result is the formation of a structure know as a *micelle* (Figure 3-10b). Polar substances that normally do not dissolve in carbon dioxide will dissolve in the center polar region of the micelle.

DeSimone was one of the founders of a dry-cleaning chain that uses liquid carbon dioxide, along with surfactants that he developed, to clean clothes. The spent liquid carbon dioxide is drained from the clothes after the wash cycle (in much the same way as the wash water in our washing machines at home is drained off after the wash cycle), and the carbon dioxide is allowed to evaporate by simply reducing the pressure. The carbon dioxide vapors are then captured, liquefied by increasing the pressure, and reused for another wash. Carbon dioxide is plentiful and inexpensive, since it can be recovered as a by-product from natural gas wells or ammonia production. Capture of carbon dioxide from these processes puts to good use this compound which would normally be released to the atmosphere and contribute to global warming (see Chapter 6). By contrast, most dry cleaners in North America presently use *perchloroethylene*, $Cl_2C{=}CCl_2$, known as PERC, as the solvent. PERC is a VOC, since it has a high vapor pressure and readily escapes into the troposphere if not carefully controlled. PERC is also a groundwater contaminant (see Chapter 14) and is a suspected human carcinogen.

 Green Chemistry: Using Ionic Liquids to Replace Organic Solvents; Cellulose, a Naturally Occurring Polymer Replacement for Petroleum-Derived Polymers

Cellulose (Figure 3-11) is a polymer of glucose that makes up about 40% of all organic matter on Earth. About 700 billion tonnes of cellulose exist on Earth, with another 40 billion tonnes produced each year as the major component of biomass by plants from atmospheric carbon dioxide and water via

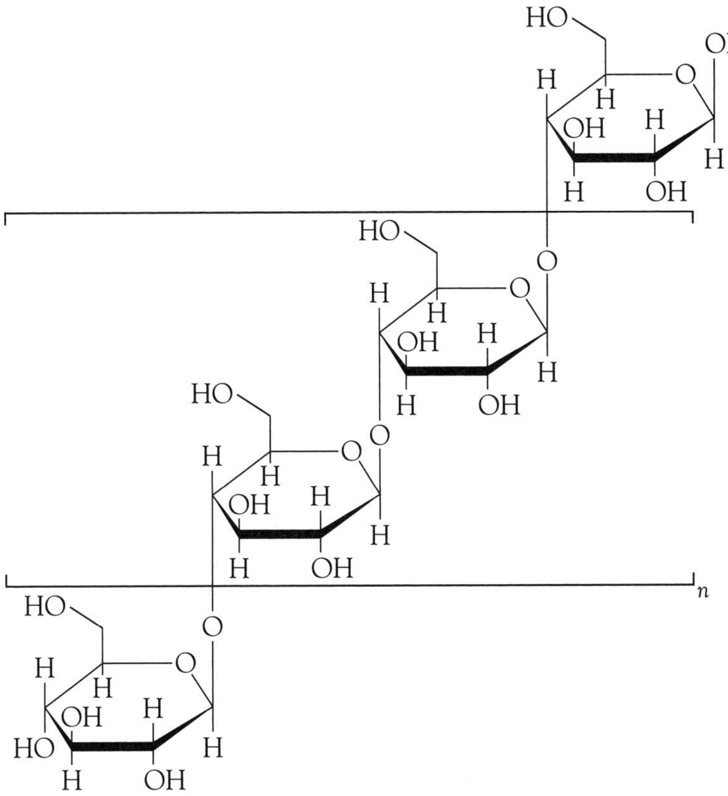

FIGURE 3-11 Structure of cellulose.

photosynthesis. This removal of carbon dioxide from the atmosphere helps to mitigate some of the global warming caused by anthropogenic emissions of this gas.

Many polymers produced from crude oil are ubiquitous in our everyday lives, including **polyethylene terephthalate** (PET), which is found in beverage bottles and polyester clothing; **polyethylene,** which is employed in making plastic bags and milk jugs; **polyvinyl chloride,** which is found as plastic pipes and shower curtains; and **polystyrene,** which we discussed in the green chemistry section in Chapter 2. Hundreds of million of kilograms of these petrochemical-based polymers are produced each year, requiring as raw material approximately 700 million barrels of crude oil. As the price of conventional crude oil increases and the supply declines, a major focus of green chemistry is the production of organic chemicals, including polymers, from biomass (see the green chemistry section in Chapter 7). An even more intriguing opportunity is to use naturally occurring polymers such as cellulose to replace the polymers synthesized from crude oil.

The use of cellulose is severely limited by its insolubility in water and in traditional organic solvents. The strong intra- and inter-chain hydrogen

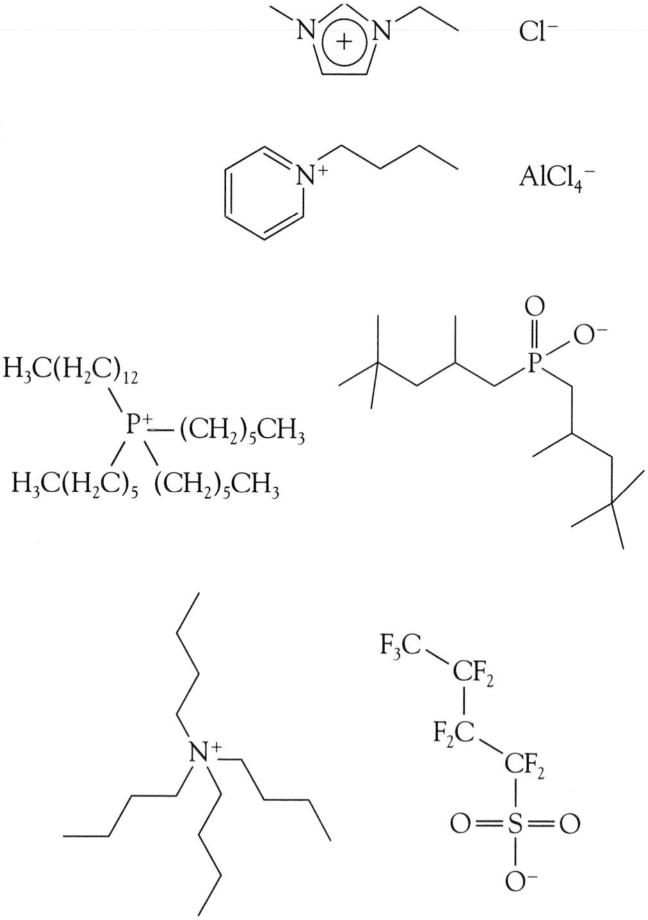

FIGURE 3-12 Ion pairs in four typical ionic liquids.

bonding between the numerous hydroxyl groups on the cellulose polymer are likely the reason for this insolubility, which results in very poor processability for cellulose. Consequently, only about 0.1 billion tonne of cellulose has been used annually as a feedstock for further processing.

In the previous green chemistry section, the replacement of traditional organic solvents with supercritical and liquid carbon dioxide was discussed. A very interesting and relatively unknown group of compounds that are of growing interest as replacements for traditional organic solvents are called *room-temperature ionic liquids*, or just **ionic liquids** (ILs). Most ionic compounds have characteristically high melting points due to their strong network of ionic bonding. For example, sodium chloride (table salt) has a melting point of 801°C. In contrast, a few ionic compounds have melting points below or moderately above ($\leq$100°C) room temperature; such compounds are known as (room-temperature) ionic liquids. ILs are generally composed of bulky ions that often have dispersed rather than localized charges and large nonpolar groups (Figure 3-12). As a consequence, their oppositely charged ions have only weak attractive interactions with one another, which results in the low melting points of these compounds.

One very beneficial characteristic of ILs is their very low vapor pressure. This is in contrast to most organic solvents, which, because of their significant vapor pressures, are VOCs and contribute to tropospheric pollution. Because ILs are ionic, they are nonvolatile and thus their potential to replace VOCs is of significant interest. Ionic liquids may also be purified and recycled, thereby adding to the green characteristics of these solvents. In addition, they are nonflammable, and many are stable up to 300°C, making them attractive for reactions and processes that require high temperatures.

Another strong interest of the green chemistry community is the use of microwave ovens to facilitate chemical processes and reactions. Conventional heat sources—such as heating mantles, Bunsen burners, and oil baths—heat materials from the outside in, transferring energy (in turn) from the heat source to the bottom of the reaction vessel, to the solvent inside the vessel, and finally to the dissolved reactants. In each step, heat energy is lost to the surroundings as it is transferred. A microwave-absorbing reactant or solvent, however, can be targeted by microwaves and therefore can be directly heated by irradiation in a microwave oven or reactor. Thus with microwave heating, the contents may be heated directly without heating the vessel. Most people have experienced this phenomenon when heating a cup of water in a household microwave oven. The water heats quite quickly while the cup remains relatively cool. Chemists have found that many reactions and processes can be accelerated in a microwave oven, whose heating efficiency has the potential to reduce energy requirements.

In order to effectively heat via a microwave source, a substance must be polar and/or ionic. ILs heat up very quickly in a microwave since by nature they are ionic. They can reach temperatures as high as 300°C in 15 seconds of microwave heating.

Robin Rogers and his group at the University of Alabama won a Presidential Green Chemistry Challenge Award in 2005 for their discovery that certain ILs readily dissolve cellulose when heated in a microwave oven. The process they developed involves the use of gentle, pulsed microwave heating in a domestic microwave oven to expedite the dissolution of cellulose in ILs. Their studies indicate that with the IL 1-butyl-3-methylimidazolium chloride, they can produce solutions with up to 25% (by mass) cellulose.

There is evidence that the chloride ion in this compound disrupts the internal hydrogen bonding in the cellulose, thereby leading to dissolution. The addition of small amounts of water solvates the chloride ions, allowing the hydrogen bonding of the cellulose to resume. The cellulose precipitates from the solution and can then be deposited as films, membranes, and fibers.

By dispersing additives in the IL either before or after the dissolution of cellulose, composite or encapsulated cellulose-based materials can be formed when the polymer is regenerated. For example, *laccase*, an enzyme found in fungi that degrades polyphenolic compounds, has been encapsulated in a cellulose support without loss of its activity. The enzyme, when supported on a cellulose film, can be immersed in an aqueous reaction environment and easily removed at the end of the reaction by removing the film, which can then be reused.

The Rogers group has also successfully suspended many other materials in cellulose. These include dyes that can be used to detect metals, such as mercury, and magnetite (Fe_3O_4) that produces a composite with uniform magnetic properties. Using this method, cellulose can be combined with other polymers to produce blends. When mixed with polypropylene, a composite that has excellent tear properties is formed. The use of this material for packaging offers significant promise. Encapsulation of medically active compounds along with magnetic materials has the potential to produce microcapsules that can be directed to specific parts of the body. In addition, titanium dioxide infusion into cellulose fibers can produce clothing and bedding with antibacterial properties. The development of cellulose-based materials has the potential to produce new materials with novel characteristics that require less use of petroleum than conventional polymers.

Sulfur Dioxide and Hydrogen Sulfide Sources and Abatement

On a global scale, most SO_2 is produced by volcanoes and by the oxidation of sulfur gases produced by the decomposition of plants. Because this natural sulfur dioxide is mainly emitted high into the atmosphere or far from populated centers, the background concentration of the gas in clean air is quite small, about 1 ppb. However, a sizable additional amount of sulfur dioxide is emitted into ground-level air, particularly over land masses in the Northern Hemisphere, due to industrial activities. The main anthropogenic source of SO_2 is the combustion of coal, a solid that, depending on the geographic area from which it is mined, contains 1 to 6% sulfur. In most countries, including the United States, the major use of coal is to generate electricity. Usually half or more of the sulfur is trapped as inclusions in the mineral content of the coal. If the coal is pulverized before combustion, this type of sulfur can be mechanically removed, as discussed below. The rest of the sulfur, which usually amounts to about 1% of the coal's mass, is bonded in the complex organic structure of the solid and cannot be removed without expensive processing that breaks covalent bonds.

Sulfur occurs to the extent of a few percent in crude oil (with higher concentrations in tar sands and shale oil), but it is reduced to the level of a few hundred ppm or less in products such as gasoline. Sulfur dioxide is emitted into air directly as SO_2 or indirectly as H_2S by the petroleum industry when oil is refined and natural gas is cleaned before delivery. Indeed, the predominant component in natural gas wells is sometimes H_2S rather than CH_4! The substantial amounts of hydrogen sulfide obtained from its removal from oil and natural gas are often converted to solid, elemental sulfur, an environmentally benign substance, using the gas-phase process called the **Claus reaction**:

$$2\ H_2S + SO_2 \longrightarrow 3\ S + 2\ H_2O$$

One-third of the molar amount of hydrogen sulfide extracted from the fossil fuel is first combusted to sulfur dioxide to provide the other reactant for this process. Huge amounts of elemental sulfur are produced by sulfur removal, especially from natural gas. Notice the analogy between the Claus reaction and the selective catalytic reduction process for nitric oxide control: They both involve the reaction together of oxidized and reduced forms of an element (N or S) to form the innocuous elemental form.

It is very important to remove hydrogen sulfide from gases before their dispersal in air because it is a highly poisonous substance, more so than sulfur dioxide. The concentration of H_2S sometimes becomes elevated in the area surrounding natural gas wells during *flaring*—the burning off of gas that cannot be immediately captured. Flaring burns only about 60% of the hydrogen sulfide content of the gas, so the remainder becomes dispersed into the surrounding air. Hydrogen sulfide is also a common pollutant in the air emissions from pulp and paper mills.

In addition to H_2S, several other smelly gases containing sulfur in a highly reduced state are emitted as air pollutants in petrochemical processes; these include CH_3SH, $(CH_3)_2S$, and CH_3SSCH_3. The term **total reduced sulfur** is used to refer to the total concentration of sulfur from H_2S and these three compounds. (The word *reduced* is used here to denote not a decrease in sulfur emissions but rather the oxidation state of the sulfur.)

Large **point sources**—individual sites that emit large amounts of a pollutant—of SO_2 are also associated with the nonferrous smelting (i.e., the conversion of ores to free metals) industry. Many valuable and useful metals, such as copper and nickel, occur in nature as ores containing the **sulfide ion,** S^{2-}. In the first stage of their conversion to the free metals they were usually "roasted" in air to remove the sulfur, which was converted to SO_2 and was traditionally released into the air. For example,

$$2\,NiS(s) + 3\,O_2(g) \longrightarrow 2\,NiO(s) + 2\,SO_2(g)$$

Ores such as copper sulfide can be smelted in a process that uses pure oxygen forced into the smelting chamber, and the very concentrated sulfur dioxide that is obtained from the reaction can be readily extracted, liquefied, and sold as a by-product. On the other hand, the SO_2 concentration in the waste gases from conventional roasting processes (such as that used for nickel) is high. Consequently, it is feasible to pass the gas over an oxidation catalyst that converts much of the SO_2 to sulfur trioxide, to which water can be added to produce commercial concentrated sulfuric acid.

$$2\,SO_2(g) + O_2(g) \longrightarrow 2\,SO_3(g)$$

$$SO_3(g) + H_2O(aq) \longrightarrow H_2SO_4(aq)$$

The latter reaction (which represents only initial reactants and end-product) is in fact accomplished in two steps (not shown) in order to ensure that none of the substances escape into the environment: First the trioxide is combined with sulfuric acid, then water is added to the resulting solution.

The Oxidation of Sulfur Dioxide in Suspended Water Droplets

Although some sulfur dioxide in air is oxidized by gas-phase reactions, most of it is converted to sulfuric acid after it has dissolved in tiny suspended water droplets present in clouds, mists, etc. The uncatalyzed oxidation of dissolved SO_2 by dissolved oxygen proceeds at a very slow rate unless a catalyst such as Fe^{3+} or the ions of other transition metals are also present in the droplets. The most important oxidizing agents in the droplets, though present only in tiny concentrations, are the dissolved atmospheric gases ozone and **hydrogen peroxide,** H_2O_2. The concentration of these two pollutants is much greater in air masses undergoing photochemical smog than in clean air.

In general, the concentration of a dissolved gas in the liquid phase can be determined by considering the equilibrium between its two forms. Thus for hydrogen peroxide, we have

$$H_2O_2(g) \rightleftharpoons H_2O_2(aq)$$

The useful form of the equilibrium constant for such processes is the **Henry's law constant,** K_H, which is equal to the concentration of the dissolved species divided by the partial pressure of the gas. For the above reaction, we have

$$K_H = \frac{[H_2O_2]}{P_{H_2O_2}}$$

If the concentration is expressed as a molarity, and the unit of pressure is atmospheres, from experimental data

$$K_H = 7.4 \times 10^4 \text{ M atm}^{-1}$$

Using this information, we can determine the molarity of H_2O_2 in a raindrop for typical clean-air conditions of 0.1 ppb, i.e., equivalent to 0.1×10^{-9} atm.

$$\begin{aligned}
[H_2O_2] &= K_H \, P_{H_2O_2} \\
&= 7.4 \times 10^4 \text{ M atm}^{-1} \times 0.1 \times 10^{-9} \text{ atm} \\
&= 7.4 \times 10^{-6} \text{ M}
\end{aligned}$$

Although a concentration of 7.4 μM seems tiny by comparison to values routinely encountered in laboratories, it is sufficient to oxidize dissolved SO_2 at an appreciable rate. The hydrogen peroxide concentration in smoggy air is an order of magnitude or more larger than in clean air, so its concentration in water droplets rises accordingly, as does the rate of oxidation of sulfur dioxide. (Recently it has been discovered that, for unknown reasons, hydrogen peroxide levels in suspended aerosol droplets in the air at several California locations are orders of magnitude larger even than predicted by Henry's law.)

The calculation of the solubility of SO_2 in raindrops is more complicated, since in the aqueous phase it exists as **sulfurous acid, H_2SO_3**:

$$SO_2(g) + H_2O(aq) \rightleftharpoons H_2SO_3(aq)$$

The Henry's law expression does not include the concentration of the solvent, water:

$$K_H = [H_2SO_3]/P_{SO_2}$$

Since $K_H = 1.0$ M atm^{-1} for SO_2, and since its concentration in a typical sample of air is about 0.1 ppm, i.e., equivalent to 0.1×10^{-6} atm, the equilibrium concentration of sulfurous acid is

$$[H_2SO_3] = K_H \, P_{SO_2}$$

$$= 1.0 \text{ M atm}^{-1} \times 0.1 \times 10^{-6} \text{ atm}$$

$$= 1.0 \times 10^{-7} \text{ M}$$

This value of about 10^{-7} M for the equilibrium concentration of H_2SO_3 is deceptive since it by no means represents *all* the sulfur dioxide that dissolves in a water droplet (see Figure 3-13). Sulfurous acid is a weak acid whose ionization to the **bisulfite ion, HSO_3^-**, must also be considered in calculating the solubility of sulfur dioxide:

$$H_2SO_3 \rightleftharpoons H^+ + HSO_3^-$$

The **acid dissociation** (or ionization) **constant** K_a for H_2SO_3 is equal to 1.7×10^{-2}, where K_a is related to concentrations by the expression

$$K_a = \frac{[H^+] \, [HSO_3^-]}{[H_2SO_3]}$$

The concentrations in such expressions are *equilibrium* values. Since the equilibrium molarity of H_2SO_3 is determined in the raindrop by its interchange with SO_2 in

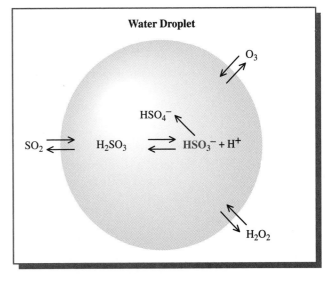

FIGURE 3-13 Dissolution of atmospheric gases SO_2, O_3, and H_2O_2 into a water droplet and their subsequent reactions.

air, we can substitute that known value into the K_a expression:

$$K_a = \frac{[H^+]\,[HSO_3^-]}{1.0 \times 10^{-7}}$$

Rearranging the equation to solve for the ion concentrations, which from stoichiometry are equal in value, we obtain

$$[HSO_3^-]^2 = 1.7 \times 10^{-2}\,M \times 1.0 \times 10^{-7}\,M$$

and hence

$$[HSO_3^-] = 4 \times 10^{-5}\,M$$

Thus the equilibrium ratio of bisulfite ion to sulfurous acid in water is about 400 : 1. Consequently, the total dissolved sulfur dioxide is about 4×10^{-5} M, rather than just the 1×10^{-7} M that represents the contribution from the un-ionized acid.

Since the concentration of hydrogen ion produced by the reaction is also 4×10^{-5} M, the pH of such raindrops is 4.4. Rain does not become much more acidic than this if no strong acids are dissolved in the droplets.

PROBLEM 3-7

Bisulfite ion can act as a weak acid and ionize further:

$$HSO_3^- \rightleftharpoons H^+ + SO_3^{2-}$$

Given that K_a for HSO_3^- is 1.2×10^{-7}, calculate the concentration of SO_3^{2-} that is present in the raindrops of pH 4.4 discussed above. [Hint: The concentrations of bisulfite and hydrogen ion will be very close to their previously established values.]

PROBLEM 3-8

Calculate the pH of rainwater in equilibrium with SO_2 in a polluted air mass for which the sulfur dioxide concentration is 1.0 ppm. [Hint: Recall the relationship between partial pressure and ppm concentration discussed earlier in the chapter.]

PROBLEM 3-9

Calculate the concentration of SO_2 that must be reached in polluted air if the dissolved gas is to produce a pH of 4.0 in raindrops without any oxidation of the sulfur.

PROBLEM 3-10

CO_2 dissolves in water to produce H_2CO_3 in the same way that SO_2 produces H_2SO_3. (a) Confirm by calculation that the pH of CO_2-saturated water at 25°C is 5.6, given that the CO_2 concentration in air is 365 ppm. For carbon dioxide, the Henry's law constant $K_H = 3.4 \times 10^{-2}\,M\,atm^{-1}$ at 25°C. The K_a for carbonic acid, H_2CO_3, is 4.5×10^{-7}. (b) Recalculate the pH for a carbon dioxide concentration of 560 ppm, i.e., double that of the preindustrial age.

Clean Coal: Reducing Sulfur Dioxide Emissions from Power Plants

When the sulfur dioxide present in exhaust gases is dilute, as in the case of power-plant emissions, its extraction by oxidation is not feasible. Instead, the SO_2 gas is removed by an acid–base reaction between it and **calcium carbonate** (limestone), $CaCO_3$, or **calcium oxide** (lime), CaO, in the form of wet, crushed solid. The emitted gases are either passed through a slurry of the wet solid or are bombarded by jets of the slurry. In some applications, fine grains of calcium oxide, rather than a slurry of calcium carbonate, are used to trap the sulfur dioxide from the emission gases. Up to 90% of the gas can be removed by such *scrubber* processes, more formally known as **flue-gas desulfurization.** In some operations, notably in Japan and Germany, the product is fully oxidized by reaction with air, and the resulting **calcium sulfate,** $CaSO_4$, is dewatered and sold as *gypsum.* Usually the product, a mixture of **calcium sulfite,** $CaSO_3$, and calcium sulfate, is a slurry—or a dry solid if granular calcium oxide was used—and is buried in a landfill. The reactions with calcium carbonate are

$$CaCO_3 + SO_2 \longrightarrow CaSO_3 + CO_2$$
$$2\,CaSO_3 + O_2 \longrightarrow 2\,CaSO_4$$

Alternatively, the sulfur dioxide can be captured by use of slurries of sodium sulfite or magnesium oxide or amine salts, and these reactant compounds and concentrated SO_2 gas are later regenerated by thermally decomposing the product.

Recently, **clean coal technologies** have been developed to use coal in ways that are cleaner and often more energy efficient than those employed in the past. In the various technologies, the cleaning can occur precombustion, during combustion, postcombustion, or by conversion of the coal to another fuel.

In **precombustion cleaning,** the coal has the sulfur associated with its mineral content—usually *pyritic sulfur,* FeS_2—removed so it cannot subsequently produce sulfur dioxide. The coal first is ground to a very small particle size, effectively into separate mineral particles and carbon particles. Since they have different densities, the two types of particles can be separated by mixing the pulverized solid in a liquid of intermediate density and allowing the fuel portion to rise to the top, where it can be skimmed off. As an alternative to such physical cleaning, biological or chemical methods can be employed. For example, bacteria cultured to eat the organic sulfur in coal can be utilized. Chemically, the sulfur can be leached from coal with a hot sodium or potassium caustic solution.

In **combustion cleaning,** the combustion conditions can be modified to reduce the formation of pollutants, and/or pollutant-absorbing substances can be injected into the fuel to capture pollutants as they form. In **fluidized-bed combustion,** pulverized coal and limestone are mixed and then suspended on jets of air (fluidized) in the combustion chamber. Virtually all the sulfur dioxide is thereby captured in solid form as calcium sulfite and sulfate before the gas can escape. This procedure allows for much-reduced combustion temperatures and therefore also greatly reduces the amount of nitrogen oxides that are formed and released.

Some of the advanced techniques used in **postcombustion cleaning**—such as the use of granular calcium oxide or sodium sulfite solutions—have already been described. In the *SNOX* process developed in Europe, cooled flue gases are mixed with ammonia gas to remove the nitric oxide by its catalytic reduction to molecular nitrogen (by the reaction discussed previously). The resulting gas is reheated and the sulfur dioxide oxidized catalytically to sulfur trioxide, which is then hydrated by water to sulfuric acid, condensed, and removed.

In the conversion of coal to other fuels, it is first gasified by reaction with steam, as will be described in Chapter 8. The gas mixture is cleaned of pollutants, and the purified gas is then burned in a gas turbine that generates electricity. The waste heat of the combustion gases is used to produce steam for a conventional turbine and thus to generate more electricity. Alternatively, the gasified coal can be converted into liquid fuels suitable for vehicular use.

Sulfur dioxide emissions from power plants can also be minimized by burning oil, natural gas, or low-sulfur coal as the fuel, though these fuels usually are more expensive than high-sulfur coal.

PROBLEM 3-11

What mass of calcium carbonate is required to react with the sulfur dioxide that is produced by burning one tonne (1000 kg) of coal that contains 5.0% sulfur by mass?

PROBLEM 3-12

Write the balanced reaction whereby sodium hydroxide can be used to scrub sulfur dioxide from exhaust gases by a reaction that produces water and sodium sulfite, Na_2SO_3. What substance would you have to react with this solution to produce calcium sulfite and regenerate the sodium hydroxide? Write the balanced equation for the latter process, and deduce the net reaction for the cycle.

The alternative to sulfur dioxide control—to simply allow the pollutant gas to be emitted into the air—can cause devastation by SO_2 to the plant life in the surrounding area unless extremely high smokestacks are used. The tallest such stacks in the world are located at Sudbury, Ontario, and reach 400 meters. However, using tall stacks simply solves a local SO_2 problem at the expense of creating a problem downwind. For example, emissions from mainland North America can sometimes be detected in Greenland.

Because of federal regulations, the amount of sulfur dioxide emitted into the air in North America has fallen substantially—by about 38% from the peak levels (in 1973) by 1998 in the United States and by 45% in Canada by 2000. The 1991 Air Quality Accord between the United States and Canada required both countries to substantially reduce their sulfur dioxide emissions beyond those of previous laws and agreements. Such emissions in the United States are restricted in accordance with the Clean Air Act, especially the 1990 amendments to it. The average concentration of sulfur dioxide in air in the United States fell by 43%, from 13.2 ppb to 7.5 ppb, from 1975 to 1991. Whereas Phase I (1995 deadline) of the Clean Air Act imposed controls on only the largest coal-fired plants, Phase II, which began in 2000, imposes more stringent requirements and applies to almost all plants. There is an SO_2 tonnage limit for each power plant that emits this gas, based upon the power it produces, and a cap on overall national emissions as well. By 2010, the SO_2 emissions from these power plants should have been reduced by 50% compared to 1980 levels.

The reductions in sulfur dioxide emissions by power plants in the U.S. Midwest have been achieved at lower-than-expected cost, due in part to the availability of cheap, low-sulfur coal (1% S, versus more than 3% S in the high-sulfur coal used previously) and inexpensive scrubbers, and in part to the implementation of a system of tradable emission permits. This permit system, in operation since the early 1980s, allows industries to buy emission allowances if they need to exceed their allowed levels or to sell excess allowances on the open market (through the Chicago Board of Trade) if they do not need their whole allowance. A similar program has been started for nitrogen oxide emissions.

The European Union issued a directive in 1988 specifying reductions from large power plants of 50–70% of SO_2 emissions from 1980 levels by 2003. The decline in SO_2 emissions in Great Britain from 1970 to 1998 amounted to 75%. This cutback was achieved mainly by switching from coal to natural gas in power stations and by the use of low-sulfur coal and the scrubbing of emissions in facilities where coal is still burned. According to the 1999 Gothenburg Protocol, Europe's sulfur dioxide emissions are to be cut beyond 1990 levels by another 63% by 2020.

Global SO_2 emissions are predicted to keep rising until about 2020, due mainly to increased releases from Asia. China has become the world leader in emitting sulfur dioxide. Because of its rapid economic expansion, the growth in the burning of coal—which supplies about two-thirds of China's energy—has recently been about 20% per year. As a consequence, its SO_2 emissions have risen rapidly—its 2005 emissions were 27% higher than those in 2000.

Because it has closed many small coal-fired power plants and either cleans the coal before combustion or scrubs emission gases, China's sulfur dioxide emissions are predicted to level off by about 2010. However, the quantity of nitric acid (from NO_X emissions) in its acid rain is predicted to increase due to the rapid rise in domestic vehicle ownership.

Japan initiated tight controls on SO_2 emissions in the 1970s, and by 1980 its power plants had almost eliminated such emissions by the widespread installation of scrubbers. The high rate of SO_2 emissions from the former Soviet bloc has declined in recent decades, due more to economic problems than to intentional controls, though in places emissions continue to make acidification a problem.

Particulates in Air Pollution

The black smoke released into the air by a diesel truck is often the most obvious form of pollution that we routinely encounter. The smoke is composed largely of particulate matter. **Particulates** are tiny solid or liquid particles—other than those of pure water—that are temporarily suspended in air and that are usually individually invisible to the naked eye. Collectively, however, such particles often form a haze that restricts visibility. Indeed, on many summer days the sky over North American and European cities is milky white rather than blue. More importantly, breathing air that contains particulates is known to be hazardous to human health. In the material that follows, we investigate the wide range of sizes of the suspended particles and their origins.

The particles that are suspended in a given mass of air are neither all of the same size or shape nor do they all have the same chemical composition. The smallest suspended particles are about 0.002 μm (i.e., 2 nm) in their dimensions; by contrast, the length of typical gaseous molecules is 0.0001 to 0.001 μm (0.1 to 1 nm). The upper limit for suspended particles corresponds to dimensions of about 100 μm (i.e., 0.1 mm). When atmospheric water droplets coalesce to form particles bigger than this, they are raindrops and fall out of the air so quickly they are not considered to be "suspended." The ranges of particle sizes for common types of suspended particulates is illustrated in Figure 3-14.

Although few of the particles suspended in air are exactly spherical in shape, it is convenient and conventional to speak of all particles as if they were so. Indeed, the **diameter** of particulates is their most important property. Qualitatively, individual particles are classified as **coarse** or as **fine** depending upon whether their diameters are greater or less than 2.5 μm, respectively. (About 100 million particles of diameter 2.5 μm would be required to cover the surface of a small coin.)

There are many common names for atmospheric particles: *dust* and *soot* refer to solids, whereas *mist* and *fog* refer to liquids, the latter denoting a high

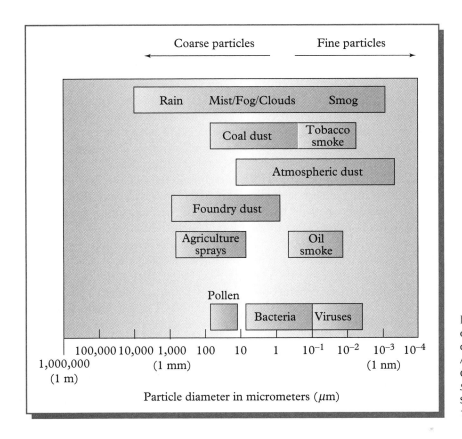

FIGURE 3-14 Sizes of common airborne fine and coarse particulates. [Source: Adapted from J. G. Henry and G. W. Heinke, *Environmental Science and Engineering* (Upper Saddle River, NJ: Prentice Hall, 1989).]

concentration of water droplets. An **aerosol** is a collection of particulates, whether solid particles or liquid droplets, dispersed in air. A true aerosol (as opposed to, say, the fairly large droplets from a hairspray dispenser) has very small particles: Their diameters are less than 100 μm.

Intuitively, one might think that all particles should settle out under the influence of gravity and be deposited onto the Earth's surface rapidly, but this is not true for the smaller ones. According to *Stokes' law*, the rate, in distance per second, at which particles settle increases with the square of their diameter. In other words, a particle half the diameter of another falls four times more slowly. The small ones fall so slowly they are suspended almost indefinitely in air (unless they stick to some object they encounter). As we shall see later, the very small ones aggregate to form larger ones, usually still in the fine size category. Fine particulates usually remain airborne for days or weeks, whereas coarse particulates settle out fairly rapidly. In addition to this sedimentation process, particles also are commonly removed naturally from air by their incorporation into falling raindrops.

Sources of Coarse Particles

The primary-versus-secondary distinction made between atmospheric gaseous pollutants is also applied to suspended particles. Most coarse particles are primary, although they often begin their existence as even coarser matter, since they originate chiefly from the disintegration of larger pieces of matter. Minerals constitute one important type of the coarse particulates in air. Because many of the large particles in atmospheric dust, particularly in rural areas, originate as soil or rock, their elemental composition is similar to that of the Earth's crust, namely high concentrations of Al, Ca, Si, and O in the form of aluminum silicates (see Chapter 16), some of which also contain the calcium ion.

Wind storms in deserts sweep large amounts of fine sand into the air. Dust storms in Asia, whose effects reach as far away as North America, are increasing due to the continuing transformation of fertile land into desert as a consequence of global warming, deforestation, and overgrazing. The wind generates coarse particles by the mechanical disintegration of leaf litter. Pollen released from plants also consists of coarse, primary particles. Wildfires and volcanic eruptions generate both fine and coarse particulate matter. Near and above oceans, the concentration of solid NaCl is very high, since sea spray leaves sodium chloride particles airborne when the water evaporates. Indeed, sea salt aerosols are by far the largest mass of primary particles in air, followed by soil dusts and debris from natural fires.

Although most coarse particulates originate with natural sources, human activities such as stone crushing in quarries and land cultivation result in particles of rock and topsoil being picked up by the wind. Coarse particles in many areas are basic, reflecting the calcium carbonate and other such salts in soils.

Sources of Fine Particles

Primary fine particles of anthropogenic origin include ones generated by the wearing of tires and vehicle brakes as well as the dust from metal smelting. The incomplete combustion of carbon-based fuels such as coal, oil, gasoline, and diesel fuel produces many fine soot particles, which are mainly crystallites (miniature crystals) of carbon. Consequently, one of the main sources of carbon-based primary atmospheric particulates, both fine and coarse, is the exhaust from vehicles, especially those having diesel engines. About half the organic content from heavy-duty diesel vehicles is elemental carbon; one can easily observe this soot as the black smoke that emanates from such equipment. Most carbon-containing emissions from gasoline-powered engines are composed of organic compounds rather than elemental carbon.

Whereas coarse particles result mainly from the breakup of larger ones, fine particles are formed mainly by chemical reactions between gases and by the coagulation of even smaller species including molecules in the vapor state, so they are mainly secondary particles. Although most of the atmospheric mass of fine particles arises from natural sources, that over urban areas often has mainly an anthropogenic origin.

The average organic content of fine particles is generally greater than that for coarse ones. In areas such as Los Angeles, up to half the organic compounds in the particulate phase are formed from the reaction of VOCs and nitrogen oxides in the photochemical smog reaction; these compounds correspond to partially oxidized hydrocarbons that have incorporated oxygen to form carboxylic acids, etc. and nitrogen to form nitro groups, etc. Aromatic hydrocarbons with at least seven carbon atoms (e.g., toluene) that enter the air of such cities from the evaporation of gasoline also form aerosols. Hydrocarbons having fewer than seven carbons give oxidation products with substantial vapor pressures and therefore remain in the gas phase.

The other important fine particles suspended in the atmosphere consist predominantly of inorganic compounds of sulfur and of nitrogen. Much of the natural sulfur in air originates as *dimethyl sulfide*, $(CH_3)_2S$, emitted from the oceans. A by-product of its oxidation in air is *carbonyl sulfide*, COS, a long-lived trace atmospheric component that also results from the atmospheric oxidation of *carbon disulfide*, CS_2, and from direct emissions from oceans and biomass. Some of the COS makes its way into the stratosphere, where it is oxidized and produces the natural sulfate aerosol found at those altitudes. Both dimethyl sulfide and hydrogen sulfide are oxidized in air mainly to sulfur dioxide, SO_2. Sulfur dioxide gas also is emitted directly in large quantities both by natural sources such as volcanoes and as pollution from power plants and smelters. It becomes oxidized over a period of hours or days to sulfuric acid and sulfates in air. Sulfuric acid, H_2SO_4, itself travels in air not as a gas but as an aerosol of fine droplets, since it has such a great affinity for water molecules. A huge volcanic eruption in Iceland in 1783 produced enough sulfuric acid particles to blanket Europe in a "great dry fog" for the entire summer, killing many people. In Iceland itself, fluoride emitted from the volcano was a worse problem since it proved fatal to crops, livestock, and people.

Another natural source of atmospheric particles has been discovered. Alkyl iodine compounds such as CH_2I_2 are emitted by seaweed into the air above coastal regions. Absorption of the ultraviolet component of light is sufficient to detach iodine atoms from such gaseous molecules. In subsequent reactions analogous to those of chlorine in the stratosphere, the iodine atoms react with ozone to form *iodine monoxide*, IO, which in turn dimerizes to form I_2O_2. The dimer and other iodine–oxygen compounds condense to form fine particles.

PROBLEM 3-13

By analogy with the reactions of atomic chlorine discussed in Chapter 2, write balanced equations for the reaction of atomic iodine with ozone and for the dimerization of IO.

Fine particles in many areas are acidic, due to their content of sulfuric and nitric acids. The nitric acid is the end-product of the oxidation of nitrogen-containing atmospheric gases such as NH_3, NO, and NO_2. Because HNO_3 has a much higher vapor pressure than does H_2SO_4, there is less condensation of nitric acid onto preexisting particles than occurs with H_2SO_4.

Both sulfuric and nitric acids in tropospheric air often eventually encounter ammonia gas that is released as a result of biological decay processes occurring at ground level. The acids undergo an acid–base reaction with the ammonia, which transforms them into the soluble salts **ammonium sulfate,** $(NH_4)_2SO_4$, and **ammonium nitrate,** NH_4NO_3. Since sulfuric acid contains two hydrogen ions, the neutralization reaction occurs in two stages, the first producing **ammonium bisulfate,** NH_4HSO_4:

$$H_2SO_4(aq) + NH_3(g) \longrightarrow NH_4HSO_4(aq)$$

$$NH_4HSO_4(aq) + NH_3(g) \longrightarrow (NH_4)_2SO_4(aq)$$

The ammonia that results from animal urine originates in the liquid as **urea,** $CO(NH_2)_2$, which subsequently hydrolyzes:

$$CO(NH_2)_2 + H_2O \longrightarrow 2\,NH_3 + CO_2$$

The neutralization of acidity by ammonia gas released into the air from live-stock and from the use of fertilizers, and by carbonate ion suspended in air from the dust raised by farming activities, is the main reason why precipitation over the central United States is not particularly acidic, and similarly for regions of China. However, some acidification results from the ionization of the **ammonium ion,** NH_4^+, a weak acid, that is produced by ammonia neutralization:

$$NH_4^+ \rightleftharpoons NH_3 + H^+$$

Although the nitrate and sulfate salts initially are formed from acids in aqueous particles, evaporation of the water can result in the production of solid particles. The predominant ions in fine particles are the anions **sulfate,** SO_4^{2-}; **bisulfate,** HSO_4^-; and **nitrate,** NO_3^-; and the cations ammonium, NH_4^+; and hydrogen ion, H^+. Aerosols dominated by oxidized sulfur compounds are called **sulfate aerosols.**

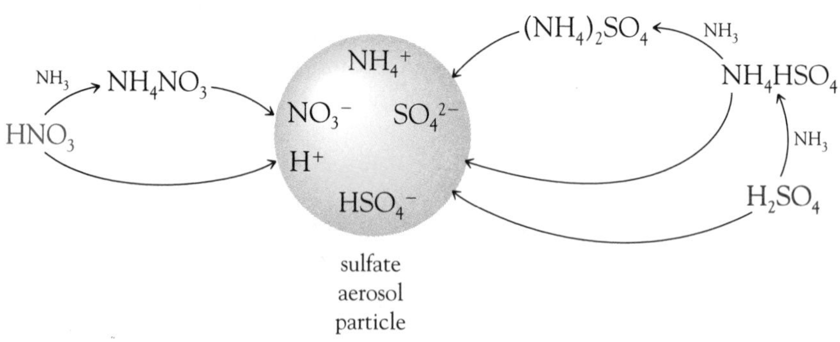

sulfate
aerosol
particle

On the west coast of North America, nitrate rather than sulfate is the predominant anion because more pollution results initially from nitrogen oxides than from sulfur dioxide, since coal mined in the western United States tends to be low in sulfur. In Great Britain, most of the fine particles in the winter months originate as soot from car exhaust and pollution from industry, whereas in the summer they arise from the oxidation of sulfur and nitrogen oxides.

If there is substantial ammonia gas in the air, nitric acid will react with it to form ammonium nitrate solid in the particulate phase. Recent simulations of smog formation in southern California indicate that although reductions in VOC concentrations without any change in NO_X would reduce ozone formation, the production of nitrate-based particulates would actually increase because more nitrogen dioxide would then react to produce nitric acid and then nitrate ion. The simultaneous control of ozone and particulates presents regulators with a formidable challenge!

In summary, coarse particles are usually either soot or inorganic (soil-like) in nature, whereas fine ones are mainly either soot, or sulfate or nitrate aerosols. Fine particles are usually acidic due to the presence of unneutralized acids, whereas coarse ones are usually basic because of their soil content.

Air Quality Indices and Size Characteristics for Particulate Matter

As we shall see in subsequent sections, the effect of particles suspended in air upon human health depends significantly upon the size of the particles involved. In the material that follows, we investigate the pollution indices used by governmental agencies to characterize the level of particulate air pollution present in an air sample as well as the effect of particle size on visibility through air masses.

The PM Indices

When air quality is monitored, the most common measure of the concentration of suspended particles is the **PM** index, which is the amount of particulate matter that is present in a given volume. Since the matter involved usually is not homogeneous, no molar mass for it can be quoted and thus concentrations are given in terms of the mass, rather than the number of moles, of particles. The usual units are *micrograms* of particulate matter *per cubic meter* of air, i.e., $\mu g/m^3$. Because smaller mass particles have a greater detrimental effect on human health than do larger ones, as we shall see later in this chapter, usually only those having a specified diameter or smaller are collected and reported. This cut-off diameter, in μm, is listed as the subscript to PM.

In recent years, government agencies in many countries, including the United States and Canada, have monitored PM_{10}, i.e., the total concentration of all particles having diameters less than 10 μm, which corresponds to all of the fine-particle range plus the smallest members of the coarse range. These are called **inhalable** particles since they can be breathed into the lungs. A typical value for PM_{10} in an urban setting is 20–30 $\mu g/m^3$. Increasingly, regulators are using the $PM_{2.5}$ index, i.e., one that includes all and only fine particles, which are also called **respirable** particles. The respirable range includes only particles that can penetrate deep into the lungs, where there are no natural mechanisms such as the cilia that line the walls of bronchial tubes to catch particles and move them up and out. Urban $PM_{2.5}$ values are usually in the 10–20 $\mu g/m^3$ range in North America, though background concentrations are only 1–5 $\mu g/m^3$. The new term **ultrafine** is applied to particles with very small diameters, usually taken to be less than 0.1 μm. Most ultrafine particles are anthropogenic in origin. In the past, the **total suspended particulates,** abbreviated TSP, which is the concentration of all particulates suspended in air, was often reported instead of a PM index.

The 1987 U.S. Air Quality Standards called for a maximum 24-hour PM_{10} level of 150 $\mu g/m^3$ and a maximum annual average of 50 $\mu g/m^3$. The United Kingdom has instituted a 24-hour PM_{10} standard of 50 $\mu g/m^3$ that cannot legally be exceeded on more than four days each year. In 1997, the U.S. EPA decided to regulate $PM_{2.5}$ levels—to an average of no more than 15 $\mu g/m^3$ annually and 65 $\mu g/m^3$ daily. The EPA estimated that the new particulate standards could prevent 15,000 premature deaths, as well as 250,000 person-days of aggravated asthma, annually. Indeed, $PM_{2.5}$ levels in the United States in 2003 were the lowest since 1999 at least, when they were first monitored by the EPA. Although 30 states met the new $PM_{2.5}$ standard in 2003, the other 20—mostly along the East Coast—did not. In 2006, the EPA proposed to lower the daily $PM_{2.5}$ level to 35 $\mu g/m^3$, which would further reduce death rates from particulate exposure by 22%. The corresponding Canadian standard is 30 $\mu g/m^3$, to be achieved by 2010. The European Union has proposed that its member states reduce their $PM_{2.5}$ levels by 20% between 2010 and 2020, with a cap of 25 $\mu g/m^3$.

PROBLEM 3-14

What would be the correct PM symbol for an index that included only ultrafine particles? What would be the PM symbol for the TSP index? Numerically, would the value for the ultrafine component of a given air mass be larger or smaller than its TSP?

As discussed in detail in Box 3-2, the distribution of particles is suspended in air peaks in the micrometer region because smaller ones coagulate to form particles of this size and further growth is slow, and because much larger ones rapidly settle out. The large increase in surface area that occurs when a large particle is split into smaller ones is explored in Problem 3-15.

BOX 3-2	Distribution of Particle Sizes in an Urban Air Sample

Because particles suspended in the atmosphere have different origins and compositions, and were formed and interacted with each other over a period of time in haphazard ways, there is a wide distribution of particle sizes present in any air mass.

One way of looking at the distribution of sizes is to plot the *number* of particles having a given diameter against the diameter; this is done in the solid curve of Figure 1 for a typical urban air sample. Notice that logarithmic scales were used in plotting both axes, in order that the details of the distribution for particles of many sizes can be seen clearly. The peak in the distribution occurs at about 0.01 μm and has a shoulder at about 0.1 μm. The net distri-

bution is probably the sum of several symmetrical (bell-shaped) distributions having peaks at different diameters. The particles of the distribution with the smallest diameter (0.01 μm) are formed by the condensation of vapors of pollutants formed by chemical reactions, such as the sulfuric acid formed by the oxidation of gaseous sulfur dioxide and the soot particles formed by combustion. The coagulation of such particles into larger ones (which can occur in minutes) and the deposition of gas molecules onto them result in the distribution having its peak at about 0.1 μm. Particles of this size also are created when water in aqueous droplets containing dissolved solids evaporates. Growth beyond this size is slow because the

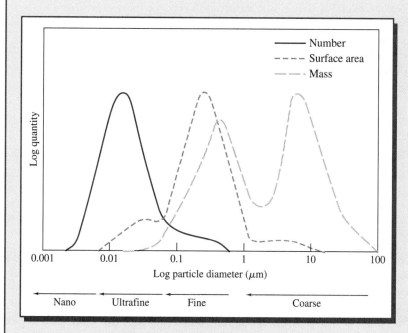

FIGURE 1 Distribution of particles by size in a typical urban environment. [Source: G. Oberdorster and M. J. Utell, *Environmental Health Perspectives* 110 (2002): A440.]

(continued on p. 140)

BOX 3-2	Distribution of Particle Sizes in an Urban Air Sample (continued)

larger the particle, the slower it moves, and thus the less likely it is to encounter and coagulate with particles of comparable size. Growth by condensation of gases is also slow for larger particles since their surface-to-mass ratio is smaller than for small particles.

The particles associated with the tail of the distribution, at almost 1 μm, are mainly soot or consist of material produced by mechanical disintegration of soil particles, etc. There are few particles of mass *larger* than a few microns in diameter because they quickly settle out of the air, although large particles that have settled on roadways often are resuspended temporarily by the action of vehicular traffic.

The plots of particle *numbers* can be misleading for some purposes because tiny particles of very small mass and surface area dominate the samples and thus the distributions. One alternative way to represent the data in a more meaningful way is to plot the total *mass* of all particles of a given size in an air sample against the diameter to see how mass is distributed among the different sizes. This type of plot is shown by the longer-dashed curve in Figure 1. The distribution function for mass is displaced to larger diameters compared to that for particle numbers: The mass

(or volume) of a particle is proportional to the cube of its diameter d (since for a sphere volume is proportional to the cube of the radius), so the height of the curve at any diameter of the distribution in the longer-dashed curve in Figure 1 corresponds to the value for the number distribution for this air mass times d^3. Consequently, in the mass distribution, the peak heights for larger particles are emphasized more than are those for smaller ones, and the whole distribution appears to shift to larger diameters. Two symmetrical distribution curves, one centered in the fine region at about 0.3 μm and the other in the coarse region at about 7 μm, appear to be superimposed to produce the final bimodal distribution. Notice that the total mass of the coarse-particle range (i.e., the sum of the area under the longer-dashed curve for $d > 2.5$ μm) in Figure 1 is greater than that for the fine region; this ratio is even larger for clean, rural air masses.

The function representing the distribution of surface area versus size is the shorter-dashed curve in Figure 1. The points on it are proportional to those for the number distribution times d^2, rather than d^3, since surface area is proportional to the square of the radius or diameter of the particle.

PROBLEM 3-15

Let k be a given measure of length; then suppose a cubic particle of dimension $3k \times 3k \times 3k$ is split up into 27 particles with size $k \times k \times k$. Calculate the relative increase in surface area when this occurs by comparing the surface area (length times width) of the six faces of the larger cube to the sum of all those of the smaller ones. From your answer, deduce whether the total surface area of a given mass of atmospheric particles is larger or smaller when it occurs as a large number of small particles rather than a small number of large ones.

Review Questions

1. In the "micrograms per cubic meter" concentration scale, to what substances do micrograms and cubic meters refer?

2. In general terms, what is meant by *photochemical smog*? What are the initial reactants in the process? Why is sunlight required?

3. What is meant by a *primary pollutant* and by a *secondary pollutant*? Give examples.

4. What is the chemical reaction by which *thermal NO* is produced? From which two sources does most urban NO arise? What is meant by the term NO_X?

5. Describe the strategies by which reduction of urban ozone levels have been attempted. What difficulties have been encountered in these efforts?

6. Describe the operation of the *three-way catalyst* in transforming emissions released by an automobile engine. Does the catalyst operate when the engine is cold? Why is it important for converters that the level of sulfur in gasoline be minimized?

7. Describe the reaction used in the *selective catalytic reduction* of nitrogen oxides.

8. What are the main anthropogenic sources of sulfur dioxide? Describe the strategies by which these emissions can be reduced. What is the *Claus reaction*?

9. What species are included in the air pollution index called *total reduced sulfur*?

10. Describe the various strategies used to produce *clean coal*.

11. Deduce the balanced reaction corresponding to the capture of sulfur dioxide gas by aqueous sodium sulfite and water to produce calcium bisulfite.

12. What is the difference between *dry* and *wet deposition*?

13. Define the term *aerosol*, and differentiate between *coarse* and *fine particulates*. What are the usual origins of these two types of atmospheric particles?

14. What are the usual chemical components of a *sulfate aerosol*?

15. Write a balanced equation illustrating the reactions that occur between one molecule of ammonia with (a) one molecule of nitric acid and with (b) one molecule of sulfuric acid.

16. What are the usual concentration units for suspended particulates? What would the designation PM_{40} mean? What do the terms *respirable* and *ultrafine* mean?

17. What is the two-step mechanism by which the hydroxyl free radical is produced in clean air?

 # Green Chemistry Questions

See the discussion of focus areas and the principles of green chemistry in the Introduction before attempting these questions.

1. *PERC* replaced gasoline and kerosene in the dry-cleaning process.

(a) Describe any environmental problems or worker hazards that would be associated with these solvents.

(b) Would these same environmental problems or worker hazards be eliminated by the use of PERC?

(c) By the use of carbon dioxide?

2. The development of surfactants for carbon dioxide by Joseph DeSimone won a Presidential Green Chemistry Challenge Award.

(a) Into which of the three focus areas for these awards does this award best fit?

(b) List two of the twelve principles of green chemistry that are addressed by the green chemistry developed by DeSimone.

3. The ions in ionic liquids (ILs) have weak ionic attractions for one another. This weak interaction is due to one or more factors including

• the presence of bulky nonpolar groups that prevent the close interaction of the charged regions of the ions, and

• delocalized and/or dispersed charges resulting in low charge density.

Inspect the ILs in Figure 3-12 and discuss the structural features of these compounds that result in weak interactions between the oppositely charged ions.

4. The discovery of the dissolution of cellulose with ionic liquids and the formation of various cellulose composites by Robin Rogers won a Presidential Green Chemistry Challenge Award.

(a) Into which of the three focus areas for these awards does this award best fit?

(b) The use of an abundant and naturally occurring polymer, a microwave heat source, and ionic liquids are three important green chemistry aspects of this study. For each of these aspects, list at least two of the twelve principles of green chemistry that are addressed in this study.

Additional Problems

1. The rate constant for the oxidation of nitric oxide by ozone is 2×10^{-14} molecule^{-1} cm^3 sec^{-1}, whereas that for the competing reaction in which it is oxidized by oxygen, that is,

$$2\,NO + O_2 \longrightarrow 2\,NO_2$$

is 2×10^{-38} molecule^{-2} cm^6 sec^{-1}. For typical concentrations encountered in morning smog episodes, namely 40 ppb for ozone and 80 ppb for nitric oxide, deduce the rates of these two reactions and decide which one is the dominant process.

2. In a particular air mass, the concentration of OH was found to be 8.7×10^6 molecules cm^{-3}, and that of carbon monoxide was 20 ppm.
(a) Calculate the rate of the reaction of OH with atmospheric CO at 30°C, given that the rate constant for the process is 5×10^{-13} e$^{-300/T}$ molecule^{-1} cm^3 sec^{-1}.
(b) Estimate the half-life of OH molecules in air at 30°C, assuming that their lifetime is determined by their reaction with CO. [Hint: Re-express the rate law as a pseudo-first-order process with the level of CO fixed at 20 ppm. Consult your introductory chemistry textbook to find the relationship between the half-life of a substance and the rate constant for its first-order decay.]

3. In the overall reaction that produces nitric oxide from N_2 and O_2, the slow step in the mechanism is the reaction between atomic oxygen and molecular nitrogen to produce nitric oxide and atomic nitrogen.
(a) Write out the chemical equation for the slow step and the rate law equation for it.
(b) Given that its rate constant at 800°C is 9.7×10^{10} L mol^{-1} sec^{-1}, and that its activation energy is 315 kJ mol^{-1}, calculate the amount by which the rate constant increases if the temperature is raised to 1100°C.

4. At combustion temperatures, the equilibrium constant for the reaction of N_2 with O_2 is about 10^{-14}. Calculate the concentration of nitric oxide that is in equilibrium with atmospheric levels of nitrogen and oxygen. Repeat the calculation for normal atmospheric temperatures, at which the

equilibrium constant is about 10^{-30}. Given that the concentration of NO that exits from the combustion zone in a vehicle is much higher than this latter equilibrium value, what does that imply about equilibrium in the reaction mixture? [Hint: Use the stoichiometry of the reaction to reduce the number of unknowns in the expression for K.]

5. The concentration of ozone in ground-level air can be determined by allowing the gas to react with an aqueous solution of potassium iodide, KI, in a redox reaction that produces molecular iodine, molecular oxygen, and potassium hydroxide.
(a) Deduce the balanced reaction for the overall process.
(b) Determine the ozone concentration, in ppb, in a 10.0-L sample of outdoor air if it required 17.0 μg of KI to react with it.

6. Perform a steady-state analysis on the 3-step reaction mechanism below. Assume that both ozone and atomic oxygen are in a steady state, and derive an expression for $[NO_2]/[NO]$.

$$NO_2 \longrightarrow NO + O$$

$$O + O_2 \longrightarrow O_3$$

$$NO + O_3 \longrightarrow NO_2 + O_2$$

7. The percentage of sulfur in coal can be determined by burning a sample of the solid and passing the resulting sulfur dioxide gas into a solution of hydrogen peroxide, which oxidizes it to sulfuric acid, and then titrating the acid. Calculate the mass percentage of sulfur in a sample if the gas from a 8.05-g sample required 44.1 mL of 0.114 M NaOH in the titration of the diprotic acid.

8. Calculate the volume, at 20°C and 1.00 atm, of SO_2 produced by the conventional roasting of 1.00 tonnes (10,000 kg) of nickel sulfide ore, NiS. What mass of pure sulfuric acid could be produced from this amount of SO_2?

9. The settling rate of particulates in air is directly proportional to the squares of their diameters (Stokes' law), provided that their densities are equal. If emitted particulates with a given diameter are found to settle out after two days, how long would it take particulates of the same material with half the diameter to settle out if they are emitted from the same tall chimney?

10. The sulfur species that is oxidized in water droplets is the bisulfite ion, HSO_3^-, so the rate of oxidation is proportional to its concentration multiplied by that of the oxidizing agent. Predict how changes in pH in the droplet will affect the rate of oxidation if (a) O_3 reacts with bisulfite ion, and if (b) hydrogen peroxide in the protonated form, $H_3O_2^+$, formed in the equilibrium

$$H_2O_2 + H^+ \rightleftharpoons H_3O_2^+$$

is the species that reacts with bisulfite.

Further Readings

1. O. Klemm, "Local and Regional Ozone: A Student Study Project," *Journal of Chemical Education* 78 (2001): 1641–1646.

2. R. J. Chironna and B. Altshuler, "Chemical Aspects of NO_X Scrubbing," *Pollution Engineering* (April 1999): 33–36; R. K. Agrawal and S. C. Wood, "Cost-Effective NO_X Reduction," *Chemical Engineering* (February 2001): 78–82.

3. A. Sheth and T. Giel, "Understanding the PM-2.5 Problem," *Pollution Engineering* (March 2000): 33–35.

4. (a) A-M. Vasic and M. Weilenmann, "Comparison of Real-World Emissions from Two-Wheelers and Passenger Cars," *Environmental Science and Technology* 40 (2006): 149–154. (b) A. Kurniawan and A. Schmidt-Ott, "Monitoring the Soot

Emissions of Passing Cars," *Environmental Science and Technology* 40 (2006): 1911–1915.

5. D. Mage et al., "Urban Air Pollution in Megacities of the World," *Atmospheric Environment* 30 (1996): 681–686.

6. R. M. Heck and R. J. Farrauto, "Automobile Exhaust Catalysts," *Applied Catalysis A: General* 221 (2001): 443–457.

7. "Fires from Hell," *New Scientist* 31 (August 2002): 34–37.

Websites of Interest

Log on to www.whfreeman.com/envchem4/ and click on Chapter 3.

CHAPTER 4

THE ENVIRONMENTAL AND HEALTH CONSEQUENCES OF POLLUTED AIR—OUTDOORS AND INDOORS

In this chapter, the following introductory chemistry topics are used:

- pH and acid–base concepts
- Balancing of redox equations

Background from Chapter 3 used in this chapter:

- Photochemical smog; thermal NO
- Coarse and fine particulates
- Aerosols
- PM_x indexes
- ppm, ppb, and $\mu g/m^3$ concentration scales for gases

Introduction

Smog, whether sulfur-based or photochemical, often has unpleasant odors due to some of its gaseous components. More seriously, the initial pollutants, intermediates, and final products of the reactions in smog affect human health and can cause damage to plants, animals, and some materials. In this chapter, we describe the detrimental effects on animals, plants, and materials of the gases and particles in polluted air—including the air we encounter indoors—and methods by which air pollution can be combated. Included in the discussions are the environmental effects of acid rain, a phenomenon that results from polluted air.

Haze

The most obvious manifestation of photochemical smog is a yellowish-brownish-gray haze that is due to the presence in air of small water droplets containing products of chemical reactions that occur among pollutants in air. This haze, familiar to most of us who live in urban areas, now extends periodically to once-pristine areas such as the Grand Canyon in Arizona.

Particles whose diameter is about that of the wavelength of visible light, i.e., 0.4–0.8 μm, can scatter light and interfere with its transmission, thereby reducing visual clarity, long-distance visibility, and the amount of sunlight reaching the ground. A high concentration in air of particles of diameters between 0.1 μm and 1 μm produces a haze. Indeed, one conventional technique of measuring the extent of particulate pollution in an air mass is to determine its haziness. The existence of smog in the air can often be determined by simply looking at buildings or hills in the distance and seeing if their appearance is partially masked by haze.

The widespread haze in the Arctic atmosphere in winter is due to sulfate aerosols that originate from the burning of coal, especially in Russia and Europe. The enhanced haziness in summertime over much of North America is due mainly to sulfate aerosols arising from industrialized areas in the United States and Canada. Fine particles also are largely responsible for the haze associated with Los Angeles and other locations subject to episodes of photochemical smog. The smog aerosols contain nitric acid that has been neutralized to salts. Also present in these aerosols are carbon-containing products that are intermediates in the photochemical smog reactions; however, intermediates formed from fuel molecules having short carbon chains usually have vapor pressures high enough that they exist as gases rather than condense onto particles. The typical composition of the fine component of an aerosol suspended over continental areas is illustrated in Figure 4-1.

Since most fine particles in urban air are secondary pollutants, their number can only be controlled by reducing emissions of the primary pollutant gases from which they are created. Thus governments have successively required more and more stringent emission controls on vehicles, power plants, etc., as discussed in Chapter 3. The switch to low-sulfur gasoline and diesel fuels should make catalytic converters on vehicles more efficient in reducing emissions.

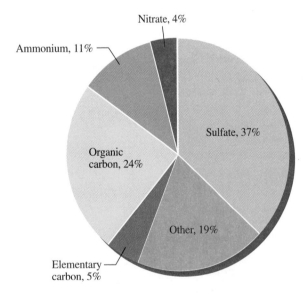

FIGURE 4-1 Typical composition of fine continental aerosol. [Adapted from J. Heintzenberg, *Tellus* 41B (1989): 149–160.]

Acid Rain

One of the most serious environmental problems facing many regions of the world is **acid rain.** This generic term covers a variety of phenomena, including *acid fog* and *acid snow,* all of which correspond to atmospheric precipitation of substantial acidity. In this section, the nature of the acids present in precipitation is discussed.

The phenomenon of acid rain was discovered by Angus Smith in Great Britain in the mid-1800s, but then it was essentially forgotten until the 1950s. It refers to precipitation that is significantly *more* acidic than "natural" (i.e., unpolluted) rain, which itself is often mildly acidic due to the presence in it of dissolved atmospheric carbon dioxide, which forms **carbonic acid,** H_2CO_3:

$$CO_2(g) + H_2O(aq) \rightleftharpoons H_2CO_3(aq)$$

The weak acid H_2CO_3 then partially ionizes to release a **hydrogen ion, H^+,** with a resultant reduction in the pH of the system:

$$H_2CO_3(aq) \rightleftharpoons H^+ + HCO_3^-$$

Because of this source of acidity, the pH of unpolluted, "natural" rain is about 5.6 (see Problem 3-10). Only rain that is appreciably more acidic than this—that is, with a pH of less than 5—is considered to be truly "acid" rain since, because of natural trace amounts of strong acids, the acidity level of rain in clean air can be a little greater than that due to carbon dioxide alone. Strong acids such as **hydrochloric acid,** HCl, produced by emissions of hydrogen chloride gas by volcanic eruptions, can produce "natural" acid rain temporarily in regions such as Alaska and New Zealand. On the other hand, the pH of unpolluted rain may be somewhat greater than 5.6 due to the presence of weakly basic substances originating with airborne soil particles that have partially dissolved in the droplets.

The two predominant acids in acid rain are **sulfuric acid,** H_2SO_4, and **nitric acid,** HNO_3, both of which are strong acids. Generally speaking, acid rain is precipitated far downwind from the source of the primary pollutants, namely **sulfur dioxide,** SO_2, and **nitric oxide,** NO. The strong acids are created during the transport of the air mass that contains the primary pollutants.

$$SO_2 \xrightarrow[H_2O]{O_2} H_2SO_4$$

$$NO_X \xrightarrow[H_2O]{O_2} HNO_3$$

Consequently, acid rain is a pollution problem that does not respect state or national boundaries because the atmospheric pollutants often undergo long-range transport. For example, most acid rain that falls in Norway, Sweden, and the Netherlands originates as sulfur and nitrogen oxides emitted in other countries in Europe. Indeed, the modern recognition of acid rain as a problem stems from observations made in Sweden in the 1950s and 1960s, which were due to emissions from outside its borders. China now has serious acid rain problems due to its high emissions of SO_2. Acidification is more serious in southern and southwestern than in northern China, where airborne alkaline dust originating in deserts neutralizes the acid. Some of the acid rain that originates in China is carried by the wind to Japan and, on occasion, all the way to North America. As the late economist-philosopher John Kenneth Galbraith noted, "Acid rain falls on the just and the unjust and also equally on the rich and poor."

The Ecological Effects of Acid Rain and of Photochemical Smog

Acid rain has a variety of ecologically damaging consequences, and the presence of acid particles in air may also have direct effects on human health. However, the effects of acid rain on soil vary dramatically from region to region. In this section, we investigate the chemical processes underlying the ecological effects of acid rain.

Nitric oxide is not especially soluble in water, and the acid (sulfurous) that sulfur dioxide produces upon dissolving in water is a weak one. Consequently, the primary pollutants NO and SO_2 themselves do not make rainwater particularly acidic. However, some of the mass of these primary pollutants is converted over a period of hours or days into the secondary pollutants sulfuric acid and nitric acid, both of which are very soluble in water and are strong acids. Indeed, virtually all the acidity in acid rain is due to the presence of these two acids. In eastern North America, sulfuric acid greatly predominates because some electrical power is generated from power plants that use high-sulfur coal. In western North America, nitric acid attributable to vehicle emissions is predominant, since the coal mined and burned there is low in sulfur.

Figure 4-2 shows a contour map of the average pH of precipitation in different regions of the world. The lowest pH ever recorded, 2.4, occurred for a rainfall in April 1974 in Scotland. Indeed, central-west Europe, including the United Kingdom, has a serious acid rain problem, as can be seen from the pH = 4.0 and 4.5 contours surrounding the area in Figure 4-2. In North America, the greatest acidity occurs in the eastern United States and in southern Ontario, since both regions lie in the path of air originally polluted by emissions from power plants in the Ohio Valley. On the other hand, much of the acidity that falls in upper New York State stems from emissions in southern Ontario.

In addition to the acids delivered to ground level during precipitation, a comparable amount is deposited on the Earth's surface by means of **dry deposition,** the process by which nonaqueous chemicals are deposited onto solid and liquid surfaces at ground level when air containing them passes over the

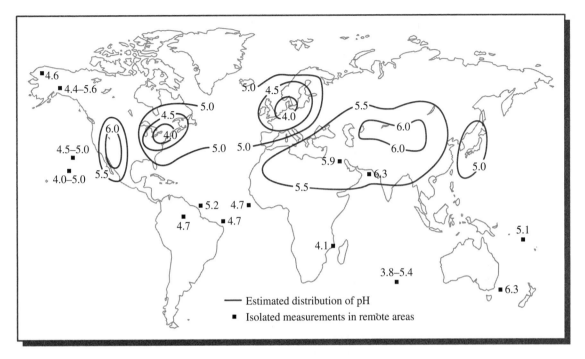

FIGURE 4-2 Global pattern of acidity of precipitation. [Source: Redrawn from J. H. Seinfeld and S. N. Pandis, *Atmospheric Chemistry and Physics* (Chichester: John Wiley, 1998).]

surfaces. Much of the original SO_2 gas is not oxidized in the air but rather is removed by dry deposition from air before reaction can occur: Oxidation and conversion to sulfuric acid occurs after deposition. **Wet deposition** processes encompass the transfer of pollutants to the Earth's surface by rain, snow, or fog—i.e., by aqueous solutions.

Neutralization of Acid Rain by Soil

The extent to which acid precipitation affects biological life in a given area depends strongly on the composition of the soil and bedrock in that area. If the bedrock is limestone or chalk, the acid can be efficiently neutralized ("buffered"), since these rocks are composed of **calcium carbonate,** $CaCO_3$, which acts as a base and reacts with acid, producing **bicarbonate ion,** HCO_3^-, as an intermediate:

$$CaCO_3(s) + H^+(aq) \longrightarrow Ca^{2+}(aq) + HCO_3^-(aq)$$

$$HCO_3^-(aq) + H^+(aq) \longrightarrow H_2CO_3(aq) \longrightarrow CO_2(g) + H_2O(aq)$$

The reactions here proceed almost to completion due to the excess of H^+ that is present. Thus the rock dissolves, producing carbon dioxide and calcium ion to replace the hydrogen ion. These same reactions are responsible for the deterioration of limestone and marble statues; fine detail, such as ears, noses, and other facial features, are gradually lost as a result of reaction with acid and with sulfur dioxide itself. Also, neutralization by calcium carbonate and similar compounds that are commonly present as suspended particles in atmospheric dust is the mechanism by which carbonic acid in normal rainfall and acid rain over some areas has a pH greater than expected.

In contrast, areas strongly affected by acid rain are those having granite or quartz bedrock, since the soil there has little capacity to neutralize the acid. Figure 4-3 shows areas of North America having low soil alkalinity, that is, low amounts of basic compounds with which acids can react. Large areas susceptible to acidity are the Precambrian Shield regions of Canada and Scandinavia. Acid rain resulting from the massive development of the tar sands to produce synthetic crude oil in northern Alberta, and the SO_2 and NO_X emissions that result, are now affecting areas in Manitoba and northern Saskatchewan that lie upwind from them, since the soils in these two areas have very little neutralizing capacity (Figure 4-3).

Acidity from precipitation leads to the deterioration of soil. When the pH of soil is lowered, plant nutrients such as the cations potassium, calcium, and magnesium are exchanged with H^+ and thereupon leached from it.

FIGURE 4-3 Regions of North America with low soil alkalinity for neutralizing acid rain. [Source: D. J. Jacob, *Introduction to Atmospheric Chemistry* (Princeton, NJ: Princeton University Press, 1999), p. 233.]

Although sulfur dioxide emission levels fell significantly in recent decades in both Europe and North America, there has not been as large a corresponding change in the pH of the precipitation, especially in northeastern North America. The lack of corresponding reduction in acidity is attributed to a decline over the same period of fly ash emissions from smokestacks and of other solid particles, all of which are alkaline and in the past neutralized a fraction of the sulfur dioxide and sulfuric acid in the same way that calcium carbonate does in soil. Thus the decline in acidity in precipitation in the northeastern United States from 1983 to 1994 amounted to 11%, although the sulfate ion molar concentrations in precipitation fell not by 5.5% (half that of H^+), but rather by 15%. The much smaller nitrate levels remained essentially unchanged in this period in this region. The change in sulfate deposition in the northeastern United States and south-central Canada

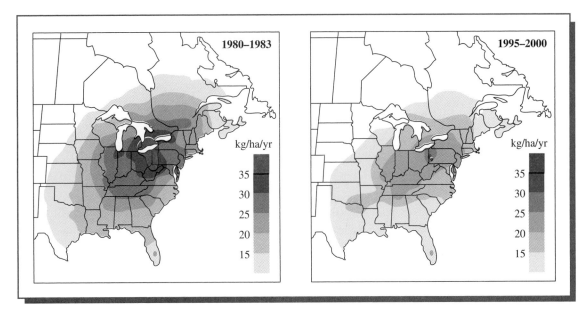

FIGURE 4-4 Wet sulfate deposition in eastern North America as a four-year mean (kilograms/hectare per year). [Source: Canadian National Atmospheric Chemistry Database, Meteorological Service of Canada, Environment Canada.]

from the early 1980s to the late 1990s is shown in Figure 4-4. In Great Britain, rainfall acidity declined by about 40% in the 1986–1997 period due to emission controls there.

Because of acid rainwater falling and draining into them, tens of thousands of lakes in the Shield regions of both Canada and Sweden have become strongly acidified, as have lesser numbers in the United States, Great Britain, and Finland. Lakes in Ontario are particularly hard hit, since they lie directly in the path of polluted air and since the soil there contains little limestone. In a few cases, attempts have been made to neutralize the acidity by adding limestone or **calcium hydroxide,** $Ca(OH)_2$, to the lakes; however, this process must be repeated every few years to sustain an acceptable pH. Adding phosphate ion to lakes can also control acidity, since it stimulates plant growth during the natural denitrification process by which **nitrate ion,** NO_3^-, is converted to reduced nitrogen with the consumption of large quantities of hydrogen ions, as shown in the reduction half-reaction

$$2\,NO_3^- + 12\,H^+ + 10\,e^- \longrightarrow N_2 + 6\,H_2O$$

In recent years, a new source of sulfuric acid in lakes has appeared—the oxidation of sulfur in shallow wetlands dried up by global warming and thereby exposed to the air.

As Problem 4-1 shows, the oxidation of **ammonium ion,** NH_4^+, to nitrate ion produces hydrogen ions. Indeed, the large emissions of ammonia into the air from manure in areas of livestock and poultry farming result in the atmospheric deposition of ammonium ion, which then is oxidized by soil microbes. The resulting H^+ contributes to the acidification of soil.

PROBLEM 4-1

Deduce the balanced redox half-reaction of conversion of ammonium ion, NH_4^+, to nitrate ion, NO_3^-, and thereby show that H^+ is also produced in this process.

In Australia, soil acidity has a completely different origin. Acidification is associated there with the removal of nitrate ion by the harvesting of plant and animal crops and by soil leaching. Presumably the loss of nitrate prevents its natural buffering of acidity by the reaction shown above. As in Canadian lakes, the effects of the acidification have been partially reversed in Australia by the addition of lime to the soil.

Until recently, acid rain in the United States was considered to be a problem for its northeastern region. Indeed, one of the hardest-hit regions is the Catskill Mountains in New York State, whose surface rocks consist of calcium-poor sandstone and from which most of the nutrients have now been leached. At the Hubbard Brook Experimental Forest in New Hampshire, half the calcium and magnesium in soil was leached by 1996 and, as a result, vegetative growth almost stopped. However, as a consequence of the reduction in SO_2 emissions, by 2003 over half the lakes in the Adirondack Mountains of New York State showed some significant recovery from acid rain. On average, the ability of lake water to neutralize acids increased by an average of 1.6 micromoles of H^+ per liter per year in the 1990s. Unfortunately, full recovery for these lakes to an acid-neutralizing capacity of 50 micromoles per liter is predicted to take another 25–100 years to achieve.

Acid rain now is also a concern in the southeastern United States. Here soils are generally thicker and thus able to neutralize more acid. However, much of that leaching ability now has been exhausted and acid levels in many waterways have increased substantially. It has been discovered that the recovery of such soils, and of those in Germany, is slowed once acid precipitation has declined, because previously stored sulfate ion is then released, causing more cation leaching and penetration of acidity deeper into the ground.

The regulatory scheme used in the United States of requiring reductions in sulfur dioxide emissions in certain geographical regions has been extended by European scientists and regulators into the concept of **critical load.** This concept recognizes that different levels of risk from acid rain are faced in different regions. Geographic areas that have buffering capacity can withstand a much greater load of acid rain before damage occurs than those without the capacity. Thus, higher sulfur dioxide emissions from a particular region can be allowed if the area in which the resulting sulfuric acid is usually deposited has a high critical load. To determine the critical loads, scientists use computer models that incorporate soil chemistry, rainfall, topography, etc. Use of the concept has had great success in Sweden, for example.

In using critical loads, pollution control becomes *effects-based* rather than *source-based*. Although the critical-loads concept has been implemented in regulations in Europe and Canada and is favored by many scientists and politicians in the United States, it has not been implemented there. Although significant progress has been made in reducing emissions of SO_2, and more reductions are scheduled both for it and for NO_X, scientists predict that these efforts will be insufficient to allow a full recovery of lakes and forests in the northeastern United States and south-central Canada.

Acidification reduces the ability of some plants to grow, including those in fresh-water systems. Because of the decrease in this productivity in lakes and streams that feed them, the amount of **dissolved organic carbon** (DOC) in the surface water has declined. The DOC contains molecules that absorb some of the ultraviolet from sunlight; thus a decline in DOC levels has allowed more penetration of UV light into the lower layers of lakes. In addition, global warming (see Chapters 6 and 7) has resulted in the drying up of some streams that supplied DOC to lakes. Furthermore, stratospheric ozone depletion has also allowed more UV to reach the Earth's surface, including lakes, in the first place. Thus fresh-water lakes have suffered a "triple whammy" from global environmental problems.

Release of Aluminum into Soil and Water Bodies by Acid Rain

Acidified lakes characteristically have elevated concentrations of dissolved **aluminum ion,** Al^{3+}, and it is now known that many of the biological effects of acid rain are due to increased levels of aluminum ion dissolved in water rather than to the hydrogen ion itself.

Aluminum ions are leached from rocks in contact with acidified water by reaction with the hydrogen ions; under normal, near-neutral pH conditions, the aluminum is immobilized in the rock by their insolubility.

$$\text{Al compounds(s)} \xrightarrow{\text{H}^+} Al^{3+}$$

Plots of dissolved aluminum concentration versus water acidity for lakes in the Adirondack Mountains of New York State and for lakes in Sweden are illustrated in Figure 4-5. (The chemistry underlying these processes is further discussed in Chapter 13, as are the reasons why natural waters have pH values of 7 or 8, rather than the 5.3 of rain.)

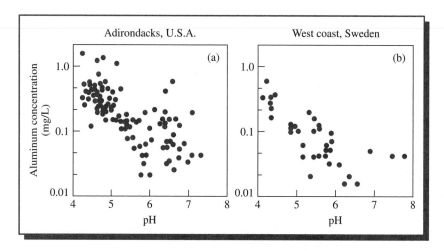

FIGURE 4-5 Aluminum concentrations versus pH of the water in different fresh-water lakes in (a) the Adirondacks and (b) western Sweden. Notice the logarithmic vertical axis. [Source: M. Havas and J. F. Jaworski, *Aluminum in the Canadian Environment* (Ottawa: National Research Council of Canada Report 24759, 1986).]

Scientists believe that both the acidity itself and the high concentrations of aluminum together are responsible for the devastating decreases in fish populations that have been observed in many acidified water systems. Different types of fish and aquatic plants vary in their tolerances for aluminum and acid, so the biological composition of a lake varies as it gradually becomes increasingly acidic. Generally speaking, fish reproduction is severely diminished even at low levels of acidity that can, however, be tolerated by adult fish. Very young fish, hatched in early spring, also are subject to the shock of very acidic water that occurs when the acidic winter snow all melts in a short time and enters the water systems.

Healthy lakes have a pH of about 7 or a little higher; few fish species survive and reproduce when the pH drops much below 5. As a result, many lakes and rivers in affected areas are now devoid of their valuable fish; for example, 30% of the salmon rivers in Nova Scotia are too acidic for Atlantic salmon to survive. The water in many acidified lakes is crystal clear due to the death of most of the flora and fauna.

However, aluminum levels draining from soils at medium-to-high elevations in the United States declined significantly over the period 1984–1998 and, if the trend continues, will not be a threat to fish by about 2012.

Effect of Air Pollution on Trees and Crops

In recent years it has become clear that air pollution can also have a severe effect on trees. The phenomenon of forest decline was first observed on a large scale in western Germany and occurs mainly at high altitudes. However, the cause-and-effect relationship behind this forest decline has been very difficult for scientists to untangle. As discussed above, acidification of the soil can leach nutrients from it and, as occurs in lakes, solubilize aluminum. This element may interfere with the uptake of nutrients by trees and other plants. Apparently both the acidity of the rain falling on affected forests and the tropospheric ozone and other oxidants in the air to which they are exposed pose significant stresses to the trees. These two stresses alone will not kill them, but when combined with drought, temperature extremes, disease, or insect attack, the trees become much more vulnerable.

Forests at high altitudes are most affected by acid precipitation, possibly because they are exposed to the base of low-level clouds, where the acidity is most concentrated. Fogs and mists are even more acidic than precipitation, since there is much less total water to dilute the acid. For example, white birch trees along the shores of Lake Superior experience dieback in regions where acid fog occurs, as it frequently does there. Deciduous trees (i.e., those that lose their leaves annually) affected by acid rain gradually die from their tops downward; the outermost leaves dry and fall prematurely and are not replenished the following spring. The trees become weakened as a result of these changes and become more susceptible to other stressors. In some regions of Europe and North America, forest soils are limed in order to combat the effects of acidity on trees.

Ground-level ozone itself has an effect on some agricultural crops due to its ability to attack plants. Apparently the ozone reacts with the gas (ethene) that the plants emit, generating free radicals that then damage plant tissue. The rate of photosynthesis is slowed, and hence the total amount of plant material is reduced, by the action of ozone. As in the case of trees, air pollution acts as a stressor to plants. The collective damage to North American crops, e.g., alfalfa in the United States and white beans in Canada, is estimated to be $3 billion a year. Other crops whose yields are adversely affected by current levels of ozone include wheat, corn, barley, soybeans, cotton, and tomatoes. The fraction of the world's cereal crops that are grown in regions of high ozone, and therefore subject to damage, is predicted to more than triple by 2025.

The Human Health Effects of Outdoor Air Pollutants

It is now well established that breathing polluted air can have a dramatic influence on human health. In this section, the most important effects of outdoor air pollutants are described, and the variation in concentration of the dominant air pollutants in different countries is discussed.

The effect that pollutants have on human health cannot be deduced from general laws of biology or physiology; they must be established by experimentation. One can imagine experiments involving animals or human volunteers in which the health effects of exposure to brief periods of artificially produced high-level pollution are studied. However, the extrapolation of information gained from short-term studies of high-level pollution to the long-term exposures at low levels is difficult. In particular, for some pollutants, there may exist a **threshold** pollutant concentration, or an exposure below which a particular health effect does not occur. In such cases, predictions obtained by assuming simple direct proportionality between exposure and effect would be unwarranted. In addition, there could be deleterious effects of chronic exposure that do not come into play when exposure, even intense exposure, to pollutants occurs only for brief periods of time.

For these reasons, the best information regarding the effects of pollutants on health comes from the large-scale "experiment" in which we are all enrolled as "test animals"—namely, living in a society in which we are routinely exposed to these pollutants for our whole lives. Because the level of exposure to any given pollutant varies considerably from place to place, scientists can collect information on health and on pollution levels in different locations, then correlate them using statistics to establish the effect of one on the other.

As would be expected, the major effects on human health from air pollution occur in and through the lungs. For example, asthmatics suffer worse episodes of their disease when the sulfur dioxide or the ozone or the particulate concentration rises in the air that they breathe. In one U.S. study, it was established that asthma attacks increased by 3% for each increase of 10 μg/m^3 in the PM$_{10}$ index (discussed in Chapter 3). A recent study in California found that asthma can be *caused* by air pollution, specifically by ozone and especially among highly active children, who naturally inhale more air into their lungs.

Another gaseous pollutant of some concern is **1,3-butadiene,** which has the structure CH$_2$=CH—CH=CH$_2$. This hydrocarbon is known as an **air toxic** since there is evidence that it causes cancer—leukemia and *non-Hodgkin's lymphoma* especially—and may also negatively affect human reproduction. It is produced as a by-product of the incomplete combustion of fuels, is produced in forest fires and wildfires, and is a component of cigarette smoke.

The Human Health Effects of Smogs

In the middle decades of the twentieth century, several Western industrialized cities experienced such serious wintertime episodes of smog from soot and sulfur pollution that the death rate increased noticeably. For example, in London, England, in December 1952, about 4000 people died within a few days—plus 8000 more in the next few months—as a result of the high concentrations of these pollutants that had built up in a stagnant, foggy air mass trapped by a temperature inversion close to the ground. Those at most risk were young children and elderly persons already suffering from bronchial problems. A ban on household coal burning, from which most of the pollutants originated, has now largely eliminated such problems. Scientists are still unsure whether the main sulfur-containing agent that caused such serious problems in London was the SO_2, the sulfuric acid droplets, or the sulfate particulates.

Today, due to pollution controls, *soot-and-sulfur smogs* are no longer a major problem in Western countries. Deaths from bronchitis have fallen by over half in the United Kingdom, the result of changes in air quality (and smoking habits). However, the quality of winter air in some areas of what was the Eastern bloc of countries, such as southern Poland, the Czech Republic, and eastern Germany, until recently was very poor on account of the burning of large amounts of high-sulfur (up to 15% S) "brown coal" for both industrial and home-heating purposes. For example, although the acceptable limit for the concentration of SO_2 in air is 80 $\mu g/m^3$ in many countries, the level of this gas in Prague surpassed 3000 $\mu g/m^3$ on occasion. Indeed, four out of five children admitted to the hospital in some areas of Czechoslovakia in the early 1990s were there for treatment of respiratory problems. However, the average SO_2 level in Prague decreased by about 50% from the early 1980s to the early 1990s, and overall in the Czech Republic SO_2 emissions are now only about 10% of 1990 levels. The tremendous improvement of air quality in eastern Germany since 1990, where mean SO_2 levels have dropped from 113 to 6 $\mu g/m^3$, has resulted in a decrease in childhood respiratory infections and an increase in lung function.

The effects of sulfur dioxide are also evident in cities such as Athens, where the death rate is found to increase by 12% when the concentrations of the gas exceed 100 $\mu g/m^3$. Detail on the ancient statues and monuments of Athens is also being seriously eroded by sulfur dioxide and its secondary pollutants. High levels of sulfur dioxide and of fine particulates, both mainly from diesel-fueled vehicles, caused about 350 premature deaths in Paris annually in the late 1980s. And the air in London is not so improved that it does not affect human health; a recent study concluded that one in every 50 heart attacks was triggered by outdoor air pollution, from a combination of smoke, CO, SO_2, and NO_2.

European cities are not the only ones affected by air pollution. Both sulfur dioxide and particulate matter levels regularly exceed World Health Organization (WHO) guidelines in Beijing, Seoul, and Mexico City. In 2002, 13 of the 20 world cities having the highest averages for airborne particulate matter were located in China; the others were Cairo, Jakarta, and five cities in India. In many large cities in the developing world, coal is still the predominant fuel and in some cases diesel-powered vehicles substantially worsen the problem. In Beijing, high SO_2 emissions from coal burners that are used to heat buildings, plus smoke from smelters on the edges of the city, plus windblown dust and sand from the Gobi Desert combine to produce poor air quality. Haze over China, produced by air pollution, so reduces sunlight intensity that it may be cutting food production by as much as 30% across a third of the country. Indeed, there are a number of cities in China in which the air quality is among the poorest in the world. According to recent projections, if no attempts are made to reduce SO_2 emissions as industrialization increases, by 2020 the concentration of the gas in Bombay and the Chinese cities of Shanghai and Chongqing will be about four times the WHO maximum safe limit.

It is a historical characteristic that once an undeveloped country starts industrial development, its outdoor air quality worsens significantly. The situation continues to deteriorate until a significant degree of affluence is attained, at which point emission controls are enacted and enforced, and the air begins to clear. Thus, although the quality of air is now improving with time in most developed countries, it is worsening in the larger cities of developing countries. Mexico City and several urban areas in China, especially Beijing, are generally considered to have the worst urban air pollution in the world at present. Half the respiratory disease in China is caused by air pollution. A report by the United Nations Environmental Program estimates that deaths worldwide from all forms of air pollution amounted to 2.7–3.0 million in 2001, a figure that may rise to 8 million by 2020.

Although acute smog episodes from soot and sulfur-based chemicals have been eliminated in the West, many residents in these countries still are chronically exposed to measurable levels of suspended particles containing sulfuric acid and sulfates due to the long-range transport of these substances from industrialized regions that still emit SO_2 into the air. For example, research has shown a positive correlation between atmospheric concentrations of ozone and oxidized sulfur and hospital admissions for respiratory problems in southern Ontario. There is some evidence that the acidity of the pollution is the main active agent in causing lung dysfunction, including wheezing and bronchitis in children. Asthmatic individuals appear to be adversely affected by acidic sulfate aerosols, even at very low concentrations.

Photochemical smog, which arises from nitrogen oxides, is now more important than sulfur-based smog in most cities, particularly those of high population and vehicle density. As discussed in Chapter 3, it consists of gases such as ozone and an aqueous phase containing water-soluble organic and inorganic compounds in the form of suspended particles. In contrast to "London smogs," which chemically were reducing in nature due to sulfur dioxide, photochemical smogs are oxidizing.

Ozone itself is a harmful air pollutant. In contrast to sulfur-based chemicals, its effect on the robust and healthy is as serious as on those with preexisting respiratory problems. Experiments with human volunteers have shown that ozone produces transient irritation in the respiratory system, giving rise to coughing, nose and throat irritation, shortness of breath, and chest pains upon deep breathing. People with respiratory problems can often tell from their symptoms—such as the tightening of their chest or the beginning of a cough—when the air quality is poor. Even healthy, young people often experience such symptoms while exercising outdoors by cycling or jogging during smog episodes. Indeed, there is evidence that the daily race times of cross-country runners increase with increasing ozone concentration in the air that they inhale. A small percentage of the day-to-day fluctuations in mortality rate in Los Angeles is explained by variations in the concentrations of air pollutants. An analysis of 95 urban centers in the United States discovered that a period of high ozone concentrations increased daily cardiovascular and respiratory mortality by about 0.5% per 10 ppb increase following a few days of continuous exposure. It is not yet clear what, if any, long-term lung dysfunction results from exposure to ozone, and indeed this is a controversial subject among scientists. Exposure to ozone produces a number of indirect health effects as well—including a decrease in sperm count.

One anticipated effect of ozone is a decreased resistance to disease from infection because of the destruction of lung tissue. Many scientists believe that chronic exposure to high levels of urban ozone leads to the premature aging of lung tissue. At the molecular level, ozone readily attacks substances containing components with $C=C$ bonds, such as occur in biological tissues of the lung. As discussed later, the fine particulates produced in the photochemical smog process can have a deleterious health effect on humans.

Most industrialized nations have enacted standards that regulate the maximum concentrations in air of sulfur dioxide, **nitrogen dioxide,** NO_2, and **carbon monoxide,** CO, as well as ozone and fine particulates (see Table 4-1), and in some cases total reduced sulfur, since all these pollutants cause health effects at sufficiently high concentrations. For example, several recent North American studies have statistically linked the rate of hospitalization for

TABLE 4-1	Air Quality Standards, in Parts per Billion, for Pollutants				
Pollutant	Time Span to Average	United States	Canada	European Union	Australia
O_3	8 hr	80	65*	60	80
CO	1 hr	35	31		
	8 hr	9	13	9	9
SO_2	1 day	140	115	48	80
	1 year	30	23		20
NO_2	1 year	53	53	21	30
$PM_{2.5}$ (in $\mu g/m^3$)	1 day	35	30*	50	25

*To be implemented by 2010.

congestive heart failure among elderly people to the daily carbon monoxide concentration in outside air. Mexico City currently has the highest levels of carbon monoxide among the world's most polluted cities. Both CO and NO_2 are usually more of a problem in indoor air and will be discussed in detail in a later section.

It has been speculated that pollution due to SO_2 and sulfates causes a decrease in resistance to colon and breast cancer in people living in northern latitudes. The suggested mechanism of this action is a reduction in the amount of available UV-B that is necessary to form vitamin D, which is a protective agent for both types of cancers. Since sulfur dioxide absorbs UV-B and sulfate particles scatter it, significant concentrations of either substance in air will reduce the amount of UV-B reaching ground level. Thus too little UV-B can have detrimental health effects, just as too much of it can—as was outlined in Chapter 2.

Finally, we note that there are some positive effects of air pollution on human health! For example, the rate of skin cancers in areas heavily polluted by ozone is probably reduced because of the ability of the gas to filter UV-B from sunlight.

Particulates as Health Risks

Particulate matter in the form of smoke from coal burning has been an air pollution problem for many hundreds of years, especially in the United Kingdom. John Evelyn wrote in his January 1684 diary that "London by reason of the excessive coldness of the air, hindering the ascent of the smoke, was so filled with the fuliginous [sooty] steam of sea-coal, that hardly could one see across the street, and this filling the lungs with its gross particles exceedingly obstructed the breast, so as one would scarce breathe." Indeed, unsuccessful attempts to control coal burning and punish offenders had begun in the thirteenth century in Britain. Perhaps Shakespeare was referring to this type of air pollution in the quotation from *Hamlet* that opened Chapter 3.

Although serious episodes of such soot-and-sulfur smogs have been largely eliminated in Western industrialized countries, the air pollution parameter that correlates most strongly with increases in the rate of disease or mortality in most such regions is the concentration of respirable (fine) particulates, $PM_{2.5}$. It appears that particulate-based air pollution has a greater effect on human health than that produced directly by pollutant gases.

Substances that dissolve into the body of a particle are said to be **absorbed** by it; those that simply stick to the surface of the particle are said to be **adsorbed** (see Figure 4-6). An important example of the latter is represented by the adsorption of large organic molecules onto the surfaces of carbon (soot) particles, as discussed later in Chapter 12. Many insoluble airborne particles are surrounded by a film of water, which can itself dissolve other substances. The adsorption of metal atoms and organic molecules on the surface of airborne particles may give rise to some of the health hazards these particles represent.

Larger particles—coarse ones, according to the definition in Chapter 3—are of less concern to human health than are small (fine) ones for several general reasons:

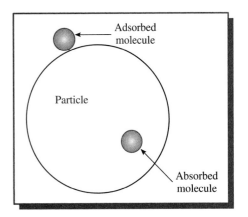

FIGURE 4-6 Contrast between adsorption and absorption of molecules on/in an airborne particle (schematic).

- Since coarse particles settle out quickly, human exposure to them via inhalation is reduced.

- When inhaled, coarse particles are efficiently filtered by the nose (including its hairs) and throat and generally do not travel as far as the lungs. In contrast, inhaled fine particles usually travel through to the lungs (which is why they are called respirable), can be adsorbed on cell surfaces there, and can consequently affect our health.

- The ratio of surface area to mass of large particles is smaller than that of small ones; thus, gram for gram, their ability to transport adsorbed gas molecules to any parts of the respiratory system and to catalyze chemical and biochemical reactions there is correspondingly smaller.

- Devices such as electrostatic precipitators, spray towers, and cyclone collectors that are used to remove particulates from air are efficient only for coarse particles. Thus, although a device may remove 95% of the total particulate mass, surface area and respirable particles are reduced by a much lower fraction; see Problem 4-2. *Baghouse filters*, which are finely woven fabric bags through which air is forced, are highly efficient in removing fine particles in the 1-μm size range, as well as all the larger ones.

The exhaust from diesel engines has been classified as "likely to be carcinogenic to humans" by the U.S. Environmental Protection Agency (EPA). Studies in California and in Seattle conclude that 70% or more of the risk to health from air toxics arises from diesel exhaust. Following court decisions, the United States will institute by 2010 a series of new regulations limiting emissions from on-road diesel vehicles.

PROBLEM 4-2

An air-filtering device is tested and is found to remove all particles larger than 1 μm in diameter, but almost none of the smaller ones. Calculate the percentage of the surface area removed by the device for a sample of particulates, 95% of the mass of which is particles of diameter 10 μm and 5% of which is particles of diameter 0.1 μm. Assume all particles are spherical and of equal density. [Hint: Recall that the surface area of a sphere is $4\pi r^2$. Calculate the surface areas of particles of each size from this formula.]

A number of studies have correlated day-to-day urban morbidity (sickness) rates, as measured by hospital admission rates, for respiratory problems against the pollution levels during the same short time period. For example, there have been several reports concerning the immediate effects on the population of southern Ontario of the pollutants—ozone gas and sulfate particulates—to which they are most exposed. In one study, the average number of hospital admissions for respiratory problems correlated best with the ozone level of the previous day, and to a slightly lesser degree with the sulfate level from the previous day, for the summers of 1983–1988. Air pollution was found to account for about 6% of summertime hospital respiratory admissions, a magnitude close to that found in previous investigations in Ontario and New York State. A recent study found that respiratory admissions correlated significantly with both PM_{10} and ozone concentrations in Spokane, Washington—an area where atmospheric sulfur dioxide is essentially nonexistent and therefore can be ruled out as the true culprit in causing the illnesses.

The strongest links between human health and exposure to airborne particulate matter are based on recent studies involving cities in the United States and are the subject of the online Case Study *The Effect of Urban Air Particulates on Human Mortality* at the website associated with this chapter.

Smoke

The burning of wood in domestic fireplaces produces large quantities of particulates, which are emitted from the chimneys into outdoor air unless catalytic converters are fitted to the smokestack. Indeed, in residential neighborhoods where wood is the predominant fuel used for heating, wood stoves contribute up to 80% of the fine particles in the air during the winter months. Outdoor wood-fired boilers, used to heat water for saunas and swimming pools, have grown so much in popularity that the particulates they emit have become a significant problem. Some newer wood stoves and boilers have catalytic converters or secondary combustion chambers in which particulates and unburned gases are more fully oxidized, thus reducing their emission to outside air.

Serious episodes of smoky haze pollution over large areas of land have occurred in recent years in Southeast Asia, especially in Malaysia and Indonesia. The smoke originates mainly from forest fires that are intentionally started in order to clear land that can be subsequently used for agriculture and to grow trees for their rubber, palm oil, or pulp content. A secondary source of the smoke is the smoldering underground fires that slowly burn in underground coal and peat deposits. Indeed, there are estimated to be a quarter million individual coal fires currently burning in Indonesia, as well as many in China and India, and there are also many peat fires in Malaysia. The fires are initiated when an outcropping of coal, or of a peat deposit that has dried after draining, is ignited, typically during one of the fires set to clear the land. Fires can also be ignited in coal, once exposed to the air, by lightning strikes and even by spontaneous combustion when the surface pyrite is oxidized and the heat released by this reaction sets the carbon ablaze. These underground fires can continue to burn for decades after the original forest fires have stopped.

A so-called *Asian brown cloud* of particles and gases from forest fires, vehicle exhausts, and domestic cookers—especially in rural areas—that burn wood, dung, and agricultural waste overhangs most of eastern and southeastern Asia annually from December to May, the main season for home heating. The brown cloud over the Indian Ocean consists mainly of smoke from the burning of dried manure in cooking fires. This haze lowers sunlight levels up to 15%, with a corresponding decline in the yield of crops such as rice and an alteration to rainfall and monsoon patterns. In contrast to the pollution aerosol over North America and Europe, to which it is comparable in magnitude, the "black carbon" content of the Asian cloud is significant. The absorption of sunlight by this elemental carbon alters the local hydrological cycle and hence the weather over the northern Indian Ocean. The lack of nitric oxide produced in the low-temperature flames of burning biomass currently limits ozone production over the area, but that will likely be reversed in the future with increased use of fossil fuels for vehicles.

Large forest fires in northern Canada produce huge quantities of carbon monoxide and volatile organic compounds (VOCs), which have been found to travel as far as the U.S. Southeast and which may well increase ozone and particulate concentrations in the air of this region.

Indoor Air Pollution

The levels of some common air pollutants often are greater indoors than outdoors, although pollutant concentrations do vary significantly from one building to another. Since most people spend more time indoors than outdoors, exposure to indoor air pollutants is an important environmental problem and may cause more problems to human health than does outdoor air. Indeed, the inadequate ventilation practices encountered in developing countries that burn coal, wood, crop residues, and other unprocessed biomass fuels create smoke and carbon monoxide pollution that produces respiratory problems and ill health among huge numbers of people in these countries. Women and young children are particularly affected since they spend more time indoors. Cooking smoke from biomass fuels increases asthma rates among elderly men and women. The particulate emissions from traditional cook stoves used indoors in developing countries can be reduced by 90% by switching from wood to charcoal. It is estimated that over 400,000 premature deaths are caused in China annually owing to exposure to dirty household fuels and to other pollutants in the air.

In the material that follows, we investigate the various indoor air pollutants that are thought to have the most serious effects on human health. To ensure that the relevant background has been covered, a discussion of several other indoor air pollutants of interest to human health is deferred until later chapters: Radon is discussed in Chapter 9, pesticides in Chapter 10, and polycyclic aromatic hydrocarbons (PAHs) in Chapter 12. Chloroform in indoor air is considered when water purification is discussed, along with indoor air contamination by chlorinated organic solvents, in Chapter 14.

Formaldehyde

The most controversial indoor organic air pollutant gas is **formaldehyde, $H_2C{=}O$**. It is a widespread trace constituent of the atmosphere since it occurs as a stable intermediate in the oxidation of methane and of other VOCs. While its concentration in clean outdoor air is too small to be important—about 10 ppb in urban areas, except during episodes of photochemical smog—the level of formaldehyde gas *indoors* is often orders of magnitude greater, in certain cases exceeding 1000 ppb (1 ppm). A survey of U.S. homes in the late 1990s found that the indoor formaldehyde concentration usually was in the 5–20 ppb range.

The chief sources of indoor exposure to this gas are emissions from cigarette smoke and from synthetic materials that contain formaldehyde resins used in urea formaldehyde foam insulation and in the adhesive employed in manufacturing plywood and particleboard (chipboard). Many useful resins (which are rigid polymeric materials) are prepared by combining formaldehyde with another organic substance. Formaldehyde itself is used in the dyeing and gluing of carpets, carpet pads, and fabrics. In the first few months and years after their manufacture, however, such materials release small amounts of free formaldehyde gas into the surrounding air. Consequently, new prefabricated structures such as mobile homes that contain chipboard generally have much higher levels of formaldehyde in their air than do older, conventional homes. Many manufacturers of pressed-wood products have now modified their production processes in order to reduce the rate at which formaldehyde is released.

The rate of formaldehyde emission from synthetic materials increases with temperature and relative humidity and declines as the materials age. Initially, formaldehyde temporarily trapped as a gas or simply adsorbed onto the materials is released into the surrounding air. There is also release of formaldehyde due to the rearrangement and dissociation of amide end-groups on resin polymers, from $R{-}NH{-}CH_2OH$ to $R{-}NH_2 + H_2CO$. Later, slow but continuous reactions of water vapor in humid air with the methylene bridges joining amide groups within the polymer backbone provide a continuing emission of formaldehyde:

$$R{-}NH{-}CH_2{-}NH{-}R + H_2O \longrightarrow 2\,R{-}NH_2 + H_2CO$$

Formaldehyde has a pungent odor, with a detection threshold in humans of about 100 parts per billion, i.e., 0.1 ppm; its odor is often noticeable in stores that sell carpets and synthetic fabrics. At somewhat higher levels, many people report irritation to their eyes, especially if they wear contact lenses, and to their noses, throats, and skin. The formaldehyde in cigarette smoke can cause eye irritation. Common symptoms of *acute* (i.e., short-term, high-level) formaldehyde exposure include coughing, wheezing, bronchitis, and chest pains. Chronic exposure to low levels of formaldehyde produces similar effects and respiratory symptoms. Formaldehyde in air may cause children to develop asthma and to have more respiratory infections and allergies and asthma attacks, although evidence for these effects is controversial. Dampness in homes, allowing the proliferation of dust mites, fungi, and bacteria, also plays a large role in increasing lower respiratory tract illnesses, especially in children.

Formaldehyde is thought to be the most important VOC in producing what is known as **sick building syndrome.** This term is used to describe situations in which the occupants of a building experience acute health effects and discomfort that seem to be linked to the time they spend in a particular building, though no specific illness or cause is apparent. Complaints commonly include

- headaches;

- irritation of eyes, nose, or throat; dry cough;

- dizziness and nausea; fatigue;

- difficulty in concentrating; and

- dry or itchy skin.

In addition to VOCs emitted from indoor sources, other factors that contribute to the syndrome include inadequate ventilation, pollutants entering from outside the building, and biological contamination of the air from bacteria, molds, pollen, and viruses that have bred in stagnant water that has accumulated in air vents, etc.

The related compounds **acetone,** $(CH_3)_2C=O$, and **4-butanone** (also called *methyl ethyl ketone*, MEK) are the ketones most commonly present in indoor air in U.S. homes, owing to their use as solvents in nail polish and paint removers, etc.

Formaldehyde is established as a **carcinogen** (a cancer-causing agent) in test animals and may also be carcinogenic to humans; it was classified as a *probable human carcinogen* by the U.S. EPA in 1987. The expected cancer sites are in the respiratory system, including the nose; cancers at these sites have been found for some people who are exposed to the gas in occupational settings. However, studies of human populations exposed to formaldehyde have led to no clear-cut conclusions concerning an increase in cancer frequency arising from nonoccupational exposure. From animal studies, an upper limit to the possible effect in humans can be estimated: It corresponds to an increase in the cancer rate of one or two cases per 10,000 people after 10 years of living in a house or trailer with high formaldehyde levels. However, the lower limit to the effect could well be a zero increase in the cancer rate. In summary, no scientific consensus has yet been reached on the dangers to human health of low-level exposure to formaldehyde.

Benzene and Other Gasoline-Related Hydrocarbons

Like formaldehyde, benzene is classified as a **hazardous air pollutant,** HAP, sometimes known as *air toxics*. **Benzene,** C_6H_6, is a stable, volatile liquid hydrocarbon that through the modern age has found a variety of uses. It is a minor constituent of gasoline and was commonly used as a solvent for many organic products, including paints and inks. The public is exposed to benzene vapor indoors from the use of solvents and gasoline, through smoking (mainly for the smoker but to a lesser extent for those inhaling second-hand smoke), and from the importing of benzene from outdoor air into the house. The levels of benzene generally are smaller outdoors and in large buildings than in individual homes, especially those with smokers living in them. A significant fraction of benzene vapor exposure occurs while riding in motor vehicles and refueling them at gas stations.

Benzene is classified by the U.S. EPA as a *known human carcinogen.* Chronic exposure at high occupational levels increases the rate of leukemia to individuals. Indeed, there were many deaths among workers in the first half of the twentieth century from exposure to benzene from petroleum-based solvents, such as those used in the rubber and glue industries; in the making of paints, adhesives, and coatings; and in dry cleaning. It also causes *aplastic anemia,* a condition in which an individual is chronically tired and is especially susceptible to infections because the bone marrow produces insufficient red blood cells. There continues to be some uncertainty, however, about whether occupational or domestic exposure to low levels of benzene vapor does indeed increase the risk for leukemia and multiple myeloma. Recently, it was estimated that benzene accounted for one-quarter of cancer deaths caused by air toxics in the United States.

Because of the serious health problems it causes, the use of benzene as a solvent has largely been phased out. In addition, its maximum allowable level in gasoline has been reduced. Benzene can be replaced in many applications by **toluene,** $C_6H_5CH_3$, which consists of molecules of benzene with one hydrogen atom replaced by a methyl group. The —CH_3 group in toluene provides liver enzymes with a site that is much easier to attack and thereby initiate metabolism than any of the very strong bonds in benzene itself. Toluene and the corresponding dimethylated benzenes called **xylenes,** the 1,2,4-trimethylated benzene, and **ethylbenzene** are all present in modern unleaded gasoline; they are very commonly detected in indoor air, as are the nonaromatic hydrocarbons *cyclohexane* and *decane*. The concentration of toluene usually greatly exceeds that of benzene itself. However, there is evidence that methylated benzenes are demethylated in catalytic converters and that, as a consequence, additional benzene is emitted into the air under some operating conditions.

Nitrogen Dioxide

Indoor concentrations of NO_2 often exceed outdoor values in homes that contain stoves, space heaters, and water heaters that are fueled by gas. The flame temperature in these appliances is sufficiently high that some nitrogen and oxygen in the air combine to form NO, which eventually is oxidized to nitrogen dioxide. In one study, it was established that NO_2 levels in homes that use gas for cooking or that have a kerosene stove average 24 ppb, compared to 9 ppb for homes that have neither. Peak concentrations near gas cooking stoves can exceed 300 ppb.

Some nitric oxide is also released from the burning of wood and other biomass fuels since these natural materials contain nitrogen. However, the flame temperature used in burning such fuels is much lower than in burning gas, so little thermal NO is produced from nitrogen in the air.

Nitrogen dioxide is soluble in biological tissue and is an oxidant, so its effects on health, if any, are expected to occur in the respiratory system. There have been many studies of the effects on respiratory illness in children owing to exposure to low levels of NO_2 emitted by gas appliances, but the results of different studies are not mutually consistent and are inadequate for establishing a cause-and-effect relationship. One study found that a 15-ppb increase in the mean NO_2 concentration in a home leads to about a 40% increase in lower respiratory system symptoms among children aged 7 to 11 years. Nitrogen dioxide is the only oxide of nitrogen that is detrimental to health at concentrations likely to be encountered in residences.

Nitrogen dioxide is probably responsible for the finding that indoor concentrations of **nitrous acid,** HNO_2, exceed those found outdoors, since the gas reacts with water to form nitrous and nitric acids:

$$2\,NO_2 + H_2O \longrightarrow HNO_2 + HNO_3$$

Indoor nitrous acid concentrations were found to correlate inversely with ozone gas concentrations, presumably because the acid is oxidized to nitric acid by the gas.

Carbon Monoxide

Carbon monoxide, CO, is a colorless, odorless gas whose concentration indoors can be greatly increased by the incomplete combustion of carbon-containing fuels such as wood, gasoline, kerosene, or gas. High indoor concentrations usually are the result of a malfunctioning combustion appliance, such as a kerosene heater. Even properly functioning kerosene or gas heaters in poorly ventilated rooms can result in CO levels in the 50–90-ppm range.

Average indoor and outdoor CO concentrations usually amount to a few parts per million, though elevated values in the 10–20-ppm range are common in parking garages due to the carbon monoxide emitted by motor vehicles. Exhaust fumes containing high levels of CO and other pollutants can enter homes having attached garages. In developing countries, carbon monoxide poisoning is a serious hazard when biomass fuels are used to heat poorly ventilated rooms in which people sleep.

The major danger from carbon monoxide arises from its ability, when inhaled, to complex strongly with the hemoglobin in blood and thus to impair its ability to transport oxygen to cells. Hemoglobin's affinity for CO is 234 times that for oxygen, and once one CO is bound to a given hemoglobin molecule, the rate of release of its remaining oxygen molecules to cells is reduced. Recent research has found that mental functioning is reduced during short-term exposure to high levels of CO and perhaps also as a result of long-term exposure to low concentrations, because the brain, like the heart, is a body organ with a high requirement for oxygen.

One important feature of the reduction in vehicular pollutant emissions over the last few decades has been the substantial decline in accidental death from acute carbon monoxide poisoning. In the United States alone, it has been estimated that more than 11,000 deaths have been avoided as a result.

Smoking tobacco is a significant source of carbon monoxide indoors. Although nonsmokers usually have less than 1% of their hemoglobin tied up as the complex with CO, the figure for smokers is many times this value because of the carbon monoxide that they inhale during smoking. Studies have shown that increased mortality from heart disease can result even if only several percent of hemoglobin is chronically tied up as the CO complex. Exposure to very high concentrations of CO results in headache, fatigue, unconsciousness, and eventually death (if such exposure is sustained for long periods).

Low-priced, easily installed carbon monoxide detectors suitable to warn residents in homes and offices when high CO levels occur are now widely available. However, scientists have begun to worry about the poorly known health effects of chronic exposure to low levels of CO and the fact that such exposure may be quite common.

Environmental Tobacco Smoke

It is well established that smoking tobacco is the leading cause of lung cancer and is one of the main contributors to heart disease. Nonsmokers are often exposed to cigarette smoke, although in lower concentrations than smokers since it is diluted by air. This **environmental tobacco smoke,** or ETS ("second-hand smoke"), has been the subject of many investigations in order to determine whether or not it is harmful to people who are exposed to it.

ETS consists of both gases and particles. The concentration of some toxic products of partial combustion is actually *higher* in sidestream smoke than in mainstream, since combustion occurs at a lower temperature—and so is less complete—in the smoldering cigarette compared to one through which air is being inhaled. Since the sidestream smoke is usually diluted by air before being inhaled, however, the concentrations of pollutants reaching the lungs of non-smokers are much lower than those reaching the lungs of smokers themselves.

The chemical constitution of tobacco smoke is complex: It contains thousands of components, several dozen of which are carcinogens. The gases in smoke include

- carbon monoxide and carbon dioxide;

- formaldehyde and several other aldehydes, ketones, and carboxylic acids;

- nitrogen oxides, hydrogen cyanide, ammonia, and a number of organic nitrogen compounds;

- methyl chloride;

- 1,3-butadiene;

- toluene, benzene, and several hundred different PAHs, to be discussed in Chapters 8 and 12; and

- cadmium and radioactive elements such as polonium (see Chapter 9).

Included in the nitrogen compounds are several **nitrosamines,** organic nitrogen compounds of formula $R_2N—N{=}O$, which, together with the PAHs, are probably the most important respiratory carcinogens in the smoke.

The particulate phase of cigarette smoke is called the **tar,** and much of it is respirable in size. The zone in a cigarette that actively burns, as occurs when a smoker inhales a puff, is quite hot (700–950°C) and produces CO and H_2 as well as the expected CO_2 and water vapor. Immediately downstream of this area is a cooler zone (200–600°C) where smoke constituents such as nicotine distill out of the tobacco. When this vapor cools farther along the cigarette path toward the smoker, much of it condenses to aerosol particles that constitute the particulate phase of the smoke.

Many people experience irritation of their eyes and airways from exposure to ETS. The gaseous components of ETS, especially formaldehyde, hydrogen cyanide, acetone, toluene, and ammonia, cause most of the odor and irritation. Exposure to ETS aggravates the symptoms of many people who suffer from asthma or from *angina pectoris,* chest pains brought on by exertion. ETS, particularly when it originates from maternal smoking, is known to induce new cases of asthma in children, especially those of preschool age. Some recent studies have established correlations between the rate of acute respiratory illness and the level of indoor $PM_{2.5}$ (which would include the total amount of respirable particulates from all sources, including tobacco smoke). **Passive smoking**—which involves inhalation of sidestream as well as already exhaled smoke—is believed by scientists to cause bronchitis, pneumonia, and other infections such as those of the ear in up to 300,000 infants, as well as several thousand instances of sudden infant death, in the United States each year. Second-hand smoke may even reduce the cognitive abilities of children, whether exposed prenatally or when young. Being exposed to second-hand smoke, whether on the job or by living with a smoker, approximately doubles a nonsmoker's chance of developing asthma.

In 1993, the U.S. EPA classified ETS as a *known human carcinogen* and estimated that it causes about 3000 lung cancer deaths annually. ETS is also considered to be responsible for killing as many as 60,000 Americans annually from heart disease. In a study of American nurses, it was found that nonsmoking women regularly exposed to ETS had a 91% greater rate of heart attacks than women who had no exposure. Apparently the smoke leads to hardening of the arteries, a main cause of heart attacks. An analysis of all recent studies on passive smoking led to the conclusion that the risks of developing lung cancer and heart disease each are increased by about one-quarter for nonsmoking spouses of smokers. Longtime workers in bars and restaurants in which smoking is permitted also have an increased rate of lung cancer, even if they themselves do not smoke. A British study estimated that ETS kills 140,000 Europeans annually through cancer and heart disease.

Asbestos

The term **asbestos** refers to a family of six naturally occurring silicate minerals that are fibrous (see Chapter 16). Structurally, they are composed of long double-stranded networks of silicon atoms connected through intervening oxygen atoms; the net negative charge of this silicate structure is neutralized by the presence of cations such as magnesium.

The most commonly used form of asbestos, **chrysotile,** has the formula $Mg_3Si_2O_5(OH)_4$. It is a white solid whose individual fibers are curly. Chrysotile, mined mainly in Russia, China, Brazil, Canada (Quebec), and Kazakhstan, is the principal type of asbestos used in North America. It has been employed in huge quantities because of its resistance to heat, its strength, and its relatively low cost. Common applications of asbestos include its use as insulation and spray-on fireproofing material in public buildings, in automobile brake-pad lining, as an additive to strengthen cement used for roofing and pipes, and as a woven fiber in fireproof cloth.

The use of asbestos has been sharply reduced since the 1970s in developed countries because it is now recognized from studies on the health of asbestos miners and other asbestos workers to be a human carcinogen. It causes *mesothelioma*, a normally rare, incurable cancer of the lining of the chest or abdomen. In addition, airborne asbestos fibers and cigarette smoke act **synergistically:** Their combined effect is greater than the sum of their individual effects (in this case, equal to the product of the two) in causing lung cancer.

There is much controversy concerning whether chrysotile should be banned outright from further use and whether or not existing asbestos insulation in buildings should be removed. Many experts feel that existing asbestos should be left in place unless it becomes damaged enough that there is a chance that its fibers will become airborne. Indeed, its removal can dramatically increase the levels of airborne asbestos in a building unless extraordinary precautions are taken. One scientist stated: "Removing asbestos is like waking up a pit bull terrier by poking a stick in its ear. We should let sleeping dogs lie." Some environmentalists, however, feel that existing asbestos is a ticking time bomb—that it should be removed as soon as possible, as one can never predict when building insulation will be damaged.

Most of the initial concern about asbestos was related to **crocidolite,** *blue asbestos,* and **amosite**, *brown asbestos.* Evidence implicating crocidolite in causing cancer in humans was already well established several decades ago. It is a material with thin, straight, and relatively short fibers that more readily penetrate lung passages, making it a more potent carcinogen than the white form. Crocidolite and amosite are mined in South Africa and Australia; they were not used much in North America but were used in many areas of Europe, including the United Kingdom.

More than 15 countries, including the European Union and Australia, have now banned all forms of asbestos. Some environmentalists and physicians worry that although workers in developed countries wear masks and overalls and handle white asbestos properly to greatly minimize their exposure to it, these practices are not yet common in developing countries. Canada, among other countries, has resisted efforts by agencies of the United Nations to place chrysotile on the list of most hazardous substances.

Review Questions

1. Discuss the relationship between atmospheric particulates and haze.

2. What is *acid rain?* What two acids predominate in it?

3. Explain why the predominant acid in acid rain differs in eastern and western North America.

4. Using chemical equations, explain how acid rain is neutralized by limestone that is present in soil.

5. Describe the effects of acid precipitation upon (a) dissolved levels of aluminum, (b) fish populations, and (c) trees.

6. What is the difference in meaning between *absorbed* and *adsorbed* when they refer to particulates?

7. Describe the major health effects of outdoor air pollutants.

8. List four important reasons why coarse particles usually are of less danger to human health than are fine particles.

9. What are the main sources of formaldehyde in indoor air? What are its effects?

10. What are the main sources of nitrogen dioxide and of carbon monoxide in indoor air? Of benzene?

11. What are the three forms of asbestos called? Why is asbestos of environmental concern?

Additional Problems

1. A sample of acidic precipitation is found to have a pH of 4.2. Upon analysis it is found to have a total sulfur concentration of 0.000010 M. Calculate the concentration of nitric acid in the sample, and from the ratio of nitric to total acid decide whether the air sample probably originated in eastern or in western North America.

2. If the pH of rainfall in upstate New York is found to be 4.0, and if the acidity is half due to nitric acid and half to the two hydrogen ions released by sulfuric acid, calculate the masses of the primary pollutants nitric oxide and sulfur dioxide that are required to acidify one liter of such rain.

3. The pH in a lake of size 3.0 km × 8.0 km and an average depth of 100 m is found to be 4.5. Calculate the mass of calcium carbonate that must be added to the lake water in order to raise its pH to 6.0.

4. The pH of a sample of rain is found to be 4.0. Calculate the percentage of HSO_4^- that is ionized in this sample, given that the acid dissociation constant for the second stage of ionization of H_2SO_4 is 1.2×10^{-2} mol L^{-1}. Repeat the calculation for a pH of 3.0. Is the trend shown by these calculations consistent with qualitative predictions made according to Le Châtelier's principle (which

states that the position of equilibrium shifts so as to minimize the effect of any stress)? [*Hint: Write the expression for the acid dissociation constant for the weak acid in terms of the concentrations of the reactants and products, and use the stoichiometry of the balanced equation to reduce the number of unknowns to one.*]

5. Calculate the mass of fine particles inhaled by an adult each year, assuming he/she inhales about 350 L of air per hour and that the average $PM_{2.5}$ index of this air is 10 $\mu g/m^3$. Assuming that each particle has a diameter of about 1 μm and that the density of the particles is about 0.5 g/mL, calculate the total surface area of this annual load of particles. [*Hint: The surface area of a spherical particle is equal to $4\pi r^2$, where r is its radius.*]

6. The detection threshold of formaldehyde by humans is about 100 ppb. Would a typical human be able to detect formaldehyde at a concentration of 250 $\mu g/m^3$ if the air temperature was 23°C and the pressure 1.00 atm?

7. What mass of formaldehyde gas must be released from building materials, carpets, etc. in order to produce a concentration of 0.50 ppm of the gas in a room having dimensions of 4 m × 5 m × 2 m?

Further Readings

1. C. T. Driscoll et al., "Acidic Deposition in the Northeastern United States: Sources and Inputs, Ecosystem Effects, and Management Strategies," *Bioscience* 51 (2001): 180–198; J. A. Lynch et al., "Acid Rain Reduced in Eastern United States," *Environmental Science and Technology* 34 (2000): 940–949; M. Heal, "Acid Rain: Is the UK Coping?" *Education in Chemistry* (July 2002): 101–104.

2. R. F. Wright et al., "Recovery of Acidified European Surface Waters," *Environmental Science and Technology* 39 (2005): 64A–72A.

3. F. Laden et al., "Association of Fine Particulate Matter from Different Sources with Daily Mortality in Six U.S. Cities," *Environmental Health Perspectives* 108 (2000): 941–947.

4. D. W. Dockery et al., "An Association Between Air Pollution and Mortality in Six U.S. Cities," *New England Journal of Medicine* 329 (1993): 1753–1759; J. Schwartz et al., "The Concentration-Response Relation Between $PM_{2.5}$ and Daily Deaths," *Environmental Health Perspectives* 110 (2002): 1025–1029; C.A. Pope III et al., "Review of Epidemiological Evidence of Health Effects of Particulate Air Pollution," *Inhalation Toxicology* 7 (1995): 1–18.

5. D. R. Gold, "Environmental Tobacco Smoke, Indoor Allergens, and Childhood Asthma." *Environmental Health Perspectives* 108, supplement 4 (2000): 643–646; "Secondhand Smoke—Is It a Hazard?" *Consumer Reports* (January 1995): 27–33.

Websites of Interest

Log on to www.whfreeman.com/envchem4/ and click on Chapter 4.

THE DETAILED CHEMISTRY OF THE ATMOSPHERE

In this chapter, the following introductory chemistry topics are used:

- Lewis structures (for nonradicals)
- Activation energy; reaction mechanisms
- Basic organic chemistry (see Appendix to this book)

Background from previous chapters used in this chapter:

- Concept of free radicals (Chapter 1)
- Atmospheric structure (Chapter 1)
- Photochemical reactions (Chapters 1 and 3)
- Stratospheric chemistry (Chapters 1 and 2)
- Photochemical air pollution, smog, NO_X (Chapter 3)

Introduction

In Chapters 1 through 4, we have discussed, in broad terms, chemical processes in the stratosphere and troposphere, emphasizing the environmental concerns that have arisen. In this chapter, the reactions that occur in clean tropospheric air and the processes encountered in the polluted air of modern cities are analyzed in more detail. In addition, the processes that deplete ozone in the stratosphere are systematized. It is only by understanding the science underlying such complicated environmental problems that we can hope to solve them.

One of the important characteristics that determines the reactivity of a species is

This Mexico City newspaper salesman wears a mask to help protect himself from air pollution during a photochemical smog episode. [Guillermo Gutierrez/AP]

| BOX 5-1 | Lewis Structures of Simple Free Radicals |

Most of the free radicals that are important in atmospheric chemistry have their unpaired electron located on a carbon, oxygen, hydrogen, or halogen atom. In a formula showing the location and position of bonds, the specific atomic location can be denoted by placing a dot above the relevant atom symbol to represent the unpaired electron, e.g., in the notation $\dot{F}$. Characteristically, such an atom forms one fewer bond than usual—its unpaired electron is not in use as a bonding electron. Thus a carbon atom on which an unpaired electron is located forms three rather than four bonds, an oxygen forms one rather than two bonds, and a halogen or hydrogen forms no bonds if it is the radical site. Usually, the unpaired electron exists as a nonbonding electron localized on one atom, not as a bonding electron shared between atoms.

For many polyatomic free radicals, the choice of atom to which the unpaired electron is to be assigned in deducing the Lewis structure is obvious from the atom–atom connections. Thus in the hydroperoxy radical HOO, the hydrogen atom cannot be the radical site, since to be part of the molecule it must form a bond to the adjacent oxygen; neither can the central oxygen be the site since it must form two bonds, one to each neighbor (or one of them would not be part of the molecule). This leaves the terminal oxygen as the radical site, and we can show the bonding network as H—O—$\dot{O}$. If desired, the unbonded electron pairs can also be shown:

$$H-\overset{\cdot\cdot}{\underset{\cdot\cdot}{O}}-\overset{\cdot}{\underset{\cdot\cdot}{O}}:$$

The procedure is more complicated in molecules that contain multiple bonds. Thus in HOCO it is not initially obvious whether it is the carbon or the terminal oxygen that carries the unpaired electron. After a little manipulation with various bonding schemes, it becomes clear that the unpaired electron could not be located on an oxygen, since to fulfill its valence requirement of four, the carbon would have to form three bonds to the other oxygen. Thus the only reasonable structure is H—O—$\dot{C}$=O.

If only a simple formula rather than a partial or complete Lewis structure is drawn for a radical, the superscript dot is placed following the formula and does not indicate which atom carries the unpaired electron. An example is HCO$^{\cdot}$, in which the actual location of the unpaired electron is at carbon, not oxygen.

For a few free radicals involving unusual bonding, such as NO$_2^{\cdot}$, these rules generate a Lewis structure that is not the dominant one; further discussion of such systems is beyond the scope of this book.

PROBLEM 1

Draw simple Lewis structures, showing the locations of the bonds and of the unpaired electron, for the following free radicals.
(a) OH$^{\cdot}$ (b) CH$_3^{\cdot}$ (c) CF$_2$Cl$^{\cdot}$
(d) H$_3$COO$^{\cdot}$ (e) H$_3$CO$^{\cdot}$ (f) ClOO$^{\cdot}$
(g) ClO$^{\cdot}$ (h) HCO$^{\cdot}$ (i) NO$^{\cdot}$

whether or not it has unpaired electrons. To emphasize that an atomic or molecular species is a free radical, in this chapter we shall place a superscript dot at the end of its molecular formula, signaling the presence of the unpaired electron. For example, the notation $OH^\cdot$ is used for the **hydroxyl free radical.** The position of the unpaired electron in the Lewis structure is often important in determining the reaction of a free radical. Box 5-1 discusses the deduction of the Lewis structure for free radicals.

Tropospheric Chemistry

The Principles of Reactivity in the Troposphere

Most gases in the troposphere are gradually oxidized by a sequence of reactions involving free radicals. For a given gas, the sequence can be predicted from the principles discussed below, which are also systematized in Figure 5-1. These tropospheric reactions are similar in many ways to those encountered in the stratosphere, which are discussed later in the chapter.

As mentioned in Chapter 3, the usual initial step in the oxidation of an atmospheric gas is its reaction with the hydroxyl free radical rather than with **molecular oxygen,** O_2. With molecules that contain a multiple bond, the hydroxyl radical usually reacts by *adding* itself to the molecule at the position of the multiple bond. Recall the general principle that radical reactions that are spontaneous are those which produce stable products, i.e., products containing strong bonds. Thus it is understandable that $OH^\cdot$ addition does *not* occur to an oxygen atom since the O—O bonds that would result are weak. Similarly, $OH^\cdot$ addition does *not* occur to CO *double* bonds since they are very strong relative to the single O—O or C—O bond that would be produced. For example, the $OH^\cdot$ radical adds to the sulfur atom, forming a strong bond, but not to an oxygen atom, in **sulfur dioxide,** SO_2:

$$O{=}S{=}O + OH^\cdot \longrightarrow O{=}\overset{\displaystyle O}{\underset{\displaystyle OH}{\overset{\|}{S}}}\!\cdot$$

(Here and elsewhere in this book we write Lewis structures that assume that *d* orbitals in atoms such as sulfur and phosphorus allow these elements to form double bonds.) Hydroxyl radical does not add to **carbon dioxide,** $O{=}C{=}O$, since the molecule contains only very strong C=O bonds. However, $OH^\cdot$ addition does occur to the carbon atom in **carbon monoxide,** CO, since the triple bond is thereby converted to the very stable double bond and a new single bond is also formed:

$$^\ominus C{\equiv}O^\oplus + OH^\cdot \longrightarrow HO{-}\overset{\cdot}{C}{=}O$$

This process is exothermic because the third C—O bond in carbon monoxide is weak relative to the other two.

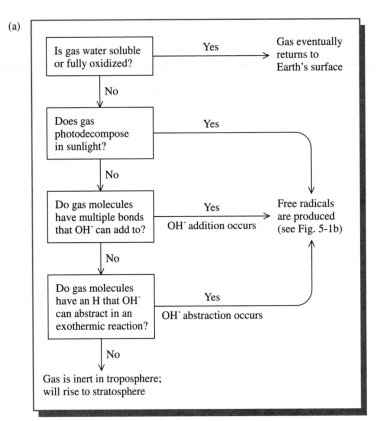

(a)

Is gas water soluble or fully oxidized? — Yes → Gas eventually returns to Earth's surface

No ↓

Does gas photodecompose in sunlight? — Yes →

No ↓

Do gas molecules have multiple bonds that OH· can add to? — Yes, OH· addition occurs → Free radicals are produced (see Fig. 5-1b)

No ↓

Do gas molecules have an H that OH· can abstract in an exothermic reaction? — Yes, OH· abstraction occurs →

No ↓

Gas is inert in troposphere; will rise to stratosphere

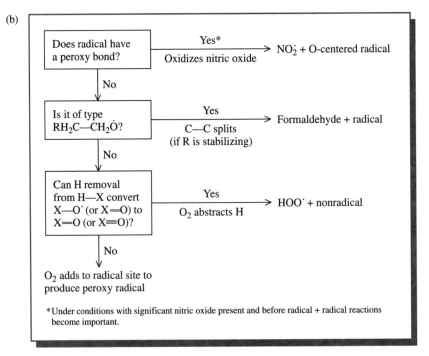

(b)

Does radical have a peroxy bond? — Yes*, Oxidizes nitric oxide → NO_2^- + O-centered radical

No ↓

Is it of type $RH_2C—CH_2\dot{O}$? — Yes, C—C splits (if R is stabilizing) → Formaldehyde + radical

No ↓

Can H removal from H—X convert X—O· (or X=O) to X=O (or X≡O)? — Yes, O_2 abstracts H → HOO· + nonradical

No ↓

O_2 adds to radical site to produce peroxy radical

*Under conditions with significant nitric oxide present and before radical + radical reactions become important.

FIGURE 5-1 (a) Decision tree illustrating the fate of gases emitted into the air. (b) Decision tree illustrating the fate of airborne free radicals.

Generally, OH˙ does not add to multiple bonds in any fully oxidized species such as CO_2, SO_3, and N_2O_5, since such processes are endothermic and therefore are very slow to occur at atmospheric temperatures. Similarly, N_2 does not react with OH˙ because the component of the nitrogen-to-nitrogen bond that would be destroyed is stronger than the N—O bond that would be formed. It doesn't react with O_2 because a high activation energy is required for this reaction to occur.

For molecules that do *not* have a reactive multiple bond but do contain hydrogen, OH˙ reacts with them by the **abstraction** of a hydrogen atom to form a water molecule and a new reactive free radical. For CH_4, NH_3, H_2S, and CH_3Cl, for instance, the reactions are

$$CH_4 + OH˙ \longrightarrow CH_3˙ + H_2O$$

$$NH_3 + OH˙ \longrightarrow NH_2˙ + H_2O$$

$$H_2S + OH˙ \longrightarrow SH˙ + H_2O$$

$$CH_3Cl + OH˙ \longrightarrow CH_2Cl˙ + H_2O$$

Because the H—OH bond formed in these reactions is very strong, the processes are all exothermic; thus only small activation energy barriers exist to impede these reactions (see Box 1-2).

PROBLEM 5-1

Why aren't gases such as CF_2Cl_2 (a CFC) readily oxidized in the troposphere? Would the same be true for CH_2Cl_2?

PROBLEM 5-2

The abstraction of the hydrogen atom in HF by OH˙ is endothermic. Comment briefly on the expected rate of this reaction: Would it be (at least potentially) fast, or necessarily very slow in the troposphere?

PROBLEM 5-3

The hydroxyl radical does not react with gaseous nitrous oxide, N_2O, even though the molecule contains multiple bonds. What can you deduce about the probable energetics (endothermic or exothermic character) of this reaction from the observed lack of reactivity?

A few gases emitted into air can absorb some of either the UV-A or the visible component of sunlight, and this input of energy is sufficient to break one of the bonds in the molecule, thereby producing two free radicals. For example, most molecules of atmospheric **formaldehyde** gas, H_2CO, react by photochemical decomposition after absorption of UV-A from sunlight:

$$H_2CO \xrightarrow{\text{UV-A } (\lambda < 338 \text{ nm})} H^{\cdot} + HCO^{\cdot}$$

In all the cases discussed, the initial reaction of a gas emitted into air produces free radicals, almost all of which are extremely reactive. The predominant fate in tropospheric air for most simple radicals is reaction with diatomic oxygen, often by an addition process: One of the oxygen atoms attaches, or "adds on," to the other reactant, usually at the site of the unpaired electron. For instance, O_2 reacts by addition with the **methyl radical**, $CH_3^{\cdot}$:

$$CH_3^{\cdot} + O_2 \longrightarrow CH_3OO^{\cdot}$$

Notice that $CH_3OO^{\cdot}$ itself is a free radical; the terminal oxygen forms only one bond and carries the unpaired electron:

$$H_3C - \overset{..}{\underset{..}{O}} - \overset{\cdot}{\underset{..}{O}}: \qquad \text{or just} \qquad H_3C - O - \overset{\cdot}{O}$$

Species such as $HOO^{\cdot}$ and $CH_3OO^{\cdot}$ are called **peroxy** radicals since they contain a peroxide-like O—O bond; recall that $HOO^{\cdot}$ is the **hydroperoxy radical.**

As radicals go, peroxy radicals are less reactive than most. They do *not* readily abstract hydrogen since the resulting peroxides would not be very stable energetically. Since the transfer of H to the peroxy radical would be endothermic and thus would possess a large activation energy, abstraction reactions for peroxy radicals are usually so slow that they are of negligible importance (in contrast to those for $OH^{\cdot}$). Peroxy radicals in the troposphere do not react with atomic oxygen because of the extremely low concentrations of the free atom in this region of the atmosphere. The most common fate of peroxy radicals in tropospheric air, except for the cleanest type of air, such as that over oceans, is reaction with **nitric oxide**, $NO^{\cdot}$, by the transfer of the "loose" oxygen atom (see the later section on stratospheric chemistry), thereby forming **nitrogen dioxide**, $NO_2^{\cdot}$, and a radical that has one fewer oxygen atoms:

$$HOO^{\cdot} + NO^{\cdot} \longrightarrow OH^{\cdot} + NO_2^{\cdot}$$
$$CH_3OO^{\cdot} + NO^{\cdot} \longrightarrow CH_3O^{\cdot} + NO_2^{\cdot}$$

It is by this type of reaction that most atmospheric $NO^{\cdot}$ is oxidized to $NO_2^{\cdot}$, at least in polluted air. Recall that this reaction also is typical of the types encountered in stratospheric chemistry (see Chapters 1 and 2) and that $NO^{\cdot}$ oxidation by ozone in sunlit conditions yields a null reaction.

For free radicals that contain nonperoxy oxygen atoms, the reaction with molecular oxygen frequently involves the abstraction of an H atom by O_2. This process occurs *provided that,* as a result, another new bond within the system is formed: A single bond involving oxygen is converted to a double one, or a double bond involving oxygen is converted to a triple one. As examples, consider the three reactions below in which a C—O single (or double) bond is converted to a double (or triple) one as a consequence of the loss of a hydrogen atom:

$$CH_3 - \dot{O} + O_2 \longrightarrow H_2C{=}O + HOO^{\cdot}$$

$$HO - \dot{C}{=}O + O_2 \longrightarrow O{=}C{=}O + HOO^{\cdot}$$

$$H - \dot{C}{=}O + O_2 \longrightarrow {}^{\ominus}C{\equiv}O^{\oplus} + HOO^{\cdot}$$

Such processes do not occur unless a new bond is created in the product free radical, since the strength of the newly created H—OO bond alone is not sufficient to compensate for the breaking of the original bond to hydrogen.

If there is no suitable hydrogen atom for O_2 to abstract, then when it collides with a radical, it instead *adds* to it at the site of the unpaired electron, as was previously discussed for simple radicals. For example, radicals of the type $R - \dot{C}{=}O$, where R is a chain of carbon atoms, add O_2 to form a peroxy radical:

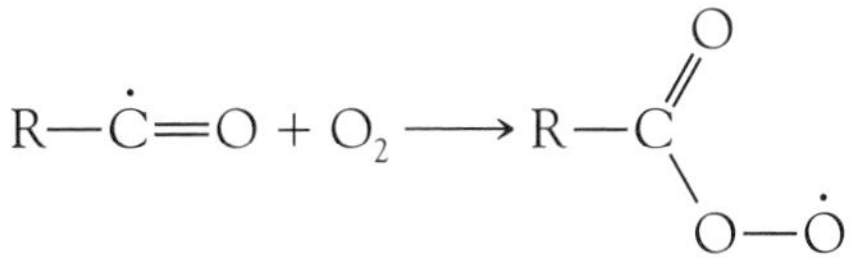

$$R - \dot{C}{=}O + O_2 \longrightarrow R - C \begin{smallmatrix} O \\ \parallel \\ \\ \diagdown \\ O{-}\dot{O} \end{smallmatrix}$$

The only exception to the generalization that oxygen-containing radicals react with O_2 occurs when the radical can decompose spontaneously in a thermoneutral or exothermic fashion. An example of this rare phenomenon is discussed later in the section on photochemical smog.

These generalizations are summarized in the form of the *decision trees* diagrammed in Figure 5-1. By using these diagrams you can deduce the sequence of reactions by which most atmospheric gases in the troposphere are oxidized.

The Tropospheric Oxidation of Methane

Gaseous **methane,** CH_4, is released into the atmosphere in large quantities as a result of **anaerobic** (i.e., O_2-free) biological decay processes and of the use of coal, oil, and, especially, natural gas. It is the predominant hydrocarbon in the atmosphere. Details concerning its production, and the effects on climate of atmospheric methane, are discussed in Chapter 6. Here, however, we shall be concerned with its conversion to carbon dioxide. A similar series of reactions is followed by other alkanes and other VOCs lacking multiple bonds.

The sequence of reactions by which methane is slowly oxidized in the atmosphere can be deduced by applying the principles outlined above and summarized in Figure 5-1, as discussed below.

Since CH_4 is not very soluble in water, does not absorb sunlight, and contains no multiple bonds, the sequence is initiated by a hydroxyl radical abstracting a hydrogen atom from a methane molecule, giving the *methyl radical*, $CH_3^{\cdot}$:

$$CH_4 + OH^{\cdot} \longrightarrow CH_3^{\cdot} + H_2O \tag{1}$$

Since the $CH_3^{\cdot}$ radical contains no oxygen, we deduce that it adds O_2, producing a peroxy radical:

$$CH_3^{\cdot} + O_2 \longrightarrow CH_3OO^{\cdot} \tag{2}$$

Further, since $CH_3OO^{\cdot}$ is a peroxy radical, we deduce that except in very clean air, it reacts with $NO^{\cdot}$ molecules in air to oxidize them by transfer of an oxygen atom:

$$CH_3OO^{\cdot} + NO^{\cdot} \longrightarrow CH_3O^{\cdot} + NO_2^{\cdot} \tag{3}$$

The radical $CH_3O^{\cdot}$ contains a C—O bond that can become C=O upon loss of one hydrogen, so we conclude from our principles that in the next step O_2 abstracts an H atom, producing the nonradical product formaldehyde, H_2CO:

$$CH_3O^{\cdot} + O_2 \longrightarrow H_2CO + HOO^{\cdot} \tag{4}$$

Thus methane is converted to formaldehyde as the first stable intermediate in its oxidation. Since formaldehyde is reactive as a gas in the atmosphere, the mechanism is not complete at this point. After several hours or days in the sunlight, most formaldehyde molecules decompose photochemically by the absorption of UV-A from sunlight, resulting in the cleavage of a C—H bond and the consequent formation of two radicals:

$$H_2CO \xrightarrow{\text{UV-A } (\lambda < 338\text{ nm})} H^{\cdot} + HCO^{\cdot} \tag{5}$$

A minority of formaldehyde molecules react with $OH^{\cdot}$ by H atom abstraction, yielding the same $HCO^{\cdot}$ radical; see Problem 5-7 for the implications of this alternative route.

The hydrogen atom from formaldehyde photolysis is itself a simple radical, and therefore it reacts by addition to O_2 to yield $HOO^{\cdot}$:

$$H^{\cdot} + O_2 \longrightarrow HOO^{\cdot} \tag{6}$$

Meanwhile, the $H\!-\!\overset{\bullet}{C}\!=\!O$ radical reacts by yielding an $H^{\bullet}$ atom to O_2, to produce carbon monoxide and $HOO^{\bullet}$, since by this route a double bond is converted to a triple one:

$$HCO^{\bullet} + O_2 \longrightarrow CO + HOO^{\bullet} \qquad (7)$$

Thus carbon monoxide also is an intermediate in the oxidation of methane. Indeed, most of the CO in a clean atmosphere is derived from this source. Since CO is not a radical and does not absorb visible or UV-A light, we deduce that it reacts ultimately by hydroxyl radical addition to its triple bond:

$$^{\ominus}C\!\equiv\!O^{\oplus} + OH^{\bullet} \longrightarrow H\!-\!O\!-\!\overset{\bullet}{C}\!=\!O \qquad (8)$$

This radical can convert its $O\!-\!C$ bond to $O\!=\!C$ by loss of H, so we deduce that O_2 readily abstracts the hydrogen:

$$H\!-\!O\!-\!\overset{\bullet}{C}\!=\!O + O_2 \longrightarrow O\!=\!C\!=\!O + HOO^{\bullet} \qquad (9)$$

Carbon in its fully oxidized form of carbon dioxide is ultimately produced from methane by this sequence of steps, which are summarized in Figure 5-2. If we add up the nine steps involved and cancel common terms, the overall reaction is seen to be

$$CH_4 + 5\,O_2 + NO^{\bullet} + 2\,OH^{\bullet} \xrightarrow{\ UV\text{-}A\ }$$
$$CO_2 + H_2O + NO_2^{\bullet} + 4\,HOO^{\bullet}$$

If to this result is added the conversion of the four $HOO^{\bullet}$ radicals back to $OH^{\bullet}$ by reaction with four $NO^{\bullet}$ molecules, the revised overall reaction is

$$CH_4 + 5\,O_2 + 5\,NO^{\bullet} \xrightarrow{\ UV\text{-}A\ } CO_2 + H_2O + 5\,NO_2^{\bullet} + 2\,OH^{\bullet}$$

We conclude that $NO^{\bullet}$ is oxidized to $NO_2^{\bullet}$ synergistically, i.e., in a mutually cooperative process, when methane is oxidized to carbon dioxide. Note also that the number of $OH^{\bullet}$ free radicals is increased as a result of the process, due to the photochemical decomposition of formaldehyde. Thus hydroxyl radical is not only a catalyst in the overall reaction but also a product of it.

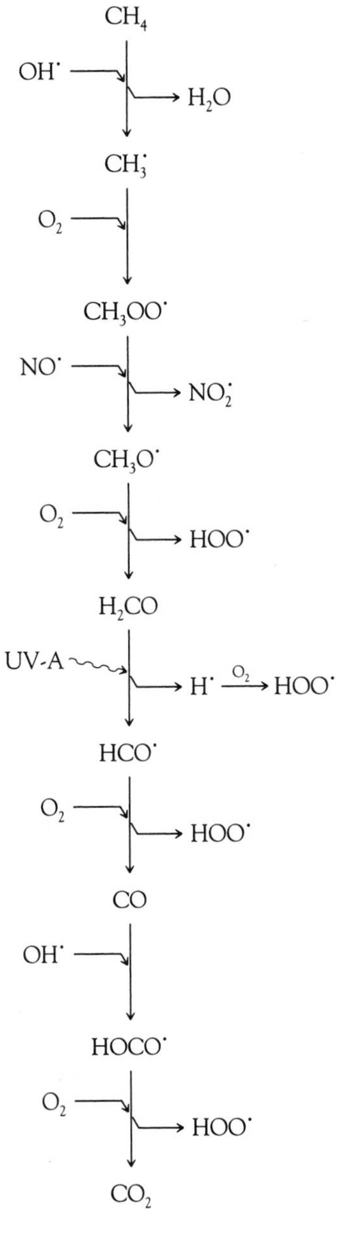

The initial step of the mechanism—the abstraction by $OH^{\bullet}$ of a hydrogen atom from methane—is a slow process, requiring about a decade to occur on average. Once this has happened, however, the subsequent steps leading to formaldehyde occur very rapidly. The slowness of the initial step in methane

FIGURE 5-2 Steps in the atmospheric oxidation of methane to carbon dioxide.

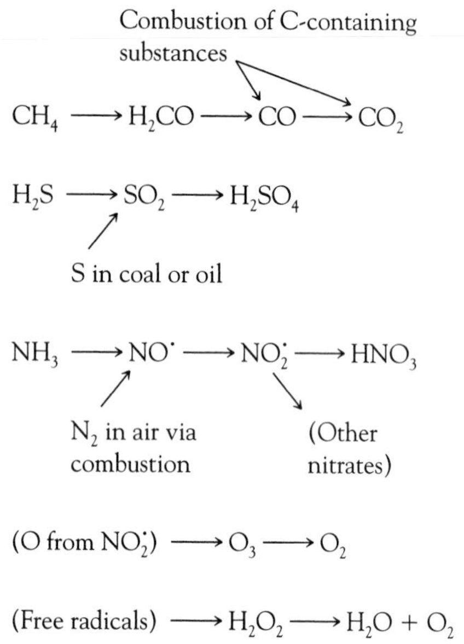

Combustion of C-containing
substances

$$CH_4 \longrightarrow H_2CO \longrightarrow CO \longrightarrow CO_2$$

$$H_2S \longrightarrow SO_2 \longrightarrow H_2SO_4$$

S in coal or oil

$$NH_3 \longrightarrow NO^{\cdot} \longrightarrow NO_2^{\cdot} \longrightarrow HNO_3$$

N$_2$ in air via   (Other
combustion   nitrates)

$$(\text{O from } NO_2^{\cdot}) \longrightarrow O_3 \longrightarrow O_2$$

$$(\text{Free radicals}) \longrightarrow H_2O_2 \longrightarrow H_2O + O_2$$

FIGURE 5-3 Stable species (i.e., nontransients) and their additional sources during sequential atmospheric oxidation processes.

oxidation, and the increasing amounts of the gas released from the surface of the Earth, have led to an increase in the atmospheric concentration of CH_4 in recent times, as discussed further in Chapter 6.

Under conditions of low nitrogen oxide concentration, such as occur over oceans, the mechanism of methane oxidation differs in some of the steps. In particular, instead of oxidizing $NO^{\cdot}$, the peroxy radicals often react with each other, combining to produce a (nonradical) peroxide:

$$2\, HO_2^{\cdot} \longrightarrow H_2O_2 + O_2$$

Under these conditions, then, the oxidation of methane in clean air decreases, rather than increases, the concentration of free radicals.

In general, during the atmospheric oxidation of any of the hydrides (simple hydrogen-containing molecules such as CH_4, H_2S, and NH_3), one or more stable species are encountered along the reaction sequence before the totally oxidized product is formed. These intermediates are also formed independently by various pollution processes. Figure 5-3 summarizes the sequences for hydrides and partially oxidized materials from the viewpoint of the stable species; close reflection will persuade you that the net result is the $OH^{\cdot}$-induced oxidation of the reduced and partially oxidized gases emitted into the air from both natural and pollution sources. In a few cases, for instance, for methane and methyl chloride, the initiation reaction is sufficiently slow that a few percent of these gases survive long enough to penetrate to the stratosphere by the upward diffusion of tropospheric air. Most hydrocarbons react much more quickly than methane (since their C—H bonds are weaker, or fast reactions with $OH^{\cdot}$ other than by hydrogen abstraction are possible) and are classified as **nonmethane hydrocarbons** (NMHC) to emphasize this distinction.

PROBLEM 5-4

Using the reaction principles developed above (the decision trees in Figure 5-1 will help here), predict the sequence of reaction steps by which atmospheric H_2 gas will be oxidized in the troposphere. What is the overall reaction?

PROBLEM 5-5

Deduce two short series of steps by which molecules of methanol, CH_3OH, are converted to formaldehyde, H_2CO, in air. The mechanisms should differ according to which hydrogen atom you decide will react first, that of CH_3 or that of OH.

PROBLEM 5-6

Write equations showing the reactions by which atmospheric carbon monoxide is oxidized to carbon dioxide. Then, by adding the process by which $HOO^{\cdot}$ is returned to $OH^{\cdot}$, deduce the overall reaction.

PROBLEM 5-7

Deduce the series of steps, and the overall reaction as well, for the oxidation of a formaldehyde molecule to CO_2, assuming that for the particular H_2CO molecule involved, the initial reaction is abstraction of H by $OH^{\cdot}$ rather than photochemical decomposition. Overall, is there any increase in the number of free radicals as a result of the oxidation if it proceeds in this manner?

Photochemical Smog: The Oxidation of Reactive Hydrocarbons

Notwithstanding the great complexity of the process, the most important features of the photochemical smog phenomenon can be understood by considering only its few main categories of reactions; these differ in speed, but not much in type, from those occurring in clean air.

We shall restrict our attention to the most reactive VOCs, namely hydrocarbons that contain a $C{=}C$ bond. (Readers unfamiliar with the basics of organic chemistry are advised to consult the Appendix, where the nature of such molecules is explored.) The simplest example is **ethene** (ethylene), C_2H_4; its structure can be written in condensed form as $H_2C{=}CH_2$; its full structural formula is

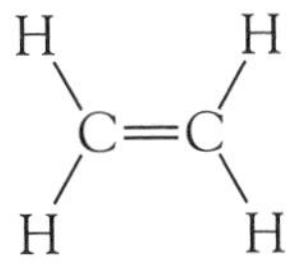

$$\begin{array}{ccc}
H & & H \\
\diagdown & & \diagup \\
& C{=}C & \\
\diagup & & \diagdown \\
H & & H
\end{array}$$

In similar hydrocarbons, one or more of the four hydrogens are replaced by other atoms or groups, often an alkyl group containing a short chain of carbon atoms such as $CH_3{-}$ or $CH_3CH_2{-}$, which will be designated simply as R, since it is generally not the chain but rather the $C{=}C$ part of the molecule that is the reactive site in atmospheric reactions.

Consider a general hydrocarbon $RHC{=}CHR$. In air it reacts with hydroxyl radical by *addition* to the $C{=}C$ bond:

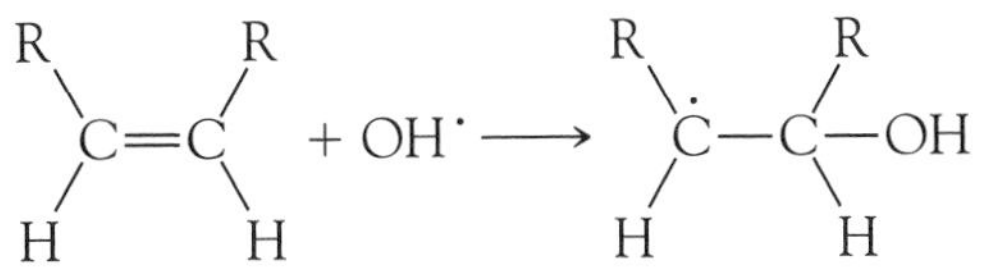

$$\begin{array}{ccccc}
R & & R & & R \qquad R \\
\diagdown & & \diagup & & \diagdown \qquad \diagup \\
& C{=}C & & +\ OH^{\cdot} \longrightarrow & C{-}C{-}OH \\
\diagup & & \diagdown & & \diagup \qquad \diagdown \\
H & & H & & H \qquad H
\end{array}$$

This addition reaction is a faster process, due to its lower activation energy, than the alternative of abstraction of hydrogen, so we can neglect the abstraction process in molecules containing a C=C link. Because the reaction of addition of $OH^\cdot$ to a multiple bond is much faster than H abstraction from methane and other alkane hydrocarbons, RHC=CHR molecules in general are much faster to react than are alkanes.

As anticipated from the reaction principles, the carbon-based radical produced from the reaction of hydroxyl radical with the hydrocarbon adds O_2 to yield a peroxy radical, which in turn oxidizes $NO^\cdot$ to $NO_2^\cdot$:

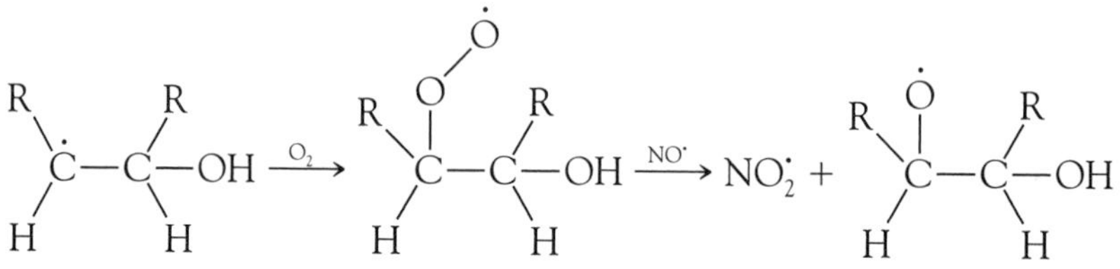

Once much of the $NO^\cdot$ has been oxidized to $NO_2^\cdot$, photochemical decomposition by sunlight of the latter gives $NO^\cdot$ plus O, which then quickly combines with molecular oxygen to give ozone, as discussed in Chapter 3. It is a characteristic of air pollution driven by photochemical processes that ozone from $NO_2^\cdot$ photodecomposition builds up to much higher levels than are found in clean air. Nitrogen dioxide is the only significant tropospheric source of the atomic oxygen from which ozone can form.

As mentioned previously, the ozone concentration does not build up substantially as a result of this sequence until most of the $NO^\cdot$ has been converted to $NO_2^\cdot$, since $NO^\cdot$ and O_3 mutually self-destruct if both are present in significant concentrations. It is only after most $NO^\cdot$ has been oxidized to $NO_2^\cdot$ as a result of reactions with peroxy free radicals that the characteristic buildup of **urban ozone** occurs, as can be seen in Figure 3-4; the transition that occurred at about 9 A.M. on a particular smoggy day in the 1960s in Los Angeles (when smog levels were higher than in more recent years) is illustrated.

One might anticipate from our reactivity principles (Figure 5-1) that the two-carbon radical mentioned above (RCHOCH(R)OH) would lose H by abstraction by O_2, but instead it decomposes spontaneously by cleavage of the C—C bond to give a nonradical molecule containing a C=C bond and another radical, $RH\overset{\cdot}{C}OH$:

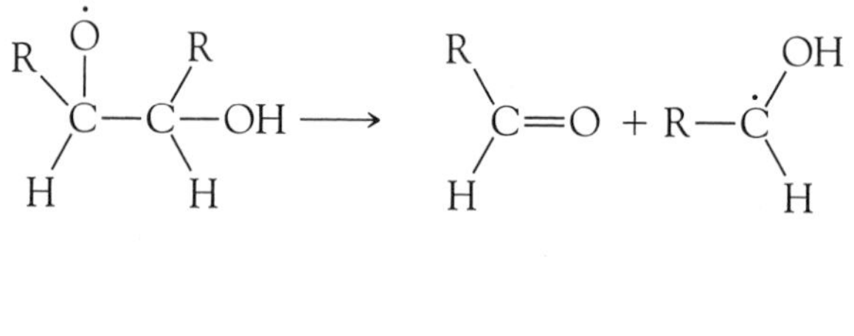

It happens that the reaction requires no energy input, i.e., ΔH is close to zero, because in this case the formation of a $C{=}O$ bond from $C{-}O$ compensates energetically for loss of the $C{-}C$ bond. Since the decomposition of this radical is not endothermic, its activation energy is small and thus the process occurs spontaneously in air.

The carbon-based radical $RH\overset{\cdot}{C}OH$ produced in the preceding reaction subsequently reacts with an O_2 molecule. Since loss of the hydroxyl hydrogen from this radical allows the $C{-}O$ bond to become $C{=}O$, the oxygen molecule abstracts the H atom and produces an aldehyde.

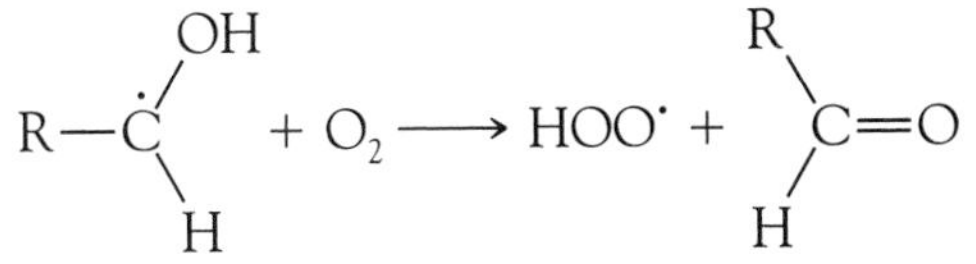

$$R{-}\overset{\displaystyle OH}{\underset{\displaystyle H}{\overset{\cdot}{C}}} + O_2 \longrightarrow HOO^{\cdot} + \overset{\displaystyle R}{\underset{\displaystyle H}{C}}{=}O$$

If we add all the above reactions, the net reaction thus far is

$$RHC{=}CHR + OH^{\cdot} + 2\,O_2 + NO^{\cdot} \longrightarrow 2\,RHC{=}O + HOO^{\cdot} + NO_2^{\cdot}$$

Thus the original $RHC{=}CHR$ pollutant molecule is converted into two aldehyde molecules, each possessing half the number of carbon atoms. Indeed, as shown in Figure 3-4, by about noon on the very smoggy day in Los Angeles, most of the reactive hydrocarbons emitted into the air by morning rush-hour traffic had been converted to aldehydes. By midafternoon, most of the aldehydes had disappeared, since they had largely been photochemically decomposed into $HCO^{\cdot}$ and $R^{\cdot}$ (alkyl) free radicals.

$$RHCO \xrightarrow{\text{sunlight}} R^{\cdot} + HCO^{\cdot}$$

The sunlight-induced decomposition of aldehydes and of ozone leads to a huge increase in the number of free radicals in the air of a city undergoing photochemical smog, although in absolute terms the concentration of radicals is still very small.

The steps in the conversion of the original $RHC{=}CHR$ molecule into aldehydes, and then of the latter to carbon dioxide (see Problem 5-8), are summarized in Figure 5-4. As indicated by the results of Problem 5-8, the net effect of the synergistic oxidation of nitric oxide and $RHC{=}CHR$ is the production of carbon dioxide, nitrogen dioxide, and more hydroxyl radicals. Thus the reaction is **autocatalytic** — its net speed will increase with time since one of its products, here $OH^{\cdot}$, catalyzes the reaction for other reactant molecules.

PROBLEM 5-8

Rewrite the net reaction shown above for $RHC{=}CHR$, assuming that R is H. Deduce the series of steps by which the formaldehyde molecules will

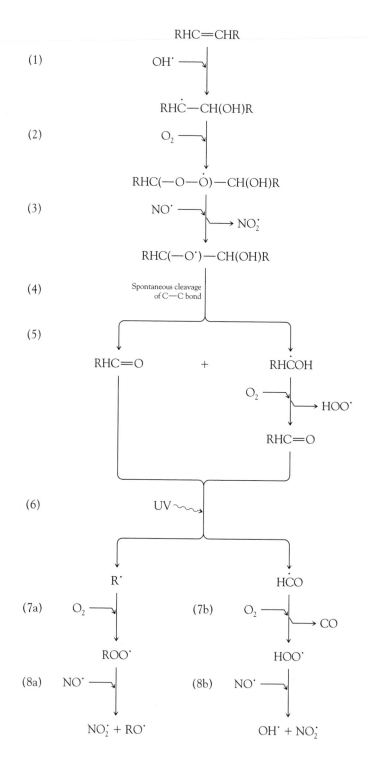

FIGURE 5-4 Mechanism of the RHC=CHR oxidation process in photochemical smog.

subsequently undergo photochemical decomposition and, by a further series of steps, be oxidized to carbon dioxide. Add these steps to the net reaction. Also add the reactions by which $HOO^{\bullet}$ oxidizes $NO^{\bullet}$. What is the final net reaction obtained by adding all these processes together?

PROBLEM 5-9

Repeat Problem 5-8, but this time assume that the alkyl group R in the aldehyde RHCO produced by photochemical smog is a simple methyl group, CH_3, and that, when the aldehyde undergoes photochemical decomposition by sunlight, the radicals $CH_3^{\bullet}$ and $HCO^{\bullet}$ are produced. Using the air reactivity principles, deduce the sequence of reactions by which these radicals are oxidized to carbon dioxide, and determine the overall reaction of conversion of RHCO to CO_2. Assume formaldehyde photolyzes.

Photochemical Smog: The Fate of the Free Radicals

In later stages of photochemical smog formation, reactions that occur between two radicals are no longer insignificant, since their concentrations have become so high. Because their rates are proportional to the *product* of two radical concentrations, these processes are important when the radical concentrations are high; i.e., they occur quickly under such conditions. Generally, the reaction of two free radicals yields a stable, nonradical product:

$$\text{radical + radical} \longrightarrow \text{nonradical molecule}$$

One important example of a radical–radical reaction is the combination of hydroxyl and nitrogen dioxide radicals to yield **nitric acid,** HNO_3, a process that, as we saw in Chapter 1, also occurs in the stratosphere:

$$OH^{\bullet} + NO_2^{\bullet} \longrightarrow HNO_3$$

This reaction is the main tropospheric sink for hydroxyl radicals. The average lifetime for an HNO_3 molecule is several days. By then it either has dissolved in water and been rained out or has been photochemically decomposed back into its components.

Similarly, combination of $OH^{\bullet}$ with $NO^{\bullet}$ gives **nitrous acid,** HONO, also written HNO_2. In sunlight the nitrous acid is almost immediately photochemically decomposed back to $OH^{\bullet}$ and $NO^{\bullet}$, but at night it is stable and therefore its concentration climbs. The observed gigantic increase by dawn in the concentration of $OH^{\bullet}$ radicals in the air of smog-ridden cities, which serves to start the oxidation of hydrocarbons, is due largely to the decomposition of the HONO that had been created the previous evening:

$$OH^\cdot + NO^\cdot \longrightarrow HONO \xrightarrow{\text{sunlight}} OH^\cdot + NO^\cdot$$

It is a characteristic of the later stages in the day of a smog episode that oxidizing agents such as nitric acid are formed in substantial quantities. The reaction of two $OH^\cdot$ radicals, or of two hydroperoxy radicals, $HOO^\cdot$, produces another atmospheric oxidizing agent, **hydrogen peroxide, H_2O_2,** which, as we have already seen, is also produced in this way in clean atmospheres devoid of nitrogen oxides:

$$2\,OH^\cdot \longrightarrow H_2O_2$$

$$2\,HOO^\cdot \longrightarrow H_2O_2 + O_2$$

The latter reaction occurs also in clean air when the concentration of NO_X is especially low and was encountered in stratospheric chemistry in Chapter 1.

The fate of the $R-\overset{\cdot}{C}=O$ radicals, produced by H atom abstraction by $OH^\cdot$ from aldehydes in the ways discussed above, is to combine with O_2 and so produce the free radical

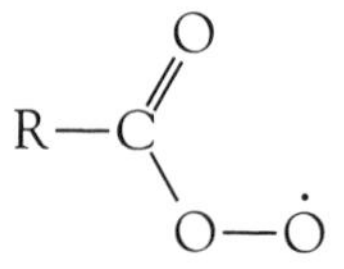

When $NO^\cdot$ is plentiful, this complex species, as expected, behaves as a peroxy radical and oxidizes nitric oxide. In the afternoon, when the concentration of $NO^\cdot$ is very low, the radical reacts instead in a radical–radical process by *adding* to $NO_2^\cdot$ to yield a nitrate. For the common case for which R is CH_3, the nitrate product formed is **peroxyacetylnitrate,** or PAN, which is a potent eye irritant in humans and is also toxic to plants.

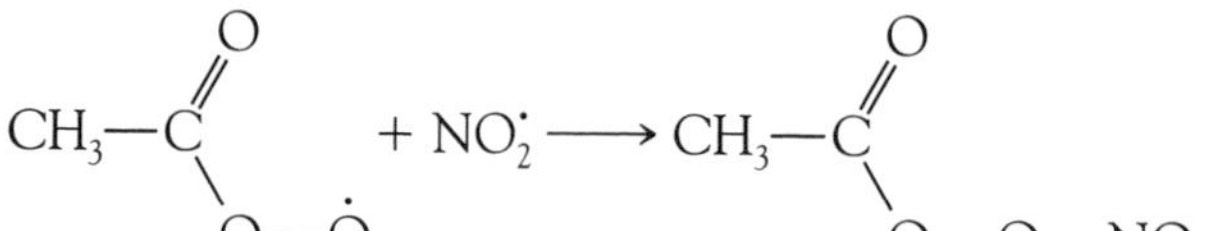

PAN

Overall then, the afternoon stage of a photochemical smog episode is characterized by a buildup of oxidizing agents such as hydrogen peroxide, nitric acid, and PAN, as well as ozone.

Another important species that is present in the later stages of smog episodes is the **nitrate radical,** $NO_3^{\cdot}$, produced when high concentrations of $NO_2^{\cdot}$ and ozone occur simultaneously:

$$NO_2^{\cdot} + O_3 \longrightarrow NO_3^{\cdot} + O_2$$

Although $NO_3^{\cdot}$ is photochemically dissociated to $NO_2^{\cdot}$ and O rapidly during the daytime, it is stable at night and plays a role similar to $OH^{\cdot}$ in attacking hydrocarbons in the hours following sundown:

$$NO_3^{\cdot} + RH \longrightarrow HNO_3 + R^{\cdot}$$

Thus at night, when the concentration of the short-lived hydroxyl radicals goes almost to zero since no new ones are being produced due to the absence of O^*, $NO_3^{\cdot}$ rather than $OH^{\cdot}$ initiates the oxidation of reduced gases in the troposphere. The similarity between $OH^{\cdot}$ and $NO_3^{\cdot}$ is not surprising since both react as $-\overset{\cdot}{O}$ radicals and form very stable $O-H$ bonds when they abstract hydrogens.

In summary, an episode of photochemical smog in a city such as Los Angeles begins at dawn, when sunlight initiates the production of hydroxyl radicals from the nitrous acid and from the ozone left over from the previous day. The initial input of nitric oxide and reactive hydrocarbons from morning rush-hour vehicle traffic reacts first to produce aldehydes (see Figure 3-4), the photolysis of which increases the concentration of free radicals and thereby speeds up the overall reaction. The increase in free radicals in the morning serves to oxidize the nitric oxide to nitrogen dioxide; photolysis of the latter causes the characteristic rise in ozone concentrations about mid-day. Oxidants such as PAN and hydrogen peroxide are also produced, especially in the afternoons due to the high free-radical concentration present at that time. The late afternoon rush-hour traffic produces more nitric oxide and hydrocarbons, which presumably react quickly under the conditions of high free-radical concentration. The smog reactions largely cease at dusk due to the lack of sunlight, but some oxidation of hydrocarbons continues due to the presence of the nitrate radical. The nitrous acid that forms after dark is stable until dawn, when its decomposition helps initiate the process for another day.

PROBLEM 5-10

Annotate (in pencil) the top of Figure 3-4 to show the *dominant* reaction occurring in the polluted air in the following time segments: (a) 5 A.M.–8 A.M.; (b) 8 A.M.–12 noon.

PROBLEM 5-11

Some formaldehyde molecules photochemically decompose to the molecular products H_2 and CO rather than to free radicals. Deduce the mechanism and overall reaction for the oxidation to CO_2 for formaldehyde molecules that initially produce the molecular products.

PROBLEM 5-12

Deduce the series of steps by which ethylene gas, $H_2C{=}CH_2$, is oxidized to CO_2 when it is released into an atmosphere undergoing a photochemical smog process. Assume in this case that aldehydes react completely by photochemical decomposition rather than by $OH^{\cdot}$ attack.

PROBLEM 5-13

Radical–radical reactions can also occur in clean air, particularly when the nitrogen oxide concentration is very low. Predict the product that will be formed when the $CH_3OO^{\cdot}$ radical intermediate of methane oxidation combines with the $HOO^{\cdot}$ free radical. Note that long oxygen chains are unstable with respect to O_2 expulsion.

Oxidation of Atmospheric SO_2: The Homogeneous Gas-Phase Mechanism

When the sky is clear or when clouds occupy only a few percent of the tropospheric volume, the predominant mechanism for the conversion of SO_2 to H_2SO_4 is a homogeneous gas-phase reaction that occurs by several sequential steps. As usual for atmospheric trace gases, the hydroxyl radical initiates the process. Since SO_2 molecules contain multiple bonds but no hydrogen, it is expected (see Figure 5-1) that the $OH^{\cdot}$ will *add* to the molecule at the sulfur atom:

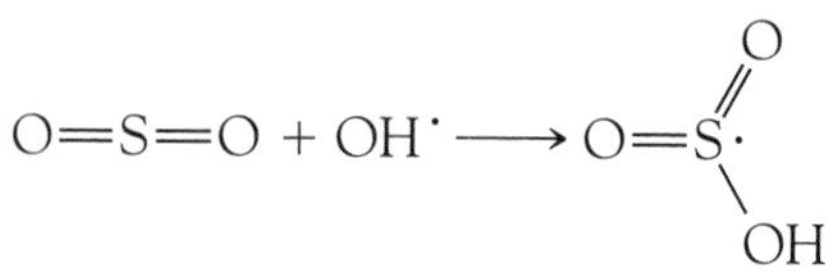

$$O{=}S{=}O + OH^{\cdot} \longrightarrow O{=}\overset{\displaystyle O}{\underset{\displaystyle OH}{\overset{\|}{S}}}{\cdot}$$

Since a stable molecule, namely **sulfur trioxide,** SO_3, can be produced from this radical by the removal of the hydrogen atom, the reaction principles predict that the next reaction in the sequence is that between the radical and an O_2 molecule to abstract H:

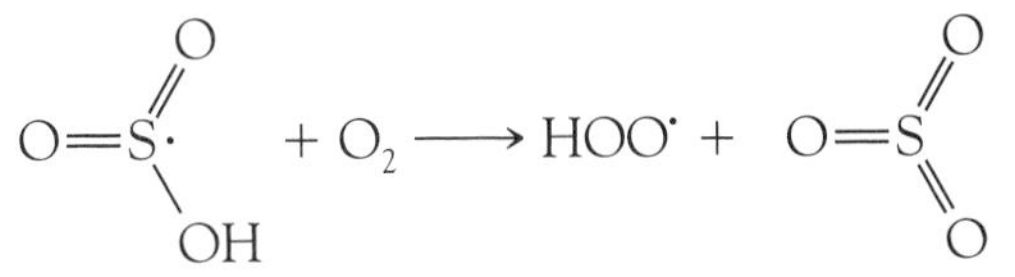

$$O=\overset{\displaystyle O}{\underset{\displaystyle OH}{S\cdot}} \;+\; O_2 \longrightarrow HOO^{\cdot} \;+\; O=\overset{\displaystyle O}{\underset{\displaystyle O}{S}}$$

The sulfur trioxide molecule rapidly combines with a gaseous water molecule to form **sulfuric acid.** Finally, the H_2SO_4 molecules react with water, whether in the form of water vapor or as a mist, to form an aerosol of droplets, each of which is an aqueous solution of sulfuric acid. The sequence of steps from gaseous SO_2 to aqueous H_2SO_4 is

$$SO_2 + OH^{\cdot} \longrightarrow HSO_3^{\cdot}$$

$$HSO_3^{\cdot} + O_2 \longrightarrow SO_3 + HOO^{\cdot}$$

$$SO_3 + H_2O \longrightarrow H_2SO_4(g)$$

$$H_2SO_4(g) + many\ H_2O \longrightarrow H_2SO_4(aq)$$

The sum of these reaction steps is

$$SO_2 + OH^{\cdot} + O_2 + many\ H_2O \longrightarrow HOO^{\cdot} + H_2SO_4(aq)$$

When we include the return of $HOO^{\cdot}$ to $OH^{\cdot}$ via reaction with $NO^{\cdot}$, the overall reaction is seen to be $OH^{\cdot}$-catalyzed co-oxidation of SO_2 and $NO^{\cdot}$:

$$SO_2 + NO^{\cdot} + O_2 + many\ H_2O \xrightarrow{\;OH^{\cdot}\ catalysis\;} NO_2^{\cdot} + H_2SO_4(aq)$$

For representative concentrations of the $OH^{\cdot}$ radical in relatively clean air, a few percent of the atmospheric SO_2 is oxidized per hour by this mechanism. The rate is much faster for air masses undergoing photochemical smog reactions since the concentration of $OH^{\cdot}$ there is much higher. However, generally only a small amount of sulfur dioxide is oxidized in cloudless air; the rest is removed by dry deposition before the reaction has time to occur.

Dissolved sulfur dioxide, SO_2, is oxidized to **sulfate ion, SO_4^{2-},** by trace amounts of the well-known oxidizing agents hydrogen peroxide, H_2O_2, and ozone, O_3, that are present in the airborne droplets, as already discussed in Chapter 3. Indeed, such reactions currently are thought to constitute the main oxidation pathways for atmospheric SO_2, except under clear sky conditions when the gas-phase homogeneous mechanism predominates. The ozone and hydrogen peroxide result mainly from sunlight-induced reactions in photochemical smog. Consequently, oxidation of SO_2 occurs most rapidly in air that has also been polluted by reactive hydrocarbons and nitrogen oxides. Since the smog reactions occur predominantly in summer, rapid oxidation of SO_2 to sulfate also is characteristic of the summer season.

Systematics of Stratospheric Chemistry

There are many similarities between the chemical reactions discussed in Chapters 1 and 2 for the stratosphere and those outlined above for the troposphere. For example, a characteristic process in both regions of the atmosphere is hydrogen atom abstraction. The stratosphere and troposphere differ, however, in which reactions are dominant. In the stratosphere $OH^{\cdot}$, O^{*}, $Cl^{\cdot}$, and $Br^{\cdot}$ are all important in abstracting a hydrogen atom from stable molecules such as methane, whereas only the hydroxyl and nitrate radicals are important in this respect in the troposphere. In the following material we systematize the Chapters 1 and 2 chemistry that is important in the stratosphere, especially in regard to processes of ozone depletion.

Processes Involving Loosely Bound Oxygen Atoms

Many of the species in the stratosphere have a **loosely bound oxygen atom,** denoted Y, which is readily detached from the rest of the molecule in several characteristic ways. In Table 5-1 we list the molecules Y—O that contain a loose oxygen. In every case, dissociation of this oxygen atom requires much less energy than is required to break any of the remaining bonds, so the resulting Y units remain intact. Notice that the Y species, except for O_2, are the free radicals that in Chapters 1 and 2 we called X when we discussed ozone destruction catalysts. In terms of electronic structure, all "loose" oxygens are joined by a single bond to another electronegative atom that possesses one or more

TABLE 5-1	Molecules Containing Loose Oxygen Atoms		
Molecule Y—O	Structure of Y—O	Y—O Bond Energy in kJ/mol	Comment
O_3	O_2—O	107	The most loose oxygen
$BrO^{\cdot}$	Br—O	235	
$HOO^{\cdot}$	HO—O	266	
$ClO^{\cdot}$	Cl—O	272	
$NO_2^{\cdot}$	ON—O	305	The least loose oxygen

nonbonding electron pairs. The interaction between the nonbonded electron pairs on this atom and those on the oxygen weakens the single bond.

The characteristic reactions involving loose oxygen are collected below.

• *Reaction with Atomic Oxygen*   Here the oxygen atom detaches the loose oxygen atom by combining with it:

$$Y—O + O \longrightarrow Y + O_2$$

These reactions are all exothermic since the $O{=}O$ bond in O_2 is much stronger than the $Y—O$ bond.

• *Photochemical Decomposition*   The $Y—O$ species absorbs UV-B, and in some cases even longer wavelength light, from sunlight and subsequently releases the loose oxygen atom:

$$Y—O + \text{ sunlight} \longrightarrow Y + O$$

• *Reaction with NO$^{\cdot}$*   Nitric oxide abstracts the loose oxygen atom:

$$Y—O + NO^{\cdot} \longrightarrow Y + NO_2^{\cdot}$$

This reaction is exothermic since the $ON—O$ bond strength (see Table 5-1) is the greatest of those involving a loose oxygen. (Recall the general principle that exothermic free-radical reactions are relatively fast.)

• *Abstraction of Oxygen from Ozone*   Abstraction of the loose oxygen atom from ozone (only) to form the $Y—O$ species is characteristic of $OH^{\cdot}$, $Cl^{\cdot}$, $Br^{\cdot}$, and $NO^{\cdot}$. Thus all these radicals act as catalytic ozone destroyers, X:

$$O_2—O + X \longrightarrow O_2 + XO$$

The reaction involving ozone is exothermic since ozone contains the weakest of the bonds involving a loose oxygen. The other YO species do not undergo this reaction to an important extent either because it is endothermic and therefore negligibly slow or because the atmospheric X species react more quickly with other chemicals.

• *Combination of Two YO Molecules*   If the concentration of YO species becomes high, they may react by the collision of two of them (identical or different species). If at least one is O_3 or $HOO^{\cdot}$, an unstable chain of three or more oxygen atoms is created when they collide and join; in these circumstances, the loose oxygens combine to form one or more molecules of O_2, which are expelled:

$$2\, O_2—O \longrightarrow 3\, O_2$$

$$2\, HO—O^{\cdot} \longrightarrow HOOH + O_2$$

$$HO—O^{\cdot} + O—O_2 \longrightarrow OH^{\cdot} + 2\, O_2$$

$$HO—O^{\cdot} + {}^{\cdot}O—Cl \longrightarrow HOCl + O_2$$

When neither is O_3 or $HOO^{\bullet}$, the two Y—O molecules combine to form a larger molecule, which subsequently often decomposes photochemically:

$$2\,NO_2^{\bullet} \longrightarrow N_2O_4$$

$$2\,ClO^{\bullet} \longrightarrow ClOOCl \xrightarrow{\text{sunlight}} \longrightarrow 2\,Cl^{\bullet} + O_2$$

$$ClO^{\bullet} + NO_2^{\bullet} \underset{\text{sunlight}}{\rightleftharpoons} ClONO_2$$

$$ClO^{\bullet} + BrO^{\bullet} \longrightarrow Cl^{\bullet} + Br^{\bullet} + O_2$$

The Y—O—O—Y molecules have little thermal stability, and even at moderate temperatures they may dissociate back to their Y—O components before light absorption and photolysis have time to occur.

PROBLEM 5-14

Which of the following species do(es) *not* contain a loose oxygen?

(a) $HOO^{\bullet}$ (b) $OH^{\bullet}$ (c) $NO^{\bullet}$ (d) O_2 (e) $ClO^{\bullet}$

PROBLEM 5-15

From which Y—O species

(a)   does $NO^{\bullet}$ abstract an oxygen atom?
(b)   does atomic oxygen abstract an oxygen atom?
(c)   does sunlight in the stratosphere detach an oxygen atom?
(d)   do the Y—O—O—Y species (with identical Y groups) form in the stratosphere?
(e)   is O_2 produced when two identical Y—O species react?

PROBLEM 5-16

Using the principles above, predict what would be the likely fate of $BrO^{\bullet}$ molecules in a region of the stratosphere that in concentrations was particularly (a) high in atomic oxygen, (b) high in $ClO^{\bullet}$, (c) high in $BrO^{\bullet}$ itself, and (d) high in sunlight intensity.

PROBLEM 5-17

Using the principles above, deduce what reaction(s) could be sources of atmospheric (a) $ClONO_2$ (b) $ClOOCl$ (c) $Cl^{\bullet}$ atoms

PROBLEM 5-18

Draw the Lewis structure for the free radical $FO^{\bullet}$. On the basis of this structure, could you predict whether it contains a loose oxygen?

PROBLEM 5-19

What is the expected product when ClO˙ reacts with NO˙? What are the possible fates of the product(s) of this reaction? Devise a mechanism incorporating (a) this reaction, (b) the reaction of Cl˙ with ozone, and (c) the photochemical decomposition of NO_2^- to NO˙ and atomic oxygen. Is the net result of this cycle, which operates in the lower stratosphere, the destruction of ozone?

Review Questions

1. Explain why OH˙ reacts more quickly than HOO˙ to abstract hydrogen from other molecules.

2. How does OH˙ react with molecules that contain hydrogen but not multiple bonds?

3. What are the two different initial steps by which atmospheric formaldehyde, H_2CO, is decomposed in air?

4. What is the reaction by which most nitric oxide molecules in the troposphere are oxidized to nitrogen dioxide?

5. What are the two common reactions by which diatomic oxygen reacts with free radicals?

6. Explain why photochemical smog is an autocatalytic process.

7. What is the fate of OH˙ radicals that react with NO˙? With NO_2^-? With other OH˙?

8. What is the fate of NO_2^- molecules that photodissociate? That react with ozone? That react with RCOO˙ radicals?

9. Why does the production of high concentrations of NO_2^- lead to an increase in ozone levels in air? Why does this not occur if much NO˙ is present?

10. What is the formula of nitrate radicals? Explain how they are similar in reactivity to hydroxyl radicals.

11. Explain how atmospheric sulfur dioxide is oxidized by gas-phase reactions in the atmosphere.

12. Explain what is meant by the term *loosely bound oxygen*. What are its four characteristic reactions in stratospheric chemistry?

Additional Problems

1. Write the two-step mechanism by which CO is oxidized to CO_2. Also include the sequence of reactions by which the hydroperoxy radical so produced oxidizes NO˙ to NO_2^-, the nitrogen dioxide is photolyzed to NO˙ and atomic oxygen, and oxygen atoms produce ozone. By adding the steps, show that the atmospheric oxidation of carbon monoxide can increase the ozone concentration by a catalytic process.

2. Using the reactivity principles developed in this chapter, deduce the series of steps and the overall reaction by which ethane, H_3C—CH_3, is oxidized in the atmosphere. Assume that the aldehydes produced in the mechanism undergo photochemical decomposition to R˙ and HCO˙.

3. When the concentration of nitrogen oxides in a region of the air is very low, peroxy radicals combine with other species rather than oxidize nitric oxide. Deduce the mechanism, including the overall equation, for the process by which carbon monoxide is oxidized to carbon dioxide under these conditions, assuming that the hydroperoxy radicals react with ozone. From your result, would you predict that ozone levels would be abnormally high or low in air masses having low nitrogen oxide concentration? *[Hint: See the generalities in the section on the systematics of stratosphere chemistry concerning reactions that produce long oxygen chains.]*

4. Predict the most likely reaction (if any) that would occur between a hydroxyl radical and each of the following atmospheric gases:

(a) $CH_3CH_2CH_3$ **(b)** $H_2C{=}CHCH_3$

(c) HCl   **(d)** H_2O

5. Draw complete Lewis structures for NO_2, HONO, and HNO_3. (Resonance structures and formal charges are not required.)

6. In Problem 1-2, the longest wavelength of light that could dissociate an O atom from O_3 was calculated and determined to occur in the IR region of the spectrum. Using the information in Table 5-1, calculate the longest wavelength of light that could photolytically cleave the loose hydrogen atom in the case of each of the remaining molecules listed in that table. What region of the electromagnetic spectrum does each correspond to?

Further Readings

1. R. Atkinson, "Atmospheric Chemistry of VOCs and NO_X," *Atmospheric Environment* 34 (2000): 2063–3101.

2. B. J. Finlayson-Pitts and J. N. Pitts, *Atmospheric Chemistry* (New York: Wiley, 1986). [A comprehensive guide to the detailed chemistry of the atmosphere.]

3. B. J. Finlayson-Pitts and J. N. Pitts, Jr., "Tropospheric Air Pollution: Ozone,

Airborne Toxics, Polycyclic Aromatic Hydrocarbons, and Particles," *Science* 276 (1997): 1045–1051.

4. J. H. Seinfeld and S. N. Pandis, *Atmospheric Chemistry and Physics* (New York: Wiley, 1998). [Another comprehensive guide to the detailed chemistry of the atmosphere.]

Websites of Interest

Log on to www.whfreeman.com/envchem4e/ and click on Chapter 5.

Environmental Instrumental Analysis I	Instrumental Determination of NO_X via Chemiluminescence

In the preceding chapters, we have seen that nitrogen oxides play a leading role in atmospheric chemistry, both in the stratosphere and at ground level. In this box, we see how the concentration of NO and NO_2 gases in environmental air samples can be determined using a sophisticated, modern method of analysis.

When chemicals react to produce light, the process is termed **chemiluminescence.** If conditions for a particular reaction are well known and can be controlled in an analytical instrument, chemiluminescence can be used as a sensitive and selective means of determining the concentration of components in the reaction. A few well-known chemical reactions that produce chemiluminescence are the basis for methods in tropospheric and stratospheric environmental analysis. One of the most common of these methods is the detection of nitric oxide (NO) and nitrogen dioxide (NO_2).

The chemiluminescence reaction that produces light in this method is the gas-phase reaction of NO with ozone (O_3):

$$NO + O_3 \longrightarrow NO_2{}^* + O_2$$

In the NO detector, this reaction takes place in a small steel reaction vessel under controlled conditions. The excited-state nitrogen dioxide created in this reaction, designated by $NO_2{}^*$, very quickly returns to ground state by giving off light in the red and infrared range of the light spectrum (600 to 2800 nanometers):

$$NO_2{}^* \longrightarrow NO_2 + light\ (\lambda = 600\ to\ 2800\ nm)$$

The amount of light produced by this reaction is dependent on pressure and temperature. A constant low pressure is maintained in the reaction vessel by use of a vacuum pump that constantly evacuates the chamber. Typical cell pressures are approximately 1 to 100 torr. The amount of light produced in the reaction is proportional to the amount of whichever reactant is *not* in excess in the reaction chamber. If ozone is provided in excess (from a steady ozone generator in this case), then the light output reflects changes in NO concentration. The light created by the reaction is detected by a photomultiplier tube (PMT) whose output is fed to a computer system. The computer software correlates the amount of light produced with the quantity of reactant by referring to the relation between light intensity and NO concentration, which is obtained from previous calibration experiments. In instruments of this kind, the PMT signal is often integrated over short time periods (e.g., 10 seconds) using what is referred to as a photon counting system, which improves sensitivity.

The schematic diagram of this instrument, shown in the figure below, includes the reaction chamber, ozone generator, PMT, computer, and

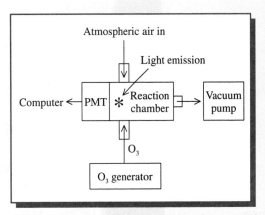

(continued on p. 200)

Environmental Instrumental Analysis I	Instrumental Determination of NO$_X$ via Chemiluminescence *(continued)*

vacuum pump. Gas from the atmosphere is sucked directly into the reaction chamber and immediately mixed with an excess of O$_3$ (i.e., more O$_3$ than NO). Typical sampling volumes are 1000 standard cubic centimeters per minute (sccm). Some instruments have gold-plated surfaces inside to prevent surface reactions and increase the light-collection abilities of the light chamber. The light produced is detected by the PMT mounted immediately adjacent to the reaction chamber and separated by a transparent window or a filter that can block out light from interfering reactions.

This same instrument can be used to determine NO$_2$ by incorporating a chemical reduction step for the incoming air, in which NO$_2$ is reduced to NO by a hot metal catalyst before entering the reaction chamber. NO is then determined as before, but now the signal includes input from the presence of the air's NO *and* NO$_2$. If alternating signals are generated in a short time period—one with the atmospheric air stream that has been reduced and the other without reduction—by means of a switching valve, then the concentration of both nitrogen oxides can be determined:

NO concentration = signal generated by unreduced air flow

NO$_2$ concentration = signal generated by reduced air flow − signal from unreduced air flow

The limits of detection and selectivity (ability to determine NO or NO$_2$ in the presence of interferants) over O$_3$, SO$_2$, and CO—all common atmospheric gases—are very good. Less than 1 ppb$_v$ NO can be routinely determined using this method (Department for Environment UK, 2004). Interferences from reduced

gas-phase components also commonly present, such as NH$_3$, can be minimized by decreasing the temperature in the catalyst chamber described above (Environmental Protection National Service, 2000).

Instruments of this kind have been used on airplane-based stratospheric and tropospheric sampling projects by the National Atmospheric and Space Administration and the National Center for Atmospheric Research. Similar instrumentation has been used for tropospheric studies of urban pollution, using laboratory-based "smog chambers," and in indoor air pollution studies.

The figure below shows the variation in concentration of NO as an airborne scientific expedition equipped with this kind of NO detector flew through cumulonimbus clouds—tall, vertically developed, and actively raining (Ridley et al., 1987).

The researchers were flying level at 9.3-km altitude while sampling the tropospheric air

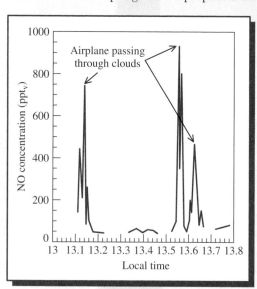

over the Pacific Ocean west of Hawaii. The regions of the flight path involving clouds are noted. The concentration of NO in parts per trillion by volume is plotted on the y-axis and local time is plotted on the x-axis. These data demonstrate the production of NO in electrically active clouds.

This method of NO and NO$_2$ measurement has also been used in experiments carried out at Antarctica at the South Pole (Davis et al., 2004). Earlier presumptions were that no nitric oxide was produced in that pristine setting since internal combustion engines, which are the major source of NO in urban environments, were absent. However, in the austral summer of 24-hour daylight, nitric oxide *is* continuously produced by sunlight photolysis of nitrate anion in the snow pack. And so, unlike diurnal variations of NO such as those seen in urban environments (see figure below), time-course studies over an entire 24-hour

period at a lab at the South Pole have shown a relatively constant NO concentration at the snow pack surface. The overall concentrations of NO at that South Polar site were also significantly higher than at other polar sites (sometimes by orders of magnitude) where similar examinations had been carried out, so it appears that the interesting meteorological conditions present at 90° S—including constant sunlight during austral summer—directly affect the production of NO. If you look carefully at the graph, you can see a dip in NO concentration that took place near midnight (0:00 h). This occurred when the shadow from a nearby building temporarily fell on the sampling site, decreasing the production of NO from the snow's nitrate ions. Plotted on the same figure is urban air NO concentration for a sampling site in Houston, Texas on August 17, 2006 (TCEQ, 2006). The common urban diurnal variation of NO is clear.

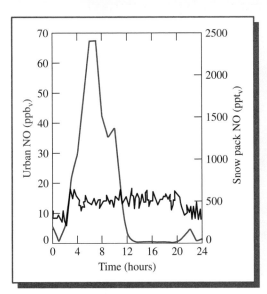

References: D. Davis, G. Chen, M. Buhr, J. Crawford, D. Lenschow, B. Lefer, R. Shetter, F. Eisele, L. Mauldin, and A. Hogan, "South Pole NO$_X$ Chemistry: An Assessment of Factors Controlling Variability and Absolute Levels," *Atmospheric Environment* 38 (2004): 5375–5388.

Department for Environment UK, "Nitrogen Dioxide in the United Kingdom, 2004," http://www.defra.gov.uk/environment/airquality/aqeg/nitrogen-dioxide/index.htm

Environmental Protection National Service UK, "Monitoring Methods for Ambient Air, 2000," http://publications.environment-agency.gov.uk/pdf/GEHO1105BJYB-e-e.pdf

B. A. Ridley, M. A. Carroll, and G. L. Gregory, "Measurements of Nitric Oxide in the Boundary Layer and Free Troposphere over the Pacific Ocean," *Journal of Geophysical Research* 92(D2) (1987): 2025–2047.

TCEQ (Texas Commission on Environmental Quality), http://www.tceq.state.tx.us.

PART II

ENERGY AND CLIMATE CHANGE

Contents of Part II

Environmental Instrumental Analysis II

- Instrumental Determination of Atmospheric Methane

Scientific American Feature Article

- A Plan to Keep Carbon in Check

CHAPTER 6

THE GREENHOUSE EFFECT

In this chapter, the following introductory chemistry topics are used:

- Combustion
- Molecular shape, bond angles and distances
- Polymers

Background from previous chapters used in this chapter:

- Sunlight wavelength regions (UV, visible, IR) (Chapter 1)
- Absorption spectra (Chapter 1)
- ppm/ppb concentration scale for gases (Chapter 1)
- CFCs and their replacements (Chapter 2)
- Tropospheric ozone; nitrous oxide (Chapter 3)
- Aerosols; sulfur dioxide (Chapter 3)

Introduction

Everyone has heard the prediction that the greenhouse effect will significantly affect climates around the world in the future. The terms *greenhouse warming* and *global warming* in ordinary usage simply mean that average global air temperatures are expected to increase by several degrees as a result of the buildup of carbon dioxide and other greenhouse gases in the atmosphere. Indeed, most atmospheric scientists

Global warming may have led to the dramatic breakup of the Larsen Ice Shelf off Antarctica in 2002. [GSFC/LaRC/JPL/MISR Team/NASA.]

believe that such **global warming** has already been under way for some time and is largely responsible for the air temperature increase of about two-thirds of a Celsius degree that has occurred since 1860.

The phenomenon of rapid global warming—with its demands for large-scale adjustments—is generally considered to be our most crucial worldwide environmental problem, although both positive and negative effects would be associated with any significant increase in the average global temperature. No one currently is sure of the extent or timing of future temperature increases, nor is it likely that reliable predictions for individual regions will ever be available much in advance of the events in question. If current models of the atmosphere are correct, however, significant warming will occur in coming decades. It is important that we understand the factors that are driving this increase so that we can, if we wish, take steps to avoid potential catastrophes caused by rapid climate change in the future.

In this chapter, the mechanism by which global warming could arise is explained, and the nature and sources of the chemicals that are responsible for the effect are analyzed. The extent of the atmospheric warming to date and other indications that change is under way are also discussed. The predictions concerning global warming in the future, and an analysis of steps that could be taken to minimize it, are presented in Chapters 7 and 8.

The Mechanism of the Greenhouse Effect

The Earth's Energy Source

The Earth's surface and atmosphere are kept warm almost exclusively by energy from the Sun, which radiates energy as light of many types. In its radiating characteristics, the Sun behaves much like a **blackbody**, i.e., an object that is 100% efficient in emitting and in absorbing light. The wavelength, λ_{peak}, in micrometers, at which the *maximum* emission of energy occurs by a radiating blackbody decreases inversely with increasing Kelvin temperature, T, according to the relationship

$$\lambda_{peak} = 2897/T$$

Since for the surface of the Sun, from which the star emits light, the temperature $T \sim 5800$ K, then from the equation it follows that λ_{peak} is about 0.50 μm, a wavelength that lies in the visible region of the spectrum (and corresponds to green light). Indeed, the maximum observed solar output (see the dashed portion of the curve in Figure 6-1) occurs in the range of visible light, i.e., that of wavelengths between 0.40 and 0.75 μm. Beyond the "red limit," the maximum wavelength for visible light, the Earth receives **infrared light** (IR) in the 0.75–4-μm region from the Sun. Of the energy received at the top of the Earth's atmosphere from the Sun, slightly over half the total is IR and most of the remainder is visible light. At the opposite end

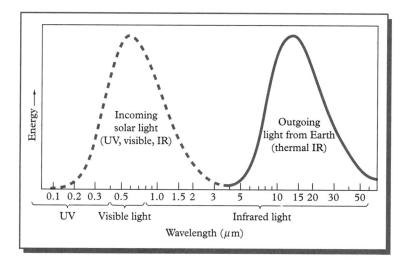

FIGURE 6-1 Wavelength distributions (using different scales) for light emitted by the Sun (dashed curve) and by Earth's surface and troposphere (solid curve). [Source: Redrawn from J. Gribbin, "Inside Science: The Greenhouse Effect," *New Scientist,* supplement (22 October 1988).]

of the visible wavelength spectrum from IR, beyond the "violet" limit, lies **ultraviolet light** (UV), which has wavelengths less than 0.4 μm and is a minor component of sunlight, as discussed in Chapter 1.

Of the total incoming sunlight of all wavelengths that impinges upon the Earth, about 50% is absorbed at its surface by water bodies, soil, vegetation, buildings, and so on. A further 20% of the incoming light is absorbed by water droplets in air (mainly in the form of clouds) and by molecular gases— the UV component by stratospheric **ozone,** O_3, and **diatomic oxygen,** O_2, and the IR by **carbon dioxide,** CO_2, and especially water vapor.

The remaining 30% of incoming sunlight is reflected back into space by clouds, suspended particles, ice, snow, sand, and other reflecting bodies, without being absorbed. The fraction of sunlight reflected back into space by an object is called its **albedo,** which therefore is about 0.30 for the Earth overall. Clouds are good reflectors, with albedos ranging from 0.4 to 0.8. Snow and ice are also highly reflecting surfaces for visible light (high albedos), whereas bare soil and bodies of water are poor reflectors (low albedos). Thus the melting of sea ice in polar regions to produce open water greatly increases the fraction of sunlight absorbed there and decreases the Earth's overall albedo. Planting trees in snow-covered forests reduces the albedo of the surface and may actually contribute to global warming.

Historical Temperature Trends

The trends in average surface temperature for the past 2000 years, as reconstructed for most of that period from indirect evidence such as tree ring growth, is shown in Figure 6-2a. (The Medieval Warm Period early in the

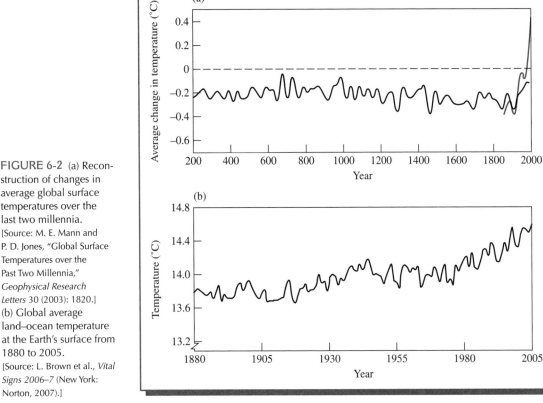

FIGURE 6-2 (a) Reconstruction of changes in average global surface temperatures over the last two millennia.
[Source: M. E. Mann and P. D. Jones, "Global Surface Temperatures over the Past Two Millennia," *Geophysical Research Letters* 30 (2003): 1820.]
(b) Global average land–ocean temperature at the Earth's surface from 1880 to 2005.
[Source: L. Brown et al., *Vital Signs 2006–7* (New York: Norton, 2007).]

millennium was apparently restricted to the North Atlantic region, so it is not very evident on the global plot.) Notice the consistently downward trend in temperature until the beginnings of the Industrial Revolution.

The warming of the climate during the twentieth century stands in stark contrast to the gradual cooling trend in the previous 900 years of the millennium, producing a "hockey stick" shape to the temperature plot in Figure 6-2a. The trends in global average surface temperature over the last century and a half are illustrated in detail in Figure 6-2b. Air temperature did not increase *continuously* throughout the twentieth century. A significant warming trend occurred in the 1910–1940 period, due to a lack of volcanic activity and a slight increase in the intensity of sunlight. This period was followed by some cooling over the next three decades, due to aerosols resulting from increased volcanic activity. These decades were succeeded in turn by a warming period that has been sustained from about 1970 to the present and that has so far amounted to a temperature increase of about 0.6°C; this is attributed almost entirely to anthropogenic influences, as discussed in detail in this chapter. Eleven of the 12 years of the 1995–2006 period are among the 12 warmest since 1850, when instrument records began. The warmest years on record were 1998 and 2005.

Earth's Energy Emissions and the Greenhouse Effect

Like any warm body, the Earth emits energy; indeed, the amount of energy that the planet absorbs and the amount that it releases must be equal over the long term if its temperature is to remain level. (Currently, the planet is absorbing slightly more than it is emitting, thereby warming the air and the oceans.) The emitted energy (see the solid portion of the curve in Figure 6-1) is neither visible nor UV light, because the Earth is not hot enough to emit light in these regions. Since the temperature of the Earth's surface is approximately 300 K, then according to the equation above for λ_{peak}, if the Earth acted like a blackbody, its wavelength of maximum emission would be about 10 μm. Indeed, the Earth's emission does peak in that general region, actually at about 13 μm, and consists of infrared light having wavelengths starting at about 5 μm and extending, albeit weakly, beyond 50 μm (Figure 6-1, solid curve). The 5–100-μm range is called the **thermal infrared** region since such energy is a form of heat, the same kind of energy a heated iron pot would radiate.

Infrared light is emitted both at the Earth's surface and by its atmosphere, though in different amounts at different altitudes since the emission rate is very temperature sensitive: In general, *the warmer a body, the more energy it emits per second*. The rate of release of energy as light by a blackbody increases in proportion to the fourth power of its Kelvin temperature:

$$\text{rate of energy release} = kT^4$$

where k is a proportionality constant. Thus doubling its absolute temperature increases sixteen-fold (2^4) the rate at which a body releases energy. More realistically, for contemporary surface conditions of planet Earth, a one-degree rise in temperature would increase the rate of energy release by 1.3%.

PROBLEM 6-1

Calculate the ratio of the rates of energy release by two otherwise identical blackbodies, one of which is at 0°C and the other at 17°C. At what temperature is the rate of energy release twice that at 0°C?

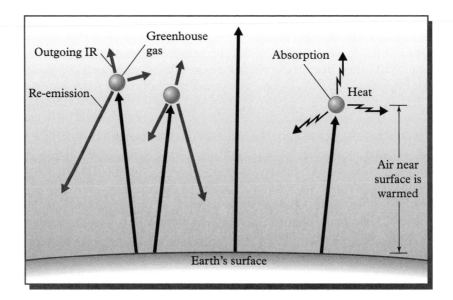

FIGURE 6-3 The greenhouse effect: Outgoing IR absorbed by greenhouse gases is either re-emitted (left side of diagram) or converted to heat (right side).

Some gases in air absorb thermal infrared light—though only at characteristic wavelengths—and therefore the IR emitted from the Earth's surface and atmosphere does not *all* escape directly to space. Very shortly after its absorption by atmospheric gases such as CO_2, the IR photon may be re-emitted. Alternatively, the absorbed energy may quickly be redistributed as heat among molecules that collide with the absorber molecule, and it may eventually be re-emitted as IR by them. Whether re-emitted immediately by the initial absorber molecule or later by others in the area, the direction of the photon is completely random (Figure 6-3). Consequently, some of this thermal IR is redirected back toward the Earth's surface and is reabsorbed there or in the air above it.

Because the air absorbs IR photons and redistributes the energy as heat to surrounding molecules, the air temperature in the region of the absorbing molecule increases. However, this air mass does not heat up without limit as its molecules trap more and more of the outgoing infrared light because there is an opposing phenomenon that prevents such a catastrophe. As explained above, the rate of energy emission increases with temperature, so the molecules that have shared the excess energy themselves emit more and more energy as infrared light as they warm up (Figure 6-3). The water droplets and vapor in clouds are also very effective in absorbing infrared light emitted from beneath them. The temperature at the tops of clouds is quite cool relative to the air beneath them, so clouds do not radiate as much energy as they absorb. Overall, air temperatures increase only enough to re-establish the planetary equality between incoming and outgoing energy.

The phenomenon of interception of outgoing IR by atmospheric constituents and its dissipation as heat to increase the temperature of the atmosphere (as illustrated in Figure 6-3) is called the **greenhouse effect.** It is responsible for the average temperature at the Earth's surface and the air close to it being about $+15°C$ rather than about $-18°C$, the temperature it would be if there were no IR-absorbing gases in the atmosphere. The surface is warmed as much by this indirect mechanism as it is by the solar energy it absorbs directly! The very fact that our planet is not entirely covered by a thick sheet of ice is due to the natural operation of the greenhouse effect, which has been in operation for billions of years.

The atmosphere operates in the same way as a blanket, retaining within the immediate region some of the heat released by a body and thereby increasing the local temperature. The phenomenon that worries environmental scientists is that increasing the concentration of the trace gases in air that absorbs thermal IR light (piling on more blankets, so to speak) would result in the conversion to heat of an even greater fraction of the outgoing thermal IR energy than occurs at present, which would thereby increase the average surface temperature well beyond 15°C. This phenomenon is sometimes referred to as the **enhanced greenhouse effect** (or *artificial global warming*) to distinguish its effects from the one that has been operating naturally for millennia.

The principal constituents of the atmosphere—N_2, O_2, and Ar—are incapable of absorbing IR light; the reasons for this will be discussed in the following section. The atmospheric gases that in the past have produced most of the greenhouse warming are water vapor (responsible for about two-thirds of the effect) and carbon dioxide (responsible for about one-quarter). Indeed, the absence of water vapor and of clouds in the dry air of desert areas leads to low nighttime temperatures there since so little of the outgoing IR is redirected back to the surface, even though the daytime temperatures are quite high on account of direct absorption of solar energy by the surface. More familiar to people living in temperate climates is the crisp chill in winter air on cloudless days and nights. Cloudy nights are usually warmer than clear ones because clouds reradiate IR that they have absorbed from surface emissions.

The greenhouse effect may be better understood by considering the following approximate model. Using physics, the temperature of an Earth that had no greenhouse gases in its air but was balanced with respect to incoming and outgoing energy would be $-18°C$, or 255 K. Since, according to the equation above, the rate of energy emission from such a planet would be $k (255)^4$, it follows that the rate of energy input from the Sun, whether or not the Earth's atmosphere contained greenhouse gases, would also be $k(255)^4$. Overall, the real Earth acts as if about 60% of the energy it emits as infrared light is eventually transmitted into space, the remainder being the fraction that is not only absorbed by greenhouse gases but that is also reradiated downward and further heats the surface and atmosphere. Thus the rate at which the Earth loses energy to space as IR is not simply kT^4, but rather $0.6\,kT^4$. Since we know that

rate of loss of energy from Earth = rate of energy input from Sun

it follows for the real Earth that

$$0.6\,kT^4 = k\,(255)^4$$

Taking the fourth root of both sides, we obtain an expression for the temperature:

$$T = (255)/0.6^{0.25}$$

so

$$T = 290\ \text{K}$$

From this model, the Earth's calculated surface temperature is 290 K, i.e., $+17°C$, an increase of 35 degrees by the operation of the natural greenhouse effect.

In reality, however, very little of the IR emitted at or near the Earth's surface escapes into space. Rather it is absorbed by the air close to the ground, then re-emitted. A simple model of the atmosphere that incorporates this effect is discussed in Box 6-1. The IR from the air close to the ground that is emitted upward is mainly absorbed by the next layer of air, which is heated by it, though to a lesser extent than is the layer underneath, and is partially re-emitted. With increasing altitude, the fraction of the IR received from the air lying below a given level is less and less likely to be absorbed, since the atmosphere becomes thinner and thinner. Thus more and more of the IR is likely to pass upward into space. Indeed, very little of the IR emitted into the upper troposphere is absorbed since the air is thin at such altitudes. Because less and less IR is absorbed with increasing altitude, less and less is degraded into heat; there is therefore a natural tendency for the air to cool the higher the altitude. In reality, other factors such as convection currents in air also play an important role in determining the decline in temperature with altitude. The temperature at the top of the troposphere, from which the emitted IR reaches outer space, is only $-18°C$, so overall the real Earth does radiate energy into space at the same temperature as was computed for the planet if it had no greenhouse gases. Thus the Earth emits the same amount of energy—equal to the amount absorbed from sunlight—into space with or without the operation of the greenhouse effect.

Earth's Energy Balance

The current energy inputs and outputs from the Earth—in watts (i.e., joules per second) per square meter of its surface and averaged over day and night, over all latitudes and longitudes, and over all seasons—are summarized

BOX 6-1	A Simple Model of the Greenhouse Effect

The calculation in the main text of the Earth's surface temperature assumed a specific value for the fraction of IR emitted from the surface that was transmitted through the atmosphere to outer space. However, this fraction—and the temperature—can be calculated from the physics of the situation. Consider a model Earth containing an atmosphere that consists of a single, uniform layer of air that is completely nontransparent to outgoing IR emitted from the surface, i.e., an atmosphere that absorbs all the IR and that converts it temporarily to heat. The layer of air itself is assumed to act as a blackbody that emits IR equally upward into space and downward back to the surface (Figure 1).

If a balance is to be achieved on Earth between incoming and outgoing energy, the air mass in the model must emit twice as much energy ($2X$) per second as is absorbed (X) from sunlight by the surface, since only half the air's energy escapes upward and is released into space. Define $rate_b$ as the total energy release rate from the air mass and $rate_a$ as the absorption rate from sunlight. Since we know that the rate of energy emission rises with the fourth power of the temperature, it follows that for any two Kelvin temperatures T_a and T_b involving the same type of blackbody, the ratio of rates of energy emission is given by

$$(rate_b/rate_a) = (T_b/T_a)^4$$

In our case, the rate ratio must be 2/1, and we know that T_a is 255 K. Thus

$$(T_b/255)^4 = 2.0$$

Taking square roots of both sides twice, we obtain

$$(T_b/255) = 2.0^{0.25} = 1.189$$

so

$$T_b = 303 \text{ K}$$

This simple model predicts the Earth's surface (and air) temperature to be 303 K, or 30°C, compared to the actual value of 15°C. The model is unrealistic and leads to an overestimation of the greenhouse effect, because it assumes that *all* the IR escaping from the surface is absorbed and that the atmosphere is completely uniform and acts exactly as a blackbody. Earth's actual surface temperature of 15°C is more consistent with a slightly more complicated model, in which about one-third of the IR emitted from the surface passes through the atmosphere unabsorbed, and about two-thirds is absorbed by the air and subsequently re-emitted. The most accurate model consists of a sequence of several layers of air, with temperatures decreasing with altitude, each layer acting as a blackbody.

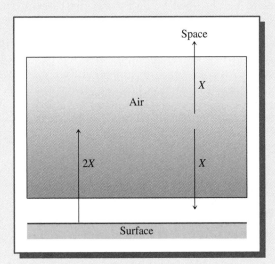

FIGURE 1 Energy released by the Earth's surface and absorbed and released by the atmosphere according to the model.

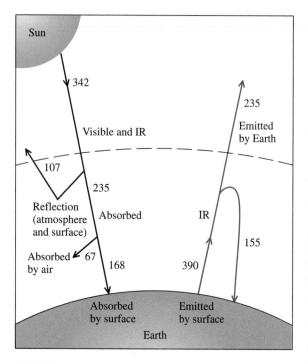

FIGURE 6-4 Globally and seasonally averaged energy fluxes to and from the Earth, in watts per square meter of surface. [Source: Data from Chapter 1 of J. T. Houghton et al., *Climate Change 1995—The Science of Climate Change* (Intergovernmental Panel on Climate Change) (Cambridge: Cambridge University Press, 1996).]

in Figure 6-4. A total of 342 watts/m^2 (W/m^2) are present in sunlight outside the Earth's atmosphere. Of this, 235 W/m^2 are absorbed by the atmosphere and the surface; this much energy must be re-emitted into space if the planet is to maintain a steady temperature. Because of the presence of greenhouse gases, however, emission of only 235 W/m^2 from the surface would not be sufficient to ensure this balance. Because absorption of IR by greenhouse gases heats the surface and lower atmosphere, the amount of IR released by them is increased. Given the current concentration of greenhouse gases in air, the balance is achieved and 235 W/m^2 escape from the top of the atmosphere into space if 390 W/m^2 are emitted from the surface, i.e., when 155 W/m^2 of IR do not escape into space.

Ironically, an increase in CO_2 concentration is predicted to cause a *cooling* of the stratosphere. This phenomenon occurs for two reasons.

• First, more outgoing thermal IR is absorbed at low altitudes (the troposphere), so less is left over to be absorbed by and warm the gases in the stratosphere.

• Second, at stratospheric temperatures, CO_2 emits more thermal IR upward to space and downward to the troposphere than it absorbs as photons—most of the absorption at these altitudes is due to water vapor and ozone—so increasing its concentration cools the stratosphere.

The observed cooling of the stratosphere has been taken to be a signal that the greenhouse effect is indeed undergoing enhancement.

Molecular Vibrations: Energy Absorption by Greenhouse Gases

Light is most likely to be absorbed by a molecule when its frequency almost exactly matches the frequency of an internal motion within the molecule. For frequencies in the infrared region, the relevant internal motions are the **vibrations** of the molecule's atoms relative to each other.

The simplest vibrational motion in a molecule is the oscillatory motion of two bonded atoms X and Y relative to each other. In this motion, called a

(a) Bond-stretching vibration

(b) Angle-bending vibration

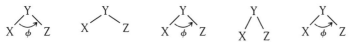

FIGURE 6-5 The two kinds of vibrations within molecules. Bond stretching (a) is illustrated for a diatomic molecule XY. The variable R represents the average value of the X–Y distance. In (b), the angle-bending vibration is shown for a triatomic molecule XYZ. The average XYZ angle is indicated by ϕ.

bond-stretching vibration, the X-to-Y distance increases beyond its average value R, then returns to R, then contracts to a lesser value, and finally returns to R, as illustrated in Figure 6-5a. Such oscillatory motion occurs in all bonds of all molecules under all temperature conditions, even at absolute zero. A huge number (about 10^{13}) of such vibrational cycles occur each second. The exact frequency of the oscillatory motion depends primarily on the type of bond—i.e., whether it is single or double or triple—and on the identity of the two atoms involved. For many bond types, e.g., the C—H bond in methane and the O—H bond in water, the stretching frequency does not fall within the thermal infrared region. The stretching frequency of carbon–fluorine bonds does, however, occur within the thermal infrared range; thus any molecules in the atmosphere with C—F bonds will absorb outgoing thermal IR light and enhance the greenhouse effect.

The other relevant type of vibration is an oscillation in the distance between two atoms X and Z bonded to a common atom Y but not bonded to each other. Such motion, called a **bending vibration,** alters the XYZ bond angle from its average value ϕ. All molecules containing three or more atoms possess bending vibrations. The oscillatory cycle of bond angle increase, followed by a decrease, and then another increase, etc., is illustrated in Figure 6-5b. The frequencies of many types of bending vibrations in most organic molecules occur within the thermal infrared region.

If infrared light is to be absorbed by a vibrating molecule, there must be a difference in the relative positions of the molecule's center of positive charge (its nuclei) and its center of negative charge (its electron "cloud") at some point during the motion. More compactly stated, in order to absorb IR light, the molecule must have a dipole moment during some stage of the vibration. Technically, there must be a *change* in the magnitude of the dipole moment during the vibration, but this is more or less guaranteed to be the case if there is a nonzero dipole moment at any point in the vibration. The positive and negative centers of charge coincide in free atoms and (by definition) in homonuclear diatomic molecules like O_2 and N_2, and the molecules have dipole moments of zero at all times in their stretching vibration. Thus argon gas, Ar, diatomic nitrogen gas, N_2, and diatomic oxygen, O_2, do not absorb IR light.

For carbon dioxide, during the vibratory motion in which both C—O bonds lengthen and shorten simultaneously, i.e., synchronously, there is at no time any difference in position between the centers of positive and negative charges, since both lie precisely at the central nucleus. Consequently, during this vibration, called the **symmetric stretch,** the molecule cannot absorb IR light. However, in the **antisymmetric stretch** vibration in CO_2, the contraction of one C—O bond occurs when the other is lengthening, or vice versa, so that during the motion the centers of charge no longer necessarily coincide. Therefore, IR light at this vibration's frequency *can* be absorbed since, at some points in the vibration, the molecule does have a dipole moment.

$$\overleftarrow{O}=C=\overrightarrow{O} \qquad\qquad \overrightarrow{O}=C=\overrightarrow{O}$$

symmetric
stretch   antisymmetric
stretch

Similarly, the bending vibration in a CO_2 molecule, in which the three atoms depart from a colinear geometry, is a vibration that can absorb IR light since the centers of positive and negative charge do not coincide when the molecule is nonlinear.

Molecules with three or more atoms generally have some vibrations that absorb IR, since even if their average shape is highly symmetric with a zero dipole moment, they undergo some vibrations that reduce this symmetry and produce a nonzero dipole moment. For example, CH_4 molecules have an average structure that is exactly tetrahedral, and hence a zero average dipole moment, because the polarities of the C—H bonds exactly cancel each other in this geometry. The zero dipole is maintained during the vibration in which all four bonds simultaneously stretch or contract. However, during the vibrational motions in which some of the bonds stretch while others contract, and those in which some H—C—H bond angles become greater than tetrahedral while others become less, the molecule has a nonzero dipole moment. Molecules of CH_4 undergoing such unsymmetrical vibrations can absorb infrared light.

PROBLEM 6-2

Deduce whether the following molecules will absorb infrared light due to internal vibrational motions:

(a) H_2 (b) CO   (c) Cl_2 (d) O_3 (e) CCl_4 (f) NO

PROBLEM 6-3

None of the four diatomic molecules listed in Problem 6-2 actually absorb much, if any, of the Earth's outbound light in the *thermal* infrared region. What does this imply about the frequencies of the bond-stretching vibrational motion of those molecules that can, in principle, absorb IR light?

The Major Greenhouse Gases

Carbon Dioxide: Absorption of Infrared Light

As stated previously, the absorption of light by a molecule occurs most efficiently when the frequencies of the light and of one of the molecule's vibrations match almost exactly. However, light of somewhat lower or higher frequency than that of the vibration is absorbed by a collection of molecules. This ability of molecules to absorb infrared light over a short range of frequencies rather than at just a single frequency occurs because it is not only the energy associated with vibration that changes when an infrared photon is absorbed; there is also a change in the energy associated with the rotation (tumbling) of the molecule about its internal axes. This **rotational energy** of a molecule can be either slightly increased or slightly decreased when IR light is absorbed to increase its **vibrational energy.** Consequently, photon absorption occurs at a slightly higher or lower frequency than that corresponding to the frequency of the vibration. Generally, the absorption tendency of a gas falls off rapidly for light frequency that lies farther and farther in either direction from the vibrational frequency.

The absorption spectrum for **carbon dioxide** in a portion of the infrared range is shown in Figure 6-6. For CO_2, the maximum absorption of light in the thermal infrared range occurs at a wavelength of 15.0 μm, which corresponds to a frequency of 2×10^{13} cycles per second (hertz). The absorption occurs at this particular frequency because it matches that of one of the vibrations in a CO_2 molecule, namely the OCO angle-bending vibration. Carbon dioxide also strongly absorbs IR light having a wavelength of 4.26 μm, which corresponds to the 7×10^{13} cycles per second (hertz) frequency of the antisymmetric OCO stretching vibration.

PROBLEM 6-4

Calculate the energy absorbed per mole and per molecule of carbon dioxide when it absorbs infrared light (a) at 15.0 μm and (b) at 4.26 μm. Express the per mole energies as fractions of that required to dissociate CO_2 into CO and atomic oxygen, given that the enthalpies of formation of the three gaseous species are -393.5, -110.5, and $+249.2$ kJ/mol, respectively. [*Hint: Recall the relationship between wavelength and energy in Chapter 1. Avogadro's constant =* 6.02×10^{23}.]

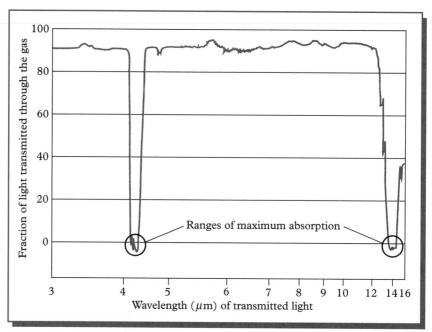

FIGURE 6-6 The infrared absorption spectrum for carbon dioxide. The scale for wavelength is linear when expressed in wavenumbers, which have units of cm^{-1}; wavenumber = 10,000/ wavelength in nm. [Source: Redrawn from A.T. Schwartz et al., *Chemistry in Context: Applying Chemistry to Society*, American Chemical Society (Dubuque, IA: Wm. C. Brown, 1994).]

The carbon dioxide molecules that are now present in air collectively absorb about half of the outgoing thermal infrared light with wavelengths in the 14–16-μm region, together with a sizable portion of that in the 12–14- and 16–18-μm regions. It is because of CO_2's absorption that the solid curve in Figure 6-7, representing the amount of IR light that actually escapes from our atmosphere, falls so steeply around 15 μm; the vertical separation between the curves is proportional to the amount of IR of a given wavelength that is being absorbed rather than escaping. Further increases in the CO_2 concentration in the atmosphere will prevent more of the remaining IR from escaping, especially in the "shoulder" regions around 15 μm, and will further warm the air. (Although carbon dioxide also absorbs IR light at 4.3 μm due to the antisymmetric stretching vibration, there is little energy emitted from the Earth at this wavelength—see Figure 6-1—so this potential absorption is not very important.)

Carbon Dioxide: Past Concentration and Emission Trends

Measurements of air trapped in ice-core samples from Antarctica indicate that the atmospheric concentration of carbon dioxide in preindustrial times (i.e., before about 1750) was about 280 ppm. The concentration had increased by one-third, to 382 ppm, by 2006. A plot of the increase in the annual atmospheric CO_2 concentration over time is shown in Figure 6-8.

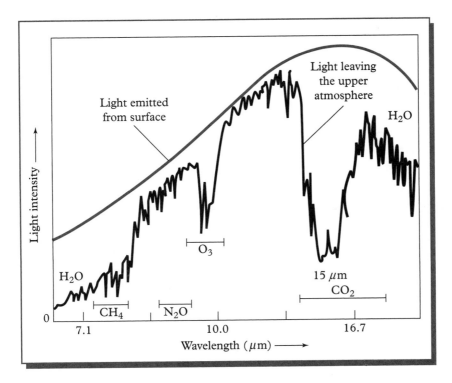

FIGURE 6-7 Experimentally measured intensity (black curve) of thermal IR light leaving the Earth's surface and lower atmosphere (above the Sahara desert) compared with the theoretical intensity (green curve) that would be expected without absorption by atmospheric greenhouse gases. The regions in which the various gases have their greatest absorption are indicated. [Source: E. S. Nesbit, *Leaving Eden* (Cambridge: Cambridge University Press, 1991).]

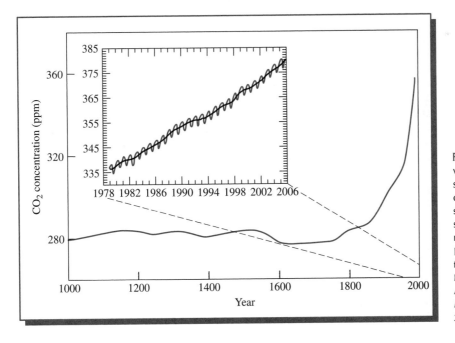

FIGURE 6-8 The historic variation in the atmospheric concentration of carbon dioxide. The insert shows the trend, with seasonal fluctuations, in recent times. [Source: Main graph: Adapted from J. L. Sarmiento and N. Gruber, "Sinks for Anthropogenic Carbon," *Physics Today* 55 (August 2002): 30; Insert: NOAA.]

The insert to the figure shows details of the increase in recent times. In the period from 1975 to 2000, the concentration grew at an average annual rate of about 0.4%, or 1.6 ppm—almost double that of the 1960s. The rate of increase in the first half-decade of the twenty-first century rose to about 2.0 ppm annually.

The seasonal fluctuations in the CO_2 concentrations are due to the spurt in the growth of vegetation in the spring and summer, which removes CO_2 from air, and the vegetation decay cycle in fall and winter, which increases it. In particular, huge quantities of CO_2 are extracted from the air each spring and summer by the process of plant photosynthesis:

$$CO_2 + H_2O \xrightarrow{\text{sunlight}} O_2 + \text{polymeric } CH_2O$$

The term *polymeric CH$_2$O* used for the product in this equation is an umbrella word for plant fiber, typically the cellulose that gives wood its mass and bulk. The CO_2 "captured" by the photosynthetic process is no longer free to function as a greenhouse gas—or as any gas—while it is packed away in this polymeric form. The carbon that is trapped in this way is called **fixed carbon.** However, the biological decay of this plant material, the reverse of the reaction, which occurs mainly in the fall and winter, frees the withdrawn carbon dioxide. Notice that the global carbon dioxide fluctuations follow the seasons of the Northern Hemisphere, since there is so much more land mass—and hence much more vegetation—there compared to the Southern Hemisphere.

Much of the considerable increase in anthropogenic contributions to the increase in carbon dioxide concentration in air is due to the combustion of **fossil fuels**—chiefly coal, oil, and natural gas—that were formed eons ago when plant and animal matter was covered by geological deposits before it could be broken down by air oxidation.

On average, each person in the industrial countries is responsible for the release of about 5 metric tons (a metric ton is 1000 kg, i.e., 2200 lb, whereas a conventional ton is 2000 lb) of CO_2 from carbon-containing fuels each year! There is considerable variation in the per capita releases among different industrialized countries; this is discussed in Chapter 7. Some of the per capita carbon dioxide output is direct, e.g., that released as gases when vehicles are driven and homes are warmed by burning a fossil fuel. The remainder is indirect, arising when energy is used to produce and transport goods; heat and cool factories, classrooms, and offices; produce and refine oil—in fact, to accomplish virtually any constructive economic purpose in an industrialized society. This topic is also discussed in greater depth in Chapter 7.

A significant amount of carbon dioxide is added to the atmosphere when forests are cleared and the wood burned to provide land for agricultural use. This sort of activity occurred on a massive scale in temperate climate zones in past centuries (consider the immense deforestation that accompanied the settlement of the United States and southern Canada) but has now shifted largely to the tropics. The greatest single amount of current deforestation occurs in Brazil and involves both rain forest and moist deciduous forest, but the annual rate of deforestation on a percentage basis is actually greater in Southeast Asia and Central America than in South America. Overall, deforestation accounts for about one-quarter of the annual anthropogenic release of CO_2, the other three-quarters originating mainly in the combustion of fossil fuels. Notwithstanding forestry harvesting operations, the total amount of carbon contained in the forests of the Northern Hemisphere (including their soils) is increasing, and in recent decades the annual increment approximately equaled the decreases in stored carbon cited above in Asia and South and Central America.

PROBLEM 6-5

Carbon dioxide is also released into the atmosphere when calcium carbonate rock (limestone) is heated to produce the quicklime, i.e., calcium oxide, used in the manufacture of cement:

$$CaCO_3(s) \longrightarrow CaO(s) + CO_2(g)$$

Calculate the mass, in metric tons, of CO_2 released per metric ton of limestone used in this process. What is the mass of carbon that the air gains for each gram of carbon dioxide that enters the atmosphere? Note that at least as much carbon dioxide is released from combustion of the fossil fuel needed to heat the limestone as is released from the limestone itself.

The growth of the total annual *emissions*, in terms of the mass of carbon, of carbon dioxide from fossil-fuel combustion and cement production since the start of the Industrial Revolution is shown by the uppermost (black) curve in Figure 6-9. The contributions to this total are shown by the other curves. Historically, the emission rate in the second half of the twentieth century grew rapidly, the rate of increase being about five times greater than that in the first half. The annual growth rate in emissions from 2000 to 2005 rose to 3%, compared to 1% in the 1990s, due mainly to a rebound in coal production (gray curve) and continuing growth in oil (green curve) and natural gas (dashed curve) usage.

Carbon Dioxide: Atmospheric Lifetime and Fate of Its Emissions

The lifetime of a carbon dioxide molecule emitted into the atmosphere is a complicated measurement since, in contrast to most gases, it is not decomposed chemically or photochemically. On average, within a few years of its release into the air, a CO_2 molecule will likely dissolve in surface seawater or be absorbed by a growing plant. However, many such carbon dioxide molecules are released back into the air a few years later on average, so this disposal is only a *temporary*

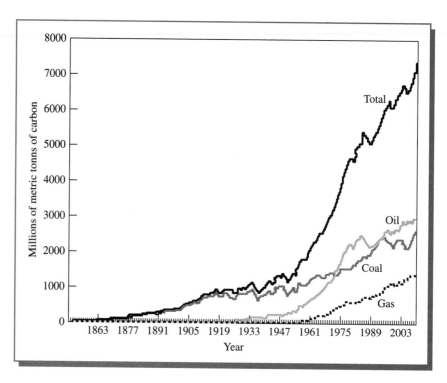

FIGURE 6-9 Annual global CO$_2$ emissions since the Industrial Revolution. The black line gives total emissions from fossil-fuel combustion and cement manufacture. The contributions from solids (mainly coal) are shown by the gray line, from liquids (mainly petroleum) by the green line, and from gases (mainly natural gas) by the dashed black line. [Source: U.S. Department of Energy Carbon Dioxide Information Analysis Center, cdiac.ornl.gov/trends/emis/glo.htm]

sink for the gas. The only *permanent* sink for it is deposition in the deep waters of the ocean and/or precipitation there as insoluble calcium carbonate. However, the top few hundred meters of seawater mix slowly with deeper waters; thus carbon dioxide that is newly dissolved in surface water requires hundreds of years to penetrate to the ocean depths. Consequently, although the oceans will ultimately dissolve much of the increased CO$_2$ now in the air, the time scale associated with this permanent sink is very long, hundreds of years.

Because the processes involving the interchange of carbon dioxide among the air, the biomass, and shallow ocean waters and between shallow and deep seawater are complicated, it is not possible to cite a meaningful average lifetime for the gas in air alone. Rather we should think of new CO$_2$ fossil-fuel emissions as being rather quickly allocated among air, the shallow ocean waters, and biomass, with interchange among these three compartments occurring continuously. Then slowly, over a period of many decades and even centuries, almost all this new carbon dioxide will eventually enter its final sink, the deep ocean. In effect, the atmosphere rids itself of almost half of any new carbon dioxide within a decade or two but requires a much longer period of time to dispose of the rest. It is commonly quoted as taking 50 to 200 years for the carbon dioxide level to adjust to its new equilibrium concentration if a source of it increases. In summary, the effective lifetime of

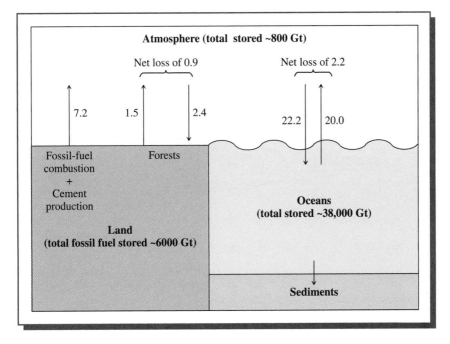

FIGURE 6-10 Annual fluxes of CO_2 to and from the atmosphere, in units of megatonnes of carbon. The total amounts stored in various locations are shown in bold. Note that the values for the air/ocean interchange include natural as well as anthropogenic carbon. [Data source: UNESCO SCOPE Policy Briefs 2006 #2. *The Global Carbon Cycle.*]

additional CO_2 in the atmosphere should be considered to be long, on the order of many decades or centuries, rather than the few years required for its initial dissolution in seawater or absorption by biomass.

The annual inputs and outputs of anthropogenic carbon dioxide to and from our atmosphere, as of the early 2000s, are summarized in Figure 6-10. (Releases and absorption by natural processes are overall in balance and are not included in the diagram.) Fossil-fuel combustion and cement production released 7.2 gigatonnes (Gt, i.e., billions of tonnes, equivalent to petagrams, 10^{15} g) of the carbon component (only) of CO_2 into the air, of which 4.4 Gt (about 60%) did not find a sink. The upper layers of the ocean absorbed about 22 Gt of carbon but released only about 20 Gt, giving a net absorption by this principal sink of almost 2 Gt. The carbon released by slash-and-burn tropical deforestation and other land-use changes amounted to about 1 Gt less than that absorbed by forest growth and soil storage. Because overall more than half the anthropogenic CO_2 emissions are quickly removed, over the short and medium term the gas continues to accumulate in the atmosphere.

The variations over the last century and a half in the various annual sources and sinks for CO_2 are summarized in Figure 6-11. Notice the year-to-year variations in the amount of the gas absorbed by the oceans and especially in that absorbed by land sinks (biomass). Although the fraction of the new emissions that remains in the atmosphere undergoes substantial variations from year to year, its average increment is increasing with time.

FIGURE 6-11 Annual
fluxes of anthropogenic
CO_2 to and from various
sources and sinks from
1850 to 2005. Note that
the unit of picograms
(10^{12} g) is equivalent to
the megatonne unit used
in Figure 6-10, since
1 megatonne = 10^6 tonnes,
1 tonne = 1000 kg, and
1 kg = 1000 g.)
[Source: M. Raupach (Global
Carbon Project), *Carbon in
the Earth System: Dynamics
and Vulnerabilities* (Beijing,
November, 2006).]

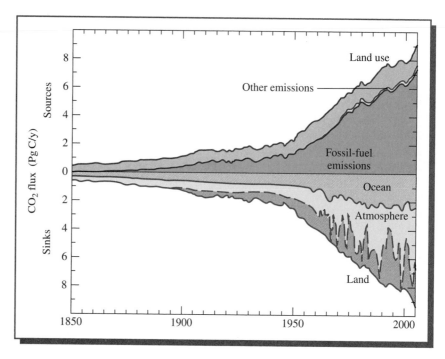

The increase in growth rate of certain types of trees due to the increased
concentration of carbon dioxide in the air is called **CO_2 fertilization.** Some
scientists suspect that the rate of photosynthesis is speeding up as the level of
CO_2 and the air temperature increase and that the formation of greater
amounts of fixed carbon represents an important sink for the gas. Indeed, an
increase in the biomass of northern temperate forests is the most likely sink
to account for the annual atmospheric CO_2 loss for which scientists had previously been unable to assign a cause. This increased activity in photosynthesis has been confirmed by satellite data for the region between 45°N and
70°N. Boreal (evergreen) forests of the Northern Hemisphere currently store
almost 1 Gt of carbon into standing biomass alone. Much of the increase in
the biomass of temperate forests at high latitudes occurs in the soil, especially
as peat. The anthropogenic releases of CO_2 amount only to about 4% of the
enormous amounts produced by nature, so a very small variation in the rate
at which carbon is absorbed into biomass could have a large effect on the
residual amount of CO_2 that accumulates in the atmosphere. Unfortunately,
scientists still do not completely understand the global carbon cycle. As Figure 6-8 indicates, however, there is no doubt that the atmospheric CO_2 concentration is increasing.

PROBLEM 6-6

(a) Given that the atmospheric burden of carbon (as CO_2) increases by about 4.7 Gt annually, calculate the increase in the ppm concentration of carbon dioxide that this brings about. (b) Given that its total concentration was 382 ppm in 2006, calculate the total mass of CO_2 that was present in the air. After converting to mass of carbon, does your answer agree with the value listed in Figure 6-10? Note that the mass of the atmosphere $= 5.1 \times 10^{21}$ g and that air's average molar mass $= 29.0$ g/mol. [*Hint: Express the amounts of CO_2 and air as moles, and recall the definition of ppm units in moles.*]

 ## Green Chemistry: Supercritical Carbon Dioxide in the Production of Computer Chips

In this example of green chemistry, we see how waste CO_2—which would normally be vented to the atmosphere—can be put to good use as a solvent. We will also see how using CO_2 as a solvent pays additional environmental dividends in terms of both energy and resource conservation and reduction of wastes.

As technology relentlessly pervades our planet, the demand for integrated circuits (ICs) and computer chips increases dramatically each year. Computer chips are used in almost any electronic device one can imagine, including telephones, televisions, radios, automobiles, trucks, computers, airplanes, rockets, smart bombs, calculators, and cameras. It is estimated that the combination of the average personal computer, keyboard, monitor, and printer has a mass of about 25 kg and contains about 9 g of silicon and metal in the ICs that are the heart of each computer.

The manufacture of computers, other electronic devices, and chips involves high-tech, high-paying, highly skilled, and highly sought-after jobs. Facilities involved in these activities are considered by most people to be "clean" industries, especially when compared to the automobile and chemical industries. It is a little-known fact, except to those who work in the field or study the chip-manufacturing process, that chip making actually creates more waste than any other process involved in the manufacture of computers and is very energy intensive! By some measures, chip manufacturing is orders of magnitude more wasteful and polluting than the production of automobiles. It is estimated that the fabrication of the chips in your computer generated about 196 kg of waste (4500 times the mass of the average chip) and used about 10,600 L of water. The ratio of the mass of the materials (chemicals and fossil fuels) needed to produce a chip to the mass of the chip is estimated at 630:1, while the analogous ratio for the production of an automobile is approximately 2:1. Consequently, there are ongoing efforts to find less resource-intensive and less wasteful methods of chip production.

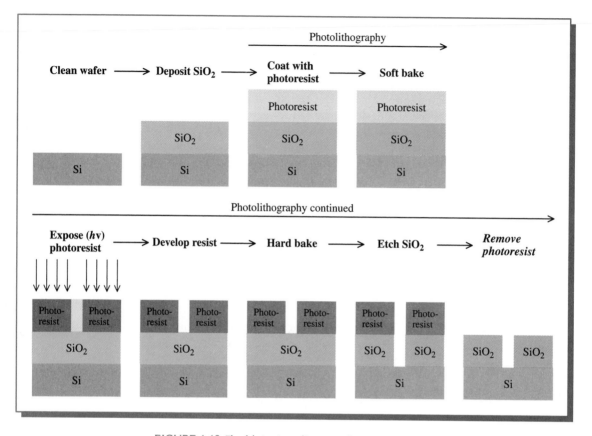

FIGURE 6-12 The fabrication of integrated circuits. [Source: L. Rothman, G. Jacobson, and C. Taylor, "Supercritical CO_2 Resist Remover–SCORR," a proposal submitted to the Presidential Green Chemistry Challenge Awards Program, 2002.]

The process of producing a 2-g computer chip entails many steps and requires 72 g of chemicals, 32 L of water (mostly for rinsing), and 700 g of process gases. For production and use over a four-year lifetime, a 2-g chip requires 1.6 kg of fossil fuels. A typical chip-manufacturing facility uses millions of liters of highly purified water per month. A few of these steps are outlined in Figure 6-12. The process begins with the mechanical or chemical cleaning of the surface of highly purified silicon, followed by deposition of silicon dioxide; then a process known as photolithography defines the shape and pattern of individual components on an IC.

Photolithography begins with the deposition of a photoresist polymer, followed by baking and exposure of selected areas of the polymer to light. The light causes the polymer to cross-link, i.e., form bonds that link the polymer chains to one another at many positions along each chain (see Figure 6-13). The chip is then developed (a process that removes the photoresist polymer from the unexposed areas) and hard baked, the SiO_2 is etched, and the

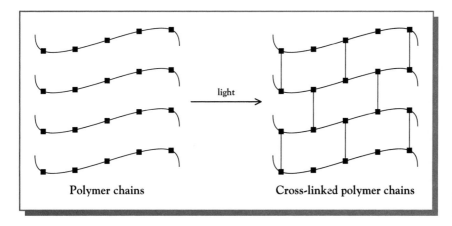

Polymer chains   **Cross-linked polymer chains**

FIGURE 6-13 The cross-linking of polymer chains.

remaining photoresist polymer is removed, creating a pattern on the surface of the chip. The removal of the photoresist is accomplished with large amounts of aqueous solutions of strong acid (sulfuric or hydrochloric) or base, or by use of organic solvents (halogenated or polycyclic aromatics). The chip is then rinsed several times with copious amounts of highly purified water and dried with alcohol. The removal of the photoresist is very resource- and energy-intensive and creates large amounts of waste. The layering, developing, and etching process is repeated several times for each chip.

Scientists at Los Alamos National Laboratories in New Mexico and SC Liquids in Nashua, New Hampshire, were awarded a Presidential Green Chemistry Challenge Award in 2002 for their development of a new process for removing photoresist in chip manufacturing, known as *SCORR* (supercritical carbon dioxide resist remover). The process employs supercritical carbon dioxide (see Box 6-2) as the solvent for removal of the photoresist (the last step of Figure 6-12). The use of SCORR offers several environmental benefits over traditional methods, including the following:

- The rinse step is no longer necessary, thus eliminating the need for millions of liters of highly purified water, the energy required to produce this water, and the associated wastewater. This also reduces the amount of fossil fuel required to produce the highly purified water and the accompanying formation of carbon dioxide.

- The need for (and waste from) hazardous and toxic chemicals such as strong acids or bases or organic solvents in the photoresist removal step is eliminated or reduced. This also enhances worker safety.

- The need for alcohols used for drying after the aqueous rinse step is eliminated.

- The carbon dioxide is recovered after each use and reused.

- The only waste left after evaporation (and recovery) of the carbon dioxide is the spent photoresist, which is unregulated.

BOX 6-2	Supercritical Carbon Dioxide

The **supercritical fluid** state of matter is produced when gases or liquids are subjected to very high pressures and, in some cases, to elevated temperatures. At pressures and temperatures at or beyond the **critical point,** separate gaseous and liquid phases of a substance no longer exist. Under these conditions, only the supercritical state, with properties that lie between those of a gas and those of a liquid, exists. For carbon dioxide, the critical pressure is 72.9 atm and the critical temperature is only 31.3°C, as illustrated in the phase diagram in Figure 1. Depending upon exactly how much pressure is applied, the physical properties of the supercritical fluid vary between those of a gas (relatively lower pressures) and those of a liquid (higher pressures); the variation of properties with P or T is particularly acute near the critical point. Thus the density of supercritical carbon dioxide varies over a considerable range, depending upon how much pressure (beyond 73 atm) is applied to it.

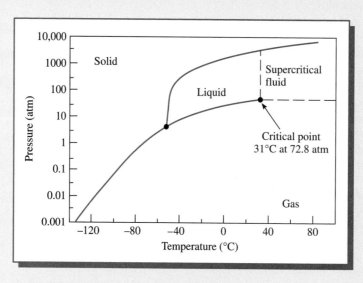

FIGURE I Phase diagram for carbon dioxide.

The carbon dioxide used in this process can be obtained as a waste by-product of other processes (as indicated in Chapter 1 during the discussion of using carbon dioxide as a blowing agent). Processes such as producing ammonia and drilling for natural gas yield large amounts of this gas, which would normally be released to the atmosphere and add to its concentration of carbon dioxide. If we can capture this "unwanted" by-product and find constructive (and environmentally sound) uses for carbon dioxide, then we have not only prevented its release into the atmosphere; we may also have reduced our reliance on valuable resources and prevented the formation of other pollutants. SCORR, the use of carbon dioxide as a blowing agent (green chemistry section, Chapter 2), and the use of carbon dioxide as a solvent for various cleaning purposes (green chemistry section, Chapter 3) are all examples of this. In general, chemists are seeking ways to find beneficial uses for by-products of other processes and reactions that would normally be considered waste and would add to the environmental burden of the planet. The use of carbon dioxide as a blowing agent and solvent for cleaning and photoresist removal offers three significant examples of this paradigm.

Employing the SCORR process yields several additional advantages. As the architecture of computer chips continually becomes smaller, water (because of its high surface tension) will no longer be able to penetrate these small spaces. Supercritical fluids have low viscosity, low surface tension, and high diffusivity. Because of these properties, they are ideal for cleaning rough, irregular surfaces with small openings. Supercritical carbon dioxide offers the answer to the cleaning problems associated with the smaller features of the new generations of chips. Other advantages of the SCORR process include the facts that (1) cleaning times are cut in half, (2) eliminating the rinse step allows for greater throughput (more chips in less time), and (3) carbon dioxide is less costly than conventional solvents.

Water Vapor: Its Infrared Absorption and Role in Feedback

Water molecules, always abundant in air, absorb thermal IR light through their H—O—H bending vibration; the peak in the spectrum for this absorption occurs at about 6.3 μm. As a consequence, almost all the relatively small amount of outgoing IR in the 5.5- to 7.5-μm region is intercepted by water vapor (see Figure 6-7). (The symmetric and antisymmetric stretching vibrations for water occur near 2.7 μm, outside the thermal IR region. The symmetric stretch in a symmetric but nonlinear molecule like H_2O does absorb IR.) Absorption of light leading to increases in the rotational energy of water molecules, without any change in vibrational energy, removes thermal infrared light of 18-μm and longer wavelengths. In fact, water is the most important greenhouse gas in the Earth's atmosphere, in the sense that it produces more greenhouse warming than does any other gas, although on a per molecule basis it is a less efficient IR absorber than is CO_2.

Although human activities, such as the burning of fossil fuels, produce water as a product, the concentration of water vapor in air is determined primarily by temperature and by other aspects of the weather. Virtually all the H_2O in the troposphere arises from the evaporation of liquid and solid water on the Earth's surface and in clouds. The rate at which water evaporates and the maximum amount of water vapor that an air mass can hold both increase sharply with increasing temperature. Indeed, the equilibrium vapor pressure of liquid water increases exponentially with temperature. The rise in air temperature that is caused by increases in the concentration of the other greenhouse gases, and by other global warming factors, heats the surface water and ice, thereby causing more evaporation to occur. Indeed, the average atmospheric content of water vapor has increased since at least the 1980s.

The consequent increase in the water vapor concentration from global warming due to increased CO_2, etc. produces an *additional* amount of global warming due to $H_2O(g)$ that is comparable in magnitude to the original amount due to the other greenhouse gases, because water vapor is a greenhouse gas! This behavior of water is an example of the general phenomenon called **positive feedback:** *The operation of a phenomenon produces a result that itself further amplifies the result*. Feedback is a reaction to change; with positive feedback, the reaction accelerates the pace of future change. On the other hand, a system whose output reduces the subsequent level of output displays **negative feedback.** An example of negative feedback from daily life is the attempt by a business to raise its profits by increasing its prices. However, the rise in price often results in a reduction in demand for the item of concern, and the rise in profits is less than anticipated. (No value judgment as to the desirability of the effect is implied by the terms *positive* and *negative*; only the increase or decrease in the pace of change is meant.)

Since it comes about as an indirect effect of increasing the levels of other gases, and since it is not within our control, the warming increment due to water is usually apportioned without further comment into the direct warming effects of the other gases. Consequently, water is not usually listed explicitly among gases whose increasing concentrations are enhancing the greenhouse effect.

Water in the form of liquid droplets in clouds also absorbs thermal IR. However, clouds also reflect some incoming sunlight, both UV and visible, back into space. It is not yet clear whether the additional cloud cover produced by increasing the atmospheric water content will have a net positive or a net negative contribution to global warming. Clouds over tropical regions are known to have a zero net effect on temperature, but those in northern latitudes produce a net cooling effect since their ability to reflect sunlight outweighs their ability to absorb IR. Thus, if increased air temperatures produce more of the latter type of cloud, the enhancement of global warming by the greenhouse effect would be damped. However, no one is sure that additional northern cloud cover would occur at the same altitudes and act in the same manner as do the current clouds. Overall, the net effect of clouds on global warming is still subject to some uncertainty.

The Atmospheric Window

As a result of absorption of IR light of other wavelengths, mainly by carbon dioxide, methane, and water, it is essentially only infrared light from 8 to 13 μm that escapes the atmosphere efficiently (see Figure 6-7). Since light of these wavelengths passes largely unimpeded, this portion of the spectrum is called the **atmospheric window.**

    The injection into the atmosphere, even in trace amounts, of gases that can absorb thermal IR light will lead to additional global warming, i.e., to enhancing the greenhouse effect. Particularly serious are pollutant gases that absorb thermal IR in the atmospheric window region, since the absorption by H_2O and CO_2 in other regions is already so great that there remains little such light for trace gases to absorb. In particular, the fraction of light absorbed by a gas as it passes through the atmosphere is logarithmically related to its concentration, C (Beer–Lambert law). Thus the additional global warming produced by carbon dioxide is logarithmically related to the increases in its concentration. However, because logarithmic functions are almost linear near $C = 0$, the warming produced by trace gases is linearly proportional to their concentration increases (see Additional Problem 5).

    In considering which potential pollutants might contribute to global warming, recall that we can dismiss free atoms and homonuclear diatomic molecules, as they cannot absorb IR light. Heteronuclear diatomic molecules such as CO and NO are also not of direct concern since their only vibration—bond stretching—has a frequency that lies outside the thermal IR region. In general, however, most long-lived gases consisting of molecules with three or more atoms are of concern since they possess many vibrations that absorb IR, one or more of which usually fall in the thermal infrared region. The important trace greenhouse gases, i.e., those whose concentration is small in absolute terms but whose ability even at these levels to warm the air is substantial, are detailed below, following a discussion of their average lifetimes in the atmosphere.

Atmospheric Residence Time

The extent to which a substance accumulates in some compartment of the environment, such as the atmosphere, depends upon the rate, R, at which it is received from the source and the mechanism by which it is eliminated, i.e., its sink. Commonly, the rate of elimination via the sink is directly proportional to the concentration, C, of the substance in the organism or environmental compartment. In chemistry, this is known as a *first-order* relationship, since the power to which the independent variable is raised is unity. If the rate constant for the elimination process is defined as k, the rate of elimination is kC:

$$\text{rate of intake} = R$$

$$\text{rate of elimination} = kC$$

In some cases, such as the atmospheric sink for methane, a reaction involving a second substance is involved, and k incorporates the steady-state concentration of this other substance.

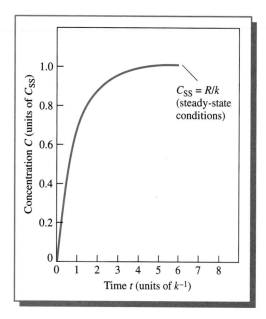

FIGURE 6-14 Increase in concentration with time to eventually reach the steady-state value, C_{SS}.

If none of the substance is initially present, that is, if $C_0 = 0$, then initially the rate of elimination must be zero. The concentration then builds up solely due to its input or ingestion, as illustrated near the origin in Figure 6-14. However, as C rises, the rate of elimination rises and increases since it is proportional to C; eventually it matches the rate of intake if R is a constant. Once this equality is achieved, C does not vary thereafter; it is in a steady state, which as we saw in Chapter 1 is defined as the state for which $dC/dt = 0$. Since under these steady-state conditions

$$\text{rate of elimination or loss} = \text{rate of intake}$$

$$kC = R$$

It follows that the steady-state value for the concentration, C_{SS}, is

$$C_{SS} = R/k$$

Often the speed of elimination or loss of a substance is discussed in terms of the **half-life period,** $t_{0.5}$, the length of time required for half of it to decay under the condition that all new input has now ceased. Under the latter condition, we know

$$dC/dt = -kC$$

Bringing to the left side of the equation all the terms involving C and putting to the right side the time dependence, we have

$$dC/C = -kdt$$

If we integrate both sides of the equation, we can find how C changes with time:

$$\int dC/C = -k \int dt$$

Performing the integration, we obtain

$$\ln C = -kt + \text{constant}$$

Thus we see that the logarithm of the concentration will decline linearly with time for the substance. At time $t = 0$, we have $\ln C = $ a constant, so we see that the (integration) constant equals the logarithm of the original concentration, C_0, the concentration at $t = 0$. Thus

$$\ln C - \ln C_0 = -kt$$

From the property of logarithms that $\log x - \log y = \log(x/y)$, we obtain the simpler equation

$$\ln(C/C_0) = -kt$$

or, in exponential form,

$$(C/C_0) = e^{-kt}$$

It is convenient to discuss the rate of decline of a substance in terms of its half-life. Substituting $C = 0.5C_0$ into the logarithmic equation above yields

$$\ln(0.5C_0/C_0) = -kt_{1/2}$$

But $\ln(0.5) = -0.693$, so we obtain

$$t_{1/2} = 0.693/k$$

We can use this final result to advantage to substitute for k in our equation for the steady-state concentration a chemical achieves when it is being both created and destroyed:

$$C_{SS} = R/k$$

We obtain

$$C_{SS} = R\, t_{0.5}/0.693 \quad \text{or} \quad C_{SS} = 1.44R\, t_{0.5}$$

Clearly, *the longer the half-life of a substance in its elimination process, the higher the steady-state accumulation level* it will achieve. The variations with time of concentrations and rates for systems of this type illustrated in Figure 6-14 are for the specific case where R and k (and hence C) are expressed in units of C_{SS}, and time t is expressed in units of k.

Every atmospheric gas that is present at or near a steady state has its own characteristic **residence time**, t_{avg}, which equals the average amount of time one of its molecules exists in air before it is removed by one means or another. Also known as the *average lifetime*, the t_{avg} of a substance is equal mathematically to the time required for its overall concentration to fall to $1/e$ times its initial value, where e is the base of natural logarithms. Since at that time $C = C_0/e$, then

$$\ln\left(\frac{C_0/e}{C_0}\right) = -kt_{avg}$$

But $\ln(1/e) = -1$, so

$$t_{avg} = 1/k$$

Substituting this expression for k into C_{SS} gives us $C_{SS} = R/t_{avg}$. Sometimes this expression is more useful in the form

$$t_{avg} = C_{SS}/R$$

since it tells us the lifetime of a substance if we know its steady-state concentration and rate of input into the environment.

In order to assess the impact of any substance on the enhanced greenhouse effect, it is necessary to know how long the substance is expected to remain in the atmosphere, since the longer its atmospheric lifetime, the greater will be its total effect. Thus, e.g., if the steady-state atmospheric concentration of a gas is 6.0 ppm and if its global rate of input, as determined by dividing the yearly amount of input by the volume of the atmosphere, is 2.0 ppm/year, then according to the preceding equation, its average residence time is 6.0 ppm/2.0 ppm/year, or 3.0 years.

The residence times of greenhouse gases such as nitrous oxide and the CFCs are many decades, so the influence of recent emissions of them will extend over long periods of time. In contrast, methane has a residence time of only about a decade.

The analysis above is applicable only to substances having a single first-order sink process. Thus it does not apply to carbon dioxide, for example, since this gas has many different sinks (dissolution in the oceans, absorption by plant matter, etc.) and sources.

PROBLEM 6-7

If the average steady-state residence time of a trace atmospheric gas is 50 years and its input rate is 2.0×10^6 kg/year, what is the total amount of it in the atmosphere?

PROBLEM 6-8

The steady-state concentration of an atmospheric gas of molar mass 42 g/mol is 7.0 μg/g of air and its residence time is 14 years. What is the annual total release of the gas into the atmosphere as a whole? See Problem 6-6 for additional data.

Other Greenhouse Gases

Methane: Absorption and Sinks

After carbon dioxide and water, **methane**, CH_4, is the next most important greenhouse gas. A methane molecule contains four C—H bonds. Although C—H bond-stretching vibrations occur well outside the thermal IR region,

HCH bond-angle-bending vibrations absorb at 7.7 μm, near the edge of the thermal IR window; consequently, atmospheric methane absorbs IR in this region.

In contrast to the century-long lifetime of carbon dioxide emissions, molecules of methane in air have an average lifetime of only about a decade. As discussed in Chapters 3 and 5, the dominant sink for atmospheric methane, accounting for almost 90% of its loss from air, is its reaction with molecules of **hydroxyl,** OH, the very reactive gas present in air in tiny concentration:

$$CH_4 + OH \longrightarrow CH_3 + H_2O$$

This reaction is the first step of a sequence that transforms methane ultimately to CO and then to CO_2 (see Chapter 5 for details).

$$CH_4 \longrightarrow \longrightarrow CH_2O \longrightarrow \longrightarrow CO \longrightarrow \longrightarrow CO_2$$

The yearly loss of methane by this reaction currently amounts to about 507 Tg (where 1 Tg, or teragram, is 10^{12} g—1 million metric tons), and the net sink from all processes amounts to about 577 Tg per year.

The other two sinks for methane gas are its reaction with soil and its loss to the stratosphere, to which a few percent of emissions eventually rise. Methane reacts there with OH, or atomic chlorine or bromine, or excited atomic oxygen; reaction with the latter produces hydroxyl radicals and eventually water molecules:

$$O^* + CH_4 \longrightarrow OH + CH_3$$

$$OH + CH_4 \longrightarrow H_2O + CH_3$$

Stratospheric water vapor acts as a significant greenhouse gas. About one-quarter of the total global warming caused by increased methane emissions is not brought about directly, but rather is due to this effect in the stratosphere whereby the region's water content is increased. Due to decreased levels of ozone and increased levels of carbon dioxide, the stratosphere has undergone cooling in recent decades; the increase in water vapor has reduced the amount of this cooling, thus contributing to the warming of the atmosphere overall.

Per molecule, increasing the amount of methane in air causes a much larger warming effect than does adding more carbon dioxide, since each CH_4 molecule is much more likely to absorb a thermal IR photon that passes through it than does a CO_2 molecule. The effect of the methane is restricted to the first decade or two after its release, however, because it is highly likely to be oxidized during this period. When considered over the century after its emission, a kilogram of methane is still about 23 times more effective in raising air temperature than the same mass of carbon dioxide; the ratio is about three times that value over the first 20 years. However, because CO_2 is so long-lived in air and its concentration has increased 80 times more, methane has been less important in heating the atmosphere. To date, methane is

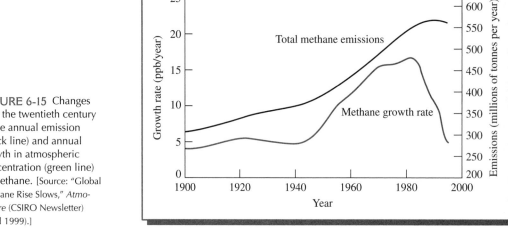

FIGURE 6-15 Changes over the twentieth century in the annual emission (black line) and annual growth in atmospheric concentration (green line) of methane. [Source: "Global Methane Rise Slows," *Atmosphere* (CSIRO Newsletter) (April 1999).]

estimated to have produced about one-third as much global warming as has carbon dioxide.

Methane: Emission Sources

About 70% of current methane emissions are anthropogenic in origin. The manner in which total methane *emissions* rose over the last century is illustrated by the black line in Figure 6-15. As in the case of carbon dioxide, post–World War II rates increased annually much more quickly than had been the case before. In the last 20 years, however, the emission rate for methane has leveled off (Figure 6-15).

Most of the methane produced from plant decay results from the process of **anaerobic decomposition,** which is decomposition of formerly living matter in the absence of air, i.e., under oxygen-starved conditions. This process converts cellulose (approximate empirical formula CH_2O) into methane and carbon dioxide:

$$2\,CH_2O \longrightarrow \longrightarrow CH_4 + CO_2$$

Anaerobic decomposition occurs on a huge scale where plant decay occurs under waterlogged conditions, e.g., in natural wetlands such as swamps and bogs and in rice paddies. Indeed, the original names for methane were *swamp gas* and *marsh gas*. Wetlands are the largest *natural* source of methane emissions, though emissions from this source have decreased sharply over the past centuries as wetlands have been drained. The huge increase in rice production over the same period has presumably led to correspondingly large increases in methane emissions from this source.

The expansion of wetlands that occurs by the deliberate flooding of land to produce more hydroelectric power adds to the total natural emissions of the gas. Deep, small reservoirs produce and emit much less methane than do shallow ones that contain large volumes of flooded biomass, such as those in the Brazilian Amazon, especially if the trees are not first removed. Indeed, the combined global warming effect of the methane and carbon dioxide produced by a large, shallow reservoir created to generate hydroelectric power can, for many years, exceed that of the carbon dioxide that would have been emitted if a coal-fired power plant were used to generate the same amount of electrical power! Hydroelectric power is not a zero-emission form of energy production if land is flooded to create it.

The anaerobic decomposition of the organic matter in garbage in landfills is another important source of methane in air. Food waste in the landfill produces the greatest amount of methane. In some communities, methane from landfills is collected and burned to generate heat, rather than being allowed to escape into the air. Although the combustion of methane produces an equal number of molecules of carbon dioxide, because the *per molecule* effect of CO_2 molecules is so much smaller than that of CH_4 molecules, the net greenhouse effect of the emission is thereby greatly reduced in relation to the amount of CO_2 absorbed from the air when the plant matter was growing.

The burning of biomass, such as forests and grasslands in tropical and semitropical areas, releases methane to the extent of about 1% of the carbon consumed, along with larger amounts of carbon monoxide (both compounds are products of incomplete—poorly ventilated—combustion).

Ruminant animals—including cattle, sheep, and certain wild animals—produce huge amounts of methane as a by-product in their stomachs when they digest the cellulose in their food. The animals subsequently emit the methane into the air by belching or flatulence. The decrease in the population of some methane-emitting wild animals (e.g., buffalo) in recent centuries has been far exceeded by the huge increase in the population of cattle and sheep. The net result has been a large increase in emissions of methane from animal sources.

It was reported by researchers in 2006 that plants, especially those growing in tropical areas, emit methane into the air as part of their aerobic metabolism, not just through the action of bacteria in anaerobic environments. The rate of emission of methane increases sharply with air temperature, approximately doubling for a 10° rise. If aerobic methane release from plants does occur, some of the observed decrease in global methane emission rates in the 1990s may have occurred because extensive tropical deforestation during that period would have greatly reduced the number of methane-emitting plants. Ironically, tropical forests cleared for livestock production may have produced as much methane as the ruminants that are now raised on that land! However, research reported by other scientists in 2007 failed to confirm that methane is produced aerobically and emitted by living plants.

Methane is released into air when natural gas pipelines leak, when coal is mined and the CH_4 trapped within it is released into the air, and when the

BOX 6-3	Determining the Emissions of "Old Carbon" Sources of Methane

The relative abundances of carbon isotopes in atmospheric carbon dioxide can be used to help deduce its origin by the following logic. The carbon in all living matter contains a small, constant fraction of a radioactive isotope, carbon-14 (^{14}C), taken in via the carbon cycle when photosynthesis captures atmospheric CO_2 and when animals in turn feed off plant matter. This fact underlies the radiocarbon dating methods used by archaeologists and anthropologists: When an organism dies, its ^{14}C decays at a known first-order rate that makes the date of its death calculable. (The assumptions justifying these methods are that biotic carbon and atmospheric carbon in CO_2 are balanced—in equilibrium with one another—and that the level of atmospheric ^{14}C is constant. The principles underlying radioactive decay are discussed in Chapter 9.)

However, in the case of atmospheric methane, the average fraction of ^{14}C is less than the value found in living tissue. This indicates that a significant fraction of the CH_4 escaping into air must be "old carbon"

that has been trapped in the ground for so long that its ^{14}C content has diminished to almost zero as a result of radioactive decay through the ages. Most methane containing old carbon is released into the air as a by-product of the mining, processing, and distribution of fossil fuels. Methane trapped in coal is released into the atmosphere when this material is mined, as is methane in oil when it is pumped from the ground. The transmission of natural gas, which is almost entirely methane, involves losses into the air due to leakage from pipelines and is the largest of the atmospheric sources of old carbon. Measurements of the methane levels in the air of various cities have indicated that much of the loss from pipelines in the past occurred in eastern Europe. Finally, there is probably a small contribution to the old carbon source from methane trapped in permafrost in far northern latitudes; the methane was formed by the decay of plant matter that lived there many thousands of years ago when the polar climate was much warmer than it is today.

gases dissolved in crude oil are released—or incompletely flared—into the air when the oil is collected or refined. Emissions from these sources have likely leveled off in the last decade. The technique by which scientists determine the component of atmospheric methane arising from fossil-fuel sources is discussed in Box 6-3.

In summary, there are six different significant sources of atmospheric methane, of which natural wetlands make the greatest contribution (~25%). The current relative importance of the five important anthropogenic sources of atmospheric methane is thought to be:

energy production/distribution ~ ruminant animal livestock
> rice production ~ biomass burning ~ landfills

Currently, the net sink for methane is thought to outweigh its sources by about 47 Tg/year, producing a slight net decrease in atmospheric methane concentration.

Methane: Concentration Trend and Possible Future Increases

Historically (i.e., before 1750), the methane concentration in air was approximately constant at about 0.75 ppm, i.e., 750 ppb. It has more than doubled since preindustrial times, to about 1.77 ppm; almost all of this increase occurred in the twentieth century because the emissions grew quickly, especially in the 1950–1980 period (see Figure 6-15). By the early 1990s, however, the rate of concentration increase had declined rapidly, and since that time it has fallen to zero in some years (see Figure 6-16b), so the methane concentration in air has been almost constant recently (Figure 6-16a). The rise in the atmospheric CH_4 level that has occurred since preindustrial times is presumed to be the consequence of such human activities as increased food production and fossil-fuel use, as discussed above.

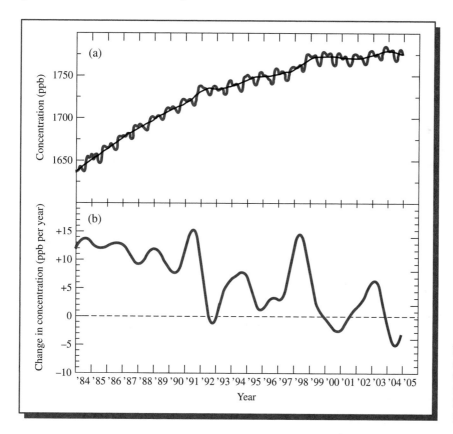

FIGURE 6-16 Atmospheric methane (a) concentration and (b) annual fluctuation in concentration in recent decades. [Source: NOAA.]

It is not known with certainty why the growth rate in methane concentration decreased recently. Since the rate of change in concentration is proportional to the difference between the rate of change in the emission rate and the rate of change in the destruction rate, change in either one or both rates could be responsible; neither can be measured directly very accurately. Natural gas pipelines carry ~90% methane, about 1.5% of which is lost to the atmosphere. Part of the decline in the emission rate of methane into the air was likely due to much-decreased emissions from pipelines in the former Soviet Union, which a few decades ago leaked much more gas than at present. However, increasing use of fossil fuels in northern Asia has probably now replaced some of these emissions. The draining, and drying out from global warming, of wetlands has resulted in decreased methane emissions from this natural source in recent decades. Some scientists have speculated that the declines in the early 1990s were related to the air temperature decreases associated with the explosion of Mount Pinatubo. The rate of oxidation of CH_4 by OH would also have increased if the concentration of hydroxyl radical increased overall.

PROBLEM 6-9

The concentration of atmospheric methane is about 1.77 ppm, and the rate constant for the reaction between CH_4 and OH is 3.6×10^{-15} cm^3 molecule^{-1} s^{-1}. Calculate the rate, in Tg per year, of methane destruction by reaction with hydroxyl radical, the concentration of which is 8.7×10^5 molecules cm^{-3}. See Problem 6-6 for additional data.

Some scientists have speculated that the rate of release of methane into air could greatly increase in the future as a *consequence* of temperature rises from the enhanced greenhouse effect. For instance, higher temperatures would accelerate the anaerobic biomass decay of plant-based matter, as occurs in a common landfill. In turn, the additional release of methane would itself cause a further rise in temperature. This is another example of positive feedback.

Methane release from biomass decay among the extensive bogs and tundra in Canada, Russia, and Scandinavia could also increase with increasing air temperature and would also constitute positive feedback. However, the rate of biomass decay and of plant growth, and thus of CH_4 production, also depends on soil moisture and therefore on rainfall, which probably would be affected by climate change in an as yet uncertain direction, so the net feedback from these sources could be positive or negative.

There is much methane currently immobilized in the permafrost of far northern regions; it was produced from the decay of plant materials during warm spells in the region but became trapped due to glaciation as temperatures became lower and lower at the start of the last ice age. Melting of the permafrost due to global warming could release large amounts of this methane. Melting would also allow the decomposition of organic matter currently present in the permafrost, with the consequent release of more methane.

In addition, there are monumental amounts of methane trapped at the bottom of the oceans, on continental shelves, in the form of *methane hydrate*. This substance has the approximate formula $CH_4 \cdot 6 H_2O$ and is an example of a **clathrate compound,** i.e., a rather remarkable structure that forms when small molecules occupy vacant spaces (holes) in a cage-like polyhedral structure formed by other molecules. In the present case, methane is caged in a 3-D ice lattice structure formed by the water molecules. The melting point of the structure is $+18°C$, somewhat higher than that of pure ice. Clathrates form under conditions of high pressure and low temperature, such as are found in cold waters and under ocean sediments. The methane was produced over thousands of years by bacteria that facilitated the anaerobic decomposition of organic matter in the sediments.

If seawater warmed by the enhanced greenhouse effect penetrates to the bottom of the oceans, the clathrate compounds could decompose and release their own methane, as well as reservoirs of pure methane currently trapped below them, to the air above. Methane trapped far below the permafrost in northern areas and in offshore areas in the Arctic also exists in the form of clathrates; it would be released eventually if the Arctic warmed sufficiently. Measurements made thus far do not indicate any significant emissions from these sources. It has been suggested by some scientists that CH_4 released from clathrates may be oxidized to CO_2 before it reaches the air, thereby greatly reducing the global warming potential.

Although the uncertainties concerning methane feedback are large, the stakes are higher than with any other gas. A few scientists believe that several positive climate feedback mechanisms, including those involving methane, could possibly combine to trigger an unstoppable warming of the globe. This worst-case scenario is called the *runaway greenhouse effect*. Such climate change would threaten all life on Earth, as the temperature would rise markedly, ocean currents would probably shift, and rainfall patterns would be very different from those we know. The possibility that the North Atlantic ocean current, which brings warm water from the south and thereby warms Europe, could cease to operate because of rapid global warming—induced by rapid increases in methane or carbon dioxide—is one of the most dramatic predictions about the possible consequences of the enhanced greenhouse effect.

PROBLEM 6-10

Calculate the mass of methane gas trapped within each kilogram of methane hydrate.

Nitrous Oxide

Another significant greenhouse trace gas is **nitrous oxide,** N_2O, also known as "laughing gas"; its molecular structure is NNO rather than the more symmetrical NON. Its bending vibration absorbs IR light in a band at 8.6 μm, i.e., within the window region, and in addition one of its bond-stretching vibrations is centered at 7.8 μm, on the shoulder of the window and at the same wavelength as one of the absorptions for methane. Per molecule, N_2O is 296 times as effective as CO_2 in causing an immediate increase in global warming. Like that of methane, the atmospheric concentration of nitrous oxide was constant until about 300 years ago, at which time it began to increase, although the level has increased from 275 ppb (preindustrial) by only 16% to 320 ppb. The yearly growth rate in the 1980s was about 0.25% but fell significantly in the early 1990s for reasons that are uncertain. The increased amounts of nitrous oxide that have accumulated in the air since preindustrial times have produced about one-third of the amount of the additional warming that methane has induced.

Less than 40% of nitrous oxide emissions currently arise from anthropogenic sources. In 1990 it was discovered that the traditional procedure, using *nitric acid*, HNO_3, of synthesizing *adipic acid* (a raw material in the preparation of nylon) resulted in the formation and release of large amounts of nitrous oxide. Since that time, nylon producers have instituted a plan to phase out N_2O emissions.

The greater part of the natural supply of nitrous oxide gas comes from release from the oceans, and most of the remainder is contributed by processes occurring in the soils of tropical regions. The gas is a by-product of the biological denitrification process in aerobic (oxygen-rich) environments and in the biological nitrification process in anaerobic (oxygen-poor) environments; the chemistry of both processes is illustrated in Figure 6-17. In **denitrification,** fully oxidized nitrogen in the form of the **nitrate ion,** NO_3^-, is reduced mostly to molecular nitrogen, N_2. In **nitrification,** reduced nitrogen in the form of ammonia or the ammonium ion is oxidized mostly to **nitrite,** NO_2^-, and nitrate ions. Chemically, the existence of the nitrous oxide by-product in both processes is simple to rationalize: Nitrification (oxidation) under oxygen-limited conditions yields some N_2O, which has less oxygen than the intended nitrite ion; denitrification (reduction) under oxygen-rich conditions yields some N_2O, which has more oxygen than the intended nitrogen molecule. Nitrification is more important than denitrification as a global source of N_2O. Normally, about 0.001 mole of nitrous oxide is emitted per mole of nitrogen oxidized, but this value increases substantially when the ammonia or ammonium concentration is high and relatively little oxygen is present. Overall, the increased use of fertilizers for agricultural purposes probably accounts for the majority of anthropogenic emissions of nitrous oxide. The decomposition of livestock-produced manure under aerobic conditions, including its use

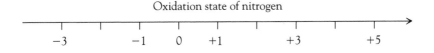

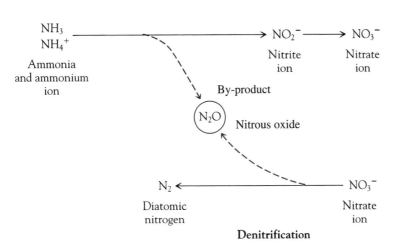

FIGURE 6-17 Nitrous oxide production as a by-product during the biological cycling of nitrogen.

as a fertilizer, contributes significantly to nitrous oxide emissions; manure produces very little N_2O if decomposed anaerobically.

Apparently, nitrous oxide released from new grasslands is particularly significant in the years following the burning of a forest. Some portion of the nitrate and ammonium fertilizers used agriculturally, particularly in tropical areas, is similarly converted (an unintended effect, to be sure) to nitrous oxide and released into the air. Tropical forests in wet areas are probably a huge natural source of the gas.

At one time it was believed that fossil-fuel combustion released nitrous oxide as a by-product of the chemical combination of the N_2 and O_2 in air, but this belief was based on faulty experiments. Only when the fuel itself contains nitrogen, as do coal and biomass (but not gasoline or natural gas), does N_2O form; apparently, N_2 from air does not enter into this process at all. However, some of the NO produced from atmospheric N_2 during fuel combustion in automobiles is unavoidably converted to N_2O rather than to N_2 in the three-way catalytic converters currently in use and is subsequently released into air. Some of the newer catalysts developed for use in automobiles do not suffer from this flaw of producing and releasing nitrous oxide during their operation.

As mentioned in Chapter 1, there are no sinks for nitrous oxide in the troposphere. Instead, all of it rises eventually to the stratosphere, where each molecule absorbs UV light and decomposes, usually to N_2 and atomic oxygen, or reacts with atomic oxygen.

CFCs and Their Replacements

Gaseous compounds consisting of molecules with carbon atoms bonded exclusively to fluorine and/or chlorine atoms have perhaps the greatest potential among trace gases to induce global warming, since they are both very persistent and absorb strongly in the 8- to 13-μm atmospheric window region. Absorption due to the C—F bond stretch is centered at 9 μm. The C—Cl bond stretch and various bond-angle-bending vibrations involving carbon atoms bonded to halogens also occur at frequencies that lie within the window region.

As discussed in Chapter 2, **chlorofluorocarbons** (CFCs) have already been released into the atmosphere in large quantities and have long residence times. Due to this persistence and to their high efficiency in absorbing thermal IR in the window region, each CFC molecule has the potential to cause the same amount of global warming as do tens of thousands of CO_2 molecules. The *net* effect of CFCs on global temperature is small, however. The heating that the CFCs produce by the redirection of thermal infrared is partially canceled by a separate effect, the cooling that they induce in the stratosphere due to their destruction of ozone. (Recall from Chapter 1 that the stratosphere is heated when oxygen atoms, recently detached photochemically from ozone molecules, collide with O_2 molecules to produce an exothermic reaction.) However, the decrease in stratospheric ozone allows more UV light to reach the lower atmosphere and the surface and to be absorbed there. The cooling and heating effects produced by CFCs occur at very different altitudes, so that their net effect on the Earth's weather may be substantial.

Ironically, the use of CFCs in insulating freezers, refrigerators, and air conditioners has reduced the energy requirements of this equipment and so has reduced CO_2 emissions resulting from electricity production.

The influence of CFCs on climate in the future will be reduced as a result of the requirements of the Montreal Protocol, which banned further production in developed countries after 1995, as discussed in Chapter 2. Most of the **HCFC** and **HFC** replacements for CFCs (with the notable exception of HFC-143a) have shorter atmospheric lifetimes and absorb less efficiently in the center of the atmospheric window region; thus on a molecule-for-molecule basis, they pose less of a greenhouse threat. However, if their levels of production and release become high in future decades because of expanding world population and increasing affluence, they will make significant contributions to global warming if they are released into the air. For this reason, many people feel that these substances must be used only in closed systems from which leakage to the atmosphere does not occur and that they must be recovered from equipment before its eventual disposal. Indeed, prevention of the chronic release of long-lived gases of all types to the atmosphere is a principle now agreed to by many scientific, business, and governmental groups. Measurements reported in 2002 indicate that the loss of the refrigerant HFC-134a (see Chapter 2) from the air-conditioning units of modern cars has a global warming impact that is about 4–5% of that from the carbon dioxide emitted from the cars.

PROBLEM 6-11

Fully fluorinated compounds such as tetrafluoromethane and hexafluoroethane are released as by-product wastes into the air in the production of aluminum. They were also briefly considered as CFC replacements. Will such molecules have a sink in the troposphere? Will they act as greenhouse gases? Would your answers be the same for monofluoromethane and monofluoroethane?

Sulfur Hexafluoride

Sulfur hexafluoride, SF_6, is a little-known greenhouse gas. It has some importance, however, because it is such a good absorber of thermal IR—23,900 times greater than CO_2 in global warming potential—and because, like other fully fluorinated compounds, it is so long-lived in the atmosphere (3200 years). It is used by electric utilities and in the semiconductor industry as an insulating gas. Formerly, it was vented to the air during routine maintenance of equipment but now is mainly recycled instead.

Tropospheric Ozone

Like methane and nitrous oxide, tropospheric **ozone,** O_3, is a "natural" greenhouse gas, but one which has a short tropospheric residence time. Although the ozone molecule is homonuclear, in its bent structure the central oxygen is not equivalent to the terminal oxygen atoms; consequently, the O—O bonds are somewhat polar. For that reason, the dipole moment of O_3 molecules changes during the symmetrical stretch vibration, which occurs in the atmospheric window region between 9 and 10 μm, and the molecules can absorb this IR light. The dip near 9 μm in the window region of the outgoing thermal IR distribution (Figure 6-7) is due to absorption by this vibration in atmospheric ozone molecules. Ozone's bending vibration occurs at 14.2 μm, near that for CO_2, and thus does not contribute much to the enhancement of the greenhouse effect, since atmospheric carbon dioxide already removes much of the outgoing light at this frequency. The antisymmetric stretching vibration of O_3 occurs at 5.7 μm, where there is very little outgoing IR.

As explained in Chapter 3, ozone is formed in the troposphere as a result of pollution from power plants and motor vehicles, from forest fires and grass fires, as well as from natural processes. As a result of these anthropogenic activities, the levels of ozone in the troposphere probably have increased since preindustrial times. The best guess is that approximately 10% of the increased global warming potential of the atmosphere results from increases in tropospheric ozone, although this value is very uncertain. The amount of thermal IR absorbed by stratospheric ozone has probably dropped slightly due to the recent decline in ozone levels.

The Climate-Modifying Effects of Aerosols

In Chapter 2, we saw that the initial neglect by scientists of the effects of atmospheric aerosol particles, specifically ice crystals in the stratosphere, led to a large underestimation of the amount of ozone that would be destroyed by chlorine. Similarly, neglect of aerosols led to misleading predictions about the extent of global warming to be expected. It is now realized that aerosols off-set and thereby mask a significant fraction of the atmospheric temperature increase that would have otherwise occurred due to the anthropogenic green-house gas emissions. The types of particulate matter of most importance in this context are particles expelled by powerful volcanic eruptions into the upper atmosphere *and* those produced by industrial processes and expelled into the lower troposphere. In order to understand how aerosols can affect global warm-ing, it is necessary to understand how they interact with light.

The Interaction of Light with Particles

All solids and liquids—including atmospheric particles—have some ability to **reflect** light. Atmospheric particles can reflect incoming sunlight, with the consequence that some of it is directed back into space and so is unavailable later for absorption at the surface (see Figure 6-18). The particles can also reflect outgoing infrared light, with the consequence that some of it is redirected back toward the Earth's surface rather than escaping from the atmosphere. The redirection of light by a particle is sometimes called **scatter-ing;** reflection backward is *backscattering*.

FIGURE 6-18 Interaction of sunlight with suspended atmospheric particles. (a) Modes of interaction. (b) Illustration of the indi-rect effect of producing increased reflection by small water droplets as compared to larger ones having the same total volume.

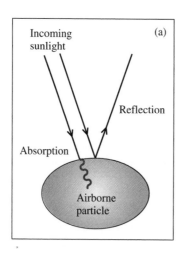

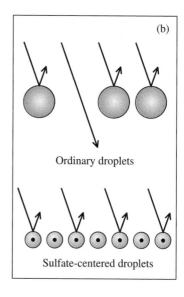

Certain types of suspended particulates in air reflect some of the sunlight that shines on them back into space and so have a significant albedo value; this reflection of sunlight by the aerosol cools the air mass and the surface below it, since none of the reflected light is subsequently absorbed and then converted to heat.

Some types of aerosol particles can **absorb** certain wavelengths of light (Figure 6-18a). Once absorbed, the energy that was associated with the light is rapidly converted into heat, which then is shared with the surrounding air molecules as a result of their collisions with the heated particle. Thus absorption of light by a particle leads to warming of the air immediately surrounding it. The absorption of sunlight, with consequent warming, is significant only for dark-colored particles such as those composed primarily of soot, often called *carbon black*, and of ash particles from volcanoes. The contribution of carbon black to global warming has only recently been fully appreciated. The emission into the atmosphere of carbon black is greatest in developing countries, where incomplete combustion of coal and biomass is widespread. Its effect globally is to increase air temperatures by its absorption of sunlight, with the subsequent export of this tropospheric air to other areas. However, its *local* effect may be cooling, since it blocks sunlight from reaching the surface. Carbon black's effects on local climate may be substantial, increasing drought in some areas and flooding in others.

Recall from Chapter 3 that the *sulfur dioxide* gas predominantly released as a pollutant from the burning of fossil fuels—especially coal—and from the smelting of nonferrous metals creates a **sulfate aerosol.** Pure sulfate aerosols do not absorb sunlight since none of their constituents—water, nitric and sulfuric acids, and the ammonium salts thereof—absorb light in the visible or the UV-A regions. The sulfate aerosols are not particularly effective in trapping outgoing thermal IR emissions. Only if tropospheric sulfate aerosols incorporate some soot will absorption of sunlight by these particles be significant.

Overall, however, anthropogenic sulfate-rich aerosols produced in abundance—especially in the Northern Hemisphere—reflect sunlight back into space much more effectively than they absorb it, so they significantly increase Earth's average albedo. As a result, less sunlight is available to be absorbed and converted to heat in the lower troposphere and at the surface. Thus the net effect of the sulfate aerosols is to cool the air near ground level and thereby to offset some of the global warming induced by greenhouse gases.

In addition to the direct effect of sulfate aerosols in reflecting sunlight, there are **indirect effects** that arise because the sulfate particles act as nuclei for the formation of small water droplets.

- Such small droplets are more effective in backscattering light than are an equal mass of larger ones (see Figure 6-18b).

- Small droplets are also less likely to coalesce into raindrops, so their clouds are longer-lived than otherwise expected and can thus reflect sunlight for longer periods.

Both these indirect effects result in more sunlight being reflected back into space, thereby cooling the Earth's surface. In addition, the "Asian brown cloud" (Chapter 4) formed by aerosol particles reduces the strength of the essential monsoon rains over India and Asia.

Some scientists have proposed that sulfate particles be injected artificially into the stratosphere, where they would reflect sunlight and thereby offset some of the effects of global warming. The lifetime of the particles in the stratosphere is several years, depending on altitude, so the sulfate would have to be replenished regularly. Although the injection of sulfate particles is considered to be a short-term solution, until controls for carbon dioxide emissions have been put in place, a few scientists have proposed injecting solid reflecting objects of macroscopic size high above the atmosphere, where they would reflect sunlight and counter global warming on a long-term basis. Such *geoengineering* of the Earth's climate is considered controversial by many scientists and policymakers because of the uncertainties involved in its potential side effects.

A short-term, dramatic example of the effects of atmospheric aerosols on climate occurred as a consequence of the massive eruption of substances into the troposphere and stratosphere by the Mount Pinatubo volcano in the Philippines in 1991. Initially, the lower stratosphere was warmed by the dominant effect of the large volcanic ash particles, which absorbed some of the incoming sunlight and subsequently converted it to heat, and by their interception of outgoing infrared from the surface. Due to their relatively large size, the ash particles were not long-lived in the stratosphere. The longer-term effect of the Pinatubo eruption was to significantly decrease air temperatures at ground level. The stratospheric aerosol that remained suspended after a few months was formed by the oxidation of the 30 million tonnes of SO_2 that the volcano had blasted directly into the lower parts of this region. The sulfate aerosol remained there for several years, during which time it efficiently reflected sunlight back into space. Many regions, including North America, experienced several cool summers in the early 1990s as a result. Due to the gradual sedimentation of the aerosol, we returned to 1990–1991 temperatures by 1995.

Aerosols and Global Warming

The cooling effect of the sulfate aerosol is concentrated almost entirely in the Northern Hemisphere because most industrial activity takes place in that half of the globe, so it is there that most emissions occur. The relatively short lifetime of such sulfate aerosols precludes their spreading to the Southern Hemisphere; consequently, the concentration of sulfate particulates is much higher over the Northern Hemisphere. The short lifetime of the sulfate particles can be understood by considering the processes of their removal from air. The average diameter of the tropospheric sulfate aerosol particles is about 0.4 μm, and their average altitude is about 0.5 km. For particles of this size and altitude, the expected atmospheric lifetime before gravitational settling to the surface is several years. However, since the sulfate aerosol droplets are also removed efficiently by rain, their actual lifetime in the lower troposphere is of the order of days rather than years.

The increase of global SO_2 emissions from fossil-fuel combustion over the last century and a half is shown in Figure 6-19. Up to 20% more sulfur dioxide is emitted by smelting, etc. Presumably the trend in anthropogenic sulfate aerosol production has approximately followed the pattern of SO_2 emissions in Figure 6-19. The initial approximately linear time increase of the sulfur dioxide emission rate changed to one with a much steeper slope after World War II, a behavior we saw previously for CO_2 and CH_4. However, the ratio of global SO_2 to CO_2 emission rates, expressed as a percentage of sulfur in the carbon of the fuel, fell from about 2.2% in the 1930s and 1940s to a constant 1.1% in recent decades, due presumably to the gradual replacement of coal by oil and natural gas.

As illustrated by the contour diagrams in Figure 6-20, the bulk of the anthropogenically produced aerosols in North America is centered above the Ohio Valley and directly reflects sunlight mostly above that area. Equal or even larger effects are observed over southern Europe and portions of China. Indeed, according to some calculations, the cooling effect from aerosols

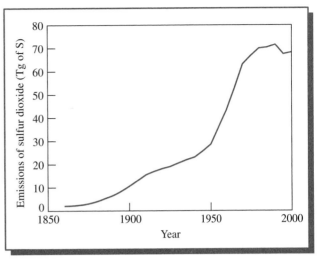

FIGURE 6-19 Estimated historical emissions of sulfur dioxide from anthropogenic sources. [Source: Adapted from S. J. Smith, H. Pitcher, and T. M. L. Wigley, "Global and Regional Anthropogenic Sulfur Dioxide Emissions," *Global and Planetary Change* 29 (2001): 99.]

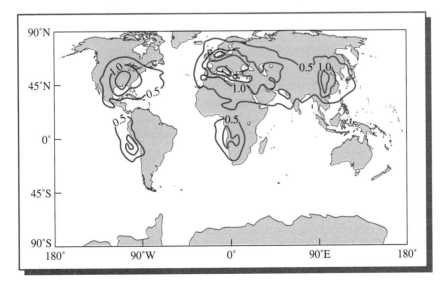

FIGURE 6-20 The amount of sunlight reflected into space by anthropogenic aerosols by the direct mechanism, in units of watts per square meter of the Earth's surface. [Source: J. T. Houghton et al., *Climate Change 1994— Radiative Forcing of Climate Change* (Intergovernmental Panel on Climate Change) (Cambridge: Cambridge University Press, 1995).]

BOX 6-4 | Cooling over China from Haze

Measurements of the amount of sunshine reaching the surface in China indicate a significant decrease over the last half-century (Figure 1). The blockage of solar intensity results from the aerosols in the air above the region produced mainly from the sulfur dioxide emitted by burning coal. As a consequence of the increasing presence of the aerosols, maximum summer temperatures in heavily polluted eastern China have fallen by about 0.6°C per decade. Similar effects are observed in the Brazilian Amazon region, due to soot and ash emitted into the local air from forest and grass fires lit to clear land, and in the African country of Zambia, due to grass fires.

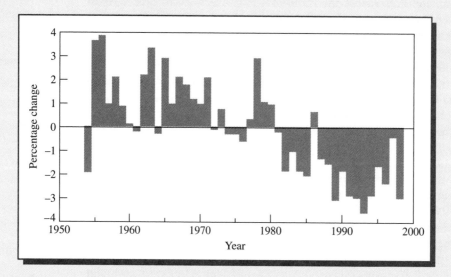

FIGURE 1 Change in the amount of sunlight reaching China relative to the average, over 50 years. [Source: F. Pearce, "Pollution Is Plunging Us into Darkness," *New Scientist* (14 December 2002):6.]

outweighs the heating effect due to greenhouse gases for some regions in these areas.

It is not clear how the amount of tropospheric sulfate aerosol will change in the future. Emissions of sulfur dioxide from power production in North America and western Europe are now more tightly controlled in order to combat acid rain, so the SO_2/CO_2 ratio in emissions from these areas should decline. However, the anthropogenic sulfate aerosol concentrations over southern Europe and parts of Russia and China are considerably higher than the current maximum values in North America and will not be affected by these legislative controls (see Box 6-4). The only substantial domestic energy source currently available to China for its rapid industrialization is coal, so the SO_2 emissions from this source and from India may well continue to rise (Chapter 3).

Aerosols also result from the oxidation of the gas **dimethyl sulfide** (DMS), $(CH_3)_2S$, which is produced by marine phytoplankton and subsequently released into the air over oceans. Once in the troposphere, DMS undergoes oxidation, some of it to SO_2, which then can oxidize to sulfuric acid, and some to *methanesulfonic acid*, CH_3SO_3H. Both of these acids form aerosol particles, which in turn lead to the formation of water droplets and hence clouds over the oceans. The particles and droplets both deflect incoming light from the Sun. Some scientists believe that increased emissions of dimethyl sulfide from the oceans will occur when seawater warms as a result of the enhancement of the greenhouse effect and that this negative feedback will temper global warming.

Although the sulfate aerosol has a short lifetime, new supplies of it are constantly being formed from the sulfur dioxide pollution that pours into the atmosphere on a daily basis. Consequently, there is a steady-state amount of the aerosol in the troposphere; sulfur dioxide emissions keep postponing the full effects of global warming induced by the rise in greenhouse gas concentrations.

Global Warming to Date

Allocation of Warming to Natural and Anthropogenic Factors

The best estimates of global warming or cooling in 2005 arising from the various factors is summarized by the bar graphs in Figure 6-21; the effect of each factor is expressed as a percentage of the total anthropogenic effect. The order of the greenhouse gases in terms of the amount of extra warming they have produced is

$$CO_2 > CH_4 > O_3 > CFCs > N_2O$$

The value in Figure 6-21 for CFCs includes the cooling of the stratosphere induced by their destruction of ozone, and that for methane includes the warming of the stratosphere produced from the additional water vapor formed there by its decomposition. The cooling labeled *surface albedo* is the net result of cooling due to changes in land use minus warming arising from the deposition of sunlight-absorbing black soot on snow and ice. Overall, the cooling from anthropogenic aerosols currently cancels about 40% of the net warming from all greenhouse gases. However, the aerosol effect—which is the sum of that from the direct and indirect effects on cloud albedo—has by far the largest uncertainty of any of the factors in Figure 6-21.

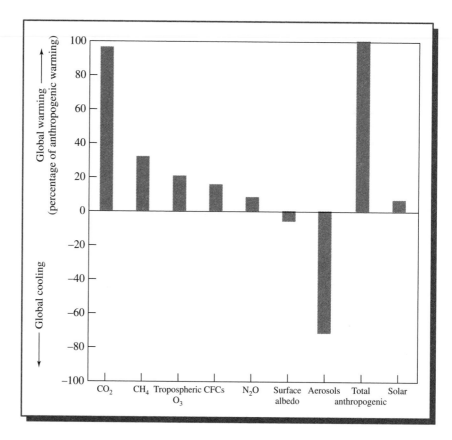

FIGURE 6-21 Contributions to global warming and cooling produced by various factors, expressed as percentages of the total anthropogenic warming. [Data source: Intergovernmental Panel on Climate Change, *Climate Change 2007: The Physical Science Basis. Summary for Policymakers* (February, 2007).]

Greenhouse gas emissions from airplanes traveling long distances high in the troposphere are particularly effective in promoting global warming. They emit, into the low-density air in which they fly, the carbon dioxide and water vapor that result from the combustion of their hydrocarbon fuel. Because the air temperature there is so low, the IR absorbed by this CO_2 and H_2O is unlikely to be re-emitted; instead, it warms the surrounding air and therefore enhances the greenhouse effect.

Global Warming: Geography

The year-to-year variations in the average worldwide surface temperature from 1880 to the present are shown in Figure 6-2. The increases in temperature were not evenly spread around the globe, however, as indicated by Figure 6-22, which shows the changes in average global temperatures in the 2001–2005 period, relative to the 1951–1980 average. In Figure 6-22, the darker the shade of green, the greater the increase in temperature; the few areas shown in gray underwent decreases in temperature. In general, air

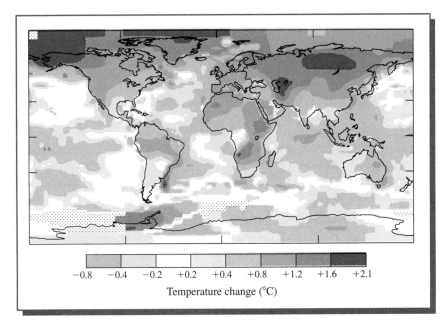

FIGURE 6-22 Changes, in degrees Celsius, in the mean surface temperature in 2001–2005 relative to the 1951–1980 mean. The dotted regions indicate areas for which data is insufficient. [Source: J. Hansen et al., "Global Temperature Change," *Proceedings of the National Academy of Science* 103 (2006): 14288.]

temperatures over land areas have increased more than those over the seas. Sulfate aerosols keep the eastern portions of the United States and Canada cooler than they would be otherwise.

The Arctic region has warmed most of all, with the consequence that its sea ice is disappearing. The melting produces a positive feedback effect: Since ice reflects sunlight more efficiently than does liquid water, the increasing amount of sunlight absorbed as the ice is replaced by open water increases surface water and surface air temperatures, thereby inducing even more melting. However, scientists have recently found that evaporation of the open water has produced more cloud cover in the region, which is a negative feedback effect that partially negates the positive one since the clouds reflect sunlight and thereby increase the albedo.

The increased lowering of air temperatures due to higher sulfate aerosol levels does not *permanently* cancel out all the warming due to greenhouse gases because of the very different atmospheric lifetimes of the particles as compared to the gases. The tropospheric aerosols last only a few days, so they do not accumulate with time and their effect is short term. In contrast, today's emissions of carbon dioxide into the atmosphere will exert effects for decades or centuries to come: CO_2 emissions are cumulative over the medium term. This is also true for CFCs and nitrous oxide, which are also important greenhouse gases, but it is less so for methane since its half-life is only about one decade. Thus, although sulfate-producing SO_2 emissions can temporarily cancel the effects of CO_2 emissions, eventually the cumulative effects of the carbon dioxide and the other greenhouse gases win out.

In summary, global warming has been experienced by most areas in the last half-century. Most of the warming is due to emissions of carbon dioxide into the atmosphere, with lesser amounts of warming from increased levels of methane, tropospheric ozone, nitrous oxide, and the introduction of CFCs. The increased water vapor in the atmosphere resulting from the warming by these gases has itself produced at least as much additional warming. The increased emissions of sulfur dioxide that accompany fossil-fuel combustion have produced aerosols that cancel out some but not all of the warming produced by the greenhouse gases.

Global Circulation Models

In continuing research that began in the 1980s, scientists have attempted by computer modeling to predict the impact of past and present increases in atmospheric gases and particles on the future climate of the planet. There are some uncertainties in such an endeavor, including the fact that we don't yet fully understand all the sources and sinks of the greenhouse gases or the net effect of aerosols. Probably the most important problem remaining with these **global circulation models** is their treatment of clouds. In particular, is the net feedback from the increased cloudiness expected for a warmed atmosphere negative, positive, or zero? Clouds operate both to cool and to heat the atmosphere. They cool it by reflecting incoming light back to space; we experience this dramatically when the Sun goes behind the clouds on a warm, sunny day and we immediately feel the air cooling. We also know that the water droplets in clouds absorb infrared light emitted from below them.

•   *Low-lying* clouds are warm, so they re-emit almost all the energy absorbed in random directions, rather than converting it to heat and thereby warming the air in their immediate surroundings. Since some of the IR emitted by these low clouds is directed downward, the surface is warmed. The fact that the IR is all re-emitted, however, means that these clouds don't warm the atmosphere much by this effect. Thus the *net* effect over a full day of low-lying clouds is to cool the Earth, because their reflection of incoming sunlight is dominant.

•   In contrast, *high-lying* clouds absorb outgoing IR but don't re-emit much of it since they are cold; all the absorbed IR is converted to heat, warming the nearby air. Thus the net effect of high-lying clouds is to warm the Earth, since their conversion of outgoing IR into heat is more important than their reflection of incoming sunlight.

Because we don't accurately know whether global warming will produce more additional low-lying or high-lying clouds, it is uncertain whether the feedback from this factor will be positive or negative.

It is not absolutely certain that all or indeed any part of the observed temperature increases in the last 100 years are attributable to the anthropogenically induced enhancement of the greenhouse effect. However, natural, century-long global warming or cooling trends as large as the half-a-degree-Celsius we have recently experienced appear only once or twice a millennium. There is an 80–90% likelihood that the increase in the twentieth century was not a wholly natural climatic fluctuation. Furthermore, on the basis of computer simulations such as that discussed above, the UN-sponsored **Intergovernmental Panel on Climate Change** (IPCC) concluded in its 2007 report that "most of the observed increase in globally averaged temperatures since the mid-twentieth-century is *very likely* due to the observed increase in anthropogenic greenhouse gas concentrations."

Some skeptics have pointed out that the evidence indicating that global warming has begun is based on air temperatures at the Earth's surface only, and that satellite data indicate that the lower troposphere as a whole has cooled rather than warmed. However, other scientists claim that this cooling is spurious and apparently arises from problems in merging data from two different satellites with slightly different calibrations and other artifacts of uncertainties in the data.

Other Signs of Global Warming

In addition to a rise in average global surface air temperature, there have been a number of other shifts that indicate that the climate indeed is changing:

- *Winters have become shorter, by about 11 days*. In the Northern Hemisphere, spring has been arriving sooner and autumn has been starting later. Over the last three decades, the advent of spring—as observed by the appearance of buds, the unfolding of leaves, and the flowering of plants—has advanced by an average of six days in Europe, while the start of autumn—as defined by the date at which leaves change color and begin to fall—has been delayed by about five days. In Alaska and northwestern Canada, average temperatures have risen one degree per decade recently, resulting in an earlier date for the last frost in the spring and significant thawing of the permafrost. Consequently, in many regions there are fewer "frost days" now than there used to be—19 fewer in the western United States and 3 fewer in the eastern part. The response of plants is driven by the increase in average daily air temperatures. Indeed, the range boundaries of some plants have increased toward the poles, and phenologies (season-dependent behavior) have shown an earlier start for spring for the great majority of plants and animals. The earlier springtime disappearance of mountain snow packs as well as higher temperatures in spring and summer have led to a substantial increase in the number, duration, and intensities of wildfires in the western United States since the mid-1980s.

- *The Earth's ice cover is shrinking fast*. Glaciers, polar icecaps, and polar sea ice are melting and disappearing at unprecedented rates as a consequence of global warming. For example, the remaining glaciers in Glacier National Park in the U.S. Rocky Mountains could disappear in 30 years if current melting rates continue. About 10% of the world's winter snow cover has disappeared since the late 1960s. Sea ice in the Arctic has not only decreased in area, by about 9% per decade recently, but it has also thinned dramatically. Warmer weather has also delayed the seasonal formation of sea ice. All these changes have caused a sharp decline in some populations of Antarctic penguins and Arctic caribou.

• *Warming water is killing much of the coral in ocean reefs and threatening sea life.* Coral reefs in tropical waters nurture and protect fish and attract scuba divers. As water warms, corals "bleach" themselves by expelling the algae that give them color and provide nutrition. Over 95% of the coral is already dead in some parts of the Seychelles Islands. Thus far, reefs in the central Pacific Ocean have escaped bleaching, and it is just beginning in the Caribbean. The death of the coral reefs not only affects the tourist trade but also threatens fishing for species that depend on the reefs for food. Beaches will erode if the reefs break up and no longer provide protection.

• *Mosquito-borne diseases have reached higher altitudes.* Because of warmer temperatures, mosquitoes are now able to survive in regions where they formerly were not viable. As a result, mosquitoes have carried malaria to higher mountain regions in parts of Africa and *dengue fever* to new regions in Central America. Outbreaks of malaria have occurred in Texas, Florida, Michigan, New York, New Jersey, and even southern Ontario in the last decade. Warmer weather and changing precipitation patterns allowed the West Nile virus, another mosquito-borne disease, to become established in the New York City area in the late 1990s and in southern Canada and virtually all the contiguous United States by the early 2000s.

• *Rising sea levels are threatening to engulf Pacific islands.* Warming of the air eventually leads to a rise in sea levels, for reasons that will be discussed in Chapter 7. The change in average sea level from 1870 to the present is illustrated in Figure 6-23. The rise since 1930 amounts to over 15 cm, a sharp rise in the rate of increase for this process compared to the past.

• *Precipitation has increased in most areas.* An important aspect of climate is the amount of precipitation—rain and snow—that falls at various locations on Earth. In the twentieth century, the total annual precipitation in the Northern Hemisphere increased, with most mid- and high-latitude regions of eastern North and South America, Europe, and Asia becoming somewhat wetter. However, many areas just north and south of the Equator, especially in Africa and parts of southern Asia, became much drier, producing more intense and longer droughts, with disastrous consequences for food production there. An overall increase in global precipitation is expected, since warming of the air warms the surface waters of lakes and oceans, and warmer water evaporates faster and thereby increases the water content of the atmosphere.

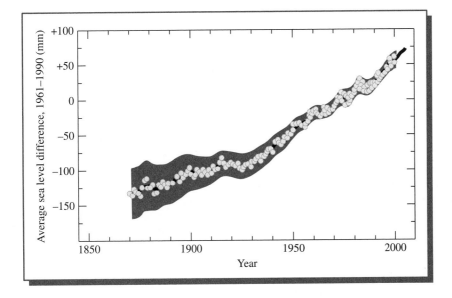

FIGURE 6-23 Changes to the global average sea level, relative to the 1961–1990 average. The smoothed curve (black) represents decadal averages, whereas the white circles show yearly values. The green shaded area represents the uncertainty interval.
[Source: Intergovernmental Panel on Climate Change, *Climate Change 2007: The Physical Science Basis. Summary for Policymakers* (February, 2007).]

- *Extreme weather is becoming more common.* The frequency of extreme and violent weather events has increased in many areas of the world. Such events include stronger blizzards and storms with heavy snow and freezing rain in northern areas but record heat waves, hurricanes, and drought in others. For example, the number of heat waves lasting three days or longer almost doubled in the United States between 1949 and 1995. The heat wave in Europe in the summer of 2003 was a tragic example of the phenomenon. Moreover, the frequency of storms with heavy or extreme precipitation increased in many non-tropical regions in the last half-century. The economic damage caused by storms, including hurricanes, has increased markedly over the last two decades.

Review Questions

1. Sketch a plot showing the main trends in global air temperature over the last century and a half.

2. What is the wavelength range, in μm, for infrared light? In what portion of this range does the Earth receive IR from the Sun? What are the wavelength limits for the *thermal IR* range?

3. Explain in terms of the mechanism involved what is meant by the *greenhouse effect*. Explain what is meant by the enhancement of the greenhouse effect.

4. Explain what is meant by the terms *symmetric* and *antisymmetric bond-stretching* vibrations and by *bending* vibrations.

5. Explain the relationship between the frequency of vibrations in a molecule and the frequencies of light it will absorb.

6. Why don't N_2 and O_2 absorb thermal IR? Why don't we consider CO and NO to be trace gases that could contribute to enhancing the greenhouse effect?

7. What are the two main anthropogenic sources of carbon dioxide in the atmosphere? What is its main sink? What is *fixed carbon*?

8. Is water vapor a greenhouse gas? If so, why is it not usually present on lists of such substances?

9. Explain what is meant by *positive* and *negative feedback*. Give an example of each as it affects global warming.

10. What is meant by the term *atmospheric window* as applied to the emission of IR from the Earth's surface? What is the range of wavelengths of this window?

11. What is meant by the *residence time* of a gas in air? How is it related to the gas's rate R of input/output and to its total concentration C?

12. What are four important trace gases that contribute to the greenhouse effect?

13. What are the six most important sources of methane?

14. What are the three most important sinks for methane in the atmosphere? Which one of them is dominant? What is meant by the term *clathrate compound*?

15. Is the enhancement of the greenhouse effect by release of methane from clathrates due to increased temperature an example of feedback? If so, is it positive or negative feedback? Would an increase in the rate and amount of photosynthesis with increasing temperatures and CO_2 levels be a case of positive or negative feedback?

16. Explain in chemical terms what is meant by *nitrification* and *denitrification*. What are the conditions under which nitrous oxide production is enhanced as a by-product of these two processes?

17. What are the main sources and sinks for N_2O in the atmosphere?

18. Are the proposed CFC replacements themselves greenhouse gases? Why is their emission considered to be less of a problem in enhancing the greenhouse effect than was that of the CFCs themselves?

19. By which two mechanisms does light interact with atmospheric particles?

20. Explain how sulfate aerosols in the troposphere affect the air temperature at the Earth's surface by both the direct and indirect mechanisms.

21. List four important signs, other than increases in average air temperature, that global warming is occurring.

 # Green Chemistry Questions

See the discussion of focus areas and the principles of green chemistry in the Introduction before attempting these questions.

1. The development of supercritical carbon dioxide for removal of photoresist (SCORR) won a Presidential Green Chemistry Challenge Award.

(a) Into which of the three focus areas for these awards does this award best fit?

(b) List four of the twelve principles of green chemistry that are addressed by the green chemistry developed by Los Alamos and SC Liquids.

2. What environmental advantages does the SCORR process offer versus conventional methods for photoresist removal?

Additional Problems

1. The common tropospheric pollutant gases SO_2 and NO_2 have molecular structures which, like that of CO_2, have the central atom connected to two oxygen atoms, but, unlike CO_2, they are nonlinear. The wavelengths for their vibrations are given in the table below. **(a)** Which of the vibrations are capable of absorbing infrared energy? **(b)** Based on the wavelengths for the IR-absorbing vibrations and the spectrum in Figure 6-7, decide which, if any, vibrations could contribute much to global warming. **(c)** What lifetime characteristic of these gases would limit their role in global warming?

Gas	Symmetric stretch	Antisymmetric stretch	Bending
SO_2	8.7 μm	7.3 μm	19.3 μm
NO_2	7.6 μm	6.2 μm	13.3 μm

2. (a) How can the fact that nitrous oxide has three vibrations that absorb infrared light be used to prove that its linear structure is NNO rather than NON? **(b)** Would methane molecules absorb IR during the vibration in which all four C—H bonds stretched or contracted in phase?

3. Anthropogenic carbon dioxide emissions into the atmosphere amounted to 178 Gt from January 1990 to December 1997. Calculate the fraction of this emitted carbon dioxide that remained in the air, given that in that same eight-year period, the carbon dioxide concentration in air rose by 11.1 ppm. Note that the molar masses of C, O, and air, respectively, are 12.0, 16.0, and 29.0 g, that the mass of the atmosphere is 5.1×10^{21} g, and that 1 Gt is 10^{15} g.

4. The total amount of methane in the atmosphere in 1992 was about 5000 Tg and was then increasing by about 0.6% annually due to the fact that the annual input rate exceeded the annual destruction rate of 530 Tg/yr. Calculate the percentage by which anthropogenic releases of methane, which

account for two-thirds of the total, had to be reduced if the atmospheric concentration of this gas was to be stabilized in 1992.

5. As mentioned in the text, the fraction F of light that is absorbed by any gas in air is logarithmically related to the concentration C of the gas and the distance d through which the light travels; this relationship is called the Beer–Lambert law:

$$\log_e(1 - F) = -KCd$$

Here K is a proportionality constant. Show by simple trial calculations that for concentrations near zero (e.g., where $KCd = 0.001$), F is related almost linearly to C, whereas for larger KCd values (e.g., near 2), doubling the concentration does not nearly double the light absorption.

6. The vapor pressure P of a liquid rises exponentially when it is heated according to the equation

$$\ln(P_2/P_1) = -\Delta H/R \, (1/T_2 - 1/T_1)$$

Here P_2 and P_1 are the vapor pressures of the liquid at the Kelvin temperatures T_2 and T_1 after and before the temperature increase, R is the gas constant 8.3 J/K mol, and ΔH is the liquid's enthalpy of vaporization, which for water is 44 kJ/mol. Calculate the percentage increase in the vapor pressure of water that occurs if the temperature is raised from 15°C to 18°C. Give several reasons why the amount of outgoing thermal infrared in water's absorption bands may not be increased by exactly the percentage you calculate if the average surface air temperature is increased to 18°C.

7. Suppose that some climatic crisis inspired the Earth's population to switch to energy systems that did not emit carbon dioxide and that the transition occurred within a decade. What would be the immediate effect of this change on the Earth's average air temperature, given that both carbon dioxide and sulfur dioxide emissions from fossil fuels would have rapidly declined?

8. Explain why CHF_2Cl (HCFC-22) has more IR absorption bands with a greater range of IR wavelengths absorbed, and absorbs IR much more efficiently on a per molecule basis, than CH_4, the hydrocarbon from which it is derived.

9. Calculate the volume of CO_2 produced at 1 atm and 20.0°C from the complete combustion of 1.00 L of gasoline. Although gasoline is a mixture of hydrocarbons (as described in Chapter 7), for the purposes of this calculation consider gasoline to have the chemical formula C_8H_{18} and the same

density as *n*-octane: 0.702 g/mL. Calculate the volume of CO_2 produced by driving 100 miles on the highway in a mid-size sedan compared to an SUV, given their highway fuel efficiencies of 33 and 19 mpg, respectively. Note that 1 gal = 3.785 L.

10. Given that the 2002 concentration of CH_4 in the atmosphere was 1.77 ppm, calculate the total mass of CH_4 in the atmosphere in 2002. Note that the total mass of the atmosphere is 5.1×10^{18} kg and the average molar mass of the atmosphere is 29.0 g/mol.

Further Readings

1. J. T. Houghton et al., *Climate Change 2001: The Scientific Basis* (Cambridge: Cambridge University Press, 2001). (Published for the Intergovernmental Panel on Climate Change.)

2. E. Claussen, ed., *Climate Change: Science, Strategies, and Solutions* (Arlington, VA: Pew Center on Global Climate Change, 2001).

3. D. Rind, "The Sun's Role in Climate Variations," *Science* 296 (2002): 673–677.

4. V. Ramanathan et al., "Aerosols, Climate, and the Hydrological Cycle," *Science* 294 (2001): 2119–2124.

5. F. W. Zwiers and A. J. Weaver, "The Causes of 20th Century Warming," *Science* 290 (2000): 2081–2137.

6. J. Hansen et al., "Global Temperature Change," *Proceedings of the National Academy of Science* 103 (2006): 14288–14293.

7. D. F. Ferretti et al., "Unexpected Changes to the Global Methane Budget over the Past 2000 Years," *Science* 309 (2005): 1714–1716.

8. P. Bousquet et al., "Contribution of Anthropogenic and Natural Sources to Atmospheric Methane Variability," *Nature* 443 (2006): 439–443.

9. C. Parmesan and G. Yohe, "A Globally Coherent Fingerprint of Climate Change Impacts Across Natural Systems," *Nature* 421 (2003): 37–42.

10. L. D. D. Harvey, *Global Warming: The Hard Science* (New York: Prentice-Hall, 2000).

11. T. Flannery, *The Weather Makers* (Toronto: Harper-Collins, 2005).

12. A. Gore, *An Inconvenient Truth* (Emmaus, PA: Rodale, 2006).

13. O. Morton, "Is This What It Takes to Save the World?" *Nature* 447 (2007): 132.

14. B. Lomberg, "The Skeptical Environmentalist" (Cambridge: Cambridge University Press, 2001), Chapter 24.

Websites of Interest

Log on to www.whfreeman.com/envchem4/ and click on Chapter 6.

CHAPTER 7

FOSSIL-FUEL ENERGY, CO$_2$ EMISSIONS, AND GLOBAL WARMING

In this chapter, the following introductory chemistry topics are used:

- Percentage composition; stoichiometry
- Combustion; heat of combustion
- Structural chemistry of hydrocarbons (see Appendix)
- Acidity; weak acids; pH
- Phase diagrams; condensation of liquids; sublimation of solids; distillation
- Catalysis
- Density
- Polymerization

Background from previous chapters used in this chapter:

- Greenhouse effect and greenhouse gases; aerosols (Chapter 6)
- Sinks (Chapter 6)
- ppm concentration scale for gases (Chapter 1)
- Clathrates (Chapter 6)
- Albedo (Chapter 6)

Introduction

As we saw in Chapter 6, the Earth's weather has probably already been affected by the enhancement of the greenhouse effect due to increasing atmospheric concentrations of carbon dioxide and other gases. A continuing buildup of CO$_2$ in the air leads to the conclusion that we are in store for further increases in global air temperatures and other changes to our climate. In this chapter, we shall inquire into predictions of the likely trends in energy usage and consequently of carbon dioxide emissions that will occur for the rest of the twenty-first century. The nature of fossil fuels and their role in generating CO_2 are then analyzed, and the prospect of burying the emissions as they are generated is then explored. We finish by considering the predictions of the consequences to the climate, and of ramifications to civilization that follow from it, if greenhouse gas emissions continue unabated or with only weak controls.

Energy Reserves and Usage

Ever since the Industrial Revolution, the worldwide use of **commercial energy**—that sold to users and usually derived on a large scale from fossil-fuel combustion, hydroelectricity, and nuclear power, as opposed to the biomass collected and used by individual families—has risen almost every year; the current annual global growth rate is about 2%. The period of the most rapid increase began after World War II, when global commercial energy consumption was only about one-tenth the current level.

Because the amounts are so huge, it is useful to discuss global quantities of energy in terms of the large energy unit EJ, an **exajoule**, which is 10^{18} joules. The total amount of commercial energy consumed by humans currently amounts to about 400 EJ annually, with the United States consuming about 100 EJ of that total.

Determinants of Growth in Energy Use

Although increases in energy usage are sometimes thought to be mainly tied to population growth, this is a dominant factor only for less developed countries, for which energy use per capita is small anyway. The usage of commercial energy by a country depends on many factors, including its population, geography, and climate, as well as on the cost of energy. However, the most important factor in total energy usage appears to be the **gross domestic product** (GDP) of the country. In industrialized societies, about 11 megajoules (11 million joules) of energy are currently needed on average to produce one (U.S.) dollar's worth of goods and services. Interestingly, the energy-to-GDP ratio for many developing countries, including China, has about the same value. The ratio for India is somewhat less than those for developed countries, and those for nondeveloped countries are lower still.

In the past, it has been found that although the energy-to-GDP ratio usually rises when a country begins to industrialize, it then drops gradually as the infrastructure becomes more substantial and efficient. For example, the ratio for the United States dropped by 44% from 1970 to 2000.

The fantastic rise in global energy usage in the second half of the twentieth century was due mainly to industrial expansion and to increases in the standard of living in the now-developed countries. The energy consumption in these countries continues to expand, though now only slowly. Per capita energy use in the United States currently amounts to 10,000 J/sec (i.e., 10,000 watts, the equivalent of one hundred 100-W light bulbs burning continuously), about twice that in the European Union and Japan, and about five times the world average.

However, economic growth in the developing countries—which contain three-quarters of the world's population—is rising more quickly and with it their total energy consumption. Thus, although per capita energy usage in China is only half the global average, it is rising. The developing countries collectively used only 30% of the world's commercial energy in 1993, but they are expected to consume more than half of it starting sometime in the next decade. According to the *International Energy Agency*, between 1994 and 2010 the rate of increase for the developing countries collectively is expected to be 4% annually, which, if compounded, would produce a doubling over that period. For developed countries, the annual rise over the same period is expected to be 1.5%, amounting to 28% if compounded over that period. The overall global increase is expected to continue to be about 2% annually over the next few decades.

PROBLEM 7-1

Any quantity V whose value increases in a time period t by a percentage of its previous value exhibits exponential growth according to the equation

$$V = V_0\, e^{kt}$$

where V_0 is the initial value and k is the fractional increase in each time period. Given that this equation will apply to the growth in energy usage if it increases by the same percentage each year, derive a general formula relating the number of years required for energy usage to double as a function of the annual fractional increase k. What is the doubling time when 4% annual growth (i.e., $k = 0.04$) is in operation? What about 3%, 1.5%, and 1.0% annual growth rates? If $k = 0.02$, how many years does it take for the energy use to increase 10-fold? If energy usage grew by a factor of 10 over 50 years, what was the annual compounded rate of growth in this period?

The usage of energy involves its transformation from one form to another, eventually resulting in its degradation to waste heat; as such, it does not pose any global environmental problem per se. However, there are usually side effects associated with energy production and/or consumption that are serious environmental issues, as we shall see in the rest of this chapter and in the next.

The most serious long-range global environmental problem associated with energy use is the release into the atmosphere of **carbon dioxide, CO_2,** when fossil fuels are combusted to produce heat. Indeed, the rest of this chapter is devoted to exploring this problem and its possible solutions, as well as to the nature and properties of fossil fuels, namely coal, petroleum, and natural gas. Carbon dioxide is produced when any carbon-containing substance undergoes complete combustion:

$$\text{C in substance} + O_2 \longrightarrow CO_2$$

As we saw in Chapter 6, CO_2 is an important greenhouse gas, and the increase in its atmospheric concentration is responsible for the largest fraction of global warming of any anthropogenic factor.

Developed countries have accounted for about three-quarters of all carbon dioxide emissions from fossil-fuel combustion and cement manufacture since the beginning of the Industrial Revolution. The emissions from these sources from various countries in the more recent period (1980–2004) are illustrated by the bands in Figure 7-1. Notice that:

• The United States was the biggest emitter country, though recent data indicate China overtook it in 2007.

• The emission from the developed countries, taken together and defined by the top of the band for "Other developed countries," has been rising slowly, as has their total energy usage, discussed previously, and now amounts to about 60% of the total.

• Emissions from the countries of the former Soviet Union decreased significantly in the early 1990s, after the fall of communism; the decrease in the

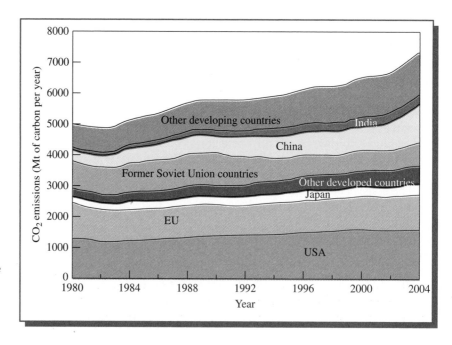

FIGURE 7-1 CO$_2$ emissions from fossil-fuel combustion for different countries and regions since 1980. [Source: M. Raupach et al., *Proceedings of the National Academy of Sciences* 104 (2007): 10288.]

1990s roughly matched the increase in emissions from developing countries over that period.

- Emissions from China have risen sharply recently, with a spike that began in 2002.

Almost all the increase in CO_2 emissions discussed above arose from increases in usage of energy derived from fossil fuels. However, the ratio of carbon dioxide to energy varies among countries and over time because of differences in the fraction of energy produced by combustion and, as discussed later in the chapter, because different fossil fuels produce very different amounts of the gas per joule. Because of the importance of CO_2, the term **carbon intensity,** defined as the ratio of CO_2 emissions per dollar of GDP, has become widely discussed by policymakers.

The global carbon intensity declined over the last half of the twentieth century, especially over its last two decades, when it declined by about one-quarter. However, as the solid green line in Figure 7-2 indicates, the global intensity remained approximately constant from 2000 to at least 2005. Currently, the production of one dollar's worth of goods and services results, on average, in the emission of about 730 g of carbon dioxide into the atmosphere. Carbon intensities are usually expressed as the carbon content alone of the emitted CO_2, or about 200 g of carbon per dollar in the present case. (Notice that the curves in Figure 7-2 are not absolute, but relative to their year 2000 values.)

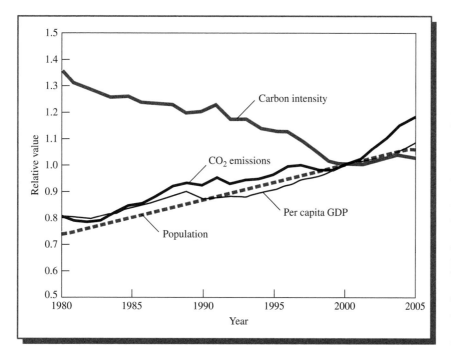

FIGURE 7-2 Global CO_2 emissions (heavy black curve; includes both fossil-fuel combustion and cement production) and important factors that influence it. Dashed green curve = population; thin black curve = per capita GDP; solid green curve = carbon intensity of GDP; heavy black curve = CO_2 emissions. Note that all parameters are normalized to year 2000 values.
[Source: M. Raupach (Global Carbon Project), *Carbon in the Earth System: Dynamics and Vulnerabilities,* Beijing, November, 2006.]

PROBLEM 7-2

Show that the carbon content of 730 g of CO$_2$ is 199 g.

Although they had very similar carbon intensity values in the early 2000s, the intensities for the world's two largest economies—the United States and China—evolved quite differently over time, as illustrated by the solid green curves in Figures 7-3a and 7-3b. American carbon intensity has fallen gradually and almost continuously, whereas that of China fell precipitously once major industrialization began, then reached a minimum in about 2000 and rose significantly at least in the half-decade that followed. Presumably, much of the decrease in the U.S. intensity is due to its continuing conversion from a manufacturing to a knowledge economy, whereas the increase in China arises from its development as an energy-intensive economy that produces large quantities of manufactured goods.

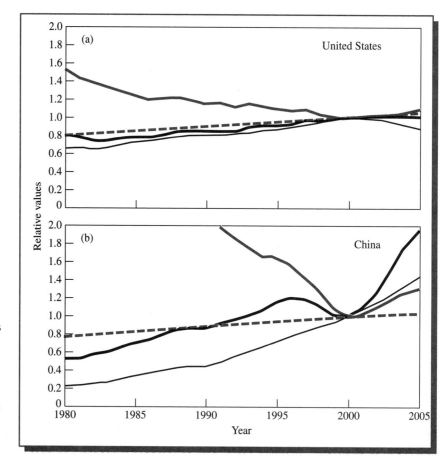

FIGURE 7-3 CO$_2$ emissions and important factors that influence it, from (a) the United States and (b) China. See Figure 7-2 for curve identification. [Source: M. Raupach (Global Carbon Project), *Carbon in the Earth System: Dynamics and Vulnerabilities*, Beijing, November, 2006.]

The driving forces behind the changes over the last few decades in global and regional CO_2 emissions can be understood by considering the various curves in Figures 7-2 and 7-3; all factors in these graphs are normalized to their year 2000 values. The continuous rise in world population over time is indicated by the dashed green line in Figure 7-2. As the world economy developed, the average GDP per person also rose, as shown by the thin black curve. The product of these two quantities is the global GDP, which rose in a faster-than-linear fashion (not shown) since each of these factors was rising more or less linearly with time. However, in the period until about 2000, the carbon intensity of the global GDP (solid green line in Figure 7-2) *declined* almost linearly with time, so the total carbon dioxide emission rate—which is a product of the three factors—rose only gradually and more or less linearly in that period (heavy black curve in Figure 7-2). From 2000 to 2005 at least, since the carbon intensity did not fall, the emission rate of carbon dioxide rose dramatically.

CO_2 emission rate = population $\times$ per capita GDP $\times$ carbon intensity

The corresponding emission curve for the United States (Figure 7-3a) involves the same combination of factors, with post-2000 emissions steady due to the continuing decline in carbon intensity. In contrast, the continuing strong rise in per capita GDP in China, combined with the increasing carbon intensity, has produced sharply increasing emissions (Figure 7-3b).

Another way of measuring the carbon dioxide emissions from different countries is to consider the per capita releases of this gas into the atmosphere. Currently the emissions of carbon dioxide amount to about 4 tonnes per person per year when averaged over the global population; this factor is usually expressed as 1 tonne of *carbon*, and it is this reference to carbon that will be used henceforth. The global average carbon emission remained remarkably steady, at about 1.1 tonnes, since rising to this value in the early 1970s, although it has increased slightly in recent years.

People in developed countries have much higher annual average emissions than do those in developing countries: 3 versus 0.5 tonnes of carbon per person. The United States leads in both total and per capita CO_2 emissions, according to the bar graph in Figure 7-4, where we have listed in order the top 20 carbon dioxide emitter countries as of 2003. The black bars indicate the country's percentage of total global emissions, and the green bars show the per capita emissions. Notice that, compared to the compact European countries and Japan, the United States, Canada, and Australia have the highest per capita CO_2 emission rates, in part due to the high transportation requirements of these vast lands. It is also true that, in these three countries, fossil-fuel energy is much cheaper than in European countries. The other developed countries have remarkably similar per capita annual carbon emissions—about 2 tonnes—which perhaps is the value currently developing countries will reach once they are fully developed. The per capita carbon emissions from the United States, the European Union (EU), and the world in total remained remarkably

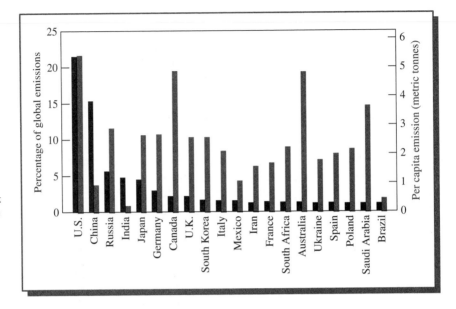

FIGURE 7-4 Total (black bars) and per capita (green bars) emissions of carbon dioxide by top 20 emitter countries in 2003. [Source: Data from Carbon Dioxide Information Analysis Center.]

constant over the last few decades of the twentieth century, the increases in GNP being matched by decreases in intensity of both energy and carbon.

Because populations of different countries vary so much, their greenhouse gas emissions per capita or per dollar of GNP are no guide to their *total* emissions. Thus in Figure 7-4, we see that China and India make substantial contributions to the total global emissions since their populations are so large, even though their per capita emission rates are still quite modest. Both countries generate most of their electricity by burning coal. China's total CO_2 emissions since the turn of the millennium have exceeded even its rapid growth of the 1980s and early 1990s. India's emissions have been growing almost linearly, currently by about 3–4% annually since 1980, though it still has the lowest per capita emissions of any of the top 20 emitter countries.

Patterns of Growth in CO_2 Concentrations

Because carbon dioxide has such a long lifetime in the atmosphere—a century or more on average—the gas *accumulates* in air. Thus almost all the CO_2 emissions from the 1990s, for example, that did not find a temporary sink will remain in the air for decades to come, adding to the bulk of the emissions from the 1980s, 1970s, and previous years. The actual carbon dioxide molecules that constitute this additional mass will change from year to year, as some CO_2 molecules leave the temporary sinks and enter the atmosphere while an equal number from the air enter one or another temporary sink.

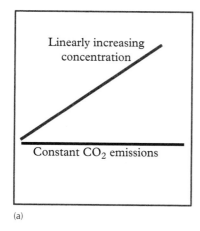
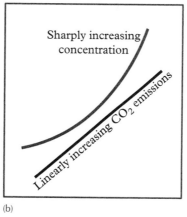

(a)   (b)

FIGURE 7-5 CO_2 concentration related to its emission level. (a) Constant emissions of CO_2 produce a linearly increasing concentration of the gas. (b) Linearly increasing emissions produce a quadratically increasing concentration of the gas.

The growth pattern of the CO_2 concentration in air is determined mainly by the pattern of CO_2 emissions. Suppose, for example, that the same amount of carbon dioxide emissions were added to the air each year and did not find a temporary sink (Figure 7-5a, black line). The total amount of CO_2 in air—and hence its concentration—would then annually increase by a constant amount. The carbon dioxide concentration increases linearly with time in this case (Figure 7-5a, green line). If the world could hold its carbon dioxide emissions constant at the year 2000 value, then the CO_2 concentration would presumably increase linearly, by the current 2-ppm yearly increment, and would consequently become about 500 ppm in 2060. Currently, global CO_2 emissions are growing slowly, increasing by only about 1% annually (see Figure 6-8), so the atmospheric carbon dioxide concentration is growing almost linearly (Figure 6-8 inset).

Another scenario, which at times has been more realistic than the situation just described, is that the CO_2 *emissions* were not the same each year but themselves *increased linearly*, that is, by a constant amount k each year. Thus, if the emissions one year amounted to A, the next year they were $A + k$, and the following year $A + 2k$, etc. (Figure 7-5b, black line). In this case, if the fraction of the gas that enters the oceanic sink each year is constant, the growth in CO_2 *concentration* will be quadratic, much sharper than linear: The resulting plot of CO_2 concentration curves upward, as illustrated in Figure 7-5b (green curve). Indeed, in the decades preceding the mid-1970s, the CO_2 concentration (Figure 6-8 inset) did increase quadratically since carbon dioxide emissions were increasing rapidly (Figure 6-9). However, since then, the CO_2 concentration has increased in an approximately linear manner with time, reflecting the slower rate of increase in emissions in this period, among other factors—including the fact that the fraction of emissions that find a temporary sink varies with time (as was discussed in Chapter 6).

Fossil Fuels

As mentioned above, most commercial energy in the world is produced currently by the burning of fossil fuels. In this section, we discuss the nature and future supplies of these fuels and compare their differing abilities to produce the greenhouse gas carbon dioxide. Later in the chapter, we shall see that the main problem of fossil-fuel usage in the present century is the CO_2 emissions that result from its combustion rather than a shortage in supply.

Fossil Fuels: Coal

The main fossil-fuel reserve is coal, which is available in abundance in many regions of the world, including developing countries, and which is cheap to mine and transport. Five countries—the United States, Russia, China, India, and Australia—have 75% of the world's coal reserves. At today's rate of consumption, coal reserves are estimated to last another 200 years, much longer than oil or gas (see below). Indeed, some observers believe the world will return to a greater reliance on coal as the major fossil fuel later in this century. The 2100 coal-fired power plants in the world are collectively responsible for about a third of all anthropogenic CO_2 emissions. Currently, coal produces about half the electric power in the United States and 80% of that in China.

Although it is a mixture, to a first approximation coal is graphitic carbon, C. It was formed from the tiny proportion of ancient plant matter that was covered over by water and could not be recycled back to CO_2 at that time. This also accounts for the buildup of O_2 in the atmosphere. Coal was formed from the highly aromatic, polymeric component of land-based woody plants called *lignin*; an approximate formula for lignin is C_3H_3O. Over long periods of time during which the material was subjected to high pressures and temperatures, both water and carbon dioxide were lost. The material polymerized further in the process to yield the very carbon-rich, hard material known as coal. Unfortunately, during its formation the coal also incorporated measurable quantities of virtually every naturally occurring element, so that when it is burned, it emits not only CO_2 and H_2O but also substantial quantities of many air pollutants—notably sulfur dioxide, fluoride, uranium and other radioactive metals, and heavy metals including mercury. Thus coal has a reputation for being a "dirty fuel." The removal of some of these impurities, especially sulfur, by various modern technologies was discussed in Chapter 3.

The burning of coal domestically in stoves and furnaces produces a great deal of soot, and it therefore has been largely discontinued in developed countries. However, coal is still used in most developed and developing countries for electric power production. When burned in such plants, the soot problem is readily solved, although emissions of sulfur and nitrogen dioxides and of mercury require more sophisticated and expensive equipment to control, also discussed in Chapter 3 and the online case study associated with Chapter 8.

The heat produced in the combustion of the fossil fuel is used to generate high-pressure steam, which in turn is used to turn turbines and thereby produce electricity. As discussed below, however, the ratio of CO_2 to energy produced from coal is substantially greater than for the other fossil fuels. Coal can also be used to produce alternative fuels—as discussed in Chapter 8—but unfortunately, the conversion processes are not very energy efficient. Although the emission of carbon dioxide is not reduced by such conversions, they do allow the removal of sulfur dioxide and other pollutants and so are "clean" ways to use coal.

To a first approximation, the amount of heat released when a carbon-containing substance burns is directly proportional to the amount of oxygen it consumes. From this principle, we can compare different fossil fuels by the amount of CO_2 released when they are burned to produce a given amount of energy.

Consider the reactions of coal (mainly carbon), oil (essentially polymers of CH_2), and natural gas (essentially CH_4) with atmospheric oxygen and written so that the amount of carbon dioxide produced is identical in each case:

$$C + O_2 \longrightarrow CO_2$$
$$CH_2 + 1.5\,O_2 \longrightarrow CO_2 + H_2O$$
$$CH_4 + 2\,O_2 \longrightarrow CO_2 + 2\,H_2O$$

It follows from the stoichiometry of these reactions that, per mole of O_2 consumed and thus approximately per joule of energy produced, natural gas generates less carbon dioxide than does oil, which in turn is superior to coal, in a ratio of 1:1.33:2. (The actual ratio is computed in Problem 7-3.)

PROBLEM 7-3

Given the following thermochemical data, determine the actual ratio of CO_2 per joule of heat released upon the combustion of methane, CH_2, and elemental carbon (graphite). ΔH_f values in kJ mol^{-1}: CH_4, -74.9; $CO_2(g)$, -393.5; $H_2O(l)$, -285.8; C(graphite), 0.0; CH_2, -20.6.

PROBLEM 7-4

The relative amounts of oxygen required to oxidize organic compounds to carbon dioxide and water can be deduced from calculating the *change* in the oxidation number (state) of the carbon atom in going from the fuel molecule to the product. Show that the ratio of oxygen required to combust C, CH_2, and CH_4 stands in the ratio of 2:3:4 according to such a calculation.

Fossil Fuels: Natural Gas

FIGURE 7-6 Releasing natural gas deposits from the Earth.

Petroleum and natural gas are essentially mixtures of hydrocarbons. They originated as the small fraction of marine organisms and plant matter that were buried and therefore cut off from the oxygen that was required for their complete oxidation. The high temperatures and pressures to which this buried material was later subjected decomposed it further, into liquid and gaseous hydrocarbons. Like petroleum, natural gas deposits are found in geological formations in which the gas mixture has been trapped by a mass of impermeable rock. Drilling a hole down through the rock releases the gas in a steady flow to the surface (Figure 7-6).

In terms of its hydrocarbon component, natural gas as it exits from the ground consists predominantly (60–90%) of **methane,** CH_4. The other component alkanes—*ethane, propane,* and the two *butane* isomers—are gases present to varying extents depending upon the geographic origin of the deposit. (See Appendix I if you are unfamiliar with the terminology and numbering systems of organic molecules.)

Methane's boiling point is so low (−164°C) that it does not readily condense into a liquid, even at moderately high pressures. In contrast, the other gaseous alkanes possess substantially higher boiling points. This makes it possible to largely remove the other alkanes from natural gas by lowering the mixture's temperature and thereby condensing these other hydrocarbons to liquids.

Sulfur compounds are also important impurities in natural gas, as previously mentioned: Some deposits contain more H_2S than CH_4! The hydrogen sulfide is removed from the gas by the Claus reaction, as discussed in Chapter 3. After processing to remove the other alkanes and the sulfur compounds, the natural gas—which now is mainly methane—is transported under pressure by pipeline to consumers.

Unfortunately, as we have discussed in Chapter 6, a small fraction of the methane being transported from its source to the consumer is lost to the atmosphere when natural gas pipelines leak; the greenhouse-enhancing effect of this methane could override some of the advantage methane has in producing less CO_2 per joule upon combustion compared to oil and especially compared to coal (see Additional Problem 3).

The enormous quantity of natural gas held in *methane hydrates* (clathrates) in ocean sediments and permafrost, as mentioned in Chapter 6, would double the fossil-fuel reserves if they could be tapped. The technology to extract the clathrates, many of which are in dilute form and mixed with sediments that lie far below the seabed, does not yet exist.

Natural Gas and Propane Fuels

In the developed world, **natural gas** is used extensively as a fuel. It consists mainly of methane but contains small amounts of ethane and propane. Normally the gas is transported by pipelines from its source to domestic consumers, who use it for cooking and heating, and to some utilities, which burn it instead of coal or oil in power plants to produce electricity.

$$CH_4(g) + 2\,O_2(g) \longrightarrow CO_2(g) + 2\,H_2O(l) \quad \Delta H = -890\,kJ\,mol^{-1}$$

Unfortunately, where pipelines do not exist, the natural gas that is produced as a by-product of petroleum production at oil wells, etc. is simply wasted by venting or flaring it off, thereby adding to the atmospheric burden of greenhouse gases.

Highly **compressed natural gas** (CNG) is used to power some vehicles, especially in Canada, Italy, Argentina, the United States, New Zealand, and Russia. Due to the cost of converting a gasoline engine to accept natural gas as the fuel, the current use of CNG in vehicles is mainly restricted to taxis and commercial trucks that are in almost constant service. For such vehicles, the additional capital cost of converting the fuel system is much less in the long run than are the savings from the lower cost of the fuel. Because the compressed gas must be maintained at very high pressures to keep its storage volume reasonable, heavy fuel tanks with thick walls are required. In order to keep the weight and size of the tank to reasonable values, the driving range (before refilling) of CNG vehicles is usually considerably shorter than that of gasoline-powered vehicles.

Compressed natural gas has both environmental advantages and disadvantages as a vehicular fuel when compared to gasoline. Since methane molecules contain no carbon chains, neither organic particulates nor reactive hydrocarbons are formed or emitted into air as a result of its combustion; however, a small amount of each pollutant type is formed from the ethane and propane component of commercial natural gas. Overall, regional air quality is improved by the use of natural gas rather than gasoline or diesel oil. However, the release of methane gas from pipelines during its transmission or from tailpipes of vehicles due to its incomplete combustion could lead to increased global warming since methane is a potent greenhouse gas. A massive conversion in North America to CNG as a vehicular fuel would be limited by supply problems for the gas, which is now used extensively for domestic heating and cooking and increasingly as the fuel in new electric power plants.

Some interesting proposals have been made recently to improve the performance of natural gas as a vehicular fuel. More efficient burning of the methane results if a small amount—about 15% by volume—of hydrogen gas is added to it. Alternatively, a smaller storage volume for methane results if it is liquefied rather than simply compressed, but more energy is expended in the process.

Similar but somewhat less serious considerations apply to **propane,** C$_3$H$_8$, also a main component of **liquified petroleum gas** (LPG), in its use as a gasoline replacement in vehicles. The heat energy produced per gram of propane combusted, 50.3 kJ, is not quite as high as that of 55.6 kJ for methane. The heat released per gram by burning gasoline depends on the composition of the particular blend under consideration, but it is generally slightly less than that of propane. Both LPG and propane are readily liquefied under pressure, so they can be stored much more efficiently than can natural gas.

Fossil Fuels: Petroleum

Petroleum, or *crude oil*, is a complex mixture of thousands of compounds, most of which are hydrocarbons; the proportions of the compounds vary from one oil field to another. The most abundant type of hydrocarbon usually is the *alkane* series, which can be generically designated by the formula C$_n$H$_{2n+2}$. In petroleum, the alkane molecules vary greatly, from the simple methane, CH$_4$ (i.e., $n = 1$), to molecules having almost 100 carbons. Most of the alkane molecules in crude oil are of two structural types: One type is simply a long, continuous chain of carbons; the other has one main chain and only short branches—e.g., 3-methylhexane.

Petroleum also contains substantial amounts of **cycloalkanes,** mainly those with five or six carbons per ring, such as the C$_6$H$_{12}$ systems *methylcyclopentane* and *cyclohexane*:

methylcyclopentane   cyclohexane

Petroleum contains some aromatic hydrocarbons, principally benzene and its simple derivatives in which one or two hydrogen atoms have been replaced by methyl or ethyl groups. Recall from Chapter 4 that toluene is benzene with one hydrogen replaced by one methyl group and that the xylenes are the three isomers having two methyl groups.

PROBLEM 7-5

Deduce the structures of all the trimethylated benzenes. *[Hint: For each of the three dimethylated benzenes, draw all the structures corresponding to placement of a third methyl group. Inspect each pair of structures you draw to eliminate duplicates.]*

It is the component of petroleum containing these aromatic hydrocarbons that is most toxic to shellfish and other fish when an oil spill occurs in an ocean, whether from an oil tanker or from an offshore oil well. Higher-molecular-weight hydrocarbons form sticky, tar-like blobs that adhere to birds, sea mammals, rocks, and other objects that the oil encounters.

Petroleum is found in certain rock formations in the ground and is pumped to the surface in oil wells. As it exits from the ground, crude oil is not a very useful substance because it is a mixture of so many compounds. To gain utility, it must first be separated into components, each of which has several particular uses.

The liquid compounds that are present in crude oil consist of hydrocarbons containing from 5 to about 20 carbon atoms each. Although no attempt is usually made to isolate individual compounds from the mixture, crude oil is separated into a number of **fractions**—different liquid solutions whose components all boil within a relatively small temperature range. This separation of oil into fractions is accomplished by a process called **distillation,** which involves the vaporization by boiling of a liquid mixture, followed by the cooling of the vapor in order to cause its condensation back to the liquid state; it is described in Box 7-1. Because of the different boiling points of the compounds, it is possible to separate the mixture into components. Each day, a total of about 10 billion liters of crude oil are distilled by this procedure in hundreds of *petroleum refineries* located around the world.

In addition to hydrocarbons, petroleum also contains some sulfur (up to 4%) in the form of compounds: **hydrogen sulfide** gas, H_2S, and organic sulfur compounds that are alcohol and ether analogs in which an S atom has replaced the oxygen. These substances are much more readily removed from oil than is the sulfur from coal, making petroleum products inherently cleaner. Diesel fuel distilled from petroleum contains a higher percentage of residual sulfur than does gasoline, and the residue from the distillation contains the highest sulfur concentration of all—as well as most of the metals vanadium and nickel from the original crude oil, usually at levels of several parts per million. Sulfur that is present in fuels generally is converted during the fuel's combustion into *sulfur dioxide*, which is a serious pollutant if released into the air (Chapter 3). Some organic compounds of nitrogen also occur in petroleum and are the source of the "fuel NO" (Chapter 3) formed when gasoline and diesel oil burn.

Petroleum and fuels made from it, such as gasoline and diesel oil, have the great advantage that they are energy-dense liquids that are convenient, relatively safe to use, and relatively cheap to produce. Virtually all nonelectrical transportation systems in both the developed and developing worlds are based on cheap petroleum fuels. The possibility of switching to alternative fuels for transportation is discussed in detail in Chapter 8. It will be much more difficult to switch from oil to other chemical feedstocks for the production of pharmaceuticals and polymers once oil runs out.

Although it took nature about half a billion years to create the world's supply of petroleum, humans will probably have used almost all of it during the 200-year period that started just before the end of the nineteenth century. Indeed, the production of petroleum in the lower 48 states of the United States has already peaked. In commerce, petroleum is measured in barrels, each of which is equivalent to 159 liters, or 42 U.S. gallons. Current annual

BOX 7-1	Petroleum Refining: Fractional Distillation

As we noted in the text, fractional distillation separates crude oil into a number of fractions having molecules of similar sizes. The crude oil mixture is first continuously fed through pipes that pass through a furnace that heats the oil to 360°–400°C. Even higher temperatures are not employed because of the tendency of the oil to decompose under such conditions. At the temperatures that are used, most of the oil is converted to a gas. The portion of the oil that is *not* vaporized is a hot liquid, called the *bottoms* or the *residuum*, which contains the heaviest molecules found in oil. It is drawn off and subsequently separated into components that are used as solid products such as waxes and asphalt, or it is used for making the form of carbon called *coke* that is employed in steel production. It is also possible, by reducing the pressure to almost a vacuum, to boil the bottoms fraction of crude oil and, by various techniques, to split the vaporized long molecules of this fraction into shorter ones, for use in gasoline and diesel fuel.

The vaporized oil is injected into a vertical distillation or *fractionating* tower, which is several meters in diameter and up to 30 m high. The temperatures in the tower *decrease* as the hot gases move higher and higher; thus the vapor cools as it rises (see Figure 1). Since they correspond to compounds with high boiling points and hence high condensation temperatures, the first gaseous molecules to recondense to liquids as the vapor rises through the tower are those with 17 or 18 or more carbon atoms. By means of a series of collection trays situated in the tower at positions where the temperature falls to an average of about about 350°C, this liquid fraction of the oil can be collected and drained off, thereby separating it from the remainder, which continues to rise in the tower. This first fraction of the petroleum, called *gas oil*, is a rather viscous liquid when cooled to room temperatures and finds commercial use as lubricating oils.

Another series of collection trays and an output pipe are located somewhat higher in the tower, where the temperature has cooled sufficiently, to about 300°C, to allow hydrocarbons in the C$_{16}$ to C$_{18}$ range to condense and be collected. This second fraction, used as diesel fuel and industrial heating oil, is called *middle distillates*. The final fraction, called *kerosene* or *heavy naphtha*, which contains primarily hydrocarbons with 12 to 16 carbons, is collected by trays near the top of the tower, where temperatures have cooled to about 150°–275°C. The kerosene fraction is used as diesel and jet fuels and as oil for home heating.

There is no particular reason why the fractions described above, with these particular boiling point ranges, should be the *only* ones collected. In fact, different petroleum distillation towers collect different fractions by using different collection temperatures, not just those we have described. The decision as to exactly which fractions are to be collected is made by considering the end uses for the various products.

At the top of the tower, the remaining uncondensed vapor contains hydrocarbons consisting primarily of molecules having 1 to 12 carbon atoms each. This vapor is cooled almost to normal outdoor temperatures in a separate unit, a procedure which condenses the molecules with 5 to 12 carbons to the liquid known as *straight-run gasoline* or *light naphtha*. This fraction, which constitutes about one-fifth of the original oil, is the basis of the gasoline used to power motor vehicles.

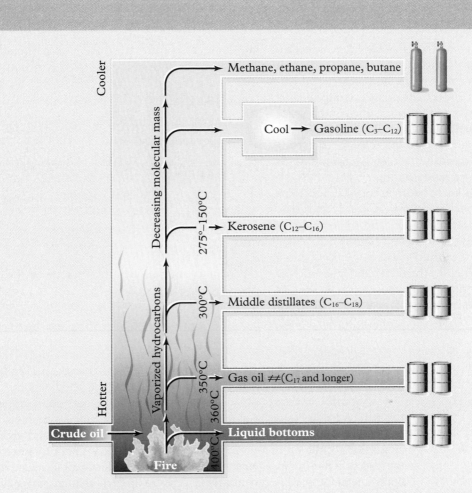

FIGURE 1 Petroleum fractions in a refining tower.

Alkanes having more than about 12 carbons cannot be used in gasoline since they do not evaporate in the engine sufficiently to burn properly.

The C_1 to C_4 gases—namely methane, ethane, propane, and butane—that remain uncondensed at the top of the tower can be collected and used for the purposes previously described for components of natural gas. The C_4 alkanes (butanes) are used as components both of gasolines and of liquefied petroleum gas. Unfortunately, the C_1 to C_3 gases are sometimes ignited and simply "flared off" into the air if facilities for their condensation or transportation do not exist at the petroleum refining site.

In summary, the fractionating tower separates crude petroleum into a number of useful

(continued on p. 278)

BOX 7-1	Petroleum Refining: Fractional Distillation *(continued)*

materials, each of which is a mixture of hydrocarbons in which the different constituents have approximately the same number of carbon atoms and which all boil within a small range relative to the larger range of the crude oil. For many applications, a further separation of a fraction into subfractions, each consisting of a smaller set of hydrocarbons, is accomplished subsequently.

In addition to hydrocarbons, crude oil also contains small quantities of compounds that contain other elements. The most predominant of these elements is sulfur, which occurs in oil to the extent of 0.5–6%, depending upon the origin of the material. Metals such as vanadium, nickel, and iron also are present, at a total concentration of more than 1000 ppm

in some cases, but they are usually found mainly in the bottoms.

PROBLEM 1

Explain the process by which crude oil is converted to usable fractions.

PROBLEM 2

How are gasoline and kerosene different? How are they alike?

PROBLEM 3

Why can both gasoline and kerosene be used as fuels?

world oil production amounts to about 4 *trillion* (4×10^{12}) liters, or 27 billion barrels. Much of the proven oil reserves are located in the Middle East.

Although it is commonly said that we are "running out" of oil and gas, this will probably not occur globally in the short-to-medium term. Improvements in the technology of petroleum extraction allow greater and greater proportions of the oil in a given deposit to be extracted. The initial extraction of oil from a reserve usually proceeds without difficulty. However, later on, the remaining oil occurs mostly in pores as droplets that are larger than the connecting necks of the porous formation; hence it will not flow because of surface tension. In *secondary extraction*, surfactants, water, or pressurized carbon dioxide can be used to lower the surface tension and drive additional oil from the reserve. In tertiary extraction, steam is injected to lower the viscosity of the remaining oil and allow it to be pumped. As the more accessible supplies dry up and the price of oil rises, such extraction processes are becoming more economically viable. Indeed, most major oilfields are now undergoing at least secondary extraction.

Some analysts believe that global oil production will peak sometime between 2005 and 2015, whereas others argue that this will not happen until at least 2030. A 2006 survey of petroleum geologists found that most believe the peak will occur between 2020 and 2040. A recent estimation of the manner in which oil production is expected to vary with time is shown as a modified bell curve in Figure 7-7. Notice that production of conventional oil

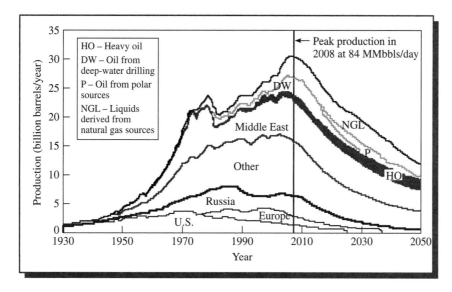

FIGURE 7-7 Historical and projected oil production for 1930–2050 from various geographic areas. [Source: C. J. Campbell, *The Coming Global Oil Crisis,* 2004. www.hubbertpeak.com/campbell]

outside the Persian Gulf area (curve above "Other") probably peaked a few years ago. Indeed, U.S. production reached a maximum in 1972. The zones at the top of the graph represent nonconventional sources of oil (including those from deep-water drilling), liquid hydrocarbons obtained from natural gas, and heavy oil from tar sands etc.; the latter sources are expected to play an increasingly important role in oil supplies in future decades.

Very-long-chain hydrocarbons from *oil shales*, a type of sedimentary rock, and *tar sands*, which is sandstone or porous rock impregnated with very viscous crude oil, could extend the petroleum supply. Indeed, the potential for oil from tar sands in Alberta, Canada, exceeds the oil reserves of Saudi Arabia! However, there are currently high economic, energy, and environmental costs for exploiting these potential reserves in a large-scale manner. For example, about 2 tonnes of the asphalt-like mixture of sand and tar in northern Alberta's heavy-oil deposits are required to produce one barrel of oil. Energy in the form of natural gas must then be expended to separate the oil from the sands and to split the long-chain hydrocarbon molecules of the tar into shorter ones for use in gasoline (see Additional Problem 4). The extraction process also consumes several times as much fresh water as the volume of oil produced, and there are already concerns that the supply of water will not be sufficient for all the planned expansions of production.

Gasoline

Regular gasoline contains predominantly C_5 to C_{11} hydrocarbons; diesel fuel contains mainly hydrocarbons with 9–11 carbon atoms. Generally speaking, the more carbon atoms in the alkane, the higher its boiling point and the lower its vapor pressure—and hence the lesser its tendency to vaporize—at a given temperature. For this reason, gasoline destined for warm summer conditions is formulated with smaller amounts of the smaller, more easily vaporized alkanes such as butanes and pentanes than is that prepared for winters in cold climates. The presence of volatile hydrocarbons in gasoline is vital in cold climates so that automobile engines can be started.

Gasoline that consists primarily of straight-chain alkanes and cycloalkanes has poor combustion characteristics when burned in internal combustion engines. A mixture of air and vaporized gasoline of this type tends to ignite spontaneously in the engine's cylinder before it is completely compressed and sparked, so the engine "knocks," with a resulting loss of power. Consequently, all gasoline is formulated to contain substances that will prevent knocking.

In contrast to unbranched alkanes, highly branched ones such as the octane isomer *2,2,4-trimethylpentane,* "isooctane" (illustrated below), have excellent burning characteristics. Unfortunately, they do not occur naturally in significant amounts in crude oil. The ability of a gasoline to generate power without engine knocking is measured by its **octane number.** To define the scale, isooctane is given the octane number of 100, and *n*-heptane is arbitrarily assigned a value of zero.

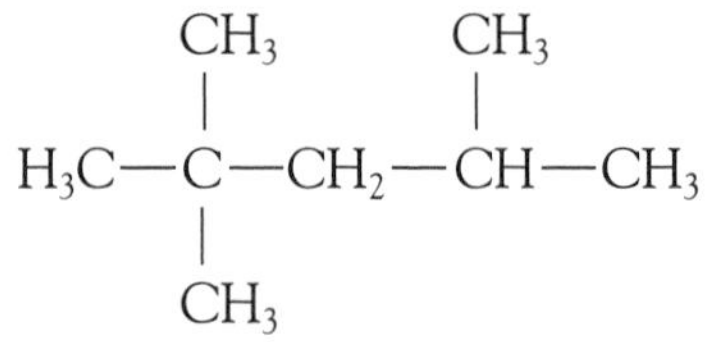

$$
\begin{array}{ccc}
\text{CH}_3 & & \text{CH}_3 \\
| & & | \\
\text{H}_3\text{C}-\text{C}-\text{CH}_2- & \text{CH}-\text{CH}_3 \\
| & & \\
\text{CH}_3 & &
\end{array}
$$

2,2,4-trimethylpentane ("isooctane")

Gasoline that has been produced simply by distillation of crude oil has an octane number of about 50, much too low for use in modern vehicles. However, when added to gasoline in small amounts, the compounds **tetramethyllead,** Pb(CH$_3$)$_4$, and its ethyl equivalent prevent engine knocking and hence greatly boost the octane number of gasoline. For decades, they were added worldwide to gasoline consisting predominantly of unbranched alkanes and cycloalkanes. However, these additives have largely been phased out now in most developed countries owing to environmental concerns about lead, a topic discussed in Chapter 15.

In some European countries and in Canada, lead compounds were replaced by small quantities of an organic compound of manganese called **MMT,** which stands for *methylcyclopentadienyl manganese tricarbonyl.* The use of MMT has been controversial for health reasons, since manganese concentrations rise in air and soil as a result, and for technological reasons, since some car manufacturers claim it degrades emissions components on vehicles. Until 1995, MMT was banned in the United States; even though this ban was revoked, its use there is very small.

The alternative to using lead or manganese additives to boost octane ratings is to blend into gasoline significant quantities of highly branched alkanes,

TABLE 7-1	Octane Numbers of Common Gasoline Additives	
Compound		**Octane Number**
Benzene		106
Toluene		118
p-Xylene (1,4-dimethylbenzene)		116
Methanol		116
Ethanol		112
MTBE		116

BTX, or other organic substances such as MTBE (discussed later), which themselves have high octane numbers. A list of the common additives is shown in Table 7-1. Collectively, the **benzene** + *toluene* + *xylene* component in gasoline is called **BTX.**

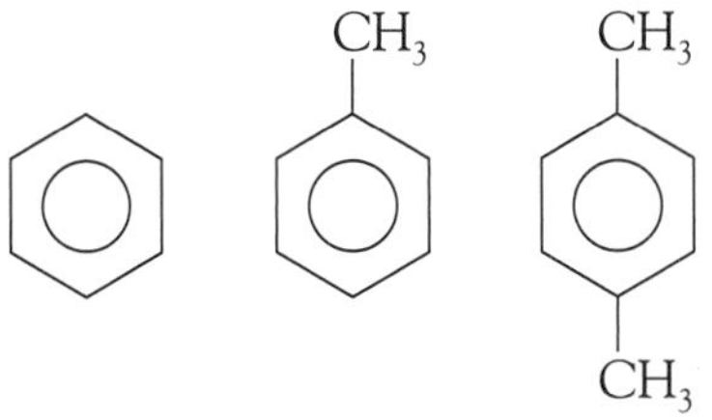

benzene   toluene   p-xylene

Gasoline often also contains some trimethylated benzenes and *ethylbenzene;* the mixture is then called BTEX. Currently, most unleaded gasoline sold in the United States contains significant quantities of BTX (as high as 40% content in the past), ethanol (especially in the Midwest), or MTBE to boost the octane number. Unfortunately, the BTX hydrocarbons are more reactive than the alkanes that they replace in causing photochemical air pollution, so in a sense the lead pollution has been reduced at the price of producing more smog. In addition, the use of BTX in unleaded gasoline in countries, such as Great Britain, where few cars were equipped with catalytic converters resulted in the past in an increase in the BTX concentrations in outdoor air. Benzene in particular is a worrisome air pollutant since at higher levels it has been linked to increases in the incidence of leukemia (see Chapter 4).

The **reformulated gasoline** used in the second half of the 1990s in North America contained a maximum of 1% benzene and 25% (volume) aromatics in total, with a minimum of 2% oxygen (by mass). The second phase of reformulated gasoline, which entered the U.S. market in 2000, reduces the benzene and BTX components even further and lowers the sulfur content to 30 ppm. The EU planned to cut the maximum level of benzene in gasoline by 75% effective 2005.

The use of alcohols and compounds derived from them as additives or as "oxygenated" motor fuels in their own right is discussed in Chapter 8.

Sequestration of CO_2

In the future, CO_2 might be removed chemically from the exhaust gases of major point sources that would otherwise release it into the atmosphere, such as power plants that burn fossil fuels and that collectively are responsible for one-quarter to one-third of total emissions. The carbon dioxide gas so recovered would then be **sequestered**—i.e., deposited in an underground or ocean location that would prevent its release into the air. For example, the CO_2 could be sequestered by burial in the deep oceans, where it would dissolve, or in very deep aquifers under land or the seas, or in empty oil and natural gas wells or coal seams. The total amount of carbon dioxide that will be produced by fossil-fuel combustion over this century will amount to more than one trillion tonnes, so vast amounts of storage would be required (Problem 7-6).

PROBLEM 7-6

Calculate the volume, and from it the length in kilometers, of each side of a cube of liquid or solid carbon dioxide whose density is the same as that of water, i.e., 1 g/cm^3, and whose mass is a trillion (10^{12}) tonnes.

Since it is not economically feasible to transport and store emission gases from power plants, the dilute carbon dioxide (usually 12–14% by volume) in the gas mixture must be captured and concentrated. The energy input required for the CO_2-concentrating phase of **carbon sequestration** schemes would represent a substantial fraction, from one-third to one-half, of the output of the power plant to which it is connected; therefore more fuel would be combusted and more air pollutants produced. The equipment required to capture and concentrate CO_2 is very large, since huge volumes of air are involved. The capture/concentration of the gas accounts for about three-quarters of the cost of the entire sequestration process and currently is estimated to cost about $100 per tonne of carbon. Nevertheless, some observers feel it would still be cheaper to sequester carbon from fossil fuels than to convert to a renewable-fuel economy.

The capture of carbon dioxide is generally accomplished by passing the cooled (to about 50°C) emission gas through an aqueous solvent containing 15–30% by mass of an *amine*, R_2NH (or RNH_2), which combines with the CO_2 to produce an anion, in a process analogous to that between the gas and water:

$$R_2NH + CO_2 \rightleftharpoons R_2N-CO_2^- + H^+$$

$$H_2O + CO_2 \rightleftharpoons HCO_3^- + H^+$$

Over time, the amine solution becomes saturated with the gas since all the amine molecules become tied up with CO_2. Heat is then used to reverse this reaction (at about 120°C), thereby producing a concentrated stream of carbon dioxide for later disposal and regenerating the amine solution after it has been cooled sufficiently. The amines commonly employed in this **chemical absorption** technology are *monoethanolamine* and *diethanolamine,* since they can absorb high loads of the gas and require relatively little heating to release it later. With these amines, over 95% recovery of CO_2 can be achieved. A strong base such as sodium hydroxide could be used to absorb the CO_2, but the bicarbonate salt it produces requires much more heat to later decompose it and release its carbon dioxide. Current research centers on finding a solvent that would bind the CO_2 efficiently but less tightly and therefore require less energy to decompose it later when concentrated.

Nitrogen and sulfur oxides must be removed from the emission gas before it interacts with the solvent, since otherwise they would react with and degrade the amine. Heating the solution to liberate the CO_2 and regenerate the amine solvent and compressing the gas once it is produced are the most energy-intensive steps of the process. In addition, enormous amounts of the solution must be transported from place to place. The decomposition of some of the amine, producing ammonia and salts, inevitably occurs during the repurification cycles.

Carbon dioxide can also be chemically absorbed by certain metal oxides, which will release the gas when heated. For example, **calcium oxide,** CaO, can quickly remove CO_2 from hot emission gases by formation of **calcium carbonate,** $CaCO_3$:

$$CaO(s) + CO_2(g) \rightleftharpoons CaCO_3(s)$$

Subsequent heating of the solid to about 900°C, once it has been largely converted to the carbonate, reverses the reaction and produces concentrated CO_2 and regenerates CaO. Unfortunately, calcium oxide deactivates relatively quickly over many absorption/deabsorption cycles, so fresh oxide must constantly be added to maintain the absorptive activity of the solid.

Three other techniques are available by which carbon dioxide can be stripped from exhaust gases:

- **Membrane separation** Polymeric membranes that allow CO_2 to pass through them, while the other gases are excluded, can be employed to recover about 85% of the carbon dioxide. This technology has been used for many decades in the oil industry, since it is more economical than chemical absorption when the concentration of carbon dioxide in the source gas is relatively high. There is currently much research and development under way in devising membranes that will be efficient and economical for power plant emissions.

- **Physical adsorption** Certain solids, such as some zeolites and activated carbon, that have large surface areas will adsorb CO_2 from the gas mixture and later release it upon heating. Methanol and glycols are also used as solvents to capture carbon dioxide from concentrated emission gases.

- **Cryogenic separation** Since CO_2 has a higher condensation temperature than nitrogen or oxygen, it can be isolated as a liquid by condensing the gas mixture at a very low temperature under high pressure. However, the energy requirement for this technique is approximately double that of chemical absorption using amines.

One way to circumvent the high expense and energy consumption required to isolate and concentrate the CO_2 from conventional power plants is by **oxycombustion.** In this technique, currently under development, the fossil fuel is burned not in air but rather in *oxygen gas*, O_2. If the stoichiometric amount of oxygen is supplied, the exhaust gas from oxycombustion will consist entirely of carbon dioxide and require no isolation step. (By contrast, since air is only 19% oxygen by volume, the maximum level of CO_2 when air is used for combustion is also only 19%.) Of course, the original isolation of oxygen from air requires energy, and the combustion facilities must be redesigned to be able to use pure oxygen. In practice, since combustion in pure O_2 produces a flame too hot (3500°C) for power plant materials, it is diluted with some of the CO_2 from the combustion to reduce the flame temperature. The output gas from oxycombustion is compressed and dried of the water vapor produced during combustion. It can then be transported by pipeline as a dense supercritical fluid (see Chapter 6).

Another scheme proposed for the future involves the conversion of a fossil fuel, either coal or natural gas, to **hydrogen gas,** H_2, which would be employed as the fuel, either in a power plant or in a vehicle, in a reaction that generates no additional carbon dioxide. In essence, the fuel value of the coal or natural gas is transferred to hydrogen by the gasification process. Such techniques for generating H_2 for use as a fuel are described in detail in Chapter 8; in general, the process corresponds to the reaction

$$\text{carbon-hydrogen fuel} + \text{water} \longrightarrow CO_2 + H_2$$

The high concentration of pressurized CO_2 in the gas mixture (in principle, 50% by volume) allows for a more economical isolation of the gas, in this case using a liquid glycol solvent, than its capture from emission gases in a conventional power plant. Alternatively, a membrane that allows only hydrogen to pass through it could be employed to produce a gas stream largely composed of CO_2. Prototype power plants in which methane is first converted to hydrogen and carbon dioxide, with the latter extracted and pumped into an underground oil field, are planned for Scotland and California. The *Future-Gen* project of the U.S. Department of Energy involves the construction of a "zero-emission" coal-gasification power plant that will capture and store all the carbon dioxide it produces.

The transfer of the energy value from a fossil fuel to hydrogen eliminates the impractical task of isolating the carbon dioxide and collecting it when the fuel itself is used to power vehicles and to heat or cool buildings, applications that currently account for more than two-thirds of its emissions. Other industrial processes in which carbon dioxide at relatively high concentration can be separated by membrane techniques include natural gas purification and fermentation plants. The concentration of CO_2 in emission gases from cement plants, which produce the gas by heating calcium carbonate to release calcium oxide, reaches 15–30% and should be susceptible to more economical capture methods than those used for the more diluted emissions from power plants.

A number of different methods and locations for storing carbon dioxide have been proposed and are under current investigation, as discussed in the following sections.

Deep Ocean Disposal of CO_2

Various schemes for delivering carbon dioxide in massive amounts to the seas and depositing it there as such are labeled *ocean acidic* in Figure 7-8 and have

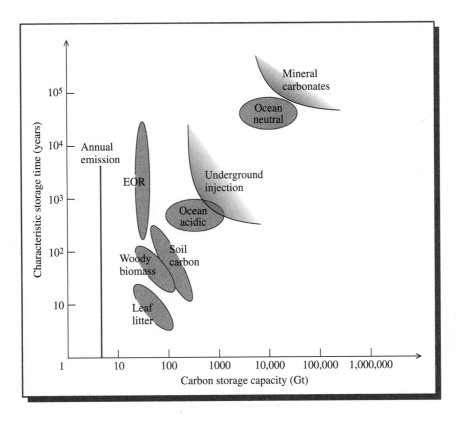

FIGURE 7-8 Capacities and storage times for various CO_2 sequestration technologies. [Source: Adapted from K. S. Lackner, "A Guide to CO_2 Sequestration," *Science* 300 (2003): 1677.]

the potential to store many hundreds of gigatonnes of carbon dioxide for many hundreds of years. The schemes are referred to as acidic because dissolving carbon dioxide gas directly in seawater produces *carbonic acid*, H_2CO_3, a weak acid that would increase the acidity of the ocean in the near vicinity:

$$CO_2(g) + H_2O(aq) \rightleftharpoons H_2CO_3(aq)$$

$$H_2CO_3(aq) \rightleftharpoons H^+ + HCO_3^-$$

Adding large amounts of carbon dioxide would lower the pH of ocean water by tenths of a unit, although much larger decreases of several pH units would occur near the points of injection.

Carbon dioxide destined for ocean storage could be transported by a pipeline, originating either from the shore or from a ship stationed above the disposal site, extending to the depth required (Figure 7-9). Even relatively shallow injection of the gas in the ocean, at 200–400-m depth, would produce a satisfactory result, provided that the seafloor there is slanted sufficiently to allow the dense, CO_2-rich water to be transported by gravity to greater depths. Simulations show that most of the gas would return to the surface and enter the atmosphere within a few decades if the CO_2-rich water was simply diluted by mixing with surrounding water, rather than sinking. Over a

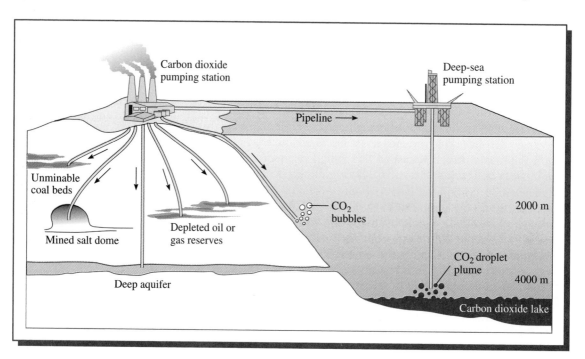

FIGURE 7-9 Potential sequestration sites for carbon dioxide. [Source: Redrawn from *Scientific American* (Feb. 2000): 72–79.]

period of centuries, excess carbon dioxide would eventually return to the atmosphere, but presumably by that time alternative energy sources would have replaced fossil fuels and the atmospheric CO_2 problem would then be less serious.

A phase diagram for CO_2 is illustrated in Figure 7-10a. Below about 500 m deep, water pressure would force pure carbon dioxide to be a compressible liquid, which above 2700 m is less dense than water and would float upward. Below that depth, it is denser than water and would sink.

However, since ocean temperatures are less than 9°C, the liquid or concentrated gas below 500 m could combine with water to form a solid, ice-like clathrate hydrate, $CO_2 \cdot 6\ H_2O$, that, if fully formed, would be denser than seawater and would sink to the deep ocean. Thus lakes of liquid and/or clathrate carbon dioxide could form on the seafloor. Figure 7-10b illustrates an experiment in which a beaker of liquid carbon dioxide was placed almost 4 km deep off Monterey Bay, California.

Direct disposal of CO_2 to the deep ocean would require a pipeline to penetrate to a depth of 3000 to 5000 m, producing a pool of liquified carbon dioxide, denser than seawater at this depth (see Figure 7-10b). Some—perhaps just the surface—or all of the liquid carbon dioxide would convert to the solid clathrate. The pool of liquid carbon dioxide would, probably over centuries, dissolve into the surrounding water. Unfortunately, sea life under

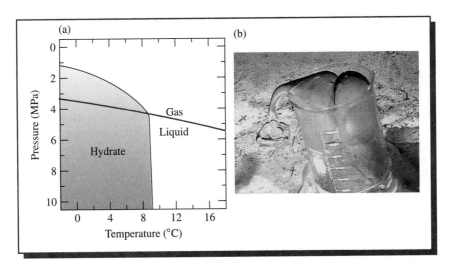

FIGURE 7-10 (a) Phase diagram for carbon dioxide. The green line shows the phase boundary between gaseous and liquid CO_2. The shaded area indicates the conditions under which the hydrate is stable if sufficient CO_2 is present. (b) Liquid carbon dioxide overflowing from a beaker placed on the seafloor at 3650-m depth. A mass of transparent hydrate formed at the upper interface, sank to the bottom of the beaker, and pushed out some of the liquid CO_2. [Source: P. G. Brewer et al., "Direct Experiments on the Ocean Disposal of Fossil Fuel CO_2," *Science* 284 (1999): 943.]

this pool would be exterminated. There is also some fear that earthquakes or asteroid impact could destabilize the pool, resulting in the release of massive amounts of carbon dioxide gas into the air above.

Near the seafloor, dissolved carbon dioxide could eventually react with the solid calcium carbonate, $CaCO_3$, in sediments formed from seashells etc. to produce soluble **calcium bicarbonate,** $Ca(HCO_3)_2$:

$$CO_2(g) + H_2O(aq) + CaCO_3(s) \longrightarrow Ca(HCO_3)_2(aq)$$

(This reaction is discussed in detail in Chapter 11.) For practical purposes, the CO_2, now chemically trapped in the bicarbonate form, would remain indefinitely in the dissolved state.

In an alternative scheme, labeled *ocean neutral* in Figure 7-8, calcium carbonate or some other suitable substance such as *calcium silicate* (a cheap, abundant mineral) would be reacted with carbon dioxide to transform it to solid *silicon dioxide*, SiO_2, and aqueous calcium bicarbonate, which could be drained into ocean depths:

$$2\ CO_2 + H_2O + CaSiO_3 \longrightarrow SiO_2\ (s) + Ca(HCO_3)_2(aq)$$

Acidity problems associated with direct carbon dioxide dissolution in seawater are avoided in this way.

Huge amounts of limestone or calcium silicate would be required for this form of sequestration, but the potential for CO_2 storage by this method is very great and the storage time of the gas is many thousands of years (Figure 7-8). In addition, it may be possible to react power plant emissions directly with a mineral, thereby avoiding the energy-expensive step of extracting and concentrating the carbon dioxide.

PROBLEM 7-7

Calculate the mass, in tonnes, of calcium carbonate that is required to react with each tonne of carbon dioxide.

Alternatively, surface rocks containing alkaline silicates could be crushed and then reacted with carbon dioxide to produce insoluble solid carbonates that could simply be buried in the ground. Unfortunately, direct carbonation reactions involving CO_2 are slow unless the mineral is heated, a step that is costly in money and energy. In one indirect scheme, magnesium silicate rock is reacted with **hydrochloric acid,** HCl, to produce silicon dioxide and **magnesium chloride,** $MgCl_2$. Reaction of this salt with carbonic acid produces insoluble **magnesium carbonate,** $MgCO_3$, and re-forms the hydrogen chloride, which, in principle, can be recycled:

$$Mg_2SiO_4 + 4\ HCl \longrightarrow 2\ MgCl_2 + 2\ H_2O + SiO_2$$
$$MgCl_2 + H_2CO_3 \longrightarrow MgCO_3 + 2\ HCl$$

There are still energy costs and additional CO_2 production associated with such procedures, however.

An alternative scheme proposed recently for carbon dioxide storage is to inject it into sediments under the seafloor. Because it would be under high pressure and at low temperature, it would exist there as a dense liquid or would combine with the water in the sediments to form the solid hydrate. Although expensive, this injection process could be useful for power plants in coastal locations.

Deep Underground Storage of CO_2

There have been suggestions that the CO_2 output from power plants could be pumped deep underground into cracks and pores in common alkaline rocks such as *calcium aluminosilicates*; there, the rocks could react with the gas, in microorganism-catalyzed processes, to produce calcium carbonate and thereby store the CO_2. Such carbonate minerals are known to be present in deep caves in Hawaii and elsewhere, so the process could well be feasible if the reactions occur quickly enough. Recently, Norway has started to pump concentrated CO_2 gas into sandstone rocks located a kilometer below the North Sea; the pores in the rock were left empty by extraction of natural gas from them in the past. The CO_2 could react with the rock and thus be immobilized.

In the short run, the easiest route to begin sequestration of carbon dioxide is probably to inject it into reservoirs containing crude oil or natural gas. However, the total capacity for carbon dioxide storage by such **enhanced oil recovery,** labeled EOR in Figure 7-8, is less than 100 Gt. This technology is already used to enhance the recovery of oil in some fields, though currently most of the CO_2 is again recovered and reused.

In an interesting international project currently under way, 5000 tonnes a day of liquefied CO_2 are sent through a 325-km-long pipeline from a coal gasification plant in North Dakota to an oilfield in Weyburn, Saskatchewan. The gas is injected through pipes 1500 m underground, allowing more oil to be extracted from the field and sequestering most of the carbon dioxide in the brine of the oil reservoir. In 2004, after the project had been under way for four years, about 3.5 million tonnes of compressed carbon dioxide gas (of a projected 20 million tonnes eventually) had been sequestered at Weyburn. The injected CO_2 that accompanies the oil extracted from the field is captured and re-injected underground. Monitoring of the site found no significant amount of the gas to be escaping through the rocks and soil of the area. Any injected carbon dioxide escaping from the area could be identified by the characteristic low $^{13}C/^{12}C$ isotope ratio of the North Dakota fossil-fuel source; artificial tracer gases were also added to the injected gas to further monitor potential emissions from the site. Analysis of the fluid that accompanies oil produced in this way revealed that the carbon dioxide has been dissolving in the reservoir brine and has been trapped by some of the minerals in the reservoir. A similar sequestration project is planned for Norway, where carbon dioxide emissions from a plant that produces methanol from natural gas would be piped to offshore oil fields and injected into undersea reservoirs to help force oil to the seabed surface.

Depleted oil and gas reservoirs could be used to store carbon dioxide (Figure 7-9). These underground caverns are known to be stable, since they have held their original materials for millions of years. Carbon dioxide storage in coal seams that lie too far underground to be mined may also be feasible. Pumping CO_2 into the coal helps it release adsorbed methane, which can then be pumped to the surface and used. Several hundred gigatonnes of carbon dioxide could be sequestered in coal.

Much larger in volume and capacity than oil and gas reservoirs are *saline aquifers*, large formations of porous rocks that are saturated with salty water and that lie far underground, well below fresh-water supplies. Carbon dioxide injected into such an aquifer initially remains a compressed gas or a supercritical liquid, but some slowly dissolves in the sometimes very alkaline brine (Figure 7-9). The brine is contained mainly in small pore spaces occupying about 10% of the volume of the porous rock. If carbon dioxide is to be stored underground, a depth of more than a kilometer is required so that its density is comparable to that of water. Even so, it is likely to rise to the top of whatever geological formation encloses it and to spread laterally. Consequently, the caprock overlying the formation must be one that is secure if significant amounts of CO_2 are not to migrate upward through the soil and into the atmosphere over time. The stability of each aquifer to potential seismic activity and gas leakage must be individually assessed before it is used.

The Norwegian energy company *Statoil* has already demonstrated the feasibility of this approach by annually storing about a million tonnes of carbon dioxide (a 9% impurity that must be removed from their crude natural gas) in a saline aquifer that lies 1000 m under the floor of the North Sea. Interestingly, Statoil found it cheaper to sequester CO_2 this way than to pay the $50 per tonne carbon tax the Norwegian government has instituted. The North Sea aquifers are sufficiently large to absorb all emissions of carbon dioxide produced by Europe for many hundreds of years. Experiments to store carbon dioxide in saline aquifers are under way in several locations in the world, including Texas and Japan. In the United States, large saline aquifers are found in the states lying just below the lower Great Lakes, in southern Florida, in northeastern Texas, and in the northern midwestern states.

Removing CO_2 from the Atmosphere

A potential technique for extracting some of the carbon dioxide that is already dispersed into the atmosphere and depositing it in ocean depths is the *iron fertilization* proposal. Experiments indicate that large portions of the seas, especially the tropical Pacific and the Southern Ocean (which surrounds Antarctica), lack plankton because they are very iron-deficient. Artificially adding iron to these areas would result in massive blooms of plankton, some of which, in the Southern Ocean at least, would descend into the deep oceans, thereby locking away the carbon dioxide used in photosynthetic activity. Experiments are under way to test the feasibility of this approach. In particular, scientists have yet to determine whether much of the additional plankton would enter the animal food chain and ultimately be converted back to carbon dioxide. In addition, some scientists have pointed out that the decomposition of the phytoplankton consumes oxygen and encourages bacteria that produce methane and nitrous oxide, thereby increasing the concentration of these greenhouse gases in air upon their release from the oceans. Other side effects of fertilization could produce additional negative environmental consequences.

Carbon dioxide can also be removed from the atmosphere by growing plants specifically for this purpose. Some utility companies and some countries have proposed a scheme by which they are given credit to offset some of their CO_2 emissions by planting forests that would absorb and temporarily sequester carbon dioxide as they grew. However, this scheme is controversial. For example, the release of CO_2 from soil into the air that occurs when ground is cleared for growing trees can exceed the total carbon dioxide absorbed by the new trees for a decade or more. In addition, the carbon stored in trees would be released back into the atmosphere if the wood burned or rotted. The conversion by pyrolysis, the decomposition of material by heat in the absence of oxygen, of wood or other biomass into charcoal or biochar produces a form of carbon that is much more durable.

In another proposal, carbon dioxide from a power plant would be used to grow vast amounts of algae, which then could be used as fuel for combustion. Several prototype facilities of this kind have been constructed. The exhaust from small power plants is led through clear tubes in which fast-growing algae are produced in an aqueous environment using sunlight to drive photosynthesis. The algae harvested from the process are dried for combustion or converted into biodiesel and ethanol fuels (see Chapter 8).

A few scientists have proposed removing carbon dioxide chemically from ambient air, for example, from high-speed winds used to turn wind turbines, using chemicals such as amines, already discussed, or some other chemical absorber. The practicality of chemically extracting carbon dioxide from air, in which its concentration is only 0.04%, compared to about 13% in power plant emissions, is yet to be determined.

Reducing CO_2 Emissions by Improving Energy Efficiency

Some writers have noted that much of the current energy expended in both industrial and domestic scenes could be saved by adopting the most *efficient* technologies for each purpose. For example, the use of low-wattage compact fluorescent light bulbs instead of incandescent bulbs would significantly reduce the amount of electrical energy used for lighting in homes; the *payback period*—until the much higher capital cost of these bulbs is more than met by savings in electrical costs—is a few years. Similarly, automobiles could be made much more energy efficient and thus use less gasoline to travel a given distance.

However, improving energy efficiency would *not* necessarily lead to a reduction in the demand for energy and a reduction in carbon dioxide emissions. The reason is that if energy-consuming equipment is made more efficient, the monetary cost for performing a given task drops, and there follows a natural tendency to use the equipment more, since it is so cheap to operate. For example, if you buy a very energy-efficient car, you will be able to afford to take more trips in it since each one would be cheaper than with a "gas guzzler." Thus some policymakers believe that because of this *rebound effect*, energy savings and CO_2 reductions from efficiency would not be achieved in the long run by making energy-consuming devices more efficient. Increased efficiency would have to be accompanied by price increases on the fuel, presumably in the form of taxes, for it to reduce overall consumption.

Reducing Methane Emissions

Although our focus so far has been on reducing emissions of carbon dioxide to the atmosphere, global warming can also be combated by decreasing the amount of methane that is released. The major opportunities for methane reduction are

- better maintenance of natural gas pipelines so as to reduce their leakage, a practice already under way in Russia;

- capture and combustion of the methane released by landfills, underground coal mines, and oil production; and

- changes in the techniques of rice production, by draining the field a few days before the plants flower, the point at which maximum emissions begin.

Energy and Carbon Dioxide Emissions in the Future

Growth in Energy Use

Since the Industrial Revolution, the rate of emission of CO_2 to the atmosphere has climbed hand-in-hand with the expansion of commercial energy use, since so much of the latter (currently 78%) is obtained from fossil-fuel sources. Barring an unforeseen, massive development of nuclear energy or renewable fuels or carbon sequestration, CO_2 emission rates are expected to match commercial energy production rate increases as the developing world undergoes industrialization and as the economies of developed countries continue to expand. Indeed, a 2003 assessment by the European Union predicts that, over the first three decades of this century, carbon dioxide emissions will rise globally by an annual average of 2.1%, due to a 1.8% annual increase in energy usage. The fraction of fossil-fuel energy obtained from coal is expected to *increase* over this period—due to higher and higher prices for oil and natural gas as they become more scarce—thereby increasing the carbon intensity of the fuel mix. The report predicts a cumulative increase in energy use for the United States of 50% and for the EU of 18%. According to the report, energy use by developing countries will triple (corresponding to 4% annual compounded growth), with the consequence that they will be responsible for 58% of CO_2 emissions by 2030, though they will still trail most industrialized countries in emissions per capita. If developing countries can implement the renewable energy technologies (discussed in Chapter 8) in constructing their economies, they would avoid the heavy fossil-fuel dependence and intense carbon dioxide emissions characteristic of all currently developed countries.

IPCC Scenarios for CO_2 Emissions and Concentrations

In its 2001 report, the U.N.'s International Panel on Climate Change (IPCC) described a number of very different scenarios for greenhouse gas emissions for the rest of the century. The magnitude of the emissions predicted at century's end vary dramatically: 5, 13.5, 20, and 29 Gt of carbon annually, the range spanning from 0.6 to 3.5 times the current value of about 8 Gt C/year. The carbon dioxide concentrations projected for 2100 for the IPCC scenarios range from 500 to more than 900 ppm, compared to today's 373- and the preindustrial 280-ppm levels.

Even with constant carbon dioxide emissions at current levels or a few percentage points lower, the carbon dioxide concentration in the atmosphere will continue to grow. Some policymakers have promoted the idea that, through international agreements or allocation schemes, the world should control future CO_2 emissions so that the atmospheric level of the gas never exceeds some specific *concentration*. Although there is no consensus on the most appropriate target, for our discussion we shall use 550 ppm. This value is twice the preindustrial value—in other words, a situation in which human actions have doubled the natural atmospheric carbon dioxide concentration.

One way in which global CO_2 *emissions* could rise and fall with time in order to eventually achieve the 550-ppm concentration target is shown by the curve in Figure 7-11a. Figure 7-11b shows how the corresponding atmospheric CO_2 *concentration* would change with time for this emission scenario. The scenario was developed assuming that international agreement on CO_2 emissions can be achieved in the relatively near future. Consequently, it assumes modest growth in CO_2 releases until about 2060, at which point a decline would set in. The temperature increase—which tracks the CO_2 concentration curve closely—by 2100 would be just under 2°C (relative to that for the year 2000). The rise in sea levels would be reduced by about

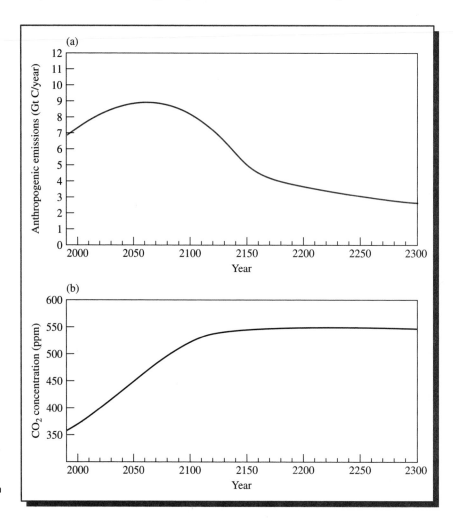

FIGURE 7-11 Approximate (a) annual CO$_2$ emission rates and (b) resultant atmospheric CO$_2$ concentrations to meet a 550-ppm stabilization target.

one-third if we embark soon on the scenario to never exceed the 550-ppm concentration of carbon dioxide.

An alternative scenario to the one shown by the curves in Figures 7-11a and 7-11b, in which effective CO$_2$ controls are not implemented until several decades later, would eventually require a sharper decline in emissions and would reach the 550-ppm limit sooner. Such alternative proposals allow more time to further develop replacement technologies, such as the solar energy techniques discussed in Chapter 8, before we begin to end our reliance on fossil fuels. Such scenarios require the world to generate more emissions-free power than today's total power consumption by about mid-century, a major challenge to achieve. By the end of the century, almost all power would have to be emissions-free. It is *not* possible to defer emission reductions indefinitely if the 550-ppm concentration target is to be achieved.

Green Chemistry: Polylactic Acid—The Production of Biodegradable Polymers from Renewable Resources; Reducing the Need for Petroleum and the Impact on the Environment

Our everyday lives are permeated by the chemicals in products such as pharmaceuticals, plastics, pesticides, personal hygiene products, cleaners, fibers, dyes, paints, clothes, building materials, computer chips, packaging, and food. The vast majority of these chemicals are ultimately made from oil, consuming approximately 2.7% of the production of this natural resource. The compounds that are isolated from oil and used to produce chemicals are known as *chemical feedstocks*. Approximately 60 billion kg of these feedstocks are employed to create 27 billion kg of polymers (many are loosely referred to as *plastics*) each year. Some of the more familiar polymers (as will be discussed in Chapter 16) that are produced from crude oil include *polyethylene terephthalate* (PET), which is used to make plastic beverage bottles and fibers for clothes; polyethylene, which is used to produce plastic grocery and trash bags; and polystyrene, which we discussed in the green chemistry section in Chapter 1. Trade names of polymers, such as *Dacron*, *Teflon*, *Styrofoam*, and *Kevlar*, represent polymers that are part of our everyday lexicon.

Approximately 2 billion kg of PET are produced each year. PET is one of the main targets for the recycling of plastics, yet less than one-quarter of this total is recycled in the United States; the rest is landfilled or incinerated. Even when PET is recycled, it generally can't be reused as beverage bottles; it is downward recycled into polyester fiber products such as carpets, T-shirts, fleece jackets, sleeping bags, and car trunk linings, or into thermoformed sheet products such as laundry scoops, nonfood containers, and containers for fruits.

When we use oil to produce items that we dispose of or incinerate (including the use of oil as a fuel), we are consuming a resource that has taken nature millions of years to produce. Petroleum is a finite, nonrenewable resource. Although there are still considerable oil reserves, at the rate of our current use we will deplete the supply of cheap, readily accessible oil within the next 30 to 40 years. We must learn to use renewable resources such as biomass rather than petroleum to produce chemical feedstocks.

Scientists at *NatureWorks LLC* (formerly Cargill Dow LLC) have developed a method for producing a polymer called **polylactic acid** (PLA) from renewable resources—such as corn (called maize in the UK and elsewhere) and sugar beets—for which they won a Presidential Green Chemistry Challenge Award in 2002. NatureWorks produces PLA at a plant in Blair, Nebraska. Ultimately, the goal is to utilize waste biomass as the source of this polymer. As in the steps shown in Figure 7-12, the corn is milled into starches, which are then reacted with water to yield glucose, which is then converted to lactic acid by natural fermentation. This naturally occurring compound is then converted to its dimer, followed by polymerization to PLA.

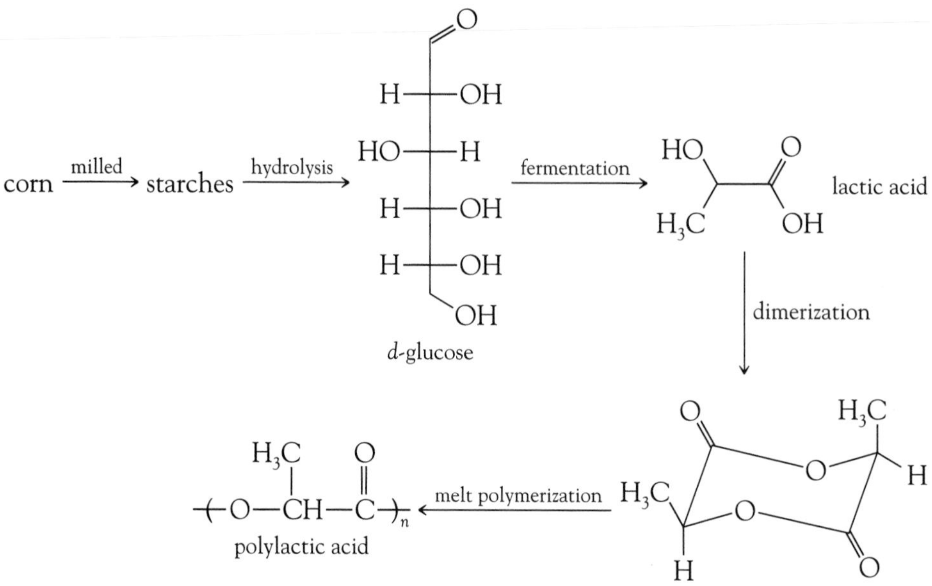

FIGURE 7-12 The synthesis of polylactic acid.

The environmental advantages of PLA over petroleum-based polymer include the following:

- It is made from annually renewable resources (corn, sugar beets, and eventually waste biomass).

- Production of PLA consumes 20–50% less fossil-fuel resources than do petroleum-based polymers.

- It uses natural fermentation to produce lactic acid. No organic solvents or other hazardous substances are used.

- It uses catalysts, resulting in reduced energy and resource consumption.

- High yields of over 95% are obtained.

- The use of recycle streams help to reduce waste.

- PLA can be recycled—converted back to its monomer via hydrolysis, then repolymerized to produce virgin polymer (i.e., closed-loop recycling).

- PLA can be composted (it is biodegradable); complete degradation occurs in a few weeks under normal composting conditions.

Another environmental consideration of PLA is that the plants, such as corn, used to produce the polymer consume atmospheric carbon dioxide, thus reducing the concentrations of this greenhouse gas. When PLA biodegrades, it releases this carbon dioxide back into the atmosphere in amounts approximately equal to the carbon dioxide absorbed by the plants used to produce it, thus making PLA in theory carbon neutral. However, fossil fuels are required during the production of PLA. Life cycle assessment studies (see Chapter 16) indicate that PLA requires 25–55% less fossil-fuel-supplied energy than do petrochemical polymers.

PLA can be used to produce products that are currently made from petroleum-based polymers such as cups; rigid food containers; food wrappers and bags; bags for refuse; furnishings for homes and offices (carpet tile, upholstery, awnings, and industrial wall panels); and fibers for clothing, pillows, and diapers. Small purveyors of natural foods have been using PLA packaging for several years. In 2005 PLA received a significant boost when Wal-Mart announced plans to use 114 million PLA containers a year. According to the company, this will save 800,000 barrels of oil annually.

Biodegradable polymers produced from renewable resources help to reduce our consumption of oil and have the potential to offer significant environmental and economic advantages over petroleum-based polymers. However, we must remember that even production of chemicals from annually renewable resources such as biomass does not offer a complete solution to energy and environmental problems. Growing crops, whether they are used to produce food or chemicals, require fertilizers and pesticides. Energy is needed to plant, cultivate, and harvest; to produce, transport, and apply fertilizers and pesticides; to make and run tractors; to transport seeds, biomass, monomers, and polymers. Use of land to produce crops for chemicals, and more significantly for bio-based fuels, also removes land that could be used to produce food and feed, and it increases the price for food.

The Extent and Potential Consequences of Future Global Warming

As we saw in Chapter 6, the Earth's weather has probably already been affected by the enhancement of the greenhouse effect due to increasing atmospheric concentrations of carbon dioxide and other gases. The continuing buildup of CO_2 in the air leads to the conclusion that we are in store for further increases in global air temperatures and other changes to our climate.

In this section, we shall summarize what projections tell us qualitatively about the climate changes to expect in the coming decades and some of their consequences for human health.

Those of us who currently suffer through severe winters each year may look forward to the warmer climate associated with the enhanced greenhouse effect. After all, in the eleventh and twelfth centuries, an increase of a few tenths of a degree in the northern temperate zone was sufficient for farming to occur on the coast of Greenland, for vineyards to flourish extensively in

England, and for the Vikings to travel the North Atlantic and settle in Newfoundland.

However, the climate changes predicted for the twenty-first century and beyond do not present a uniformly pleasant prospect. The *rate* of change in our climate, which to date has been modest, will be dramatic by the middle of the century. Indeed, the rapid rate of global change will probably be the greatest problem with which humanity will have to contend. A more gradual transition, even to the same end result, would be much easier to handle, not only for humans but for all living organisms on the planet.

It is very difficult for scientists to model the climate—even with the assistance of the fastest computers in the world—in order to make definitive statements about what changes will occur in particular regions in the future. We know that there will be substantial changes in the climate, but we are unable to specify exactly what they will be.

Predictions for Climate Change by 2100

The significant changes in the Earth's climate that have occurred in the past half-century, that are predicted to continue over the twenty-first century, and that are likely to have been at least partially caused by anthropogenic effects, as judged by the IPCC in its 2007 report, are summarized in Table 7-2.

According to sophisticated computer simulations of the future climate reported by the IPCC, the rise in the average global air temperature by 2100 (compared to 1990) could be as small as 1.4°C or as large as 4.0°C. As in the twentieth century, more warming will occur at night than during the day. The magnitude of the temperature increase will depend greatly on whether emissions (including those of sulfur dioxide) are controlled or not. At a minimum, however, the world will warm more than twice as fast in this century as it did in the last. Part of the rather large range of the prediction is due to uncertainty about exactly how sensitive the climate is to carbon dioxide. Indeed, research reported in 2003 indicates that aerosols have accounted for more greenhouse warming in the past than previously thought; consequently, scientists may have significantly underestimated the sensitivity of temperature to CO_2, in which case the predicted increases will have to be revised upward.

An increase of a few degrees in temperature may seem small, but our current average air temperature is less than 6°C warmer than that in the coldest periods of the ice ages! Snow cover and sea-ice area will continue to decline. There may well be enough melting of ice in the Arctic region for the Northwest Passage to be used for commercial transport, since the warming of all Arctic regions in winter is projected to be much greater than the global average. Indeed, the Arctic regions of Alaska and western Canada warmed at the rate of 0.3–0.4°C per decade in the 1961–2004 period. The Arctic (Southern) Ocean will probably become ice-free in the summer, a situation that has not

| TABLE 7-2 | Recent Trends and Projections for Extreme Weather Events That Have Undergone Recent Change |

Phenomenon and Direction of Trend	Likelihood That Trend Occurred in Late 20th Century (Typically Post 1960)	Likelihood of a Human Contribution to Observed Trend	Likelihood of Future Trends Based on Projections for 21st Century
Warmer days and fewer cold days and nights over most land areas	Very likely	Likely	Virtually certain
Warmer and more frequent hot days and nights over most land areas	Very likely	Likely (nights)	Virtually certain
Warm spells/heat waves. Frequency increases over most land areas	Likely	More likely than not	Very likely
Heavy precipitation events. Frequency (or proportion of total rainfall from heavy falls) increases over most areas	Likely	More likely than not	Very likely
Area affected by droughts increases	Likely in many regions since 1970s	More likely than not	Likely
Intense tropical cyclone activity increases	Likely in some regions since 1970	More likely than not	Likely
Increased incidence of extremely high sea level (excludes tsunamis)	Likely	More likely than not	Likely

Source: IPCC, *Climate Change 2007: The Physical Science Basis. Summary for Policymakers:* www.ipcc.ch

occurred for at least a million years. A similar situation is occurring over the land; because of global warming, the land is snow-covered for a shorter duration in winter and has a much higher albedo than the soil and vegetation exposed in the spring. In addition, most of the region of permafrost—land in northern Canada, Alaska, Siberia, and northern Scandinavia that stays frozen year-round—will likely melt to a depth of 3 m or more during this century.

The total amount of global rainfall is projected to increase, since more water will evaporate at the higher surface temperatures. The global average precipitation increases by about 2% for every Centigrade degree rise in temperature. Although the world overall will become more humid, some areas will become drier. To make matters worse, most areas of the world that currently suffer from drought are predicted to become even drier. Continental interiors at mid-latitudes will have continuing risk of drought in summer due to continued drying of the soil, the increased rate of evaporation from higher air temperatures being greater than the increase in the rate of precipitation. Subtropical areas will experience less precipitation, and equatorial and high-latitude regions will experience more, continuing the twentieth-century trends.

An increase in the average atmospheric temperature means that the air and water at the Earth's surface would contain more energy and that more extreme weather disturbances could result, the effect of global warming that will affect many of us the most. The number of days having intense rain showers or very high temperatures will both increase. Storm wind intensities and heavy downpours will increase in some tropical areas.

Predictions About Sea Levels

Although air and land surfaces are quick to warm with an increase in average global temperature, the same is not true of seawater. It takes many centuries for an increase in air temperature to gradually make its way down deeper and deeper into the ocean. For this reason, the rise in sea levels resulting from any particular amount of global warming is largely delayed for many years. Consequently, even if atmospheric carbon dioxide levels did not increase at all beyond today's values and no further global warming occurred, sea levels would *continue* to rise for centuries to come, as lower layers of the oceans became heated—and expanded—by air that has already been warmed.

Sea levels are predicted to rise by about half a meter by 2100—in addition to the 10–25-cm rise experienced over the last 100 years—though there is a large uncertainty in this value. About half of the predicted rise in sea level is due to the melting of glaciers, and the other half arises from the **thermal expansion** of seawater. The expansion occurs because the density of water *decreases* gradually as water warms beyond 4°C, the temperature at which it reaches its maximum density, as illustrated in Figure 7-13. Since density is mass divided by volume, and since the mass of a given sample of water cannot change, the volume it occupies must increase if its density decreases. Thus, as seawater warms (beyond 4°C), the volume occupied by a gram or kilogram of it increases; the only way that this can occur is if the top of the water—the sea level—increases.

Scientists predict that the Greenland Ice Sheet may eventually melt entirely due to global warming, raising ocean levels by about seven meters, but they are uncertain as to the fate of the Antarctic Ice Sheet. However, complete melting will require millennia to complete. Parts of both ice sheets sit above sea level on land. Consequently, the transfer, by the melting of their surface ice and the resulting draining of liquid water into the oceans, causes an increase in sea levels, as does the transfer into the oceans of icebergs broken off from the ice sheets. The collapse of ice shelves, which are the extensions into the sea of glaciers, such as the Larsen-B shelf in the Antarctic that collapsed in the early 2000s, allows glaciers that had been blocked to migrate more quickly toward open water.

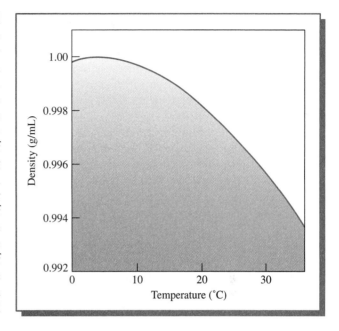

FIGURE 7-13 Density of liquid water versus temperature.

Although an increase of half a meter in sea level does not seem very large, there are countries, like Bangladesh and the small island nation of Tuvalu in the South Pacific, in which much of the population currently lives on land that would be flooded by a rise in sea level of this amount. Damage from tropical storms would increase because of these higher sea levels.

In the long term, the most dramatic—though unlikely—effect of substantial global warming would be a change in the circulation patterns of water in the Atlantic Ocean. Currently, warm surface waters flow northward from the tropics into the North Atlantic, bringing heat to Europe and to a lesser extent to eastern North America. Some scientists have speculated that a rapid rise in temperature and rainfall levels could weaken or even eliminate this circulation pattern, as historical geological records indicate has happened in the past.

Climate Predictions for Specific Regions

It is much harder for scientists to make specific, reliable predictions for individual regions than for the globe as a whole. The climate changes that seem likely for the various continents, according to the 2007 IPCC report, are summarized in Table 7-3.

The higher latitudes of the Northern Hemisphere are expected to experience temperature increases substantially greater than the global average. Warming over some land areas, including the United States and Canada, should be noticeably faster than the average rate for the globe.

TABLE 7-3	Significant Impacts of Climate Change That Will Likely Occur in the Continents in the Twenty-First Century
Arctic	Significant retreat of ice; disrupted habitats of polar megafauna; accelerated loss of ice from Greenland Ice Sheet and mountain glaciers; shifting of fisheries; replacement of most tundra by boreal forest; greater exposure to UV radiation
North America	Reduced springtime snowpack; changing river flows; shifting ecosystems, with loss of niche environments; rising sea level and increased intensity and energy of Atlantic hurricanes increase coastal flooding and storm damage; more frequent and intense heat waves and wildfires; improved agriculture and forest productivity for a few decades
Europe	More intense winter precipitation, river flooding, and other hazards; increased summer heat waves and melting of mountain glaciers; greater water stress in southern regions; intensifying regional climatic differences; greater biotic stress, causing shifts in flora; tourism shift from Mediterranean region
Central and Northern Asia	Widespread melting of permafrost, disrupting transportation and buildings; greater swampiness and ecosystem stress from warming; increased release of methane; coastal erosion due to sea ice retreat
Central America and West Indies	Greater likelihood of intense rainfall and more powerful hurricanes; increased coral bleaching; some inundation from sea-level rise; biodiversity loss
Southern Asia	Sea-level rise and more intense cyclones increase flooding of deltas and coastal plains; major loss of mangroves and coral reefs; melting of mountain glaciers reduces vital river flows; increased pressure on water resources with rising population and need for irrigation; possible monsoon perturbation
Pacific and small islands	Inundation of low-lying coral islands as sea level rises; salinization of aquifers; widespread coral bleaching; more powerful typhoons and possible intensification of ENSC extremes
Global oceans	Made more acidic by increasing CO_2 concentration; deep overturning circulation possibly reduced by warming and freshening in North Atlantic
Africa	Declining agricultural yields and diminished food security; increased occurrence of drought and stresses on water supplies; disruption of ecosystems and loss of biodiversity, including some major species; some coastal inundation
South America	Disruption of tropical forests and significant loss of biodiversity; melting glaciers reduce water supplies; increased moisture stress in agricultural regions; more frequent occurrence of intense periods of rain, leading to more flash floods
Australia and New Zealand	Substantial loss of coral along Great Barrier Reef; significant diminishment of water resources; coastal inundation of some settled areas; increased fire risk; some early benefits to agriculture
Antarctica and Southern Ocean	Increasing risk of significant ice loss from West Antarctic Ice Sheet, risking much higher sea level in centuries ahead; accelerating loss of sea ice, disrupting marine life and penguins

Source: Scientific Expert Group Report on Climate Change and Sustainable Development, *Confronting Climate Change,*
United Nations Foundation (2007): www.confrontingclimatechange.org.

In the U.S. Midwest and the area just north of it in Canada, as well as in southern Europe, the soil will probably become much less moist because of increased rates of evaporation in the warmer air and ground. This could affect the continued suitability of these areas for the growing of grain. High-latitude regions, however, could experience increased productivity, at least where the soil is suitable for agriculture. In areas that become drier, the positive CO_2 fertilization effect on plants will cancel some of the negative effects of decreased rainfall. There will be longer frost-free growing seasons at northern latitudes but increased chances that heat stress will affect crops grown there. Food production in temperate areas will probably also be affected by the attack of insects that in the past have been largely killed off during the winters but that could survive and flourish under warmer conditions.

Temperature and moisture changes will occur quickly compared to those that have taken place in the past, and consequently some ecosystems will be destabilized. Coastal ecosystems such as coral reefs are particularly at risk. The species composition of forests is likely to change, especially in regions far removed from the Equator. For example, the hardwood forest in eastern North America may be at risk of extinction if climate zones shift more quickly than their migration can keep up with. The boreal forest of central Canada could be eliminated by fire by 2050; indeed, the frequency of fires in these woodlands is already climbing.

Predicted Effects on Human Health

Many scientists have concluded that human health will be affected adversely by global warming. There will probably be more extreme heat waves in summers but fewer prolonged cold snaps in winters. The expected doubling in the annual number of very hot days in temperate zones will affect people who are especially vulnerable to extreme heat, such as the very young, the very old, and those with chronic respiratory diseases, heart disease, or high blood pressure. This will be particularly acute for poor people, who have less access to air conditioning. The heat wave in the summer of 2003 in Europe resulted in the death of at least 10,000 people in France alone.

Domestic violence and civil disturbances could also increase, as they tend to occur more frequently in hot weather. On the other hand, there would probably be a decrease in cold-related illnesses because of the milder winters. Air quality in summer will probably degrade further, as the background concentration of ground-level ozone will probably increase substantially. Higher CO_2 concentrations and warmer weather will increase the production and release of pollen by plants such as ragweed, thus exacerbating allergic responses.

Less directly, global warming may extend the range of insects carrying diseases such as malaria into regions where people have developed no immunity, and it may intensify transmission in regions where such diseases are already prevalent. It has been predicted that malaria could claim an additional million victims annually if the temperature rise is sufficient to allow parasite-bearing mosquitoes to spread into areas not now affected. In North America, suitable habitat for the mosquitoes is being lost, so fortunately it should not be as much of a problem for Americans and Canadians. A different type of mosquito carries the dengue and yellow fever viruses, and its range could increase with warming, thereby spreading these diseases. There is also evidence that cholera rates increase with warming of ocean surface waters, because coastal blooms of algae are breeding grounds for the disease, and they increase with water temperature. Some experts in disease control discount these predictions, arguing that other effects such as increased rainfall could well negate or even reverse the effects on disease rates of increases in air temperature.

Animal health could also be affected by the spreading of disease by parasites. In addition, some species, such as polar bears and caribou, that live in very cold regions could be at risk of extinction from habitat changes that threaten their hunting practices.

International Agreements on Greenhouse Gas Emissions

Faced with the prospect that increased CO_2 emissions over the next century could result in a significant increase in global air temperature, with its resultant modifications of climate, some governments and organizations have been debating how future emissions can be minimized while still allowing economic growth to occur.

The first agreement on greenhouse gas emissions was reached at the Rio Environmental Summit meeting in 1992; each developed country was to ensure that its CO_2 emission rate in 2000 would be no greater than that in 1990. This target was met, in fact, by very few countries; by 2000, most were emitting at levels well above their targets.

The second agreement was reached in negotiations held at Kyoto, Japan, in 1997. Thirty-nine industrialized nations agreed to decrease their collective CO_2-equivalent emissions by 5.2%, compared to 1990 levels, by 2008–2012. The greenhouse gases affected by the Kyoto Accord are carbon dioxide, methane, nitrous oxide, hydrofluorocarbons, perfluorocarbons, and sulfur hexafluoride.

Under the agreement, the United States was due to cut its emissions to 7% less than its 1990 level, Canada and Japan by 6%, and the European Union collectively by 8% (with wide variations for individual countries within this unit). Some countries, such as Australia, were permitted to increase their emissions beyond 1990 levels. Emissions by developing countries were not controlled by the Kyoto Accord, since they were not significant players in emitting greenhouse gases in the past and therefore did not contribute much to current global warming.

If countries were to achieve their Kyoto agreement levels, the annual per capita CO_2 emissions in 2010 in developed countries would have decreased from 3.1 tonnes of carbon in 1997 to 2.8 tonnes, whereas, because of economic development, emissions from developing countries would probably have risen from 0.5 to 0.7 tonne. The CO_2 concentration in air would have been a little over 1 ppm less than would have otherwise resulted. However, the United States and Australia subsequently withdrew from the agreement. As an alternative to cutting greenhouse gas emissions in line with the Kyoto treaty, the U.S. government proposed in 2003 to reduce the carbon intensity of the U.S. economy by 18% by 2012. Nevertheless, some U.S. states—California and several of those in New England—have decided on their own to limit greenhouse gas emissions. However, the extent to which most other countries will meet their targets by 2012 is in doubt. Canada, for example, had increased its emissions by 30% in 2006 rather than decreased them compared to 1990 levels.

The existing increase, by one-third, of the atmospheric CO_2 level, as well as the temperature increase and climate modification that this probably caused, resulted in large part from the industrialization and increased standard of living in developed countries. Without a significant change in the methods by which energy is produced and stored, and/or implementation of carbon sequestration on a massive scale, these same nations will continue to require about the same rate of CO_2 emissions in the future in order to maintain their economic growth.

Rather than utilizing a procedure in which countries have CO_2 emission targets that are negotiated at international meetings, schemes have been discussed that are based on allocations that could be traded between countries on the open market. In a manner similar to the way in which SO_2 emission rights currently are traded in the United States, countries that need to emit more than their collective CO_2 allocations could purchase unused allocations from countries with an excess. A bonus of this scheme is that it provides an incentive to develop and invest in cleaner technologies, since avoiding CO_2 emissions could be cheaper than purchasing additional rights—especially in the future, when few nations will have excess emission capacity and the price of emission rights will rise.

The question of how CO_2 allocations can be made fairly in order to initiate the free-market CO_2 emission-trading scheme is a perplexing one. In the simplest scheme, each country would be assigned an allocation based strictly upon its (current) population. For example, if it was concluded that the current average annual emission of 1 tonne of carbon as CO_2 per capita could be sustained indefinitely, then this quantity would be allocated to a country for each of its residents. If it was decided to cut back current global emission levels, e.g., by one-quarter, then only 0.75 tonne per capita per year would be allocated etc.

An immediate consequence of the per capita allocation method would be the annual transfer of substantial funds from all developed countries to developing and undeveloped ones, since, according to the data in Figure 7-4, the former all exceed the 1-tonne average, by factors ranging from 2 to 5. Although this method would provide external funding so that developing countries could establish efficient energy infrastructures, it would likely not prove popular in developed countries, since their energy costs would rise.

One alternative allocation scheme is based on how much energy is required for industrial production by a country and how efficiently it uses energy. Thus a country's carbon dioxide allocation would be directly proportional to its GDP. This allocation method rewards compact, energy-efficient developed countries at the expense of those—both developed and developing—that emit more CO_2 per unit of GDP. However, such a scheme would permit continued economic growth by developing countries, since their CO_2 allocations would track their economic growth. The global ratio of allowed carbon dioxide to dollar of GDP would have to decline with time if global emissions are to be controlled, since global GDP rises by several percent per year. Interestingly, the CO_2/$ GNP ratio is more independent of the level of economic development than is the ratio based upon population; e.g., China emits about 1.0 kg of carbon dioxide for each dollar of production, compared to 0.9 kg for the United States, 0.5 kg for Japan, 1.0 kg for Germany, and 0.7 kg for India.

Some policymakers believe that **carbon taxes,** i.e., taxes based on the amount of carbon contained in a fuel rather than upon its total mass, should be instituted as a disincentive to use fossil fuels, especially coal, since it generates more CO_2 per joule of energy produced than does natural gas. In fact, the hydrogen-to-carbon ratio of the average global fuel mix has been continuously increasing over the last century and a half, as we moved from economies whose energy source was dominated by wood (H/C ratio of about 0.1) to coal (1.0 ratio) to oil (about 2.0) and now to natural gas (4.0); this is the same direction as moving to lower CO_2/energy ratios, as implied above. Carbon taxes could be phased in over a period of decades, starting at a low price that would gradually increase, thereby giving time for low-carbon-emission technologies to be further developed and implemented.

We conclude by commenting on the paradox that faces humanity today concerning the enhancement of the greenhouse effect. On the one hand, there exists the slight possibility that doubling or quadrupling the CO_2 concentration will have no measurable effect on climate and that efforts to prevent such an increase not only would be an economic burden for both the developed and the developing worlds, but would perhaps be unnecessary. On the other hand, if the predictions of scientists who model the Earth's climate turn out to be realistic, but we do nothing to prevent further buildup of the gases, both present and future generations will collectively suffer from rapid and perhaps cataclysmic changes to the Earth's climate.

Review Questions

1. Define the term *commercial energy*. On what factors does the magnitude of its use in a country depend?

2. What is the equation relating exponential growth to the annual increase in a quantity?

3. Define the term *carbon intensity*. Describe how the carbon intensity has changed over the last few decades **(a)** globally, **(b)** for the United States, and **(c)** for China.

4. How does the rate of CO_2 emissions by a country depend on its population, its carbon intensity, and its GDP?

5. Explain why the concentration of CO_2 in air is expected to rise linearly with time if its rate of emission remains a constant.

6. What are the ultimate origins of coal, oil, and natural gas? Which fuel is in greatest reserve abundance?

7. What is the most important class of hydrocarbons present in crude oil?

8. What is meant by the *BTX fraction* of gasoline? Is it toxic?

9. What is meant by *engine knocking?*

10. How is the *octane number* rating scale for fuels defined?

11. List several ways in which the octane number of fuels can be increased by the addition of other compounds to straight-chain alkane mixtures.

12. What is the main component of *natural gas?* Write out the balanced chemical equation illustrating its combustion.

13. Why is natural gas considered to be an environmentally superior fuel to oil or coal? What phenomenon involved in its transmission by pipeline might offset this advantage?

14. What is meant by the term *CNG?* What are the advantages and disadvantages to fueling vehicles with CNG?

15. What is meant by the term *carbon sequestration?*

16. Name three techniques whereby carbon dioxide could be stripped from power plant emission gases.

17. Define *oxycombustion* and state its advantages.

18. Explain the difference between "ocean acidic" and "ocean neutral" techniques of storing CO_2 in oceans.

19. Define *enhanced oil recovery* and explain its relationship to the underground storage of carbon dioxide.

20. Explain why sea levels are expected to rise as global air temperatures increase.

21. List some of the consequences, including those affecting human health, that may occur as a consequence of global warming in the future. Why might soil in some areas be too dry for agriculture even though more rain falls on it?

22. What is the *Kyoto Accord?* What gases are limited in their emissions under it? Would the Kyoto agreement have halted global warming?

23. Describe the scheme whereby a nation's allocation of carbon dioxide emission could be traded on a market. Describe two schemes by which initial allocations could be made.

24. What is a *carbon tax* and what are the arguments in favor of it? Why do you think many people oppose it?

Green Chemistry Questions

See the discussion of focus areas and the principles of green chemistry in the Introduction before attempting these questions.

1. The formation of polylactic acid (PLA) from biomass developed by NatureWorks LLC won a Presidential Green Chemistry Challenge Award.

(a) Into which of the three focus areas for these awards does this award best fit?

(b) List three of the twelve principles of green chemistry that are addressed by the green chemistry developed by NatureWorks LLC.

2. What are the environmental advantages of using PLA in place of petroleum-based polymers?

3. Why does the use of biodegradable polymers (to replace petroleum-based polymers) not offer complete solutions to energy and environmental problems?

Additional Problems

1. Using a ruler and calculator, estimate from Figure 7-1 the fraction of CO_2 emissions in 2004 compared to 1990 **(a)** from the United States and **(b)** collectively from China, India, and other developing countries. Did the fraction of emissions from EU countries increase or decrease over that period?

2. List several reasons that you think proponents of carbon dioxide allocations based strictly on a country's population would advance in support of their position. What objections can be raised to their position? Repeat this exercise for an allocation scheme based on current GDP.

3. The replacement by natural gas of oil or coal used in power plants has been proposed as a mechanism by which CO_2 emissions can be reduced. However, much of the advantage of switching to gas can be offset since methane escaping into the atmosphere from gas pipelines is 23 times as effective, on a molecule-per-molecule basis, in causing global warming as is carbon dioxide. Calculate the maximum percentage of CH_4 that can escape if replacement of oil by natural gas is to reduce the rate of global warming. *[Hint: Recall that the heat energy outputs of the fuels are proportional to the amount of O_2 they consume. Assume the empirical formula for oil is CH_2.]*

4. Canada has massive supplies of heavy oil in tar sands, which are being used to make gasoline by combining them with natural gas. Assume that the empirical formulas of these three fuels are CH, CH_2, and CH_4, respectively, and that gasoline is made by hydrogenating the tar with hydrogen produced from natural gas in its reaction with water to produce CO_2 and H_2. Combine the hydrogenation and hydrogen production equations so as to use all the H_2, thereby deducing the overall reaction of CH and CH_4 with steam to produce gasoline and carbon dioxide.

5. Given that the density of dry ice (solid CO_2) is 1.56 g/cm^3, calculate the diameter in meters of a dry-ice ball that would be produced from the 5 metric tonnes of CO_2 produced on average by each person in developed countries each year.

6. A sign was spotted outside a farmers' market with the slogan: "Help Stop Climate Change: Buy Local Produce." Explain whether taking the advice of this sign could indeed "help stop climate change."

7. Using a ruler and calculator, estimate from Figure 7-7 the fractions of petroleum that are predicted by 2050 to originate **(a)** from heavy oil, and **(b)** from natural gas liquids. What is the fraction of oil of all kinds produced in 2050 relative to that of 2008?

Further Readings

1. A. Witze, "That's Oil, Folks," *Nature* 445 (2007): 14–17.

2. J. L. Sarmiento and N. Gruber, "Sinks for Anthropogenic Carbon," *Physics Today* 55 (Aug. 2002): 30–36.

3. H. H. Khoo and R. B. H. Tan, "Life Cycle Investigation of CO_2 Recovery and Sequestration," *Environmental Science and Technology* 40 (2006): 4016–4024.

4. E. Rubin et al., *IPCC Special Report: Carbon Dioxide Capture and Storage:* www.ipcc.ch/activity/srccs/index.htm

5. R. H. Socolow, "Can We Bury Global Warming?" *Scientific American* (July 2005): 49–55.

6. H. Inhaber and H. Saunders, "Road to Nowhere," *The Sciences* (November/December 1994): 20–25. (Argues that energy conservation leads to increased consumption.)

Websites of Interest

Log on to www.whfreeman.com/envchem4/ and click on Chapter 7.

CHAPTER 8

RENEWABLE ENERGY, ALTERNATIVE FUELS, AND THE HYDROGEN ECONOMY

In this chapter, the following introductory chemistry topics are used:

- Ideal gas law
- Thermochemistry calculations
- Electronic structure of atoms
- Electrochemistry: oxidation numbers; redox half-reactions; batteries; electrolysis
- Crystalline versus amorphous solids
- Basic organic chemistry structures: alcohols, ethers, carboxylic acids, esters, sugars, carbohydrates
- Vapor pressure of liquids
- Distillation

Background from previous chapters used in this chapter:

- Greenhouse gases (carbon dioxide, methane, nitrous oxide) (Chapter 6)
- Anaerobic decomposition; clathrates (Chapter 6)
- Fossil fuels (Chapter 7)
- Light absorption as energy; photons (Chapter 1)
- Air pollution: photochemical smog, particulates, gaseous pollutants (Chapters 3 and 5)
- Catalytic converters; thermal NO (Chapter 3)

A wind farm in Scotland.
(Image State)

Introduction

In Chapters 3–7, we have seen how the atmosphere has been affected by emissions into it of pollutant gases such as sulfur and nitrogen oxides and greenhouse gases such as carbon dioxide and methane. The emphasis in this chapter is on alternative technologies under development that could reduce the anthropogenic production of such gases in the future while still allowing economic growth to occur. We begin by considering some possible solutions to the further buildup of atmospheric CO_2 by a partial switchover from fossil fuels to renewable energy, especially solar power. We then make an extensive survey of the various alternative fuels, including biofuels and hydrogen, that may be more "greenhouse friendly" than those used at present and that also would be effective in reducing air pollution. The generation of energy by nuclear power is discussed in Chapter 9.

Renewable Energy

The Sun sends enough energy to the Earth to supply all of our conceivable energy requirements, about 10,000 times more than we use now and will in the future, if only we could trap it efficiently. In addition to being plentiful and reliable, it is **renewable energy**—energy that will not run out *and* whose capture and use do not result in the direct emission of greenhouse gases or other pollutants.

The world currently uses about 12 **terawatts** (TW, 10^{12} watts) of power, about 85% of which is generated by the burning of fossil fuels. Since 1 watt is 1 joule per second, and since there are 3.2×10^7 seconds in a year, our annual power consumption is about 3.8×10^{20} J, 380 EJ. Given that an average light bulb is rated at 60 W, we are using the equivalent of 200 billion light bulbs at a time, an average of about 35 per person, nonstop. Of course, this figure is an average for people in developed and developing countries; if we redo the calculation for North Americans, we are using about 200 60-W light bulbs for every man, woman, and child!

The pie chart in Figure 8-1 illustrates the sources of the world's commercial energy in 2004; the percentages for energy used to generate electricity are shown in parentheses. Clearly, most renewable energy currently is generated by burning biomass and by hydroelectricity, with the latter used to generate electricity. An assessment in 2003 by the European Union (EU) for energy use in the year 2030 predicts that renewable energy, including wind, geothermal, and the direct forms of solar energy, will not keep pace with rising energy demand. Because people from rural regions of Asia and Africa will burn less firewood, due both to their migration to cities and to disappearing forests, renewable energy collectively will drop from the current 13% (only 2% of which is not from biomass) to only 8% of the global supply.

Hydroelectric Power

Of all the forms of renewable energy, hydroelectric power is by far the most important. Worldwide, it constitutes over 80% of renewable energy (other than that based upon biomass) and 2% of global commercial energy.

Hydroelectric power is an indirect form of solar energy. In the hydrological cycle, the Sun's energy evaporates water from oceans, lakes, rivers, and the soil and transports the H_2O molecules upward in the atmosphere via winds. After the water molecules condense to raindrops, they still possess considerable potential energy owing to their elevation, only a portion of which is dissipated if they fall onto land or a water body that lies above sea level. We can harness some of its remaining potential energy by forcing the downward-flowing water to turn turbines and thereby generate electricity.

Although there are small hydroelectric installations that use the flow of a river, most large-scale facilities use dams and waterfalls, where the water pressure—and hence the power yield—is much greater. In particular, the energy imparted to a turbine by falling water is directly proportional not only to the volume of the water but also to the height from which it falls. For this reason, new hydroelectric projects usually involve the construction of a high dam along the path of a flowing river. Water then collects behind the dam and its level rises to a considerable height. The water that is allowed to flow over the top of the dam falls a considerable distance before encountering the turbines positioned near the bottom. Unfortunately, the collection of water behind the dam floods considerable areas of land, creating a lake with environmental problems such as those discussed below.

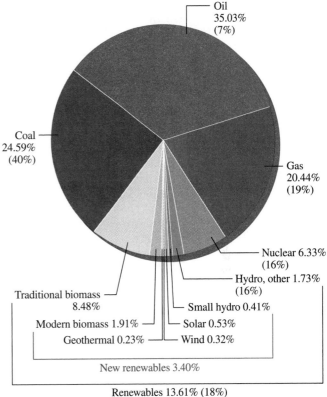

FIGURE 8-1 Sources of primary world energy supply in 2004.
[Source: J. Goldemberg, "Ethanol for a Sustainable Future," *Science* 315 (2007): 808–810.]

If all sites worldwide were exploited, the total amount of energy obtainable from hydroelectric sources would be about 100 EJ per year; about 20% of this total is obtained currently. Most of the sites that require little modification to use, and that are located within a reasonable distance of centers that use considerable electric power, have already been exploited; to use a common expression, most of the "low-hanging fruit" has already been picked. However, there are many river systems in developing countries, especially Africa, where considerable new hydroelectric power is currently being developed by the construction of dams.

Although hydropower is often thought of as pollution-free, there are environmental and social costs associated with it, especially ones resulting from the creation of the reservoirs behind dams. The most important of these costs include

- the displacement of human populations from lands flooded to create reservoirs;

- the eutrophication of water in reservoirs;

- the release of greenhouse gases, especially methane, from flooded areas;

- the release of mercury into reservoir water and consequently into the fish that swim in the water and human populations that eat the fish (this topic is discussed in more detail in the online Case Study *Mercury Pollution and the James Bay Hydroelectric Project (Canada)* and in Chapter 15);

- the devastation to fish populations, such as salmon, from the blockage of their migratory routes by dams; and

- the buildup of silt behind dams, with the result that less silt is carried to locations farther along the waterway.

Unfortunately, the construction of new hydroelectric projects involving the damming of river systems, especially in developing countries, often proceeds without adequate environmental assessment and planning ahead of time. The World Bank and several other large financiers of such hydroelectric facilities do insist on an independent assessment of the project's impacts before they provide financial assistance.

The largest hydroelectric project in the world is the 26-turbine Three Gorges Dam in China, which, when completed in 2009, will provide 18 MW of power—equivalent to five large coal-fired power plants—and will have cost $25 billion to construct. Although about a million people had to be relocated to avoid being flooded by the artificial lake, the dam also controls flooding on the Yangtze River and thereby saves thousand of lives.

The expansion of wetlands that occurs by the deliberate flooding of land to produce a large, deep reservoir of water generally creates a long lake covering hundreds or thousands of square kilometers. The Three Gorges Dam will result in a lake that is 600 km long! The deep water in such lakes is largely anaerobic, especially if the flooded land was not first cleared of vegetation. The anaerobic decomposition of the original trees, bushes, etc. on the land produces carbon dioxide and methane in almost equal volumes, both of which escape from the surface and enter the atmosphere. The emissions from such reservoirs are significant. This is particularly true for methane, since it is such a powerful greenhouse gas (Chapter 6). Deep, small reservoirs produce and emit much less methane than do shallow ones that contain large areas of flooded biomass, such as those in the Brazilian Amazon. Indeed, the combined global warming effect of the methane and carbon dioxide produced by a large, shallow reservoir created to generate hydroelectric power can, for many years, exceed that of the carbon dioxide that would be emitted if a coal-fired power plant were used instead to generate the same amount of electrical power! Even after the original vegetation has decayed, new plants that have grown on the shores of the lakes during the dry season, when water levels recede, are later engulfed by rising water in the wet season; they eventually decompose, releasing more methane.

Wind Energy

Winds are air flows that result from the tendency of air masses that have undergone different amounts of heating, and that therefore have developed unequal pressures, to equalize those pressures. The air flows from regions of high pressure to those of low pressure. The heating of air results directly or indirectly from the absorption of sunlight; indeed, about 2% of the Sun's energy received on Earth is transformed into wind energy. A large quantity of such indirect solar energy, about 300 EJ annually, is potentially available as **wind power,** although only 0.05% of it currently is being tapped.

Polar areas receive less sunlight and therefore less heat than do the tropics. To compensate for the resulting temperature difference between tropical and polar regions, winds arise in the air as do currents in the oceans. Warm air and water are carried toward the poles, whereas cold air and water are transported in the opposite direction, toward the Equator. However, these flows do not follow simple trajectories, owing to factors such as the spinning motion of the Earth around its axis and the effects of local terrain.

The force of the wind can be exploited to do useful work or to generate electrical energy in the same way that the force of flowing water is used in hydroelectric power plants. Historically, the strong, sustained winds in central North America were exploited by windmills to pump water and later to generate small amounts of electricity on individual farms until the middle of the twentieth century. Of course, windmills have been in use in Europe— especially in Holland—for centuries.

In recent decades, the large-scale generation of electricity by huge, high-tech windmills gathered in "wind farms" has become feasible. Wind power is currently the world's fastest-growing source of energy. Figure 8-2 (green curve) illustrates the increasing rate of annual wind-power installations in recent years. In 2001 alone, global wind-power capacity grew by one-third and

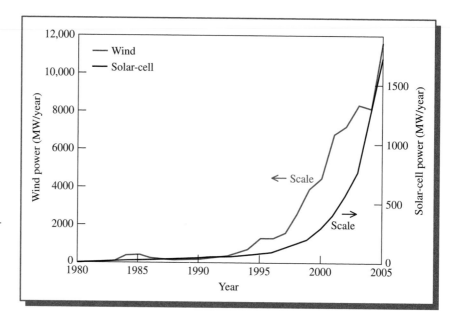

FIGURE 8-2 Annual production of wind energy and solar-cell energy. [Source: Redrawn from L. R. Brown et al., *Vital Signs 2006–2007* (New York: W. W. Norton, 2007).]

has been rising at about 25% per year, compounded, since then. As a consequence, overall in the period 1995–2005, it rose 12-fold. As of 2005, about 60,000 megawatts (MW) of wind power had been installed. Large growth in wind-power installations has occurred in Germany (currently the world leader in wind power, with 40% of installed power), Spain, the United States (which was the leader in the 1990s), India, and Denmark (which generates more than one-fifth of its electricity this way). This technology could be useful in many other parts of the world as well. A 2003 EU report predicts that 4% of the world's energy will be produced by wind power in 2030.

Technically, six times the 2001 world electricity output could be produced from wind, but only 0.5% was actually produced globally in this way in 2005, although 3% of Europe's electricity was produced in this way. A landmass the size of China would be needed to satisfy world electricity demand from wind alone. More realistically, wind power could be expanded to provide up to perhaps one-fifth of the world's electricity.

If price is not taken into account, then the country with the highest potential for wind power is the United States. About 90% of the U.S. potential for wind power lies in 12 states in the Midwest, ranging from North Dakota to northern Texas (see Figure 8-3), though the demand for electricity is centered far from most of these areas. Indeed, the United States has enough potential wind power to supply all its electricity now and in the foreseeable future. The world's largest wind farm covers 130 km^2 in Oregon and Washington and will eventually involve 460 turbines.

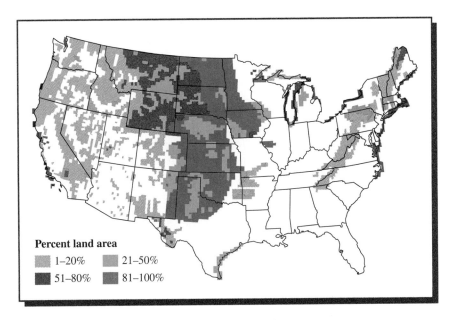

FIGURE 8-3 Percentage of land area estimated to have class 3 or higher wind power in the contiguous United States. [Source: "Wind Energy Resource Atlas of the United States," Chapter 2: http://rredc.nrel.gov/wind/pubs/atlas/maps/chap2/2-10m.html.]

Wind Speed and Windmill Size

As one would intuitively expect, the greater the velocity, v, of the wind, the greater the amount of energy a windmill or wind turbine will produce. In fact, the energy increases very sharply with wind speed. The energy yield from wind is proportional to v^3, i.e., to the third power of the wind speed. Consequently, a small improvement in velocity produces a large increase in yield; e.g., an increase from 22 to 26 mph improves the energy yield by two-thirds!

The cubic dependence of energy on wind speed is the result of two factors. First, the *kinetic energy* of the motion of the air mass in the direction of the wind is proportional to the square of the air speed, since from physics we know that for any moving body, its kinetic energy is given by $mv^2/2$. Second, the amount of wind passing over the blades per unit of time increases linearly in direct proportion to the wind speed. The energy available to the wind turbine is equal to the product of these two factors, so it is proportional to v^3. Hence, much of the energy available to windmills occurs in short bursts of high velocity due to this strong dependence of energy on wind speed.

The energy the windmill can gather is proportional to the square of its blade length, since the area the blade sweeps out is proportional to the length squared. Since wind speeds increase with height above the ground, a tall turbine is also more efficient for this reason.

Each windmill in a wind farm extracts energy from the flow of air on it, so the individual windmills must be physically separated from each other to some extent. For technical reasons, no more than about one-third of the energy passing by a windmill can be extracted from the flow of air around it.

Potential Wind-Energy Sites

As a consequence of the local terrain, some geographical regions experience almost constant windy conditions. Geographical areas are commonly divided into seven classes of potential wind-power density, with class 7 having the highest potential. Ideal locations for wind farms are those with an almost constant flow of nonturbulent winds in all seasons. Although wind energy does rise steeply with wind velocity, locations with sudden gusts of high-speed wind are not considered favorable. Locations at less than 2-km altitude, with average wind speeds of at least 5 m/sec, corresponding to 18 km/hr (11 mph), are generally required for a location to be considered economically feasible. Some authors use the criterion of annual mean wind speeds ≥ 6.9 m/sec (25 km/hr or 15 mph) measured at 80 m, the tip of the blade height of modern windmills, as suitable for low-cost wind-power generation. Such sites are considered to be of class 3 or higher potential wind-energy sites. The regions of high wind-power potential at reasonable cost are the United States, Canada, South America, the European countries that are members of the Organization for Economic Cooperation and Development (OECD), and the former Soviet Union. The areas with the lowest potential are in Africa, eastern Europe, and Southeast Asia. In most areas, the potential exceeds the current electricity usage.

Economic and Environmental Considerations

The most efficient and largest commercial wind turbines currently are the 2-MW units, three times larger than the models of the mid-1990s. Behemoths with huge, 120-m blades are under development; they will deliver 5 MW. About 660 North American homes can be supplied with electricity from a 2-MW system on typical hot afternoons, when the power draw peaks due to air conditioner usage. By contrast, modern coal-fired power plants generate from 125 MW to 1000 MW, so hundreds of windmills would be required to replace the power generated by one coal-fired plant.

In terms of **energy payback**—the amount of time required to generate the energy used in constructing the unit—that for wind is only 3–4 months. Carbon dioxide emissions from wind power are the smallest for any power source (see Figure 8-4).

Of all the forms of renewable energy, wind power is the most economical. The cost of generating electricity using modern windmill technology—and feeding it into existing power grids—is now almost competitive with conventional energy sources. In a 2004 report, the lowest-cost wind energy was quoted as 5¢ (U.S.) per kilowatt hour (kWh, the energy of 1 kilowatt of power used continuously for 1 hour). This is about the same as that from new coal-fired plants, almost as low as that generated by natural gas, and less than one-tenth of its cost 20 years earlier. As mentioned above, however, there would also be significant transmission costs if wind energy were to be expanded in the midwestern states. If the world switches eventually to a hydrogen economy, hydrogen generation powered by wind in this area could generate much of the U.S. supply. Currently, the electrolyzers required for the process are expensive.

Some individual buildings, including households that are too remote to be connected to power lines, generate their own electricity using rooftop-mounted wind turbines. When all the power generated is not needed to run the 12-V appliances etc. within the building, the excess is stored in 12-V batteries, to be drawn upon in times of low or no wind.

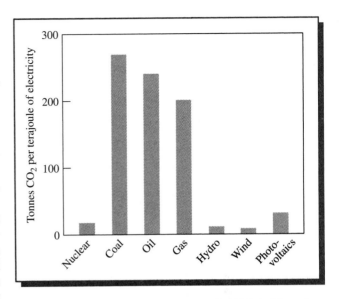

FIGURE 8-4 CO_2 emissions associated with different energy sources. [Source: "The Power to Choose," *New Scientist,* (6 September 1997): 18.]

There is often public resistance to placing wind turbines in populated areas because they are unattractive. For this reason, placing them on agricultural land or even offshore is becoming popular. A huge new 1000-MW wind farm, involving more than 300 turbines, is planned for central Labrador, in eastern Canada. One of the advantages of this project is its remote location, hundreds of kilometers from habitation, so any perceived unsightliness is not an issue. The pros and cons of wind power are summarized in Table 8-1.

Coastal daytime summer breezes arise because of the difference in density between the air over the water and that over the adjacent land. Since sunlight heats dry land more quickly than it does water, the air over the land also becomes warmer than that over the lake or sea. Since warm air rises—due to its lower density (according to the gas law, density is inversely proportional to Kelvin temperature)—and high above the surface moves out to sea, the remaining air over the land surface has a lower density and pressure than that over the sea. Consequently, to equalize pressures, surface air flows from over the sea to the land mass, creating a cooling sea breeze. At night, the situation is reversed, since the land cools more quickly than does the water, producing an outward breeze by the same mechanism.

Consistently breezy offshore shallow areas, such as the sandbanks off the coasts of Denmark and Ireland, are ideal sites and are now used extensively for wind farms. Indeed, offshore locations are popular in Europe, and most are anchored in water 8–10 m deep. A recent study indicated that locations in New England, on Lake Erie, and off the coast of mid-Atlantic states alone

TABLE 8-1	Pros and Cons of Wind Power
Arguments Against Wind Power	**Arguments in Favor of Wind Power**
Many sites—including offshore ones—are far from centers of demand, requiring long transmission lines to be built.	This is also true of many potential new hydroelectric projects.
Wind power needs some tax incentives to compete with traditional forms of electricity production.	Conventional and nuclear power plants receive much larger, though indirect, subsidies.
The construction of windmills at some remote sites requires roads, forest clearing, and other destructive infrastructure.	
Windmills kill wildlife, especially bats and birds of prey.	Studies show that very few birds are killed by wind turbines, especially compared to the number killed by cars, cats, etc.
Huge areas of land, and therefore of habitat, are required to construct enough windmills to have a substantial effect on electricity supply.	
The continuous motion of the blades produces low-grade noise pollution nearby.	Noise level is comparable to traffic.
On-shore wind farms are a form of "visual pollution."	Sites remote from areas of dense population can be used.
Wind power is usually intermittent, with a low annual load factor, and requires backup facilities using traditional resources to remain constantly on-call.	Excess wind energy can be stored mechanically by pumping water to elevated storage facilities or in batteries and then used when needed to produce electricity.
	Very little greenhouse gas emissions are associated with wind energy compared to fossil-fuel combustion. There is no nuclear waste to store or potential radiation problems compared to nuclear power.

could generate up to 20% of the U.S. electricity supply. However, the physical conditions at some potential offshore locations are rather harsh, and it is difficult to service broken turbines in open water. West-coast waters off North America are too deep for such placements, and those in the southeastern United States are too prone to hurricanes.

In summary, there exists considerable potential for wind energy to supply a significant fraction of the future electricity supply of many countries at lower cost to the environment than with any other feasible alternative. The price for new wind-power supplies is comparable to that of newly constructed coal- and nuclear-fired power plants and would probably be lower if any realistic costs associated with the environmental impact of conventional sources is assessed in the future. However, there are unresolved problems of efficient energy storage for many locations having intermittent winds that prevent the adoption of wind as the main source of electrical energy generation. Indeed, many power grids are unwilling to rely on wind for more than a small fraction of their energy supply due to its intermittent nature. However, the development of *flow batteries*, in which the chemicals formed when they are being charged—e.g., by excess wind capacity—can be removed from the batteries, stored in containers, and later resupplied during discharge, may help overcome storage problems.

Biomass

The **biomass** produced by the worldwide process of photosynthesis constitutes a form of solar energy. The annual amount of energy currently produced from this source is about 55 EJ; a much larger amount is potentially available. The use of wood, crop residues, and dung (dried excrement from plant-consuming animals) has been a traditional source of energy in undeveloped countries, but its domestic and small-scale use is very polluting to the air and inefficient. Small-scale, polluting biomass burning is generally phased out in favor of commercial energy such as fossil fuels and electricity as a country's economy develops. Nevertheless, biomass was second only to hydroelectric power in the production of renewable energy in the United States in the late twentieth century.

Recently, technology has been developed to use biomass in large-scale installations that do not pollute the air. For example, wood-chip waste can be burned to produce steam. Alternatively, wood can be gasified, or digested by bacteria, and converted into alcohol fuels (as discussed later in this chapter). Fast-growing trees on plantations could be used for this purpose, using land not suitable or needed for agriculture. Currently, crops such as corn and sugarcane are grown to produce ethanol for fuel, but often these facilities consume so much fossil fuel in their operation that little is saved overall in CO_2 emissions (also discussed later in the chapter).

Overall, the power density of photosynthesis (about 0.6 W/m^2) is too low for it to supply most of the world's energy needs. The density is low because the efficiency of conversion of sunlight to chemical energy in photosynthesis is very low, no more than 1–2%, even in the most productive areas. At today's consumption levels, the amount of land required to supply our energy needs by biomass equals that of all agricultural land currently developed, more than 10% of Earth's land surface.

Geothermal Energy

Geothermal energy, though not solar-based, is another form of renewable energy. It has proven particularly useful in countries that have no fossil-fuel resources. Geothermal energy is heat that emanates from beneath the Earth's surface and results from the radioactive decay of elements and from conduction from the molten core (>5000°C) of the Earth. Because of the movement of crustal (tectonic) plates, there are volcanic zones in which the heat is brought closer than usual to the surface. An example of the heat gradient with increasing depth for a geothermal zone is compared to that in a non-geothermal area in Figure 8-5. When deep groundwater circulates within a geothermal zone, the water is heated by contact with the hot rocks and is sometimes vaporized. If the hot fluids are trapped in porous rocks under a layer of impermeable rock, a geothermal reservoir can form.

Geothermal energy is available as steam and/or hot water at temperatures ranging from 50 to 350°C from reservoirs of hot groundwater. The fluid generally has to be piped 200–3000 m to the surface to be usable, though in a few places it exits spontaneously from the ground as "hot springs." Production of hot fluid generally declines with time once a reservoir is tapped.

Geothermal energy in the form of moderately hot water (50–150°C) is most often used directly for space heating of buildings, including greenhouses, and for aquaculture. High-temperature geothermal energy (>220°C) in the form of steam or superheated water is usually found only in volcanic regions and island chains and is used to generate electrical energy. Hot water of intermediate temperature is used both for heating and for generating electricity.

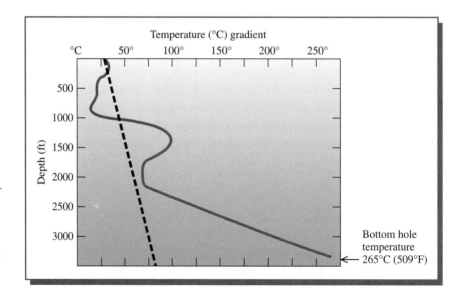

FIGURE 8-5 Underground temperature gradients in a typical area (dashed line) and in one having geothermal potential (solid green curve). [Source: Geothermal Education Office, at http://geothermal.marin.org.]

The generation of electricity by geothermal energy can be accomplished directly if superheated steam is available, since it will drive turbines. The efficiency of heat-to-electricity conversion is not great, however, since the temperature difference between the steam and the cooling water used to condense the steam is not large, in accordance with the second law of thermodynamics discussed later in this chapter. The condensed steam is pumped down an injection well back into the source to sustain production. If the geothermal energy is available only as moderately hot water, its energy can be first transferred to an organic fluid—such as *isopentane*—whose boiling point is lower than 100°C, and the hot organic vapor can be used to turn the turbine. The overall efficiency of such power conversion is less than 12%. Alternatively, in "flash steam" power plants, high-pressure hot water from the reservoir undergoes a large drop in pressure in a flash tank, at which point some of it "flashes" to steam, which is used to drive turbines.

Currently, over 8000 MW of geothermal electrical power has been installed globally, providing some developing countries with a significant fraction of their electricity. For example, both the Philippines and Indonesia generate a significant fraction of their electricity geothermally, in power plants that are as large as 100 MW. The United States (in California and Hawaii especially), Mexico, and Indonesia are other leading producers of geothermal electricity. A map showing the location of potential geothermal power sites in the United States is presented in Figure 8-6. A recent report

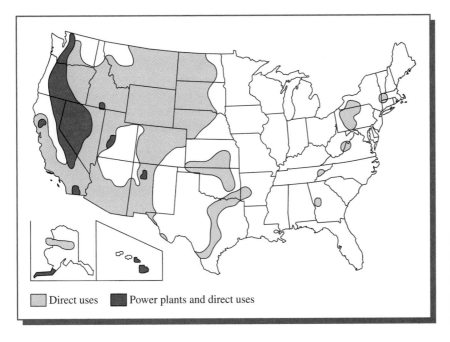

Direct uses ■ Power plants and direct uses

FIGURE 8-6 Regions of potential sites for geothermal energy production in the United States. [Source: Geothermal Education Office, at http://geothermal.marin.org.]

from MIT concluded that enough geothermal energy could be obtained to meet the entire electricity needs of the United States by piping water far underground to be vaporized by contact with hot rocks and using the resultant steam to turn turbines.

As expected, the main costs associated with the production of geothermal electrical power are capital ones; operating costs are low. High capital investment is needed for exploring, drilling the wells, and installing electrical generating equipment. The direct capital cost per kilowatt of installed capacity is about $1500 for large power plants that have a high-quality steam resource available.

The use of geothermal energy—including the wastewater from geothermal power generation—for space heating is widespread; 58 countries worldwide employ it to some extent, for a total usage of over 12,000 MW. Per capita, Iceland leads the world in geothermal energy. As a result of the mid-Atlantic meeting of two tectonic plates deep under the island, geothermal energy is plentiful. For some years, over 80% of the hot water and heating in its major city, Reykjavik, has been supplied by hot water tapped from sources a few hundred meters below the ground and distributed by pipelines. Deeper geothermal reservoirs, from which steam can be obtained, are tapped to generate some of the country's electricity. Eventually, the plan is to tap more reservoirs and use the excess electricity to produce hydrogen gas by electrolysis of water. Iceland hopes to become the world's first economy based on hydrogen and to have its economy totally free of fossil fuels by about 2030.

One drawback to geothermal energy is the large quantity of hydrogen sulfide gas that is often released when the hot fluid is tapped below the Earth's surface, especially at deep sites. Emissions of the gas into air are minimized by scrubbing the fluid. Some carbon dioxide that accompanies the hot fluid from the ground is released into the air, but the amount is tiny compared to what would be emitted if the same amount of energy were generated by burning fossil fuels. At some sites, the hot groundwater contains significant amounts of dissolved, sometimes acidic, minerals, which can corrode equipment and deposit scale. In some cases, the minerals contain sellable substances such as silica and zinc.

The impact of the wastewater on local aquatic life is eliminated if it is reinjected into the ground after its heat has been extracted. There is also some danger of land subsidence if reinjection does not occur.

In the past, geothermal energy has been exploited only where unusual geological features result in the existence of steam or hot water close to the Earth's surface. A project under way in Achen, Germany, is attempting to overcome this limitation by sinking a borehole 2.5 km (1.1 miles) into the Earth's crust, where rock temperatures are about 80°C. Cold water will flow down the outer component of the pipe, being warmed as it descends, and then be pumped back to the surface through an inner pipe, where it will be used.

Another application of geothermal energy is the use of *heat pumps*. These devices draw heat energy from the soil under the ground or from a shallow underground river and pump it to the surface to supplement heating of buildings in winter. In summer, the direction of heat flow is reversed, with energy being transferred from buildings above the ground to the underground location.

Wave and Tidal Power

Wave power and **tidal power** can be obtained in many coastal regions of the world and is competitive economically in niche markets. It is estimated that about 20 EJ of power is potentially recoverable from waves and tides.

The source of the energy of tides is the gravitational influence of the Sun and Moon on the water's mass. In some locations, coastal currents generated by tides can be exploited to turn submerged turbines mounted on pipes that fit into holes drilled in the seafloor. Because water is so much denser than air, slow currents—about 10 km/hour is best—driving such "submerged windmills" efficiently generate electricity.

Tides cause large masses of water to be lifted and then lowered twice a day. If the tides in a coastal basin are generally high, a gate that can be opened or closed can be built across the basin. When the tide is coming in, the gate is left open so the water behind it rises. At high tide, the gate is closed. The dammed water leaving the basin turns a turbine, generating electricity.

Three tidal power plants are currently in operation, located in France, Nova Scotia, and Russia. These installations had high capital costs and can operate only twice daily. Although the energy produced is renewable and pollution-free, sedimentation occurs behind the dam gates, and tidal mudflats are often destroyed as a result of the operation.

Wave power at the sea's surface can be exploited. The machines, based on an *oscillating water column*, consist of a chamber that contains trapped air located just above the water surface. Wave power is generated by using the up-and-down motion of water that results from waves, which are caused by winds and thus are an indirect form of wind energy. The rising wave compresses air trapped in the chamber. The high-pressure air is then released through a valve, turning a turbine to produce electricity. As the wave recedes, air rushes back in through another valve, also spinning the turbine. Currently there are thousands of oceanic navigation buoys whose 60-W light bulbs are powered by this mechanism. Large-scale wave-power facilities are still in the future.

Types of Direct Solar Energy

The direct absorption of energy from sunlight and its subsequent conversion to more useful forms of energy such as electricity can occur by two mechanisms:

Thermal conversion Sunlight (especially its infrared component, which accounts for half its energy content) is captured as heat energy by some absorbing material. (An everday example of such a material is a shiny metal surface, which becomes very hot when left in sunlight.) Solar energy is an excellent source of heat at temperatures near or below the boiling point of water, a category that accounts for up to half of total solar energy usage.

Photoconversion   The absorption of photons associated with the ultraviolet, visible, and near-infrared components of sunlight brings about the excitation to higher energy levels of electrons in the absorbing material. The excitation subsequently causes a physical or chemical change (rather than a simple degradation to heat).

An example of *passive* solar technology—systems that use no continuous active intervention or additional energy source to operate them—is the use of solar box cookers in developing countries. In temperate climates, the design of buildings to absorb and retain (by insulation) a maximum fraction of the solar energy that falls on them in winter is another example.

Solar water heaters are used extensively in Australia, Israel, the southern United States, and other hot areas that receive lots of sunshine. They are also used extensively in China, Germany, Turkey, and Japan. The water heaters represent the biggest use of *active* solar technologies, which are defined as those that employ an additional energy source to operate them. Solar collectors located on the rooftops of private homes and apartment buildings, as well as some commercial establishments such as car washes, contain water that is circulated around a closed system by an electrically driven pump. Sunlight is absorbed by a black flat-plate collector, which transfers the heat to the water that flows over it and is bounded on the outside by glass or a plastic window. The hot water is pumped to an insulated storage tank until it is required for bathing or laundry or to help heat swimming pool water.

In more elaborate installations, the hot water is passed through a **heat exchanger,** which is a system of pipes over which air is passed and thereby heated by thermal transfer. The hot air can be used immediately in winter to heat the rooms of the building. If not needed immediately, the heat can be stored in other materials such as rocks. Usually a backup system, in which water can be heated electrically or by burning fossil fuel, is incorporated to provide heat on cloudy days or in high-demand situations.

Using Thermal Conversion to Produce Electricity

By focusing the sunlight reflected by mirrors onto a receiver that contains a solid or a fluid, very high temperatures can be achieved. The hot fluid can be used to generate electricity by turning turbines.

As discussed in the next section, the fraction of thermal energy that can be extracted and converted to electricity from a mass of hot fluid at an absolute temperature T_h is limited by the *second law of thermodynamics* to be no greater than $(T_h - T_c)/T_h$, where T_c is the final absolute temperature of the cooling water. Consequently, it is advantageous to use a gas that has been heated to the highest possible temperature to maximize the amount of energy that is transformed to electricity rather than just degraded to waste heat. Indeed, temperatures of 1500°C have been achieved in steam heated by focusing sunlight. Generally, power plants need gas heated to 1200–1350°C at a pressure of 10–30 atm to operate. Simply focusing sunlight on tubes of air cannot achieve more than 700°C and 1 atm pressure. In one promising design, sunlight is focused by mirrors onto ceramic pins, which absorb the solar energy and heat up to 1800°C. Because they have a large surface area, the pins efficiently transfer the heat to air that flows around them. The hot, pressurized air (rather than steam) is then used to turn turbines and produce electricity.

The **solar thermal electricity** that results from power plants of this type may become competitive in price with conventional sources. This is particularly true if the waste heat, e.g., steam near the boiling point of water, can also be used for some purpose. This technique of using the waste heat from a heat-to-electricity conversion for a constructive purpose is called the **cogeneration** of energy. (It is a common feature of new power plants fueled by natural gas.) Unfortunately, power plants based on steam require large amounts of cooling water to condense the steam back into the liquid state as part of the system's cycle (see Figure 8-7); in many areas that have abundant land and sunlight, there is little water available for this purpose. Scientists have also pointed out that if the absorption of sunlight by such systems occurred on a massive scale, the Earth's albedo would be altered, with consequent effects on the climate that are difficult to predict.

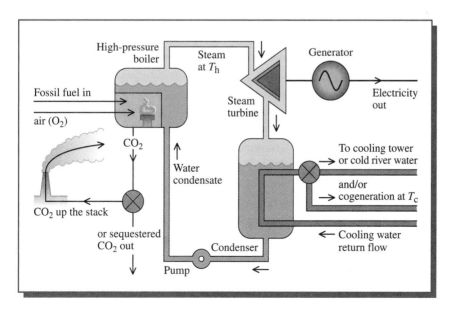

FIGURE 8-7 The generation of electricity from a steam turbine cycle. In solar thermal electricity generation, the water is heated by the Sun's rays rather than by burning a fossil fuel. T_h and T_c refer to steam and water temperatures, as discussed in the text. [Source: Modified from M. I. Hoffert et al., "Advanced Technology Paths to Global Climate Stability: Energy for a Greenhouse Planet," *Science* 298 (2002): 981.]

Another way to use the very-high-temperature heat energy is to drive a thermochemical process such as the reduction of a metal oxide to the free metal (and oxygen gas). The metal could then be used to generate electricity in batteries or to react with water to form hydrogen fuel. In either case, the product is the metal oxide, which can be recycled for reuse. An example under development is the dissociation of *zinc oxide*, ZnO, into metallic zinc and oxygen at temperatures above 1700°C. Alternatively, the heat could be used to produce a combination of carbon monoxide and hydrogen from carbon dioxide and methane.

$$CO_2 + CH_4 \longrightarrow 2\,H_2 + 2\,CO \qquad \Delta H = +248\ \text{kJ/mol}$$

Reversal of this highly endothermic reaction produces heat that can be used to generate electricity etc. without the net emission of greenhouse gases, since the methane and carbon dioxide products are collected and reused.

Limitations on the Conversion of Energy: The Second Law of Thermodynamics

In all processes that convert high-temperature heat into electricity, a portion of the original heat energy is inevitably lost as waste heat at a lower temperature. This loss is partially unavoidable as a consequence of the **second law of thermodynamics** and applies to the production of solar thermal electricity as well as to other energy conversion processes.

According to the second law, **entropy** (or disorder) *must increase*—or at the least be unchanged—*when one type of energy is converted into another.* The law tells us that for any body at absolute temperature T that possesses an amount q of heat, the entropy S is a positive quantity given by the formula

$$S = q/T$$

Since high-quality (low-disorder) energy such as electricity has essentially zero entropy, clearly one cannot convert 100% of the heat into electricity, since the change ΔS in entropy associated with the conversion would be negative. However, if *some* of the initial heat energy q_h at the initial high temperature T_h is degraded to a smaller quantity q_c at a lower temperature T_c, then the entropy *change* ΔS for the process could be positive or zero:

$$\Delta S = \text{entropy of energy after conversion} - \text{entropy of energy before conversion}$$
$$= q_c/T_c + 0 - q_h/T_h$$

For the most complete conversion to electricity possible, $\Delta S = 0$; for this situation the equation can be rearranged to give the new relationship

$$q_c/T_c = q_h/T_h \qquad \text{or} \qquad q_c = q_h\,T_c/T_h$$

The amount of heat converted to electricity is $q_h - q_c$, which upon substitution for q_c is equal to

$$
\begin{aligned}
\text{heat converted} &= q_h - q_h \, T_c/T_h \\
&= q_h \, (1 - T_c/T_h) \\
&= q_h \, (T_h - T_c)/T_h
\end{aligned}
$$

Thus the maximum fraction of the original heat that can be converted to electricity is

$$\text{heat converted/initial heat} = (T_h - T_c)/T_h$$

In other words, the *maximum* yield of electricity increases as the *difference* in temperatures between the original heat source and the waste heat increases. Thus if sunlight can be converted to heat at 1500°C (1783 K), and if the temperature of the waste heat could be held to 27°C (300 K), the fraction of energy that could be converted to electricity would be

$$(1783 - 300)/1783 = 0.83$$

In fact, the efficiencies calculated by the formula are somewhat overestimated when more than one physical phase is involved. For example, associated with the conversion cycle in traditional power plants is a step that condenses steam back into liquid water (Figure 8-7), a process in which entropy is decreased. Consequently, additional energy beyond that calculated must be degraded to waste heat to compensate.

PROBLEM 8-1

What is the maximum percentage of heat at 900°C that could be converted into electricity if the waste heat was produced as steam at 100°C?

PROBLEM 8-2

Electricity could be obtained by exploiting the thermal gradient between the surface and the deep waters of the ocean. The maximum gradient, about 20°, is achieved in tropical waters. What is the maximum percentage of the energy associated with this gradient that could be converted to electricity if the surface (cooling) water temperature is 25°C?

PROBLEM 8-3

In order to reach a conversion efficiency of 50%, to what minimum Celsius temperature must a heat source be raised if the waste heat has a temperature of 57°C?

Solar Cells

Electricity can be produced directly from solar energy by the photoconversion mechanism. This application exploits the **photovoltaic effect,** which is the creation of separated positive and negative charges in a material as a result of the excitation by light of an electron within the solid from its normal energy level to a higher, excited state. Both the excited electron and the location of the site of positive charge (the "hole") are mobile within the solid, so an electrical current could be made to flow in the material. The hole "moves" by means of the transfer of a *bonding* electron from an atom adjacent to the initial hole to the atom on which the hole is now located, thereby moving the position of the positive charge. Successive bonding-electron transfers of this type allow the hole to continue moving.

The material used for photovoltaic or solar cells is a semiconductor, which is a solid that has a conducting behavior intermediate between that of a metal (freely conducting) and an insulator (nonconducting). In semiconductors, the bonds linking the atoms are relatively weak, so the separation in energy between the bonding and antibonding levels is relatively small (compared to that of an insulator). Consequently, the energy required to excite an electron from the least stable of the filled, bonding levels to the most stable of the empty, antibonding levels is small but finite. The most common semiconductor used in solar cells is elemental **silicon,** for which this *band gap* separating the energy levels is 124 kJ/mol, which corresponds to infrared light.

Silicon's light absorption ability extends from the band-gap energy of 124 kJ/mol through to energies associated with the visible region, so most of the photons of sunlight are absorbed. However, all the photon energy in excess of the 124 kJ/mol band gap is wasted by being converted into heat rather than promoting current flow. When this loss of energy is combined with that wasted by recombination of electrons and holes even in the purest single crystal silicon, only a maximum of 28% of sunlight's energy can be converted into electricity. Commercial cells now have efficiencies of 15–20%. A higher proportion of sunlight energy can be absorbed and utilized if cell wafers having slightly different absorption characteristics are combined in a single cell. A triple-junction cell, combining layers of gallium indium phosphide, gallium arsenide, and germanium, can absorb sunlight with 40% efficiency but is very expensive to manufacture. Amorphous silicon has a maximum efficiency of only slightly more than half the value for pure silicon—because electron–hole combination occurs more readily, converting more of the energy into heat—but is used extensively because it is so much less expensive to manufacture and can be produced in thin films.

Each solar cell measures about 10 cm × 10 cm × 200 μm thick and produces only about 1 W of electricity, so to generate electricity in useful quantities, many are joined together in a *solar array*. One problem with the electricity generated using solar cells is that it is *direct current* (dc) rather than the *alternating current* (ac) that is used in power grids and by most equipment and appliances. The dc electricity can be converted to ac, although with the loss of some power (as waste heat).

The cost of producing the solar cells and the problem of storing the electricity for use at night and on cloudy days are the greatest barriers to their increased use. As with other applications of solar energy, the capital cost in creating the infrastructure required to capture and use the "free" energy of the Sun is substantial. The cost of the encapsulation of cells, the wiring, and the construction of supporting structures is relatively high since each cell is inefficient; collectively, this adds about as much to the total cost of a power system as the cost of the cells themselves. Currently, crystalline solar cells cost about \$4/W to manufacture, so the system itself costs about \$8/W; since the average home can be supplied by a 4-kW system, the cost of using cells to completely power a house is about \$32,000. If solar cells could be made 20% efficient, this power demand could be met by about 20–25 m^2 of solar panels, i.e., an area about 5 m $\times$ 5 m.

Although the solar cells do not generate any carbon dioxide during their operation, their manufacture does consume significant amounts of energy and therefore causes substantial CO_2 emissions. Indeed, according to Figure 8-4, photovoltaics are the most CO_2-intensive of the various renewable energy forms. The energy and carbon dioxide payback periods for solar cells and their infrastructure currently are about three years but are expected to fall to one to two years when manufacturing techniques improve further. Once the cells are manufactured, however, their use to generate power in a home saves about as much carbon dioxide per year as is emitted from the family car. The Japanese government subsidized solar cells for tens of thousands of rooftops and, indeed, Japan is now the world's largest producer of such cells.

The cost of manufacturing solar cells has continued to fall with time, but electricity generated in this way is still not close to being competitive with that produced by conventional methods of power generation. Currently, 90% of cells are made from crystalline silicon and the remaining 10% from thin-film amorphous silicon. To date, the silicon used for solar cells has been castoff or excess silicon from the semiconductor electronics industry. However, this supply will rapidly become inadequate if and when the computer industry revives from its recession.

Photovoltaic power may become attractive in hot, sunny locations such as the southwestern United States, where the peak power demand, driven by the need for air conditioning, coincides in time (summer afternoons) with the peak solar energy availability. Already, solar-cell power (plus storage) is cheaper than extending power grid lines a kilometer or more away from an existing network into a remote region, and it is competitive in cost with the use of diesel generators for this purpose. Portions of this textbook were written at a seaside location that is within sight of an offshore lighthouse powered by solar cells. The technology is also commonly used in water pumps, remote roadside telephones and signs, and satellites.

The use of solar cells in developing countries, most of which have sunshine in abundance, could obviate the need for the creation of power grids to carry electricity over long distances from source to user and represents the greatest potential market for expansion of photovoltaic power. Indeed, the majority of solar cells made in the United States are exported. Solar-cell electricity is already used to power water pumps, lights, refrigerators, and TVs in some developing countries. A mid-1990s survey found that 45% of new solar cells were used to electrify homes, villages, and water pumps; 36% were used for communications and other industrial applications; and 14% were used for the generation of grid-bound electricity. However, by 2000, more than 50% of new solar cells were connected to grid systems.

As with wind power, world production of solar cells for power production has soared since the mid-1990s (see black curve in Figure 8-2; note the difference in scales for the two types of power). Although almost 1700 MW of solar-cell capacity was shipped in 2005, it represents only 15% of the wind-turbine power installed that year. Although Germany installed almost half the global new solar-cell capacity in 2005, the United States may shortly resume its leadership in this area, since California plans to install 3000 MW of solar power in the next decade. Japan expects to have 4800 MW installed by 2010.

Could solar cells be used to supply *all* of the electricity needs of developed countries? Using today's efficiencies and materials, the energy needs of the United States could be met by a continuous array of solar cells covering a square of land about 160 km × 160 km (100 miles × 100 miles). Although this area of 26,000 km^2 does not seem extraordinarily large, it is about a thousand times larger than the *total* area covered by all the solar cells manufactured through 1998 and would cost trillions of dollars to construct! An area about 10 times larger—about 0.1% of the Earth's surface—could supply all the world's energy needs.

Conclusions About Solar Energy

In these discussions, some general features concerning the use of solar energy, as opposed to fossil-fuel and nuclear energy, have emerged, and others have also been reached by energy analysts. Many of these conclusions apply to all forms of renewable energy.

The **advantages** of solar energy appear to be that it

- is **free** and fantastically **abundant;**

- has **low environmental impact;**

- has **low operating costs;**

- does not require imported oil or large, centralized suppliers and expensive distribution networks; and

- has **high public acceptance** as a "natural" form of energy.

  The **disadvantages** of solar energy appear to be that it

- is **intermittent** in its availability and thus requires that efficient storage or backup systems be constructed so that power can be supplied continuously;

- is **diffuse**—it provides a low density of energy per unit of surface collection area, so large areas of solar collectors are required to harvest the energy (one kilowatt requires about one square meter, on average);

- requires **high capital costs** to construct the energy collection and storage systems, offsetting the free nature of the energy itself for many years until the investment is paid off; and

- receives little or **no economic (tax) or regulatory credit** from governments in recognition of the low amount of air pollution and greenhouse gas emissions it causes relative to fossil-fuel usage.

Alternative Fuels: Alcohols, Ethers, and Esters

For environmental and supply reasons, attention is turning to the development of cleaner-burning alternatives to hydrocarbon fuels, especially to power vehicles. Some of these alternatives are, at least in principle, renewable in the sense that their production could be sustained indefinitely into the future without resulting in the accumulation of additional carbon dioxide. In the material that follows, we discuss the nature and properties of the major contenders for alternative fuels. In a later section, we take a longer-range viewpoint and consider hydrogen, the ultimate "fuel of the future."

The organic fuels considered here have the inherent advantage over hydrogen—and even over natural gas—that they are liquids under normal temperatures and pressures that burn easily in air to produce considerable heat; like the gasoline and diesel fuels with which they can be blended or which they can replace, they are *energy-dense* fuels. Since they all contain carbon, however, their combustion releases carbon dioxide.

The alternative fuels for vehicles fall into three classes: alcohols, ethers, and esters. Because they all contain some oxygen, they generally produce a little less energy per liter than do gasoline and diesel fuel. However, their oxygen content results in low emissions of many air pollutants. NO_X emissions from these organic liquids are also lower than from pure gasoline because the flame temperature is lower and thus less thermal NO (Chapter 3) is formed.

Ethanol as a Fuel

Ethanol, C_2H_5OH, also called *ethyl alcohol* or *grain alcohol*, is a colorless liquid that has been used as an automobile fuel as far back as the late 1800s; indeed, Henry Ford designed his original cars to run on ethanol.

As a fuel for vehicles, ethanol can be used "neat," i.e., in pure form, or as a component in a solution that includes gasoline. Often these fuels are referred to by the letter E (for ethanol) followed by a subscript that indicates the percentage of alcohol in the gasoline–ethanol mix. In North America, the "gasohol" currently sold is about 10% ethanol and 90% gasoline, i.e., E_{10}. Ethanol and gasoline are freely soluble in each other, so all possible combinations can be produced. Currently in Brazil, E_{23} is used by all gasoline-powered vehicles. Pure ethanol, E_{100}, is used mainly in Brazil, where about one-eighth of car engines have been designed to use it.

One attractive feature of "oxygenated" transportation fuels such as ethanol is that they result in lower emissions of many pollutants—specifically carbon monoxide, alkenes, aromatics, and particulates—compared to emissions from combustion of pure gasoline or diesel fuel, particularly from older vehicles that do not have catalytic converters. In North America, however, the turnover of vehicle fleets means that very few cars still on the road emit much CO. The reduction in urban ozone that would result from the lowered emissions of carbon monoxide and reactive hydrocarbons would be countered by increases due to the higher amounts of acetaldehyde (Chapter 5) and vaporized ethanol that would be emitted. This is particularly true for urban areas in which ozone formation is NO_X-limited rather than hydrocarbon-limited. However, NO_X emissions from engines burning ethanol are lower than from those burning gasoline. Studies in Rio de Janeiro indicate that the concentration in air of the important pollutant *peroxyacetylnitrate*, PAN (see Chapter 5), which is readily formed from the acetaldehyde emissions, has increased due to the use of ethanol fuel; since Brazilian cars are not equipped with catalytic converters, this finding is not directly relevant to the North American situation. However, measurements in Albuquerque, New Mexico, have established that the use of ethanol as a gasoline additive increased the concentrations of pollutants such as PAN in the air of that city. It is curious that some proponents of ethanol as a fuel point to its ability to reduce CO emissions, which in fact is important only for cars without catalytic converters, but downplay the effects of acetaldehyde emissions by stating that they can always be minimized by use of catalytic converters!

One of the difficulties in using pure ethanol (or pure methanol) as a vehicular fuel is its low vapor pressure: See Figure 8-8, in which the vapor pressure of gasoline–ethanol mixtures is plotted against its composition, with pure gasoline at the left side of the horizontal axis and pure ethanol at the right. Thus, in cold climates, there is very little vaporized fuel available with which to start a cold automobile engine. However, a blend of

85% ethanol and 15% gasoline has a high enough vapor pressure (Figure 8-8) to overcome the *cold start* problem; in North America and Europe, E_{85}-fueled vehicles are now available.

The E_{10} blend of ethanol in gasoline is now widely available in North America. In order to reduce the evaporation of volatile organic compounds (VOCs) from gasoline—since they contribute to the ozone problem—the United States regulates the maximum vapor pressure of gasoline sold during the summer months. To achieve the lower overall volatility, the amount of (highly volatile) *butane* in gasoline is being reduced and replaced by substances that have low volatility. Unfortunately, as a minor additive, ethanol is quite volatile and actually increases the vapor pressure of gasoline. (The same phenomenon occurs for methanol–gasoline blends.) This behavior can be understood by reference to Figure 8-8. As ethanol is added to gasoline, the vapor pressure of the mixture rises sharply since the C_2 compound behaves in this hydrocarbon environment much like a low-molecular-weight—and therefore volatile—hydrocarbon. In contrast, as a pure liquid, ethanol has a lower vapor pressure than gasoline (see Figure 8-8) since the extensive hydrogen bonding between ethanol molecules in this situation provides a "glue" that makes it difficult to break the molecules apart and vaporize them.

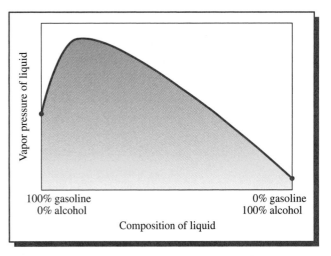

FIGURE 8-8 Variation in vapor pressure with composition for typical alcohol–gasoline mixtures.

Another disadvantage of ethanol (which is even more valid for methanol) as a fuel is that the energy it produces per liter combusted is somewhat less than is generated by an equal quantity of gasoline; to travel the same distance, fuel tanks for alcohols would have to be larger. In principle, about 1.25 gal of ethanol are needed to generate the same amount of energy as is obtained with 1 gal of gasoline. In practice, however, the efficiency of the combustion is greater with the alcohols, so the volume penalty is not this large.

Ethanol Production

Industrially, ethanol is made by catalytically adding water to petroleum-based *ethene*, $CH_2{=}CH_2$, to produce CH_3CH_2OH. In contrast, ethanol for fuel is produced on a massive scale by the fermentation of carbohydrates in plants. Such **bioethanol** is produced by the yeast-driven fermentation principally of **glucose,** $C_6H_{12}O_6$. In North America, most carbohydrate used for ethanol production is derived from the starch in kernels of corn, although some wheat and other grains are also used. In Brazil and some other semitropical countries, sucrose from sugarcane is used. A number of developing countries, including Thailand and China, are producing ethanol from *cassava*, a woody shrub that produces a tuberous root high in starch content. In the fermentation process, the sunlight-derived energy of the glucose becomes more concentrated in the ethanol product, since some of the carbon is released as carbon dioxide gas:

$$C_6H_{12}O_6 \xrightarrow{\text{yeast}} 2\,CO_2 + 2\,C_2H_5OH$$

PROBLEM 8-4

By comparing oxidation numbers, show that the carbon atoms in ethanol are more reduced—and therefore serve as better fuels when oxidized—than would the carbon atoms in the glucose molecules from which they originated before fermentation. Show also that there is no net change in oxidation number of carbon when going from reactants to products in the fermentation reaction.

PROBLEM 8-5

Using the enthalpies of formation listed below, calculate the enthalpy change for the fermentation reaction of glucose into ethanol and carbon dioxide. Is the process exothermic or endothermic? From your answers, decide whether the fuel value of the ethanol product is slightly greater or slightly less than that of the glucose from which it is created.

ΔH_f values in kJ mol^{-1}:

$C_6H_{12}O_6(s)$	-1273.2
$C_2H_5OH(l)$	-277.8
$CO_2(g)$	-393.5

As with all biofuels, the attractions of using ethanol to replace some of the hydrocarbon component in gasoline are that

- the producing country becomes less reliant on imported petroleum;

- air pollution is reduced; and

- the net amount of CO_2 emitted into the air is reduced.

Using biofuels such as ethanol is thought to counterbalance much of the resulting greenhouse gas emissions because the plants absorb the carbon, used in photosynthesis, that accounts for much of their mass; the atmosphere is thereby depleted of some of its CO_2. The harvested plant biomass is converted by fermentation into a fuel, which is combusted, and the carbon is released back into the air as carbon dioxide in the same amount that the plant had absorbed to grow. The resulting net change to atmospheric CO_2 from growing and then burning the fuel would be zero. Since the process can be repeated the next season by growing more biomass in the same field, the fuel would be renewable. In the case of ethanol, the photosynthesis reaction is

$$6\,CO_2(g) + 6\,H_2O(g) \longrightarrow C_6H_{12}O_6 + 6\,O_2$$

The reverse of this reaction is the combustion process for glucose.

Unfortunately, a large quantity of water must be used in fermentation in order to solubilize the starch from which the glucose is obtained; otherwise, the yeast dies if it is present in concentrated alcohol. Indeed, the greater the percentage of alcohol in the mixture, the slower the rate of conversion. A total inhibition of fermentation occurs when the alcohol solution reaches about 8–11% ethanol by volume (i.e., about one-tenth of the aqueous solution). For this reason, only dilute solutions of alcohol can be produced by fermentation. However, a dilute solution of ethanol in water (equivalent in alcohol content to that in wine) will not burn.

To be used as a vehicular fuel, almost all the water must be removed from the ethanol solution produced by fermentation. Consequently, the solution is distilled to separate the alcohol from the water. Distillation is a very energy-intensive process, since the watery mixture must be constantly kept at a boil. What ultimately results from repeated distillations is not pure ethanol but a solution of 95.6% ethanol and 4.4% water (by volume). The last vestiges of water cannot be removed by more distillation; however, this removal can be accomplished by a process involving a molecular sieve that also uses heat energy when the sieve is dried so that it can be reused. Many of Brazil's vehicles operate on *hydrous ethanol,* i.e., 95% C_2H_5OH.

The crux of the controversy about whether or not bioethanol is a renewable fuel is that heat generated by burning a large amount of fuel is needed to distill the ethanol from the water. In the modern production of ethanol from corn in the United States and Canada, a nonrenewable fuel—either coal or natural gas—is burned to generate the heat required in the distillation process. As a result of this combustion, a large amount of carbon dioxide—a significant fraction of that produced when the alcohol is later burned as a fuel—is released into the atmosphere at this stage. However, if, as is done in Brazil, biomass crop residues (called *bagasse* in the case of sugarcane) rather than a fossil fuel are used to power the distillation, the carbon dioxide that they release upon combustion is reabsorbed by growth of such material the next season, so there is very little net addition of CO_2 into the air from this step. However, the particulate air pollution from the smoke that can accompany biomass combustion restricts its use.

Many scientists and policymakers have attempted to add up the pluses and minuses of greenhouse gas release and absorption, as well as of energy production and consumption, in generating ethanol for fuel in North America; they have compared these findings to the corresponding values for gasoline in order to determine whether or not ethanol is truly a renewable fuel. Their conclusions about whether the overall balances for ethanol relative to gasoline are positive or negative depend largely on the assumptions they make, although all agree that the size of the difference is relatively small. The analyses are complicated by the fact that commercial materials such as corn gluten feed, corn oil, and dried distiller grains, obtained from the nonstarch component of the corn kernels, are co-produced with ethanol in the distillation step of the corn mash. Presumably, some of the fossil-energy usage and greenhouse gas emissions in the process ought to be associated with the *co-products*, rather than assigning it all to the alcohol, since the co-products displace other substances on the market that would require energy to produce. Most analyses conclude that modern North American ethanol production from corn requires about two-thirds the amount of fossil fuel that would be required to generate the same amount of energy in the form of gasoline produced from conventional petroleum sources.

We can conclude, then, that the production and use of ethanol derived from carbohydrate biomass in North America reduces by about one-third the amount of fossil fuel per se that is required to produce energy for vehicles. In essence, the energy of ethanol is derived from a combination of two parts fossil fuel and one part captured solar energy; thus the production of ethanol from corn in North America is largely the conversion of natural gas or coal into a convenient vehicular fuel. Ethanol causes about 86% of the greenhouse effect enhancement of the gasoline that it displaces. This greenhouse gas percentage exceeds that for the fossil fuel it consumes in production mainly because it includes the contribution from the nitrous oxide gas that is emitted as a by-product when fertilizers are used to grow the corn; another contributor is the carbon dioxide released when nitrogenous fertilizers are made synthetically.

Although bioethanol produced from sugarcane has a better energy balance than that produced from corn, sugarcane requires considerable water to grow. Although irrigation is not required in the parts of Brazil where it currently is produced, growing sugarcane in some other countries is placing a burden on their water resources.

The rapid recent growth in world production of ethanol is illustrated by the green curve in Figure 8-9. Currently, massive amounts of bioethanol—over 16 billion liters annually—for use as vehicular fuel are produced from sugarcane in Brazil; unfortunately, the resulting air and water pollution and soil erosion are massive. Smaller quantities of ethanol are obtained from cane in Zimbabwe and from corn and grain in some midwestern American states and, recently, in Canada. As of 2005, more than 16 billion liters of ethanol fuel were also being produced annually from the starch of corn and grain in the midwestern United States; that amount is expected to more than double by 2009. About 15% of the corn crop in the United States and about half the sugarcane crop in Brazil were used to produce bioethanol in 2005. Many farmers in the United States and Canada have provided major political

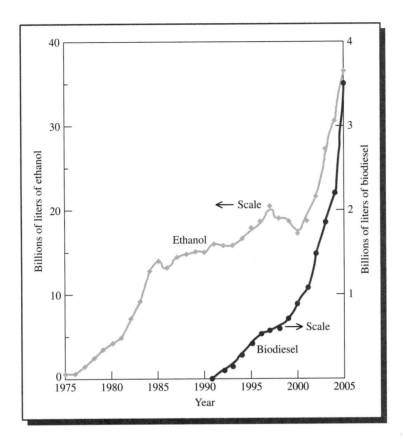

FIGURE 8-9 Annual global ethanol (light green curve) and biodiesel (dark green curve) production. [Source: Data from L. R. Brown et al., *Vital Signs 2006–2007* (New York: W.W. Norton, 2007).]

support for the production and use of ethanol in gasoline, in particular for the government subsidies required to make it economically competitive with petroleum.

Can ethanol produced from corn ever replace petroleum worldwide? The growing area required to do this would be about twice the arable land used for all food crops today, so clearly the answer is "no." Indeed, in developed countries, current energy consumption rates exceed the energy generated in food crops. Thus alcohol from fermentation of corn is unlikely to become a major fuel replacement for gasoline.

An emerging technology for biomass ethanol production uses cellulose and hemicellulose components of plants, rather than starch, as the abundant feedstock from which sugars are produced and fermented. The hope is that such *cellulostic ethanol* can be produced in the future in larger amounts and at cheaper price in terms of energy and dollars than that obtained from starch sources. The main components in the woody plants being considered for cellulostic ethanol are *cellulose* (35–50%), *hemicellulose* (25–30%), and *lignin* (15–30%). Cellulose is a long polymer of the C_6 sugar glucose (see Figure 3-11), approximate formula $(CH_2O)_n$, whereas hemicellulose consists of shorter polymer chains consisting mainly of sugars such as xylose that contain five, rather than six, carbon atoms. Lignin is an unfermentable component of the biomass.

In order to be converted into alcohol, the woody biomass must first be *pretreated* in order to break the seal of lignin and to disrupt the crystalline cellulose structure, to enable enzymes to reach and react with the cellulose and hemicellulose within. A number of different pretreatment techniques, some of them physical and some chemical (such as treatment with steam or dilute acid, or ammonia), are available, but all those developed to date are relatively expensive. Once the pretreatment has liberated the cellulose, enzymes are available to depolymerize it via hydrolysis to glucose, which then is fermented to ethanol:

$$\text{biomass} \xrightarrow[\text{pretreatment}]{} \underset{\text{cellulose}}{(CH_2O)_n} \xrightarrow[\substack{\text{enzyme} \\ \text{hydrolysis}}]{} \underset{\text{glucose}}{(CH_2O)_6} \xrightarrow[\text{fermentation}]{} \text{ethanol}$$

Until recently, the enzymes used to hydrolyze cellulose were expensive to produce, but this problem has now been overcome, leaving pretreatment as the most expensive step in the sequence. Unfortunately, the C_5 sugars that are the chief product of the depolymerization of hemicellulose are not fermented to ethanol by naturally occurring enzymes, though genetically engineered yeasts have now been produced that ferment both C_5 and C_6 sugars, albeit producing a very dilute ($<5\%$) alcohol solution.

The biomass for bioethanol production could, for example, be *switchgrass*—a perennial wild grass that was once widespread in the Great Plains of the United States—grown for this purpose; agricultural residue, such as the stalks and other nonkernel parts of corn plants; or wood, waste paper, or municipal solid waste. However, nitrogen fertilization—with its accompanying nitrous oxide release—would be required to maintain switchgrass production. Short-rotation hardwood crops—grown on land that is currently out of production or on marginal cropland—require substantially fewer fertilizers and pesticides than do corn and switchgrass; together with waste from other forestry and agricultural operations, these crops can produce much more biomass per unit area than corn.

A real advantage to the production of cellulose for bioethanol is the combustion of the mechanically dewatered lignin component of the biomass to fuel the distillation process. This greatly reduces the amount of fossil fuel required to about 8%, and the net greenhouse gas emissions to 12%, of those involved in the energy-equivalent gasoline cycle. Presumably, substantial fossil-fuel reductions would also be achieved in producing ethanol from corn itself if the stalks, cobs, etc., rather than coal and/or natural gas, were used to fuel the distillation.

Methanol

Methanol, CH_3OH, is a colorless liquid that, like ethanol, is somewhat less dense than water. Although methanol was produced in the past from the destructive distillation of wood, giving rise to its historical name *wood alcohol*, it is now produced mainly from a fossil fuel.

Methanol can be blended with gasoline to produce a fuel that burns more cleanly than gasoline. In a labeling scheme analogous to that used for ethanol–gasoline mixtures, blends of methanol are designated by an M rating; thus M_5 corresponds to 5% methanol and 95% gasoline.

One disadvantage to methanol blends is that the pure alcohol is only soluble to the extent of about 15% in gasoline, corresponding to M_{15}; greater amounts of methanol form a second layer rather than dissolve. The inadvertent presence of water causes this unacceptable phase separation to occur at an even smaller percentage of methanol. Additives such as *tertiary-butyl alcohol* (2-methyl-2-propanol) that are soluble in both methanol and gasoline prevent such separations from occurring. Looking at things from the other direction, gasoline is moderately soluble in methanol, so fuel blends such as M_{85} have been tested and are now on sale in limited quantities. Another difficulty is that methanol cannot be used in conventional automobile engines because it reacts with and corrodes some engine and fuel tank components.

Some concern has been expressed about the safety of methanol for use as a vehicular fuel, given its toxicity. Methanol–water solutions have been widely used as windshield washer liquids in northern climates for many years without much environmental impact. The use of methanol as a fuel may be more dangerous, as it would involve a very high concentration of the alcohol. Ethanol is much less toxic than either methanol or gasoline.

However, alcohols also possess some advantages: They are inherently high-octane fuels, and indeed, methanol is used to power all the cars at the *Indy 500* races. Methanol has the added advantage that it does not produce a fireball when a tank-rupturing crash of racing cars occurs: It vaporizes less rapidly than does gasoline, and, once formed, the vapor disperses more quickly.

PROBLEM 8-6

Given that the enthalpies of combustion, per mole, of methanol and ethanol are −726 and −1367 kJ and that the density of each is 0.79 g/mL, calculate the heat released by methanol and by ethanol (a) per gram and (b) per milliliter. From your results, comment on the superiority of one alcohol or the other with respect to energy intensity based on weight and volume. Are these alcohols superior or inferior to methane as fuels in terms of energy intensity per gram? (See Problem 7-4 for data.) How do they compare to gasoline, for which about 43 kJ are released per gram?

The conventional conversion of either coal or natural gas into methanol begins with the reaction of the fossil fuel with steam to produce a mixture of CO and H_2, often called **synthesis gas:**

$$C(s) + H_2O(g) \longrightarrow CO(g) + H_2(g)$$

$$CH_4(g) + H_2O(g) \longrightarrow CO(g) + 3\ H_2(g)$$

In the first process, steam is blown over white-hot coal; in the second, methane gas is combined with steam that has been heated to about 1000°C. These methods produce synthesis gas from nonrenewable raw materials. An analogous mixture of hydrogen and carbon monoxide can also be obtained by heating renewable sources of biomass such as wood or the cellulosic component of garbage. The wood is first chipped and then gasified. The gaseous product is a mixture of CO, CO_2, and H_2. The use of wood gasification to generate methanol or electricity out of biomass is highly promising, for both produce useful energy without significant greenhouse gas emissions. The various processes for synthesis gas and methanol production are summarized in schematic form in Figure 8-10.

Methanol is synthesized from a 2:1 molar ratio of H_2 to CO in the presence of a catalyst:

$$2\ H_2 + CO \xrightarrow{\text{catalyst (Cu/ZnO)}} CH_3OH$$

Unfortunately, existing catalysts allow only a partial conversion (about one-fifth) of the gases into methanol for each pass of the gas mixture over the catalyst, and the processes are energy-intensive and require relatively high temperatures. Research is under way to develop catalysts that will operate at lower temperatures and thereby allow higher yields.

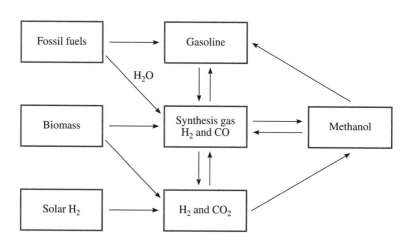

FIGURE 8-10 Scheme for the production of fuels in a hydrogen economy.

PROBLEM 8-7

The enthalpies of formation of $CO(g)$ and $CH_3OH(l)$, respectively, are -110.5 and -239.1 kJ/mol. Calculate the enthalpy of the reaction that forms methanol from synthesis gas. From your answer, predict whether the equilibrium amount of methanol obtained will increase or decrease as the temperature is lowered. Given your result, comment on the interest in developing low-temperature catalysts.

The correct 2:1 molar ratio of H_2 to CO required for the methanol synthesis reaction above is rarely obtained initially from the raw materials. For example, the reaction of steam with coal instead gives synthesis gas with a 1:1 ratio, and with natural gas a 3:1 ratio is obtained. The ratio can be adjusted to the required 2:1 by subjecting the mixture to the **water-gas shift reaction,** which is an equilibrium that can be written as

$$CO_2 + H_2 \xrightleftharpoons{\text{catalyst}} CO + H_2O$$

or as its reverse. Since running the reaction in the direction shown consumes H_2 and produces CO, and the opposite result is obtained by running the reaction in reverse, the initial 3:1 or 1:1 ratio of H_2 to CO can be altered to 2:1 by the partial conversion of the excess material, whether it is H_2 or CO, into the other, deficient material.

For example, consider the adjustment of the 3:1 ratio produced by the reaction of methane with steam to the required 2:1 ratio. Call the initial molar amount of CO produced a; then the initial amount of H_2 is $3a$. Since the hydrogen is initially in excess, some of it must be converted to CO; thus the appropriate direction for the shift reaction is indeed the forward direction written above. When this reaction achieves equilibrium, a molar amount x of hydrogen will have been consumed and an additional molar amount x of carbon monoxide will have been produced:

$$CO_2 + H_2 \longrightarrow CO + H_2O$$

from initial reaction:	$3a$	a
at new equilibrium:	$3a - x$	$a + x$

The value of x is obtained by requiring that the new, equilibrium ratio of H_2 to CO be 2:1:

$$\frac{3a - x}{a + x} = \frac{2}{1}$$

By algebraic manipulation of this equation, the ratio of x to a can be obtained:

$$x/a = 1/3$$

Thus the fraction $x/3a$ of the initial amount of H_2 from the natural gas that must be converted to CO is 1/9.

The two chemical reactions that when combined correspond to the conversion of methane in the correct 2:1 ratio are shown and added together below; it has been assumed for simplicity that $a = 1$:

$$CH_4 + H_2O \longrightarrow CO + 3\,H_2$$

$$\underline{1/3\,H_2 + 1/3\,CO_2 \longrightarrow 1/3\,CO + 1/3\,H_2O}$$

sum $CH_4 + 2/3\,H_2O + 1/3\,CO_2 \longrightarrow 4/3\,CO + 8/3\,H_2$

When combined with a catalyst, the CO and H_2 from the sum of these reactions will yield 4/3 mole of CH_3OH, giving the net overall reaction

$$CH_4 + 2/3\,H_2O + 1/3\,CO_2 \longrightarrow 4/3\,CH_3OH$$

PROBLEM 8-8

In order to synthesize a compound with the empirical formula CH_3O (and no other products) starting from methane and steam, what ratio of H_2 to CO would be required? What fraction of the hydrogen gas produced from the reaction of methane and steam would have to be converted to carbon monoxide using the water-gas shift reaction to accomplish the transformation?

Research is currently under way to find how to directly convert methane into methanol in a much more efficient manner than that described above. Most of the difficulty stems from the fact that methane is a very unreactive substance: Its C—H bond dissociation energy is highest of all the alkanes. Once one C—H bond is broken, however, the molecule becomes highly reactive because the other C—H bonds are weakened, and in the presence of oxygen it tends to oxidize completely to carbon dioxide rather than partially to a useful intermediate stage such as methanol.

Methanol can also be produced by combining carbon dioxide and hydrogen gases (see Figure 8-10) in the presence of a suitable catalyst:

$$CO_2(g) + 3\,H_2(g) \xrightarrow{\text{catalyst}} CH_3OH(l) + H_2O(l)$$

Since this reaction is only slightly exothermic, most of the fuel energy of the hydrogen is present in the methanol product. Based on equilibrium considerations, methanol formation would be favored by low temperatures and high pressures; research has centered on finding a catalyst that will operate efficiently under such conditions without being deactivated. Low temperatures also prevent the formation of carbon monoxide rather than methanol.

Some of the massive quantities of CO_2 that are released annually into the atmosphere could be used as reactants in this process. Indeed, methanol produced in this manner could be considered a renewable fuel *provided* that the hydrogen is produced without the consumption of a fossil fuel, e.g., by solar energy (see below).

Although methanol can be used as a vehicular fuel on its own, there are chemical reactions by which it can be converted to gasoline. Similarly, synthesis gas itself can be converted to gasoline, thereby allowing the production of this fuel from either natural gas or coal (Figure 8-10). Currently, neither of these processes, nor the production of fuel methanol itself, is sufficiently efficient that it can compete economically with gasoline produced from crude oil.

Ethers

Methanol can be used to produce **dimethyl ether,** CH_3—O—CH_3, which has been tested as a replacement for diesel fuel in trucks and buses:

$$2\ CH_3OH \longrightarrow CH_3OCH_3 + H_2O$$

This ether is nontoxic and degrades easily in the atmosphere—in fact, it is used as a propellant in spray cans. Since its molecules contain no C—C bonds, soot-based particulate matter is produced in its combustion in only very small quantities (see Chapters 3 and 5) compared to those obtained from diesel fuel. The NO_X emissions from dimethyl ether combustion are also lower than usually found for diesel engines.

Methanol is also used to produce the oxygenated gasoline additive **MTBE,** which stands for *methyl tertiary-butyl* ether, the structure of which is illustrated below:

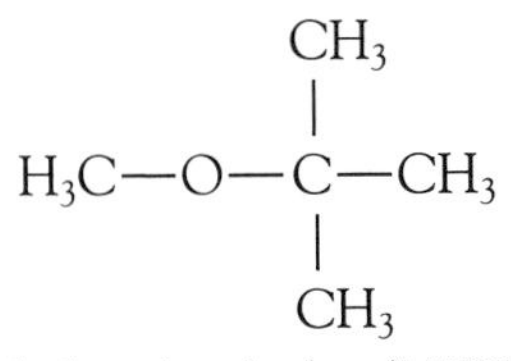

$$CH_3$$
$$|$$
$$H_3C-O-C-CH_3$$
$$|$$
$$CH_3$$

methyl *tert*-butyl ether (MTBE)

MTBE, octane number 116, is used in some North American and European unleaded gasoline blends—up to 15%—to increase their overall octane number and to reduce carbon monoxide (and unburned hydrocarbon) air pollution; the reason is that, like the alcohols, it is an "oxygenated" fuel that generates less CO during its combustion than would the hydrocarbons it replaces. The advantages to using MTBE rather than ethanol as an additive are that it has a higher octane number and does not evaporate as readily. However, as with alcohols, its combustion can also produce more aldehydes and other oxygen-containing air pollutants than result from the hydrocarbons it replaces.

The usage of MTBE has become controversial. It has an objectionable odor resembling turpentine and ether. Another problem associated with MTBE is its contamination of well water, which has now occurred at many sites in the United States. The sources of MTBE in well water are leaking underground fuel tanks; leaking pipelines; and spillage of gasoline at gas stations, in vehicle accidents, and by homeowners. In contrast to the hydrocarbon components of gasoline, MTBE is rather soluble in water and therefore is quite mobile in soil and groundwater. It is also quite resistant to biological degradation because its carbon chains are very short; its half-life is on the order of years. Various U.S. states and the U.S. EPA have set action levels for MTBE in drinking water at a few tens of parts per billion, values that are exceeded in some supplies and at which the odor of the additive is apparent. Because of concerns about well-water contamination, California and several other states have now banned the use of MTBE in gasoline, and its use as a gasoline additive has dropped sharply, having been replaced by ethanol, isooctane, and other high-octane substances.

Biodiesel

Another biofuel that has found some application, especially in the United States and Europe, is the mixture of fatty acid methyl esters, $R—COOCH_3$, called **biodiesel.** This material usually corresponds to an oil—usually derived from a plant source such as soybeans or rapeseed (canola)—that has been esterified and can be used in diesel engines. The rapid rise in annual global biodiesel production began in the late 1990s, as illustrated by the dark green curve in Figure 8-9; note the difference by a factor of 10 in the scales for bioethanol and biodiesel.

In principle, the raw vegetable oils could be blended with diesel oil—or even used neat—as a fuel. Indeed, when diesel engines were introduced in the early twentieth century, they were fueled with pure peanut oil. However, due to its high viscosity and impurities—such as free fatty acids, water, and odorous substances—the raw oil cannot be used in modern diesel engines. Even refined vegetable oil cannot be used as a general fuel, because of its viscosity and because the polymerization of the unsaturated hydrocarbon components of the oils that occurs during combustion produces gums that result in carbon deposits and thickening of lubricating oil in the engine. One solution to the viscosity problem, employed by so-called *Grease cars* that are fueled with grease from deep-frying, is to heat the oil onboard the vehicle.

More commonly, the virgin vegetable oils are transformed into a less viscous, less corrosive liquid that is used as the fuel. The main component in the original oil is **triglycerides,** which are triesters of various fatty acids with **glycerin,** CH_2OH—$CHOH$—CH_2OH. The transformation converts each triglyceride molecule into three methyl esters of long-chain fatty acid molecules, which then constitute the fuel, and a molecule of glycerin (also called *glycerol*), which is removed from the mixture of fatty acids and sold separately for other uses. To accomplish the transformation, the triglyceride is reacted using base or acid catalysis with methanol obtained from natural gas, as described in detail in the green chemistry section that follows. The use of methanol, most commercial supplies of which involve its synthesis from natural gas, makes biodiesel less than 100% renewable, though the great majority of the carbon atoms in the fuel esters—and hence in its fuel value— originate with the vegetable oil.

Overall, soybean-derived biodiesel generates over 90% more energy than is used to produce it, compared to about 25% for corn-based ethanol. Biodiesel blends produce less carbon monoxide, particulate matter (PM_{10}), and sulfur dioxide emissions when combusted than does the 100% diesel fuel that they replace; the reduction in soot and CO arises because it is an oxygen-containing fuel. There is controversy as to whether biodiesel blends produce more or less NO_X than does pure diesel. Although energy and fossil-fuel-derived methanol are used in its production, and nitrous oxide emissions are associated with fertilizers to grow the plants, biodiesel produced from soybeans on existing agricultural land overall reduces the CO_2 equivalent emissions by about 40%. The much greater decrease in CO_2 from biodiesel, compared to corn-based ethanol, is due primarily to the much lower amount of energy required: Soybeans create oil that can be obtained readily from seed by physical methods, whereas ethanol requires fuel-intensive distillation. Soybean production also uses much less fertilizer and releases much less nitrogen, phosphorus, and harmful pesticides into the environment than does corn production for ethanol. Of course, the two biofuels are used in different types of vehicles, so a comparison between them is of limited significance.

The fraction of biodiesel in diesel fuel is designated by a system analogous to that used for alcohols in gasoline. Thus B_5 symbolizes diesel fuel containing 5% biodiesel by volume, and B_{100} is pure biodiesel. In the past, the most common blend was B_{20}—indeed, the U.S. Navy, the largest single user of biodiesel in the world, uses that blend in all nontactical vehicles—but more dilute blends such as B_2 and B_5 are becoming popular. Currently, the largest manufacturer of biodiesel is Germany, which produced more in 2005 than the rest of the world combined. The European Union has mandated that all fuels contain 5.75% biofuels by 2010, which will mean a tripling of their consumption there compared to 2005 levels.

Almost all the biodiesel produced in the United States uses domestic soybeans (which are about 20% oil) as its raw material. The oil yield, about 40%, is even higher for rapeseed (canola). In tropical areas, massive plantations of palm trees are being planted to produce palm oil for biodiesel, since the yield of oil per square kilometer greatly exceeds that from soybean or rapeseed crops. Unfortunately, in the rush to produce more palm oil destined to become biodiesel in Europe, huge areas of tropical rain forest in Malaysia, Brazil, and Borneo and peatland in Indonesia have been burned and cleared and thereby destroyed; the result is large amounts of greenhouse gas emissions, totaling one-twelfth of all global CO_2 in the case of Indonesia. The advantage of biofuels in producing lower greenhouse gases than conventional gasoline or diesel will be overcome by these emissions for many years.

Anther concern about biofuels is the effect they have on the price of food. In 2007, the Food and Agriculture Organization of the United Nations noted that the rapidly increasing demand for biofuels is transforming agriculture worldwide and contributing to the increase in food prices. The inflation in prices applies not only to the corn, sugar, and vegetable oil sources but also indirectly to prices for the livestock that are usually fed these crops.

In the future, species of algae that contain as much as 50% oil may be grown as raw material from which biodiesel fuel would be obtained, since the yield per square kilometer could greatly exceed that of even tropical palm oil. However, the practical mass production of single-species algae has not yet been achieved.

Green Chemistry: Valuable Chemical Feedstocks from Glycerin, a Waste By-Product in the Production of Biodiesel

As mentioned above, biodiesel is produced from a transesterification reaction of the triglycerides that make up animal fats and vegetable oils (Figure 8-11). This reaction not only produces the methyl esters of the fatty acids, which comprise biodiesel, but also yields glycerin as a by-product. For every 9 L of biodiesel produced, about 1 L of glycerin is formed. The market for the glycerin by-product has not kept pace with biodiesel production; consequently, there is now a glut of glycerin on the market. Chemists and engineers have been searching for new uses for glycerin and for processes to convert glycerin to other valuable and useful chemicals.

Professor Galen Suppes and his group at the University of Missouri won a 2006 Presidential Green Chemistry Challenge Award for their discovery of a process to convert glycerin to **propylene glycol** and **acetol** (Figure 8-12). Although other methods exist for the conversion of glycerin to propylene glycol, they require high temperatures (200–400°C) and high pressures (1450–4700 psi). As a result of the stringent reaction conditions, these other

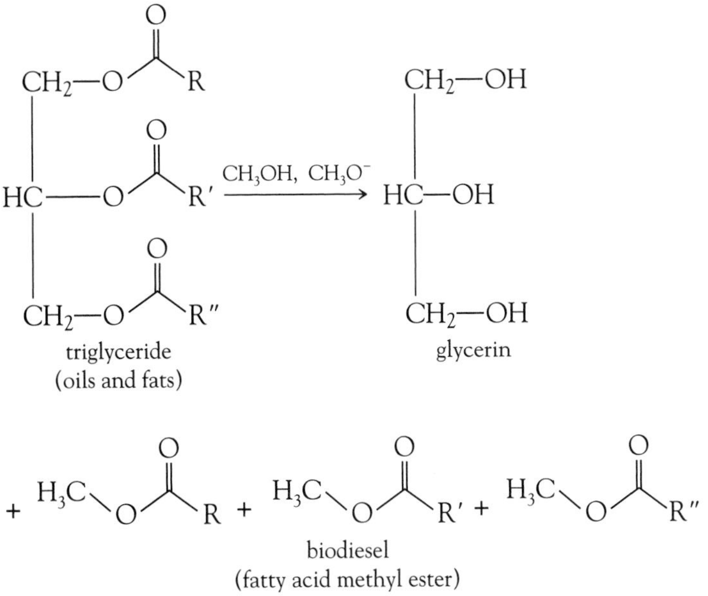

FIGURE 8-11 Preparation of biodiesel.

triglyceride
(oils and fats)

glycerin

biodiesel
(fatty acid methyl ester)

methods require specialized costly equipment that prohibits their commercialization. These methods also suffered from poor yields and low selectivity.

In the Suppes synthesis (Figure 8-12), propylene glycol is formed by reaction of glycerin with hydrogen in the presence of *copper chromite* catalyst. The advantage of this synthesis is that it takes place at relatively low temperatures and pressures, with high yields and high selectivity, making it commercially feasible. Since the present commercial synthesis of propylene glycol is from petroleum-based feedstocks, the Suppes synthesis offers a way to reduce our dependence on fossil fuels by producing propylene glycol from biomass (waste glycerin from biodiesel production). Propylene glycol has a relatively large market, over a billion kilograms per year; finding a commercially viable use for the glycerin by-product of biodiesel production will help lower the cost of biodiesel and in turn encourage its production and use.

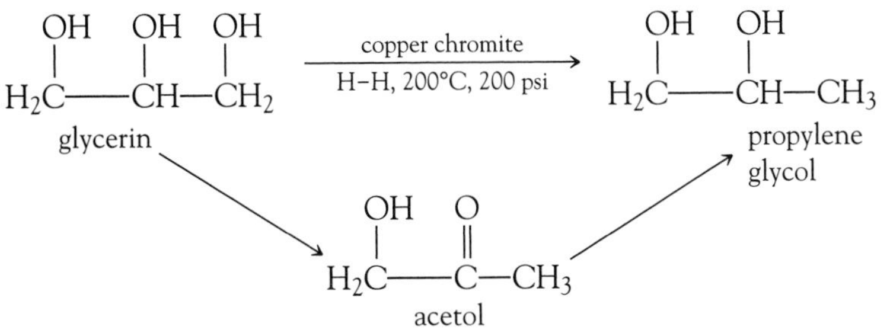

FIGURE 8-12 Preparation of propylene glycol and acetol from glycerin. In the absence of H_2, acetol can be isolated as the major product by distillation as it is formed.

glycerin

propylene glycol

acetol

Propylene glycol has many desirable properties, including its low toxicity, which allow it to be used in a number of different products. Propylene glycol is the only glycol approved by the U.S. Food and Drug Administration for use in products meant for human consumption. It can be found in alcoholic beverages, confections and frostings, ice cream, nuts and nut products, and seasoning and flavorings. This compound may also be present in cosmetics, pharmaceuticals, pet food, tobacco, paints, detergents, fragrances, resins, and antifreeze. Propylene glycol's many functions include preservative, moisturizer, wetting agent, coolant, and solvent. Currently, the bulk of automobile antifreeze is composed of ethylene glycol, which is acutely toxic. The economical production of large quantities of propylene glycol from the glycerin waste product of biodiesel production offers the potential to replace ethylene glycol in antifreeze with a significantly less toxic substance.

The Suppes group also discovered that glycerin could be reacted with copper chromite in the absence of hydrogen to produce acetol (Figure 8-12). They believe that this discovery could be even more important than the formation of propylene glycol, since acetol can act as precursor to many other organic compounds. Thus these other organic compounds could ultimately be formed from biomass rather than the petroleum-based precursors from which they are presently made. This would not only decrease the dependence on crude oil but would also lower the price of biodiesel by making the glycerin by-product more valuable.

Hydrogen—Fuel of the Future?

Hydrogen gas can be used as a fuel in the same way as carbon-containing compounds; some futurists believe that the world will eventually have a hydrogen-based economy. Hydrogen gas combines with oxygen gas to produce water, and in the process it releases a substantial quantity of energy:

$$H_2(g) + 1/2\ O_2(g) \xrightarrow{\text{spark}} H_2O(g) \qquad \Delta H = -242\ \text{kJ/mol}$$

The idea that hydrogen would be the ultimate fuel of the future goes back at least as far as 1874, when it was mentioned by a character in the novel *Mysterious Island* by Jules Verne. Indeed, hydrogen has already found use as fuel in applications for which lightness is an important factor, namely, in powering the Saturn rockets to the moon and the U.S. space shuttles.

Hydrogen is superior even to electricity in some ways, since its transmission by pipelines over long distances consumes less energy than the transmission through wires of the same amount of energy as electricity and since batteries are not required for local storage of energy.

However, as we shall see in the material that follows, the substantial technical problems in the production, storage, transportation, and usage of hydrogen—the need to create a new infrastructure for it—mean that a hydrogen economy is probably still many decades away.

Combusting Hydrogen

Hydrogen gas can be combined with oxygen to produce heat by conventional flame combustion or by low-temperature combustion in catalytic heaters. The combustion efficiency, i.e., the fraction of energy converted to useful energy rather than to waste heat, is approximately 25%, about the same as for gasoline. The main advantages to using hydrogen as a combustion fuel are its low mass per unit of energy produced and the smaller (but not zero) quantity of polluting gases its combustion produces, when compared to other fuels. BMW, Ford, and Mazda may begin marketing hydrogen-fueled combustion-engine cars by 2010.

Although it is sometimes stated that hydrogen combustion produces only water vapor and no pollutants, this in fact is not true. Of course, no carbon-containing pollutants, including carbon dioxide, are emitted. Since combustion involves a flame, however, some of the nitrogen from the air that is used as the source of oxygen reacts to form nitrogen oxides, NO_X. Some *hydrogen peroxide*, H_2O_2, is released as well. Thus hydrogen-burning vehicles are not really zero-emission systems. It is true that the lower flame temperature for the $H_2 + O_2$ combustion, compared to that for fossil fuels with oxygen, inherently produces less NO_X, perhaps two-thirds less. The nitrogen oxide release can be eliminated by using pure oxygen rather than air to burn the hydrogen; alternatively, it can be reduced even further by passing the emission gases over a catalytic converter or by lowering the flame temperature as much as possible, e.g., by reducing the H_2/O_2 ratio to half the stoichiometric amount.

Generating Electricity by Powering Fuel Cells with Hydrogen

Hydrogen and oxygen can be combined in **fuel cells** in order to produce electricity (a hydrogen technology also used in space vehicles). Fuel cells are similar in operation to batteries except that the reactants are supplied *continuously*. In the hydrogen–oxygen fuel cell, the two gases are passed over separate electrodes that are connected by an external electrical connection through which electrons travel and also by an electrolyte through which ions travel.

The components of a fuel cell, then, are the same as those of an electrolysis operation in which water would be split into hydrogen and oxygen, but the chemical reaction that occurs is exactly the opposite. Instead of electricity being used to drive the electrolysis reaction, it is produced. Fuel cells have the advantage over combustion since a more useful form of energy is produced (electricity rather than heat) and the process produces no polluting gases as by-products. In principle, the only product of the reaction is water. At the catalytic surface of the first electrode, the H_2 gas produces H^+ ions and electrons, which travel around the external circuit to the second electrode, across which O_2 gas is bubbled (see Figure 8-13). Meanwhile, the H^+ ions

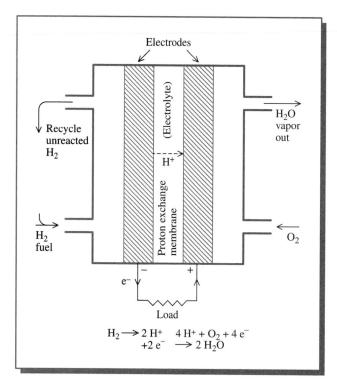

$$H_2 \rightarrow 2\,H^+ + 2\,e^-$$
$$4\,H^+ + O_2 + 4\,e^- \rightarrow 2\,H_2O$$

FIGURE 8-13 Schematic diagram of a hydrogen–oxygen fuel cell (PEMFC version).

travel through the electrolyte and recombine with the electrons and O_2 to produce water at the electrode. Although some of the reaction energy is necessarily released as heat—about 20%, due to requirements of the second law of thermodynamics—most of it is converted to electrical energy associated with the current that flows between the electrodes. Electric motors, whether in a fuel-cell or battery vehicle, are 80–85% efficient in converting electrical to mechanical energy. Real fuel cells overall are now about 50–55% efficient; 70% efficiency may be attained eventually. By contrast, internal combustion engines using gasoline are 15–25% efficient, diesels 30–35%.

Many auto manufacturers are currently pursuing the development of electric cars that use fuel cells. Prototype buses running in Vancouver and Chicago use innovative fuel cells for their power source. The electrolyte used in the fuel cells of these vehicles is a hair-thin (about 100 μm) synthetic polymer that acts as a proton exchange membrane. The membrane, when moist, conducts protons well since it incorporates sulfonate groups. It also keeps the hydrogen and oxygen gases from mixing. The electrodes of such **polymeric-electrolyte-membrane fuel cells,** labeled PEMFC in Figure 8-14, are graphite with a small amount of platinum dispersed as nanoparticles in thin layers (about 50 μm thick) on its surface. Each cell, which operates at about 80°C, generates about 0.8 V of electricity, so many must be stacked together in order to provide sufficient power for the vehicle. In the current versions of the buses, compressed hydrogen is stored in tanks under the roof of the bus.

Some incentives to develop vehicles that use fuel cells powered by hydrogen are

- to reduce urban smog, which is partially produced from emissions from gasoline and diesel engines;

- to reduce energy consumption, since fuel cells are much more efficient in producing motive power than are combustion engines; and

- to reduce carbon dioxide emissions, since fuel cells powered by hydrogen are carbon-free.

Some analysts point out that the cost of improving air quality and CO_2 emissions by switching the transportation system to such vehicles over the next few decades is much higher than alternative strategies, such as scrapping old cars, improving vehicle fuel efficiency, reducing NO_X emissions from power plants, and capturing and sequestering CO_2 emissions from power plants. In addition, hydrogen leaked into the air acts indirectly as a greenhouse gas, since it reacts with and decreases the concentration of OH; this result will slightly increase the atmospheric lifetime of methane and hence its concentration, since the main sink for CH_4 is its reaction with OH, as discussed in Chapters 5 and 6.

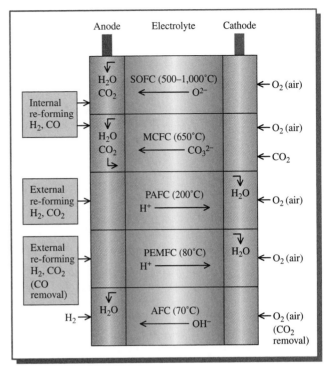

FIGURE 8-14 Operating characteristics of various fuel cells. [Source: B. C. H. Steele and A. Heinzel, "Materials for Fuel Cell Technologies," *Nature* 414 (2001): 345.]

Obtaining Fuel-Cell Hydrogen from Liquid Fuels

Because of the limited practicality of transporting hydrogen in individual cars and trucks, there is active research in designing systems that allow it to be extracted as needed from liquid fuels, which are much more convenient to transport. For example, in the near future, the hydrogen may instead be obtained as needed from liquid methanol by onboard decomposition of the latter to hydrogen gas using the reverse of the methanol-formation reaction discussed earlier:

$$CH_3OH \longrightarrow 2\,H_2 + CO$$

In the General Motors version of this process, the *re-former* unit operates at 275°C and uses a copper oxide/zinc oxide catalyst to promote the reaction. The water-gas shift reaction is subsequently used to react the CO in the synthesis gas with steam and provide additional H_2 gas, giving the overall reaction

$$CH_3OH + H_2O \longrightarrow 3\,H_2 + CO_2$$

Similar processes have been developed that convert gasoline, diesel fuel, octane, or aqueous ethanol into carbon dioxide and hydrogen. Unfortunately, the current PEMFC and alkaline fuel cells (see Problem 8-9), as well as one based on phosphoric acid, all require relatively pure hydrogen, free especially of carbon monoxide—a gas that is formed in the re-former process and is difficult to eliminate completely. Carbon monoxide bonds to the sites of the catalyst (e.g., platinum) intended to promote the fuel-cell reaction and blocks the catalytic activity there. Concentrations of CO greater than 20 ppm in the hydrogen gas slow down most fuel cells appreciably. Perhaps a CO-tolerant electrode catalyst, possibly one incorporating a second metal or a metal oxide as well as platinum, will be developed in the future to overcome this problem by oxidizing the adsorbed CO to carbon dioxide. Hydrogen that is virtually free of carbon monoxide could be produced from methanol by an oxidative steam re-forming process at 230°C:

$$4\,CH_3OH + O_2 + 2\,H_2O \longrightarrow 10\,H_2 + 4\,CO_2$$

Since oxygen is involved, however, not all the fuel value of methanol is captured in the hydrogen product.

PROBLEM 8-9

An alkaline electrolyte can be used in the H_2-O_2 fuel cell to replace the acidic environment mentioned previously (see Figure 8-13). Assuming that the reaction of O_2 with water and electrons produces hydroxide ions, OH^-, and that these ions travel to the other electrode, where they react with hydrogen to give up electrons and produce more water, deduce the two balanced half-reactions and the balanced overall reaction for such a fuel cell. (The *alkaline fuel cell*, labeled AFC in Figure 8-14, is used in space shuttles and Apollo spacecraft to provide electricity.)

Fuel cells may also be used in the near future in small electric power plants, partly because pollutant emissions from them are so small compared to those from fossil-fuel combustion (e.g., only about 1% of the NO_X). Indeed, **phosphoric acid**–based fuel cells (PAFC in Figure 8-14) have been operating since the early 1990s in some hospitals and hotels to generate power. The most promising fuel cells for power plants involve a **molten carbonate** salt—e.g., potassium and lithium carbonates, plus additives, at 650°C—as the electrolyte. (The *molten-carbonate fuel cell* is labeled MCFC in Figure 8-14.) The hydrogen gas reacts with **carbonate ions,** CO_3^{2-}, to produce carbon dioxide, water, and electrons at the anode, while the carbon dioxide reacts with electrons and oxygen from air to re-form the carbonate ions at the cathode. The carbon dioxide must be recycled from the anode back to the cathode during the cell's operation. The hydrogen is produced on-site by reaction of methane with steam, so CH_4 is the actual source of the fuel energy here. By-product waste heat can be recovered and used in a cogeneration mode, and the dc electricity produced by the fuel cell is converted to ac for distribution. MCFC efficiency approaches 50%.

Like the molten-carbonate fuel cell, a cell based on **solid oxides** (see SOFC, Figure 8-14) is much more tolerant of carbon monoxide impurities in the fuel hydrogen gas than are the other fuel cells. The electrolyte in the solid-oxide fuel cell is a ceramic mixture of oxides of zirconium and yttrium. The **oxide anion,** O^{2-}, produced from oxygen gas carries the charge between the electrodes, travels through the solid from cathode to anode as the fuel is consumed, and forms water at the anode. The high operating temperature (up to 1000°C) of the solid-oxide cell allows fuel to be re-formed internally and to form hydrogen ions without the use of expensive catalysts, so methane or other hydrocarbons can be used as the fuel instead of hydrogen. However, carbon deposits tend to form at the anode and stick to it at the high operating temperatures that are involved with these units.

The solid-oxide fuel cell, like the molten-carbonate one, is practical for central power plants, and its fuel efficiency is also high. Both types suffer from practical problems with electrodes, such as carbon deposition. The latter problem for solid-oxide fuel cells can be overcome by operating at a lower temperature.

PROBLEM 8-10

For the molten-carbonate fuel cell, obtain balanced half-reactions for the processes at the two electrodes and add them together to determine the overall reaction.

Electric Cars Powered by Batteries

An alternative to vehicles that use fuel cells are those powered by batteries. Some *electric cars* have already been produced, and most use a number of the same sort of lead-acid batteries that gasoline-powered vehicles have traditionally employed singly to operate the starter motor. In the future, electric cars will probably use nickel-cadmium, nickel-metal hydride, and lithium-based batteries. The practical difficulties that discourage widespread adoption of such vehicles are their high cost, the low driving range between battery charges, the length of the battery recharge period, and the weight of the batteries. Like fuel-cell systems, they have the attraction of zero emissions during operation, little operating noise, and low maintenance costs. Of course, pollution is emitted into the environment when the electricity required for these cars is generated in the first place. Some researchers have predicted that lead pollution stemming from the manufacture, handling, disposal, and recycling of lead-acid batteries would raise lead emissions into the environment by an amount that would exceed the levels that were associated with leaded gasoline. Critics of this analysis have pointed out that the data used are flawed and that not all the lead lost in the processing steps will be emitted into the environment rather than disposed of properly.

Even electric vehicles are not really pollution-free if a fossil fuel is burned to generate the electricity to charge the battery, since the fossil fuel's combustion in any power plant yields NO_x that is released into the atmosphere.

The turn of the century saw the introduction of **hybrid** combustion/electric powered vehicles, such as the Toyota Prius, into the market. Such vehicles overcome the requirement of long and frequent recharging of all-battery systems, since the battery is recharged continuously when the (small) gasoline engine is in operation and is producing excess power. During braking, recovered kinetic energy is channeled back to the battery. Motive power for the vehicle is supplied by both the gasoline and electric motors, the proportion depending on the driving situation. Hybrid vehicles are highly fuel efficient, so they emit much less carbon dioxide and much less nitrogen oxides, carbon monoxide, and VOCs than conventional vehicles. The batteries involved are nickel-metal hydride, which are lighter and more compact than lead batteries of equivalent potential.

Other Uses for Fuel Cells

The first wave of consumer products powered by fuel cells will likely be on the market in the next year or two. Laptop computers will be powered by fuel cells rather than rechargeable batteries. These fuel cells will use methanol rather than hydrogen directly as fuel and will contain removable cartridges containing the alcohol. The advantage for the user of fuel cells over battery power is a much longer working time before the power runs out. Using methanol or natural gas rather than H_2 directly as the fuel in fuel cells does avoid the problem of the generation and storage of hydrogen. One problem with methanol is the generation of by-product carbon monoxide, which poisons the catalyst, as discussed previously. Diluting the methanol with water lessens this problem but cuts the power output of the cell. In addition, neither methanol nor any other liquid fuel that could in principle be used directly instead of H_2 in a fuel cell reacts fast enough to produce the required electrical current in a vehicle, although recent research on the use of dilute solutions of methanol indicates that these problems may be overcome eventually.

Storing Hydrogen

In rocketry applications, hydrogen is stored as a liquid, as is oxygen. Since hydrogen's boiling point of only 20 K ($-253°C$) at 1 atm pressure is so low, a large amount of energy must be expended in keeping it very cold, in addition to the energy used to liquefy it. This drawback effectively limits the applications of liquid hydrogen to a few specialized situations in which its lightness (low density) is the most important factor.

Hydrogen could be stored as a compressed gas, in much the same way as is done for methane in the form of natural gas. However, compared to CH_4, hydrogen has a drawback: A much greater amount of H_2 gas needs to be stored in order to release the same amount of energy. Compared with methane, the combustion of one mole of hydrogen consumes only one-quarter of the oxygen and consequently generates about one-quarter of the energy, even though both occupy equal volumes under the same pressure (ideal gas law). Thus the "bulky" nature of hydrogen gas limits its applications (see Problem 8-13).

It is instructive to compare the volumes of hydrogen under different conditions required to fuel a hydrogen-fuel-cell car (assuming 50% efficiency) to travel 400 km (240 miles), approximately the distance one can obtain in an efficient gasoline-powered car with a tank capacity of 40–50 L. The amount of hydrogen required is 4 kg, which occupies

- 45,000 L, or 45 m^3—e.g., a balloon having a 5-m diameter or a cube 3.6 m on each side, if it exists as a gas at normal atmospheric pressure; or

- 225 L (about 60 gal, equal to about five normal-sized gasoline tanks) as a gas compressed to about 200 atm (routinely achievable); or

- 56 L as a liquid (or solid) maintained at −252°C (at 1 atm pressure); or

- 35–75 L if stored as a metal hydride, if effective systems can be developed, as discussed below.

A practical and safe way to store hydrogen for use in small vehicles may be in the form of a **metal hydride.** Many metals, including alloys, absorb large amounts of hydrogen gas reversibly—as a sponge absorbs water. The molecular form of hydrogen becomes dissociated into atoms at the surface of the metal as it is absorbed and forms metal hydrides by incorporating the small atoms of hydrogen in "holes" in the crystalline structure of the metal. Thus the hydrogen exists as atoms, not molecules, within the lattice, which expands slightly to incorporate them. For example, titanium metal absorbs hydrogen to form the hydride of formula TiH_2, a compound in which the density of hydrogen is twice that of liquid H_2! Heating the solid gradually releases the hydrogen as a molecular gas, which then can be burned in air or oxygen to power the vehicle.

Research continues to find a light metal alloy that can efficiently store hydrogen without making the vehicle excessively heavy. Even existing metal hydride systems are lighter than the pressurized tanks needed to store liquid hydrogen. Most industrial research now centers on metal systems. Practical considerations require that an alloy to store hydrogen

- be capable of quickly and reversibly absorbing hydrogen,

- not become brittle after many repeated cycles of absorption and desorption,

- operate in the pressure and temperature ranges of 1–10 atm and 0–100°C,

- not be so dense that it weighs down the vehicle excessively (a concentration of hydrogen of at least 6.5 mass % is the U.S. Department of Energy target), and

- not require a huge volume (at least 62 kg H/m^3, equivalent to 4 kg in 65 L, is the target).

Lanthanum–nickel alloys derived from LaNi$_5$ have all the above characteristics except one: They are too heavy (mass % < 2), a deficiency shared by all known metal hydrides that operate near ambient temperature. Many lighter hydrides and alloys, such as MgH$_2$ and Mg$_2$NiH$_4$, are known, but they do not operate reversibly under moderate conditions. Research on systems formed by the lighter metals continues but has not yet been successful in producing alloys that fulfil all five conditions listed.

One of the practical difficulties in using hydrogen as a fuel is its tendency to react over time with the metal in pipelines or storage containers in which it is used. This reaction embrittles the metal, eventually deteriorating it to form a powder. Recent progress has been made in overcoming this difficulty by using composite materials rather than simple metals as the structural materials for storage and transport facilities.

Some research in the past reported that tiny fibers made of graphite, a light material, can store up to three times their weight in hydrogen between the graphite layers and would be a safe, lightweight storage mechanism for hydrogen. However, research in the late 1990s involving extensive hydrogen storage in carbon nanotubes has not proven to be reproducible. One of the difficulties is the very tiny (milligram) samples of carbon nanotubes that are available for experimentation. Some researchers believe that, if the nanotubes are broken so that they have an open end, hydrogen can enter the tube. Other experiments indicate that only a single layer of H$_2$ gas conventionally adsorbed to the outside of the tubes is actually stored, a concentration too small ($<2\%$ by mass) to be useful.

Overall, the problem of devising a practical, economical, and safe way of storing hydrogen has not yet been achieved, and in the eyes of some analysts, "no breakthrough is yet in sight," despite much interest and research activity. It may be that the weight requirements associated with all practical methods of storing hydrogen will limit its use to large vehicles such as buses and airplanes.

As mentioned in the discussion of fuel cells, in some applications it may be more feasible to transport and store hydrogen in the form of an energy-dense liquid such as methanol and to use it as required for power generation. *Toluene*, C$_7$H$_8$, has also been proposed as a long-distance hydrogen carrier; it could then be dehydrogenated when the hydrogen is required.

Another way to temporarily store hydrogen is by means of the alkali salt (lithium or sodium) of the **borohydride ion,** BH$_4^-$. For their prototype mini-van that runs on fuel cells, Chrysler uses a 20% solution of sodium borohydride in water to store hydrogen, which is released when the solution is pumped over a ruthenium catalyst, prompting the redox reaction of the H$^-$ in borohydride with the H$^+$ in water to produce H$_2$:

$$BH_4^- + 3\,H_2O \longrightarrow 4\,H_2 + H_2BO_3^-$$

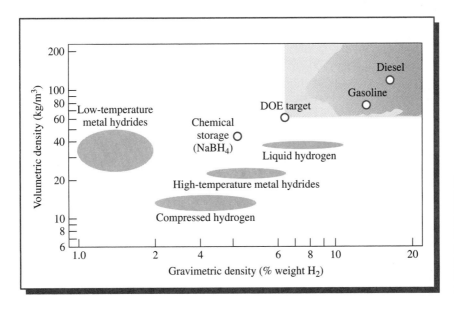

FIGURE 8-15 The performance of various hydrogen storage techniques. Note that both horizontal and vertical scales are logarithmic. [Source: R. F. Service, "The Hydrogen Backlash," *Science* 305 (2004): 958–961.]

The density of hydrogen in the borohydride solution is comparable to that in liquid hydrogen. The analogous *alanate ion*, AlH_4^-, in the form of its sodium salt is also a candidate for hydrogen storage in vehicular fuel cells:

$$2\,NaAlH_4 \longrightarrow 2\,NaH + 2\,Al + 3\,H_2$$

Several molecular boron compounds, including BH_3NH_3 and an organoboron–phosphorus system, have recently been proposed as hydrogen carriers.

The performance of metal systems for storing hydrogen is compared to that of the compressed and liquefied element and that of gasoline and diesel fuel in Figure 8-15. No practical system discovered to date has reached the target of the U.S. Department of Energy (Figure 8-15) in terms of combining high density with a high percentage of hydrogen in its mass ("gravimetric density" of at least 6%).

The possibility of storing hydrogen in clathrates in water—much like methane in water clathrates (Chapter 6)—may be feasible. Although hydrogen molecules are too small to be efficiently trapped at low pressures, it has been discovered recently that two or four H_2 molecules can be stored in each ice clathrate at room temperature at enormous pressures, about 2000 atm. Once formed, though, the clathrate can be stored at liquid nitrogen temperatures at reduced pressure.

PROBLEM 8-11

Calculate the mass of titanium metal required to absorb each kilogram of hydrogen and form TiH_2 in a "tankful" of hydrogen. Repeat the calculation for magnesium if the hydride has the formula MgH_2. Which metal is superior for storage of hydrogen from a weight standpoint?

PROBLEM 8-12

Assuming that the energy released by combustion of H_2 is proportional to the amount of oxygen it consumes, estimate the ratio of heat released by one mole of methane compared to one mole of hydrogen gas.

PROBLEM 8-13

Using the thermochemical information in the H_2 combustion equation, calculate the enthalpy (heat) of combustion of hydrogen per gram, and by comparing it to that of methane (see Problem 7-4), decide which fuel is superior on a weight basis. By comparing the actual energy released by combustion per mole of gas—and hence per molar volume—decide which fuel is superior on a volume basis.

Producing Hydrogen

It is important to realize that hydrogen is not an energy *source*, since it does not occur as the free element in the Earth's crust. Hydrogen gas is an **energy vector,** or carrier, only; it must be produced, usually from water and/or methane, with the consumption of large amounts of energy and/or other fuels. The industrial infrastructure that would be required to produce enough hydrogen to fuel all the vehicles in the United States is enormous, since it would require about as much energy as the current electric power capacity.

The most expensive commercial way to produce hydrogen is by electrolysis of water, using electricity generated by some energy source:

$$2\ H_2O(l) \xrightarrow{\text{electricity}} 2\ H_2(g) + O_2(g)$$

Unfortunately, about half the electrical energy is inadvertently converted to heat and therefore wasted in this process.

A hope for the future is that wind power or solar energy from photovoltaic collectors will become economically efficient in providing electricity to generate hydrogen. Currently, there are prototype plants in Saudi Arabia and Germany that use electricity from solar energy to produce hydrogen, a process about 7% efficient. The stored energy is later recovered by reacting the hydrogen with oxygen. Excess electricity from hydroelectric or nuclear power or wind-power installations—i.e., power generated but not required immediately for use—could be used to produce hydrogen by electrolysis of water.

Even better than the use of solar electricity to electrolyze water would be the direct decomposition of water into hydrogen and oxygen by absorbed sunlight, but no practical, efficient method has yet been devised to effect this transformation. One of the difficulties in using sunlight to decompose water is that H_2O does not absorb light in the visible or UV-A regions; thus some substance must be found that can absorb sunlight, transfer the energy to the decomposition process, and finally be regenerated. The substances proposed to date for this purpose are very inefficient in converting sunlight into energy. In addition, since the light-absorbing substances and others required are not 100% recoverable at the end of the cycle, they must be continuously resupplied—thus the hydrogen that is produced is not really a renewable fuel.

One catalyst that has been found to convert sunlight into hydrogen by electrolyzing water is **titanium dioxide,** TiO_2. A small potential is applied to the electrode in the cell's operation. Titanium dioxide is stable to sunlight (unlike many other potential light-absorbing materials) and cheap, but pure TiO_2 absorbs only ultraviolet light. By blending carbon into TiO_2 so that C replaces some of the oxide ions, the efficiency in producing hydrogen gas is increased eightfold, to more than 8% of the Sun's energy, because the addition of carbon extends absorption into the visible region (to 535 nm).

PROBLEM 8-14

Determine the longest wavelength of light that has photons capable of decomposing liquid water into H_2 and O_2 gases, given that for this process $\Delta H = +285.8$ kJ/mol of water. In which region of the spectrum does this wavelength lie? [Hint: Recall from Chapter 1 the relationship between reaction enthalpy and light wavelength.]

In principle, the thermal conversion of sunlight into heat can produce temperatures hot enough to decompose water into hydrogen and oxygen. Research in Israel, using a *solar tower* of mirrors to concentrate sunlight by a factor of 10,000 and thereby produce temperatures of about 2200°C in a reactor, has succeeded in splitting about one-quarter of water vapor at low pressures into H_2 and O_2.

Various thermochemical cycles by which water can indirectly be decomposed by heat into hydrogen and oxygen have been proposed. Ideally, such cycles should operate at moderate temperatures, be efficient in conversion of heat into hydrogen, and not degrade the reactants so they can be recycled. Perhaps the most practical is the **sulfur–iodine cycle,** in which elemental iodine is first reduced by sulfur dioxide to hydrogen iodide and sulfuric acid:

$$I_2 + SO_2 + 2\ H_2O \longrightarrow 2\ HI + H_2SO_4 \qquad (at\ 120°C)$$

The hydrogen iodide is then thermally decomposed into hydrogen gas, recovering the elemental iodine, and the sulfuric acid is thermally decomposed into oxygen gas, recovering the sulfur dioxide:

$$2\ HI + heat\ (320°C) \longrightarrow H_2 + I_2$$

$$H_2SO_4 + heat\ (830°C) \longrightarrow SO_2 + H_2O + 1/2\ O_2$$

Since the reactants HI and SO_2 are recovered in high yield, the cycle can be repeated over and over. Heat produced by nuclear reactors could drive this cycle, which has a conversion efficiency of about 50%.

Hydrogen gas can be produced by reacting a fossil fuel such as coal or petroleum or natural gas with steam to form hydrogen and carbon dioxide. The energy value of the fuel is transferred from carbon to the hydrogen atoms of water; chemically speaking, the reduced status of the carbon is transferred to the hydrogen. The net reactions, assuming coal to be mainly graphite, are:

$$C + 2\ H_2O \longrightarrow 2\ H_2 + CO_2$$

$$CH_4 + 2\ H_2O \longrightarrow 4\ H_2 + CO_2$$

Notice that as much carbon dioxide is produced in this way as would be obtained by combustion of the fossil fuels in oxygen. As discussed previously, the actual conversions occur in two steps: First the fossil fuel reacts with steam to yield carbon monoxide and some hydrogen (Figure 8-10). Then the CO/H_2 synthesis gas mixture and additional steam are passed over a suitable catalyst to obtain additional hydrogen and complete the oxidation of the carbon by the water-gas shift reaction driven in the direction shown here:

$$CO + H_2O \xrightarrow{\text{catalyst}} CO_2 + H_2$$

It is interesting to note that at the turn of the twentieth century and for several decades thereafter, the synthesis gas produced when coal reacts with steam was itself used as the fuel in many municipal street-lighting systems around the world.

Hydrogen gas could be produced in a renewable way from biomass grown for this purpose. Some research indicates that aqueous solutions of both glucose and glycerol can be decomposed at moderate temperatures (225–265°C) and pressures (27–54 atm) with a platinum-based catalyst to produce hydrogen and carbon dioxide (Figure 8-10).

Associated with every conversion of one fuel to another are energy losses, mainly to waste heat, some of which are dictated by the second law of thermodynamics and therefore cannot be avoided. The energy of natural gas can be transferred to hydrogen with an efficiency of about 72%; transfer from coal is 55–60% efficient. Thus, if the resulting fuel is used only to generate heat, significantly less CO_2 is emitted if the original fossil fuel is burned rather than being first converted to hydrogen.

Finally, it should be mentioned that hydrogen is considered to be a dangerous fuel due to its high flammability and explosiveness; it ignites more easily than do most conventional fuels. On the positive side, however, spills of liquid hydrogen rapidly evaporate and rise high into the air. (Some of the fears surrounding hydrogen stem from the 1930s incident in which the airship *Hindenburg* was destroyed in a catastrophic fire. However, it was the thin layer of aluminum encasing the hydrogen gas that initially was ignited, not the H_2 itself.)

Review Questions

1. Define the term *renewable energy*, and list several forms of it. Which form is growing the fastest?

2. Name four environmental/social problems associated with the expansion of hydroelectric power.

3. What is the mathematical relationship between the energy generated by a windmill and **(a)** the wind's speed and **(b)** the length of the windmill blades?

4. Explain the origin of coastal winds.

5. List four pros and cons of wind power.

6. Define *energy payback*, and state which form of renewable energy has the lowest payback period and the lowest cost at present.

7. What is meant by *geothermal energy*? Give some examples of how and where it is tapped.

8. Describe the difference between the two methods of absorbing energy from sunlight. What is the difference between active and passive systems?

9. What is meant by *solar thermal electricity*, and how is it generated? What is meant by the term *cogeneration*?

10. State the *second law of thermodynamics*. According to this law, what formula gives the maximum fraction of heat that can be transformed into electricity?

11. Define the *photovoltaic effect*. What is the chief difficulty preventing the widespread use of solar cells?

12. List four advantages and four disadvantages of solar energy.

13. What are the advantages and disadvantages of using alcohol fuels in regard to air pollution? What is meant by E_{10} fuel?

14. Describe the method used in producing ethanol in high volume for use as a fuel. What are the potential feedstocks for this process?

15. What are the highly energy-intensive steps involved in production of fuel ethanol? Why isn't ethanol a fully "renewable" fuel? What is meant by the term *cellulostic ethanol*?

16. What is the *water-gas shift reaction*? Describe the methods by which methanol can be produced in volume for use as a fuel. What does M_{85} mean?

17. Chemically speaking, what is *biodiesel* and how is it produced?

18. Describe the three ways in which hydrogen can be stored in vehicles for use as a fuel, and discuss briefly the disadvantages of each method.

19. Does the burning of hydrogen really produce no pollutants? Under what conditions do no pollutants form?

20. What is the difference between an *energy source* and an *energy carrier* (vector)? Into which category does H_2 fall?

21. Describe how a hydrogen fuel cell works, and write balanced half-reactions for its operation in acidic media. What other types of fuel cells exist?

22. Describe the production of hydrogen gas by electrolysis. Can solar energy be used for this purpose? Why isn't water decomposed directly by absorption of sunlight?

Green Chemistry Questions

See the discussion of focus areas and the principles of green chemistry in the Introduction before attempting these questions.

1. The first reaction below is the Suppes synthesis of propylene glycol from glycerin (which won a Presidential Green Chemistry Challenge Award); the second reaction is the commercial synthesis of propylene glycol from propene. Glycerin (as was discussed in this chapter) can be obtained from biomass while propene is a petrochemical. Another aspect of these two preparations of propylene glycol to consider when assessing their environmental impact is their atom economy. Calculate the atom economy of these syntheses. To aid you, for each synthesis, the atoms of the reactants that are incorporated into propylene glycol are given in green while those that are wasted are in black.

$$
\underset{\text{glycerin}}{\overset{\overset{\displaystyle OH \quad OH \quad OH}{|\qquad|\qquad|}}{H_2C\!-\!\!-\!\!-\!CH\!-\!CH_2}} \xrightarrow[\text{H—H, 200°C, 200 psi}]{\text{copper chromite}} \underset{\text{propylene glycol}}{\overset{\overset{\displaystyle OH \quad OH}{|\qquad|}}{H_2C\!-\!\!-\!\!-\!CH\!-\!CH_3}}
$$

$$
H_3C\!-\!CH\!=\!CH_2 \xrightarrow[H_2O]{Cl_2} \underset{}{\overset{\overset{\displaystyle H\!-\!O \quad Cl}{|\qquad|}}{H_3C\!-\!CH\!-\!CH_2}} \xrightarrow{NaOH} H_3C\!-\!\overset{\displaystyle O}{\overset{/\backslash}{HC\!-\!CH_2}} \xrightarrow[\substack{120-190°C \\ \text{high pressure}}]{H_2O} \underset{}{\overset{\overset{\displaystyle OH \ OH}{|\quad|}}{H_3C\!-\!HC\!-\!CH_2}}
$$

2. The development of the preparation of propylene glycol and acetol from glycerin by Suppes won a Presidential Green Chemistry Challenge Award.

(a) Into which of the three focus areas for these awards does this award best fit?

(b) List at least two of the twelve principles of green chemistry that are addressed by the green chemistry developed by Suppes.

3. If you have had a course in organic chemistry, try to deduce a reaction mechanism for the conversion of fats and oils to biodiesel and glycerin as shown in Figure 8-11.

4. Where does diesel fuel come from? How is it produced? What compounds make up diesel fuel?

5. Ethanol and methanol are being used as automobile fuels. What do you think about the use of the alcohols propylene glycol and glycerin as automobile fuels? [Hint: It many be useful to look up the boiling points of ethanol, methanol, ethylene glycol, and glycerin.]

Additional Problems

1. Given that an average of 342 W of sunlight energy falls on each square meter of Earth, that the surface area of a sphere is $4\pi r^2$, and that the Earth's radius r is about 6400 km, calculate in joules the total amount of sunlight received annually by the Earth. What percentage of this quantity needs to be captured in order to provide our current commercial energy needs?

2. Consider the use of methanol, CH_3OH, as an oxygenated liquid fuel for suitably modified cars. **(a)** By writing the balanced chemical equation for its combustion in air, determine whether it is more similar to coal, oil, or natural gas in terms of the joules of energy released per mole of CO_2 produced. **(b)** Determine the balanced equation by which methanol can be produced by reacting elemental carbon (coal) with water vapor, given that CO_2 is the only other product in the reaction. **(c)** Does the combined scheme of parts (a) and (b) represent a way of using coal but producing less carbon dioxide per joule than by its direct combustion? Explain your answer.

3. Deduce the fraction of the CO or H_2 produced by the reaction of coal with steam that must be converted to H_2 or CO, respectively, by the water-gas shift reaction in order to obtain the 2:1 ratio of hydrogen to carbon monoxide that is required to synthesize methanol. Deduce also the net reactions of conversion of coal to methanol.

4. Deduce the balanced reaction in which synthesis gas is formed by combining equal volumes of methane and carbon dioxide. From enthalpy of formation data given in Problems 7-4 and 8-7, deduce the enthalpy change for this reaction. By applying Le Châtelier's principle, deduce whether the conversion of the gases to carbon monoxide and hydrogen will be favored by low or by high pressures, and by low or by high temperatures. Combining these results with those obtained in Additional Problem 3, determine the fraction of

the total carbon dioxide resulting from the production and combustion of methanol synthesized from this synthesis gas that would be renewable, i.e., recycled from the consumption of carbon dioxide in the process.

5. Contact several new car dealerships in your area to discover which vehicles currently on sale can use one or more of the alternative fuels CNG, LPG (propane; see Chapter 7), M_{85}, E_{100} or E_{85}, hydrogen, or electricity. For each fuel and vehicle, inquire about the average kilometers or miles per liter or gallon of fuel; from this information and using fuel prices obtained from local service stations, estimate the driving cost per kilometer or mile for each vehicle and fuel combination. Are any of the combinations competitive in cost with gasoline?

6. Contact a garage in your area that converts existing gasoline-fueled vehicles into those that accept CNG or LPG (propane; see Chapter 7). Determine the common conversion price and what the likely kilometers-per-liter or miles-per-gallon performance for the new fuel in a common vehicle will be. From the cost of the fuels in your area, estimate the distance that the vehicle must be driven before the cost of the conversion has been met by savings on the fuel.

7. The Bay of Fundy, located between the Canadian provinces of Nova Scotia and New Brunswick, has the highest tides in the world. The difference between high and low tides can be as much as 16 m. A total of 14 billion tonnes of seawater flow into and out of Minas Basin, a part of the Bay of Fundy, during each tide. The energy tapped as tidal power comes from the change in potential energy of this water as it falls in the Earth's gravitation field. Given that the potential energy of a mass m at height h in a gravitational field is given by $m \times g \times h$, where g is the gravitational constant, 9.807 m/s^2, calculate the amount of energy that corresponds to a tidal drop of 16 m in the Minas Basin.

8. One way of considering the direct environmental impact of the combustion of various fuels is to look at the amount of heat produced per unit of CO_2 generated. Determine and compare the values of kilojoules of heat per mole of CO_2 produced in the case of the combustion of methanol, ethanol, and n-octane. Use the following molar enthalpies of combustion (ΔH_c, in kJ/mol): methanol, -726; ethanol, -1367; and n-octane, -5450. Comment on the values obtained.

9. The use of biomass fuels, including scavenged wood and wood-chip waste, has been proposed as a way of reducing CO_2 emissions, even though such fuels would produce a large amount of CO_2 per unit of heat produced. Explain the rationale behind this idea.

10. Assuming that MTBE reacts in the atmosphere at the methyl group attached to the oxygen, use the principles of atmospheric reactivity developed in Chapter 5 to show that the first stable product in its decomposition sequence in air is an ester.

Further Readings

1. S. Pacala and R. Socolow, "Stabilization Wedges: Solving the Climate Problem for the Next 50 Years with Current Technologies," *Science* 305 (2004): 968–972.

2. M. Hoogwijk et al., "Assessment of the Global and Regional Geographical, Technical, and Economic Potential of Onshore Wind Energy," *Energy Economics* 26 (2004): 889–919.

3. N. Fell, "Deep Heat," *New Scientist* (22 February 2003): 40–42.

4. R. Gomez and J. L. Segura, "Plastic Solar Cells," *Journal of Chemical Education* 84 (2007): 253–258.

5. P. Hoffmann, *Tomorrow's Energy: Hydrogen, Fuel Cells, and the Prospects for a Cleaner Planet* (Cambridge, MA: MIT Press, 2001).

6. R. F. Service, "The Hydrogen Backlash," *Science* 305 (2004): 958–961 (the August 13, 2004, issue of *Science* contains many articles on the hydrogen economy).

Websites of Interest

Log on to www.whfreeman.com/envchem4/ and click on Chapter 8.

CHAPTER 9

RADIOACTIVITY, RADON, AND NUCLEAR ENERGY

In this chapter, the following introductory chemistry topics are used:

- Mass and atomic numbers; isotope symbolism; elementary particles
- Half-lives; exponential decay and first-order processes

Background from previous chapters used in this chapter:

- Concepts of free radicals (Chapter 1) and synergism (Chapter 4)
- Half-life equation (Chapter 6)
- Second law of thermodynamics (Chapter 8)

Introduction

In all other chapters of this book, we are concerned with chemicals and chemical processes. In this chapter, we consider *nuclear* processes and how they affect the environment, our health, and our energy supply. These concerns all center on the effects of radioactivity, and it is with this topic that we begin. This allows us to discuss **radon,** the most important radioactive indoor air pollutant, and depleted uranium. We then switch to nuclear energy and explore the ways in which electricity can be produced from it and the environmental consequences of the radioactive waste that these processes generate.

The San Onofre nuclear power generating station at San Diego, California. All current nuclear energy is generated by fission, though fusion plants may be viable in the future. (Corbis Images)

Radioactivity and Radon Gas

The Nature of Radioactivity

Although most atomic nuclei are stable indefinitely, some are not. The unstable, or **radioactive,** nuclei spontaneously decompose by emitting a small particle that is very fast moving and therefore carries with it a great deal of energy. In some types of nuclear decomposition processes, atoms are converted from those of one element to those of another as a consequence of this emission. Very heavy elements are particularly prone to this type of decomposition, which occurs by the emission of a small particle. The nuclei produced by emission of the particle may or may not themselves be radioactive; if they are, they will undergo another decomposition at a later time.

Recall from introductory chemistry that the **mass number** is the *number of heavy particles*—protons and neutrons—and not the actual mass of the nucleus. An **alpha (α) particle** is a radioactively emitted particle that has a charge of $+2$ and a mass number of 4—it has two neutrons and two protons—and it is identical to a common **helium** nucleus. Thus an α particle is written as $_2^4\text{He}$, where 4 is its mass number and 2 refers to its nuclear charge (i.e., number of protons). The nucleus that remains behind after an atom has lost an α particle has a nuclear charge that is 2 units *less* than the original, and it is 4 units *lighter*. For example, when a $_{88}^{226}\text{Ra}$ (radium-226) nucleus emits an α particle, the resulting nucleus has a mass number of $226 - 4 = 222$ units and a nuclear charge of $88 - 2 = 86$; this is a wholly new element that is an isotope of the element radon. The process can be written as a *nuclear reaction:*

$$_{88}^{226}\text{Ra} \longrightarrow {_{86}^{222}}\text{Rn} + {_2^4}\text{He}$$

Notice that both the total mass number and the total nuclear charge individually balance in such equations.

A **beta (β) particle** is an electron. It is formed when a neutron splits into a proton and an electron in the nucleus. Since the proton remains behind in the nucleus when the electron leaves it, the nuclear charge (or atomic number) *increases* by 1 unit (you may imagine this effect as "subtracting a negative particle"). There is no change in mass number of the nucleus, since the total number of neutrons plus protons remains the same. For example, when an atom of the lead isotope $_{82}^{214}\text{Pb}$ (lead-214) decays radioactively by the emission of a β particle, the nuclear charge of the product is $82 + 1 = 83$, corresponding to the element bismuth; the mass number remains 214:

$$_{82}^{214}\text{Pb} \longrightarrow {_{83}^{214}}\text{Bi} + {_{-1}^0}\text{e}$$

Notice that the symbol $_{-1}^0\text{e}$ used here for the electron shows its mass number (zero) and its charge; in the equation the total mass numbers and nuclear charge numbers each balance.

TABLE 9-1	Summary of Small Particles Produced by Radioactivity		
Particle Symbol and Name	Chemical Symbol	Comment	Effect on Nucleus of Particle Emission
α (alpha)	^4_2He	Nucleus of a helium atom	Atomic number reduced by 2
β (beta)	$^0_{-1}\text{e}$	Fast-moving electron	Atomic number increased by 1
γ (gamma)	None	High-energy photon	None

One other important type of radioactivity is the emission of a **gamma (γ) particle** (also called a *ray*) by a nucleus. This is a huge amount of energy concentrated in one photon and possesses no particle mass. Neither the nuclear mass number nor the nuclear charge changes when a γ particle is emitted. The emission of a γ ray often accompanies the emission of an α or β particle from a radioactive nucleus. The properties of all three types of nuclear radiation are summarized in Table 9-1.

PROBLEM 9-1

Deduce the nature of the species that belongs in the blank for each of the following nuclear reactions:

(a) $^{222}_{86}\text{Rn} \longrightarrow ^4_2\text{He} + \underline{\hspace{1cm}}$

(b) $^{214}_{83}\text{Bi} \longrightarrow \beta + \underline{\hspace{1cm}}$

(c) $^{214}\text{Po} \longrightarrow ^{214}\text{Pb} + \underline{\hspace{1cm}}$

(d) $\underline{\hspace{1cm}} \longrightarrow ^{234}_{90}\text{Th} + \alpha$

The Health Effects of Ionizing Radiation

The α and β particles that are produced in the radioactive decay of a nucleus are not in themselves harmful chemicals, since they are simply the nucleus of a helium atom and an electron. However, they are ejected from the nucleus with an incredible amount of energy of motion. When this energy is absorbed by the matter encountered by the particle, it often ionizes atoms or molecules; for that reason, it is called **ionizing radiation,** or just *radiation*. This radiation is potentially dangerous if we absorb it, since the molecular components of our bodies can be ionized or otherwise damaged.

Although α and β particles are energetic, they cannot travel far within the human body, since they lose more and more of their energy—and consequently slow down—as they collide with more and more atoms.

Alpha particles can travel only a few thousandths of a centimeter within the body, so they are not penetrating. This is true because they are relatively massive, and when they interact with matter they slow down, capture electrons from it, and are converted into harmless atoms of helium gas. If an α particle is emitted outside the body, it will usually be absorbed in the air or by the layer of dead skin, so it will do you no harm. However, inhaled or ingested radioactive atoms can cause serious internal damage when they emit α particles. The damage is particularly severe with α particles since their energy is concentrated in a small area of absorption located within about 0.05 mm of the point of emission. In their interaction with matter, α particles are highly damaging—the most highly damaging of all particles—since they can knock atoms out of molecules or ions out of crystal sites. If the molecules affected are DNA or its associated enzymes, cell death can result. A more serious consequence for the individual can be the creation of mutations that could lead to cancer.

Beta particles move much faster than α particles since they are much lighter and can travel about 1 m in air or about 3 cm in water or biological tissue before losing their excess energy. Like α particles, they can cause considerable damage to cells if they are emitted from particles that have been inhaled or ingested and the radioactive nucleus is consequently close to the cell when it decays.

Gamma rays easily pass through concrete walls—and our skin. A few centimeters of lead are required to shield us from γ rays. Gamma particles are the most penetrating and therefore the most damaging of the three, traveling a few dozen centimeters into our bodies or even right through them. They are generally the most dangerous type of radioactivity, since they can penetrate matter efficiently and do not have to be inhaled or ingested. Although they can pass through our bodies, γ rays lose some of their energy in the process, and cells can be damaged by this transferred energy, since it can ionize molecules. Ionized DNA and protein molecules cannot carry out their normal functions, potentially resulting in radiation sickness and cancer.

The ions produced by radiation when its energy is transferred to molecules are free radicals; hence they are highly reactive (Chapters 1, 3, and 5). For example, a water molecule can be ionized by an α, β, or γ ray or by an X-ray. The resulting H_2O^+ free-radical ion subsequently dissociates into a hydrogen ion and the *hydroxyl free radical*, OH:

$$H_2O + radiation \longrightarrow e^- + H_2O^+$$

$$H_2O^+ \longrightarrow H^+ + OH$$

If the affected water molecule is contained in a cell, the hydroxyl radical can engage in harmful reactions with biological molecules in the cell, such as DNA and proteins. In some cases, radiation damage is sufficient to kill cells of living organisms. This is the basis of food irradiation, where the death of microorganisms helps prevent subsequent spoilage of the food.

If human beings are exposed to substantial, though sublethal, amounts of ionizing radiation, they can develop *radiation sickness*. The earliest effects of this malady to be observed occur in tissues containing cells that divide rapidly, because damage to the cell's DNA or protein can affect cell division. Such rapidly dividing cells are found in bone marrow, where white blood cells are produced, and in the lining of the stomach. It is not surprising, then, to find that early symptoms of radiation sickness include nausea and a drop in white blood cell count. Children are more susceptible to radiation than adults because their tissues involve more cell division. On the other hand, radiation can be effectively used to kill cancer cells since they are dividing rapidly. Unfortunately, radiation therapy cannot be completely selective in terms of the cells it affects, so it has side effects such as nausea.

Long-term effects from radiation may show up in genetic damage, because chromosomes may have undergone damage or their DNA may have mutated. Such damage may lead to cancer in the person exposed or to effects in her or his offspring if the changes occurred in the ovaries or testes.

Quantifying the Amount of Radiation Energy Absorbed

The amount of radiation absorbed by the human body is measured in **rad** (radiation absorbed dose) units, where 1 rad is the quantity of radiation that deposits 0.01 joule of energy to 1 kilogram of body tissue. The rad is not a particularly useful quantity, however, since the damage inflicted by 1 rad of α particles is 10 to 20 times greater than that inflicted by 1 rad of β particles or γ rays. The scale of absorbed radiation that incorporates this biological effectiveness factor is the **rem** (roentgen equivalent man). A more modern unit than the rem is the sievert, Sv, which equals 100 rem.

On average, we each receive about 0.3 rem, i.e., 300 mrem or 3000 μSv, of radiation annually. The origin on average is about

- 55% from radon in indoor and outdoor air;

- 8% from cosmic rays from outer space;

- 8% from rocks and soil;

- 11% from natural radioactive isotopes (e.g., ^{40}K, ^{14}C) of elements that are present in our own bodies; and

- 18% from anthropogenic sources, chiefly medical X-rays.

The average contribution from nuclear power production is negligible at present.

An *acute* exposure of more than 25 rem results in a measurable decrease in a person's white blood cell count; over 100 rem produces nausea and hair loss; and an exposure of over 500 rem results in a 50% chance of death within a few weeks.

Radioactive Nucleus Decay

The radioactive decay of the atoms in an isotope sample does not occur all at once. For example, in a sample of **uranium-238**, ^{238}U, just large enough to be visible, there are about 10^{20} atoms. Only about 10^7 of the ^{238}U nuclei in the sample decompose in a given second, so it requires billions of years for the decomposition process to be complete for the sample as a whole.

Since all radioactive nuclei disintegrate by processes that are kinetically first-order, it is convenient to express the decomposition rate as the time period required for half the nuclei in a sample to disintegrate—its **half-life,** $t_{1/2}$. (This is the same property that was used in earlier chapters to discuss the decomposition of substances by chemical reactions.) For example, the half-life of ^{238}U is about 4.5 billion years. Thus about half of this isotope of uranium existing when the Earth was formed (about 4.5 billion years ago) has now disintegrated; half the *remaining* ^{238}U, amounting to one-quarter of the original, will disintegrate over the next 4.5 billion years, leaving one-quarter of the original still intact. After three half-lives have passed, only one-eighth of the original will remain, and only one-sixteenth will be there after four half-lives.

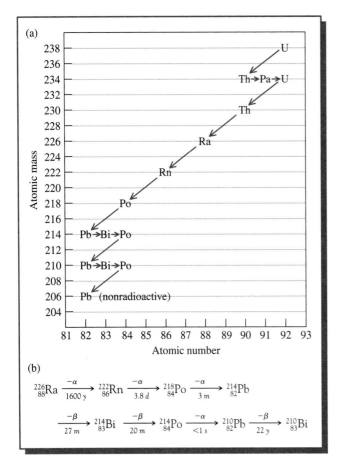

FIGURE 9-1 (a) The ^{238}U radioactive decay series. (b) The radium–radon portion of the ^{238}U radioactive decay series. The symbol above the arrow signifies the type of particle (α or β) emitted during the transition. The time period indicated below an arrow is the half-life of the unstable isotope.

PROBLEM 9-2

Recall from introductory chemistry that for first-order processes, the fraction F of reactant left after time t of reaction has elapsed is given by $F = e^{-kt}$, where k is the rate constant. Calculate the time required for 99% of a sample of ^{238}U to disintegrate. [*Hint: The relationship between k and $t_{1/2}$ is $k = 0.693/t_{1/2}$.*]

Radon from the Uranium-238 Decay Sequence

Many rocks and granite soils contain uranium, so the radioactive decay process takes place under our feet each day, with radon gas being one of its unwelcome products. Each $^{238}_{92}U$ nucleus eventually emits an α particle, and an atom of the *thorium* isotope $^{234}_{90}\text{Th}$ is formed:

$$^{238}_{92}\text{U} \longrightarrow {}^{234}_{90}\text{Th} + {}^{4}_{2}\text{He}$$

This is the first of 14 sequential radioactive decay processes that a ^{238}U nucleus undergoes, as illustrated in Figure 9-1a. The last of these reactions produces $^{206}_{82}\text{Pb}$, a nonradioactive (stable) isotope of *lead*; therefore, the sequence stops. The concentration of each member of the series can be determined using steady-state principles (see Box 9-1).

BOX 9-1 | **Steady-State Analysis of the Radioactive Decay Series**

The various members, except the first and last, of the 14-step radioactive series of Figure 9-1a in a body of undisturbed uranium ore are each in a steady state (Chapter 1) with respect to their concentrations. Since all nuclear disintegration processes are kinetically first-order, the rate of production of each species, C, is proportional to the concentration of B, the species that precedes it in the chain $A \rightarrow B \rightarrow C \rightarrow D \rightarrow \cdots$. The rate of disintegration of C is proportional to its own concentration. Thus

$$\text{d}[C]/\text{d}t = k_{p}[B] - k_{d}[C] = 0$$

at steady state

so

$$[C]_{ss}/[B]_{ss} = k_{p}/k_{d}$$

Since each first-order rate constant is inversely proportional to the half-life period for the process, it follows that

$$[C]_{ss}/[B]_{ss} = t_{C}/t_{B}$$

where t_{C} and t_{B} correspond to the half-life periods for decay by disintegration of C and B. Thus the steady-state concentration of any species along the series relative to the one that precedes it is directly proportional to its half-life decay period and inversely proportional

to the half-life decay period of the species that produces it. For example, the half-life of ^{222}Rn is 3.8 days and that of the ^{226}Ra that produces it is 1600 years, or 1.5×10^{5} times as long, so the steady-state ratio of radon to radium is $1/1.5 \times 10^{5}$ or 6.5×10^{-6}. In qualitative terms, the radon concentration never comes close to reaching that of the radium since it decays so quickly after it is formed.

PROBLEM 1

Show that the ratio of the steady-state concentration of C relative to the first member, A, of the radioactive series $A \rightarrow B \rightarrow C \rightarrow D \rightarrow \cdots$ is equal to the ratio of their half-lives for disintegration, and thus that the steady-state concentration of each member of the series is directly proportional to its disintegration half-life.

PROBLEM 2

Using the equations developed in Box 14-4, deduce how long it will take for radon-222 to reach 80% of its steady-state concentration after a sample of radium-226 is purified of contamination by any other substance. [*Hint: Recall that for first-order reactions, $t_{0.5} = 0.693/k$.*]

Of particular interest is the portion of the 14-step sequence of ^{238}U radioactive decay that involves radon, since this element is the only one, other than the helium produced from the α particles, that is gaseous and therefore is mobile. Details concerning this portion of the radioactive decay series are shown in Figure 9-1b. The immediate precursor of the radon is radium-226, which has a half-life of 1600 years and decays by emission of an α particle:

$$^{226}_{88}\text{Ra} \longrightarrow {}^{222}_{86}\text{Rn} + {}^{4}_{2}\text{He}$$

The ^{222}Rn isotope has a half-life of 3.8 days, which can be long enough for it to diffuse through the solid rock or soil in which it is initially formed. Most radon escapes directly into outdoor air when the surface of the Earth where it appears is not covered, e.g., by a building. The very small background concentration of radon in air that this produces nevertheless yields about half of our exposure to radioactivity, as listed above. Although the radon decays in a few days, it is constantly replaced by the decay of more radium.

Some scientists have pointed out that radon gas accumulates to unhealthy levels in caves, including some that are often used for recreational purposes. However, it is in certain homes that radon becomes an important indoor air pollutant. Most radon that seeps into homes comes from the top meter of soil below and around the foundation; radon produced much deeper than this will probably decay to a nongaseous and therefore immobile element before it reaches the surface. Loose, sandy soil allows the maximum diffusion of radon gas, whereas frozen, compacted, or clay soil inhibits its flow. Radon enters the basements of homes through holes and cracks in their concrete foundations. The intake is increased significantly if the air pressure in the basement is low. The material used to construct the homes and water from artesian wells are other potential sources of radon in homes. Groundwater systems serving up to a few hundred people often have radon levels almost 10 times those of surface waters. When well water is heated and exposed to air, as occurs when it exits from a showerhead, radon is released to the air. However, radon from water usually represents only a small fraction of that arising from soil, although it represents a greater health hazard than that contributed from water disinfection by-products and other dissolved chemicals.

Measuring the Rate of Disintegration and Health Threat from Environmental Radiation

The rate of radioactive disintegrations in a sample of matter is usually measured in **bequerels,** Bq, where 1 Bq corresponds to the disintegration of one atomic nucleus per second. The other unit used is the **curie,** Ci, which equals 3.7×10^{10} Bq and is the radioactivity produced by one gram of ^{226}Ra. Environmental regulations are usually expressed in the number of bequerels per unit volume or, in the United States, in terms of the number of picocuries, where 1 pCi $= 10^{-12}$ Ci. For example, the U.S. EPA uses 4 pCi per liter of air as the upper limit for a safe radon level in houses.

We can calculate the amount of energy from radiation that is absorbed by a person's lungs in a year if he or she breathes air containing radon at the 4 pCi/L level, since the energy of each α particle emitted by a radon atom is known by measurement to be 9.0×10^{-13} J. Since 4 pCi/L is equivalent to $4 \times 10^{-12} \times 3.7 \times 10^{10} = 0.15$ disintegrations per liter per second, and since there are $60 \times 60 \times 24 \times 365$ sec in a year, the total annual number of disintegrations in 1 L of air is 4.7×10^{6}. Thus the total amount of energy liberated annually in the process is $4.7 \times 10^{6} \times 9.0 \times 10^{-13}$ J $= 4.2 \times 10^{-6}$ J. If we assume that all this energy is absorbed by a person's lung tissues (rather than by the air in the lungs), that the lung volume is about 1 L, and that the lung mass is about 3 kg, then since 1 rad $= 0.01$ J/kg, the energy absorbed is 1.4×10^{-4} rad or 0.14 mrad. Using a factor of 10 to convert rads to rems for α-particle radiation, we find that the annual radiation dose is about 1.4 mrem, i.e., about 0.5% of background exposure.

The Daughters of Radon

Radon, the heaviest member of the noble gas group, is chemically inert under ambient conditions and remains a monatomic gas. As such, it becomes part of the air that we breathe when it enters our homes. Because of its inertness, physical state, and low solubility in body fluids, radon *itself* does not pose much of a danger; the chance that it will disintegrate during the short time it is present in our lungs is small and, as discussed above, the range of α particles in air before they lose most of their energy is less than 10 cm.

The danger arises instead from the radioactivity of the next three elements in the disintegration sequence of radon—namely, *polonium, lead,* and *bismuth* (see Figure 9-1b). These *descendants* are termed **daughters** of radon, which in turn is called the *parent* element. In macroscopic amounts, these particular daughter elements are solids, and when formed in the air from radon they all quickly adhere to dust particles. Some dust particles adhere to lung surfaces when inhaled, and it is under these conditions that the elements pose a health threat. In particular, both the ^{218}Po, which is formed directly from ^{222}Rn, and the ^{214}Po, which is formed later in the sequence (Figure 9-1b), emit energetic α particles that can cause radiation damage to the bronchial cells near which the dust particles reside. This damage can eventually lead to lung cancer. Indeed, as will be discussed, radon (or rather its daughters) is the second leading cause of such cancers, although it follows smoking by a wide margin.

Although some radon daughters in the sequence disintegrate by β-particle emission, the deleterious health effects of these particles are considered negligible because the α particles carry much more energy and, as discussed, it is the disruption of cell molecules by the burst of high energy that initiates cancer.

Notice that the sequence (see Figure 9-1b) of radon decay to ^{210}Pb formation takes less than a week on average. In contrast, disintegration of ^{210}Pb to ^{210}Bi has a half-life of 22 years, and, in fact, most of the lead will have been cleared from the body before this process occurs.

Measuring the Health Danger from Radon and Its Daughters

The greatest exposure to α particles from radon disintegration is experienced by miners who work in poorly ventilated underground uranium mines. Their rate of lung cancer is indeed higher than that of the general public, even after corrections to the data have been made for the effects of smoking. From statistical data relating their excess incidence of lung cancer to their cumulative level of exposure to radiation, a mathematical relationship between cancer incidence and radon exposure has been developed. Scientists have extrapolated this relationship to determine the risk to the general population from the generally lower levels of radon to which the public is exposed.

Based on linear extrapolation from the miners' data and other sources, the U.S. EPA estimates that radon currently causes about 21,000 excess lung cancer deaths annually; the estimate for the United Kingdom is 2000 cases per year. Most of the excess deaths are associated with smokers, since radon and cigarette smoke are synergistic (see Chapter 4) in causing lung cancer. In particular, the risk of lung cancer by age 75 is 4 in 1000 for a nonsmoker living in a house with zero radon and is only increased to 7 in 1000 if he or she has constant exposure by inhalation to 400 Bq/m^3 of radioactivity from the gas, for a net increase of 3 in 1000. However, a smoker's chances of lung cancer rise from 100 to 160 in 1000, an increase of 60, by exposure to the same level of radioactivity. It is estimated that radon causes about 10% of all lung cancers, which is about half the mortality rate from automobile accidents, for example.

The level of radioactivity in air is stated in units of becquerels (Bq) per cubic meter. The average indoor radon concentration globally is about 39 Bq/m^3, compared to the usual outdoor level of about 10 Bq/m^3. The radioactivity "action level" for indoor air, beyond which mitigation measures should be taken, is 150 Bq/m^3 (4 pCi/L) in the United States; that in Great Britain, Norway, and Sweden—and that proposed for Canada—is 200 Bq/m^3. Because radon dissolved in drinking water can escape into the air when it comes out of the tap, a maximum contaminant level of 150 Bq/L has been established by the U.S. EPA; the World Health Organization (WHO) guideline is 100 Bq/L. Radon concentrations at these levels are associated with groundwater that has passed through rock formations that contain natural uranium and radium.

Those skeptical of using uranium miners' data point out that the calculated estimates of radon-caused lung cancer may be too high, since miners work in much dustier conditions than are found in homes and their breathing during hard labor is much deeper than normal. Consequently, there is a much greater chance that radon daughters will find their way deep into the lungs of miners in comparison with the general population. The miners' exposure to arsenic and diesel exhaust may also contribute to an increase in the lung cancer rate that would have been counted as due to radon.

In order to establish whether or not radon gas buildup in homes causes lung cancer, several epidemiological studies were undertaken in the 1990s. These analyses, one from Sweden, one from Canada, and one from the United States, reached contradictory conclusions about the risk of radon to householders. In the Swedish report the rate of lung cancer in nonsmokers and especially in smokers was found to increase with increasing levels of radon in their homes. The Canadian study focused on residents of Winnipeg, Manitoba, which has the highest average radon levels in Canada; no linkage between radon levels and lung cancer incidence was found. The U.S. study, conducted among non-smoking women in Missouri, found little evidence for a trend of increasing lung cancer with increasing indoor radon concentration.

The environmental radon problem has received the greatest attention in the United States, where there are currently programs to test the air in the basements of a large number of homes for significantly elevated levels of the gas. Once radon is identified, the owners can then alter the air circulation patterns to reduce radon levels in living areas, thereby reducing the additional risk of contracting lung cancer. Gaps and cracks in basement walls are sealed, and venting pipes through basement floor slabs are installed. About 800,000 U.S. homes have undergone mitigation to reduce high radon levels, at an average cost of $1200 per house, since the 1980s. It has been pointed out, however, that the normally high mobility of the U.S. population means that, on average, a given individual who happens to live in a house with a high level of radon will be there for only a few years and likely spend most of his or her life in a house with a lower level (since only about 7% of houses have high levels). Consequently, the estimate of increased lung cancer death mentioned at the beginning of this section is likely much too high.

A minority of scientists do not believe that radioactivity at very low levels causes harm to humans. They are skeptical about the *linear-no-threshold* (LNT) assumption—that the observed effects of high doses of radioactivity can be extrapolated linearly to very low doses and that there is no threshold below which radioactivity causes no harmful effects such as cancer. Some point to evidence of a threshold near about 1000 Bq/m^3 for the effects of residential radon in causing lung cancer, and others point to the lack of cancer incidence in many areas having high natural radon levels. However, reviews published in 2005 and 2006 of epidemiological data from North America and Europe find no threshold for increase in lung cancer with residential radon.

Some scientists believe in a theory called *hormesis*, which states that exposure to radioactivity (and some chemicals) at very low doses for short periods of time can be *positive* to human health. Although there are some animal and cell studies that support this idea, it is not widely accepted among scientists. Indeed, the LNT theory is supported by recent evidence concerning cancer incidence among Russian citizens who were exposed inadvertently to very low doses of radioactivity from nuclear weapons production.

Depleted Uranium

Depleted uranium is what remains of natural uranium once most of the ^{235}U isotope, used in power reactors and for nuclear weapons, and the ^{234}U have been extracted from it. Indeed, about 200 kg of depleted uranium is produced for every kilogram of the highly enriched element, since uranium is only 0.7% ^{238}U. Because uranium is so dense (70% denser than lead), it is useful for making armor-piercing weapons, especially projectiles used against tanks. When the shells hit hard targets, the uranium ignites and the combustion creates clouds of dust containing *uranium oxide*. This dust settles and contaminates the soil in the area, but before that, it can be inhaled by people in the vicinity. Since most of the ^{235}U has been extracted from the uranium (for use in bombs and power production), and since ^{238}U has such a very long half-life, depleted uranium is less radioactive in terms of α-particle emissions (by almost half) than the naturally occurring element. However, the β emission from daughters of ^{238}U is still present. Concerns have been expressed about the effect of the residual radioactivity from depleted uranium on troops and civilians exposed to it during wartime.

Dirty Bombs

A **dirty bomb** is a conventional, chemical-based explosive mixed with radioactive material that would be distributed over a wide area as a result of the bomb's explosion. Dirty bombs could be made by terrorists, using radioactive material stolen from either hospitals or research institutes, or purchased on the black market from supplies originating in the former Soviet Union or other countries that have undergone disruption with a consequent loss of security at their nuclear energy facilities. Although the actual danger to human health from a dispersal of dirty bomb materials into the environment may well be quite small, the fear it would generate in the public and the expense of decontamination of wide areas would be substantial.

A similar terrorist threat would be the deliberate crashing of hijacked aircraft into nuclear power plants or into containers of nuclear fuel. However, most nuclear energy industry executives deny that the power plants or containers could be breached, and radioactivity released, by such an event.

Nuclear Energy

Although most of the energy we use originates as heat generated by the combustion of carbon-containing fuels, heat in commercial quantities can also be produced indirectly when certain processes involving atomic nuclei occur; this power source is called **nuclear energy,** used mainly to produce electricity. Since nuclear forces are much stronger than chemical bond forces, the energy released per atom in nuclear reactions is immense compared to that obtained in combustion reactions. One of the attractions of nuclear power is that it does not generate *carbon dioxide* or other greenhouse gases during its operation. Some policymakers have promoted expansion of nuclear power as a way to combat global warming in the future.

There are two processes by which energy is obtained from atomic nuclei: fission and fusion.

- In **fission,** the collision of certain types of heavy nuclei (all of which have many neutrons and protons) with a neutron results in the splitting of the nucleus into two similarly sized fragments. Since the separated fragments are more stable energetically than was the original heavy nucleus, energy is released by the process.

- The combination of two very light nuclei to form one combined nucleus is called **fusion.** It also results in the release of huge amounts of energy, since the combined nucleus is more stable than the original, lighter ones.

Fission Reactors

Currently, there are 440 fission-based nuclear power plants in operation in more than 30 countries in the world. Collectively, they generate 17% of global electricity demand, including 23% of that in developed countries, 16% of that in the former Soviet Union, but only 2% of that in developing countries. Global production more than tripled between 1980 and 2000. The fraction of electricity produced by nuclear energy in various countries is listed in Table 9-2.

TABLE 9-2	Nuclear Power Around the World (2005)	
Country	Number of Power Reactors	Proportion of Electricity Generated by Nuclear Power
United States	103	19%
France	59	79%
Japan	55	29%
Russia	31	16%
Great Britain	23	20%
South Korea	20	45%
Canada	18	15%
Germany	17	31%
India	16	3%
Ukraine	15	49%
Sweden	10	45%
China	10	2%
Others	65	

The most economically useful example of fission, and the one mainly used by power plants, is induced by the collision of a ^{235}U nucleus with a neutron. The combination of these two particles is unstable. When it decomposes, the products vary but are typically a nucleus of *barium*, ^{142}Ba, one of *krypton*, ^{91}Kr, and three neutrons:

$$\,_0^1 n \;+\; \,_{92}^{235}U \longrightarrow \,_{56}^{142}Ba \;+\; \,_{36}^{91}Kr \;+\; 3\,_0^1 n$$

$$\text{(n)} + \left(\,_{92}^{235}U\right) \longrightarrow \left(\,_{56}^{142}Ba\right) + \left(\,_{36}^{91}Kr\right) + \begin{array}{c} \text{(n)} \\ \text{(n)} \\ \text{(n)} \end{array}$$

Not all the uranium nuclei that absorb a neutron form exactly the same products, but the process always produces two nuclei of about the sizes of Ba and Kr, together with several neutrons.

The two new nuclei produced in fission reactions are very fast-moving, as are the neutrons. It is heat energy from this excess kinetic energy that is used to produce electrical power. Indeed, the generation of electricity by nuclear energy and by the burning of fossil fuels both involve using the energy source to produce steam, which is then used to turn large turbines that produce the electricity.

An average of about three neutrons are produced per ^{235}U nucleus that reacts; one of these neutrons can produce the fission of another ^{235}U nucleus, and so on, yielding a **chain reaction.** In atomic bombs, the extra neutrons are used to induce a very rapid fission of all the uranium that is constrained to stay in a small volume, so energy is released explosively. In contrast, the energy is released gradually in a nuclear power reactor by ensuring that, on average, only one neutron released from each ^{235}U fission event initiates the fission of another nucleus. The extra neutrons produced in fission are absorbed by the **control rods** in a nuclear power reactor. These are bars made from neutron-absorbing elements such as *cadmium*. The position of the rods in the reactor can be varied to control the rate of fission.

The neutrons that leave ^{235}U nuclei upon fission are too fast-moving to be efficiently absorbed by other nuclei and cause further fission, so they must be slowed down if they are to be useful. This is accomplished by the **moderator,** which, depending upon the type of reactor, can be regular water, heavy water (i.e., water enriched in deuterium), or *graphite*. The **coolant** material that is used to carry off the heat energy produced by fission in most reactor types is water, but gaseous carbon dioxide is used in some. Ultimately, high-pressure, high-temperature steam is created—just as in a fossil-fuel power generating plant—and used to turn turbines and create electricity. The operating efficiency of nuclear power reactors is significantly lower than that of fossil-fuel ones since the core of the nuclear reactor, from which heat is drawn, is only a little above 300°C, whereas that of fossil-fuel plants may be as high as 550°C. The various steps in the production of electricity by a fission-based nuclear power plant are illustrated in Figure 9-2.

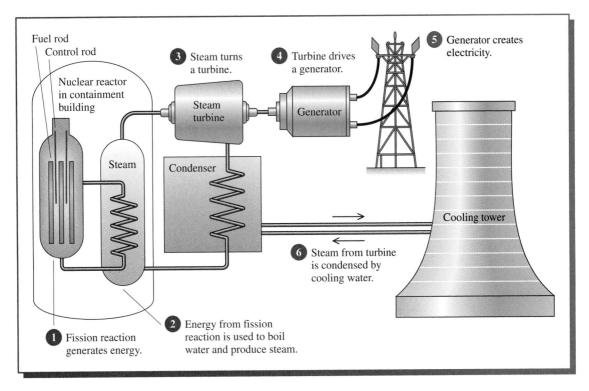

FIGURE 9-2 Step-by-step schematic diagram of the production of electrical power by a nuclear fission reactor.

Calculate the maximum percentage efficiency in converting heat at 300°C into electricity according to the second law of thermodynamics (see Chapter 8), assuming the cooling water is at 17°C. Repeat the calculation for 550°C heat. Do your calculated values lie above the average efficiencies of 30% and 40% observed for nuclear and coal-fired power plants, respectively?

The uranium in reactors is contained in a series of enclosed bars called **fuel rods.** When the uranium fuel in a rod is "spent," i.e., when its ^{235}U content is too low for it to be useful as a fuel since too few neutrons are produced per second, it is removed from the reactor.

The only naturally occurring uranium isotope that can undergo fission is ^{235}U, which constitutes only 0.7% of the native element. The remaining uranium is ^{238}U (99.3% natural abundance). A neutron produced by the fission of a ^{235}U nucleus can be absorbed on collision by a ^{238}U nucleus. The resulting ^{239}U nucleus is radioactive and emits a β particle, as does the heavy product (^{239}Np) of this process; consequently, a nucleus of **plutonium, ^{239}Pu**, is produced:

$$^{1}_{0}n + {}^{238}_{92}U \longrightarrow {}^{239}_{92}U \xrightarrow{-\beta} {}^{239}_{93}Np \xrightarrow{-\beta} {}^{239}_{94}Pu$$

$$\boxed{n} + \boxed{{}^{238}_{92}U} \longrightarrow \boxed{{}^{239}_{92}U} \longrightarrow \beta + \boxed{{}^{239}_{93}Np} \longrightarrow \beta + \boxed{{}^{239}_{94}Pu}$$

Thus, ^{239}Pu is produced as a by-product of the operation of nuclear power reactors.

Unfortunately, both the nuclei into which the ^{235}U nucleus fissions and the ^{239}Pu by-product are highly radioactive substances. As a consequence, the material in spent fuel rods is much more radioactive than was the original uranium. The fission products each consist of about half the original uranium nucleus (minus a few neutrons), with atomic numbers in the 30s to 50s rather than 92. Since the optimum neutron/proton ratio of nuclei increases with mass number, the fission products are *neutron-rich*. The excess neutrons split into a proton and an electron, with the latter expelled from the nucleus as an energetic β particle, thereby increasing the atomic number by 1 and lowering the neutron/proton ratio. Successive neutron decay continues until the product is no longer neutron-rich. For example, the krypton produced in the prototypical fission reaction previously discussed has a neutron/proton ratio of 1.53, typical of that of a nucleus the size of uranium. It has a half-life of only 10 seconds before it emits a β particle and becomes rubidium, which in turn (1-minute half-life) emits a β particle to become strontium, which in turn (10-hour half-life) emits a β particle to become yttrium, which in turn (59-day half-life) emits a β particle to become a stable, nonradioactive isotope of zirconium with a normal neutron/proton ratio (1.28) for elements of this size.

$$^{91}_{36}Kr \longrightarrow {}^{91}_{37}Rb \longrightarrow {}^{91}_{38}Sr \longrightarrow {}^{91}_{39}Y \longrightarrow {}^{91}_{40}Zr$$
$$\quad +\beta \qquad\quad +\beta \qquad\quad +\beta \qquad\quad +\beta$$

The spent fuel rods, in which heat is produced by the energy associated with the β-particle emissions, are immersed in ponds of water for a few months until most of such decay processes have occurred and the rods have cooled down.

A few of the ^{235}U fission products have longer half-lives than those discussed above. After 10 years, most of the radioactivity from spent fuel rods is due to strontium-90, ^{90}Sr (half-life of 29 years), and to cesium-137, ^{137}Cs (half-life of 30 years), both of which are β-particle emitters. The dispersal of radioactive strontium and cesium into the environment would constitute a serious environmental problem. Ions of both metals can be readily incorporated into the body because strontium and cesium readily replace chemically similar elements that are integral parts of animal bodies, including humans.

Strontium, a Group II metal, concentrates in the bones and teeth, replacing ions of *calcium*, also a Group II metal that forms +2 ions. Ions of *cesium*, a Group I metal, can replace those of *potassium*, also from Group I and also +1 in charge, which is widespread in all body cells. Once these elements are in place in the body, their radioactive disintegration produces β particles that can damage the cells in which they are present or near. Radioactive waste from the spent fuel rods of nuclear power plants must therefore be carefully monitored and will eventually have to be deposited in a secure environment from which it cannot escape.

Environmental Problems of Uranium Fuel

Several of the steps in the *nuclear fuel cycle*, illustrated in Figure 9-3, generate environmental waste.

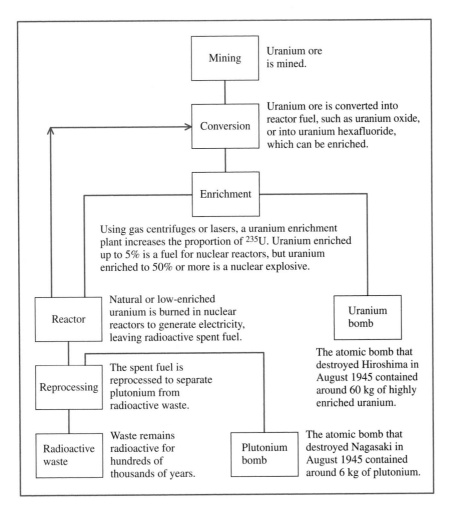

FIGURE 9-3 The nuclear fuel cycle. [Source: R. Edwards, "A Struggle for Nuclear Power," *New Scientist* (22 March 2003): 8.]

Major suppliers and processors of uranium ore are Canada, Australia, Russia, Niger, Ukraine, and Kazakstan. During the mining of uranium, contamination of the environment by radioactive substances commonly occurs. Since naturally occurring uranium slowly decays into other substances that are also radioactive, uranium ore contains a variety of radioactive elements. As a result, the very large volume of waste material that remains after the uranium is chemically extracted from the ore is itself radioactive.

The radioactive material that was immobilized in the original rock ore is in the form of liquid and powder *tailings* after mining. Gaseous emissions also escape from the wastes. As in other mining operations, the liquid tailings are normally held in special ponds until the solids separate. Pollution of the local groundwater can occur if these ponds leak or overflow. In addition, when the solid tailings are exposed to the weather and are partially dissolved by rainfall, they can contaminate local water supplies.

The use of the solid tailings as landfill on which buildings are constructed can also lead to problems, since the radon produced by the radioactive decay of the radium in the tailings is quite mobile. As previously noted, radon is a particular hazard for uranium ore miners, since the radioactive gas is always present in the ore and is released into the air of the mine. Indeed, the incidence of lung cancer among such miners was particularly high until mine ventilation was improved to permit more frequent changes of air and thus a more efficient clearing out of accumulated radon gas.

In most but not all nuclear power reactors (the Canadian *CANDU* system being the main exception), the uranium fuel must be enriched in the fissionable ^{235}U isotope; its abundance is increased to 3.0% from the 0.7% in the naturally occurring element. The extent of enrichment required for use in bombs is much greater. Uranium sufficiently enriched for this purpose, called *weapons-grade* material, is 90% or more ^{235}U. Enrichment is a very expensive, energy-intensive process since it requires physical rather than chemical means of separation, given the fact that all isotopes of a given element behave identically chemically. For separation, the uranium is temporarily converted to the gaseous compound *uranium hexafluoride*, UF_6.

The CANDU reactor is able to use nonenriched uranium by employing **heavy water** rather than normal water as a moderator, since doing so decreases the probability that neutrons will be absorbed and therefore unavailable to continue the chain reaction. Heavy water contains a much-higher-than-normal fraction of **deuterium**, ^2H, the naturally occurring (0.02% natural abundance) nonradioactive isotope of hydrogen. Some of the neutrons produced by fission are absorbed by deuterium nuclei, producing **tritium**, ^3H.

$$^2H + {}^1n \longrightarrow {}^3H$$

Indeed, CANDU reactors produce about 30 times as much tritium as do light-water reactors. Tritium levels in moderator and coolant water increase with time, as the product of this reaction accumulates. Small losses of coolant from reactors result in the release of tritium into the aqueous environment.

Tritium is radioactive, a low-energy β-particle emitter with a half-life of 12.3 years. Less than 1% of the tritium in the present environment occurs naturally; the remainder is due to fallout from nuclear weapons testing in past decades and to releases from nuclear power reactors. The U.S. EPA classifies tritium as a human carcinogen; there is evidence that it is mutagenic and teratogenic (causes birth defects) as well. The EPA's maximum limit for tritium in drinking water is 740 Bq/L; Canada's limit is almost 10 times as high. The waters of Lake Huron and Lake Ontario, along whose shores several CANDU reactors are located, contain about 7 Bq/L of tritium, compared to 2 Bq/L for Lake Superior, which has no nuclear reactors nearby.

The Future of Fission-Based Nuclear Power

In North America and much of Europe, public opinion about fission-based nuclear power shifted from positive to negative in recent decades, partly as a consequence of the accidents at the *Three Mile Island* power plant in Harrisburg, Pennsylvania, in 1979 and at the Chernobyl, Ukraine, power station in 1986. No new reactors have been ordered in the United States since the Three Mile Island incident, and several power plants in both the United States and Canada have been shut down. The last new plant opened in Great Britain in 1995, and there are no plans for new ones.

Because of the increased need for electricity in developing countries and the desire to produce electricity with fewer greenhouse gas emissions, nuclear power has made a comeback in the twenty-first century. About 168 new reactors are expected to start up by 2020, the majority of them in China, India, Russia and elsewhere in eastern Europe, South Korea, Taiwan, and South Africa; some new ones are also expected to be built in the United States, Japan, Europe, and Canada. In addition, a nuclear energy plant may be built to generate the heat needed to extract oil from the tar sands in Alberta, Canada. Many existing reactors in these countries are approaching the ends of their useful lives, giving rise to some of the impetus for new units. The only new plant actually under construction in Europe is in Finland. The main advantages and disadvantages of nuclear power production are summarized in Table 9-3.

The Catastrophe at Chernobyl

A more realistic fear than a nuclear power plant running out of control and blowing up like an atomic bomb is that the highly radioactive fission products contained in operating fuel rods could be spread into the surrounding

TABLE 9-3	Inherent Advantages and Disadvantages of Nuclear Power Production from Fission	
Advantages	**Disadvantages**	
Minimal air and water pollution	Production of radioactive wastes that require special handling	
Efficient use of fuel resources	Possibility of an accident producing serious human health problems	
Relatively low operating cost	Long-term storage of waste and decommissioning of plants will likely involve high costs	
	Requirement of an international security system to prevent diversion of nuclear materials to weapons use	

countryside if a nonnuclear explosion occurs in a power plant. This in fact did occur at one of the nuclear power plants in Chernobyl, Ukraine, in 1985. During a routine test, engineers at the plant lost control of the reactor by overriding the plant's safety mechanisms and withdrew most of the control rods from the reactor core. As a consequence, the reactor overheated and a fire broke out in the graphite moderator. This produced a huge explosion, which blew off the heavy plate covering the building, and several hundred million curies of radioactivity were released into the air and spread over a wide area. Several dozen people, mainly plant operators and firefighters, died immediately from high doses of radiation. The dispersed radioactivity was mainly concentrated in the noble gas isotopes ^{131}Xe and ^{85}Kr, in ^{131}I, and in the cesium isotopes ^{134}Cs and ^{137}Cs, all of which are fission products.

The major chronic health consequence to the public of the explosion at Chernobyl has been the great increase in thyroid cancer among children in the area. The thyroid cancer was initiated presumably by β radiation from radioactive iodine, ^{131}I, which has a half-life of only 8 days. The human body concentrates iodine in the thyroid gland. Thankfully, thyroid cancer has a high cure rate. In the most contaminated areas of Ukraine, about 4 times as many children as normal have been diagnosed with thyroid cancer, with very young children the most susceptible to the radioactivity. The probability of a Ukrainian child having thyroid cancer was found to increase in direct proportion to the dose of radioactive iodine ingested after the accident. Most of the iodine in the children's systems resulted from drinking milk from cows that had grazed on contaminated plants and from eating leafy vegetables, since the radioactive iodine was deposited from the air onto plant surfaces. The soil in the affected areas is chronically deficient in iodine, so the substance would readily have been taken up by the thyroid in children chronically deficient in the element. Indeed, tablets of *potassium iodide*, KI, designed to flood the body with iodine and thereby dilute the radioactive form, were distributed to the inhabitants around Chernobyl—but not until a week after the explosion.

Radiation experts predict that about 4000–9000 premature deaths among people who were living in the heavily contaminated regions—Ukraine, Belarus, and Russia—have or will eventually be associated with radioactivity from the Chernobyl event, including not only ^{131}I but also the longer-lived cesium isotopes ^{134}Cs and ^{137}Cs (the latter having a half-life of 30 years). Cancers other than that of the thyroid have much longer development times. There is some fear that females who were lactating or pubescent just after the accident may be at higher risk of breast cancer. A much larger number of Europeans—perhaps as many as 60,000—living outside the most affected areas who ingested smaller doses may also be affected eventually if the dose–response relationship for radiation does not have a threshold. However, according to the World Health Organization, the largest public health problem for the people who were living in the highly contaminated regions at the time of the Chernobyl disaster is a long-term decrease in their mental health, stemming from the traumas of perceived health risks and from relocation out of areas declared uninhabitable, as was the case for a third of a million people.

The Accident at Three Mile Island

The other nuclear power plant accident was much less serious than the one at Chernobyl. It occurred in 1979 at the Three Mile Island plant in Pennsylvania. The problem originated with failures in the water-based cooling systems in the reactor. Although control rods were inserted into the core and the fission process stopped, the reactor kept heating. The source of the heat was the radioactive decay by γ-ray emission, not of uranium, but of fission products that had naturally built up over time in the fuel rods. Because of mechanical malfunction and operator error, the core became partially uncovered; it was therefore heated even further (>2200°C), and half of it melted. Hydrogen and oxygen gases were produced from the decomposition of the superheated water.

Fortunately, there was no large explosion at Three Mile Island: The containment facility was not breached. In contrast to the Chernobyl accident, the liquid and solid radioactive materials that escaped from the reactor did not escape from the containment building; the release of ^{131}I was about a million times less. Significant amounts of radioactive noble gas isotopes ^{131}Xe and ^{85}Kr did escape into the air but, as at Chernobyl, they were quickly diluted in the atmosphere.

Plutonium

The plutonium-239 isotope that is produced during uranium fission is an α-particle emitter and has a long half-life of 24,000 years. After 1000 years, the main sources of radioactivity from spent fuel rods will be plutonium and other very heavy elements, since the medium-sized nuclei produced in fission, having much shorter half-lives than 1000 years, will have largely decayed by that time. Thus the long-term radioactivity of the spent fuel rods can be greatly reduced by chemically removing the very heavy elements from it.

PROBLEM 9-4

Given that the half-life of ^{239}Pu is 24,000 years, how many years will it take for the level of radioactivity from plutonium in a sample to decrease to 1/128 (i.e., about 1%) of its original value?

The ^{239}Pu that forms in fuel rods is itself fissionable; once its concentration in the rods becomes high enough, some of it also undergoes fission and contributes to the power output of the reactor. However, many of the other fission by-products are efficient neutron absorbers, and their concentration eventually builds up to the point where the nuclear chain reaction cannot occur. The plutonium that accumulated over time, and the remaining uranium, in such fuel rods can be chemically removed from the rest of the spent fuel by **reprocessing.** In this procedure, the rods are first dissolved in concentrated **nitric acid,** HNO_3. The actinide elements—uranium, plutonium, and minor amounts of several others produced by fission and subsequent radioactive decay—are separated from the non-actinide fission by-products such as barium. This is done by selectively complexing the actinides with a ligand that makes them much more soluble in an organic medium than in an aqueous one. In the PUREX (*plutonium and uranium recovery by extraction*) process, 30% **tri-*n*-butyl phosphate,** $O{=}P(O{-}CH_2CH_2CH_2CH_3)_3$, in a hydrocarbon solvent such as kerosene is mixed with the nitric acid solution. The actinides complex with the phosphate and migrate into the kerosene component of the two-phase kerosene–nitric acid system. Unfortunately, some of the fission by-products dissolve in the organic phase, since the phosphate compound becomes decomposed to some extent by the radioactivity present and forms a product that can complex the fission metals. For this reason, several solvent extraction cycles are required to completely separate actinides from fission by-products. The aqueous layer is a highly radioactive form of *high-level waste* (HLW).

In some situations, it is desirable to separate the uranium from the plutonium before or after it has been extracted from spent fuel. For example, the uranium still contains about 0.3% ^{235}U, which can be converted again into reactor fuel. Alternatively, removing the uranium from the rods greatly reduces the volume of material that must be stored. If the uranium is extracted from the dissolved fuel rods while the plutonium remains mixed with the highly radioactive fission by-products, there is less reason to worry about the latter being diverted into the making of bombs. In the UREX (*uranium recovery by extraction*) process, plutonium in the dissolved rods is selectively reduced from Pu^{4+} to Pu^{3+} by addition of a mild reducing agent such as *acetohydroxamic acid* (AHA), CH_3—$C(=O)$—NHOH. Uranium as U^{4+} (or as U(VI) in UO_2^{2+}) is then isolated by solvent extraction with tributyl phosphate, producing a two-phase system, as in the PUREX process, but with the plutonium remaining in the aqueous phase since Pu^{3+} preferentially exists in water rather than in an organic solvent.

Since reprocessing uses chemical procedures, it is much less energy-intensive than the isotope separation processes. However, the liquid that remains after the valuable heavy elements have been removed is still highly radioactive due to the presence of fission products. No method of disposal for this waste has yet been approved or implemented; it is still stored in metal tanks. Unfortunately, many of the older tanks in which the liquid is stored have begun to leak at installations such as Hanford, Washington, where plutonium for nuclear weapons was manufactured by reprocessing. Reprocessing is costly, in part because the radioactive waste it produces must be stored.

Spent fuel rods from civilian power reactors in a number of countries have been reprocessed in France, the United Kingdom, India, Russia, and Japan, though not in the United States or Canada. This reprocessing has resulted in a buildup of hundreds of tonnes of plutonium. Comparable quantities of the element are also available from the dismantling of nuclear weapons by the United States and Russia (see Box 9-2).

Aside from questions of health and safety, a major problem associated with the handling of plutonium from both civilian and military sources involves the security measures needed to prevent the material from slipping into the hands of terrorists and rogue governments who wish to fashion their own bombs. Only a few kilograms of weapons-grade plutonium, which consists of 93% or more of ^{239}Pu, is required to make an atomic bomb. A somewhat larger quantity of *reactor-grade* plutonium, which contains more of the other isotopes of plutonium and similar elements, is needed for a bomb. The world's current stockpile of plutonium exceeds 1000 metric tons and continues to grow.

Breeder reactors are nuclear power reactors that are designed specifically to maximize the production of by-product plutonium; such reactors actually produce more fissionable material than they consume. The special **fast breeder** reactors start with a small amount of ^{238}Pu (initially obtained from conventional uranium-fueled reactors) and produce more of it from ^{238}U from the reaction previously discussed. The reactors are cooled by liquid sodium, since water is not adequate for the purpose and would absorb neutrons. They use fast neutrons, rather than those moderated to produce fission, since fast neutrons are more likely to be captured by ^{238}U and produce its

---

BOX 9-2 | Radioactive Contamination by Plutonium Production

Plutonium has been deliberately produced for more than 50 years to provide the readily fissionable material needed for nuclear weapons. In the United States, most plutonium production and processing were carried out at Hanford, Washington. Because huge quantities of radioactive waste were produced, stored, and disposed of at this facility, the surrounding environment is now so heavily polluted that it has been called the "dirtiest place on Earth." About 190,000 m^3 of highly radioactive solid waste and 760 million liters of moderately radioactive liquid waste and toxic chemicals were deposited in the ground at this site. As much as a metric ton of plutonium may be contained within the masses of solid wastes buried there. The cleanup of the wastes will cost between $50 and 200 billion and

will not be completed earlier than 2020. One proposed—though expensive—technique for immobilizing the radioactive wastes involves passing a strong electrical current through the contaminated soil for a period of days; the electricity would fuse the soil and sand into a glassy rock from which the contaminants could not escape.

Plutonium was used as the actual explosive material in some atomic bombs and as a "trigger" for hydrogen (fusion) bombs, forcing the reactants together and thereby initiating the thermonuclear explosion. Consequently, about 100 metric tons of plutonium must be removed from nuclear weapons as tens of thousands of them are dismantled by the United States and Russia over the next few decades.

---

daughters, which in turn quickly decay to plutonium. The plutonium must be separated from the residual uranium by reprocessing before it can be used as a fuel. Unfortunately, in an accident involving failure of the cooling system, the reactor core could melt since plutonium's melting point is only 640°C. Another potential problem is the violent explosion that would occur if liquid sodium were to come into contact with water or air.

Alternatively, the **thermal breeder reactor** converts ^{232}Th by neutron absorption and β decay into ^{233}U, which can also be used as fuel in nuclear reactors since it will undergo fission; India has experimented with this design since it has substantial natural reserves of thorium.

Because there have been operational problems with the very technologically sophisticated fast breeder reactors, the programs were abandoned in the United States, the United Kingdom, Germany, and France. Russia and Japan still operate demonstrator breeder reactors, and both India and China are building demonstration units. Overall, the future of breeder reactors is in doubt.

The radiation from plutonium is weak and becomes a concern only if the material is ingested or inhaled, since its radiation (α particles) cannot pass through dead layers of skin or clothing or indeed through even a few centimeters of air. Elemental, metallic plutonium cannot be easily absorbed by the body. However, when exposed to air, plutonium forms the oxide PuO_2, a powdery dust that disperses readily and can be inhaled. Even microscopic amounts of plutonium oxide lodged in the lungs can induce lung cancer.

Two methods have been proposed to dispose of excess plutonium:

- Mix it with other highly radioactive liquid waste and then **vitrify** the mixture into durable glass logs that would be buried far underground in metal canisters, as will be discussed. **Vitrification** chemically bonds liquid waste into a stable and durable borosilicate glass in which fission product oxides make up about 20% of the mass.

- Convert it to *plutonium dioxide*, PuO_2, and mix it with *uranium oxide* to produce a **mixed oxide fuel,** MOX (containing a few percent plutonium), that could be used in existing nuclear power plants. Indeed, some MOX is now employed in reactors in France, Germany, and Switzerland and in the future will probably be used in the United States, Canada, Belgium, and Japan. However, producing MOX fuel is much more expensive at present than is producing low-enriched uranium, so the incentive to use it is not economic. The issue of using mixed oxide fuel is very controversial in Great Britain, where one-third of the world's nonmilitary plutonium is presently stockpiled.

Nuclear Waste

Although nuclear energy has been used to generate electricity for many decades, there is still no consensus among scientists and policymakers concerning the best procedure for the long-term storage of the radioactive wastes generated by these plants. Initially, spent fuel rods are simply stored above ground—often under cooled water—for several years or decades until the level of radioactivity has been reduced and the rods have therefore cooled somewhat. At this stage, the rods can be transferred to dry storage, e.g., in concrete canisters. If the fuel rods are reprocessed to remove the uranium and plutonium, the remaining highly radioactive liquid waste is subsequently solidified by heating to dehydrate it and decompose the nitrate salts. The resulting powder, composed mainly of oxides of the fission product metals, is then incorporated into borosilicate glass (at about the 20% level) for storage in metal cylinders.

Whether or not the plutonium is removed, most nuclear waste disposal plans assume that the solid material would be encapsulated and immobilized in a glass or ceramic form and then buried far below the Earth's surface. The container for this ultimate disposal would likely be made of a metal, such as *copper* or *titanium*, that is highly resistant to corrosion. The canisters are designed to last for at least several hundred years before leakage could occur; by that time the level of radioactivity would have declined substantially, though almost all of the plutonium would remain. In Sweden, canisters are being designed to last for 100,000 years, at which time the level of radioactivity inside would be no greater than that of uranium ore.

Canisters will ultimately be buried in vaults 300–1000 m below the surface. The geological features of the burial sites should include high stability (from disruption by earthquakes or volcanic activity) and low permeability to assure minimal interactions with groundwater and the biosphere. Deep geological disposal is the only method by which safety requirements can be met without burdening future generations with monitoring and management responsibilities. However, some governments do not accept the idea of *permanent* disposal of such wastes. They want to be able to eventually recover the plutonium from spent fuel rods if it is needed in the future for nuclear power.

Originally, the U.S. plan for the storage of high-level nuclear waste was to use about 10,000 stainless-steel containers, each weighing several tonnes. However, a nickel–chromium–molybdenum alloy of steel (C-22) is now proposed for this role since it is more resistant to corrosion. The United States plans to use a depository to be created 300 m below the surface at Yucca Mountain, Nevada. In contrast to the plans of other countries, the depository will be in the unsaturated zone (see Chapter 14), 300 m above the water table, where air is still present in the soil. Conditions in the unsaturated zone are oxidizing. The original idea was to have a dry repository, since water would have been the main agent for release and transport of radioactive nuclei. However, some evidence now exists that there is rapid transport of water through the area even at this depth. Consequently, a so-called *drip shield* of titanium would eventually be placed over each canister to keep water away from it. Other countries, including Canada, plan to use deeper sites, which will be in the saturated zone so that ambient conditions will be reducing.

After the engineered barriers have eventually failed, the release of radioactive nuclei will depend on the chemical durability of the fuel. Since the uranium in the spent nuclear fuel occurs as U(IV) in the form of UO_2, even small amounts of moisture will result in its oxidation to U(VI), as UO_2^{2+}, which is quite soluble and mobile, at a much faster rate than would occur under reducing conditions. The presence of Fe^{2+}, resulting from the rusting of the steel canister, would reverse the oxidation and precipitate the uranium oxide of any U(VI) that came into contact with it.

As of the mid-2000s, no country had as yet implemented permanent geological storage for nuclear waste, all such schemes having been delayed by technical or political setbacks. As a consequence, about a quarter of a million tonnes of spent fuel rods are currently stored in water pools. In some sites in the United States—home to 70,000 tonnes of waste—the rods are so crowded together that neutron-absorbing boron panels must be inserted between them in order to prevent chain reactions from occurring.

Perhaps the most advanced research on nuclear waste disposal is under way in Sweden, where large iron canisters coated with copper have been buried in a chamber half a kilometer underground, covered with clay, and the chamber sealed off with concrete. Electric heaters are used to mimic the effects of radioactive decay. Sensors are used to monitor the temperature and water movement in the chamber walls.

Fusion Reactors

The optimum energetic stability per nuclear particle (protons and neutrons) occurs for nuclei of intermediate size, such as iron. That is why the fission of a heavy nucleus into two fragments of intermediate size releases energy. Similarly, the fusion of two very light nuclei to produce a heavier one also releases substantial quantities of energy. Indeed, fusion reactions are the source of the energy in stars, including our own Sun, and in hydrogen bombs.

Unfortunately, fusion reactions all have extremely large activation energies due to the huge electrostatic repulsion that exists between the positively charged nuclei when they are brought very close together, which must happen before fusion can occur. Consequently, it is difficult to initiate and sustain a controlled fusion reaction that provides more energy than it consumes.

The fusion reactions that have the greatest potential as producers of useful commercial energy involve the nuclei of the heavier isotopes of hydrogen, namely deuterium, ^{2}H, and tritium, ^{3}H. Note that two different sets of products can be produced from these sample reactions:

$$\underset{\text{deuterium}}{^2_1\text{H}} + \underset{\text{deuterium}}{^2_1\text{H}} \Bigg\langle \begin{array}{l} \nearrow\ ^3_2\text{He} + \ ^1_0\text{n} \\ \qquad\quad \text{or} \\ \searrow\ ^3_1\text{H} + \ ^1_1\text{H} \end{array}$$

The energy that is released when one of these reactions occurs is about 4×10^8 kJ/mol, which is about 1 million times the energy produced in a typical exothermic *chemical* reaction. An abundant supply of deuterium is available, since it is a nonradioactive, naturally occurring isotope (constituting 0.015% of hydrogen) and thus is a natural component of all water.

A somewhat lower activation energy is required for the reaction of deuterium with tritium:

$$\underset{\text{deuterium}}{^2_1\text{H}} + \underset{\text{tritium}}{^3_1\text{H}} \rightarrow\ ^4_2\text{He} + \ ^1_0\text{n}$$

However, because tritium is a radioactive element (a β emitter) with a short half-life (12 years), it is not a significant component of naturally occurring hydrogen and would have to be synthesized by the fission of the relatively scarce element *lithium*.

The environmental consequences of generating electrical power from fusion reactors should be less serious than those associated with fission reactors. The only radioactive waste produced *directly* in quantity would be tritium, although the neutrons emitted in the process could produce radioactive substances when they are absorbed by other atoms. Although the β particle that tritium emits is not sufficiently energetic to penetrate the outer layer of human skin, tritium is nevertheless dangerous since biological systems incorporate it as readily as they do normal hydrogen (^{1}H or ^{2}H)—by inhalation, absorption through the skin, or ingestion of water or food. Currently, tritium in drinking water (some of which results from artificial sources) constitutes the source of about 3% of our exposure to radioactivity.

About 80% of the energy emitted in the deuterium–tritium reaction is associated with the neutrons. The energy, captured by using neutron-heated coolants, would be used to create superheated steam to drive turbines. Unfortunately, the intense neutron bombardment will cause severe degradation of the structure used to confine the fusion reactants and will generate large amounts of radioactive nuclei. These problems would be largely overcome if so-called *advanced fuels* were to be used as reactants. Thus the fusion reaction of deuterium with helium-3 ($^3_2$He) releases only a few percent of its reaction energy as neutrons (the products of the dominant process being protons and ^{4}He nuclei), and that of two ^{3}He nuclei (to produce two protons and ^{4}He) is essentially neutron-free. These processes could operate with much higher energy conversion efficiencies but require even higher initiation temperatures and confinement conditions. The other serious problem is the lack of ^{3}He on Earth: The Moon is the best source for this material!

PROBLEM 9-5

Write and balance the two fusion reactions involving ^{3}He mentioned above.

In 2006, a consortium consisting of the European Union, the United States, China, India, Russia, Japan, and South Korea agreed to finance the ITER (International Thermonuclear Experimental Reactor) project, the world's first nuclear fusion reactor, to be built by about 2016 in Provence, France, at a cost of $13 billion. The reactor will not produce usable energy, but it is being constructed to demonstrate that it is possible to build a reactor that will generate more power than it consumes. The reactor design is based on the *tokamak*, a doughnut-shaped machine having a series of overlapping magnetic fields that can hold the hot plasma within reactor walls. A demonstration power plant, which could use the excess heat to boil water and turn turbines to generate electricity, is slated for about 2040; it will use the experience gained in the ITER project in dealing with the helium and tritium generated by the fusion reactions. Hopefully, the problem of radioactive reactor materials generated by the neutrons emitted by the fusion reaction will also have been overcome by that time.

The Energy Released in Nuclear Processes

The energy, E, released in fission and fusion processes comes from the conversion of a tiny fraction, m, of the masses of the atoms and other particles involved. According to Einstein's famous equation, the energy released is

$$E = mc^2$$

where c is the speed of light.

For example, in the conversion of one mole of deuterium atoms (mass 2.0140 g) and one mole of tritium atoms (mass 3.01605 g) into one mole of ^{4}He atoms (mass 4.00260 g) and one mole of neutrons (mass 1.008665 g), a total of 0.0188 g of matter is lost by conversion to energy. Since $m = 0.0188 \times 10^{-3}$ kg and $c = 2.99792 \times 10^8$ m/s, the energy released is

$$
\begin{aligned}
E = mc^2 &= (0.0188 \times 10^{-3} \text{ kg}) \times (2.99792 \times 10^8 \text{ m/s})^2 \\
&= 1.69 \times 10^{12} \text{ kg m}^2/\text{s}^2 \\
&= 1.69 \times 10^{12} \text{ J}
\end{aligned}
$$

This energy, 1690 million kilojoules per mole, is about 7 million times the energy released when the same quantity of hydrogen is burned in oxygen to produce water.

PROBLEM 9-6

Calculate the amount of energy released in the fission process described earlier in the chapter in which uranium-235 and a neutron are fissioned to barium-142, krypton-91, and 3 neutrons. Assume 1 mole of ^{235}U reacts and that the atomic masses of ^{235}U, ^{142}Ba, ^{91}Kr, and a neutron are, respectively, 235.044, 141.926, 91.923, and 1.008665. Is the energy released much greater than, much less than, or about the same per nucleus as that for the fusion reaction between deuterium and tritium?

Review Questions

1. What is the particulate nature of the radioactive emissions α and β particles? What is a γ ray?

2. Why are α particles dangerous to health only if ingested or inhaled?

3. What is meant by the terms *rad* and *rem*?

4. Explain the origin of radon gas in buildings.

5. Explain what is meant by the term *daughters of radon*. Why they are more dangerous to health than radon itself?

6. What is *depleted uranium*? Is it radioactive at all?

7. Explain what is meant by a *dirty bomb*.

8. Define *fission* and write the reaction in which a ^{235}U nucleus is fissioned into typical products.

9. What are the functions in a fission power reactor of **(a)** the moderator and **(b)** the coolant?

10. Explain why the spent fuel rods from fission reactors are more radioactive than the initial fuel.

11. Why would radioactive strontium and cesium be particularly harmful to human health?

12. Write the nuclear reaction that produces plutonium from ^{238}U in a fission reactor.

13. Describe why the mining of uranium ore often pollutes the local environment.

14. What is a *breeder reactor*? Why is it useful to breed fissionable fuel? What is meant by *reprocessing* and why is it done?

15. Write the nuclear reaction that produces ^{233}U from ^{232}Th following the absorption of a neutron.

16. Describe the two main methods that have been proposed to dispose of excess plutonium.

17. Define *fusion* and give two examples of fusion processes (i.e., reactions) that may be used in power reactors of the future.

18. Describe the nature of any radioactive by-products of the operation of fusion reactors. What damage could the neutrons do?

Additional Problems

1. Another possible radioactive decay mechanism involves the emission of a positron, a particle with the same mass as an electron but the opposite (i.e., positive) charge. The net nuclear effect of positron emission is the change of a proton into a neutron. Explain how the result of positron emission differs from β emission in terms of the periodic table. Deduce the symbol for a positron to be used in balanced nuclear equations, by analogy with the symbol $_{-1}^{0}e$ used for a β particle (the symbol e for an electron is also used here to represent a positron). Predict the decay products and write the balanced equation for the positron decay of the radioactive isotopes $_{11}^{22}Na$ and $_{7}^{13}N$.

2. Tablets of KI were distributed to people living around Chernobyl a week after the nuclear reactor explosion occurred, to help flush the radioactive ^{131}I isotope out of their bodies. What percentage of the ^{131}I released by the explosion would have remained by that time? How long would it have taken for 99% of the released ^{131}I to undergo

radioactive decay? Use the kinetics equation for the first-order decay of a species A as a function of time: $[A] = [A]_o\, e^{-kt}$, where k is the first-order rate constant for the decay. (The rate constant can be obtained from the value of the half-life given in the chapter: $t_{1/2} = 0.693/k$).

3. One physical property that differs for $^{235}UF_6$ and $^{238}UF_6$ is the rate of effusion of these gases through porous membranes. The lighter $^{235}UF_6$ will effuse faster, resulting in enrichment of this desired isotope after passing through the membrane. The effusion rate of gases can be described by Graham's law, which states that the effusion rate of a particle is inversely proportional to the square of its mass. Use this law to determine relative effusion rates of $^{235}UF_6$ and $^{238}UF_6$. Based on your result, explain why it is necessary in practice to perform this process through hundreds of membrane barriers to achieve useful enrichment.

Further Readings

1. J. H. Lubin et al., "Lung Cancer in Radon-Exposed Miners and Estimation of Risk from Indoor Exposure," *Journal of the National Cancer Institute* 87 (1995): 817.

2. D. Williams, "Return to the Inferno: Chernobyl After 20 Years," *Science* 312 (2006): 180.

3. D. Butler, "Energy: Nuclear Power's New Dawn," *Nature* 429 (2004): 238; E. Marris, "Nuclear Reincarnation," *Nature* 441 (2006): 796.

4. K. Becker, "Residential Radon and the LNT Hypothesis," *International Congress Series* 1225 (2002): 259.

5. D. Krewski et al., "Residential Radon and the Risk of Lung Cancer," *Epidemiology* 16 (2005): 137.

6. N. Zevos, "Radioactivity, Radiation, and the Chemistry of Nuclear Waste," *Journal of Chemical Education* 79 (2002): 692.

7. R. C. Ewing, "Less Geology in the Geological Disposal of Nuclear Waste," *Science* 286 (1999): 415.

8. K. D. Crowley and J. F. Ahearne, "Managing the Environmental Legacy of U.S. Nuclear-Weapons Production," *American Scientist* 90 (2002): 514.

9. G. Brumfiel, "Just Around the Corner" [fusion power], *Nature* 436 (2005): 318.

Websites of Interest

Log on to www.whfreeman.com/envchem4/ and click on Chapter 9.

Environmental Instrumental Analysis II	Instrumental Determination of Atmospheric Methane

In the preceding chapters, we have seen that methane is an important gas, not only in chemical reactions in the stratosphere and at ground level, but also in producing global warming. In this box, the quantification of methane in environmental air samples is described.

The determination of methane concentration in the atmosphere can be accomplished by return of atmospheric gas samples to the laboratory for analysis or by measurement using more rugged field instrumentation for direct analysis where the sample is taken— e.g., in a rain forest, on a research vessel at sea, or from an airplane. Both methods usually rely on **gas chromatography** (GC) as a means of separating the complex mixtures of atmospheric components and a very sensitive instrument called the **flame ionization detector** (FID) to actually measure each of those separated analytes.

The power of gas chromatography stems from its ability to separate individual components of a mixture injected into the chromatographic column and to use the FID to identify the analytes as they exit the column one by one. The GC separation process occurs as the gaseous compounds, under the influence of the column's temperature and chromatographic surface and a flowing gas called the carrier gas or mobile phase, undergo a series of nondestructive interactions with the chromatographic surface as they pass through the column. Since each compound to be separated—methane and other atmospheric components in this case— interacts to a slightly different extent with the column's chromatographic surface, the result is a different transit time through the column for each compound and, therefore, an individual *exit time* or *retention time*.

High-resolution capillary GC columns, designed to separate literally hundreds of compounds from a single mixture, have a very small inside diameter: less than 0.53 mm to as small as 0.05 mm. Many different internal chromatographic surfaces are available; each is designed to separate different families of analytes. For the paraffin family, of which methane is the first member, nonpolar chromatographic surfaces or phases are used. In a sample of air collected near a rain forest slash-and-burn site, many other hydrocarbons are routinely detected, e.g., isoprene (C_5H_8), ethane, propane, and β-pinene. The table below details the concentration of these compounds 30 meters above the ground in the biosphere near Manaus, Brazil, in July and August of 1985. Tethered balloons were raised and lowered to predetermined heights and a radio-controlled valve system was used to collect samples of biospheric gases in Teflon bags. These gas samples were transferred to stainless steel canisters and then returned for GC/FID analysis to the National Center for Atmospheric Research (Zimmerman et al., 1988).

Compound	Concentration (ppb) at 30 m above ground
Methane	1657
Ethane	1.17
Isoprene	2.56
β-pinene	0.03

Carbon dioxide sampled from the atmosphere can also be determined very sensitively with this method if the carbon dioxide is reduced to methane before FID analysis. This is easily accomplished by flowing the sample across a hot metal catalyst such as nickel.

The GC/FID system shown in the diagram below contains three basic parts: the injector, chromatographic column, and detector. The chromatographic column is plumbed into the GC's injector, where the mixture is initially introduced into the column. The column—whose temperature is controlled by a small oven—terminates in the base of the FID, where each exiting analyte is detected.

The FID response to hydrocarbons is based on the detection of carbon-containing ions created in the combustion processes of the FID's flame, which is created by burning a hydrogen/air mixture. This extremely sensitive system can detect CH_4 in the picogram (10^{-12} g) range, which translates into parts per billion by volume in atmospheric gas samples.

The result of all chromatographic methods is, in one form or another, a graphical representation of the data called a **chromatogram.** This is most often a plot of retention time versus detector signal. Careful analysis of the chromatogram yields the

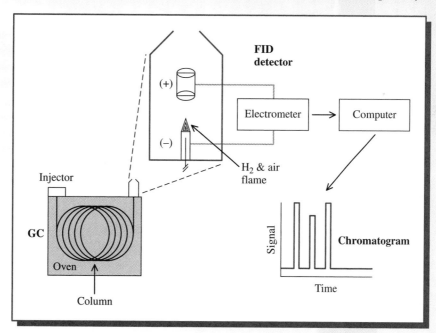

(continued on p. 400)

Instrumental Determination
of Atmospheric Methane *(continued)*

identity, based on known standards, and quantity of each of the components in the original atmospheric mixture. The figure below shows a schematic of five stages in the chromatographic process, with the injector—where the sample is introduced—at the top of the column and the detector at the bottom. In reality, the GC column is 10 m or longer and is curled up inside the GC oven. A chromatogram "in

progress" for each of the stages is presented in the lower panels.

Methane production and emission from biospheric sources is an important field of study because of the importance of methane in the enhanced greenhouse effect. As the inset in Figure 6-8 shows, atmospheric carbon dioxide concentration varies in a seasonal "breathing pattern," with a decrease in atmospheric

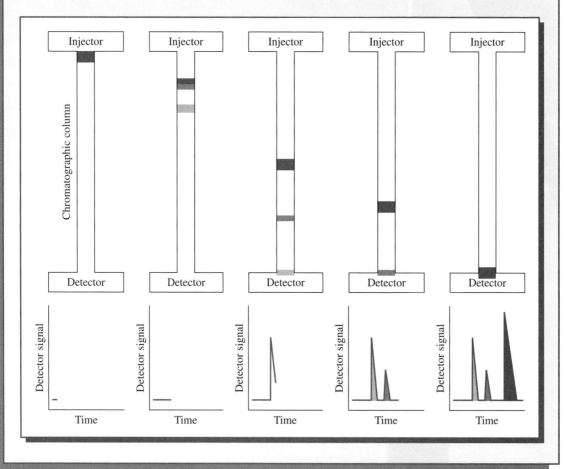

CO_2 in the spring as it is taken up by plants and a CO_2 increase in the fall as it is released from decaying organic matter. Since CH_4 is also produced and consumed in the biosphere, scientists are interested in following its production and consumption in various biomes such as lakes or permafrost. Recent work in Asia using this GC/FID method has shown that the methane flux from a subtropical lake in southeastern China varies greatly with the season, as does that of CO_2 (Xing et al., 2005). But while CO_2 flux is positive (to the atmosphere) in autumn and winter and negative in spring and summer, methane flux is positive year-round in this setting. Using a small gas-collection chamber at the surface of this lake (lake area: 27.9 km^2), researchers found an average flux of 23 mg CH_4 per m^2 per day. This works out to an emission of approximately 2.4 H 10^5 kg CH_4 per year from that lake. And overall, this investigation showed that annually the lake contributed 7.5×10^5 kg carbon to the atmosphere, taking into account methane and negative and positive CO_2 fluxes. Interestingly, workers performing these same kinds of studies with soils taken from the Arctic tundra and examined in the lab found that normally moist soils in central Siberia are a net atmospheric sink for methane year-round (Rodionow et al., 2005).

References: Chemistry-Based Animations, 2006: http://www.shsu.edu/~chm_tgc/sounds/sound.html.

A. Rodionow, H. Flessa, O. Kazansky, and G. Guggenberger, "Organic Matter Composition and Potential Trace Gas Production of Permafrost Soils in the Forest Tundra in Northern Siberia," *Geoderma* (2006).

Y. Xing, P. Xie, H. Yang, L. Ni, Y. Wang, and K. Rong, "Methane and Carbon Dioxide Fluxes from a Shallow Hypereutrophic Subtropical Lake in China," *Atmospheric Environment* 39 (2005): 5532–5540.

P. R. Zimmerman, J. P. Greenberg, and C. E. Westberg, "Measurements of Atmospheric Hydrocarbons and Biogenic Emission Fluxes in the Amazon Boundary Layer," *Journal of Geophysical Research* 93(D2) (1988): 1407–1416.

▶ Humanity faces a choice between two futures: doing nothing to curb emissions (which poses huge climate risks) and bringing them under control (which has costs but also benefits).

A Plan to Keep *Carbon* in Check

Getting a grip on greenhouse gases is daunting but doable. The technologies already exist. But there is no time to lose

BY ROBERT H. SOCOLOW AND STEPHEN W. PACALA

OVERVIEW

❋ Humanity can emit only so much carbon dioxide into the atmosphere before the climate enters a state unknown in recent geologic history and goes haywire. Climate scientists typically see the risks growing rapidly as CO_2 levels approach a doubling of their pre-18th-century value.

❋ To make the problem manageable, the required reduction in emissions can be broken down into "wedges"—an incremental reduction of a size that matches available technology.

Retreating glaciers, stronger hurricanes, hotter summers, thinner polar bears: the ominous harbingers of global warming are driving companies and governments to work toward an unprecedented change in the historical pattern of fossil-fuel use. Faster and faster, year after year for two centuries, human beings have been transferring carbon to the atmosphere from below the surface of the earth. Today the world's coal, oil and natural gas industries dig up and pump out about seven billion tons of carbon a year, and society burns nearly all of it, releasing carbon dioxide (CO_2). Ever more people are convinced that prudence dictates a reversal of the present course of rising CO_2 emissions.

The boundary separating the truly dangerous consequences of emissions from the merely unwise is probably located near (but below) a doubling of the concentration of CO_2 that was in the atmosphere in the 18th century, before the Industrial Revolution began. Every increase in concentration carries new risks, but avoiding that danger zone would reduce the likelihood of triggering major, irreversible climate changes, such as the disappearance of the Greenland ice cap. Two years ago the two of us provided a simple framework to relate future CO_2 emissions to this goal.

We contrasted two 50-year futures. In one future, the emissions rate continues to grow at the pace of the past 30 years for the next 50 years, reaching 14 billion tons of carbon a year in 2056. (Higher or lower rates are, of course, plausible.) At that point, a tripling of preindustrial carbon concentrations would be very difficult to avoid, even with concerted efforts to decarbonize the world's energy systems

Robert H. Socolow and Stephen W. Parcala, "A Plan to Keep Carbon in Check," *Scientific American*, September 2006, 50–57.

MANAGING THE CLIMATE PROBLEM

At the present rate of growth, emissions of carbon dioxide will double by 2056 (*below left*). Even if the world then takes action to level them off, the atmospheric concentration of the gas will be headed above 560 parts per million, double the preindustrial value (*below right*)—a level widely regarded as capable of triggering severe climate changes. But if the world flattens out emissions beginning now and later ramps them down, it should be able to keep concentration substantially below 560 ppm.

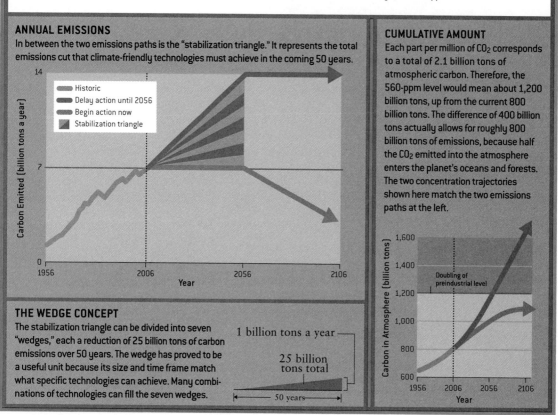

ANNUAL EMISSIONS

In between the two emissions paths is the "stabilization triangle." It represents the total emissions cut that climate-friendly technologies must achieve in the coming 50 years.

Legend:
- Historic
- Delay action until 2056
- Begin action now
- Stabilization triangle

Carbon Emitted (billion tons a year)

14 / 7 / 0

1956 · 2006 · 2056 · 2106

Year

THE WEDGE CONCEPT

The stabilization triangle can be divided into seven "wedges," each a reduction of 25 billion tons of carbon emissions over 50 years. The wedge has proved to be a useful unit because its size and time frame match what specific technologies can achieve. Many combinations of technologies can fill the seven wedges.

1 billion tons a year
25 billion tons total
← 50 years →

CUMULATIVE AMOUNT

Each part per million of CO_2 corresponds to a total of 2.1 billion tons of atmospheric carbon. Therefore, the 560-ppm level would mean about 1,200 billion tons, up from the current 800 billion tons. The difference of 400 billion tons actually allows for roughly 800 billion tons of emissions, because half the CO_2 emitted into the atmosphere enters the planet's oceans and forests. The two concentration trajectories shown here match the two emissions paths at the left.

Carbon in Atmosphere (billion tons)

1,600 / 1,400 / 1,200 / 1,000 / 800 / 600

Doubling of preindustrial level

1956 · 2006 · 2056 · 2106

Year

over the following 100 years. In the other future, emissions are frozen at the present value of seven billion tons a year for the next 50 years and then reduced by about half over the following 50 years. In this way, a doubling of CO_2 levels can be avoided. The difference between these 50-year emission paths—one ramping up and one flattening out—we called the stabilization triangle [*see box on this page*].

To hold global emissions constant while the world's economy continues to grow is a daunting task. Over the past 30 years, as the gross world product of goods and services grew at close to 3 percent a year on average, carbon emissions rose half as fast. Thus, the ratio of emissions to dollars of gross world product, known as the carbon intensity of the global economy, fell about 1.5 percent a year. For global emissions to be the same

in 2056 as today, the carbon intensity will need to fall not half as fast but fully as fast as the global economy grows.

Two long-term trends are certain to continue and will help. First, as societies get richer, the services sector—education, health, leisure, banking and so on—grows in importance relative to energy-intensive activities, such as steel production. All by itself, this shift lowers the carbon intensity of an economy.

Second, deeply ingrained in the patterns of technology evolution is the substitution of cleverness for energy. Hundreds of power plants are not needed today because the world has invested in much more efficient refrigerators, air conditioners and motors than were available two decades ago. Hundreds of oil and gas fields have been developed more slowly because aircraft engines consume less fuel and the windows in gas-heated homes leak less heat.

The task of holding global emissions constant would be out of reach, were it not for the fact that all the driving and flying in 2056 will be in vehicles not yet designed, most of the buildings that will be around then are not yet built, the locations of many of the communities that will contain these buildings and determine their inhabitants' commuting patterns have not yet been chosen, and utility owners are only now beginning to plan for the power plants that will be needed to light up those communities. Today's notoriously inefficient energy system can be replaced if the world gives unprecedented attention to energy efficiency. Dramatic changes are plausible over the next 50 years because so much of the energy canvas is still blank.

To make the task of reducing emissions vivid, we sliced the stabilization triangle into seven equal pieces, or "wedges," each representing one billion tons a year of averted emissions 50 years from now (starting from zero today). For example, a car driven 10,000 miles a year with a fuel efficiency of 30 miles per gallon (mpg) emits close to one ton of carbon annually. Transport experts predict that two billion cars will be zipping along the world's roads in 2056, each driven an average of 10,000 miles a year. If their average fuel efficiency were 30 mpg, their tailpipes would spew two billion

> Holding carbon dioxide emissions constant for 50 years, without choking off economic growth, is within our grasp.

tons of carbon that year. At 60 mpg, they would give off a billion tons. The latter scenario would therefore yield one wedge.

Wedges

IN OUR FRAMEWORK, you are allowed to count as wedges only those differences in two 2056 worlds that result from deliberate carbon policy. The current pace of emissions growth already includes some steady reduction in carbon intensity. The goal is to reduce it even more. For instance, those who believe that cars will average 60 mpg in 2056 even in a world that pays no attention to carbon cannot count this improvement as a wedge, because it is already implicit in the baseline projection.

Moreover, you are allowed to count only strategies that involve the scaling up of technologies already commercialized somewhere in the world. You are not allowed to count pie in the sky. Our goal in developing the wedge framework was to be pragmatic and realistic—to propose engineering our way out of the problem and not waiting for the cavalry to come over the hill. We argued that even with these two counting rules, the world can fill all seven wedges, and in several different ways [see box on page 407]. Individual countries—operating within a framework of international cooperation—will decide which wedges to pursue, depending on their institutional and economic capacities, natural resource endowments and political predilections.

To be sure, achieving nearly every one of the wedges requires new science and engineering to squeeze down costs and address the problems that inevitably accompany widespread deployment of new technologies. But holding CO_2 emissions in 2056 to their present rate, without choking off economic growth, is a desirable outcome within our grasp.

Ending the era of conventional coal-fired power plants is at the very top of the decarbonization agenda. Coal has become more competitive as a source of power and fuel because of energy security concerns and because of an increase in the cost of oil and gas. That is a problem because a coal power plant burns twice as much carbon per unit of electricity as a natural gas plant. In the absence of a concern about carbon, the world's coal utilities could build a few thousand large (1,000-megawatt) conventional coal plants in the next 50 years. Seven hundred such plants emit one wedge's worth of carbon. Therefore, the world could take some big steps toward the target of freezing emissions by not building those plants. The time to start is now. Facilities built in this decade could easily be around in 2056.

Efficiency in electricity use is the most obvious substitute for coal. Of the 14 billion tons of carbon emissions projected for 2056, perhaps six billion will come from producing power, mostly from coal. Residential and commercial buildings account for 60 percent of global electricity demand today (70 percent in the U.S.) and will consume most of the new power.

So cutting buildings' electricity use in half—by equipping them with superefficient lighting and appliances—could lead to two wedges. Another wedge would be achieved if industry finds additional ways to use electricity more efficiently.

Decarbonizing the Supply

EVEN AFTER energy-efficient technology has penetrated deeply, the world will still need power plants. They can be coal plants but they will need to be carbon-smart ones that capture the CO_2 and pump it into the ground [see "Can We Bury Global Warming?" by Robert H. Socolow; SCIENTIFIC AMERICAN, July 2005]. Today's high oil prices are lowering the cost of the transition to this technology, because captured CO_2 can often be sold to an oil company that injects it into oil fields to squeeze out more oil; thus, the higher the price of oil, the more valuable the captured CO_2. To achieve one wedge, utilities need to equip 800 large coal plants to capture and store nearly all the CO_2 otherwise emitted. Even

in a carbon-constrained world, coal mining and coal power can stay in business, thanks to carbon capture and storage.

The large natural gas power plants operating in 2056 could capture and store their CO_2, too, perhaps accounting for yet another wedge. Renewable and nuclear energy can contribute as well. Renewable power can be produced from sunlight directly, either to energize photovoltaic cells or, using focusing mirrors, to heat a fluid and drive a turbine. Or the route can be indirect, harnessing hydropower and wind power, both of which rely on sun-driven weather patterns. The intermittency of renewable power does not diminish its capacity to contribute wedges; even if coal and natural gas plants provide the backup power, they run only part-time (in tandem with energy storage) and use less carbon than if they ran all year. Not strictly renewable, but also usually included in the family, is geothermal energy, obtained by mining the heat in the earth's interior. Any of these sources, scaled up from its current contribution, could produce a wedge. One must be careful not to double-count the possibilities; the same coal plant can be left unbuilt only once.

Nuclear power is probably the most controversial of all the wedge strategies. If the fleet of nuclear power plants were to expand by a factor of five by 2056, displacing conventional coal plants, it would provide two wedges. If the current

THE AUTHORS

ROBERT H. SOCOLOW and STEPHEN W. PACALA lead the Carbon Mitigation Initiative at Princeton University, where Socolow is a mechanical engineering professor and Pacala an ecology professor. The initiative is funded by BP and Ford. Socolow specializes in energy-efficient technology, global carbon management and carbon sequestration. He was co-editor (with John Harte) of *Patient Earth,* published in 1971 as one of the first college-level presentations of environmental studies. He is the recipient of the 2003 Leo Szilard Lectureship Award from the American Physical Society. Pacala investigates the interaction of the biosphere, atmosphere and hydrosphere on global scales, with an emphasis on the carbon cycle. He is director of the Princeton Environmental Institute.

15 WAYS TO MAKE A WEDGE

An overall carbon strategy for the next half a century produces seven wedges' worth of emissions reductions. Here are 15 technologies from which those seven can be chosen (taking care to avoid double-counting). Each of these measures, when phased in over 50 years, prevents the release of 25 billion tons of carbon. Leaving one wedge blank symbolizes that this list is by no means exhaustive.

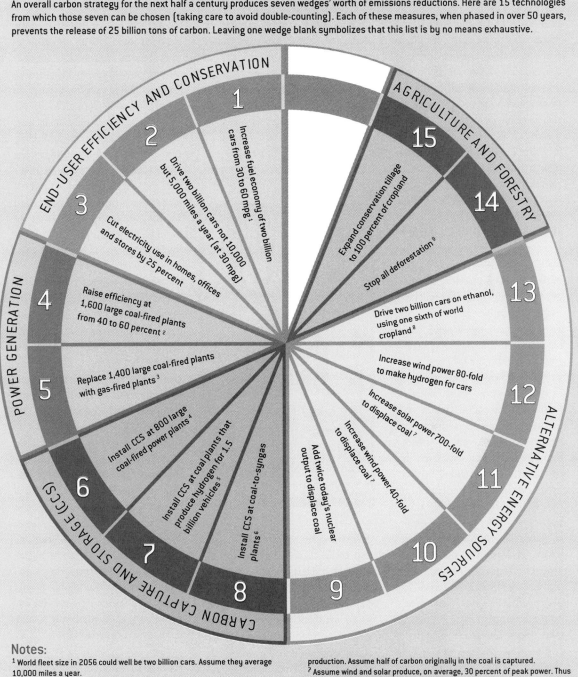

END-USER EFFICIENCY AND CONSERVATION

1. Increase fuel economy of two billion cars from 30 to 60 mpg[1]
2. Drive two billion cars not 10,000 but 5,000 miles a year (at 30 mpg)
3. Cut electricity use in homes, offices and stores by 25 percent

POWER GENERATION

4. Raise efficiency at 1,600 large coal-fired plants from 40 to 60 percent[2]
5. Replace 1,400 large coal-fired plants with gas-fired plants[3]

CARBON CAPTURE AND STORAGE (CCS)

6. Install CCS at 800 large coal-fired power plants[4]
7. Install CCS at coal plants that produce hydrogen for 1.5 billion vehicles[5]
8. Install CCS at coal-to-syngas plants[6]

ALTERNATIVE ENERGY SOURCES

9. Add twice today's nuclear output to displace coal
10. Increase wind power 40-fold to displace coal[7]
11. Increase solar power 700-fold to displace coal[7]
12. Increase wind power 80-fold to make hydrogen for cars

AGRICULTURE AND FORESTRY

13. Drive two billion cars on ethanol, using one sixth of world cropland[8]
14. Stop all deforestation[9]
15. Expand conservation tillage to 100 percent of cropland

Notes:

[1] World fleet size in 2056 could well be two billion cars. Assume they average 10,000 miles a year.

[2] "Large" is one-gigawatt (GW) capacity. Plants run 90 percent of the time.

[3] Here and below, assume coal plants run 90 percent of the time at 50 percent efficiency. Present coal power output is equivalent to 800 such plants.

[4] Assume 90 percent of CO_2 is captured.

[5] Assume a car (10,000 miles a year, 60 miles per gallon equivalent) requires 170 kilograms of hydrogen a year.

[6] Assume 30 million barrels of synfuels a day, about a third of today's total oil production. Assume half of carbon originally in the coal is captured.

[7] Assume wind and solar produce, on average, 30 percent of peak power. Thus replace 2,100 GW of 90-percent-time coal power with 2,100 GW (peak) wind or solar plus 1,400 GW of load-following coal power, for net displacement of 700 GW.

[8] Assume 60-mpg cars, 10,000 miles a year, biomass yield of 15 tons a hectare, and negligible fossil-fuel inputs. World cropland is 1,500 million hectares.

[9] Carbon emissions from deforestation are currently about two billion tons a year. Assume that by 2056 the rate falls by half in the business-as-usual projection and to zero in the flat path.

JANET CHAO

fleet were to be shut down and replaced with modern coal plants without carbon capture and storage, the result would be minus one-half wedge. Whether nuclear power will be scaled up or down will depend on whether governments can find political solutions to waste disposal and on whether plants can run without accidents. (Nuclear plants are mutual hostages: the world's least well-run plant can imperil the future of all the others.) Also critical will be strict rules that prevent civilian nuclear technology from becoming a stimulus for nuclear weapons development. These rules will have to be uniform across all countries, so as to remove the sense of a double standard that has long been a spur to clandestine facilities.

Oil accounted for 43 percent of global carbon emissions from fossil fuels in 2002, while coal accounted for 37 percent; natural gas made up the remainder. More than half the oil was used for transport. So smartening up electricity production alone cannot fill the stabilization triangle; transportation, too, must be decarbonized. As with coal-fired electricity, at least a wedge may be available from each of three complementary options: reduced use, improved efficiency and decarbonized energy sources. People can take fewer unwanted trips (telecommuting instead of vehicle commuting) and pursue the travel they cherish (adventure, family visits) in fuel-efficient vehicles running on low-carbon fuel. The fuel can

be a product of crop residues or dedicated crops, hydrogen made from low-carbon electricity, or low-carbon electricity itself, charging an onboard battery. Sources of the low-carbon electricity could include wind, nuclear power, or coal with capture and storage.

Looming over this task is the prospect that, in the interest of

39 percent
U.S. share of global carbon emissions in 1952

23 percent
U.S. share in 2002

The U.S. share of global emissions can be expected to continue to drop.

energy security, the transport system could become more carbon-intensive. That will happen if transport fuels are derived from coal instead of petroleum. Coal-based synthetic fuels, known as synfuels, provide a way to reduce global demand for oil, lowering its cost and decreasing global dependence on Middle East petroleum. But it is a decidedly climate-unfriendly strategy. A synfuel-

powered car emits the same amount of CO_2 as a gasoline-powered car, but synfuel fabrication from coal spews out far more carbon than does refining gasoline from crude oil—enough to double the emissions per mile of driving. From the perspective of mitigating climate change, it is fortunate that the emissions at a synfuels plant can be captured and stored. If business-as-usual trends did lead to the widespread adoption of synfuel, then capturing CO_2 at synfuels plants might well produce a wedge.

Not all wedges involve new energy technology. If all the farmers in the world practiced no-till agriculture rather than conventional plowing, they would contribute a wedge. Eliminating deforestation would result in two wedges, if the alternative were for deforestation to continue at current rates. Curtailing emissions of methane, which today contribute about half as much to greenhouse warming as CO_2, may provide more than one wedge: needed is a deeper understanding of the anaerobic biological emissions from cattle, rice paddies and irrigated land. Lower birth rates can produce a wedge, too—for example, if they hold the global population in 2056 near eight billion people when it otherwise would have grown to nine billion.

Action Plan

WHAT SET OF POLICIES will yield seven wedges? To be sure, the dramatic changes we anticipate

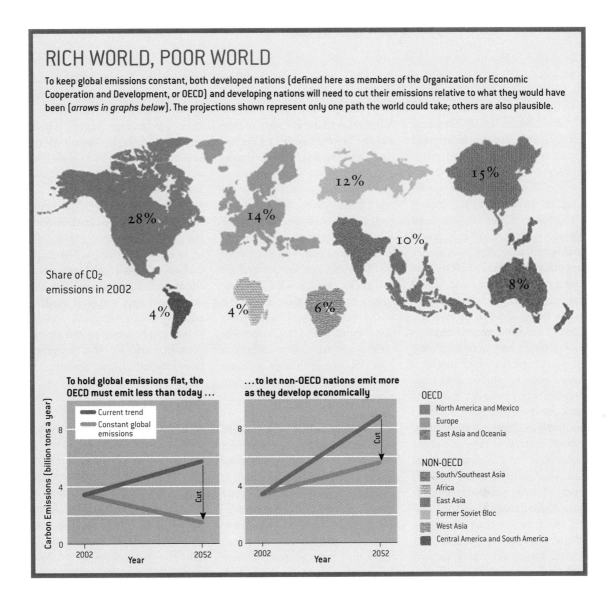

RICH WORLD, POOR WORLD

To keep global emissions constant, both developed nations (defined here as members of the Organization for Economic Cooperation and Development, or OECD) and developing nations will need to cut their emissions relative to what they would have been (*arrows in graphs below*). The projections shown represent only one path the world could take; others are also plausible.

Share of CO_2 emissions in 2002

28% 14% 12% 15% 10% 8% 4% 4% 6%

To hold global emissions flat, the OECD must emit less than today ...

...to let non-OECD nations emit more as they develop economically

Carbon Emissions (billion tons a year)

— Current trend
— Constant global emissions

Cut

Cut

2002 Year 2052

2002 Year 2052

OECD
- North America and Mexico
- Europe
- East Asia and Oceania

NON-OECD
- South/Southeast Asia
- Africa
- East Asia
- Former Soviet Bloc
- West Asia
- Central America and South America

JEN CHRISTIANSEN; SOURCE: "GLOBAL, REGIONAL, AND NATIONAL FOSSIL FUEL CO₂ EMISSIONS," BY G. MARLAND, T. A. BODEN AND R. J. ANDRES, IN *TRENDS: A COMPENDIUM OF DATA ON GLOBAL CHANGE.* CARBON DIOXIDE INFORMATION ANALYSIS CENTER, OAK RIDGE NATIONAL LABORATORY, 2006

in the fossil-fuel system, including routine use of CO_2 capture and storage, will require institutions that reliably communicate a price for present and future carbon emissions. We estimate that the price needed to jumpstart this transition is in the ballpark of $100 to $200 per ton of carbon—the range that would make it cheaper for owners of coal plants to capture and store CO_2 rather than vent it. The price might fall as technologies climb the learning curve. A carbon emissions price of $100 per ton is comparable to the current U.S. production credit for new renewable and nuclear energy relative to coal, and it is about half the current U.S. subsidy of ethanol relative to gasoline. It also was the price of CO_2 emissions in the European Union's emissions trading system for nearly a year, spanning 2005 and 2006. (One ton of carbon is carried in 3.7 tons of carbon dioxide, so this price is also $27 per ton of CO_2.) Based on carbon content, $100 per ton of carbon is $12 per barrel of oil and $60 per ton of coal. It is

25 cents per gallon of gasoline and two cents per kilowatt-hour of electricity from coal.

But a price on CO_2 emissions, on its own, may not be enough. Governments may need to stimulate the commercialization of low-carbon technologies to increase the number of competitive options available in the future. Examples include wind, photovoltaic power and hybrid cars. Also appropriate are policies designed to prevent the construction of long-lived capital facilities that are mismatched to future policy. Utilities, for instance, need to be encouraged to invest in CO_2 capture and storage for new coal power plants, which would be very costly to retrofit later. Still another set of policies can harness the capacity of energy producers to promote efficiency—motivating power utilities to care about the installation and maintenance of efficient appliances, natural gas companies to care about the buildings where their gas is burned, and oil companies to care about the engines that run on their fuel.

To freeze emissions at the current level, if one category of emissions goes up, another must come down. If emissions from natural gas increase, the combined emissions from oil and coal must decrease. If emissions from air travel climb, those from some other economic sector must fall. And if today's poor countries are to emit more, today's richer countries must emit less.

How much less? It is easy to bracket the answer. Currently the industrial nations—the members of the Organization for Economic Cooperation and Development (OECD)—account for almost exactly half the planet's CO_2 emissions, and the developing countries plus the nations formerly part of the Soviet Union account for the other half. In a world of constant total carbon emissions, keeping the OECD's share at 50 percent seems impossible to justify in the face of the enormous pent-up demand for energy in the non-OECD countries, where more than 80 percent of the world's people live. On the other hand, the OECD member states must emit *some* carbon in 2056. Simple arithmetic indicates that to hold global emissions rates steady, non-OECD emissions cannot even double.

One intermediate value results if all OECD countries were to meet the emissions-reduction target for the U.K. that was articulated in 2003 by Prime Minister Tony Blair—namely, a 60 percent reduction by 2050, relative to recent levels. The non-OECD countries could then emit 60 percent more CO_2. On average, by midcentury they would have one half the per capita emissions of the OECD countries. The CO_2 output of every country, rich or poor today, would be well below what it is generally projected to be in the absence of climate policy. In the case of the U.S., it would be about four times less.

Blair's goal would leave the average American emitting twice as much as the world average, as opposed to five times as much today. The U.S. could meet this goal in many ways [*see illustration on facing page*]. These strategies will be followed by most other countries as well. The resultant cross-pollination will lower every country's costs.

Fortunately, the goal of decarbonization does not conflict with the goal of eliminating the world's most extreme poverty. The extra carbon emissions produced when the world's nations accelerate the delivery of electricity and modern cooking fuel to the earth's poorest people can be compensated for by, at most, one fifth of a wedge of emissions reductions elsewhere.

Beyond 2056

THE STABILIZATION triangle deals only with the first 50-year leg of the future. One can imagine a relay race made of 50-year segments, in which the first runner passes a baton to the second in 2056. Intergenerational equity requires that the two runners have roughly equally difficult tasks. It seems to us that the task we have given the second runner (to cut the 2056 emissions rate in half between 2056 and 2106) will not be harder than the task of the first runner (to keep global emissions in 2056 at present levels)—provided that between now and 2056 the world invests in research and development to get ready. A vigorous effort can prepare the revolutionary technologies that will give the second half of the century a running start. Those options could include scrubbing CO_2 directly from the air, carbon storage in minerals, nuclear fusion, nuclear thermal hydrogen, and artificial photosynthesis. Conceivably, one or more of these technologies may arrive in time to help the first runner, although, as we have argued, the world should not count on it.

As we look back from 2056, if global emissions of CO_2 are indeed no larger than today's, what will have been accomplished? The world will have confronted energy production and energy efficiency at the consumer level, in all economic sectors and in economies at all levels of development. Buildings and lights and refrigerators, cars and trucks and planes, will be transformed. Transformed, also, will be the ways we use them.

The world will have a fossil-fuel energy system about as large as today's but one that is infused with modern controls and advanced materials and that is almost unrecognizably cleaner. There will be integrated production of power, fuels and heat; greatly reduced air and water pollution; and extensive carbon capture and storage. Alongside the fossil energy system will be a non-fossil energy system approximately as large. Extensive direct and indirect harvesting of renewable energy will have brought about the revitalization of rural areas and the reclamation of degraded lands. If nuclear power is playing a large role, strong international enforcement mechanisms will

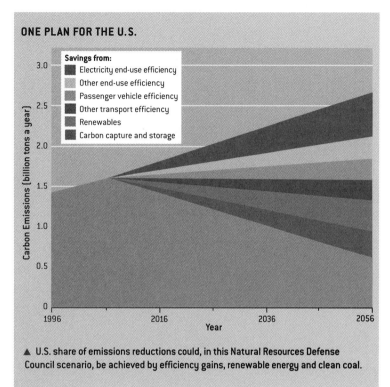

ONE PLAN FOR THE U.S.

Savings from:
- Electricity end-use efficiency
- Other end-use efficiency
- Passenger vehicle efficiency
- Other transport efficiency
- Renewables
- Carbon capture and storage

Carbon Emissions (billion tons a year) — Year: 1996, 2016, 2036, 2056

▲ U.S. share of emissions reductions could, in this Natural Resources Defense Council scenario, be achieved by efficiency gains, renewable energy and clean coal.

have come into being to control the spread of nuclear technology from energy to weapons. Economic growth will have been maintained; the poor and the rich will both be richer. And our descendants will not be forced to exhaust so much

treasure, innovation and energy to ward off rising sea level, heat, hurricanes and drought.

Critically, a planetary consciousness will have grown. Humanity will have learned to address its collective destiny—and to share the planet. ∎SA

MORE TO EXPLORE

Stabilization Wedges: Solving the Climate Problem for the Next 50 Years with Current Technologies. S. Pacala and R. Socolow in *Science*, Vol. 305, pages 968–972; August 13, 2004.

The calculations behind the individual wedges are available at www.princeton.edu/cmi

Energy statistics are available at www.eia.doe.gov, www.iea.org and www.bp.com; carbon emissions data can also be found at cdiac.esd.ornl.gov

JEN CHRISTIANSEN; SOURCE: DANIEL A. LASHOF AND DAVID G. HAWKINS *Natural Resources Defense Council*

PART III

TOXIC ORGANIC COMPOUNDS

Contents of Part III

Environmental Instrumental Analysis III

- Electron Capture Detection of Pesticides

Environmental Instrumental Analysis IV

- Gas Chromatography/Mass Spectrometry (GC/MS)

Scientific American **Feature Article**

- Tackling Malaria

CHAPTER 10

PESTICIDES

In this chapter, the following introductory chemistry topics are used:

- Elementary organic chemistry (as in the Appendix in this book)
- Concepts of vapor pressure; solubility; half-life; chemical versus physical change; enzymes; acids and bases
- Molarity

Background from previous chapters used in this chapter:

- Photochemical reactions (Chapters 1, 3, 5)
- Adsorption (Chapter 4)

Introduction

The term **synthetic chemical** is used to describe substances that generally do not occur in nature but have been synthesized by chemists from simpler substances. The great majority of commercial synthetic chemicals are organic compounds, and most use petroleum or natural gas as the original source of their carbon.

In Chapters 10, 11, and 12, the environmental consequences of the widespread use of synthetic organic chemicals are discussed. Our emphasis will be on those substances whose toxicity has led to concerns regarding human health, as well as the well-being of lower organisms.

In this chapter, we discuss pesticides, considering the environmental problems associated with their use and some general principles of toxicology that apply to their effects on human health. In Chapters 11 and 12, nonpesticidal organic compounds of environmental concern are described.

Types of Pesticides

Pesticides are substances that kill or otherwise control an unwanted organism. The most common categories of pesticides are listed in Table 10-1. All

TABLE 10-1	Pesticides and Their Targets
Pesticide Type	**Target Organism**
Acaricide	Mites
Algicide	Algae
Avicide	Birds
Bactericide	Bacteria
Disinfectant	Microorganisms
Fungicide	Fungi
Herbicide	Plants
Insecticide	Insects
Larvicide	Insect larvae
Molluscicide	Snails, slugs
Nematicide	Nematodes
Piscicide	Fish
Rodenticide	Rodents
Termiticides	Termites

chemical pesticides share the common property of blocking a vital metabolic process of the organism to which they are toxic. In this chapter, we discuss first **insecticides**—substances that kill insects—and then **herbicides**—compounds that kill plants. Also mentioned in passing are some **fungicides,** substances that are used to control the growth of various types of fungus, especially to protect stored seeds before planting. Collectively, these three categories represent the great bulk of the *1 billion* kilograms of pesticides that are used annually in North America.

About half of the usage of pesticides in North America involves agriculture; worldwide the figure rises to 85%. Almost all commercial food crops are now produced with the use of synthetic insecticides, herbicides, and fungicides, except of course in organic farming, where natural pesticides are used. Indeed, the current ability to produce and harvest large amounts of food on relatively small amounts of land with a relatively small input of human labor has been made possible by the use of pesticides. Currently, the greatest U.S. use of insecticides occurs in the growing of cotton, whereas the majority of herbicides are used in the growing of corn and soybeans. Recently the application of insecticides to cotton has been significantly reduced by the introduction of cotton that has been genetically modified to incorporate resistance to insects, in particular the bollworm. Depending upon the main crops they produce, countries vary in what types of pesticides they consume in large amounts. For example, herbicides account for three-quarters of pesticide use in Malaysia, whereas insecticides are the most widely used pesticides in India and the Philippines, as are fungicides in Colombia.

Some 80–90% of American domestic households contain at least one synthetic pesticide; typical examples are weed killers for the lawn and garden, algae controls for the swimming pool, flea powders for use on pets, and sprays to kill insects such as cockroaches.

Almost since their introduction, synthetic pesticides have been a concern because of the potential impact on human health of eating food contaminated with these chemicals. Half the foods eaten in the United States contain measurable levels of at least one pesticide. For that reason, many have been banned or restricted in their use. Nevertheless, the U.S. National Academy of Science has pointed out that pesticide regulation to date has not paid enough attention to the protection of human health, especially that of infants and children, whose growth and development are at stake. Children, kilogram for kilogram, eat more food than adults and tend to eat more foods (such as apples, grapes, and carrots) with higher pesticide levels than do adults, and their internal organs—including the brain—are still developing and maturing, thereby making them more vulnerable to any negative effects these chemicals may have. In addition, small children play on floors and lawns and put many objects in their mouths, increasing their exposure to pesticides used in homes and yards.

Some scientists dismiss these concerns by emphasizing that living plants themselves manufacture insecticides in order to discourage insects and fungi from consuming them, and consequently we are exposed in our food supply to much higher concentrations of these natural pesticides than to synthetic ones. Natural pesticides are not necessarily less toxic than are synthetic ones, as we shall see.

Traditional Insecticides

Insecticides of one type or another have been used by society for thousands of years. One principal motivation for using insecticides is to control disease: Human deaths due to insect-borne diseases through the ages have greatly exceeded those attributable to warfare. The use of insecticides has greatly reduced the incidence of diseases transmitted by insects and the rodents which bear them: malaria, yellow fever, bubonic plague, sleeping sickness, and, recently, illness from the *West Nile virus* scarcely exhaust the list of these scourges. The other principal motivation for insecticide usage is to prevent insects from attacking food crops; nevertheless, even with extensive use of pesticides, about one-third of the world's total crop yield is destroyed by pests or weeds during growth, harvesting, and storage. People also try to control insects such as the mosquito and the common fly simply because their presence is annoying.

The earliest recorded usage of pesticides was the burning of sulfur to fumigate Greek homes around 1000 B.C. **Fumigants** are pesticides that enter the insect as an inhaled gas. The use of **sulfur dioxide,** SO_2, from the burning of solid *sulfur,* sometimes by incorporating the element in candles, continued at least into the nineteenth century. Sulfur itself, in the form of dusts and sprays, was also used as an insecticide and as a fungicide; it is still employed as a fumigant against *powdery mildew* on plants. Inorganic fluorides, such as *sodium fluoride,* NaF, were used domestically to control ant populations. Both sodium fluoride and *boric acid* were used to kill cockroaches in infested buildings. Various oils, whether derived from petroleum or from living sources such as fish and whales, have been utilized for hundreds of years as insecticides and as *dormant sprays* to kill insect eggs.

The use of *arsenic* and its compounds to control insects dates back to Roman times. It was employed by the Chinese in the sixteenth century and became quite widespread from the late nineteenth century until World War II. Arsenic-containing salts have also been used as stomach poisons, and they kill insects that ingest them. Arsenic compounds, which continued to be heavily used as insecticides into the early 1950s, are discussed in more detail in Chapter 15.

Unfortunately, most inorganic and metal-containing organic pesticides are quite toxic to humans and other mammals, especially at the dosage levels that are required to make them effective as pesticides. Mass poisonings have occurred as a result of the use of some mercury-based fungicides, as will be discussed in Chapter 15. In addition, toxic metals and semimetals, such as the arsenic commonly used in pesticides, are not biodegradable: Once released into the environment, they will remain indefinitely in the water, wildlife, soil, or sediments and may enter the food supply if liberated from these sites.

Organochlorine Insecticides

Organic insecticides developed during and after World War II have largely displaced these inorganic and metal-organic substances. Usually only small amounts of the organic compounds are required to be effective against the target pests, so smaller amounts of chemicals enter the environment. Given a dose of any of these compounds large enough to act as a pesticide, the organic substances are generally much less toxic to humans than are the inorganic and metal-organic pesticides. Pesticide formulations containing only organic compounds were initially thought to be biodegradable, though as we shall see, this has certainly not been found to be true in many cases.

In the 1940s and the 1950s, the chemical industries in North America and western Europe produced large quantities of many new pesticides, especially insecticides. The active ingredients in most of these pesticides were

organochlorines, which are organic compounds that contain chlorine. Many organochlorines share several notable properties:

- stability against decomposition or degradation in the environment;

- very low solubility in water, unless oxygen or nitrogen is also present in the molecules;

- high solubility in hydrocarbon-like environments, such as the fatty material in living matter;

- relatively high toxicity to insects, but low toxicity to humans.

As an example of an organochlorine pesticide, consider **hexachlorobenzene** (HCB), C_6Cl_6. The compound is stable, easy to prepare from chlorine and benzene, and for several decades after World War II it found use as an agricultural fungicide for cereal crops. Since it is extremely persistent and is still emitted as a by-product in the chemical industry and in combustion processes, it remains a widespread environmental contaminant. It is of concern because it causes liver cancer in laboratory rodents and therefore perhaps also in humans. Our current daily exposure to HCB is not sufficient to pose a significant health hazard, even though it is estimated that 99% of Americans have detectable levels of the chemical in their body fat.

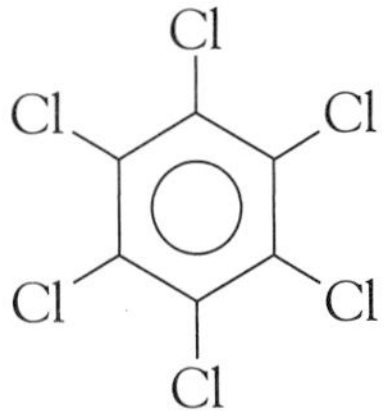

hexachlorobenzene (HCB)

Hexachlorobenzene is one of the 12 chemicals, the "dirty dozen," listed by the United Nations Environmental Programme as **persistent organic pollutants,** or POPs, which are being banned or phased out by international agreement (see Table 10-2). Each country signing the Stockholm Treaty that bans these POPs must develop its own implementation plan. Other compounds with the same negative characteristics as the dirty dozen may be added later to the list.

Another chlorinated benzene, namely the 1,4 isomer of **dichlorobenzene,** is used as an insecticidal fumigant. It is used in one type of domestic *mothball* and as a deodorizer in restroom urinals, garbage pails, and so on. Although it is a crystalline solid, it has an appreciable vapor pressure. Consequently, enough of it will vaporize to act as an effective insecticide in the immediate area around the solid. The same compound has also been used as a soil fumigant. However, it is an animal carcinogen and accumulates to some extent in the environment. It may well be the chemical responsible for the

TABLE 10-2	U.N.'s Persistent Organic Pollutants (POP) "Dirty Dozen" and Their Status in Several Countries						
POP	**U.S.**	**Canada**	**U.K.**	**Mexico**	**China**	**India**	**Use**
DDT	X	X	X	R	R	R	Mosquitoes
Aldrin	X	X	X	X	OK	OK	Termites
Dieldrin	X	X	X	X	OK	R	Crops
Endrin	X	X	X	X	OK	X	Rodents
Chlordane	R	X	X	OK	R	OK	Termites
Heptachlor	R	X	X	X	OK	OK	Soil insects
Hexachloro-benzene	X	X	X				Fungicide
Mirex	X	X		R	R		Ants, termites
Toxaphene	X	X	X	X	OK	X	Ticks, mites
PCBs*	X	R	R	OK			Many uses
Dioxins*	BP	BP	BP	BP	BP	BP	
Furans*	BP	BP	BP	BP	BP	BP	

Codes: *not pesticides; X = banned or no registered uses; R = severely restricted uses only; OK = not restricted; BP = by-product, not produced deliberately.

greatest carcinogenic risk of all indoor VOCs (volatile organic compounds; see Chapter 3). Recent research indicates that it is present in the blood of most U.S. residents; its presence at high levels is associated with reduced pulmonary function.

Pesticides in Water

The solubilities of trace substances such as HCB in liquids and solids are often expressed on a "parts per" scale, rather than on a mass or moles per unit volume basis. However, the "parts per" scales for condensed (nongaseous) media express the ratio of the *mass* of the solute to the *mass* of the solution, *not* the ratio of moles or molecules that is used for gases. Since the mass of 1 liter of a natural water sample is very close to 1 kilogram, the HCB solubility of 0.0062 milligrams solute per liter corresponds to 0.0062 mg solute per 1000 grams of solution. Multiplying both numerator and denominator by 1000, we conclude that 0.0062 mg/L is equivalent to 0.0062 grams of solute per 1 million grams of solution, i.e., to 0.0062 parts per million. *In general, the value for the ppm solubility of any trace substance in water is the same as its value in units of milligrams per liter or micrograms per gram.* Similarly, the ppb scale in aqueous solutions is equivalent to micrograms per liter, or nanograms per gram.

Another commonly used set of units for trace contaminants—particularly for substances dissolved in a medium such as soil or a biological specimen—is *micrograms* (of the contaminant) *per gram* (of the medium), μg/g. For example, the concentration of a pesticide in human fat could be listed as 2.0 μg/g. However, this scale is equivalent to that of parts per million, which can be seen by multiplying the numerator and denominator of μg/g by 1 million, giving grams per million grams. Similarly, the nanograms per gram (ng/g) scale is equivalent to parts per billion.

PROBLEM 10-1

For aqueous solutions, (a) convert 0.04 μg/L to the ppm and ppb scales, (b) convert 3 ppb to the μg/L scale, and (c) convert 0.30 μg/g to the ppb scale.

The pollution of aquatic environments is not merely a question of the concentration of pollutant actually dissolved in *solution,* and the small values for the water solubilities of organochlorines may be deceptive on this score. Most organochlorine compounds are much more soluble in organic media than in water. In bodies of water such as rivers and lakes, organochlorines are much more likely to be adsorbed on the surfaces of organic particulate matter suspended in the water and on the muddy sediments at the bottom than to be dissolved in the water itself. From these surfaces, they enter living organisms such as fish. For reasons discussed in detail later, their concentration in fish is often thousands or millions of times greater than in polluted drinking water. It is due to this phenomenon that concentrations of organochlorines have often reached dangerous levels in many species. Many organochlorine insecticides have been removed from use as a consequence. For humans, the amount of organochlorines ingested by eating a single lake fish is generally greater than the total organochlorine acquired in a lifetime of drinking water from the same lake!

DDT

DDT, or **para-dichlorodiphenyltrichloroethane,** has had a tumultuous history as an insecticide, as discussed in detail in the online Case Study *To Ban or Not to Ban DDT? Its History and Future.*

Unfortunately, following its introduction during World War II, DDT was widely overused in the 1950s and 1960s, particularly in agriculture, which consumed 70–80% of its production in the United States, and in forestry. As a result, its environmental concentration rose rapidly and it began to affect the reproductive abilities of birds that indirectly incorporated it into their bodies. By 1962, DDT was being called an "elixir of death" by the writer Rachel Carson in her influential book *Silent Spring* because of its role in decreasing the populations of birds such as the bald eagle, whose dietary intake of the chemical was very high.

DDT's Structure

Structurally, a DDT molecule is a substituted *ethane*. At one carbon, all three hydrogens are replaced by chlorine atoms, while at the other, two of the three hydrogens are replaced by a benzene ring. Each ring contains a chlorine atom at the para position, i.e., directly opposite the ring carbon that is joined to the (shaded) ethane unit:

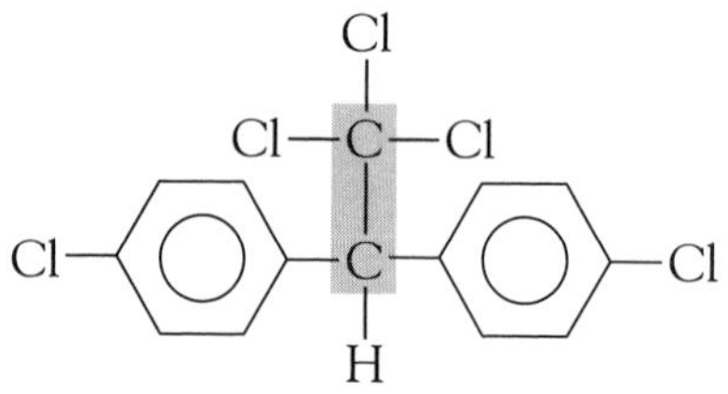

(DDT): *para*-dichlorodiphenyltrichloroethane

Many animal species metabolize (convert by biochemical reactions into other substances) DDT by the elimination of HCl; a hydrogen atom is removed from one ethane carbon and a chlorine atom from the other, thereby creating a derivative of ethene called **DDE, dichlorodiphenyldichloroethene:**

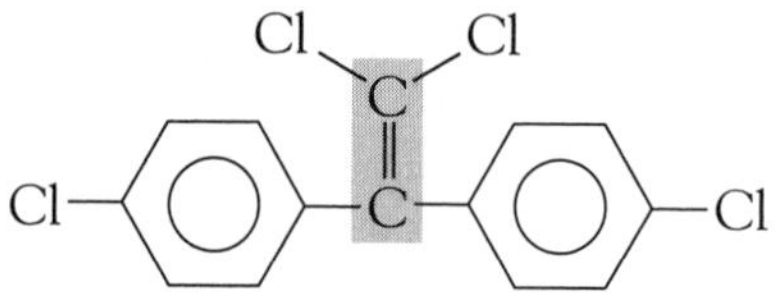

DDE

Substances that are produced by the metabolism of a chemical are called **metabolites;** thus DDE is a metabolite of DDT. The chemical DDE is also produced slowly in the environment by the degradation of DDT under alkaline conditions and by DDT-resistant insects that detoxify DDT by this transformation. Unfortunately, in some birds DDE interferes with the enzyme that regulates the distribution of calcium, so contaminated birds produce eggs that have shells (calcium carbonate) too thin to withstand the weight of the parents who sit on them to hatch them.

PROBLEM 10-2

The structures shown for DDT and DDE have both the ring chlorines in the para position, and are sometimes labeled p,p'-DDT and p,p'-DDE, where the p and p' prefixes refer to the para chlorine positions in the first and second rings, respectively. Deduce the structures and the appropriate labels for all the other unique isomers of both DDT and DDE in which there is one chlorine atom on each ring. Note that the two rings are equivalent, so that, e.g., o,m'-DDT is the same compound as is m,o'-DDT.

DDE in Body Fat

DDT's persistence made it an ideal insecticide: one spraying gave protection from insects for weeks to years, depending upon the method of application. Its persistence is due to

- its low vapor pressure and its consequent slow rate of evaporation;

- its low reactivity with respect to light and to chemicals and microorganisms in the environment; and

- its very low solubility in water.

Its rate of evaporation from the upper layer of soil in southern Canada, for example, is so slow that its volatilization half-life is about 200 years. Like other organochlorine insecticides, DDT is soluble in organic solvents and therefore in the fat of animal tissue. DDT and/or its degradation products have been found in all birds and fish that have been analyzed, even those living in deserts or ocean depths.

We all have some "DDT" (to the extent of about 3 ppm for North American adults) stored in our body fat. In humans, most ingested DDT is slowly but eventually eliminated. Most of the "DDT" stored in human fat is actually the DDE that was present in the food we have eaten, having previously been converted from DDT that was originally in the environment. Unfortunately DDE is almost nondegradable biologically and is very fat-soluble, so it remains in our bodies for a long time.

DDT Levels in Modern Times

For environmental reasons, DDT is now banned from use in most Western industrialized countries. Its use had been declining anyway as resistant insect populations evolved that metabolized DDT to the noninsecticidal DDE and thus rendered it inactive. The United Nations includes DDT on its list of twelve persistent organic pollutants (Table 10-2). Of the dozen, only DDT will not be totally banned. One permitted use is in the production of **dicofol,** a miticide whose molecular structure is identical except that DDT's remaining hydrogen atom on the ethane unit is replaced by a hydroxyl group, —OH. A controversial exception to the ban on DDT is for its use in controlling mosquitoes involved in the transmission of malaria. The issues in this debate are explored in the online Case Study *To Ban or Not to Ban DDT? Its History and Future.*

DDT and DDE still enter the environment everywhere as a result of long-range air transport—a topic discussed at length in Chapter 12—from developing countries where DDT is still in use to control malaria and typhus and for some agricultural purposes. DDT also continues to be degassed from soils in developed countries where it was used for agricultural purposes decades ago.

Thanks to restrictions and bans, the environmental concentrations of DDT and DDE in developed countries dropped substantially in the early and middle years of the 1970s and have now become stabilized at low levels, that of DDT itself being very small. The concentration of DDT itself declined more rapidly than that of its metabolites such as DDE. Nevertheless, as a result of the substantial decline in DDE levels that has occurred, bald eagles have made a comeback around Lake Erie and elsewhere. In fact, in the summer of 2007 bald eagles were removed from the list of endangered species in the United States. Similarly, the population of Arctic peregrine falcons, a bird that was driven to near-extinction due to the effects of DDE, has now recovered to such an extent that it has also been removed from the endangered species list.

The concentration of DDT in humans has also declined drastically; for example, its 1997 level in human breast milk of Swedish women was only 1% of what it was in 1967. As illustrated in Figure 10-1a, the DDT levels

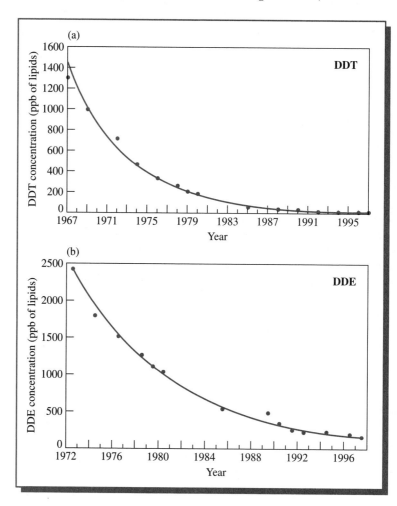

FIGURE 10-1 Trends in (a) DDT and (b) DDE concentrations in breast milk of Swedish women. [Source: K. Noren and D. Meironyte, "Certain Organochlorine and Organobromine Contaminants in Swedish Human Milk in Perspective of Past 20–30 Years," *Chemosphere* 40 (2000): 1111.]

in these women also began to decline exponentially beginning in the late 1960s, since DDT has a half-life of about 4 years. Data for women in North America show similar declines over this period. The concentration of DDE in breast milk did not drop as quickly as did that of DDT, its half-life being somewhat longer, about 6 years (Figure 10-1b). However, DDE levels remain high in regions such as Central and South America, Mexico, and Africa where DDT has been used more recently.

The Accumulation of Organochlorines in Biological Systems

Many organochlorine compounds are found in the tissues of fish in concentrations that are orders of magnitude higher than are those in the waters in which they swim. *Hydrophobic* (water-hating) substances like DDT are particularly liable to exhibit this phenomenon. There are several reasons for this **bioaccumulation** of chemicals in biological systems.

Bioconcentration

In the first place, many organochlorines are inherently much more soluble in hydrocarbon-like media, such as the fatty tissue in fish, than they are in water. Thus, when water passes through a fish's gills, the compounds selectively diffuse from the water into the fish's fatty flesh and become more concentrated there: This process (which also affects other organisms besides fish) is called **bioconcentration.** The **bioconcentration factor,** BCF, represents the equilibrium ratio of the concentration of a specific chemical in a fish relative to that in the surrounding water, provided that the diffusion mechanism represents the only source of the substance to the fish. BCF values occur over a very wide range and vary not only from chemical to chemical but also, to a certain extent, from one type of fish to another, particularly because of variations in the abilities of different fish to metabolize a given substance.

The BCF of a chemical can be predicted, to within about a factor of 10, for a typical fish from a simple laboratory experiment: The chemical is allowed to equilibrate between the liquid layers in a two-phase system made up of water and **1-octanol,** $CH_3(CH_2)_6CH_2OH$, an alcohol that has been found experimentally to be a suitable surrogate for the fatty portions of fish. The **partition coefficient,** K_{ow}, for a substance S is defined as

$$K_{ow} = [S]_{octanol}/[S]_{water}$$

where the square brackets denote concentrations in molarity units. (Since water and fat have approximately the same densities, the ratio of molarities

TABLE 10-3	Selected Data for Some Pesticides	
Pesticide	Solubility in H_2O (ppm)	log K_{ow}
HCB	0.0062	5.5–6.2
DDT	0.0034	6.2
Toxaphene	3	5.3
Dieldrin	0.1	6.2
Mirex	0.20	6.9–7.5
Malathion	145	2.9
Parathion	24	3.8
Atrazine	35–70	2.2–2.7

Data from K. Verschueren, *Handbook of Environmental Data on Organic Chemicals* (New York: Van Nostrand Reinhold, 1996).

in the two phases is identical to the ratio of their masses; consequently, K_{ow} can also be taken as the ratio of ppm or ppb concentrations.) Largely as a matter of convenience, the value of K_{ow} is often reported as its base 10 logarithm, since its magnitude sometimes is quite large. For example, for DDT (see Table 10-3), K_{ow} is about 1,000,000, i.e., 10^6, and so log K_{ow} = 6. Experimentally, the bioconcentration factor for DDT lies in the range of about 20,000 to 400,000, depending upon the type of fish. The K_{ow} value of a compound is a fairly reliable approximation to the BCF values found for fish. The approximation that K_{ow} = S typically breaks down when the molecules are too large to diffuse into the fish.

In general, the higher its octanol–water partition coefficient K_{ow}, the more likely a chemical is to be adsorbed on organic matter in soils and sediment and ultimately to migrate to fat tissues of living organisms. However, log K_{ow} values of 7 or 8 or higher are indicative of chemicals with such strong adsorption to sediments that they are actually unlikely to be mobile enough to enter living tissue. Thus, it is chemicals with log K_{ow} values in the 4–7 range that bioconcentrate to the greatest degree.

PROBLEM 10-3

For HCB, log K_{ow} = 5.7. What would be the predicted concentration of HCB due to bioconcentration in the fat of fish that swim in waters containing 0.000010 ppm of the chemical?

Biomagnification

Fish also accumulate organic chemicals from the food they eat and from their intake of particulates in water and sediments onto which the chemicals have been adsorbed. In many such cases, the chemicals are not metabolized by the fish: The substance simply accumulates in the fatty tissue of the fish, where its concentration there increases with time. For example, the concentration of DDT in trout from Lake Ontario increases almost linearly as the fish ages, as illustrated in Figure 10-2. The average concentration of many chemicals also increases dramatically as one proceeds up a **food chain,** which is a sequence of species, each one of which feeds mainly upon the one preceding it in the chain. The **food web,** incorporating interlocking food chains, for the Great Lakes is illustrated in Figure 10-3. Over a lifetime, a fish eats many times its weight in food from the lower levels of the food chain but retains rather than eliminates or metabolizes most organochlorine chemicals from these meals.

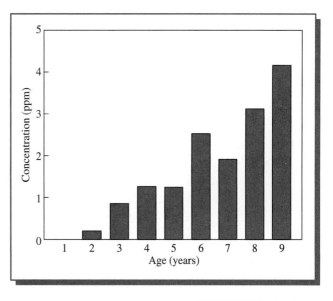

FIGURE 10-2 Variation with age of DDT concentration in Lake Ontario trout caught in the same year. [Source: "Toxic Chemicals in the Great Lakes and Associated Effects" (Ottawa: Government of Canada, 1991).]

A chemical whose concentration increases along a food chain is said to be **biomagnified.** In essence, the biomagnification results from a sequence of bioaccumulation steps that occur along the chain. The difference between bioconcentration from water and biomagnification along a food chain is illustrated symbolically in Figure 10-4. The biomagnification of DDT along some of the Great Lakes food chains is shown in Figure 10-3. Notice, for example, the herring gull's higher level of DDT compared with the fish below it in the chains. Fish at the top of the aquatic part of the chain bioaccumulate DDT rather effectively, so that even higher concentrations are found in the birds of prey that feed on them.

As an example of biomagnification, consider that the DDT/DDE concentration in seawater in Long Island Sound and the protected waters of its southern shore at one time was as high as 0.000003 ppm, but it reached 0.04 ppm in the plankton, 0.5 ppm in the fat of minnows, 2 ppm in the needlefish that swim in these waters, and 25 ppm in the fat of the cormorants and osprey that feed on the fish, for a total biomagnification factor of about 10 million. It is by such mechanisms that DDE levels in some birds of prey became so great that their ability to reproduce successfully was impaired. The bioaccumulation of organochlorines in fish and other animals is the reason that most of the human daily intake of such chemicals enters via our food supply rather than from the water we drink.

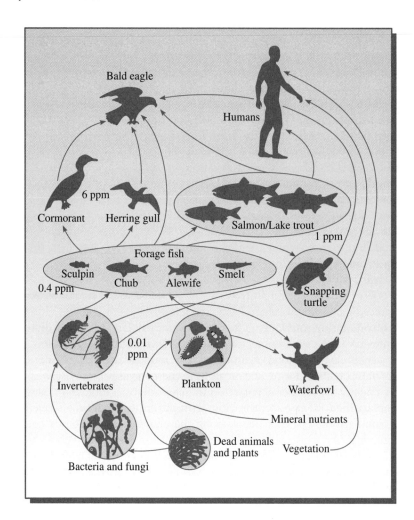

FIGURE 10-3 Simplified food web for the Great Lakes with typical DDT concentrations for some species. [Source: "Toxic Chemicals in the Great Lakes and Associated Effects" (Ottawa: Government of Canada, 1991).]

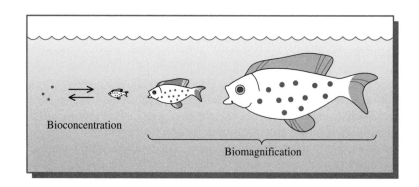

FIGURE 10-4 Schematic representation of the two modes of bioaccumulation that operate in biological matter present in a body of water.

Less Persistent Analogs of DDT

A number of compounds having the same general molecular structure as DDT are found to display similar insecticidal properties. This similarity arises from the mechanism of DDT action, which is due more to its molecular *shape* than to its chemical interactions. The shape of a DDT molecule is determined by the two tetrahedral carbons in the ethane unit and by the two flat benzene rings. In insects, DDT and other molecules with the same general size and 3-D shape become wedged in the nerve channel that leads out from the nerve cell. Normally, this channel transmits impulses only as needed via sodium ions. But a continuous series of Na^+-initiated nerve impulses is produced when the DDT molecule holds open the channel. As a consequence, the muscles of the insect twitch constantly, eventually exhausting it with convulsions that lead to death. The same process does not occur in humans and other warm-blooded animals since DDT molecules do not exhibit such binding action in nerve channels.

Examples of other molecules with DDT-like action include **DDD** (sometimes called TDE), *para*-**dichlorodiphenyldichloroethane,** which is an environmental degradation product of DDT: It differs only in that one chlorine from the —CCl_3 group is replaced by a hydrogen. Since the overall shapes and sizes of DDT and DDD are similar, their toxicity to insects is as well. In the past, DDD was itself sold as an insecticide, but it has also been discontinued because it bioaccumulates. Notice that DDE, unlike DDT and DDD, is based upon a *planar* C=C unit rather than a C—C linkage which has tetrahedral groups at each end. Thus, whereas DDD is a DDT-like insecticide, DDE is not, since its three-dimensional shape is very different: DDE is flat rather than propeller-shaped, so it does not become wedged in the insect's nerve channels.

Scientists have devised analogs to DDT that have its same general size and shape; consequently, they possess the same insecticidal properties but are more biodegradeable and thus do not present the bioaccumulation problem associated with DDT. The most important of these analogs is **methoxychlor:**

methoxychlor

The *para*-chlorines of DDT are replaced in methoxychlor by *methoxy* groups, —OCH_3, which are approximately the same size as chlorine but which react much more readily. In particular, under reducing conditions, the O—CH_3 bonds in methoxychlor are subject to attack by hydrogen ions in water, converting the O—CH_3 units to O—H bonds and free methane molecules, CH_4. The hydroxylated products are water-soluble products that not only degrade in the environment but are excreted rather than accumulated by organisms. Methoxychlor is still used both domestically and agriculturally to control flies and mosquitoes.

PROBLEM 10-4

Draw the molecular structure of DDD.

PROBLEM 10-5

Methyl groups are approximately the same size as chlorine atoms, but hydrogen atoms are significantly smaller. Would you expect insecticidal properties for DDT molecules in which (a) the —CCl_3 group is replaced by —$(CH_3)_3$, and (b) the *para*-chlorines are replaced by hydrogens?

Other Organochlorine Insecticides

Toxaphene

During the 1970s, after DDT had been banned, the insecticide that replaced it in many agricultural applications, such as the growing of cotton and soybeans, was **toxaphene.** It is a mixture of hundreds of similar substances, all of which are produced when the hydrocarbon *camphene*, produced from chemicals extracted from pine trees, is partially chlorinated.

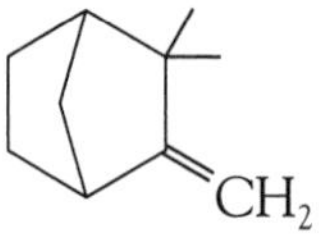

camphene

Toxaphene became the most heavily used insecticide (1966–1976) in the United States before restrictions were placed on its use in 1982 and a total ban imposed in 1990. More than 85% of toxaphene use in the United States occurred in the southeastern cotton-growing states, though some was used as a herbicide in the growing of peanuts and soybeans.

Toxaphene is extremely toxic to fish. Indeed, it was first used in the 1950s in North America to rid lakes of undesirable fish. However, it was found to be so persistent that the lakes could not be successfully restocked for years thereafter!

Toxaphene bioaccumulates in fatty tissues and causes cancer in test rodents; consequently it is one of the dirty dozen persistent organic pollutants (Table 10-2) and is now banned in developed countries as well as some developing ones. It still is being deposited in bodies of water remote from its point of usage because of long-range transport by air (see Chapter 12) from countries that still make restricted use of it.

Hexachlorinated Cyclohexane

During World War II, the derivative of cyclohexane having one of the two hydrogens on each carbon replaced by chlorine, namely, **1,2,3,4,5,6-hexachlorocyclohexane,** was discovered to be an effective insecticide against a wide variety of insects. In fact, there exist eight isomers having this formula;

they differ only in the relative orientations of the chlorine atoms bonded to different carbons. (The diagrammatic formula below is not intended to illustrate the chlorines' orientations—only their points of attachment.)

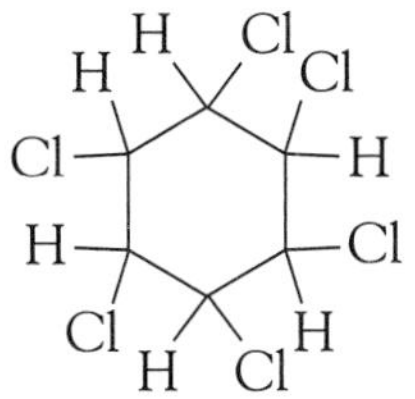

1, 2, 3, 4, 5, 6-hexachlorocyclohexane

(This compound is also known as *benzene hexachloride* or BHC, not to be confused with hexachlorobenzene)

A commercial mixture of most of the hexachlorocyclohexane isomers was used to control mosquitoes and in agricultural applications after World War II. Its use has been restricted since the 1970s due to its toxicity and tendency to bioaccumulate. Only one of the eight isomers, the gamma isomer, actually kills insects; now sold separately under the name *Lindane*, it was the active ingredient in several commercial medical preparations used to rid children of lice and scabies and agriculturally to treat seeds and seedlings.

Chlorinated Cyclopentadiene Insecticides

Cyclopentadiene, shown at left below, is an abundant by-product of petroleum refining. As its name implies, there are two double bonds in each molecule. When fully chlorinated (diagram at right), it can be combined with one of several other organic molecules (Diels–Alder reaction) to produce a series of insecticidal compounds with properties, including environmental persistence, that at first made them superficially attractive.

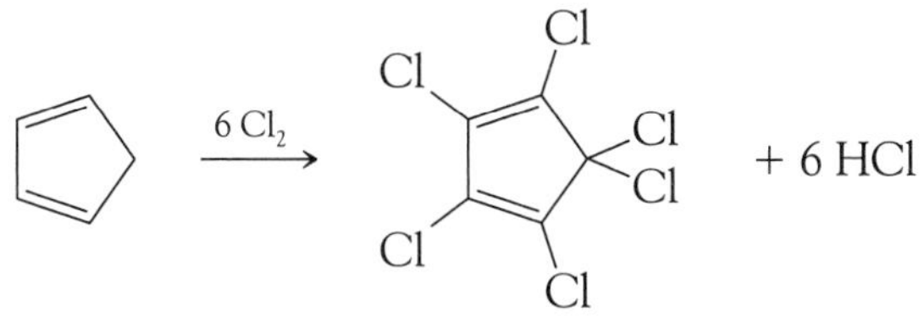

cyclopentadiene   perchlorocyclopentadiene

All the cyclodiene pesticides contain the hexachlorinated five-membered ring bonded to at least two other carbon atoms and containing a residue of one double bond.

Most cyclodiene insecticides that were commercially important have now been branded as persistent organic pollutants by the United Nations Environmental Programme and are listed in Table 10-2. They were used to control soil insects, fire ants, cockroaches, termites, grasshoppers, locusts, and other insect pests. In many such applications, their persistence was an advantage since they did not have to be reapplied frequently.

The cyclodiene pesticides, starting with **aldrin** and **dieldrin**—which are structurally identical except that the latter corresponds to the former with one of its C=C bonds epoxidized (see Figure 12-2)—arrived on the market in about 1950. Given their persistence, their potential toxicity, their tendency to accumulate in fatty tissues, and the suspicion that dieldrin was causing excess mortality of adult bald eagles, the use of almost all of these compounds has now either been banned or severely restricted in North America and most western European countries. Nonetheless, dieldrin and DDT were the most common POPs still detectable in food in 2002. Some of the compounds are still in use elsewhere (see Table 10-2).

Agricultural uses of dieldrin, mainly to combat soil insects, and its use in buildings to control termites, were largely prohibited in North America by the mid-1980s but continue in many developing countries. Dieldrin was used extensively in tropical countries to control the tsetse fly and is still used in some countries to kill termites. It continues to enter water systems even in developed countries by percolating from waste disposal sites. Danish studies have found a correlation between blood dieldrin levels in women and their increasing risk of developing, and dying from, breast cancer. Dieldrin and another cyclodiene insecticide containing a three-membered epoxide ring, *heptachlor epoxide*, have been associated with increased risk of non-Hodgkin's lymphoma.

If two perchlorocyclopentadiene molecules are chemically combined, the resultant molecule, known commercially as as **mirex,** also acts as an insecticide and is particularly effective against the fire ant found in the southeastern United States.

mirex

(All 10 carbon atoms are bonded to chlorine, but for clarity the individual chlorine atoms are not shown; only the total is displayed in the formula.) Mirex was also sold as a flame retardant additive for synthetic and natural materials. Most mirex use occurred in the 1960s, although it is still produced and used in China and is used to fight giant termites in parts of Australia. Mirex is classified as a persistent organic pollutant by the U.N. and has been banned in many countries since the mid-1970s.

For the most part, the chlorinated cyclodiene pesticides are chemical products of the past. Their use has been phased out or at least severely restricted because of environmental and human health considerations. The only cyclodiene insecticide still in widespread use is **endosulfan,** which is discussed in detail in Box 10-1.

BOX 10-1	The Controversial Insecticide Endosulfan

Endosulfan does not appear on the U.N. list of POPs; indeed, it is one of the few cyclopentadiene pesticides still widely available on the market. Structurally, endosulfan molecules consist of a perchlorinated five-membered ring attached by methylene groups to an $O\!=\!S(O\!-\!)_2$ unit.

endosulfan

It is used as an insecticide and an acaricide in agricultural applications, its domestic applications having been phased out—at least in the United States. Both its environmental persistence and its tendency to bioconcentrate are much lower than those of the other cyclodienes because it is much more reactive, owing to the presence of the sulfur–oxygen group. In the environment, some of it is converted to the sulfate, which has an additional oxygen doubly bonded to the sulfur, thereby producing S(VI) from S(IV) in the original pesticide. Unfortunately, **endosulfan sulfate** is just as toxic as endosulfan and is more persistent. The remaining fraction of endosulfan in the environment is converted to the sulfur-free diol in which the $-CH_2-$ groups are each bonded to $-OH$, and which itself undergoes further degradation reactions.

Although endosulfan is classified as *moderately toxic* by the WHO, it is considered to be *highly hazardous* by the U.S. EPA. It is readily absorbed by the stomach and lungs and through the skin. Although acutely toxic, endosulfan does not persist in mammals because it is degraded into water-soluble compounds and eliminated from the body within a few days or weeks. Because of its short lifetime, the risk from consuming it in drinking water or in food is usually not a concern. Consequently, most of the health concerns about endosulfan relate to its ability to act as an acute poison to workers who handle and apply it in agricultural settings. It is also of environmental concern due to its toxicity to nontarget fish, birds, and other animals. Indeed, there have been massive fish kills in the United States (Alabama) and in other countries in waters that were inadvertently contaminated with the insecticide.

Owing to its toxicity, endosulfan has been banned in many countries and its use restricted in others. The bans followed the accidental deaths of farmworkers and nearby residents as well as its use in committing suicide. For example, it was introduced into Sri Lanka to replace category I insecticides that previously had been used in suicides, but it soon became employed for the same purpose. Since it was banned there in 1998, other category II insecticides that replaced it have been employed instead by would-be suicides.

In addition to its use on some food and grain crops as well as on tea and coffee plantations, endosulfan is also used in the growing of cotton in both developed countries, such as the United States and Australia, and developing countries. For example, endosulfan was introduced late in the twentieth century into French-speaking countries in West Africa when bollworm caterpillars (boll weevils)—which

(continued on p. 434)

| BOX 10-1 | The Controversial Insecticide Endosulfan *(continued)* |

enter cotton flower buds and bolls and destroy them—became resistant to the lower-toxicity insecticides (pyrethroids) that previously had been employed. Unfortunately, most farmers in certain countries in this region, including Benin, cannot afford the protective clothing—goggles, gloves, and respirators—that should be used when handling and spraying endo-sulfan. In addition, the cotton farmers some-times use leftover endosulfan inappropriately as an insecticide on their vegetable crops. Since endosulfan is almost insoluble in water, washing food sprayed with it is largely ineffective.

Principles of Toxicology

Toxicology is the study of the harmful effects on living organisms of substances that are foreign to them. The substances of interest include both synthetic chemicals and those that exist naturally in the environment. In toxicology, the effects are normally determined by injecting or feeding animals with the substance of interest and observing how the health of the animal is affected. By contrast, in **epidemiology,** scientists do not run experiments in a lab but instead determine the health history of a selected group of human beings and attempt to relate differences in disease rates and so on to differences in sub-stances to which they have been accidentally exposed.

Toxicological data concerning the harmfulness of a substance to an organism, such as an organochlorine pesticide or a heavy metal, are gathered most easily by determining its **acute toxicity,** which is the rapid onset of symptoms—including death at the extreme limit—immediately following the intake of a dose of the substance. For example, experiments show that it takes only a few tenths of 1 microgram of the most acutely toxic synthetic compound—the "dioxin" to be discussed in Chapter 11—to kill most rodents within a few hours after it is administered orally to them.

Although the acute toxicity of a substance is of interest when we are exposed accidentally to pure chemicals, in **environmental toxicology** we are usually more concerned about **chronic** (continuous, long-term) **exposures** at relatively low individual doses of a toxic chemical that is present in the air that we breathe, the water we drink, or the food we eat. Generally speaking, any effects of such continuing exposures are also long-lasting and therefore also classified as chronic.

The same chemical may give rise to both acute and chronic effects in the same organism, although usually by different physiological mechanisms. For example, a symptom of acute toxicity in humans of exposure to many organochlorines is a skin irritation that leads to **chloracne,** a persistent, disfig-uring, and painful analog to common acne, and there is the fear that persistent exposure to much lower individual doses than those which produce the skin disease could eventually lead to cancer.

The three types of substances that produce detrimental effects on human health that are of most concern are

- **mutagens,** substances that cause mutations in DNA, most of which are harmful and can produce inheritable traits if they occur in DNA of cells present in sperm or eggs;

- **carcinogens,** substances that cause cancer; and

- **teratogens,** substances in the mother that cause birth defects in the fetus.

Some carcinogens operate in an initiation step, in which the substance—sometimes after itself having been transformed in the body—reacts directly with a strand of DNA. This alteration in DNA can lead to the growth of cancer cells. Others, called *promoters*, act only after cancer has been initiated, but they speed up the process of tumor formation.

Dose–Response Relationships

Most of the quantitative information concerning the toxicity of substances is obtained from experiments performed by administering doses of the substances to small animals, although recently tests on certain bacteria have been used to determine whether a substance is likely to be a carcinogen or not. Owing to practical considerations including cost and time, most experiments involve acute rather than chronic toxicity, even though it is the latter that usually is of primary interest in environmental science. To determine directly the effects of continuous, low-level exposures over long periods would require a very large number of test animals and a long project time. The practical alternative is to evaluate the effects using high doses—at which point the effects are substantial and clear-cut—and then extrapolate the results down to environmental exposures. Unfortunately, there is no assurance that such extrapolations are always reliable, since the cellular mechanisms that produce the effects at high and low doses could differ.

The **dose** of the substance administered in toxicity tests is usually expressed as the mass of the chemical, usually in milligrams, per unit of the test animal's body weight, usually expressed in kilograms, thus giving units of milligrams per kilogram, mg/kg. The division by body weight is necessary because the toxicity of a given amount of a substance usually decreases as the size of the individual increases. (Recall that the maximum recommended doses for medicines such as headache remedies are smaller for children than for adults, primarily because of the difference in body masses.) It is also assumed that toxicity values obtained from experiments on small test animals are approximately transferable to humans provided the differences in body weight are taken into account. Normally the toxicity of a substance increases with increasing dose, although exceptions are known.

If a dose of a few tenths of a microgram of a certain substance is sufficient to kill a mouse, approximately what mass of the substance would be fatal to you? What average level of the substance would have to be present in the water you drink if you were to receive a fatal dose from this source in a week? Note that your weight in kilograms is that in pounds divided by 2.2.

Individuals differ significantly in their susceptibility to a given chemical: Some respond to it even at very low doses, whereas others require a much higher dose before they respond. It is for this reason that scientists created **dose–response relationships** for toxic substances, including environmental agents. A typical dose–response curve for acute toxicity is illustrated in Figure 10-5a. The dose is plotted on the (horizontal) x-axis, and the cumulative

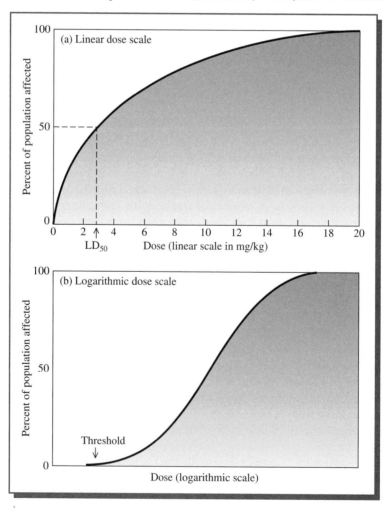

FIGURE 10-5 Dose–response curves for (a) linear dose scale; (b) logarithmic dose scale.

percentage of test animals that display the measured effect (e.g., death) when administered a particular dose is shown on the (vertical) y-axis. For example, in Figure 10-5a, about 60% of the test animals were affected by a dose of about 4 mg/kg.

Because the range of doses on such graphs often exceeds an order of magnitude, and because the effects at the low end of the concentration scale are often important in environmental decision making and cannot be seen clearly using linear scales, the dose–response plot is often recast by using a logarithmic scale for doses. Usually an S-shaped or sigmoidal-shaped curve results from this transformation—see Figure 10-5b.

Most often, the response effect on test animals that is used to construct dose–response curves is death. *The dose that proves to be lethal to 50% of the population of test animals is called the* LD_{50} *value of the substance*; its determination from a dose–response curve is illustrated by the dashed lines in Figure 10-5a. The *smaller* the value of LD_{50}, the *more potent* (i.e., more toxic) is the chemical, since less of it is required to affect the animal. A chemical much less toxic than that illustrated in Figure 10-5b would have a sigmoidal curve shifted to the right of the one shown.

Many sources quote values for the LOD_{50}, the *lethal oral dose*, when the chemical has been administered orally to the test animals, as opposed to dermal or some other means of delivery. For example, the LOD_{50} value for DDT for rats is about 110 mg/kg. As mentioned previously, the presumption is usually made that LD_{50} and LOD_{50} values are approximately transferable between species. In the case of DDT, for example, humans are known to have survived doses of about 10 mg/kg, so presumably the LOD_{50} value for humans is greater than 10 mg/kg. However, we have no *direct* evidence that the 110 mg/kg value for rats is also valid for humans.

Of much more concern than the acute toxicity of DDT is its ability to cause chronic effects in humans, such as cancer. Although DDT is not traditionally considered to be a human carcinogen, some small-scale epidemiological studies found that the higher the concentration of DDE in a woman's blood, the more likely she was to have contracted breast cancer. However, later full-scale studies in the United States have failed to confirm this association between breast cancer and DDE. Thus it does not seem that lifetime exposure to DDT is an important cause of breast cancer, but these studies do not address the issue raised recently of whether exposure during the teenage years, when breasts are developing rapidly, might be a factor in developing breast cancer decades later.

The range of LD_{50} and LOD_{50} values for acute toxicity of various chemical and biological substances is enormous, and spans about ten powers of 10. Indeed, all substances are toxic at sufficiently high doses; as the Renaissance-era German philosopher Paracelsus observed, all things are poison, and it is the dose that differentiates a poison from a remedy. The *World Health Organization* (WHO) has devised descriptors for four broad levels of toxicity for substances, especially pesticides; the range and descriptor for each class is shown in

| TABLE 10-4 | WHO and U.S. EPA Pesticide Hazard Categorization | | | | |

WHO Category Number	U.S. EPA Category*	WHO Description	LOD_{50}† (mg/kg)	Examples	
				Synthetic Pesticide	"Natural" Pesticide
Ia	I	Extremely hazardous	<5	aldicarb; parathion; methyl parathion; turbufos	
Ib	I	Highly hazardous	5–50	azinphos-methyl; carbofuran; dichlorvos	nicotine
II	II	Moderately hazardous	50–500	carbaryl; chlorpyrifos; diazinon; dimethoate; **endosulfan;** fenitrothion; lindane; paraquat; propoxur	permethrin; pyrethrins; rotenone
III	III	Slightly hazardous	500–5000	alachlor; malathion; metolachlor; 2,4-D family	allethrin
III	IV		>5000		

* The United States EPA does not distinguish between WHO classes Ia and Ib but uses a single category I. The EPA also defines a fourth category, IV, for substances with LOD_{50} values greater than 5000 mg/kg.
†The LOD_{50} values quoted are for the solid form and are based upon experiments with rats; LD_{50} ranges are a factor of 2 higher. Lethal dose ranges for liquids are a factor of 4 larger than their respective LD_{50} and LOD_{50} ranges.

Table 10-4, along with some examples of pesticides that fall in each. All the pesticides in the *extremely toxic* category Ia are synthetic, but nicotine—which as a solution of its sulfate salt has been used as an organic insecticide in gardens for many decades—falls in category Ib, *highly hazardous*. The U.S. EPA classifies pesticides in a similar fashion to WHO but does not distinguish between the two sublevels of WHO's category I. Category II substances, *moderately hazardous*, include many pesticides—both synthetic and organic—still on the market. WHO's category III, *slightly hazardous*, is subdivided into III and IV by the U.S. EPA.

In the dose–response curves for some substances, there exists a dose below which none of the animals are affected; this is called the **threshold,** and it is illustrated in Figure 10-5b. The highest dose at which no effects are seen lies slightly below it and is called the **no observable effects level** (NOEL), although sometimes the two terms are used interchangeably. It is difficult to determine the threshold or NOEL level: It may be that if more animals were involved in a particular study, effects at low doses might be uncovered that are not apparent with only a small number of test animals. Most toxicologists believe that for toxic effects other than carcinogenesis there is probably a nonzero threshold for each chemical. A few scientists hold the controversial view that for some substances the curve in Figure 10-5b actually falls *below* the zero or NOEL value for very low concentrations before returning to zero at zero dose, indicating that tiny amounts of these substances could have a positive rather than a negative effect on health.

Experiments involving test animals are also used to determine how carcinogenic a compound is. However, the simple **Ames test,** which uses bacteria, can be used fairly reliably to distinguish compounds likely to be human carcinogens from those that are not.

A parameter that is useful in judging whether a specific chemical is present in an environmental sample in dangerous amounts or not is the **lethal concentration, LC,** of the substance. Usually this is listed as its LC_{50}, the concentration of the substance that is lethal to 50% of a specified organism within a fixed exposure period. LC_{50} values may refer to the concentration of a substance in air or in aqueous solution to which the organism is exposed and usually have units of milligrams per liter. For example, the LC_{50} for rainbow trout for a four-day exposure to endosulfan in water is only 0.001 mg/L; indeed, this insecticide is "supertoxic" to many fish species. The LC_{50} for a shorter exposure would be a value greater than 0.001.

Risk Assessment

Once toxicological and/or epidemiological information concerning a chemical is available, a **risk assessment** analysis can be performed. This analysis tries to answer quantitatively the questions "What are the likely types of toxicity expected for the human population exposed to a chemical?" and "What is the probability of each effect occurring in the population?" Where necessary, risk assessment also tries to determine permissible exposures to the substance in question.

In order to perform a risk assessment on a chemical, it is necessary to know

- **hazard evaluation** information; i.e., the types of toxicity (acute? cancer? birth defects?) that are expected from it;

- quantitative **dose–response** information concerning the various possible modes of exposure (oral, dermal, inhalation) for it; and

- an estimation of the potential **human exposure** to the chemical.

For chronic exposures, threshold or NOEL dose information is normally expressed as milligrams of the chemical per kilogram of body weight per day. In determining the threshold level for the most sensitive members of the human population, it is common to divide the NOEL from animal studies by a *safety factor*, typically 100. The resulting value is called the maximum **acceptable daily intake,** ADI, or maximum daily dose; the U.S. EPA instead refers to it as the **toxicity reference dose,** RfD. Note that the ADI or RfD value does *not* represent a sharp dividing line separating absolutely safe from absolutely unsafe exposures, since the transferability of toxicity information from animals to humans is not exact, and in most cases the safety factor presumably is quite generous. Some scientists have suggested dividing the NOEL by a further factor of 10 in order to protect very susceptible groups such as children. Indeed, the Food Quality Protection Act in the United States requires that the EPA set limits for residues of pesticides on foods 10 times lower than what is considered to be safe for an adult.

PROBLEM 10-7

The NOEL for a chemical is found to be 0.010 mg/kg/day. What would its ADI or RfD value for adults be set at? What mass of the compound is the maximum that a 55-kg woman should ingest daily?

As mentioned, in a risk assessment, an attempt is made to estimate the exposure of the affected population. For example, for chemicals whose mode of exposure is primarily through drinking water, regulatory agencies such as the U.S. EPA consider a hypothetical average person who drinks about 2 L of water daily and whose body weight averages 70 kg (154 lb) through life. If the ADI (or RfD) of a substance is 0.0020 mg/kg/day, then for the 70-kg person, the mass of it that can be consumed per day is $0.0020 \times 70 = 0.14$ mg/day. Thus the maximum allowable concentration of the chemical in water would be 0.14 mg/2 L = 0.07 mg/L = 0.07 ppm. Of course, if there are other significant sources of the substance, they must be taken into account in determining the drinking-water standard. Also, exposure to several chemicals of the same type (e.g., several organochlorines) could produce additive effects, so the standard for each one should presumably be lowered to take this into consideration.

The EPA regulates exposure to carcinogens by assuming that the dose–response relationship has no threshold and can be linearly extrapolated from zero dose to the area for which experimental results are available. The maximum daily doses are then determined by assuming that each person receives the dose every day over his or her lifetime and that this exposure should not increase the likelihood of cancer to more than one person in every million.

In determining regulations to control risk, usually no consideration is given to the economic costs involved. Many economists believe that because regulations cost money to implement, and because society may decide that it has only limited resources that it is willing to spend overall on regulation, a cost–benefit analysis should be used to help decide which substances to regulate. Associated with this line of thinking is the idea that regulations should show a positive payback: The benefits from regulation should exceed the costs. Of course, it is difficult to place a specific monetary value on environmental benefits in many cases. For example, is the $200,000 average cost for saving a life by regulating the chloroform content in water (see Chapter 14) a reasonable amount to pay? If so, would it still be reasonable if the cost instead was 10 or 100 or 1000 times this amount?

Organophosphate and Carbamate Insecticides

Organochlorine insecticides have now largely been replaced by those that are less persistent and less subject to bioaccumulation. The most important classes of these newer insecticides are organophosphates and carbamates, both of which are discussed below.

Organophosphate Insecticides

Structurally, all **organophosphate** (OP) pesticide molecules can be considered as derivatives of **phosphoric acid,** $O{=}P(OH)_3$, and consist of a central, pentavalent phosphorus atom to which are connected

- an oxygen or sulfur atom doubly bonded to the P atom,

- two *methoxy* ($-OCH_3$) or *ethoxy* ($-OCH_2CH_3$) groups singly bonded to the P atom, and

- a longer, more complicated, characteristic R group singly bonded to phosphorus, usually through an oxygen or sulfur atom, that differentiates one organophosphate insecticide from another.

The organophosphates are toxic to insects because they inhibit enzymes in the nervous system; thus they function as nerve poisons. In particular, organophosphates disrupt the communication that is carried between cells by the **acetylcholine** molecule. This cell-to-cell transmission cannot operate properly unless the bound acetylcholine molecule is destroyed after it has executed its function. Organophosphates block the action of the enzymes whose job it is to destroy the acetylcholine, by selectively bonding to them. (At the atomic level, it is the phosphorus atom of the organophosphate molecule that attaches to the enzyme and stays bound to it for many hours.) The presence of the insecticide molecule has the effect of suppressing the dissociation of the bound acetylcholine. Consequently, continuous stimulation of the receptor cell and its target muscle occurs. Normally the enzyme occurs in large excess and therefore some exposure to OPs occurs without immediate effects, but symptoms begin to appear if a majority of them

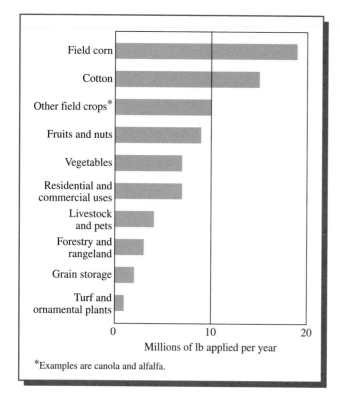

Field corn
Cotton
Other field crops*
Fruits and nuts
Vegetables
Residential and commercial uses
Livestock and pets
Forestry and rangeland
Grain storage
Turf and ornamental plants

0   10   20

Millions of lb applied per year

*Examples are canola and alfalfa.

FIGURE 10-6 Consumption of organophosphate pesticides by various crops in the United States. [Source: B. Hileman, "Reexamining Pesticide Risk," *Chemical and Engineering News* (17 July 2000): 34].

are inactivated. Since the affected nerves include those controlling gastrointestinal activities and bronchial secretions, massive gastric and respiratory secretions occur along with involuntary motions. Death ensues if 80–90% or more of the enzyme sites are inactivated.

The largest use of organophosphate insecticides is in agriculture (see Figure 10-6), but they also find many uses domestically. Organophosphates generally are nonpersistent; in this respect, they represent an environmental advance over organochlorines since they do not bioaccumulate in the food chain and present a chronic exposure and health problem to us later. However, they are generally much more acutely toxic to humans and other mammals than are organochlorines. Many organophosphates represent an acute danger to the health of those who apply them and to others who may come into contact with them. Exposure to these chemicals by inhalation, swallowing, or absorption through the skin can lead to immediate health problems. However, organophosphates metabolize relatively quickly and are excreted in the urine.

After application, most organophosphates decompose within days or weeks and thus are seldom found to bioconcentrate appreciably in food chains. However, because they have a wide range of uses in homes, on lawns, in commercial buildings, and in agriculture, almost everyone is regularly exposed to them. There is some evidence that OPs cause chronic as well as acute health problems. Children in the United States generally have their greatest exposure to organophosphates through the food they eat, with exposure from drinking water only 1–10% of that from food. A study reported in 2003 concerning a group of preschool children in Seattle, Washington, who consumed mainly organic fruits, vegetables, and juices found that they had much lower exposure to organophosphates, as measured by metabolites in their urine, than children with conventional diets. A number of recent studies, none of them large enough to be statistically definitive, have found links between indoor use of insecticides—organophosphates especially—and childhood leukemia and brain cancer. The U.S. EPA has placed OPs in its highest priority group in its current re-examination of pesticides.

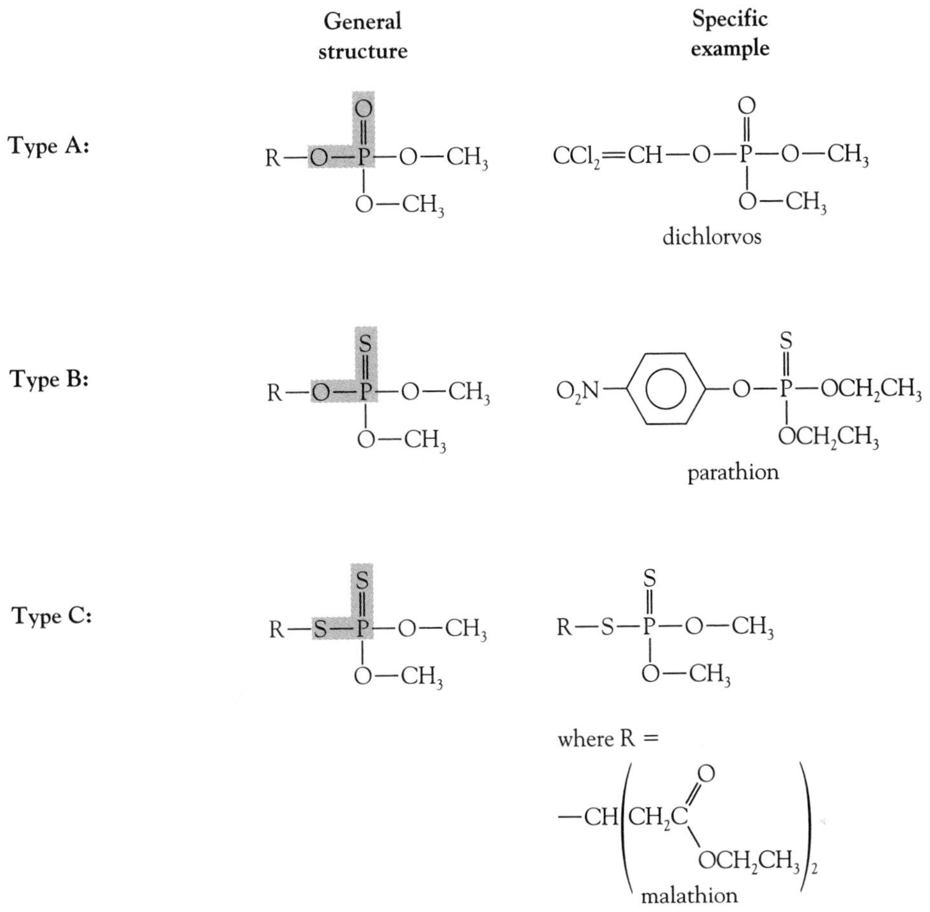

FIGURE 10-7 General structures and examples of organophosphate insecticides. In some molecules (e.g., parathion) the methoxy groups are replaced by ethoxy groups.

The three main subclasses of the organophosphate insecticides (for convenience called Types A, B, and C) are illustrated in Figure 10-7. Those containing a P=S unit are converted within the insect into the corresponding molecules with a P=O unit, producing a more toxic substance. The P=S form is used initially because it penetrates the insect more readily and is more stable than is the corresponding P=O compound. Such organophosphates decompose fairly rapidly in the environment because oxygen in air alters P=S bonds to P=O.

Most organophosphate insecticides decompose in the environment by *hydrolysis* reactions, i.e., reaction with H_2O. Water molecules split P—S and P—O bonds in organophosphates—by adding H to the sulfur or oxygen atom and OH to the phosphorus—and also add to the C—O bonds in the phosphoric acid esters formed as intermediates, ultimately yielding nontoxic substances such as phosphoric acid and alcohols and thiols: e.g.,

$$R'\!-\!S\!-\!\overset{\overset{\displaystyle O}{\|}}{P}(OR)_2 + H\!-\!OH \longrightarrow R'\!-\!S\!-\!H + HO\!-\!\overset{\overset{\displaystyle O}{\|}}{P}(OR)_2$$

$$\longrightarrow \longrightarrow R'SH + 2\,ROH + O\!=\!P(OH)_3$$

Dichlorvos is an example of a Type A organophosphate and so contains no sulfur. It is a relatively volatile insecticide and is used as a domestic fumigant released from impregnated fly strips hung from ceilings and light fixtures. The chemical slowly evaporates, and its vapor kills flies in the room. Plastic is impregnated with dichlorvos for use in flea collars. It is also used to treat worm infections in animals and as an agricultural insecticide. It is classed as highly hazardous (Table 10-4) to mammals; its LOD_{50} is 25 mg/kg in rats. (Note the *-os* ending of the name of this insecticide. The commercial names of organophosphate insecticides often have this or an *-fos* ending, signaling their nature.)

Parathion (Figure 10-7) is an example of a Type B organophosphate, i.e., one in which the doubly bonded oxygen (only) has been replaced by sulfur. It is described as extremely hazardous ($LOD_{50} = 3$ mg/kg) and has probably been responsible for more deaths of agricultural field workers than any other pesticide. Since it is nonspecific to insects, its use can inadvertently kill birds and other nontarget organisms. Bees, too, which are often economically valuable, are indiscriminately destroyed by parathion. It is now banned in some Western industrialized countries but is still widely used in developing countries. The background levels of parathion in homes in the agricultural community in the United States have dropped by an order of magnitude since it was banned in 1991. The structure of *fenitrothion* is very similar to parathion, but it is much less toxic ($LOD_{50} = 250$ mg/kg) because it is less effective in deactivating bound acetylcholine in mammals as compared to insects. Fenitrothion has been used extensively as a spray against spruce budworm in evergreen forests in eastern Canada, though not without controversy.

Diazinon, another Type B $O_3P\!=\!S$ organophosphate, was widely used for insect control in homes (against ants and roaches), gardens and lawns (including grub control), shrubs, and on pets, since it was thought to be relatively safe (moderately hazardous; $LOD_{50} = 300$ mg/kg), although its usage did have some restrictions since it is toxic to birds. However, there now is evidence that children treated for diazinon poisoning suffer persistent neurobehavioral problems. Consequently, the residential uses of diazinon in North America have been phased out, though it is still used in agriculture.

Chlorpyrifos (moderately hazardous; $LOD_{50} = 135$ mg/kg), another Type B insecticide, was commonly used in households to control cockroaches, ants, termites, and other insects. Indeed, it was the insecticide most commonly sprayed by exterminators to kill cockroaches. However, it has been withdrawn in the United States for home and garden use and banned for certain crops due to potential health concerns, especially involving fetal and childhood exposure, following experiments using rats, and due to unintentional poisonings. It was detected in 93% of the 1300 Americans whose blood was analyzed in a recent survey of common body contaminants. Its use had been restricted as of the late 1990s by the EPA, as had that of **methyl parathion** (which is identical to parathion except that it has methyl rather than ethyl groups), which is heavily used in the growing of cotton.

Most indoor uses of organophosphates have been eliminated by regulatory action in recent years. A recent survey linked the use of indoor insecticides, but not herbicides or outdoor insecticides, especially during pregnancy, to an increase in leukemia among the children born following fetal exposures.

Malathion (slightly hazardous) is the most important example of Type C organophosphates, which are those in which two oxygens have been replaced by sulfur, resulting in a $P{=}S$ and one $P{-}S$ unit. Upon exposure to oxygen, malathion is converted to *malaoxon*, which has the doubly bonded sulfur atom replaced by oxygen and is more toxic than malathion. Introduced in 1950, malathion is not particularly toxic to mammals ($LOD_{50} = 885$ mg/kg) since they metabolize it faster than their livers convert the $P{=}S$ form to the activated $P{=}O$ form, but insects do not have the requisite enzyme to deactivate it and hence it is fatal to them. However, if improperly stored, malathion can be converted to an isomer that is almost 100 times as toxic; it was responsible for the death of five spray workers and the sickness of thousands more during a malaria eradication program in Pakistan in 1976.

Malathion is still used in domestic fly sprays and to protect agricultural crops. It and chlorpyrifos were the organophosphorus insecticides commonly detected in the urine of the 23 schoolchildren from Seattle, Washington, who participated in the experiment in 2003. When they switched to an all-organic-food diet for a week, the insecticides in their urine samples fell immediately to nondetectable levels. The insecticides were again detected once the children in the survey switched back to conventional diets.

In combination with a protein bait, low concentrations of malathion have been sprayed from helicopters over several areas in the United States (California, Florida, Texas) to combat infestations of the Mediterranean fruit fly, a dangerously destructive pest. As in California, aerial spraying of malathion in Chile to combat the fruit fly has also proven to be controversial. It has also been sprayed on parts of New York City (1999) and Florida (1990) to protect against *St. Louis encephalitis*, carried by mosquitoes. For decades, the entire city of Winnipeg, Manitoba, has been sprayed with malathion several times each summer to keep down the mosquito population and now also to protect against the West Nile virus. A recent analysis concluded that the health risks from West Nile virus exceed those from the malathion used to control the mosquitoes that carry it.

Dimethoate (moderately hazardous; LOD_{50} = 250 mg/kg) belongs to the same $O_2SP{=}S$ structural Type C as malathion; it is widely used for insect control on food crops, including those grown in backyards. Like malathion, it is converted into its oxygen analog, which is the metabolite responsible for its toxic action. Some of its uses in the United States have been terminated.

Yet another Type C member, **azinphos-methyl,** is classed as a highly hazardous insecticide (LOD = 5 mg/kg) but has been widely used by professionals on fruits and vegetables. However, all its uses in the United States and Canada are slowly being phased out, since it can pose acute risks to agricultural workers. According to a survey in the mid-1990s, azinphos-methyl and another Type C organophosphate, *phosalone*, together with the fungicide *diphenylamine*, were the most common pesticides present in residues in and on apples grown commercially in Canada.

Carbamate Insecticides

The mode of action of **carbamate** insecticides is similar to that of the organophosphates; they differ in that in carbamates, it is a carbon atom rather than a phosphorus that attacks the acetylcholine-destroying enzyme. They are more attractive for some applications since their dermal toxicity is rather low. The carbamates, introduced as insecticides in 1951, are derivatives of **carbamic acid,** H_2NCOOH. One of the hydrogens attached to the nitrogen is replaced by an alkyl group, usually methyl, and the hydrogen attached to the oxygen is replaced by a longer, more complicated organic group symbolized below simply as R:

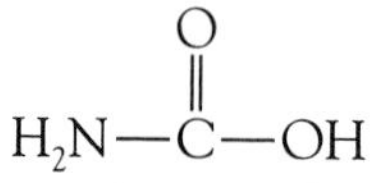

$$H_2N-\overset{\overset{\textstyle O}{\|}}{C}-OH \qquad\qquad CH_3-\overset{\overset{\textstyle H}{|}}{N}-\overset{\overset{\textstyle O}{\|}}{C}-O-R$$

carbamic acid 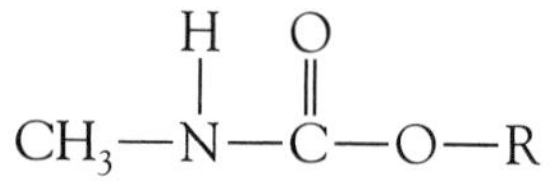   the general formula of a carbamate

Like the organophosphates, the carbamates are short-lived in the environment since they undergo hydrolysis reactions and decompose to simple, nontoxic products. The reaction with water involves the addition of H—OH to the N—C bond; the species HO—C—OR decomposes to release CO_2 and the alcohol R—OH.

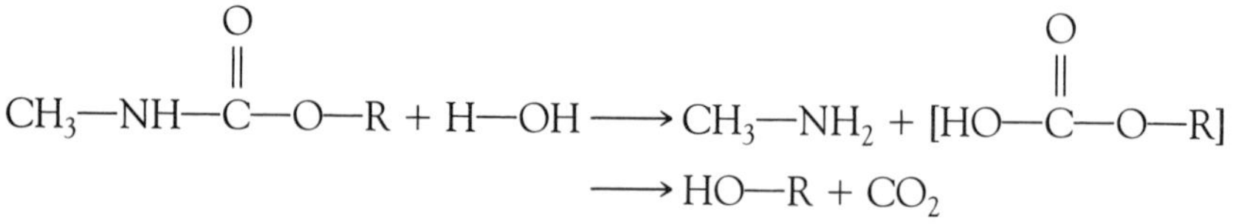

$$CH_3-NH-\overset{\overset{\textstyle O}{\|}}{C}-O-R + H-OH \longrightarrow CH_3-NH_2 + [HO-\overset{\overset{\textstyle O}{\|}}{C}-O-R]$$
$$\longrightarrow HO-R + CO_2$$

Important examples of the carbamate pesticides are **carbofuran** (LOD_{50} = 8 mg/kg), **carbaryl** (LOD_{50} = 307), and **aldicarb** (LOD_{50} = 0.9). Although carbaryl, a widely used lawn and garden insecticide, has a low toxicity to mammals, it is particularly toxic to honeybees. Aldicarb is not only highly toxic to mammals—including humans—it is also somewhat water-soluble (>1 ppm) and persistent, so although it does not bioaccumulate, it can build up in groundwater supplies—and in crops irrigated with the water.

Health Problems of Organophosphates and Carbamates

The organophosphates and carbamates solved the problem of environmental persistence and accumulation associated with organochlorine insecticides, but sometimes at the expense of dramatically increased acute toxicity to the humans and animals who encounter them while the chemicals are still in the active form. These less persistent insecticides—together with the pyrethroids mentioned below—largely replaced organochlorines in residential uses. Organophosphates and carbamates are a particular problem in developing countries, where widespread ignorance about their hazards and failure to use protective clothing—due to lack of information or to the heat—has led to many deaths among agricultural workers. The types of pesticides used in developing countries are also more likely to be highly toxic, even banned elsewhere for that reason. Estimates by the United Nations and the World Health Organization put the number of persons who suffer acute illnesses from short-term exposure to pesticides in the millions annually; 10,000–40,000 die each year from the poisoning, about three-quarters of these in developing countries. Although most of the deaths from pesticide poisonings occur in developing countries, about 20,000 people receive emergency medical care in the United States annually for actual or suspected poisoning from pesticides, and about 30 Americans annually die from it.

Natural and Green Insecticides, and Integrated Pest Management

Pesticides from Natural Sources

As pointed out earlier, many plants themselves manufacture certain molecules for their own self-protection that either kill or disable insects. Chemists have isolated some of these compounds so that they can be used to control insects in other contexts. Examples are nicotine, rotenone, the pheromones, and juvenile hormones.

One group of natural pesticides that has been used by humans for centuries is the **pyrethrins.** The original compounds, the general structure for which is illustrated on the next page, were obtained from the flowers of a species of chrysanthemum.

general pyrethrin structure

In the form of dried, ground-up flower heads, pyrethrins were used in Napoleonic times to control body lice; they are still used in flea sprays for animals. They are generally considered to be safe to use. Like organophosphates, they paralyze insects, though they usually do not kill them. Unfortunately, these compounds are unstable in sunlight. For that reason, several synthetic pyrethrin-like insecticides that are stable outdoors—and therefore can be used in agricultural and garden applications—have been developed by chemists. Semisynthetic pyrethrin derivatives, called **pyrethroids,** are usually given names ending in *-thrin* to denote their nature (e.g., *permethrin*). Pyrethroids are now also a common ingredient in domestic insecticides, as any visit to a garden center will testify. They have also been used in Mexico to spray houses in which malaria control is still necessary and to spray neighborhoods in New York City to reduce mosquito populations carrying the West Nile virus. In order to make them more effective as insecticides, pyrethroid formulations are usually mixed with **piperonyl butoxide,** a semisynthetic derivative of the natural product *safrole*, which can be extracted from sassafras plants. The piperonyl butoxide interferes with the enzyme that insects use to detoxify the pyrethroids, thus making them more potent in destroying the insects.

Pyrethroids are now so widely used that their metabolites were detected in the urine of most of the elementary schoolchildren in the Seattle study group mentioned previously. Interestingly, the pyrethroid levels did not decrease when the children adopted an organic diet. However, the most important contribution to high pyrethroid levels was not dietary but correlated with the use of pyrethrin insecticides by their parents.

Rotenone, a complex natural product derived from the roots of certain tropical bean plants, has been used as a crop insecticide for over 150 years and for centuries to paralyze and/or kill fish. The compound enters fish via their gills and disrupts their respiratory system. It is also highly efficient against insects but is decomposed by sunlight. Rotenone is widely use in hundreds of commercial products, including flea-and-tick powders and sprays for tomato plants. Notice in Table 10-4 that the "natural" insecticides pyrethrins and rotenone are classified as moderately hazardous, since they have about the same acute toxicity as some synthetic ones such as malathion, even though they are often marketed as safer, natural pesticides. There is epidemiological and toxicological evidence that chronic exposure to rotenone can contribute to the onset of Parkinson's disease. Indeed, there is some evidence that exposure to pesticides in general contributes to its incidence.

Integrated Pest Management

In recent years, **integrated pest management** (IPM) strategies have been developed. They combine the best features of various feasible methods of pest control—not just the use of chemicals—into a long-range, ecologically sound plan to control pests so that they do not cause economic injury. Generally speaking, a unique plan is developed for each area and crop, with chemicals used only as a last resort and when the monetary cost of their use will be more than recovered by increased crop yield. The six pest control methods that can be combined are

- chemical control—the use of chemical pesticides, both synthetic and natural;

- biological control—reducing pest populations by the introduction of predators, parasites, or pathogens;

- cultural control—introducing farm practices that prevent pests from flourishing;

- host-plant resistance—using plants that are resistant to attack, including plants adapted by genetic engineering to have greater resistance;

- physical control—using nonchemical methods to reduce pest population; and

- regulatory control—preventing the invasion of an area by new species.

 ## Green Chemistry: Insecticides That Target Only Certain Insects

Insecticides such as the organophosphates and carbamates interrupt the function of specific enzymes that are common to most insects (and to humans). They are thus toxic to a wide array of insect species and are known as **broad-spectrum insecticides.** Although it may be an advantage to kill more than one species with a single pesticide, this often leads to the demise of beneficial insects such as pollinators (bees) and natural enemies (lady bugs and praying mantises) of insects that are pests.

An approach to limiting the environmental effect of an insecticide is to develop insecticides that are toxic to only certain species, i.e., the target organism. One way to accomplish this is to find a biological function that is unique to the target organism and develop an insecticide that interrupts only that function. The Rohm and Haas Company of Philadelphia, Pennsylvania, won a Presidential Green Chemistry Challenge Award in 1998 for the development of *Confirm, Mach 2,* and *Intrepid.* The active ingredients in these insecticides are members of the **diacylhydrazine** family of compounds (Figure 10-8a) and are effective in controlling caterpillars.

Caterpillars are the larval stage of insects such as moths and butterflies, and during the larval stage they must shed their cuticle to grow. The concentration

(a)

(b)

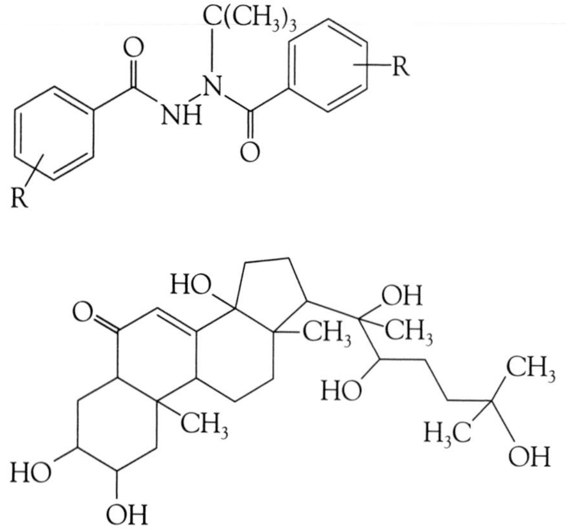

FIGURE 10-8 (a) Diacyl-hydrazine pesticides; (b) 20-hydroxyecdysone.

of *20-hydroxyecdysone* (Figure 10-8b), which is produced by the caterpillar and is a member of the steroid family of compounds, increases during the molting process. As a result of its presence, the caterpillar ceases to feed and sheds its cuticle. The concentration of this natural compound then decreases and the caterpillar resumes feeding. The diacylhydrazines present in the commercial products Confirm, Mach 2, and Intrepid mimic 20-hydroxyecdysone; however, their concentrations do not decline, and consequently the insect never resumes feeding. The insect thus dies of starvation or dehydration. These insecticides target only insects that go through molting stages during their growth; most other insects will be unaffected.

Confirm and Intrepid are classified as **reduced-risk pesticides** by the U.S. EPA. This classification program was started in 1993. To be placed in this category, a pesticide must meet one or more of the following requirements:

1. it reduces pesticide risks to human health;

2. it reduces pesticide risks to nontarget organisms;

3. it reduces the potential for contamination of valued environmental resources; or

4. it broadens the adoption of IPM (integrated pest management, discussed in the preceding section) or makes it more effective.

The diacylhydrazene insecticides certainly meet the first two of these requirements. In order to encourage the development of lower-risk pesticides, the EPA rewards the developers of pesticides that contain active ingredients that meet the EPA reduced-risk criteria with expedited review. (http://www.epa.gov/opppmsd1/PR_Notices/pr97-3.html)

 Green Chemistry: A New Method for Controlling Termites

Termites invade over 1.5 million homes in the United States annually and cause about $1.5 billion in damage. Traditional treatments for termites involve treating the soil around the affected structure with 100–200 gal of pesticide solution to create an impenetrable barrier. This process may result in groundwater contamination, accidental worker exposure, and detrimental effects to beneficial insects.

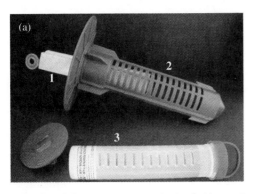

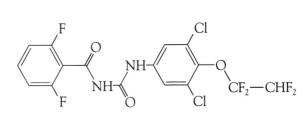

FIGURE 10-9 (a) Sentricon monitoring/baiting station; (b) hexaflumuron structure.
[Source: Photo by Michael Cann.]

Dow AgroSciences in Indianapolis won a Presidential Green Chemistry Challenge Award in 2000 for its development of *Sentricon*. In contrast to the traditional control of termites, Sentricon employs monitoring stations to first detect the presence of termites prior to the use of any insecticide. The monitoring stations (Figure 10-9a) consist of pieces of wood (1) contained in perforated plastic tubes (2), which are placed in the ground around the structure. If termites are detected in any of the monitoring stations, the wood pieces are then replaced by a perforated plastic tube (3) containing the bait. Bait stations may also be placed in the structure. The bait consists of a mixture of cellulose and the pesticide **hexaflumuron** (Figure 10-9b). Hexaflumuron interrupts the molting process of termites and thus is not harmful to most beneficial insects. Termites that have ingested the bait return to their nests and share the bait by trophallaxis, thus spreading the insecticide throughout the colony. Once the colony has been decimated, the bait is replaced with wood and monitoring resumes.

Hexaflumuron was the first pesticide to be classified as a reduced-risk pesticide by the U.S. EPA. Hexaflumuron is significantly less toxic and is used in quantities 100 to 1000 times smaller than traditional pesticides (see Table 10-5) employed for termite control.

TABLE 10-5	Toxicities of Traditional Termiticides vs. Hexaflumuron*	
Pesticide (Compound Type)	**Acute Oral LD$_{50}$ (mg/kg)**	**Acute Dermal LD$_{50}$ (mg/kg)**
Chlorpyrifos (organophosphate)	135–163	2000
Permethrin (pyrethroid)	430–4000	>4000
Imidacloprid (chloronicotinyl)	424–475	>5000
Fipronil (pyrazole)	100	>2000
Hexaflumuron	>5000	>2000

* Typical quantity applied: traditional pesticides, 750–7000 g; hexaflumuron, 2–5 g.

Herbicides

Herbicides are chemicals that destroy plants. They are usually employed to kill weeds without causing injury to desirable vegetation; e.g., to eliminate broad-leaf weeds from lawns without killing the grass. The agricultural use of herbicides has replaced human and mechanical weeding in developed countries and has thereby sharply reduced the number of people employed in agriculture. Herbicides also are used to eliminate undesirable plants from roadsides, railway and powerline rights-of-way, etc., and sometimes to defoliate entire regions. Ever since the late 1960s, herbicides have been the most widely used type of pesticide in North America. As of the early 1990s, about half the U.S. herbicides used were applied to corn, soybean, and cotton crops.

In ancient times, armies sometimes used salt or a mixture of brine and ashes to sterilize land that they had conquered, intending to make it uninhabitable by future generations of the enemy. In the first half of the twentieth century, several inorganic compounds were used as weed killers—principally *sodium arsenite*, Na_3AsO_3; *sodium chlorate*, $NaClO_3$; and *copper sulfate*, $CuSO_4$. The latter two belong to a large group of salts formerly used as herbicidal sprays that kill plants by the rather primitive action of extracting the water from them, while at the same time leaving the land treated in this way still capable of supporting agriculture.

Organic derivatives of arsenic gradually replaced inorganic compounds as agricultural herbicides since they are less toxic to mammals (see Chapter 15). However, both inorganic and metal-organic herbicides have been largely phased out because of their persistence in soil. Completely organic herbicides now dominate the market; their utility is based partially on the fact that they are much more toxic to certain types of plants than to others, so they can be used to eradicate the former while leaving the latter unharmed.

Atrazine and Other Triazines

One modern class of herbicides is the **triazines,** which are based upon the symmetric, aromatic structure shown below, which has alternating carbon and nitrogen atoms in a six-membered benzene-like ring:

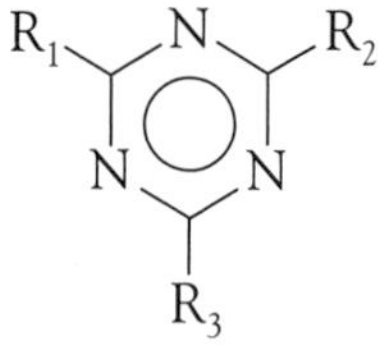

the general formula of the triazines

In triazines that are useful as herbicides, $R_1 = Cl$ and R_2 and $R_3 =$ amino groups, which are nitrogen atoms singly bonded to hydrogens and/or carbon chains.

The best-known member of this group is **atrazine,** which was introduced in 1958 and has been used since that time in huge quantities to destroy weeds in corn fields. Indeed, atrazine is the most heavily used herbicide in the United States (accounting for 40% of all weedkillers applied in the country, including use on 75% of corn crops) and probably the world. In atrazine, R_2 is $-NH-CH_2CH_3$ and R_3 is $-NH-CH(CH_3)_2$.

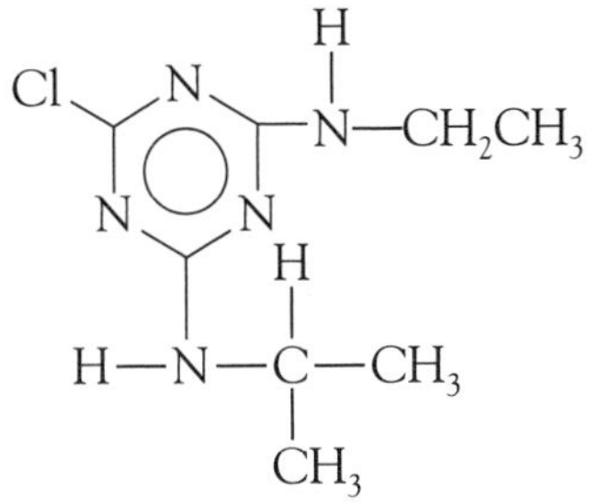

atrazine

It is usually applied to cultivated soils, at the rate of a few kilograms per hectare or one kilogram per acre, in order to kill grassy weeds, mainly in support of corn and soybean cultivation.

Biochemically, atrazine acts as a herbicide by blocking photosynthesis in the plant in the photochemical stage that initiates the reduction of atmospheric carbon dioxide to carbohydrate. Higher plants, including corn, tolerate triazines better than do weeds since they rapidly degrade them to nontoxic metabolites. However, if the triazine concentration builds up in a soil—e.g., because of lack of moisture to degrade it—a stage can be reached where no plants will grow. In high concentrations, atrazine has been used to eliminate all plant life, e.g., to create parking lots.

Some weeds are becoming atrazine-tolerant. The main ecological risk from its widespread use is the death of sensitive plants in water systems close to agricultural fields. Canada has set 2 ppb as the maximum concentration in water for protection of aquatic life. Some controversial recent research regarding the effects of low levels of atrazine on wildlife is discussed in Chapter 12.

While in soil, atrazine is degraded by microbes. One such biochemical reaction results in the replacement of chlorine by a hydroxyl group, $-OH$, yielding a metabolite that is not toxic to plants. The other microbial pathways involve the loss of either the ethyl group or the isopropyl group from an amino unit, with its replacement by hydrogen; these metabolites are toxic to plants.

Although it only persists in most soils for a few months, once it or its metabolites enter waterways, atrazine's half-life is several years. For example, in the Great Lakes, its half-life is about 2–5 years, whereas it is less than half a year in Chesapeake Bay, presumably because the water is warmer so metabolism is faster there.

Atrazine is moderately soluble (30 ppm) in water. During rainstorms, it is readily desorbed from soil particles and dissolved in the water moving through the soil. In waterways that drain agricultural land on which atrazine is used, its concentration typically is found to be a few parts per billion. Usually atrazine is detectable in well water in such regions. The possible risks of atrazine to amphibians are discussed in Chapter 12.

Unfortunately, atrazine is not removed by typical treatments of drinking water unless carbon filtration is used. However, less than 0.25% of the population in the U.S. cornbelt states consume atrazine at greater than 3 ppb, its **maximum contaminant level,** MCL. The MCL values for substances are the maximum permissible concentrations of substances dissolved in the water of any public system in the United States and are based on average annual concentrations. Currently the U.S. EPA is re-evaluating the potential risk of atrazine to humans and the environment. Several European countries have banned atrazine as they would any pesticide that exceeds a level of 0.1 ppb in drinking water, regardless of whether or not it has been proven to be a human health risk.

Since atrazine's measured BCF is less than 10, bioaccumulation does not represent a significant problem. Atrazine is not a very acutely toxic compound (its LOD_{50} is about 2000 mg/kg). However, some surveys on the health of farmers and other individuals exposed to it in high concentrations show disturbing links to higher cancer rates and a higher incidence of birth defects. No definitive studies linking atrazine use to human health problems have as yet been reported. Nevertheless, the U.S. EPA has listed it as a *possible human carcinogen* and has directed states to devise plans to protect groundwater from herbicide contamination.

In certain American agricultural regions, atrazine usage has been banned outright. For a few years, the triazine called **cyanazine**—which has the same chemical formula as atrazine except that one hydrogen in the isopropyl group is replaced by a cyanide group—became quite popular as an agricultural herbicide, but its manufacturer now has voluntarily phased it out of production because of questions about its effect on human health. Other triazines with similar uses are *simazine* and *metribuzin*.

Chloroacetamides and the Occurrence of Pesticides in Groundwater

In some regions where soybeans and corn are grown intensively, atrazine has yielded its status as the herbicide of choice to one of the **chloroacetamides,** which are derivatives of **chloroacetic acid,** $ClCH_2COOH$, in which the —OH group is replaced by an amino group —NR_1R_2. The most prominent herbicides of this type are **alachlor, metolachlor,** and **acetochlor.** These three compounds differ only in minor variations in the complicated organic groups R_1 and R_2 attached to the amino nitrogen. Alachlor is a carcinogen in animals, and metolachlor is suspected of being one as well. The EPA has proposed that the use of alachlor, metolachlor, atrazine, and simazine be carefully managed in areas where they are used intensively since they represent a significant risk to groundwater.

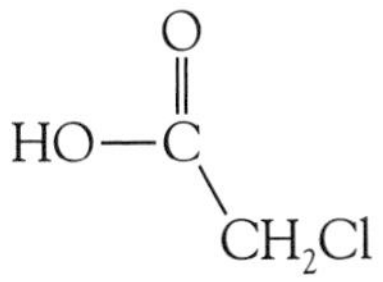

$$HO-\overset{\overset{\displaystyle O}{\|}}{C}\diagdown_{CH_2Cl} \qquad\qquad R_1R_2N-\overset{\overset{\displaystyle O}{\|}}{C}\diagdown_{CH_2Cl}$$

chloroacetic acid 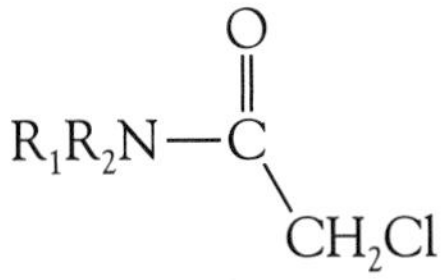   general structure of chloroacetamide herbicides

Generally, the concentrations of these herbicides in waterways that drain agricultural land peak in May and are nondetectable by the end of summer; however, all are somewhat toxic to fish. Alachlor and its degradation products have been detected in groundwater that lies under corn fields. Metolachlor is known to degrade in the environment by the action of sunlight and of water. The chloroacetamides degrade by reaction with water since their amide unit undergoes hydrolysis, producing an amine and chloracetic acid:

$$R_1R_2N-\overset{\overset{\displaystyle O}{\|}}{C}-CH_2Cl + H-OH \longrightarrow R_1R_2N-H + HO-\overset{\overset{\displaystyle O}{\|}}{C}-CH_2Cl$$

Atrazine and its metabolite and metolachlor were the agricultural herbicides most often detected in streams and shallow groundwater in both urban and agricultural areas according to an investigation by the U.S. Geologic Survey in the 1990s. Domestic herbicides found most often were the triazines simazine and prometon. Insecticides found in highest concentrations—principally carbaryl and the organophosphates diazinon, malathion, and chlorpyrifos—were higher in urban than rural regions, presumably because of domestic usage. More than 95% of the streams, and 50% of the groundwater samples, were found to contain at least one pesticide at detectable levels. Research in Switzerland has found levels of atrazine, alachlor, and other agricultural pesticides in rainwater that exceed drinking water standards. Presumably the pesticides evaporated from farm fields.

Glyphosate

Glyphosate is an example of a *phosphonate,* a class of compounds that are structurally similar to organophosphates except that one oxygen of the four that surround phosphorus is missing and is replaced by an organic group, in this case a methylene group, —CH$_2$—, attached to the simple amino acid *glycine*.

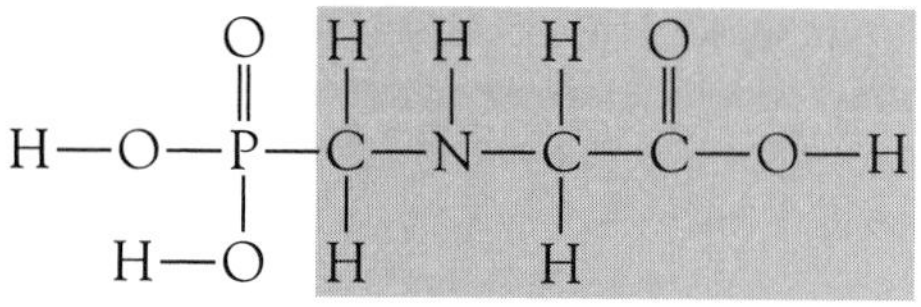

$$\underset{\displaystyle \text{glyphosate}}{\text{H}-\text{O}-\overset{\displaystyle \overset{\text{O}}{\|}}{\underset{\displaystyle \underset{\text{H}-\text{O}}{|}}{\text{P}}}-\overset{\displaystyle \overset{\text{H}}{|}}{\underset{\displaystyle \underset{\text{H}}{|}}{\text{C}}}-\overset{\displaystyle \overset{\text{H}}{|}}{\underset{\displaystyle \underset{\text{H}}{|}}{\text{N}}}-\overset{\displaystyle \overset{\text{H}}{|}}{\underset{\displaystyle \underset{\text{H}}{|}}{\text{C}}}-\overset{\displaystyle \overset{\text{O}}{\|}}{\text{C}}-\text{O}-\text{H}}$$

glyphosate

Glyphosate is widely used as a herbicide, e.g., as the commercial product *Roundup*. It is rather nontoxic: its LD$_{50}$ values are high for both oral and dermal routes of exposure, although acute ingestion of or exposure to large quantities of it is fatal. Dermal and oral absorption of it is small, and it is eliminated essentially unmetabolized. Glyphosate is nonresidual, and there is no evidence that it bioaccumulates in animal tissue or is carcinogenic or teratogenic. The same is true of its initial breakdown product, the substance corresponding to cleavage of the rightmost NH—CH$_2$ bond in the structure above.

Glyphosate operates by inhibiting the synthesis of amino acids containing the aromatic benzene ring, which in turn prevents protein synthesis from occurring. Although it kills almost all plants, some strains of soybeans have been genetically altered using biotechnology so that they are resistant to glyphosate; consequently, it can be used as a weed killer in the growth of the crop (see Box 10-2). Its advantages in growing soybeans are that it replaces several different herbicides and that only one application is required, though the total volume of herbicide used is not reduced substantially. Its greater tendency to stay adsorbed on soil means that it has a lesser tendency to occur in runoff and subsequently in water supplies than do the herbicides atrazine and alachlor, which it replaces. The evidence gathered so far indicates that glyphosate is a relatively benign herbicide.

Phenoxy Herbicides

Phenoxy weedkillers were introduced at the end of World War II. Environmentally, the by-products contained in commercial products of such herbicides are often of greater concern than the herbicides themselves, as we shall see in Chapter 11. For that reason, we begin by discussing the chemistry of **phenol,** the fundamental component of these compounds.

Phenols are mildly acidic; in the presence of concentrated solutions of a strong base like NaOH, the hydrogen of the OH group is lost as H$^+$ (as occurs

BOX 10-2 | Genetically Engineered Plants

In 1940 the world population was 2.3 billion people; by 1985 it had more than doubled, and it now exceeds 6.5 billion. Fortunately, beginning in the 1940s, a "green revolution" in agriculture took place that allowed the world to feed this burgeoning population. Extensive development and use of pesticides (many of which have been mentioned in this chapter) and fertilizers, along with irrigation and plant breeding programs, led to dramatic increases in the yield per acre of crops. Total worldwide grain production increased from 600 million metric tons in 1950 to more than 1600 million metric tons by 1985. Since 1995, production has leveled off at 1800–2000 metric tons. However, the human population continues to grow and is expected to reach 9 billion by 2050.

Since the 1980s, talk of a second green revolution has centered about genetically engineered plants. Traditional crossbreeding of wheat plants over many years has resulted in plants that yield two to three times more grain than previously existing varieties and are more resistant to pests and diseases. Genetic engineering of plants offers the possibility of doing these same things and additional feats in much less time and with more selectivity than traditional crossbreeding.

Genetic engineering involves taking a portion of the DNA from one species and inserting it into the DNA of another, unlike species. One striking example of this technique has been to take the human DNA (gene) that codes for the synthesis of the protein insulin and insert it into that of bacteria, thus allowing the bacteria to produce insulin. This results in the production of human insulin to be used for medical purposes.

Transgenic plants have been produced which have enhanced resistance to herbicides,

drought, pests, salinity, and frost, as well as improved taste and nutritional value. The best-known examples of herbicide-resistant plants that have been developed are known as *Roundup Ready*. Roundup, as was previously mentioned, is a commonly used broad-spectrum herbicide. Monsanto, its manufacturer, has developed and patented genetically altered seeds for soy, corn, alfalfa, sorghum, canola, and cotton which grow into plants that are resistant to destruction by Roundup. Fields planted with these crops can be sprayed indiscriminately to destroy weeds, with little concern for destruction of the crop.

The use of transgenic plants has been widely adopted in the United States. In 2005, 87% of all soybean acreage in the United States was planted with transgenic crops, followed by cotton at 79% and corn at 52%. The top five countries in growing transgenic crops in 2005 were the United States, Argentina, Brazil, China, and Canada.

Although transgenic plants offer the possibility of improving upon what nature has provided us, there are significant concerns about these organisms, especially in Europe. Concerns include:

- the use of greater quantities of herbicides, since there is less fear of destroying a crop from the indiscriminate application of the herbicide;

- the spread of herbicide resistance to related plants that become "super weeds"; and

- the decrease in genetic diversity of crops as farmers all use the same seeds.

In addition to these concerns, genetically engineered grains have not resulted in substantial increases in crop yields.

with any common acid) and the *phenoxide anion*, $C_6H_5O^-$, is produced in the form of its sodium salt:

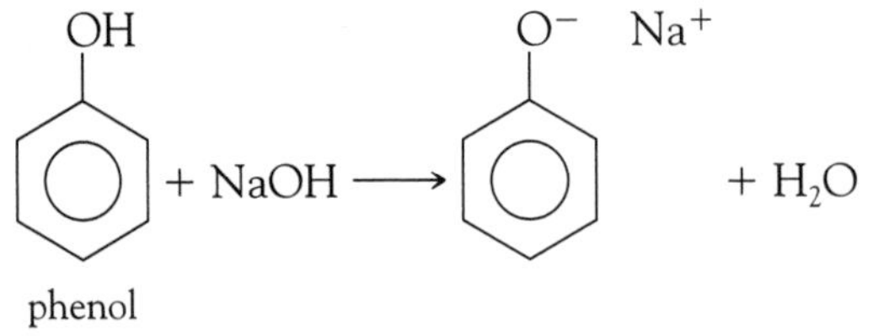

phenol

The O^-Na^+ group is a reactive one, and this property can be exploited in order to prepare molecules containing the C—O—C linkage. Thus if an R—Cl molecule is heated together with a salt containing the phenoxide ion, NaCl is eliminated and the phenoxy oxygen links the benzene ring to the R group:

$$C_6H_5O^-Na^+ + Cl—R \longrightarrow C_6H_5—O—R + NaCl$$

Such a reaction is the most direct commercial route to the large-scale preparation of the herbicide, introduced in 1944, whose well-known commercial name is **2,4,5-T.** Here (in the reaction immediately above) the R group is **acetic acid,** CH_3COOH, minus one of its methyl group hydrogens, so that R = $—CH_2COOH$, and the Cl—R reactant is Cl—CH_2COOH. Then, according to the reaction, $C_6H_5—O—CH_2COOH$, called **phenoxyacetic acid,** is obtained as an intermediate in the production of the actual herbicides.

In the commercial herbicides, some of the five remaining hydrogen atoms of the benzene ring in phenoxyacetic acid are replaced by chlorine atoms.

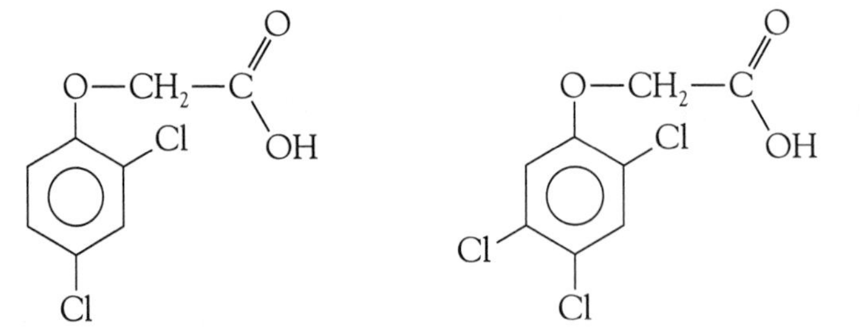

2,4-D
2,4-dichlorophenoxyacetic acid

2,4,5-T
2,4,5-trichlorophenoxyacetic acid

Note that the numbering scheme for the benzene ring begins at the carbon attached to the oxygen.

The **2,4-D** compound (**2,4-dichlorophenoxy acetic acid**) is used to kill broad-leaf weeds in lawns, golf course fairways and greens, and agricultural fields. In contrast, 2,4,5-T (**2,4,5-trichlorophenoxy acetic acid**) is effective in clearing brush, for instance, on roadsides and powerline corridors. Like the P—O—C bonds in organophosphates, the O—C bond to the —CH$_2$— group in 2,4-D and analogous phenoxy herbicides undergoes a hydrolysis reaction in the environment, degrading the compound to a phenol

$$R—O—CH_2—COOH + H—OH \longrightarrow R—O—H + HOCH_2—COOH$$

The herbicide MCPA is 2,4-D with the chlorine in the 2 position replaced by a methyl group, —CH$_3$. The herbicides called *dichlorprop*, *silvex*, and *mecoprop* are identical to 2,4-D, to 2,4,5-T, and to MCPA, respectively, except that their molecules have a methyl group replacing one hydrogen atom of the —CH$_2$— group in the acid chain; thus they are phenoxy herbicides that are based on **propionic acid,** CH$_3$—CH$_2$—COOH, rather than on acetic acid. The herbicide *dicamba* is the same as 2,4-D with a methoxy group at the 5 position of the benzene ring; it is often used as a weedkiller in corn fields.

Huge quantities of 2,4-D and its closely related analogs described above are used in developed countries for the control of weeds in both agricultural and domestic settings. In some communities, their continued use on lawns has become a controversial practice because of their suspected effects on human health. In particular, farmers in the midwestern United States who mix and apply large quantities of 2,4-D to their crops are found to have an increased incidence of the cancer known as non-Hodgkin's lymphoma.

The Degradation of Pesticides

Although some pesticides such as DDT are very long-lived in the environment, most undergo chemical or biochemical reactions within a few days or months, producing other compounds. Based upon their typical half-lives in the environment, the U.S. EPA classifies pesticides as being:

- *nonpersistent*, if they last less than 30 days;

- *moderately persistent* for those lasting 30–100 days; and

- *persistent* for lifetimes greater than 100 days.

Like most organic compounds, pesticides in the environment—whether present in air, water, or soil—degrade to other compounds, which in turn decompose further. The complete eventual breakdown of organic compounds to CO$_2$, H$_2$O, and stable inorganic forms of its other elements is called *mineralization*.

In air, the degradation process usually begins either with attack on the organic molecule by the hydroxyl radical, OH, or with a photochemical reaction if the substance absorbs light with wavelength greater than about 285 nm, in accordance with the principles discussed in Chapter 5.

Photochemical decomposition is possible also for pesticides present in water or adsorbed onto soil resident at the Earth's surface. In some instances, adsorption onto soil particles increases the maximum wavelength of light the substance absorbs into the range in sunlight—an example is the herbicide **paraquat,** which undergoes photolysis more rapidly when adsorbed on clay than it does in solution. Complexation of organic molecules by metal ions also usually increases their maximum wavelength of absorption, thereby activating them for photochemical decomposition by sunlight in some cases.

As we already have discussed several times in this chapter, pesticides in water and in soil can undergo hydrolysis reactions, especially when the water is somewhat acidic or somewhat basic, since catalysis by H^+ or OH^- can then speed up the processes significantly. Organophosphate insecticides, for example, hydrolyze in alkaline water and soil owing to attack by OH^- on the P—O—C link. Even in quite dry soils, hydrated aluminum ions produce hydrogen ions that in the existing moisture can catalyze hydrolysis.

$$Al(H_2O)_6{}^{3+} \longrightarrow Al(H_2O)_5OH^{2+} + H^+$$

For example, in triazine herbicides, hydrolysis can convert their C—Cl bonds to C—OH ones, thereby eliminating their herbicidal activity. Organic compounds, including pesticides, can also be transformed in water or soil by oxidation or reduction reactions. Although dissolved O_2 itself can oxidize, its reactions are often accelerated by the presence of dissolved or adsorbed transition metal ions, which oxidize the pesticide and whose reduced form is subsequently re-oxidized by O_2. For example, Fe^{3+} is a good oxidizing agent for many organic compounds; the Fe^{2+} state to which it is reduced in the process is subsequently oxidized by oxygen back to Fe^{3+}, thereby completing the cycle.

Reducing agents are commonly found in anaerobic waters and soil; they include Fe^{2+} and *sulfide ion,* S^{2-}. For example, pesticides that contain a C—Cl unit are dechlorinated by iron when it abstracts an electron from the C—Cl bond, thereby releasing Cl^- and forming Fe^{3+} and a reactive carbon-based free radical.

Even more important than the chemical processes described above are degradation reactions facilitated by microbial action in water and soil. *Chemheterotrophs* are microorganisms that derive the energy they require from redox reactions and their carbon from organic compounds. The metabolic reactions proceed in stepwise fashion, the individual steps usually being oxidation, reduction, or hydrolysis. However, the rates of degradation vary over a very wide range, depending on the molecular structure of the pesticide and the properties of the soil. Compounds containing functional groups such as —OH, —NO_2, —NH_2, and carboxylate degrade most readily in soils since they contain a site for enzymatic attack and are relatively soluble in water, whereas highly chlorinated hydrocarbons are much more resistant since there is no reactive site and their water solubility is very low.

A common example of a microbial oxidation step is enzyme-catalyzed **epoxidation,** a process in which an oxygen atom from an O_2 molecule is added to a C=C bond, even one contained within an aromatic benzene ring system:

Following epoxidation, the adduct can undergo further reactions, for example,

- rearrangement to a hydroxylated compound, thereby reestablishing the highly stable aromatic ring; or

- hydrolysis to produce an ortho-dihydroxyl compound; or

- the addition of further oxygen and water to other double bonds within the aromatic system.

Subsequent reactions at an adjacent pair of carbons having —OH groups often lead to ring cleavage at that site, yielding a dicarboxylic acid.

Summary

In general, there is no pesticide that is completely "safe." However, the elimination of all synthetic pesticides would lead to an increase in the transmission of disease by insects and an increase in the cost of food, both of which would affect human health adversely. Any decision about discontinuing the production and use of a given pesticide must consider whether cheap, safer alternatives are available and, if not, what the consequences are of both action and inaction. The quandary about whether to ban the use of DDT in tropical developing countries is an excellent illustration.

When a new pesticide, or indeed any other synthetic chemical, is about to be introduced into the market, many environmental groups and some government agencies have proposed that we should err on the side of being too cautious and only allow its introduction if there are no signs that problems could arise. They propose that in such situations, to prevent possible harm to the health of humans and other organisms, we should employ what is now known as the **precautionary principle.** One definition of this principle was given at the 1992 U.N. Conference in Rio on Environment and Development: "Where there are threats of serious or irreversible damage, lack of full scientific certainty shall not be used as a reason for postponing cost-effective measures to prevent environmental degradation." Opponents of the use of this principle point out that it is impossible to anticipate all possible consequences, positive or negative, of introducing a new substance and that consequently we could become frozen into inaction. The best technique for predicting where a given pesticide will ultimately end up in the environment is through the calculations described in Box 10-3.

BOX 10-3 | The Environmental Distribution of Pollutants

When a persistent chemical, such as DDT, is released into the environment, we find that later some of it has dissolved in natural bodies of water, some is in the air, some is present in soil and sediments, and some is located in living matter. A constant interchange of the chemical occurs among these various physical phases. It is possible to estimate the amount and concentration of the chemical in each phase once the release of the chemical has stopped and sufficient time has passed that equilibrium among the phases has been achieved. Even when equilibrium conditions are not yet in place, it is of value to determine the phases where the chemical will ultimately be concentrated.

Recall from your previous background in chemistry that in calculations involving substances participating in *chemical* reactions, we algebraically combine experimental values of equilibrium constants with information concerning initial concentrations in order to determine equilibrium concentrations. A somewhat analogous procedure can be applied to determine the distribution of a substance when by *physical* processes it has achieved equilibrium between several phases. The condition that equilibrium has been achieved in its distribution is that the **fugacity,** f, of the substance, which is defined as *its tendency to escape from a given physical phase*, is equal for all phases. Fugacity has units of pressure, e.g., atmospheres or kilopascals. Thus, for example, when all the DDT in the environment has distributed itself among air, water, sediment, biota, etc., the concentrations in each phase are such that its tendency to escape from any phase (and enter any other) has the same value for all phases.

As you might expect, the fugacity of a substance in a given phase is proportional to its concentration, C, in that phase:

$$f = C/Z$$

where Z is the *fugacity capacity constant* for the substance and the phase. Generally, the higher value of Z, the greater the tendency of a chemical to concentrate in that phase. (These capacity constants are analogous to the equilibrium constants used in chemical reaction calculations.) If we use x to denote the phase of interest, then

$$f_x = C_x/Z_x$$

We can determine the concentration in each phase by rearranging the equation, to give

$$C_x = f_x Z_x$$

At equilibrium, the f_x values for all phases are identical, equal to f, say. Thus, if we know f, we can determine the concentration in each phase from the simplified equation

$$C_x = f Z_x$$

As in chemical equilibrium problems, we usually know the total number of moles, n_{total}, of the material. As in chemical problems, it is useful to state the mass conservation condition: the sum of the equilibrium number of moles, n_x, present in each phase x must add up to n_{total}. By definition, each n_x is equal to the concentration C_x times the volume V_x for the phase:

$$n_x = C_x V_x$$

Substitution of the next-to-last equation into the last one gives

$$n_x = f Z_x V_x$$

When we sum the n_x values over all phases x of interest, we must obtain the total number of moles. Thus

$$n_{total} = f \Sigma Z_x V_x$$

Rearrangement of this equation allows us to calculate the value of the system fugacity:

$$f = n_{total}/\Sigma Z_x V_x$$

An Example of a Fugacity Calculation

As an example of how fugacity calculations are carried out in practice, consider the distribution of 1 mole of DDT among three phases: air, water, and sediment in a model compartment of Earth (Figure 1). As discussed later, we take the volume of air to be 10^{10} m^3, the water volume to be 7×10^6 m^3, and the volume of accessible sediment to be 2×10^4 m^3. The values of the Z_x constants for DDT, in units of mol/atm m^3, are determined from experimental data to be

for the air phase, 40.3
for the water phase, 3.92×10^4
for the sediment phase, 2.25×10^9

In the evaluations of Z_x values from experimental data, a temperature of 25°C is usually assumed for simplicity. The Z_x values for sediment (and biota) are assumed to be proportional to the octanol–water partition coefficients K_{ow} discussed earlier in the chapter.

After substitution of the values for Z_x and V_x, the value of the fugacity in this case is

$$f = 1.0/(40.3 \times 10^{10} + 3.92 \times 10^4 \times 7 \times 10^6 + 2.25 \times 10^9 \times 2 \times 10^4)$$
$$= 1.0/(4.03 \times 10^{11} + 2.74 \times 10^{11} + 4.5 \times 10^{13})$$
$$= 2.19 \times 10^{-14} \text{ atm}$$

The concentration of the chemical can now be computed for each phase:

$$C_x = f Z_x$$

(continued on p. 464)

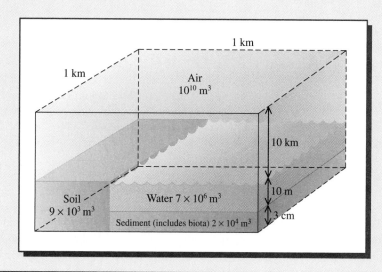

FIGURE 1 Model world parameters used in fugacity calculations.

BOX 10-3 | The Environmental Distribution of Pollutants *(continued)*

so

DDT concentration in air = $2.19 \times 10^{-14} \times 40.3 = 8.8 \times 10^{-13}$ mol/m^3

DDT concentration in water = $2.19 \times 10^{-14} \times 3.92 \times 10^4 = 8.6 \times 10^{-10}$ mol/m^3

DDT concentration in sediment = $2.19 \times 10^{-14} \times 2.25 \times 10^9 = 4.9 \times 10^{-5}$ mol/m^3

Notice the preferential concentration of DDT in sediment, which is hydrophobic due to its carbon content.

The *amounts* in each phase are given by the *fZV* values, i.e., the concentrations multiplied by the respective volumes. Then the number of moles of DDT

in air = $8.8 \times 10^{-13} \times 1 \times 10^{10} = 0.0088$ mol

in water = $8.6 \times 10^{-10} \times 7 \times 10^6 = 0.0060$ mol

in sediment = $4.9 \times 10^{-5} \times 2 \times 10^4 = 0.98$ mol

Thus we see that, with air, water, and sediment accessible to it, 98% of the DDT will be found in sediment, and about 1% in air and in water. Notice that the concentration of DDT in water is greater than in air, but the total amount of it in air exceeds that in water because the air volume is so much larger. This sort of interchange in ordering between amount and concentration in different phases is common for pollutant chemicals.

The Parameters for the Model World in Fugacity Calculations Are Estimates

The volumes for the various phases used in the above calculation are based upon a model "world" (Figure 1) whose components are able to be in equilibrium with each other. Since only concentrations are obtained in the calculations, it is important only that the *relative* volumes, not their absolute values, be used. The model world has an area of 1 kilometer by 1 kilometer, whose characteristics are assumed

to be average for the real Earth. The atmosphere is taken to be 10 kilometers high, which is a reasonable approximation to the troposphere. The air volume then is (1000 m × 1000 m) × (10,000 m) = 10^{10} m^3. The 1-km square is assumed to be 70% covered by water and 30% by soil. The average water depth is taken to be 10 meters, which is relatively shallow since we are interested only in the part that achieves equilibrium with the air. Thus the water volume is 0.7 × (1000 m × 1000 m) × 10 m = 7×10^6 m^3. The sediment in equilibrium with this water is assumed to be only 3 centimeters deep, giving it a volume of 0.7 × 1000 m × 1000 m × 0.03 m = 2.1×10^4 m^3. In addition to air, water, and sediment, the model usually also includes soil, whose effective volume is 9×10^3 m^3, plus 35 m^3 of solids suspended in the water, and about 3.5 m^3 of biota such as fish. The Z values for biota are usually of the same order of magnitude as those for sediment, so the concentration of a given chemical in biota is close to that in sediment.

PROBLEM 1

The Z values for hexachlorobenzene are 4×10^{-4} in air, 9.5×10^{-5} in water, and 2.3 in sediment (and biota). Using the model world volumes above, calculate the equilibrium concentrations when 1 mole of hexachlorobenzene is distributed among air, water, and sediment.

PROBLEM 2

In fugacity calculations, the Z values for dieldrin are 4×10^{-4} in air, 2.0 in water, and 2×10^{-5} in sediment (and biota). Using the model world volumes above, calculate the equilibrium concentrations when 1 mole of dieldrin is distributed among air, water, and sediment.

Review Questions

1. What are the three main categories of pesticides? What types of organisms are killed by each category?

2. What is meant by the term *fumigant*?

3. Name three important properties shared by organochlorine pesticides.

4. Draw the structure of DDT, and state what the initials stand for.

5. What units are usually used to state the concentrations of trace contaminants in water?

6. What were the main uses of DDT? Explain why it is no longer used in many developed countries and why some developing countries wish to continue using it.

7. Explain how DDT functions as an insecticide.

8. Draw the structure of DDE. Is it a pesticide or not? Explain.

9. Explain what is meant by the terms *bioconcentration* and *bioconcentration factor* (BCF).

10. Explain what is meant by the term *biomagnification*, and how it differs from bioconcentration.

11. Write the defining equation for the partition coefficient K_{ow}. How is it related to a compound's BCF? What is octanol supposed to be a surrogate for in this experiment?

12. Describe one analog of DDT that works in the same fashion but does not bioaccumulate.

13. In general terms, explain what toxaphene is and why it is no longer in use.

14. Draw the structure of cyclopentadiene. Name at least three insecticides produced from it.

15. Define the terms *acute toxicity* and *dose*.

16. Sketch a typical dose–response curve relationship for a toxic chemical using **(a)** a linear, and **(b)** a logarithmic scale for doses.

17. Define the terms LD_{50} and LOD_{50}.

18. What are the general structures of the three main subclasses of organophosphate insecticides? Give the name of one insecticide in each subclass. Explain how organophosphates function as insecticides.

19. In what way are organophosphate insecticides considered superior to organochlorines as pesticides? In what way are they more dangerous?

20. What is the general structure of carbamate insecticides? Name one example.

21. What are five of the pest control methods that are used in pest control management?

22. What is the function of an herbicide? Name a few "old-fashioned" insecticides.

23. What is the general structure of a triazine herbicide? Name two commercial examples.

24. What is the general structure of chloroacetamide herbicides? Name one example.

25. What is the formula of gyphosate? What are its advantages over other herbicides?

26. What is phenol? Draw its structure and that of 2,4-dichlorophenol.

27. Draw the structures, and write out the names, of the two most important phenoxy herbicides.

28. What is meant by the *precautionary principle*?

29. Write out three examples of hydrolysis reactions by which pesticides are degraded in the environment.

Green Chemistry Questions

See the discussion of focus areas and the principles of green chemistry in the Introduction before attempting these questions.

1. The development of the insecticides Confirm, Mach2, and Intrepid won a Presidential Green Chemistry Challenge Award.
(a) Into which of the three focus areas for these awards does this award best fit?
(b) List one of the twelve principles of green chemistry that are addressed by these new pesticides.

2. What environmental advantage do Confirm, Mach2, and Intrepid offer compared to conventional pesticides?

3. **(a)** What is a U.S. EPA *reduced-risk pesticide?*

(b) Which categories under the reduced-risk criteria do Confirm, Mach2, and Intrepid meet?

4. How do Confirm, Mach2, and Intrepid act to only target specific insects?

5. The development of the Sentricon system won a Presidential Green Chemistry Challenge Award.
(a) Into which of the three focus areas for these awards does this award best fit?
(b) List two of the twelve principles of green chemistry that are addressed by the hexaflumuron/Sentricon system?

6. What environmental advantages does the hexaflumuron/Sentricon system offer compared to conventional termite control pesticide methods?

7. Which categories under the reduced-risk criteria does the hexaflumuron/Sentricon system meet?

Additional Problems

1. The threshold/NOEL level found for a particular chemical from animal studies is 0.004 mg/kg body weight per day. The only source for the chemical is freshwater fish, where it occurs at an average level of 0.2 ppm. What is the maximum average daily consumption of such fish that would keep exposure level below the ADI or RfD for the compound?

2. An approximate mathematical fit to the form of the dose–response curve of Figure 10-5a is
$$R = 1 - e^{-d}$$
where R is the fractional response and d is the dose.
(a) Plot R versus d for values of d ranging from 0 to 5 on both linear and logarithmic scales for d. (Be sure to include some small values of d, from 0.01 to 0.10, in the logarithmic plot to ensure that the form of the curve near zero is displayed.) Do the forms of the curves resemble those in Figures 10-5a and 10-5b, respectively?

(b) Both from your graphs and by solving the equation above analytically, find the dose corresponding to LD_{50}.
(c) Does the function R have a nonzero threshold at low doses? Can you confidently predict the answer to this from inspecting your logarithmic dose–response curve?

3. The fat (lipid) content of breast milk averages about 4.2 g/100 mL. Based on Figure 10-1b, calculate the mass of DDE that would have been ingested by a typical breast-fed Swedish infant in 1972 consuming 250 mL of breast milk.

4. The BCF for a substance in a particular aquatic species (not just its fat tissues) can be estimated as the K_{ow} value for the substance times the fraction of body fat in the species of interest. Rainbow trout, which average 5.0% body fat, taken from a particular lake were found to contain 22 ppb in their tissues.

Use the information in Table 10-3 to determine the concentration of parathion in the lake.

5. The pesticide azinphos-methyl has a 96-hour LC_{50} value of 3 ppb for rainbow trout. In an unfortunate incident, 200 g of this pesticide was sprayed on a field, and a subsequent heavy rainfall washed 35% of it into a nearby lake having a surface area of 30,000 m^2 and an average depth of 0.5 m. Would the pesticide concentration in the lake water have been sufficient to kill a significant fraction of the rainbow trout in the lake?

Further Readings

1. V. Turusov et al., "DDT: Ubiquity, Persistence, and Risks," *Environmental Health Perspectives* 110 (2002): 125.

2. B. Hileman, "Reexamining Pesticide Risk," *Chemical and Engineering News* (17 July 2000): 34.

3. G. Santaoro, "Silent Summer," *Discover* (July 2000): 76.

4. M. Lopez-Cervantes et al., "Dichlorodiphenyl-trichlorethane Burden and Breast Cancer Risk: A Meta-Analysis of the Epidemiological Evidence," *Environmental Health Perspectives* 112 (2004): 207.

5. K. Noren and D. Meironyte, "Certain Organochlorine and Organobromine Contaminants in Swedish Milk in Perspective of Past 20–30 Years," *Chemosphere* 40 (2000): 1111.

6. (a) C. Lu et al., "A Longitudinal Approach to Assessing Urban and Suburban Children's Exposure to Pyrethroid Pesticides," *Environmental Health Perspectives* 114 (2006): 1419. (b) C. Lu et al., "Organic Diets Significantly Lower Children's Dietary Exposure to Organophosphate Pesticides," *Environmental Health Perspectives* 114 (2006): 260.

7. G. M. Williams et al., "Safety Evaluation and Risk Assessment of the Herbicide Roundup and Its Active Ingredient, Glyphosate, for Humans," *Regulatory Toxicology and Pharmacology* 31 (2000): 117.

8. T. P. Brown et al., "Pesticides and Parkinson's Disease—Is There a Link ?" *Environmental Health Perspectives* 114 (2006): 156.

Websites of Interest

Log on to www.whfreeman.com/envchem4/ and click on Chapter 10.

CHAPTER 11

DIOXINS, FURANS, AND PCBs

In this chapter, the following introductory chemistry topics are used:

- Elementary organic chemistry (as in the Appendix in this book)
- First-order kinetics rate law
- Concept of vapor pressure

Background from previous chapters used in this chapter:

- Structure of phenols (Chapter 10)
- Carcinogens and toxicology concepts, including LD_{50} (Chapter 10)
- Concept of adsorption (Chapter 4)

Introduction

As we have seen in Chapter 10, compounds used as pesticides are somewhat toxic to humans, and can bioaccumulate and cause environmental problems. But sometimes it is the highly toxic trace impurities in commercial lots of such substances that are the principal concern regarding human health. In this chapter, we shall analyze how such hazardous by-products, especially dioxins, arise in the environment, not only from pesticide manufacture but from other anthropogenic processes as well. Also considered are PCBs, industrial chemicals of widespread environmental concern with respect to both their own properties and those of their contaminants. As we shall see, the toxicity mechanisms by which these contaminants, the PCBs themselves, and dioxins operate all are quite similar.

Dioxins

Dioxins are not articles of commerce, nor are they normally produced deliberately for purposes other than scientific investigation. They arise as by-products in the production of certain herbicides and in some other processes, as we shall see below.

Dioxin Production in the Preparation of 2,4,5-T

Traditionally, the industrial synthesis of the herbicide called *2,4,5-T* (discussed in Chapter 10) started with *2,4,5-trichlorophenol*, which itself was produced by reacting NaOH with the appropriate *tetrachlorobenzene*. The OH group replaces one chlorine atom in the process. Unfortunately, during this synthesis there occurs an additional reaction that converts a very small portion of the trichlorophenol product into "dioxin." In this side reaction, two *trichlorophenoxy anions* react with each other, resulting in the elimination of two chloride ions:

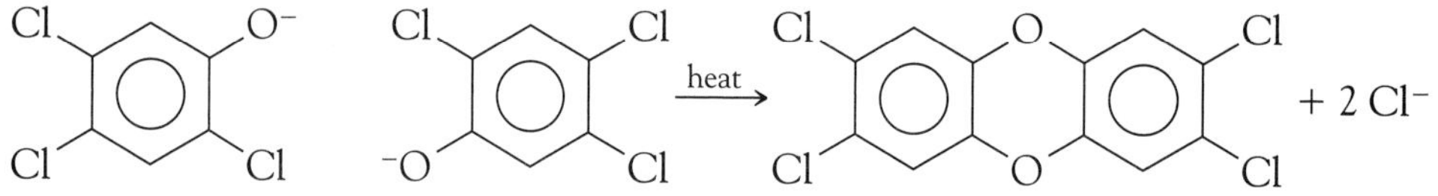

"dioxin"
(tetrachlorodibenzo-*p*-dioxin)

In this process a new six-membered ring is formed which links the two chlorinated benzene rings. This central ring has two oxygen atoms located *para* (i.e., opposite) to each other, as is found in the simple molecule **1,4-dioxin** or **para-dioxin** (*p*-dioxin).

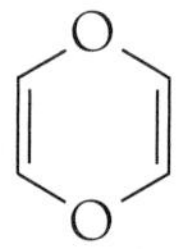

1,4-dioxin

Although the molecule labeled "dioxin" above is correctly known as a *tetrachlorodibenzo-p-dioxin*, it has become popularly known simply as "dioxin," with the understanding that it is the most toxic of a class of related compounds.

The side reaction that produces dioxin as a by-product is kinetically second-order in chlorophenoxide. In other words, the rate of the reaction depends on the square (or second power) of the ion's concentration. Consequently, the rate of dioxin production increases dramatically as the initial chlorophenoxide ion concentration increases. Also, the rate of this side reaction increases more rapidly with increasing reaction temperature than does the main reaction. Therefore, the extent to which the trichlorophenol and consequently the commercial herbicide become contaminated with the dioxin by-product can be minimized by controlling concentration and temperature in the preparation of the original trichlorophenol. Today, the contamination of commercial 2,4,5-T by this dioxin can be kept to less than 0.1 ppm by keeping both the phenoxide concentration and the temperature low. Nevertheless, its manufacture and use in North America were phased out in the mid-1980s because of concerns about its dioxin content, however small.

A 1:1 mixture of the herbicides 2,4-D and 2,4,5-T called *Agent Orange* was used extensively as a defoliant during the Vietnam War. Since the mixture contained dioxin levels of about 10 ppm, it is clear that the reaction used to produce the trichlorophenol used for 2,4,5-T preparation was not carefully controlled so as to minimize contamination. As a result, the soil in southern Vietnam is contaminated by dioxins. The consequences of this contamination for the residents, and for the American troops who were exposed while spraying was underway, are still controversial. There is some evidence that the rate of melanoma has increased in Air Force personnel who were involved in the spraying. It was realized recently that the defoliation potential of Agent Orange was originally tested in the 1960s by the U.S. armed forces near Gagetown, New Brunswick, Canada, and that the soil in the area is polluted by the substance.

Environmental contamination by dioxin also occurred as the result of an explosion in a chemical factory in Seveso, Italy, in 1976. The factory produced 2,4,5-trichlorophenol from tetrachlorobenzene, as described above. On one occasion the reaction was not brought to a complete halt before the workers left for the weekend. The reaction continued unmonitored, and the heat subsequently released by the reaction eventually resulted in an explosion. Since the trichlorophenol had been heated to a high temperature, a considerable amount of dioxin—probably several kilograms—was produced. The explosion distributed the toxin into the environment, and many wildlife deaths resulted from the contamination. Although a large number of humans, both adults and children, were also exposed to the chemical as a result of this explosion, no serious health effects to humans were found for many years. Recent studies, however, have established that the rates of several types of cancers are elevated in people who lived in the zones most exposed to dioxin from the explosion. Specifically, the risk of contracting breast cancer increased in proportion to the dioxin exposure, as measured by the level of the substance in women's blood samples taken soon after the explosion.

Dioxin Numbering System

The nomenclature and numbering system used for ring systems like the dioxins is a little unusual. Since the central dioxin ring is connected on either side to benzene rings, the three-ring unit is properly known as **dibenzo-*p*-dioxin.**

The chlorine substitution on the outer rings also should be recognized, so the dioxin shown below is a **tetrachlorodibenzo-*p*-dioxin,** or **TCDD.**

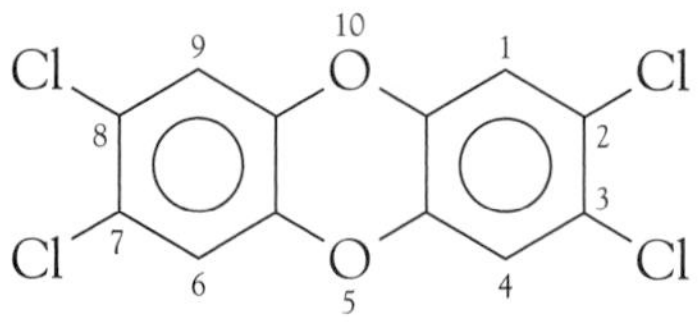

2,3,7,8-tetrachlorodibenzo-*p*-dioxin
(2,3,7,8-TCDD)

The numbering scheme for the ring carbons in dioxins takes into account the fact that the carbons shared between two rings carry no hydrogen atoms and so need not be numbered. Thus C-1 is the carbon next to one joining the rings, and the numbering follows a direct path from there. By convention, the oxygen atoms are also part of the numbered sequence in this scheme, although their locations are not used in naming any of this family of compounds. The initial (C-1) position for the numbering system is chosen to give the lowest possible value to the first substituent; if there is a choice after this criterion has been applied, then that which gives the lowest number to the second substituent is used, etc. Applying these rules, the dioxin shown above is named 2,3,7,8-TCDD, or to give it its full title **2,3,7,8-tetrachlorodibenzo-*p*-dioxin.** No wonder it is simply called "dioxin" in the press!

There are actually 75 different chlorinated dibenzo-*p*-dioxin compounds, when one includes all the possibilities having between one and eight chlorines, given that a number of isomers exist for most of these eight types. Different members of a chemical family that differ only in the number and position of the same substituent are called **congeners.**

All dioxin congeners are planar: All carbon, oxygen, hydrogen, and chlorine atoms lie in the same plane. For convenience, we refer to the carbon atoms closest to the central dioxin ring as alpha carbons, and the outer ones as beta carbons:

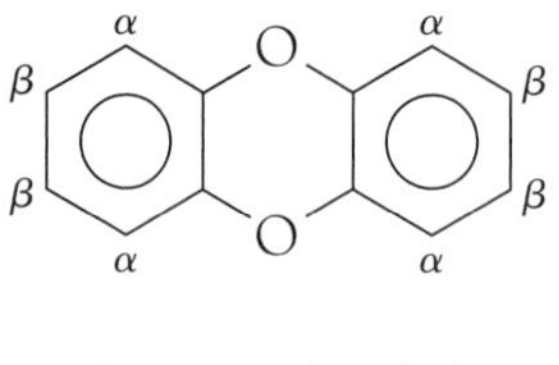

The unsubstituted dibenzo-*p*-dioxin molecule has two types of symmetry that are useful when considering substitution patterns. First there is lateral, or *left–right*, symmetry: The carbon atom labeled β at the top of the left-side ring is equivalent to the β carbon at the top of the ring at the right side, and similarly for the two β carbons at the bottom. The dioxin ring also has *up–down*

symmetry: The carbon atom labeled β at the top of the left-side ring is equivalent to the β carbon at the bottom of that ring, and similarly for the two β carbons of the right-side ring. Thus all the four β carbons are equivalent in the unsubstituted dioxin. Similarly, the four α carbons all denote equivalent positions. Consequently, for example, there are only two unique monochlorodibenzo-p-dioxins; due to the equivalence of the four α positions, those that would otherwise be numbered 4-, 6-, and 9-chlorodibenzo-p-dioxin are all equivalent to the 1- molecule. Similarly, 3-, 7-, and 8-chlorodibenzo-p-dioxins are all equivalent to the 2- molecule due to the equivalence of the β positions. Some or all of the equivalences can be lost when multiple substitution occurs.

PROBLEM 11-1

By drawing the structures and comparing them, decide whether 1,3-, 2,4-, 6,8-, and 7,9-dichlorodibenzo-p-dioxins are all unique compounds or whether they are all really the same compound. Are 1,2- and 1,8-dichlorodibenzo-p-dioxins unique compounds? Using a systematic procedure, deduce the structures of all unique dichlorodibenzo-p-dioxins, keeping in mind that before substitution the two rings are equivalent and that the molecule has up–down symmetry.

Chlorophenols as Pesticides

In addition to their use as starting materials in the production of herbicides, chlorophenols find use as wood preservatives (fungicides) and as slimicides. The most common preservative, in use since 1936, is **pentachlorophenol** (PCP, though not the "angel dust" compound known by the same initials); all the benzene's six hydrogens have been substituted in this compound:

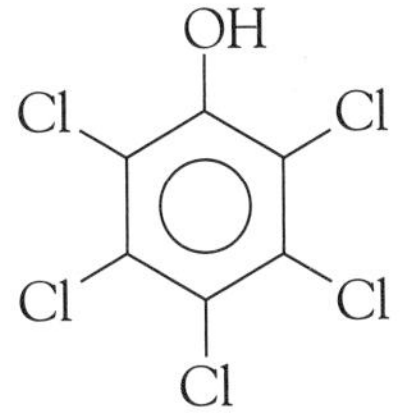

pentachlorophenol (PCP)

Commercial PCP is not pure pentachlorophenol but is significantly contaminated with *2,3,4,6-tetrachlorophenol*. This mixture has many pesticidal applications: It is used as a herbicide (e.g., as a preharvest defoliant), an insecticide (termite control), a fungicide (wood preservation and seed treatment), and a molluscicide (snail control). Some trichlorophenol isomers and some tetrachlorophenol isomers are also sold as wood preservatives.

Unfortunately, if wood treated with such preservatives is eventually burned, a fraction of the chlorophenols can react to eliminate HCl, thereby producing members of the chlorinated dioxin family. Thus **octachlorodibenzo-p-dioxin**, OCDD, is produced as an unwanted by-product in the incomplete combustion of pentachlorophenol products:

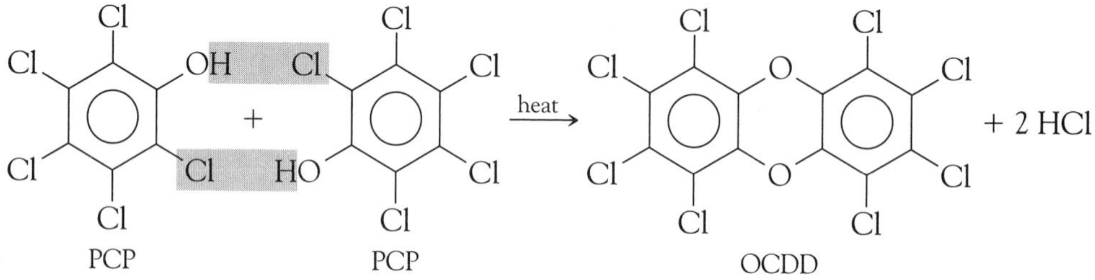

OCDD is the most prevalent dioxin congener found in human fat and in many environmental samples.

Indeed, pentachlorophenols are one of the largest chemical sources of dioxins to the environment; however, the main dioxin they contain, OCDD, is not particularly toxic, as discussed in a later section. Commercial supplies of chlorinated phenols themselves are contaminated with various dioxins.

PROBLEM 11-2

In naming OCDD and pentachlorophenol, no numbers are used to specify the positions of the chlorine substituents. Why is that not necessary here, whereas it is required in, e.g., 2,3,7,8-TCDD?

In general, any two phenol molecules that each have a chlorine on a carbon atom that is next to that having an OH group can combine to produce a dibenzo-*p*-dioxin molecule. The two phenols that combine need not be identical but simply need to make contact when they have been heated sufficiently to facilitate HCl elimination and dioxin formation. Similarly, coupling of phenoxide anions can occur with Cl⁻ elimination, as discussed previously in the case of 2,4,5-T synthesis. The methodology of problem solving that can be used to deduce the chlorophenolic origin of environmental dioxins is discussed in Box 11-1.

PROBLEM 11-3

(a) Deduce the structures and the correct numbering for the two tetra-chlorophenol isomers that exist in addition to the 2,3,4,6 isomer mentioned in the text. (b) For each of these two isomers, deduce the structure and names of the dioxin(s) that would result if two molecules of that isomer were to react together.

BOX 11-1 | Deducing the Probable Chlorophenolic Origins of a Dioxin

The chlorophenolic source of dioxins found in environmental samples can be deduced by reversing the logic used in the text to deduce which dioxin would be produced by the coupling of two specific chlorophenols.

Consider the congener 1,2,7,8-tetra-chlorodibenzo-p-dioxin; it could have been formed by elimination of two HCl molecules from two chlorophenol molecules in the following two ways (here T stands for trichlorophenol).

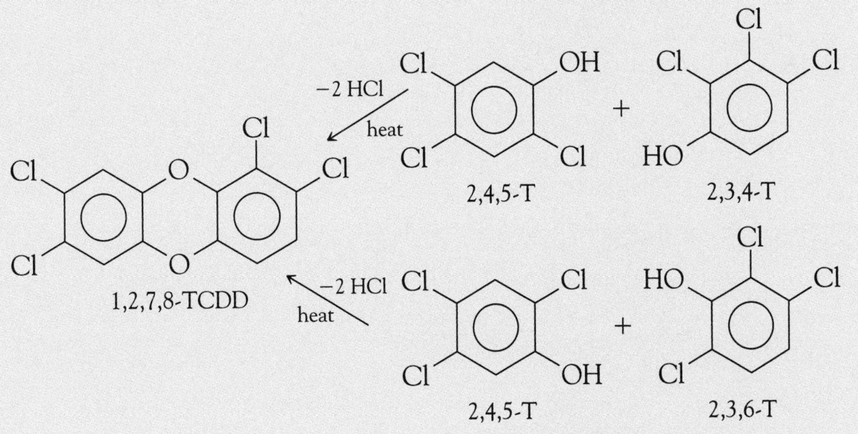

dioxin structure came from the chlorophenol on the right side of the molecule and the bottom oxygen from the chlorophenol on the left, leads to the possibility that the trichlorophenol molecules that combined were the 2,4,5 and the 2,3,6 congeners. Thus a 1,2,7,8-tetrachlorodibenzo-p-dioxin molecule in the environment could have arisen by combination of a 2,4,5-trichlorophenol molecule with either a 2,3,4- or a 2,3,6-substituted congener.

If it is assumed that the oxygen atom at the top of the dioxin congener originates with the chlorophenol congener on the left side of the dioxin molecule, then the bottom oxygen must come from the chlorophenol on the right side of the dioxin molecule; with this set of assumptions, the original reactants must have been 2,4,5- and 2,3,4-trichlorophenol. (Notice in the diagram above that the chlorine atoms eliminated must have arisen from positions adjacent to the oxygen atoms.) The alternative possibility, that the oxygen atom at the top of the

Unfortunately, some dioxins undergo rearrangement of substituents during their formation, so such a "retrosynthesis" approach is not an infallible guide to the origin of dioxins discovered in the environment.

PROBLEM 1

Deduce the two possible combinations of polychlorophenol molecules that, when coupled together through loss of two HCl molecules, would produce a molecule of 1,2,9-trichlorodibenzo-p-dioxin.

When both carbon atoms adjacent to the one bonded to —OH or O⁻ bear chlorine atoms on one (or both) of the chlorophenol molecules or chlorophenoxy ions, several possible dioxins can be formed in some cases. Consider, for example, the possible couplings of *2,3,6-trichlorophenol* with *2-chlorophenol*. The two possible orientations of the trichlorophenol with respect to the chlorophenol are illustrated below. The lower one corresponds to the upper one rotated by 180° about the axis (dashed green line) running through the O atom and the carbon atom that is para to it. Thus we see that both 1,4- and 1,2-dichlorodibenzo-*p*-dioxins can be produced, depending upon orientation. In practice, an almost equal mixture of the two isomers will be formed. (See Additional Problem 2.)

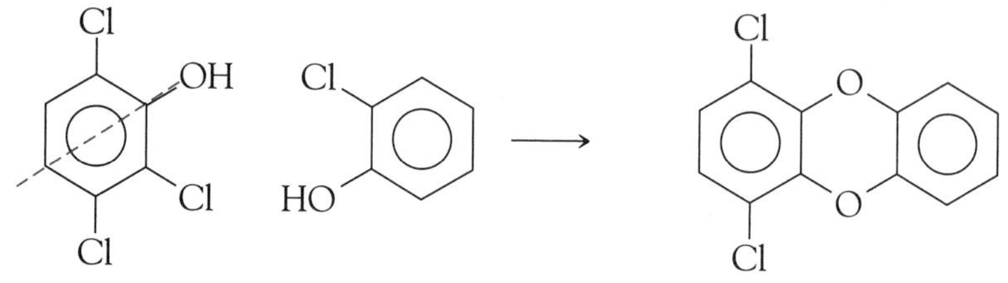

1,4-dichloro isomer

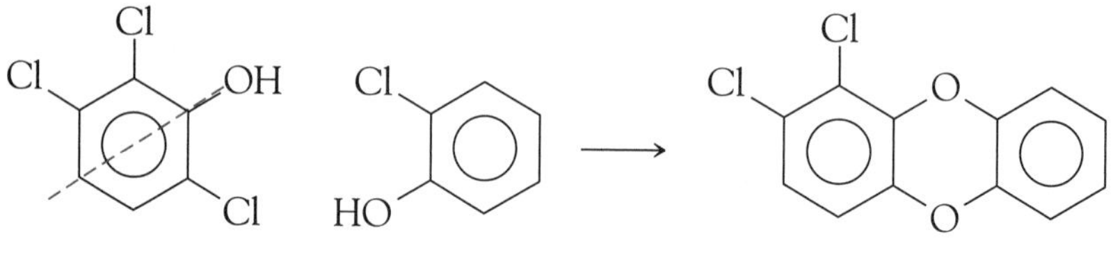

1,2-dichloro isomer

PROBLEM 11-4

Deduce what dioxin(s) would be produced in side reactions if 2,4-D were to be synthesized from 2,4-dichlorophenol.

Detecting Dioxins in Food and Water

As a consequence of their widespread occurrence in the environment and their tendency to dissolve in fatty matter, dioxins bioaccumulate in the food chain. More than 90% of human exposure to dioxins is attributable to the food we eat, particularly meat, fish, and dairy products. Typically, dioxins and furans (a group of compounds resembling the dioxins in structure, which we'll discuss later) are present in fish and meat at levels of tens or hundreds of picograms (pg, or 10^{-12} gram) per gram of the food; in other words, they occur at levels of tens or hundreds of parts per trillion. However, the bulk of dioxins and furans in nature are not present in biological systems: Attachments to soil and to sediments of rivers, lakes, and oceans are their most common sinks.

The ability of chemists to detect TCDD and other organochlorines in environmental samples has improved by orders of magnitude over the past few decades. In the early 1960s, when Carson's *Silent Spring* was published, the lower limit for analysis of DDT and other such compounds was the parts-per-million level. Ten years later, detecting such substances at the parts-per-billion level was possible but not common or easy. By 1990, parts-per-trillion detection was possible in soil or biota samples, and parts-per-quadrillion was possible for water samples. Today, a few labs detect some substances at limits up to 1000 times lower than these! At these latter levels, many organochlorines are found in *every* environmental sample, no matter how "clean" an environment it came from. By the late 1990s, researchers at the Centers for Disease Control in Atlanta were able to detect as little as 10^{-16} grams of TCDD in human serum samples.

The potential impact on human health of exposure to dioxins is documented later, following a discussion of the properties of PCBs and furans, two types of chemicals with which dioxins share many properties.

PROBLEM 11-5

Given its formula and Avogadro's constant (6.02×10^{23} molecules/mol), deduce how many molecules are present in 10^{-16} grams of TCDD.

PCBs

The well-known acronym **PCBs** stands for **polychlorinated biphenyls,** a group of industrial organochlorine chemicals that became a major environmental concern in the 1980s and 1990s. Although not pesticides, they found a wide variety of applications in modern society because of certain other properties they possess. Since the late 1950s, over 1 million metric tons of PCBs have been produced, about half in the United States and the rest mainly in France, Japan, and the former Eastern bloc nations. Like many other organochlorines, they are very persistent in the environment and they bioaccumulate in living systems. As a result of careless disposal practices, they have become a major environmental pollutant in many areas of the world. More than 95% of the entire U.S. population has detectable concentrations of PCBs in their bodies. Due both to their own toxicity and to that of their "furan" contaminants, PCBs in the environment have become a cause for concern because of their potential impact on human health, particularly with regard to growth and development.

In the following sections, we consider what PCBs are, how they are made, what they are used for, and how they become contaminated and released into the environment.

The Structure of PCB Molecules

Biphenyl molecules consist of two benzene rings linked by a single bond formed between two carbons that have each lost their hydrogen atom:

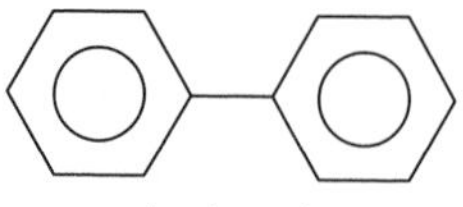

biphenyl

Like benzene, if biphenyl reacts with Cl_2 in the presence of a *ferric chloride* ($FeCl_3$) catalyst, some of its hydrogen atoms are replaced by chlorine atoms. The more chlorine initially present and the longer the reaction is allowed to proceed, the greater the extent (on average) of chlorination of the biphenyl molecule. The products are polychlorinated biphenyls, PCBs. The reaction of biphenyl with chlorine produces a mixture of many of the 209 congeners of the PCB family; the exact proportions depend upon the ratio of chlorine to biphenyl, the reaction time, and the reaction temperature. An example of a PCB molecule is shown below:

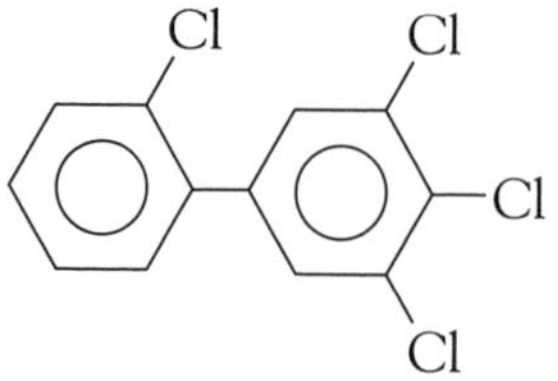

2,3',4',5'-tetrachlorobiphenyl

Although many individual PCB compounds are solids, the mixtures are liquids or are solids with low melting points. Commercially, individual PCB compounds were not isolated; rather they were sold as partially separated mixtures, with the average chlorine content in different products ranging from 21% to 68%.

PROBLEM 11-6

The general formula for any PCB congener is $C_{12}H_{10-n}Cl_n$, where n ranges from 1 to 10. Calculate the average number of chlorine atoms per PCB molecule in a mixture of congeners that is 60% chlorine by mass, a common value for commercial samples.

The Numbering Systems for PCBs

The numbering scheme used for individual PCB congeners begins with the carbon that is joined to a carbon in the other ring; it is given the number 1, and the other carbons around the ring are numbered sequentially. As illustrated below, the positions in the second ring are also numbered 1 through 6, starting with the ring-joining carbon, but are distinguished by primes. By convention, the 2′ position in the second ring lies on the *same* side of the C—C bond joining the rings as does the 2 position in the first ring, and so on.

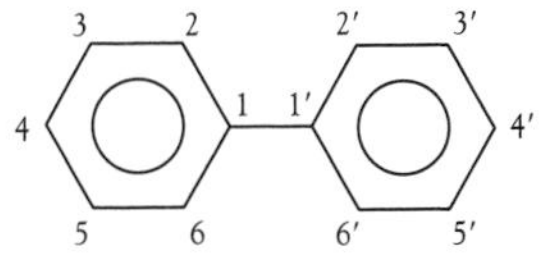

In most instances, the two rings in a chlorinated biphenyl molecule are not equivalent since the patterns of substitution differ. The unprimed ring is chosen to be the one that will give a substituent with the lowest-numbered carbon. Using all these rules, we can deduce that the name of the PCB molecule shown on page 478 is *2,3′,4′,5′-tetrachlorobiphenyl*.

Very rapid rotation occurs around carbon–carbon single bonds in most organic molecules, including the C—C link joining the two rings in biphenyl and in most PCBs. Thus it is not normally possible to isolate compounds corresponding to different relative orientations of the two rings ("rotamers") in a PCB. For example, *3,3′-* and *3,5′-dichlorobiphenyl* are not individually isolatable compounds, since one form is constantly being converted into the other and back again by rapid rotation about the C—C bond linking the rings:

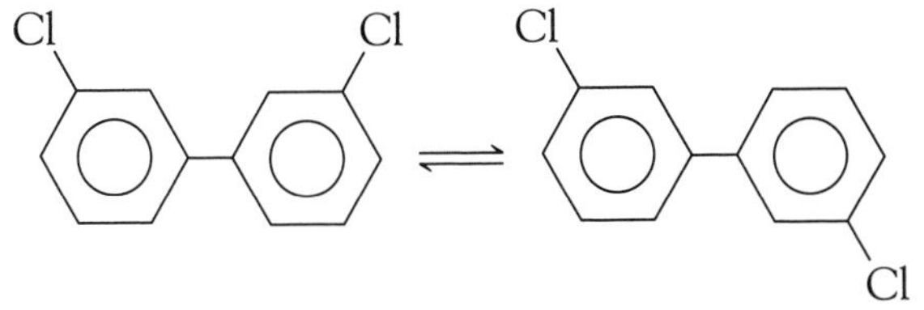

3,3′-dichlorobiphenyl   3,5′-dichlorobiphenyl

The name used for such a compound is that which has the lowest number for the second chlorine, so the system shown above is called the 3,3′ isomer. Although the rings rotate rapidly with respect to each other, the energetically optimum orientation is the one having the rings coplanar or close to it, except, as we shall see later, when large atoms or groups occupy the 2 and 6 positions.

PROBLEM 11-7

Using a systematic procedure, draw the structures of all unique dichlorobiphenyls, assuming first that free rotation about the bond joining the rings does *not* occur. Then deduce which pairs of structures become identical if free rotation does occur.

Commercial Uses of PCBs

All PCBs are practically insoluble in water but are soluble in hydrophobic media, such as fatty or oily substances. Commercially, they were attractive because they

- are chemically inert liquids and are difficult to burn,

- have low vapor pressures,

- are inexpensive to produce, and

- are excellent electrical insulators.

As a result of these properties, they were used extensively as the coolant fluids in power transformers and capacitors. Later, they were also employed as *plasticizers*, i.e., agents used to make plastic materials such as PVC products more flexible, in carbonless copy paper, as de-inking solvents for recycling newsprint, as heat transfer fluids in machinery, as waterproofing agents, and so on.

Because of their stability and extensive usage, together with careless disposal practices, PCBs became widespread and persistent environmental contaminants. When their accumulation and harmful effects became recognized, **open uses**—i.e., those for which their disposal could not be controlled—were terminated. Although North American production of PCBs was halted in 1977, the substances remain in use in some electrical transformers currently in service. As these units are gradually decommissioned, their PCB content usually is stored in order to prevent further contamination of the environment. In the United States, the EPA expected a 90% reduction of PCB use in electrical equipment by 2006. Canada has proposed a phase-out of all PCB uses by 2008. In some locales, stored PCBs are destroyed by incineration, using techniques discussed in Chapter 16. Previously, PCB-containing transformers and capacitors were often just dumped into landfills, and their PCB content was allowed to leak into the ground. Indeed, PCBs were inadvertently released into the environment during their production, their use, their storage, and their disposal.

PCBs Cycling Among Air, Water, and Sediments

If released into the environment, PCBs persist for many years because they are so resistant to breakdown by chemical or biological agents. Although their solubility in water is very slight—indeed they are more likely to be adsorbed onto suspended particles in the water than dissolved in it—the tiny amounts of PCBs in surface waters are constantly being volatilized and subsequently redeposited on land or in water after traveling in air for a few days. By such mechanisms, PCBs have been transported worldwide. There are measurable background levels of PCBs even in polar regions and at the bottom of oceans. Indeed, the ultimate sinks for PCBs that are mobile are in the deep sediments of oceans and large lakes. This environmental load of PCBs will continue to be recycled among air, land, and water, including the biosphere, for decades to come, as analyzed in greater detail in Chapter 12. Only a minority of PCBs manufactured in the past are currently found in the environment or have been destroyed; much of the production lingers in storage or old electrical equipment and may ultimately be released. Recent research has found that PCB releases from older consumer products into indoor air, which then is eventually vented outside, is a major source of PCBs in urban air.

A quantitative measure of the recycling of substances within a water body is provided by the **mass balance** of its current annual inputs and outputs of the compounds. The PCB mass balance for a very large, relatively clean water body—Lake Superior—is illustrated in Figure 11-1. Although it is the least polluted of the Great Lakes, Lake Superior's burden of PCBs in its water and its sediments is substantial. Currently, almost all the input of PCBs occurs from the air, with relatively little added from industries or via tributary rivers (Figure 11-1). Overall, Lake Superior is now gradually "exhaling" its historical load of PCBs into the air, the output to air being much greater than the annual input from the atmosphere. Little of Lake Superior's PCB content

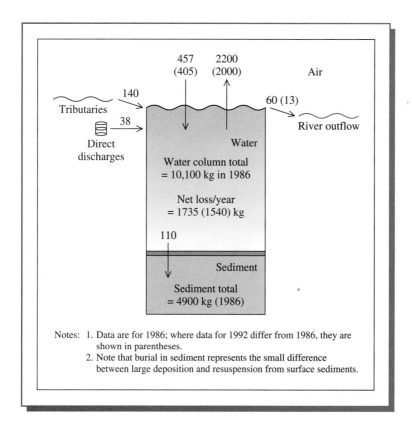

FIGURE 11-1 Mass balance of PCBs in Lake Superior, in kilograms per year. [Source: *The State of Canada's Environment 1996* (Ottawa: Government of Canada, 1996).]

is now being lost to sediments; about as much is redissolved from them as is deposited onto them each year.

By contrast, the mass balance of PCBs in Lake Ontario, another of the Great Lakes, is quite different from that of Lake Superior. The PCB concentration in Lake Ontario water substantially exceeds that of Lake Superior, since it is located in a much more industrialized area. In Lake Ontario, the greatest current input comes from land-based sources such as waste dumps that still leach PCBs into the lake and its tributaries. About the same quantity of PCBs are present in the water flowing out of the lake. Approximately equal amounts are lost annually to sediments and to the atmosphere; about one-third of such losses are canceled by new inputs from the sediments and the air.

PROBLEM 11-8

The PCB concentration in Lake Michigan is declining according to a first-order rate law having a rate constant of 0.078/year. If the PCB concentration in Lake Michigan averaged 0.047 ppt in 1994, what will it be in 2010? In what year will the concentration fall to 0.010 ppt? What is the half-life period of PCBs in this lake? [*Hint: Recall that for first-order processes, the fraction f of any sample that still remains after time t has passed is $f = e^{-kt}$.*]

Because of their persistence and their solubility in fatty tissue, PCBs in food chains undergo biomagnification; an example is shown in Figure 11-2.

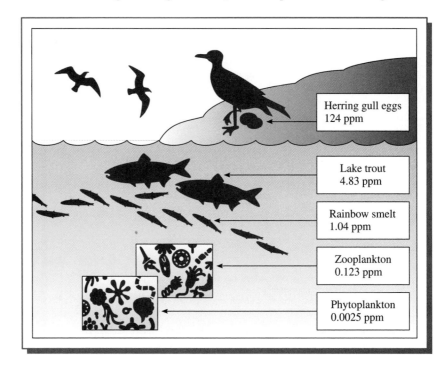

FIGURE 11-2 The bio-magnification of PCBs in the Great Lakes aquatic food chain. [Source: *The State of Canada's Environment 1996* (Ottawa: Government of Canada, 1991).]

Notice that the ratio of PCBs in the eggs of herring gulls in the Great Lakes was 50,000 times that in the phytoplankton in the water at the time of these measurements. The good news is that the average level of PCBs in such eggs has fallen with time in many locations, as the data in Figure 11-3a illustrate for

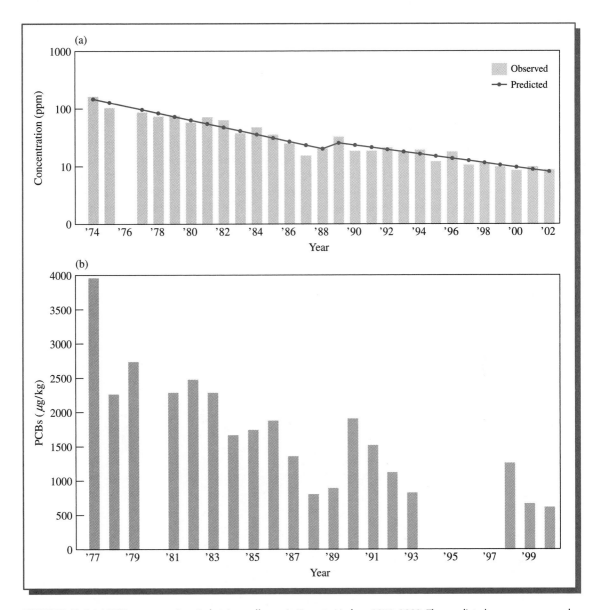

FIGURE 11-3 (a) PCB concentrations in herring gull eggs in Toronto Harbor, 1974–2002. The predicted curves correspond to exponential decay in those time periods. [Source: Dr. Chip Weseloh, Environment Canada.] (b) PCB concentrations in 65-cm coho salmon from Lake Ontario. [Source: Ontario Ministry of the Environment.]

gull colonies in Lake Ontario around Toronto. The concentrations are plotted in the figure on a logarithmic scale, and the data seem to fit two intersecting straight lines, corresponding to first-order decay sequences with half-lives of first five years and more recently seven years. This complicated behavior may arise because of continuing sources of PCBs to the system. PCB levels in fish at the top of the Lake Ontario food chain have also declined since the 1970s, but the current rate of decrease is slow and erratic (see Figure 11-3b).

The relative concentrations of the congeners of a PCB mixture begin to change once they enter the environment. Microorganisms in soils and sediments and large organisms such as fish both preferentially metabolize congeners having relatively few chlorine atoms. Thus the relative concentrations of the more heavily chlorinated congeners increases with time since they are degraded much more slowly. Thus, for example, between 1977 and 1993, the proportion of PCB molecules with four or five chlorine atoms decreased by 6% each in trout in Lake Ontario, whereas those with seven or eight chlorines increased by 7% and 4%, respectively. However, PCBs present in anaerobic soil are eventually dechlorinated microbially at their meta and para positions, leaving congeners that are only chlorinated at ortho positions. Aerobic degradation occurs with congeners having adjacent carbons (ortho + meta, or meta + para) chlorine-free.

PCB Contamination by Furans

Strong heating of PCBs in the presence of a source of oxygen can result in the production of small amounts of *furans*. These compounds are structurally similar to dioxins; they differ only in that the molecules are missing one oxygen atom in the central ring. The furan ring contains five atoms, one of which is oxygen and the other four of which are carbon atoms that participate in double bonds:

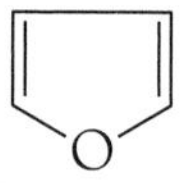

furan

The **dibenzofurans** (DFs) have a benzene ring fused to opposite sides of the furan ring:

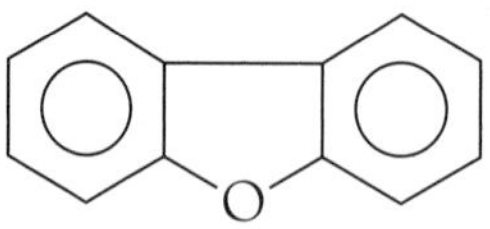

dibenzofuran

As with dioxins, all chlorinated dibenzofuran congeners are planar; i.e., all C, O, H, and Cl atoms lie in the same plane. They are formed from

PCBs by the elimination of the atoms X and Y bonded to two carbons that are ortho in position to those that link the rings *and* that lie on the same side of the C—C link between the rings:

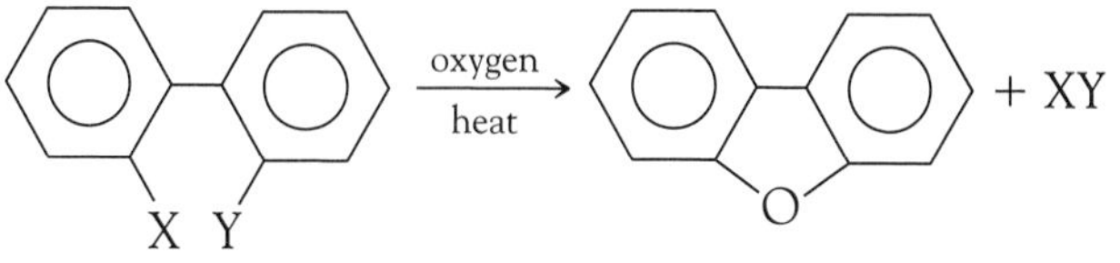

The atoms X and Y can both be chlorine, or one can be hydrogen and the other one chlorine, so the molecule eliminated can be Cl_2 or ClH (i.e., HCl), respectively. A more detailed analysis of the nature of the specific furans that result from particular PCB congeners is given in Box 11-2.

Most of the chlorine in the original PCB molecule is still present in the dibenzofuran; *polychlorinated dibenzofurans* are known commonly as **PCDFs.** The numbering scheme for substituents is the same as that for dioxins (PCDDs); note, however, that by convention the numbering starts next to a carbon that forms the single C—C bond *opposite* the oxygen.

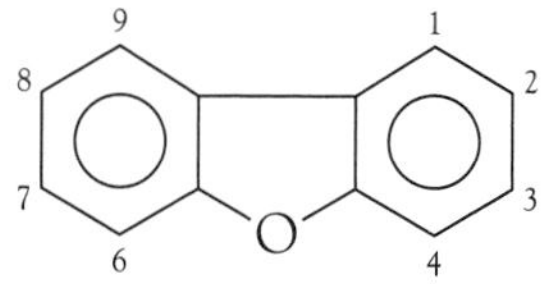

While there exist 75 different chlorine-substituted dibenzo-*p*-dioxins, there are 135 dibenzofuran congeners, since the symmetry of the ring system is lower for furans. In particular, although the furans have the same left–right symmetry as dioxins, they do not have their up–down symmetry.

PROBLEM 11-9

Draw the structures of all the 16 unique dichlorodibenzofurans, and deduce the numbering required in their names. [Hint: Use a systematic procedure to generate all, but include only those congeners that correspond to unique molecules; i.e., be careful to eliminate duplicates. For example, start by placing one chlorine at C-1 and then generate all the possible isomers corresponding to different positions for the second chlorine. Then place the first chlorine at C-2 and repeat the procedure, noting that the 1,2-dichloro isomer is generated both times. Continue the procedure with the first chlorine at C-3, etc.]

BOX 11-2   Predicting the Furans That Will Form from a Given PCB

In deducing the nature of the polychlorinated dibenzofuran (PCDF) that would
be formed from a particular PCB, it should be
remembered that free rotation occurs about
the single bond joining the two rings in the
original biphenyl in all PCBs at the elevated
temperatures of the reaction. Thus HCl elimination in 2,3'-dichlorobiphenyl gives both
4- and 2-chlorodibenzofuran.

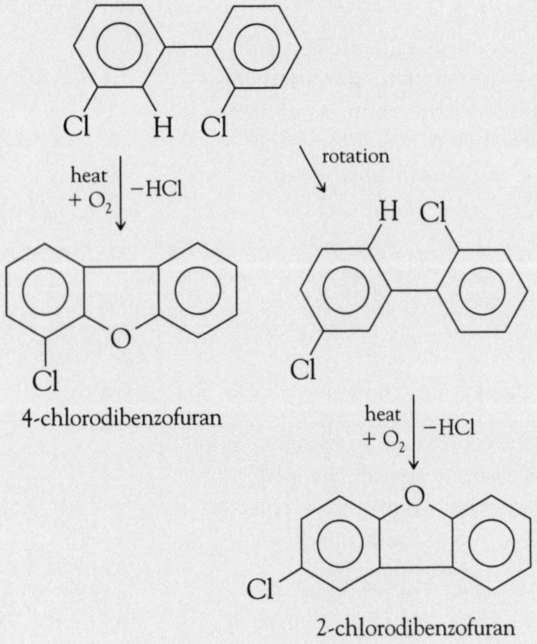

At the high temperatures of this reaction,
some interchange of the adjacent substituents
in the 2 and 3 positions (ortho and meta)
of any given ring can occur as a prelude to
HCl elimination; in particular *chlorine can
move from an ortho to a meta position, and
hydrogen from meta to ortho, preceding HCl
elimination.* For example, when 2,6,2',6'-
tetrachlorobiphenyl (see next page) is heated
in air, some of its molecules lose a pair of
ortho chlorines to give a dichlorodibenzofu-
ran, and some first interchange Cl and H in
one ring to eliminate HCl and produce a
trichlorodibenzofuran. Free rotation about the
C—C bond does *not* occur *after* the interchange, as presumably the elimination occurs
immediately.

4-chlorodibenzofuran

2-chlorodibenzofuran

Almost all commercial PCB samples are contaminated with some
PCDFs, but this usually amounts to only a few ppm in the originally manufactured liquids. However, if the PCBs are heated to high temperatures and if
some oxygen is present, conversion of PCBs to PCDFs increases the level of
contamination by orders of magnitude. The furan concentration in used PCB
cooling fluids is found to be greater than in the virgin materials, presumably
due to the moderate heating that the fluid undergoes during its normal use.

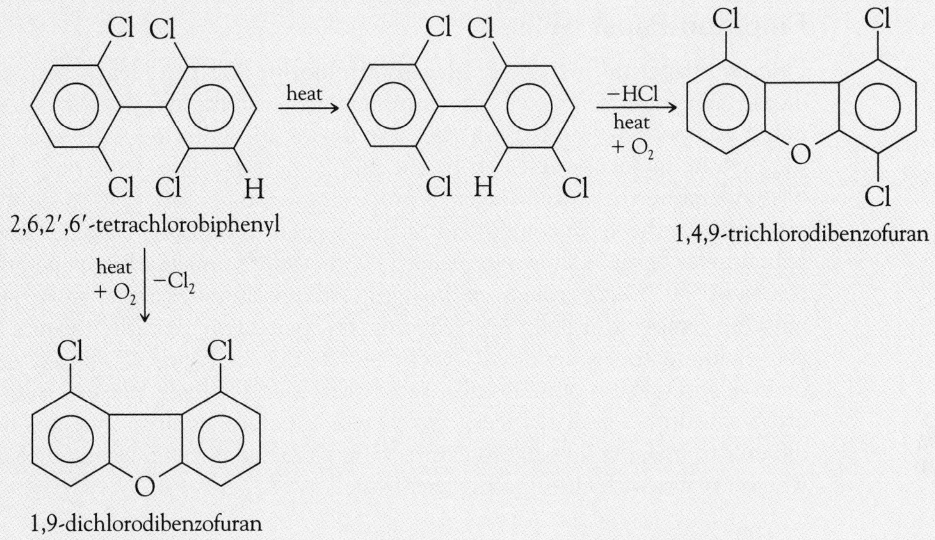

2,6,2′,6′-tetrachlorobiphenyl

1,4,9-trichlorodibenzofuran

1,9-dichlorodibenzofuran

PROBLEM 1

For each PCB shown below, deduce which furans would be expected to be produced by Cl$_2$ or HCl elimination when the PCB is heated in air. Write the correct name for each PCDF.

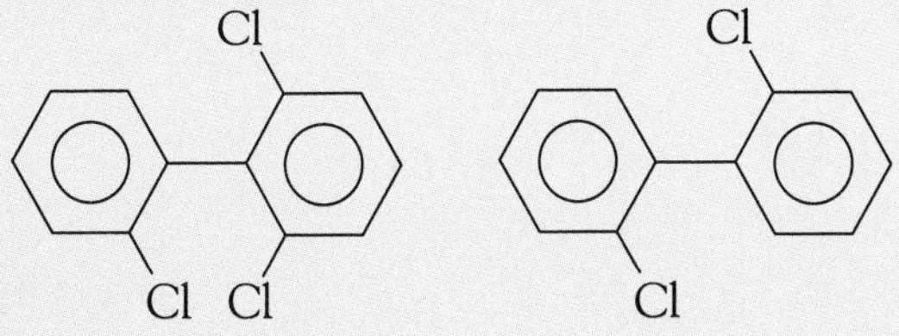

PROBLEM 2

Recently it has been discovered that upon strong heating in air PCBs can also react by elimination of two ortho hydrogen atoms (one on each ring) as H$_2$. Decide which, if any, additional PCDFs will be produced if the PCBs in Problem 1 can eliminate H$_2$.

Furan production also occurs if one attempts to burn PCBs with anything but an unusually hot flame.

Other Sources of Dioxins and Furans

In addition to the sources discussed above, polychlorinated dibenzofurans and dibenzo-*p*-dioxins are also produced as by-products in a myriad of processes, including the bleaching of pulp, the incineration of garbage and hospital waste, the recycling of metals and sintering of iron ore, and the production of common solvents such as *tri-* and *perchloroethene*.

Pulp-and-Paper Mills

Pulp-and-paper mills that use **elemental chlorine,** Cl_2, to bleach pulp are dioxin and furan sources. These contaminants, among many other chlorinated compounds, result from the reaction of the chlorine with some of the organic molecules released by the pulp. The tan color of the pulp that has undergone the initial stages of processing is due to the light-absorbing properties of the *lignin* component of the original wood fibers. A generalized structure for lignin is shown in Figure 11-4. In order to make white paper, the residual (~10%) component of the lignin still present after initial processing must be removed, usually by bleaching the pulp with oxidizing agents. If you examine the generalized structure of lignin in Figure 11-4, you can observe several sites of monosubstituted *phenols* and *phenolic ethers* as well as ortho-substituted *phenyl diethers*. From these structural components, it is not difficult to imagine how lignin can serve as a precursor to furans and dioxins when it reacts with chlorinating agents such as Cl_2.

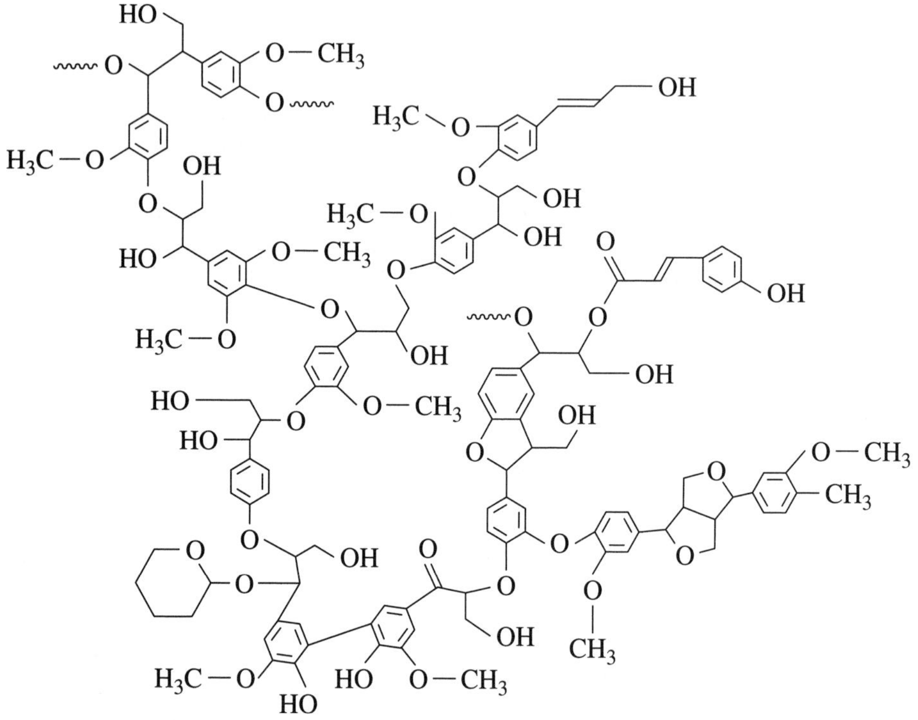

FIGURE 11-4 Generalized structure of lignin. [Source: M. C. Cann and M. E. Connelly, *Real-World Cases in Green Chemistry* (Washington D.C.: American Chemical Society, 2000).]

More furans than dioxins are formed in the bleaching of pulp by elemental chlorine. The furan congeners of highest concentrations in the pulp are *1,2,7,8-TCDF* and the more toxic *2,3,7,8-TCDF*. Unfortunately, the most abundant dioxin produced by the pulp-and-paper bleaching process is the highly toxic 2,3,7,8-TCDD congener. The paper and effluent contain dioxins at parts-per-trillion levels, which resulted in total releases in the past, in North America, of several hundred grams of 2,3,7,8-TCDD annually.

Because of the problems of producing dioxins and furans, the use of elemental chlorine as a bleaching agent for paper was banned in the United States as of April 2001. Most pulp-and-paper mills there and in other developed countries switched their bleaching agent from elemental chlorine to **chlorine dioxide,** ClO_2, from which the furan and dioxin output is much smaller, even undetectable in many cases. The difference is due to the mechanism by which the compounds attack the pulp's residual lignin. Elemental chlorine reacts to insert chlorine as a substituent on the aromatic rings in lignin, yielding products that are soluble in alkali and that can then be washed away. Experiments suggest that during the oxidation of the lignin, two of the component benzene rings can couple together to form a dibenzofuran or dibenzo-*p*-dioxin system that subsequently is chlorinated and in the process becomes detached from the lignin system. In contrast, chlorine dioxide destroys the aromaticity of the benzene rings by free-radical processes and therefore produces fewer chlorinated products that contain six-membered rings.

Some mills now produce paper pulp without any use of chlorine compounds. Ozone, hydrogen peroxide, and even high-pressure oxygen are the alternative bleaching agents used in these **totally chlorine free** (TCF) pulp mills. Mills that still use chlorine to bleach now remove contaminants from wastewater by treatments such as reverse osmosis (see Chapter 14).

The use of chlorine to disinfect drinking water and the chlorinated by-products that are formed in the process are discussed in Chapter 14.

 ## Green Chemistry: H_2O_2, an Environmentally Benign Bleaching Agent for the Production of Paper

TCF bleaching agents for paper such as **hydrogen peroxide** (H_2O_2), *ozone*, and *diatomic oxygen* have been developed. While TCF agents eliminate the formation of dioxins and furans, these methods in general are problematic because these oxidizing agents are not as strong as elemental chlorine or chlorine dioxide. Thus they generally require longer reaction times and higher temperatures (more energy input), and they lead to significant breakdown of the cellulose fibers, which weakens the paper, requiring more wood to produce the same amount of paper.

Terry Collins of Carnegie Mellon University earned a Presidential Green Chemistry Challenge Award in 1999 for his development of compounds

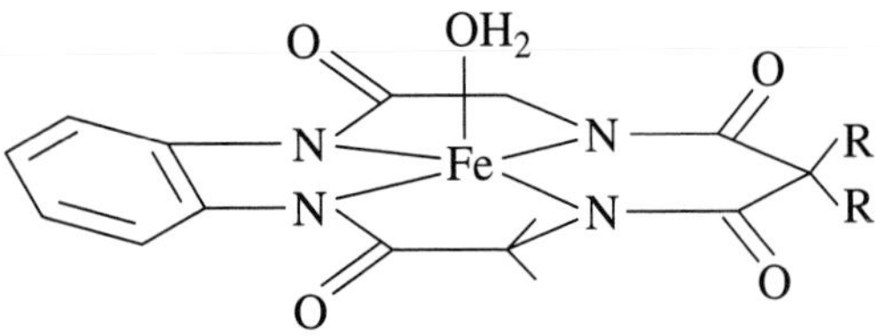

FIGURE 11-5 Tetraamido-macrocyclic ligands (TAML): activators for hydrogen peroxide.

[Source: M. C. Cann and M. E. Connelly, *Real-World Cases in Green Chemistry* (Washington D.C.: American Chemical Society, 2000).]

known as **tetraamido-macrocyclic ligands** (TAMLs, Figure 11-5), which enhance the oxidizing strength of hydrogen peroxide. Hydrogen peroxide is a particularly enticing oxidizing reagent since its by-products are water and oxygen, which are environmentally benign. The use of TAML in conjunction with hydrogen peroxide reduces the temperature and reaction times normally required for bleaching paper with hydrogen peroxide, thus making hydrogen peroxide a viable alternative for this process. See also the feature article "Little Green Molecules" immediately following the Introduction to this book.

The TAMLs can be modified by varying the alkyl groups (R) on the right side of the structure in Figure 11-5. Changing these groups influences the lifetime of these catalysts. In uses such as the bleaching of paper it is important for the TAML catalysts to decompose in a relatively short period of time so they do not become a burden to the environment. However, they must last long enough to fulfill their role as catalysts for hydrogen peroxide. TAML catalysts not only offer significant promise for the bleaching of paper, they are also being considered for use in laundry applications, the disinfection of water, and the decontamination of biological warfare agents such as anthrax.

Fires and Incineration as Sources of Dioxins and Furans

Fires of many kinds, including forest fires and those in incinerators, release various congeners of the dioxin and furan families into the environment; these chemicals are produced as minor by-products from the chlorine and organic matter in the fuel. Dioxin and furan production seem unavoidable whenever combustion of organic matter occurs in the presence of chlorine, unless steps are taken to ensure complete combustion by using very high flame temperatures. Some environmentalists worry particularly about the dioxin emissions when the chlorine-containing plastic PVC is incinerated or involved in other fires. Indeed, research on the combustion of newspapers indicates that chlorinated dioxin and furan production rises as the amount of salt or PVC present also increases. In many environmental samples of combustion products, several dozen different dioxin congeners are found, all in comparable amounts. Congeners with relatively high numbers of chlorine substituents usually are the most prevalent.

Incinerators now are the largest anthropogenic source of dioxins in the environment. Dioxins and furans are formed in the postcombustion zone of incinerators, where the temperature is much lower (250–500°C) than in the flame itself (see Chapter 16). They are formed during the oxidative degradation of the graphite-like structures in the soot particles that were produced during the incomplete combustion of the waste. Trace metal ions in the original waste probably catalyze the process. The small amounts of chlorine in the waste provide more than enough of this element required to partially chlorinate the furans and dioxins. The dibenzofuran and dibenzo-*p*-dioxin ring systems are formed at high temperatures (> 650°C); chlorination progressively occurs when the temperature cools below 650°C and gradually is reduced to 200°C.

Characteristically, incineration produces a greater mass of furans than dioxins. The yields of specific dioxin congeners increase with the degree of chlorination through to OCDD, whereas the peak production of furans occurs with four to six chlorines. In contrast to waste incineration, industrial coal combustion generates little dioxin because it burns much more completely, generating little soot to decompose later into dioxins and furans.

Chlorine Content of Dioxin and Furan Emissions

The profile of estimated annual global PCDF and PCDD emissions for the congeners with four to eight chlorines, i.e., those believed to be toxic, is shown in Figure 11-6a. As discussed previously, furans outnumber dioxins, and furans peak with congeners having four chlorines, whereas the dioxin

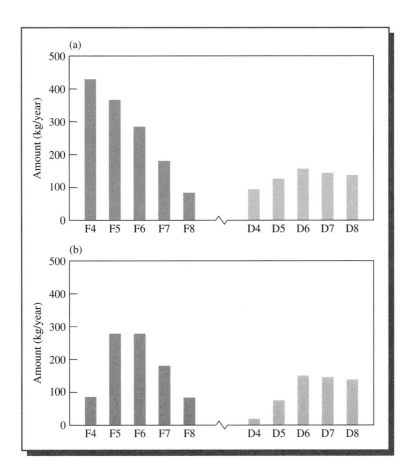

FIGURE 11-6 Annual PCDD and PCDF (a) emissions and (b) deposition rates after reactions with the hydroxyl radical OH. The letters F and D represent furans and dioxins; the numbers indicate the number of chlorine atoms per molecule. [Source: J. I. Baker and R. A. Hites, "Is Combustion a Major Source of Polychlorinated Dibenzo-*p*-Dioxins and Dibenzofurans to the Environment?" *Environmental Science and Technology* 34 (2000): 2879.]

peak is less pronounced and occurs with about six chlorines. Furans with these intermediate amounts of chlorine have toxicities similar to that of 2,3,7,8-TCDD, whereas fully chlorinated dioxin molecules have low toxicities. Consequently, the threat to human health from furans in the environment may exceed that from dioxins.

The mass of the dioxin and furan compounds that are eventually deposited from air onto soil and sediments, the primary mechanism by which dioxins eventually enter the food chain, is shown in Figure 11-6b. The loss in mass between emission and deposition is greater the *fewer* the number of chlorines present; hence the significant difference between amounts emitted and deposited for the tetrachloro and pentachloro congeners, but not those more heavily chlorinated, in Figure 11-6. This differentiation occurs because the principal loss mechanism is attacked at an unsubstituted carbon by the **hydroxyl free radical,** OH (followed by atmospheric oxidation of the resulting radical, as expected from the principles discussed in Chapters 3 and 5), and the rate of this initial reaction is greater the fewer the chlorines present. The amount of OCDD, the octachloro congener of dioxin, that is found to be deposited greatly exceeds the estimate from this figure, which is based mainly on combustion sources. Some scientists believe that the discrepancy arises because much additional OCDD is created in water droplets in air by the sunlight-initiated photochemical decomposition of PCP, pentachlorophenol, which eventually results in coupling of two PCPs to produce OCDD.

Very small concentrations of dioxins—particularly highly chlorinated ones—were present in the environment in the preindustrial era, presumably as a result of forest fires, volcanoes, etc. Indeed, forest fires are still probably Canada's largest source of dioxins. According to the analysis of soils and the sediments in lakes, the greatest anthropogenic input of dioxins and furans to the environment in developed countries began in the 1930s and 1940s, and peaked in the 1960s and 1970s. The principal sources were combustion/incineration, the smelting and processing of metals, the chemical industry, and existing environmental reservoirs. Inadvertent production of dioxins continues today, but at a slower rate—about half of the maximum, according to some sediment samples. The decrease in emissions resulted from deliberate steps taken by industrialized nations to reduce the production and dispersal of these toxic by-products. In particular, dioxin emissions from large sources in the United States declined by 75% from 1987 to 1995 alone, primarily due to reductions in air emissions from municipal and medical waste incinerators. New regulations should increase the reduction to 95%. However, uncontrolled combustion of nonpoint sources such as rural backyard trash burning in barrels—especially when plastics such as PVC are included in the mix—has not yet been brought under control.

Once created, dioxins and furans are transported from place to place mainly via the atmosphere (Chapter 12). Eventually they are deposited and can enter the food chain, becoming bioaccumulated in plants and animals. As previously mentioned, our exposure to them arises almost entirely through the foods that we eat. In the next section, we try to answer the question of what effects, if any, this exposure has on our health.

The Health Effects of Dioxins, Furans, and PCBs

Over a billion dollars has been spent on research to determine the extent to which dioxins, furans, and PCBs cause toxic reactions in humans. Nevertheless, conclusions about this issue are still tentative and controversial. Evidence about toxicity is derived from two sources:

- toxicological experiments on animals that have been deliberately exposed to the chemicals; and

- epidemiological studies of humans who have been accidentally exposed.

It is generally agreed that most PCBs are *not* acutely toxic to humans: The LD_{50} values of most congeners are large. In high doses, PCBs cause cancer in test animals, and consequently they are listed as a "probable human carcinogen" by the U.S. EPA. However, studies of humans exposed to them have produced inconsistent results. Most groups of people who have been exposed to relatively high concentrations of PCBs—e.g., as a result of their employment in electrical capacitor plants—have not experienced a higher overall death rate. (See, however, the comments about human cancer in Chapter 12.) The most common reaction to exposure to PCBs is *chloracne*, a biological response by humans to exposure to many types of organochlorine compounds.

Inadvertent PCB Poisonings

The most dramatic effects observed to human health from exposure to PCB mixtures occurred when two sets of people, one in Japan in 1968 and the other in Taiwan in 1979, unintentionally consumed PCBs that had accidentally been mixed with cooking oil. In the Japanese incident, and probably in the Taiwanese case as well, the PCBs had been used as a heat exchanger fluid in the deodorization process for the oil. Since the PCB-contaminated oils had been heated, their level of PCDF contamination was much greater than occurs in freshly prepared commercial PCBs. The thousands of Japanese and the Taiwanese people who consumed the contaminated oils suffered health effects far worse than has been found for workers at PCB manufacturing and handling plants, even though the resulting PCB levels in their bodies were about the same. From this difference, it has been concluded that the main toxic agents in the poisonings were the PCDFs and that they and dioxins were collectively responsible for about two-thirds of the health effects, with the PCBs themselves responsible for the remainder. Indeed, studies on laboratory animals indicate that the furans involved in these incidents are more than 500 times as toxic on a gram-for-gram basis than are pure PCBs. Cognitive development, as measured by IQ scores, of children born to the most highly exposed Taiwanese mothers—even if births occurred long after the consumption of the contaminated oil—was found to be significantly lower than that of their siblings born before the accident occurred and for children of unexposed mothers. Interestingly, children whose fathers but not their mothers had consumed the oil showed no detrimental effects. Further effects of this incident are discussed in Chapter 12 (Environmental Estrogens section).

An incident comparable to the Asian cooking oil poisonings occurred in early 1999 in Belgium. Several kilograms of a mixture of PCBs that had previously been heated to a high temperature—converting a tiny fraction of the PCBs to furans—were put into an 80,000-kg batch of animal fat, which was then mixed with animal feed and shipped to about 1000 farmers. Poultry producers subsequently noticed a sudden drop in egg production and in egg hatchability, and high concentrations of dioxins were subsequently found in chicken meat. The contaminated food was withdrawn from the market and destroyed.

Effects of *in Utero* Exposure to PCBs

Are humans who have been exposed to PCBs through their diets especially susceptible to reproductive problems? To answer this question, Sandra and Joseph Jacobson and their co-workers at Wayne State University in Detroit have spent more than two decades studying the offspring of people from the Lake Michigan area, including children whose mothers regularly eat fish from the lake and who, as a result, are expected to have elevated levels of PCBs in their bodies. They have discovered statistically significant differences in children born to women who have high levels of PCBs; these differences were present not only at birth but persisted to the age of at least eleven years.

The *prenatal* (i.e., prebirth) exposure of the infants to PCBs was determined by analyzing the blood of their umbilical cords for these chemicals after birth. Because analytical techniques in the early 1980s were not sufficiently sensitive to detect the amounts of PCBs in all the umbilical cord samples, exposure for many infants was estimated from the PCB levels in their mother's blood and breast milk. The *postnatal* exposure of the children was assessed by analyzing their mother's breast milk and also by analyzing blood samples from the children at the age of four years.

The Jacobsons discovered that, at birth, the children of mothers who had transmitted the highest amounts of PCBs to the children *before* birth had, on average, a slightly lower birth weight and a slightly smaller head circumference; they were also on average slightly more premature than those born to women who passed along lower amounts. The severity of these deficits was larger the greater their prenatal exposure to PCBs. When tested at the age of seven months, many of the affected children displayed small difficulties with visual recognition memory, again with the extent of the problem increasing with prenatal transmission of PCBs. At the age of four years, the lower body weight observed at birth for highly exposed infants still lingered. More serious was the observation that the four-year-olds displayed progressively lower scores on several tests of mental functioning (with respect to verbal and memory abilities) the greater their prenatal PCB exposure. A comparable study in Holland, using more modern analytical instrumentation, found similar results (see Figure 11-7a).

At eleven years of age, the effects of the prenatal PCBs were still apparent: the IQ scores of the part of the group that had been the most highly exposed before birth averaged 6 points below the others; the most affected mental functions were memory and attention span. However, prenatal PCB exposure at any but the highest levels did not appear to have affected the IQ of the eleven-year-olds (see Figure 11-7b).

Interestingly, at both four and eleven years of age, the children's total body content of PCBs, which is determined mainly from the breast milk they consumed as infants and to a lesser extent by the fish in their diet rather than from any prenatal transmission of the chemicals, was *not* the relevant factor in determining these physical and mental deficits. Rather, it was the smaller amounts of PCBs that had been transmitted from mother to fetus that were important. Thus PCBs appear to interfere with the proper prenatal development of the brain and with the mechanisms that ultimately determine physical size.

The study by the Jacobsons is one of the clearest examples available concerning the influence of toxic environmental chemicals on the health of human beings. It is important to realize that the highest levels of PCBs to which the children in this group were exposed before birth are not much greater than those to which the majority of unborn children in the general population were subjected a few decades ago. And while the chemicals did not produce gross birth defects in the children, they did result in small and consistent deficits of several kinds.

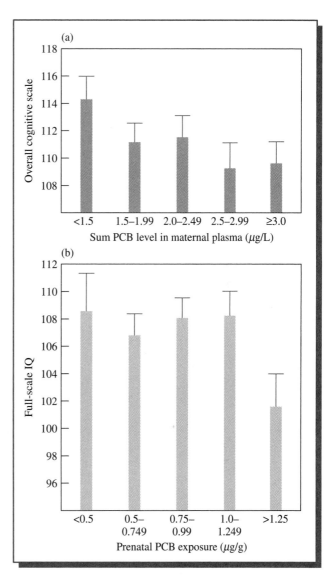

FIGURE 11-7 The effect on the intelligence of children of receiving PCBs prenatally: (a) overall cognitive abilities at 3½ years of age, (b) full-scale IQ at 11 years. [Sources: (a) S. Patandin et al., "Effects of Environmental Exposure to Polychlorinated Biphenyls and Dioxins on Cognative Abilities in Dutch Children at 42 Months of Age," *Journal of Pediatrics* 134 (1999): 33. (b) J. L. Jacobson et al., "A Benchmark Dose Analysis of Prenatal Exposure to Polychlorinated Biphenyls," *Environmental Health Perspectives* 110 (2002): 393.]

Studies of infants in North Carolina and in upper New York State have produced similar results to those found by the Jacobsons. Nevertheless, some doubt has been cast on the Jacobsons' results by an analysis that pointed out the difficulties in establishing the original *in utero* exposures. A more recent study in the Netherlands overcame these analytical difficulties, since the ability to detect very low levels of chemicals such as PCBs in blood serum has improved dramatically. The ongoing Dutch study has supported the Jacobsons' findings. It found that prenatal exposure to PCBs is more important than postnatal exposure and produces lower birth and growth weight and lesser cognitive abilities in young children, although the latter deficit was found for all but the least exposed fraction of children, rather than only the most highly exposed fraction. Negative effects on the IQs of three-year-old children by postnatal exposure to PCBs, as measured by blood levels, has recently been discovered by researchers in Germany. Subtle cognitive deficits have also been found in children in northern Quebec whose PCB concentrations are high due to the long-range air transport and subsequent deposition and entry into the food chain of the compounds (as discussed in Chapter 12). Dutch researchers also found that PCBs and dioxins transmitted to babies during gestation and via breast milk weakened their immune systems, contributing to more infections in the first few years of life.

The Toxicity Patterns of Dioxins, Furans, and PCBs

Research has shown that single doses of 2,3,7,8-TCDD administered to pregnant laboratory animals cause reproductive effects in their offspring. These results have raised some alarm about the potential effects of dioxins on human reproduction. In this connection, many scientists are worried about the dangers posed by environmental chemicals such as dioxins, furans, PCBs, and other organochlorines that can affect sex hormones, as discussed in Chapter 12.

Test results from studies of animals indicate that the acute toxicity of dioxins, furans, and PCBs depends to an extraordinary degree on the extent and pattern of chlorine substitution. The following generalization can be made: the very toxic dioxins are those with four beta chlorine atoms, and few if any alpha chlorines (see diagram on page 472 for definitions of the alpha and beta positions). Thus the most toxic is 2,3,7,8-TCDD, which has the maximum number (four) of beta chlorines and no alpha chlorines.

2,3,7,8-TCDD

Dioxin congeners that have three beta chlorines, but no (or only one) alpha chlorine, are appreciably toxic, but less so than the 2,3,7,8 compound. Fully chlorinated dioxin, that is, octachlorodibenzo-*p*-dioxin (OCDD), has a very low toxicity since all the alpha positions are also occupied by chlorine.

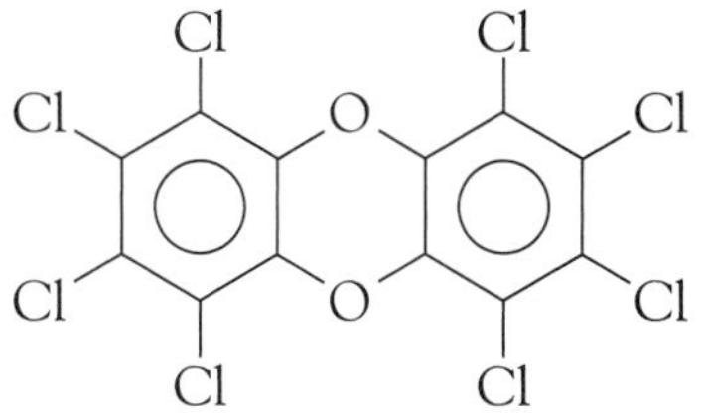

OCDD

Similarly, mono- and dichloro dioxins are usually not considered highly toxic, even if the chlorines are present in beta positions.

PROBLEM 11-10

Predict the order of relative toxicities of the following three dioxin congeners, given that, for systems not too dissimilar to TCDD, the presence of an alpha chlorine reduces the toxicity less than does the absence of a beta chlorine:

2,3,7-trichlorodibenzo-*p*-dioxin

1,2,3-trichlorodibenzo-*p*-dioxin

1,2,3,7,8-pentachlorodibenzo-*p*-dioxin

The toxicity pattern for furans is similar though not identical to that for dioxins in that the most toxic congeners have chlorines in all the beta positions. However, the most toxic furan, the 2,3,4,7,8 congener, does have one chlorine atom in an alpha position.

According to animal tests, the most acutely toxic PCBs are those having *no* chlorine atoms (or at most one) in the positions that are ortho to the carbons that join the rings, that is, on the 2, 2′, 6, and 6′ carbons. Without ortho chlorines, the two benzene rings can easily adopt an almost coplanar configuration, and rotation about the C—C bond joining the rings is rapid. However, because of the large size of chlorine atoms, they get in each other's way if they are present in both ortho positions on the same side of the two rings; this interaction forces the rings to twist away from each other, preventing such rings from adopting the coplanar geometry:

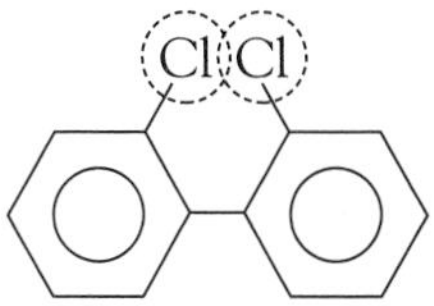

Consequently, PCB molecules with chlorines at three or four of the ortho positions cannot adopt a coplanar geometry.

If the rings are not kept from coplanarity by interference between chlorine atoms, and if the hydrogen atoms at certain meta and para carbons are replaced by chlorine, then the PCB molecule can readily attain a coplanar geometry that is similar in size and shape to 2,3,7,8-TCDD. Such PCB molecules are found to be highly toxic. Apparently 2,3,7,8-TCDD and other molecules of its size and shape readily fit into the same cavity in a specific biological receptor; the complex of the molecule and the receptor can pass through cell membranes and thereby initiate toxic action. By comparing molecular models, it is not difficult to see, e.g., that the most toxic PCB, namely *3,3′,4,4′,5′-pentachlorobiphenyl*, is almost the same size and shape as 2,3,7,8-TCDD. It is believed that some of the toxicity of the cooking-oil incidents arose from coplanar PCBs.

Only a very small fraction of commercial PCB mixtures corresponds to coplanar PCBs having no ortho chlorines. Although individually less toxic, PCBs with one ortho chlorine, and with chlorines in both para and at least one meta position, contribute substantially to the overall toxicity of PCB mixtures since they are far more prevalent than those having no ortho chlorines.

In humans, the more highly chlorinated furans, dioxins, and PCBs are stored in fatty tissues and are neither readily metabolized nor excreted. This persistence is a consequence of their structure: Few of them contain hydrogen atoms on adjacent pairs of carbons at which hydroxyl groups, OH, can readily be added in the biochemical reactions that are necessary for their elimination. In contrast, those compounds with few chlorines always contain one or more such adjacent pairs of hydrogens and tend to be excreted after hydroxylation, rather than stored for a long time.

The TEQ Scale

Since most organisms, including humans, have a mixture of many dioxins, furans, and PCBs stored in their body fat, and since all these compounds act qualitatively in the same way, it is useful to have a measure of the *net* toxicity of the mixture. To this end, scientists often report concentrations of these organochlorines in terms of the equivalent amount of 2,3,7,8-TCDD that, if present alone, would produce the same toxic effect. An international **toxicity equivalency factor,** or **TEQ,** has been devised that rates each dioxin, furan, and PCB congener's toxicity relative to that of 2,3,7,8-TCDD, which is arbitrarily assigned a value of 1.0. Recently, *polybrominated biphenyls* have also been added to this scale.

A summary of the TEQ values for some of the more toxic dioxins, furans, and PCBs is given in Table 11-1. As an example, consider an individual who ingests 30 pg (picograms) of 2,3,7,8-TCDD, 60 pg of 1,2,3,7,8-PCDF, and

TABLE 11-1	Toxicity Equivalence Factors (TEQ) for Some Important Dioxins, Furans, and PCBs	
Dioxin or Furan or PCB	**Toxicity Equivalency Factor**	
2,3,7,8-Tetrachlorodibenzo-*p*-dioxin	1	
1,2,3,7,8-Pentachlorodibenzo-*p*-dioxin	0.5	
1,2,3,4,7,8-Hexachlorodibenzo-*p*-dioxin		
1,2,3,7,8,9-Hexachlorodibenzo-*p*-dioxin	0.1	
1,2,3,6,7,8-Hexachlorodibenzo-*p*-dioxin		
1,2,3,4,6,7,8-Heptachlorodibenzo-*p*-dioxin	0.01	
Octachlorodibenzo-*p*-dioxin	0.001	
2,3,7,8-Tetrachlorodibenzofuran	0.1	
2,3,4,7,8-Pentachlorodibenzofuran	0.5	
1,2,3,7,8-Pentachlorodibenzofuran	0.05	
1,2,3,4,7,8-Hexachlorodibenzofuran		
1,2,3,7,8,9-Hexachlorodibenzofuran	0.1	
1,2,3,6,7,8-Hexachlorodibenzofuran		
2,3,4,6,7,8-Hexachlorodibenzofuran		
1,2,3,4,6,7,8-Heptachlorodibenzofuran	0.01	
1,2,3,4,7,8,9-Heptachlorodibenzofuran		
Octachlorodibenzofuran	0.001	
3,3′,4,4′,5-Pentachlorobiphenyl	0.1	
3,3′,4,4′,5,5′-Hexachlorobiphenyl	0.01	

200 pg of OCDD. Since the TEQ factors for these three substances are, respectively, 1.0, 0.05, and 0.001, the intake is equivalent to

$$(30 \text{ pg} \times 1.0) + (60 \text{ pg} \times 0.05) + (200 \text{ pg} \times 0.001) = 33.2 \text{ pg}$$

Thus, even though a total of 290 pg of dioxins and furans were ingested by this person, the mixture is equivalent in its toxicity to an intake of only 33.2 pg of 2,3,7,8-TCDD.

PROBLEM 11-11

Using the TEQ values in Table 11-1, calculate the number of equivalent picograms of 2,3,7,8-TCDD that corresponds to an intake of a mixture of 24 pg of 1,2,3,7,8,9-hexachlorodibenzo-*p*-dioxin, 52 pg of 2,3,4,7,8-pentachlorodibenzofuran, and 200 pg of octachlorodibenzofuran.

The TEQ values for environmental samples are sometimes reported in the media as if they represent the concentration of 2,3,7,8-TCDD itself. However, this compound often is not even the dominant contributor to the TEQ toxicity. As discussed previously, combustion of organic matter produces relatively few toxic dioxins; the TEQ from such sources is often dominated by penta- and hexachlorinated furans. Similarly, the TEQ arising from the use of chlorination in bleaching paper is dominated by toxicity from tetrachlorinated furans.

Dioxins, Furans, and PCBs in Food

About 95% of human exposure to dioxins and furans arises from the presence of the compounds in food. A bar graph showing the TEQ values for contamination of various types of foods purchased in U.S. supermarkets in the 1990s is shown in Figure 11-8. Notice that fresh-water fish contained the highest levels of both PCB and furan toxicity. Recently, average TEQ levels of several ppt have been found for farmed salmon; these high levels originate with the fishmeal and fish oil the growing salmon had been fed. However, the TEQ levels in young chickens and turkeys and in hogs sold in the United States has declined significantly—by 20 to 80%—in recent years. The composite vegan diet (i.e., all vegetable, fruit, and grain, with no animal products at all) has a very low TEQ compared to that from animal-based components.

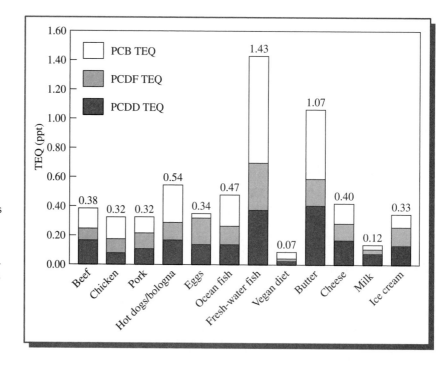

FIGURE 11-8 TEQ values for foods collected from U.S. supermarkets. [Source: A. Schecter et al., "Levels of Dioxins, Dibenzofurans, PCB and DDE Congeners in Pooled Food Samples Collected in 1995 at Supermarkets Across the United States," *Chemosphere* 34 (1997): 1437.]

A 2003 report by the U.S. Institute of Medicine recommended that girls should reduce their consumption of animal products in order to reduce the amount of dioxins that build up in their body fat and that could subsequently affect any children they might have in the future.

PROBLEM 11-12

Given that the average TEQ of animal-based foods was about 0.4 pg of TCDD equivalent per gram when the data in Figure 11-8 were collected, and that the LD_{50} for 2,3,7,8-TCDD is about 0.001 mg/kg body weight, what mass of animal-based food would you have had to consume to ingest a fatal dose of it?

In the mid-1980s, the average total concentration of all dioxins and furans in the fat tissue of adult North Americans was about 1000 ppt. However, because highly chlorinated and therefore less toxic congeners predominated, the TEQ value was much less: about 40 ppt of 2,3,7,8-TCDD equivalent. By the 1990s, the TEQ in human fat had fallen to about 15 ppt. The historical maximum TEQ levels, about 75 ppt, were reached in the 1970s. The variation with time of stored dioxins and furans fits a model in which the daily TEQ dose amounted to about 0.5 pg/kg body weight in the early decades of the twentieth century, rose to over 6 pg/kg in the 1940s to 1970s, and has now declined again to 0.5 pg/kg.

The average North American adult's body contains about 15 kg of fat, so his or her total body burden of 2,3,7,8-TCDD equivalents now amounts to about 0.2 μg. Given that the average residence time t_{avg} of dioxins and furans in the human body is about seven years, and using the relationship from Chapter 6 connecting t_{avg} to the total amount C and the input rate R, i.e.,

$$t_{avg} = C/R$$

then the average human rate of intake of 2,3,7,8-TCDD equivalents is calculated to be

$$R = C/t_{avg} = 0.2\ \mu g/7\ y = 0.03\ \mu g/y$$

This value, which corresponds to about 1 pg per kilogram body weight per day, is close to the intake estimated over the last few years from the model discussed previously.

Dioxins as Probable Human Carcinogens

Although there is little argument as to the *relative* acute toxicities of various dioxin and furan congeners, their *absolute* risk to humans is *very* controversial. The amount of 2,3,7,8-TCDD per kilogram body weight required to kill a guinea pig is extraordinarily small—about 1 μg—making it the most toxic synthetic chemical known for that species. However, the LD_{50} required to kill many other types of animals is hundreds or thousands of times this amount—e.g., the LOD_{50} for hamsters 1200 $\mu g/kg$; for frogs, 1000 $\mu g/kg$; for rabbits and mice, 115 $\mu g/kg$; for monkeys, 70 $\mu g/kg$; and for dogs, it may be as low as 30 $\mu g/kg$. A 2000 EPA report on dioxin suggests that humans fall in the middle range for acute susceptibility to dioxins; humans given doses of 100 $\mu g/kg$ suffered no apparent ill effects beyond chloroacne.

Another widespread exposure of humans to 2,3,7,8-TCDD occurred in the early 1970s in and around Times Beach, Missouri. Waste oil containing PCBs and 2,3,7,8-TCDD from the manufacture of 2,4,5-trichlorophenol was used for dust control on gravel roads. Some horses died due to exposure in an area where the dioxin contamination was particularly high, and some children became ill. A decade later, widespread contamination of the soil in the town was discovered. In 1997, over 200,000 tonnes of soil from this town and 26 other affected sites in eastern Missouri that had soil 2,3,7,8-TCDD levels of 30–200 ppb were excavated and incinerated in order to remediate the problem. (The town of Times Beach was abandoned in 1982 due in part to a severe flash flood.) Although exposure to the chemical seems to have negatively affected their immune systems, a major study in 1986 did not find evidence of increased disease prevalence in the exposed group of Times Beach residents. Less formal studies and anecdotal evidence, however, indicate problems with seizures and congenital abnormalities, and so the issue remains controversial.

Scientists are more concerned about the long-term effects of exposure to dioxins than about their acute toxicities. The Seveso study discussed previously was the first to show an increased rate of cancer among people exposed accidentally to TCDD. A recent analysis of the health of American workers who were employed in industries that produced chemicals contaminated with 2,3,7,8-TCDD indicates that exposure to it at relatively high levels may cause cancer. The current theory concerning the action of dioxin predicts that there should be a threshold below which no toxic effects will occur, and recent studies of workers exposed to 2,3,7,8-TCDD supports this hypothesis.

In animal studies, a threshold of about 1000 pg (i.e., 1 ng) of 2,3,7,8-TCDD equivalent per kilogram of body weight per day is observed with respect to the cancer-causing ability of dioxins and furans. In determining the maximum tolerable human exposure to such compounds, many governments apply a safety factor of 100, resulting in a guideline for maximum exposure of 10 pg/kg/day averaged over a lifetime. Currently, the average American ingests only about one-tenth this amount from animal fats in his or her food supply. Exposure levels near the guideline limit are expected for persons consuming large amounts of fish that have elevated dioxin and furan levels.

The U.S. EPA draft report of 2000 concerning the health risks of dioxins concluded that 2,3,7,8-TCDD is a (known) human carcinogen—although this characterization was a point of controversy among members of the expert committee that reviewed the report—and that the mixtures of dioxins to which people are exposed is a "likely human carcinogen." The *International Agency for Research on Cancer* of the *World Health Organization* had previously classified TCDD as a known human carcinogen. Experiments had indicated that 2,3,7,8-TCDD was the most potent multisite carcinogen known in test animals. The EPA estimates that the most sensitive and most highly exposed Americans stand at least a 1-in-1000 chance of developing cancer from dioxin. The report notes that noncancer effects of dioxin are at least as important as cancer.

However, there is currently no clear indication that the level of cancer has risen in the general American population due to dioxins, though this could well be due to the inability to relate effects to exposure at current levels. In addition to cancer, the report concludes that dioxins adversely affect the endocrine and immune systems and the development of fetuses, as will be discussed in Chapter 12.

For furans, direct evidence of human susceptibility is available from the incidents of PCB-contaminated cooking-oil consumption mentioned above. The most common symptoms observed in these groups were chloracne and other skin problems. Unusual pigmentation occurred in the skin of babies born to some of the mothers who had been exposed. The children also often had low birth weight and experienced a rather high infant mortality rate. Children who directly consumed the oil showed retarded growth and abnormal tooth development. Many of the victims also reported aching or numbness in various parts of their bodies and frequent bronchial problems. Other than chloracne, such symptoms are not observed in workers who are occupationally exposed to PCBs and whose body burdens of PCBs are comparable to those who consumed the contaminated oil, but in which the PCDF concentration is orders of magnitude lower. However, some children of these workers had mild cases of the less serious problems seen in the poisoned group.

Ukrainian president Viktor Yushchenko was apparently the recent victim of deliberate dioxin poisoning. Since September 2004, he has suffered from ulcers in his stomach and intestines, problems with his liver and spleen, and disfiguring facial cysts that have left him looking far older than he is (Figure 11-9).

Human Exposure to Dioxins, Furans, and PCBs

There continues to be vigorous debate in scientific, industrial, and medical communities regarding the environmental dangers of dioxins, furans, and PCBs. In one camp are those who feel that the dangers from these chemicals have been wildly overstated in the media and by some special-interest groups. They point to the very low concentrations of these substances that exist in the environment, to the possibility that there is a threshold below which

FIGURE 11-9 Ukranian president Viktor Yushchenko before and after he was poisoned with dioxin. [Source: AP Photo/Efrem Lukatsky.]

these compounds have no effect on human health, to the lack of known human fatalities resulting from them, and to the enormous economic costs associated with instituting further controls and cleanup measures. At the other extreme are persons who point to the substantial biomagnification and high toxicity per molecule of these substances and to their presence in almost all environments. They consider the detrimental effects such as cancer and birth deformities caused by these chemicals in wildlife to be "warning canaries" that signal potential ill effects in humans. Discovering where the truth lies between these opposing viewpoints presents a challenge even for environmental science students, to say nothing of the public at large!

Review Questions

1. Using structural diagrams, write the reaction by which 2,3,7,8-TCDD is produced from 2,4,5-trichlorophenol.

2. Draw the structure of 1,2,7,8-TCDD. What is the full name for this dioxin?

3. What, chemically speaking, was *Agent Orange*, and how was it used?

4. Draw the structure of *pentachlorophenol*. What is its main use as a compound? What is the main dioxin congener that it could produce?

5. What does *PCB* stand for? Draw the structural diagram of the 3,4′,5′-trichloro PCB molecule.

6. What were the main uses for PCBs? What is meant by an *open use*?

7. Draw the structure of a representative polychlorinated dibenzofuran congener.

8. Other than the chlorophenols and PCBs, what are some of the other sources of dioxins and furans in the environment? What is currently the biggest anthropogenic source of dioxins?

9. From what medium—air, food, or water—does most human exposure to dioxin come about? Why is this so?

10. What molecules can be eliminated by a PCB molecule when it is heated to moderately high temperatures?

11. Are PCBs acutely toxic to humans? What is the basis for health concerns about them? Recount recent evidence that shows that PCBs can affect human development.

12. Are all dioxin congeners equally toxic? If not, what pattern of chlorine substitution leads to the greatest toxicity? Which is the most toxic dioxin?

13. What is meant by a *coplanar PCB*? What structural features give rise to noncoplanarity?

14. What does *TEQ* stand for? Why is it used?

15. Is dioxin carcinogenic to humans or not? Discuss the evidence for and against.

 ## Green Chemistry Questions

See the discussion of focus areas and the principles of green chemistry in the Introduction before attempting these questions.

1. The development of TAML catalyst for hydrogen peroxide oxidation by Terry Collins won a Presidential Green Chemistry Challenge Award.

(a) Into which of the three focus areas for these awards does this award best fit?

(b) List at least three of the twelve principles of green chemistry that are addressed by the green chemistry developed by Collins.

2. What environmental advantages does the TAML/hydrogen peroxide method of bleaching pulp have over the use of elemental chlorine?

Additional Problems

1. Deduce which combination(s) of two different tetrachlorophenol isomers would produce the following hexachlorodibenzo-*p*-dioxins:

(a) the 1,2,3,7,8,9 isomer, **(b)** the 1,2,4,6,8,9 isomer, and **(c)** the 1,2,3,6,7,9 isomer.
[*Hint: See Box 11-1.*]

2. Deduce which dioxins would likely result from the low-temperature combustion of a commercial sample of PCP.

3. In the purification of wastewater contaminated by pentachlorophenol and 2,3,5,6-tetrachlorophenol using ultraviolet light, it was noticed that OCDD and 1,2,3,4,6,7,8-heptachlorodibenzo-*p*-dioxin were formed. Deduce whether the latter was formed by the coupling of a molecule of each of the phenols or must have been formed by photochemical dechlorination of OCDD. What potential flaw exists in treating water by UV light, given the nature of the products that are formed?

4. Using mechanical ball-and-stick or computer-generated molecular models, construct structures for **(a)** 2,3,7,8-TCDD, **(b)** dibenzofuran, and **(c)** biphenyl. Place chlorines onto the dibenzofuran and biphenyl models such that the space filled by the carbon, oxygen, and chlorine atoms overlaps that of the dioxin as much as possible without occupying much of the space associated with the alpha positions. Do the resulting congeners represent the most toxic furan and biphenyl according to TEQ values?

5. Consider the PCDF on the next page. Deduce which PCBs could produce this furan if they are moderately heated in air, given that PCDFs can result from HCl elimination with or without 2,3 interchange, or from Cl_2 elimination. [Hint: See Box 11-2.]

6. By comparing the average dioxin levels in humans to those in our food, decide whether or not dioxin is biomagnified in the transition. Given the food typically eaten by domestic animals such as cattle and chickens compared to that of the fresh-water fish we eat, can you explain why dioxin TEQ levels in Figure 11-8 for the fish exceed those for the animals? Why is the TEQ value for butter so much higher than that for milk, and that for hot dogs greater than the general levels for meat? Why is the TEQ value for a vegan diet so very low? Given the vegan-diet TEQ and that for the domestic animals, can you predict whether biomagnification occurs for the latter relative to its diet?

7. Predict the order of toxicity of the three tetra-chlorobiphenyls with the following numbering: **(a)** 2,4,3',4'; **(b)** 3,4,5,4'; **(c)** 2,4,2',6'. [Hint: Will these PCB molecules be coplanar?]

8. In the potential reactions of chlorinated phenols undergoing incineration in separate facilities, what dioxins could be formed from **(a)** 2,5-dichlorophenol, and **(b)** 2,4,6-trichlorophenol? Name the two dioxins, and predict which would be the more toxic.

9. The partitioning of PCBs among air, water, and sediments can be estimated by the fugacity model discussed in Chapter 10. The Z values for a typical environmental PCB are 4×10^{-4} in air, 0.03 in water, and about 10,000 in sediment (and biota). Using the model world volumes in Chapter 10, calculate the equilibrium concentrations when 1 mole of PCBs is distributed among air, water, and sediment.

Websites of Interest

Log on to www.whfreeman.com/envchem4/ and click on Chapter 11.

CHAPTER 12

OTHER TOXIC ORGANIC COMPOUNDS OF ENVIRONMENTAL CONCERN

In this chapter, the following introductory chemistry topics are used:

- Elementary organic chemistry (as in the Appendix in this book)
- Concept of vapor pressure

Background from previous chapters used in this chapter:

- Maximum Contaminant Level (MCL) (Chapter 10)
- Structures of dioxins, furans, and PCBs; congeners (Chapter 11)
- Structures of DDT, DDE, and atrazine (Chapter 10)
- Concepts of bioaccumulation; K_{ow} (Chapter 10)
- Concepts of free-radical and chain reactions (Chapters 1–5)
- LD_{50} (Chapter 10)
- Adsorption (Chapter 4)

Introduction

In this chapter, we shall look at a series of toxic organic compounds that do not contain chlorine but that have become common air and water pollutants. We begin by considering PAHs, pollutants that accompany the combustion of most natural organic materials, and then consider a wide range of environmental chemicals—most of which have been encountered in other contexts—that may have disruptive effects on our reproductive systems. We then consider the surprising mechanism by which persistent substances become distributed around the world by air currents. Finally, we survey two new classes of commercial products that are causing environmental concern and are being transported to remote regions by this mechanism. The topic of pharmaceutical contamination of drinking water will be discussed in Chapter 16.

Polynuclear Aromatic Hydrocarbons (PAHs)

One of the most common and ancient types of environmental pollutant, whether in air, water, or soil, is a large series of hydrocarbons known as PAHs. We first discuss the molecular structure of these compounds, then relate their occurrence in the environment and their health effects on humans.

The Molecular Structure of PAHs

There is a series of hydrocarbons whose molecules contain several six-membered benzene-like rings connected by the sharing of a pair of adjacent carbon atoms between two adjoining **fused** rings. The simplest example is **naphthalene,** $C_{10}H_8$:

naphthalene

Notice that there are ten, not twelve, carbon atoms in total and that there are only eight hydrogen atoms, since the shared carbons have no attached hydrogen atoms. As a compound, naphthalene is a volatile solid whose vapor is toxic to some insects. It has found use as one form of "moth balls," the other being *1,4-dichlorobenzene.*

Conceptually, there are two ways to fuse a third benzene ring to two carbons in naphthalene; one results in a linear arrangement for the centers (the "nuclei") of the rings while the other is a "branched" arrangement:

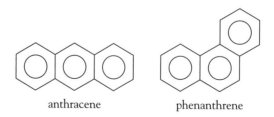

anthracene   phenanthrene

The resulting molecules, **anthracene** and **phenanthrene,** are PAH pollutants arising from incomplete combustion, especially of wood and coal. They are also released into the environment from the dumpsites of industrial plants that convert coal into gaseous fuel and from the refining of petroleum and shale. In rivers and lakes they are found mainly attached to sediments rather than dissolved in the water; both are subsequently partially incorporated into fresh-water mussels.

PROBLEM 12-1

Draw the full structural diagram for phenanthrene, showing all atoms and bonds explicitly.

PROBLEM 12-2

By determining their molecular formulas, show that the molecules below are not additional isomers of $C_{14}H_{10}$:

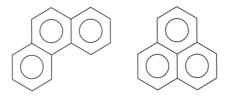

PROBLEM 12-3

Using a systematic procedure, deduce the structural formulas for the five unique isomers of $C_{18}H_{12}$, which contains four fused benzene rings.

In general, hydrocarbons that display benzene-like properties are called **aromatic;** those that contain fused benzene rings are called **polynuclear** (or polycyclic) **aromatic hydrocarbons,** or **PAHs** for short. Like benzene itself, most PAHs have a planar geometry and contain alternating single and double bonds producing unusually high stability. A few PAHs also contain one or more five-membered rings. Other than naphthalene, they are not manufactured commercially. However, some PAHs are extracted from coal tar and used in commerce.

PAHs as Air Pollutants

PAHs are common air pollutants and are strongly implicated in the degradation of human health. Typically, the concentration of PAHs in urban outdoor air amounts to a few nanograms per cubic meter, although it reaches 10 times this amount in very polluted environments. PAHs are formed when carbon-containing materials are incompletely burned. Elevated PAH concentrations in domestic indoor air are typically due to the smoking of tobacco and the burning of wood and coal.

PAHs containing four or fewer rings usually remain as gases if they are released into air, since the vapor pressures of their liquid forms are relatively high. After spending on average less than a day in outside air, such PAHs are degraded by a sequence of free-radical reactions that begin, as expected from our previous analysis of air chemistry (Chapters 3 and 5), by the addition of the OH radical to a double bond.

In contrast to their smaller analogs, PAHs with more than four benzene rings do not exist for long in air as gaseous molecules. Owing to their low vapor pressure, they condense and become adsorbed onto the surfaces of suspended soot and ash particles. PAHs are found mainly on particles of submicron, i.e., respirable, size; consequently, the PAHs can be transported into the lungs by breathing.

Figure 12-1 illustrates the concentrations of the most abundant PAHs in an air sample taken in Sweden. The leftmost four PAHs in the figure have three or four fused rings and, as expected, occur mainly as gases. The remaining ones, which in contrast are found mainly associated with fine particulate

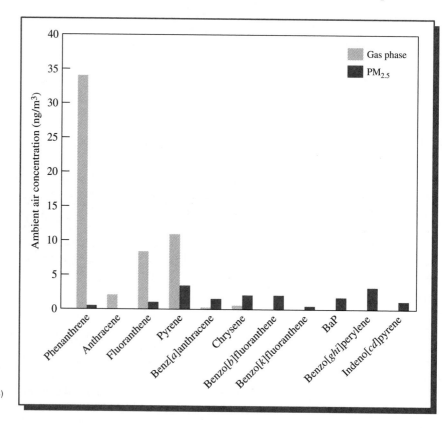

FIGURE 12-1 The distribution of individual PAHs attached to fine particles (PM$_{2.5}$) and in the gas phase in an ambient air sample from Sweden.
[Source: C.-E. Bostrom et al., "Cancer Risk Assessment for Polycyclic Aromatic Hydrocarbons," *Environmental Health Perspectives* 110 (supplement 3) (2002): 451.]

matter, are larger. Even PAHs with two to four rings are adsorbed onto particles in the wintertime, since their vapor pressure decreases sharply at lower temperatures. The role of vapor pressure in determining the long-range air transport of pollutants such as PAHs is discussed later in this chapter.

By 2001, the concentrations of PAHs in rural air in eastern North America (Nova Scotia) were found to have declined by an order of magnitude since reaching their peak in 1985, presumably due to air pollution abatement programs in Canada and the United States. In contrast, the concentrations of PAHs in sediments laid down in urban lakes increased substantially over that period.

Soot itself is mainly graphite-like carbon; it consists of a collection of tiny crystals (crystallites), each composed of stacks of planar layers of carbon atoms, all of which occur in fused benzene rings. **Graphite** is the ultimate PAH—its parallel planes of fused benzene rings each contain a vast number of carbon atoms.

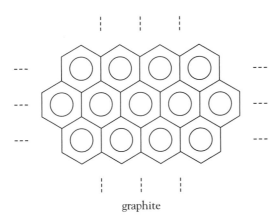

graphite

There are no hydrogen atoms in graphite except at the periphery of the layers. The surfaces of soot particles are excellent adsorbers of gaseous molecules, especially organic ones.

PAHs are introduced into the environment from a number of sources: the exhaust of gasoline and especially diesel combustion engines, the "tar" of cigarette smoke, the surface of charred or burned food, the smoke from burning wood or coal, and other combustion processes in which the carbon of the fuel is not completely converted to CO or CO_2. Although PAHs constitute only about 0.1% of airborne particulate matter, their existence as air pollutants is of concern since many them are carcinogenic, at least in test animals. Vehicle exhaust, especially from diesel engines, older gasoline-powered cars, and all vehicles in which the engine has not warmed up, is the major contributor to PAH levels in cities. Aluminum smelters are a source of PAHs since their heated graphite anodes deteriorate over time, releasing the hydrocarbons.

A survey of PAH emissions in Taiwan from exhaust stacks found the following trends for different types of restaurants, the differences presumably arising from the style of cooking employed:

Chinese >> Western > fast-food > Japanese

PAH levels in the air were particularly high near the site of World Trade Center buildings in New York for many days after they had been destroyed in September 2001. However, the heat generated by the flames carried most of the smoke aloft. In addition to PAHs, the high temperatures and anaerobic conditions within the piles of debris resulted in the oxidation by chlorine of metals and organic substances, producing a wide variety of contaminants.

PAHs as Water Pollutants

Polycyclic aromatic hydrocarbons also are serious water pollutants. PAHs enter the aquatic environment as a result of spills of oil from tankers, refineries, and offshore oil-drilling sites. In drinking water, the PAH level typically amounts to a few parts per trillion and usually is an unimportant source of these compounds to humans. The larger PAHs bioaccumulate in the fatty tissues of some marine organisms and have been linked to the production of liver lesions and tumors in some fish. PAHs, PCBs, and the insecticide mirex are thought to play a role in the devastation of the populations of beluga whales in the St. Lawrence River; since discharges of these pollutants to the river have decreased substantially in recent years, the health of the belugas may eventually improve.

PAHs are generated in substantial quantity in the production of such coal-tar derivatives as *creosote*, a wood preservative used especially on railway ties. Up to 85% of the 200 compounds in creosote are PAHs, including some carcinogenic ones. In addition to concerns about PAH exposure to workers installing the ties, gardeners who buy used railway ties for landscape design and people burning discarded ties are at risk. The leaching of PAHs from the creosote used to preserve the immersed lumber of fishing docks and the like represents a significant source of pollution to crustaceans such as lobsters.

Formation of PAHs During Incomplete Combustion

The mechanism of PAH formation during combustion of organic materials is complex but is due primarily to the repolymerization of hydrocarbon fragments that are formed during the **cracking** (i.e., the splitting into several parts) of larger fuel molecules in the flame. Fragments containing two carbon atoms are particularly prevalent after cracking and partial combustion have occurred. Two C_2 fragments can combine to form a C_4 free-radical chain, which could add another C_2 to form a six-membered ring. Such reactions occur quickly if one of the original C_2 fragments is itself a free radical.

The repolymerization reaction occurs particularly under oxygen-deficient conditions. Generally the PAH formation rate increases as the oxygen-to-fuel ratio decreases. The fragments often lose some hydrogen, which forms water after combining with oxygen during the reaction steps. The carbon-rich fragments combine to form the polynuclear aromatic hydrocarbons, which are the most stable molecules that have a high C-to-H ratio.

Since neither methane, the main component of natural gas, nor methanol molecules contain any C—C bonds with which to form C_2 units, their combustion as fuels produces very little PAH or other soot-based particulates.

Carcinogenic Properties of PAHs

The most notorious and common carcinogenic PAH is **benzo[a]pyrene,** BaP, which contains five fused benzene rings:

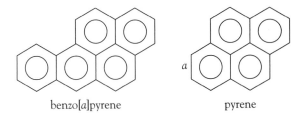

benzo[a]pyrene   pyrene

The molecule is named as a derivative of **pyrene,** which has the structure shown at the right. Conceptually, if an additional benzene ring is added at the pyrene bond labeled a, the benzo[a]pyrene molecule is obtained.

Benzo[a]pyrene is a common by-product of the incomplete combustion of fossil fuels, of organic matter (including garbage), and of wood. It is a carcinogen in test animals and a probable human carcinogen. BaP is worrisome since it bioaccumulates in the food chain—its log K_{ow} value is 6.3, comparable to that of many organochlorine insecticides (see Table 10-3). It is considered to be one of the top 15 organic carcinogens in drinking water, as listed in Table 12-1 along with the other 14 compounds in this category, most of which are pesticides.

A second example of a carcinogenic PAH is the four-ring hydrocarbon **benz[a]anthracene,** which is anthracene with an additional benzene ring fused to the a bond:

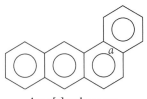

benz[a]anthracene

The relative positions in space of the fused rings in PAHs play a major role in determining their level of carcinogenic behavior in animals. The

TABLE 12-1	Top Organic Carcinogens in U.S. Drinking Water*		
Chemical		**MCL in ppb**	**Cancer Risk per 100,000 People**
Ethylene dibromide		0.05	12.5
Toxaphene		3	9.6
Vinyl chloride		2	8.4
Heptachlor		0.4	5.2
Heptachlor epoxide		0.2	5.2
Hexachlorobenzene		1	4.6
Benz[a]pyrene		0.2	4.2
Chlordane		2	2.0
Carbon tetrachloride		5	1.9
1,2-Dichloroethane		5	1.3
PCBs		0.5	0.5
Pentachlorophenol		1	0.3
Di(2-ethylhexyl) phthalate		6	0.2
Dichloromethane (methylene dichloride)		5	0.1

Source: A. H. Smith et al., "Arsenic Epidemiology and Drinking Water Standards," *Science* 296 (2002): 2146.

*Chemicals ordered by their cancer risk, were they to be consumed at their MCL (maximum contaminant level).

PAHs that are the most potent carcinogens each possess a **bay region** formed by the branching in the benzene ring sequence: The organization of carbon atoms as an open bay indirectly imparts a high degree of biochemical reactivity to the PAH, as explained in Box 12-1.

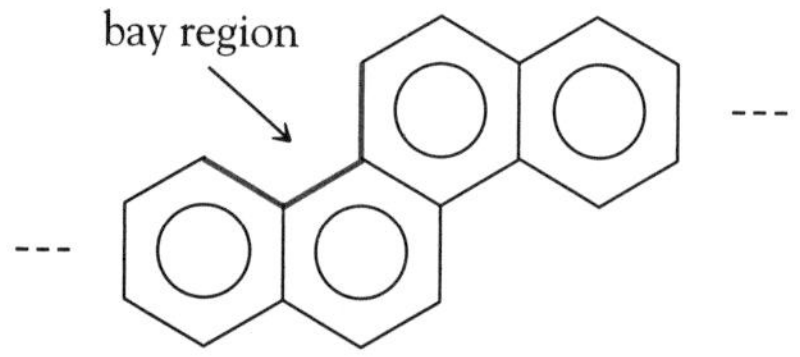

bay region

PROBLEM 12-4

Based upon the bay region theory, would you expect naphthalene to be a carcinogen? How about anthracene or phenanthrene? What about the PAH called benzo[ghi]perylene, whose structure is shown on the next page?

BOX 12-1 | More on the Mechanism of PAH Carcinogenesis

Research has established that the PAH molecules themselves are not carcinogenic agents; rather they must be transformed by several metabolic reactions in the body before the actual cancer-causing species is produced.

The first chemical transformation that occurs in the body is the formation of an epoxide ring across one C=C bond in the PAH. The specific epoxide of interest to the carcinogenic behavior of benzo[a]pyrene is

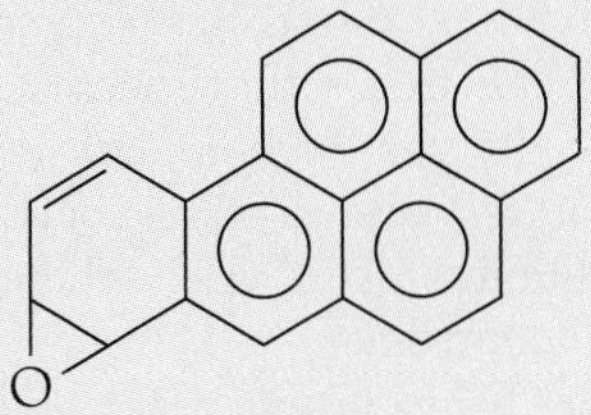

A fraction of these epoxide molecules subsequently add H_2O, to yield two —OH groups on adjacent carbons:

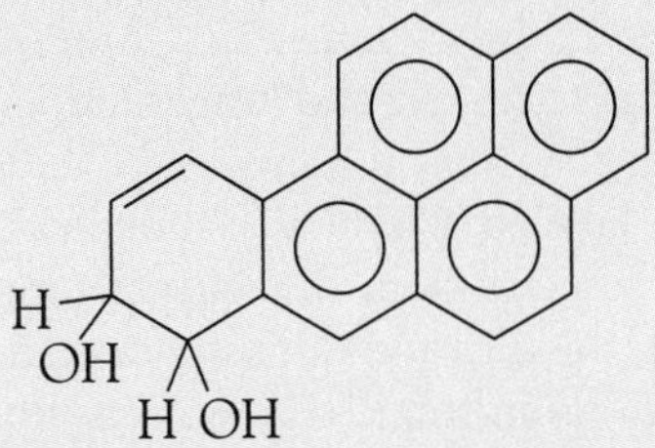

The double bond (shown in the structure) that remains in the same ring as the two —OH groups subsequently undergoes epoxidation, yielding the molecule that is the active carcinogen:

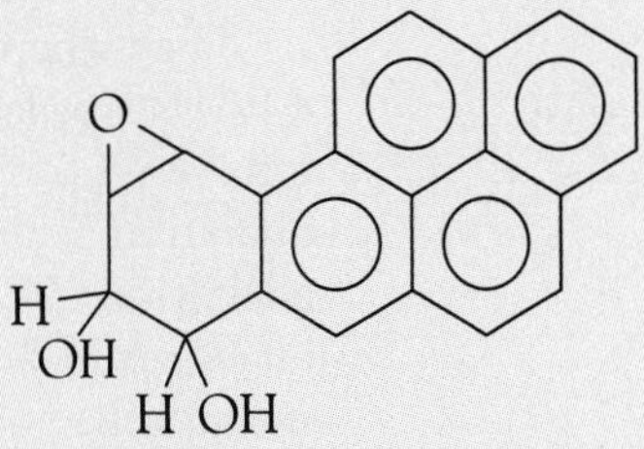

By adding H^+, this molecule can form a particularly stable cation that can bind to molecules such as DNA, thereby inducing mutations and cancer.

The metabolic reactions of epoxide formation and H_2O addition are part of the body's attempt to introduce —OH groups into hydrophobic molecules like PAHs to make them more easily dissolved in water and then eliminated. For BaP and other PAHs that possess a bay region, one of the intermediate products in this multistep process can be diverted instead into the formation of a very stable cation that induces cancer.

benzo[ghi]perylene

Some PAHs with certain of their hydrogen atoms replaced by methyl groups are even more potent carcinogens than are the parent hydrocarbons.

Environmental Levels of PAHs and Human Cancer

Has exposure to PAHs been demonstrated to produce cancer in humans? The answer is both yes and no. For over 200 years, it has been known that prolonged exposure in occupational settings to very high levels of coal tar, the principal toxic ingredient of which is benzo[*a*]pyrene, leads to cancer in humans. In 1775 the occurrence of scrotal cancer in chimney sweeps was associated with the soot lodged in the crevices of the skin of their genitalia. Modern workers in coke-oven and gas-production plants likewise experience increased levels of lung and kidney cancer due to this PAH.

The evidence for cancer induction in the general public, whose exposure to PAHs is at levels that are orders of magnitude lower than in these occupational environments, is less clear-cut. The main cause of lung cancer is the inhalation of cigarette smoke, which contains many carcinogenic compounds in addition to PAHs; the deduction from health statistics of the much smaller influences of air pollutants such as PAHs from other sources is difficult to accomplish. Some scientists speculate that the higher death rate from lung cancer in cities as compared to rural areas found in many countries is due in part to breathing carcinogenic air pollutants like PAHs, although other factors such as a higher smoking rate also contribute.

Many cities in developing countries have chronic problems with carbon-based particulate air pollution. For example, the serious indoor and outdoor air pollution, which arises primarily from the unvented burning of coal and biomass for cooking and heating and consists primarily of PAHs, sulfur dioxide, and particulate matter, is reputed to be responsible for over 1 million deaths annually in China. The rate of lung cancer in Chinese women is higher than that for men, possibly due to higher and longer indoor exposures to PAHs from coal burning and from cooking-oil fumes.

Diesel-engine exhaust has been labeled a "probable human carcinogen." It contains not only PAHs but also some of their derivatives containing the nitro group, $-NO_2$, as a substituent; these nitrated substances are even more active carcinogens than are the corresponding PAHs. For example, *nitropyrene* and *dinitropyrene* are responsible for much of the mutagenic character of diesel exhaust, i.e., its ability to cause mutations that could ultimately produce cancer. These compounds are formed within the engines by the reaction of pyrene with NO_2 and N_2O_4. There is also evidence that PAHs are nitrated by some of the constituents of photochemical smog.

As discussed in Chapter 3, the emission of particulates from heavy-duty diesel engines can be controlled by means of filter traps in the exhaust stream. These devices temporarily retain the solids, including soot, in the exhaust and eventually oxidize them further. Some scientists have worried that, during their residence in the traps, PAHs could undergo reaction to produce even greater quantities of nitrated PAHs. While this apparently does occur, tests indicate that the total mutagenic activity of a given amount of engine exhaust is actually decreased by the devices, since most of it is oxidized in the trap.

For most nonsmokers in developed countries, by far the greatest exposure to carcinogenic PAHs arises from their diet, rather than directly from polluted air, water, or soil. As expected from their mode of preparation, charcoal-broiled and smoked meat and fish contain some of the highest levels of PAHs found in food. However, leafy vegetables such as lettuce and spinach can constitute an even greater source of carcinogenic PAHs due to the deposition of these substances from the air onto the leaves of the vegetables while they are growing. Unrefined grains also contribute significantly to the total amount of PAHs ingested from food.

Environmental Estrogens

In the last two decades, a new threat to the health of wildlife, and possibly of humans, exposed to synthetic organic chemicals in the environment has been identified. It has been established that certain compounds can affect the reproductive and immune health of higher organisms and may also increase the rate of cancer in reproductive organs. Much of the public interest in this issue was stirred by the 1996 publication of the book *Our Stolen Future* by Theo Colburn and her associates. However, there remains great uncertainty and debate in the scientific community about whether or not there are significant risks to human health from environmental levels of these compounds.

Mechanism of Action of Environmental Estrogens

The chemicals in question interfere with the system in the organism that operates by transmitting extremely low concentrations, on the parts-per-trillion level, of chemical messenger molecules called **hormones.** The hormones flow through the bloodstream from the point of production and storage to their target organs, including those involved in sexual reproduction in both females and males. The arrival of the hormone at a receptor is a signal for the cell to initiate action of some type. Much of the concern about humans centers upon interference with **estrogens,** the female sex hormones (which also are present in males, but at a smaller concentration.) There also exist environmental substances that interfere with **androgens,** the male sex hormones, and with thyroid hormones. Sex hormones, including estrogens and androgens, contain the characteristic four-ring steroid structure (see top of Figure 12-2). They are produced from cholesterol in the ovaries of females and the testes of males in response to signals from the brain and other organs.

Hormone actions are initiated by the binding of the hormone to a specific receptor within a cell. The resulting hormone–receptor complex binds to specific regions of DNA in the cell nucleus, an action that in turn determines the action of the genes. Certain environmental chemicals can also bind to the hormone receptor and thereby either mimic or block the action of the hormone itself. In particular, the "promiscuous" estrogen receptor will bind to a number of compounds, even ones that bear little structural

FIGURE 12-2 The structures of estradiol and some environmental estrogens.

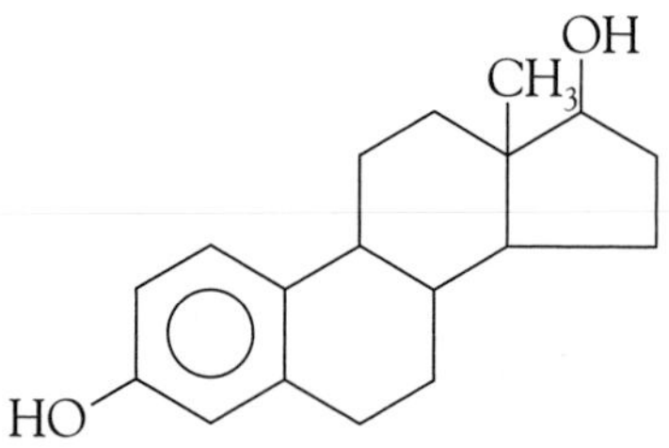

estradiol, the main estrogen

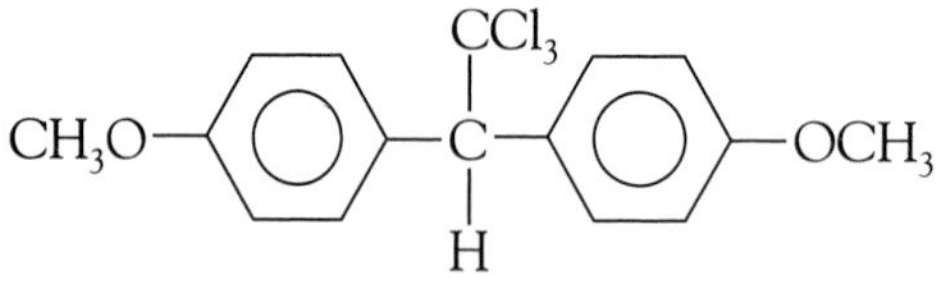

methoxychlor

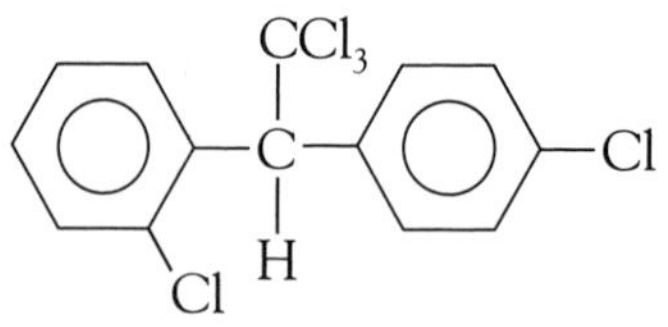

o,p'-DDT

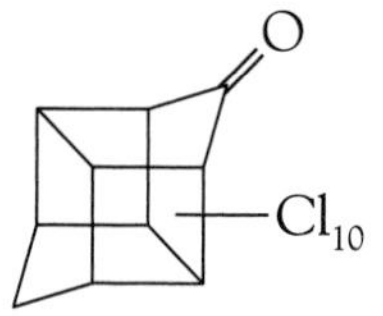

kepone

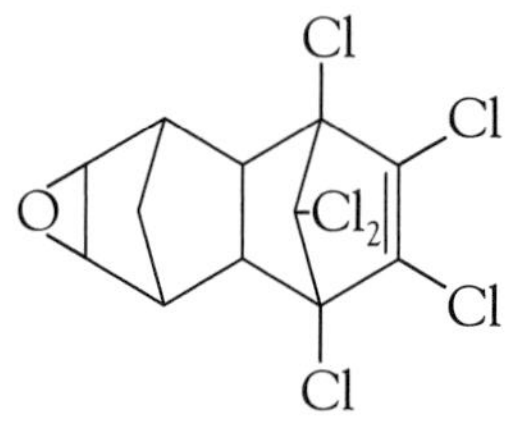

dieldrin

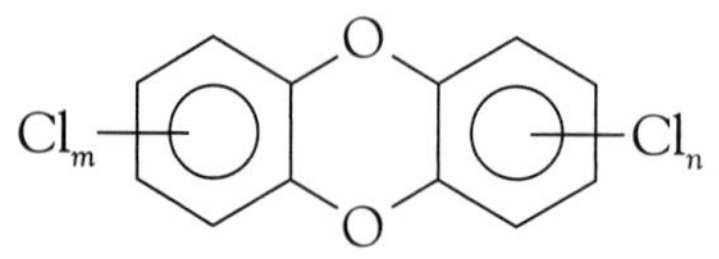

dioxins

PCBs

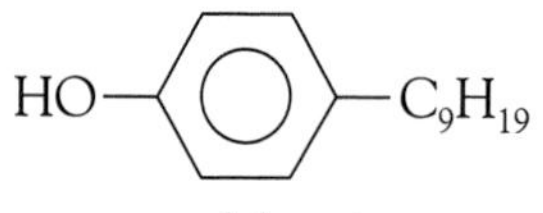

nonylphenol

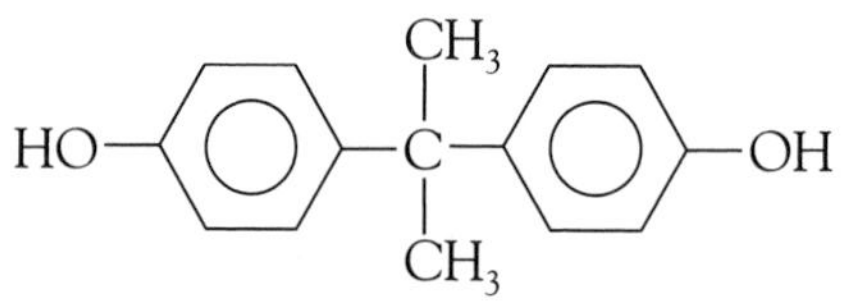

bisphenol-A

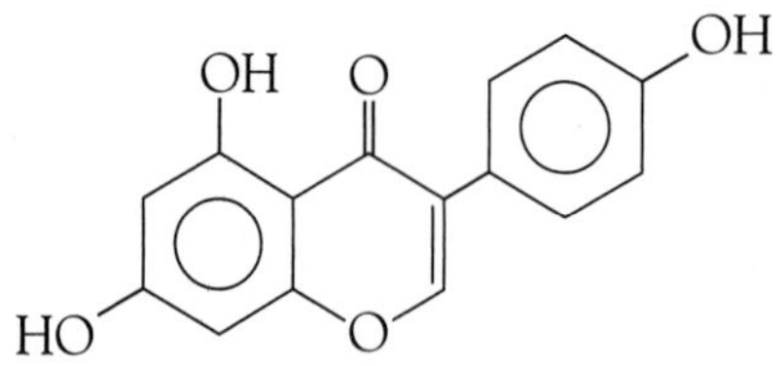

genistein

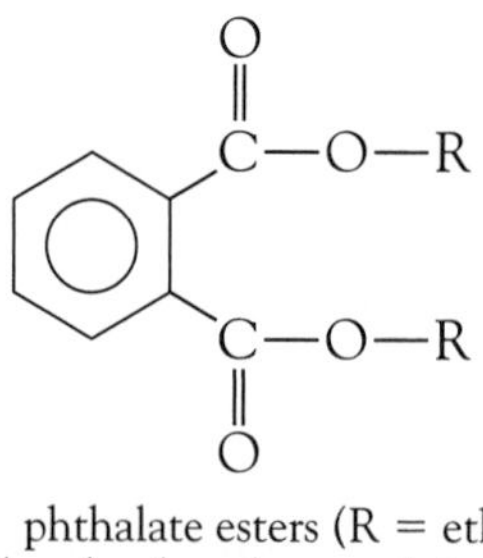

phthalate esters (R = ethyl, n-butyl, n-hexyl, n-octyl, isononyl, isodecyl, benzyl butyl, 2-ethylhexyl)

resemblance to estrogen itself. However, most such compounds bind to the receptor with only a small fraction of the strength of estrogen itself. Estrogen itself binds to its main receptor by hydrogen bonding from its two —OH groups and by attractive van der Waals forces from its ring system to amino acid side chains. A second estrogen receptor has recently been discovered. In addition to the interfering mechanisms discussed above, some compounds can accelerate the breakdown of natural hormones, and it has recently been established that some promote the conversion of male hormones into female ones.

As a class, substances that interfere with the **endocrine system** of hormone production and transmission are often referred to as **environmental estrogens** (or *ecoestrogens*). More general names for this class of substances have also been suggested: synthetic *hormonally active agents* (HAAs), *endocrine disrupting chemicals* (EDCs), *environmental hormones*, and *xenoestrogens*.

The Chemicals That Operate as Environmental Estrogens

Although most ecoestrogens operate by attaching themselves to, or by blocking access to, the hormone receptor, there is not a strong resemblance in overall structure between synthetic substances that have been identified as hormonally active and the natural sex hormones, nor is there much structural similarity of the synthetic ones to each other. It is true that many molecules identified as environmental estrogens contain one or more hydroxyl groups, as does estrogen. Oxygen-containing organochlorine insecticides that are known to act hormonally include *methoxychlor* and *kepone* (Figure 12-2). However, other environmental estrogens are organochlorine molecules, including some *PCBs*, *dioxins*, and insecticides (Figure 12-2) that contain no oxygen atoms. In metabolizing them, however, the body itself attaches hydroxyl groups to some of the nonchlorinated carbon atoms in such molecules. Some of the resulting hydroxylated organochlorines are more hormonally active agents than the original compounds. The same is true with polycyclic aromatic hydrocarbons when they are hydroxylated.

Most nonchlorinated environmental estrogens are alcohols, often with the phenolic ring that is present in the steroidal hormones. **Nonylphenol** (Figure 12-2) is one important example. **Octylphenol** (which has a C_8 rather than a C_9 chain attached to the phenol ring) is even more hormonally active. Both these alkylphenols occur in the environment—including drinking water—as a result of the breakdown, e.g., in sewage treatment plants, of larger *ethoxylate* molecules. The ethoxylates are used in detergents, spermicides, paints, and some plastics. They are also commonly used as emulsifying agents in pesticide formulations, and the nonylphenols produced by their decomposition enter the food chain via sprayed fruits and vegetables. The

European Union is moving toward banning the use of ethoxylates, as Norway has done already for applications for which alternatives exist.

Another phenolic environmental estrogen of concern is **bisphenol-A** (Figure 12-2); the prefix *bis* means "two" and is used here to signify that two phenol rings are connected together. It is a widely used substance that is polymerized industrially into the polycarbonate plastics employed as linings in most food and beverage cans and in some epoxy resins. Bisphenol-A is a controversial chemical, with very different views regarding its potential estrogenic effects in humans being taken by the plastics industry and some of the scientists whose research they support financially compared to those of some academic scientists who have performed tests with the compound on animals.

Some bisphenol-A is leached if food containers made using the polymerized resin are heated for sterilization purposes. The plastics industry states that migration of bisphenol-A from plastic food containers does not occur under normal cleaning and use conditions, although some consumer groups claim evidence that some, albeit small, amount of leaching from polycarbonate plastics such as baby bottles does occur. Although resins containing the compound are used to line aluminum beer and soda cans, bisphenol-A apparently does not leach from these containers. Bisphenol-A also could potentially leach from dental sealants that are made from resins prepared from it, though the initial evidence that this occurs is now in doubt. Urine testing indicates that the great majority of Americans have been exposed to bisphenol-A, though most of it ingested by humans is metabolized.

Another hormonally active diol is **genistein,** a molecule whose structure (Figure 12-2) has similarities to that of estrogen. Genistein is produced naturally by plants, rather than being a synthetic product. It is a member of a group of natural chemicals called *flavonoids*. Genistein is found in significant concentrations in wood products and in soybeans and soy-based food products. Researchers have discovered that genistein from pulp-mill effluent may cause feminizing and other reproductive effects in fish that swim in the effluent-fed waters. Both nonylphenol and genistein have been found to produce effects at very low doses in test animals. It is not clear yet whether genistein overall has a positive or a negative effect on human health. However, some scientists are worried about the levels of genistein ingested by babies fed with soy milk.

The PCB congeners commonly found in environmental samples and identified recently by screening as showing some, albeit weak, estrogenic activity all have at least one ortho chlorine atom.

Phthalate esters (Figure 12-2) are widely used as plasticizers in common plastics such as polyvinyl chloride, PVC, from which they can leach into the environment, since they are not chemically bonded to the polymer. They have antiandrogenic action. An important example is **di-2-ethylhexyl phthalate,** DEHP, which is present in many plastics found around the home, including some used by children, where phalate plasticizers can constitute up

to 45% of the weight of the object. The presence of DEHP in plastic medical devices such as intravenous PVC bags is also of concern since leakage of small amounts the plasticizer into the patient occurs during medical procedures. In the United States, neither DEHP nor other phthalates are used in food wrap or food packaging.

The European Commission has banned the use of phthalate softeners in PVC toys meant for children under three years of age, since such children tend to suck and chew on these toys, particularly rattles and teethers, and would thereby extract and ingest some of the phthalate esters. However, a scientific panel convened by the U.S. Consumer Product Safety Commission concluded that the most common plasticizer used in PVC toys, **diisononyl phthalate,** does not pose a risk to humans.

The compounds discussed above are thought to be the most significant environmental estrogens uncovered to date. However, the estrogenic properties of many synthetic substances is not yet known. The U.S. EPA in 1999 began an extensive process of screening potential endocrine disruptors.

Effects of Environmental Estrogens on Wildlife

The most devastating consequences of environmental estrogens commonly are not observed in the mammals that originally ingest them. Rather, they result from their transfer from the mother to the fetus or egg. Their presence disrupts the hormone balance in the recipient and causes reproductive abnormalities or produces changes that can result in cancer when the offspring grows to adulthood. During its development, the fetus is particularly sensitive to fluctuations in hormone levels. For that reason, exposure to low levels of natural or environmental hormones can result in physiological changes that do not occur in adults who are exposed at the same levels.

The most famous example of the environmental effects of hormone-like chemicals upon wildlife involves alligators in Lake Apopka, Florida. In 1980, massive amounts of DDT and its analogs were spilled into the lake. In the mid-1980s, Professor Louis Guilette, Jr., of the University of Florida at Gainesville found that very few alligator eggs were hatching, and few hatchlings survived of those that were born, thereby threatening the future population of the colony. Furthermore, the eggs that did hatch produced alligators with abnormal reproductive systems, which therefore were unlikely themselves to be able to reproduce. The ratio of natural estrogen to the male sex hormone testosterone was greatly elevated in the young alligators. Presumably as a consequence, the penises of male alligators were reduced in size compared to the norm. Apparently these effects were caused by DDE, the metabolite of DDT, which has been found by research to inhibit binding of male hormones to their receptor.

Another Florida-based example involves the nearly extinct Florida panthers that live in the Everglades. Their reproductive difficulties may result from the consumption of raccoons whose levels of DDE and other endocrine disruptors are high as a consequence of their consumption of contaminated fish. Some researchers have linked reproductive problems of birds in the Great Lakes area, such as embryo mortality and deformities, to the hormonal activity of pollutants such as PCBs and dioxins. Abnormalities in the reproduction and/or development of frogs, seals, polar bears, mollusks, and several types of birds have been linked to exposure of the fetuses to endocrine disrupting chemicals.

Test have revealed that the most estrogenic component of commercial DDT is not the main (~75%) ingredient, the p,p'-DDT isomer (Chapter 10), but is rather the minor (15–20%) o,p'-DDT isomer (Figure 12-2), which has one of the ring chlorines in the ortho position and one in the para. Some scientists have speculated that women who are directly exposed to o,p'-DDT by spraying may be at much higher risk of subsequently developing breast cancer than those in the developed world, whose main exposure now is through DDE in their diets.

Research reported in 2002 and 2003 indicated that environmental concentrations, ppb levels, of the herbicide *atrazine* (Chapter 10) could modify the balance of hormones in just-hatched frogs and thereby affect their sexual development. The effect of atrazine was to increase the levels of an enzyme that converts the male hormone testosterone to the female one, estrogen. About 20% of the male tadpoles developed into hermaphrodites, having both testes and ovaries. Because of this, atrazine may be contributing to the worldwide decline in the amphibian population. This hormonal action indicates yet another way, in addition to attaching to or blocking a hormone receptor, that environmental chemicals could upset hormonal balances. However, research reported from some other groups has failed to duplicate the findings at very low atrazine concentrations, so the issue of whether or not this is a significant environmental problem is as yet unresolved.

Researchers have also found that both natural estrogen, secreted by women as part of their monthly cycles, and the synthetic derivative used in birth control pills are present in wastewater and sewage effluent, and can cause feminization of males in some species of fish. Such feminization was encountered in the 1990s in some British waterways and was initially blamed on the presence of nonylphenol discharges in the water. A survey of estrogens in coastal waters found much higher levels in shallow bays that receive input from sewage than in the open oceans.

Effects of Environmental Estrogens on Humans

The areas of human health that are considered to be at potential risk from exposure to environmental hormones are reproduction, neurobehavior, immune function, and cancer.

Much of the human evidence concerning the possible effects of estrogen mimics on developing fetuses was obtained from the experience of women who took the synthetic estrogen *DES* (diethylstilbestrol) in the 1948–1971 period to prevent miscarriage. Many of the daughters of these women are sterile, and a small fraction of them have developed a rare vaginal cancer, as a consequence of their prenatal exposure to DES. The male offspring have an increased incidence of abnormalities in their sexual organs, have decreased average sperm counts, and may have an increased risk of testicular cancer as a consequence of this exposure, but their fertility is not affected.

Although there is good evidence that high concentrations of environmental estrogens have caused reproductive problems in wildlife and laboratory animals, it is not certain that comparable effects occur in humans at levels to which we are exposed. In the early 1990s, a connection between the rise in environmental contamination by endocrine disruptors and an apparent increase in certain male disorders of the reproductive system was postulated. In particular, the decline in male sperm counts and quality, the increase in the rate of testicular cancer, and the increase in incidence of male reproductive problems in newborns were cited. However, the changes in the frequency of these problems found in some locations may not, in fact, be general phenomena. Both sperm counts and testicular cancer rates vary significantly between geographical regions, and the variations are neither worldwide nor apparently closely linked to differences in pollution levels. Similarly, no clear picture has emerged relating trends in exposure to environmental estrogens to human fertility or the rates of spontaneous abortion, though an association has been found between delayed conception and exposure to high concentrations of environmental contaminants.

Perhaps the most dramatic effects of an environmental hormone on human reproduction is the effect of dioxins and furans in influencing the male:female sex ratio, i.e., the ratio—normally about 0.51—of boys to girls at birth. Epidemiological evidence from the dioxin-exposed group in Seveso, Italy, and from the Taiwanese PCB-poisoned-oil group (Chapter 11) indicates that males who were exposed to high concentrations of dioxins, or to furans, dioxins, and PCBs, respectively, during adolescence are in adulthood much less likely to father boys than girls. Most studies of men exposed *as adults* to high levels of dioxins did not show a similar result, and the same is largely true for men who were over age 20 and for all females at the time of the Taiwanese incident. However, research from Austria indicates that workers occupationally exposed to TCDD when under 20 years of age later fathered many more girls than boys, whereas initial dioxin exposure at a later age had no effect on the sex ratio.

Thus it appears that exposure during puberty to environmental hormones from the dioxin-furan-PCB family of compounds can have measurable lifelong effects on subsequent reproduction characteristics of human males. Interestingly, higher-than-normal exposure of males to industrial PCB mixtures through the diet can result in a *higher* male:female sex ratio, perhaps because some PCB congeners exhibit estrogenic whereas others exhibit antiestrogenic or androgenic behavior, the dominant effect depending upon the ratio of one to the other. In addition to PCB congeners, particularly those having one ortho and two para chlorines, other environmental compounds that have been found to interfere with the androgen receptor include the common DDT metabolite *p,p'-DDE* and the fungicide *vinclozolin*. Recent research indicates that environmental estrogens in humans can affect sexual characteristics: Girls who were exposed prenatally to high levels of DDE reach puberty on average almost a full year before those with the lowest exposures. Boys were not affected in this way.

The effects on neurological development of prenatal exposure to PCBs cited in Chapter 11 may arise from disruption of thyroid hormones, but further research is required to establish this connection more definitely.

Concern about human health effects from exposure to phthalates has increased over the past few years. Of particular interest—and controversy—has been research linking phthalate metabolite levels in new mothers to the incomplete development of the sex organs in male infants born to them, possibly due to prenatal exposure to phthalates. An expert panel convened by the U.S. government found this research to be inconclusive and recommended that it be repeated and expanded. A subsequent report linked phthalate levels in human breast milk to abnormal levels of reproductive hormones in infant boys.

Some researchers have postulated a connection between the increasing exposure of the general population to endocrine disrupting chemicals and the rising incidence of cancers at hormonally sensitive sites, including the breast, the uterus, the testes, and the prostate gland. A recent review of the medical literature concluded that "the overall strength of a causal association is weak" between the two, although "there is not enough information to completely reject the hypothesis" that endocrine disruptors could play a role.

Some scientists discount *any* adverse health effects of synthetic chemicals acting as environmental estrogens by pointing out that we all ingest much greater quantities of plant-based estrogen mimics called **phytoestrogens,** including genistein. Common sources of these natural chemicals include all soy products, broccoli, wheat, apples, and cherries. Indeed, there is some evidence that phytoestrogens have a *protective* effect against some types of cancers. It is true that phytoestrogens are quickly metabolized by the body and perhaps do not survive long enough to exert effects on a developing fetus, for example, whereas many synthetic environmental estrogens are stored in body fat rather than being metabolized. However, the current average dioxin level in humans is less than an order of magnitude greater than the

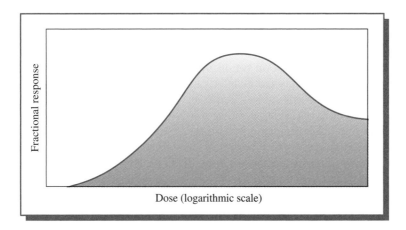

FIGURE 12-3
Dose–response
curve (schematic) for
estrogen and its mimics.

lowest dose found by experiments to cause reproductive problems in the off-spring of rats.

The greatest difficulty that scientists face in trying to discover whether environmental estrogens affect human health significantly is the lack of exposure data from the suspected substances. Thus, the jury is still out regarding whether environmental estrogens pose a substantial threat to humans or not.

Another puzzle in the environmental estrogen story relates to their dose–response behavior. In contrast to most toxic substances for which the response increases with the dose, eventually leveling off, the response curves for action by estrogen and estrogen mimics have an inverted-U shape (see Figure 12-3). The greatest effects are produced at low dosages; large dosages shut down the responding system to some extent. For example, the effects of atrazine on frogs are found by some researchers to operate only at very low concentrations, not at higher levels.

As an added complication, some estrogenic compounds are found to increase estrogenic activity in some tissues and block it in others! The relevance of the findings discussed above and many other unanswered questions in the environmental estrogen story will only be sorted out once much more research is done in this fascinating new chapter concerning the effects of low environmental levels of toxic organic chemicals upon living organisms.

The Long-Range Transport of Atmospheric Pollutants

At first glance, it seems amazing to discover that relatively nonvolatile organochlorines and PAHs can eventually travel thousands of kilometers by air from their point of release and end up contaminating relatively pristine areas of the world such as the Arctic. Some quantitative understanding of this **long-range transport of atmospheric pollutants** (LRTAP) has been made using principles of physical chemistry.

By a global fractionation (or distillation) process, pollutants travel at different rates and are deposited in different geographical regions depending upon their physical properties. Most persistent organic pollutants have sufficient volatility to evaporate—often rather slowly—at normal environmental temperatures from their temporary locations at the surface of soil or water bodies. However, because the vapor pressure of any chemical increases exponentially with temperature, evaporation is favored in tropical and semitropical areas, so these geographic regions are rarely the *final* resting places for pollutants. In contrast, cold air temperatures favor the condensation and adsorption of gaseous compounds onto suspended atmospheric particles, most of which are subsequently deposited onto the Earth's surface. Thus the Arctic and Antarctic regions are the final resting places for relatively mobile pollutants that are not deposited at lower latitudes because of their high volatility. Unfortunately, these compounds degrade even more slowly in these regions because temperatures there are so cold.

Example of pollutants that migrate to polar regions are the highly *chlorinated benzenes*; PAHs having three rings; and PCBs, dioxins, and furans that have only a few chlorines (see Table 12-2). Substances with even greater volatility, such as naphthalene and the less chlorinated benzenes, are not deposited even at the cold temperatures of polar regions; consequently, they continue their worldwide travels more or less indefinitely until they are chemically destroyed, usually by reaction initiated by the hydroxyl radical.

As implied in Table 12-2, the mobility of a chemical increases as the vapor pressure of its condensed form (as measured by that of the supercooled liquid at 25°C) increases. In addition, mobility increases as the temperature of condensation of the vapor form of the pollutant gas decreases. Thus, substances that do not condense until the temperature drops to −30°C or lower eventually accumulate in polar regions, where such air temperatures are common. Substances having condensation temperatures below −50°C remain airborne indefinitely, since not even polar regions sustain such temperatures for long.

DDT is an intermediate case on these transport scales. It does evaporate sufficiently rapidly (supercooled liquid vapor pressure is 0.005 pascals), but its relatively high condensation temperature of 13°C (55°F) means that much of it becomes permanently deposited at mid-latitudes (especially in the winter) and only a small fraction of it migrates to the Arctic.

Although PCBs are predicted by the model to deposit mainly in temperate areas rather than migrating *en masse* to the Arctic, the migration that does occur is sufficient that animals there are quite contaminated by these

TABLE 12-2	Predicted Mobilities of Persistent Airborne Pollutants			
Global Transport Behavior	Low Mobility	Relatively Low Mobility	Relatively High Mobility	High Mobility
Property				
Vapor pressure of liquid at 25°C in pascals*	10^{-4}		10^{-2}	1
Condensation temperature	30°C		−10°C	−50°C
Examples				
PAHs	>4 rings	4 rings	3 rings	1–2 rings
Chlorobenzenes	—	—	5–6 Cl	0–4 Cl
PCBs	9–8 Cl	4–8 Cl	1–4 Cl	0–1 Cl
PCDDs/PCDFs	4–8 Cl	2–4 Cl	0–1 Cl	—
Pesticide examples	Mirex	DDT Toxaphene Chlordane	HCB Dieldrin Hexachloro-cyclohexane	Moth balls

Source: Adapted mainly from F. Wania and D. Mackay, "Tracking the Distribution of Persistent Organic Pollutants," *Environmental Science and Technology* 30 (1996): 390A–396A.
*For the supercooled liquid.

chemicals. The world record for PCB contamination, 90 ppm, is held by polar bears in Spitsbergen, Norway. Even breast milk is higher in PCBs for women who live in far northern areas than in more temperate ones, a result partially of their high-fat diet since organochlorines are known to accumulate in such a medium.

PROBLEM 12-5

DDE has a 25°C vapor pressure (for its supercooled liquid) of 0.0032 Pa and a condensation temperature of −2°C. Is DDE more or less volatile than DDT? Predict whether a larger or smaller fraction of the fraction that does vaporize will be deposited at polar latitudes compared to DDT itself.

Owing to the variations in air temperature during their transport, most molecules of mobile pollutants experience several successive cycles of evaporation and condensation as they migrate gradually toward colder climates.

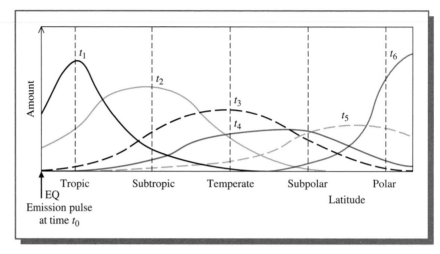

FIGURE 12-4 Calculated variation with time in the geographic distribution of an airborne pollutant released at the Equator (EQ). [Source: F. Wania and D. Mackay, "Tracking the Distribution of an Airborne Pollutant," *Environmental Science and Technology* 30 (1996): 390A.]

This "grasshopper effect" is illustrated in Figure 12-4 for a pulse of a relatively mobile pollutant that was emitted near the Equator at time t_0. At a later time t_1, the majority of the pollutant mass is still present in tropical regions, but at a subsequent time t_2 it has moved mainly to the subtropics. Whether it eventually ever moves ("hops") from temperate and subpolar regions to polar ones (at a later time t_6) depends upon whether or not its mobility is sufficiently high.

Brominated Fire Retardants

Highly brominated organic compounds are common commercial fire retardants. Large amounts of these fire retardants are used worldwide, and because of their persistence, they now are accumulating in the environment and have even been detected in the Arctic, to which they presumably migrated by the LRTAP mechanism discussed above. Based upon animal studies, they may have potential for liver toxicity, thyroid hormone-level disruption, and reproduction and development effects.

Many brominated organic compounds function as fire retardants because, when heated to 200–300°C—the approximate temperature range in which many polymers begin to decompose—they release free bromine atoms, which react with the free radicals of combustion and thereby quench any fire. For example, when a molecule of such a compound has absorbed sufficient energy from the fire and released Br atoms from one or more of its C—Br bonds, the atom may react with one of the free H atoms associated with the free-radical mechanism of combustion:

$$H + Br \longrightarrow HBr$$

The **hydrogen bromide,** HBr, molecule so formed may subsequently react with a free hydroxyl radical, which otherwise would also have continued to propagate the combustion:

$$OH + HBr \longrightarrow H_2O + Br$$

The net reaction of these two steps is the formation of a water molecule by the reaction of H and OH, thereby reducing the concentration of highly reactive free radicals.

In this way, energy is withdrawn from the propagation mechanism of the combustion process, and the fire is quenched. The decomposition temperature of brominated fire retardants lies just below those of the polymers that they protect.

Iodine organic compounds would be even more effective in withdrawing energy, but since the C—I bond is so weak, they decompose at too low a temperature. Fluorinated compounds are generally unsuitable as fire retardants because most C—F bonds are so strong that free fluorine atoms would not be released and because, once formed, HF molecules are so stable that they would not participate in further reaction.

PBDEs: A New Persistent Pollutant

Brominated diphenyl ethers, especially at levels of 5–30%, are incorporated into polyurethane foam, textiles, ready-made plastic products, and certain electronic equipment to prevent them from ever catching on fire. They are found in common domestic products such as carpet padding, mattresses, curtains, and upholstered couches and chairs.

From a conceptual viewpoint, the molecular structure of **diphenyl ether** is analogous to biphenyl, except that an ether oxygen atom joins the two benzene rings. Bromine atoms can occupy any of the 10 other positions on the rings, analogous to the chlorine atoms in PCBs, again giving 209 possible congeners. Molecules with fewer than four bromines are generally not present in commercial mixtures.

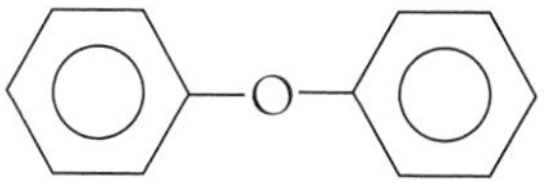

diphenyl ether

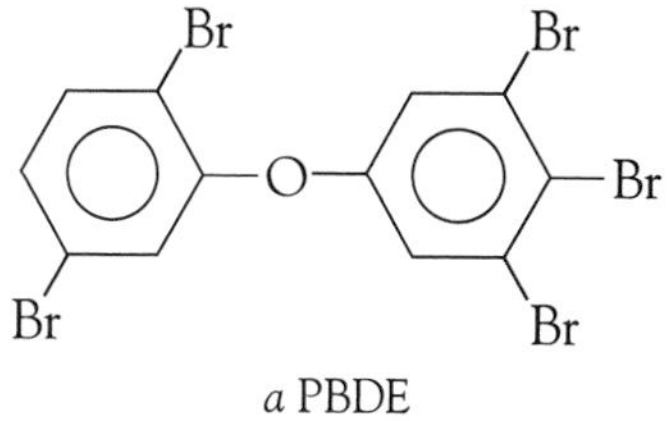

a PBDE

As a class, these compounds are known as **polybrominated diphenyl ethers, PBDEs.**

PBDE molecules are of particular environmental concern because some migration of them has occurred from their commercial products into the environment, where they now are widely distributed. Like PCBs, they are persistent and lipophilic, they bioaccumulate, and some of them are toxic. PBDEs have been detected in U.S. sewage sludge (much of it destined to be spread on agricultural land), in some fish caught in the wild, and even in sperm whales, which normally feed only in deep ocean waters. Also like PCBs, the commercial products are mixtures of congeners, not pure compounds, although the number of congeners present in each product is relatively small. Unlike most PCBs, PBDEs are solids under ambient conditions rather than liquids.

The acute toxicity of PBDEs decreases as the number of bromines per molecule increases. Consequently, the least toxic PBDE is the fully brominated congener **decabromodiphenyl ether.** It is the almost exclusive ingredient ($> 97\%$) in *Deca*, the commercial mixture that is the predominant PBDE product on the market and that is used as the flame retardant in plastic components in computers and TV housings. Some scientists suspect that the PBDEs in Deca, while not themselves highly toxic, may degrade in the environment by loss of some bromine, thereby increasing the toxicity of the mixture, since PBDEs having intermediate bromine content are more toxic than is decabromodiphenyl ether. Indeed, there is growing evidence that Deca can undergo debromination by photochemical decomposition in sunlight; by reduction with elemental iron; and by metabolic processes in fish, such as carp and rainbow trout, and in rats. It is also debrominated anaerobically in sewage sludge if certain other chemicals are also present.

PROBLEM 12-6

Deduce the structures of (a) the three unique PBDEs formed by loss of one bromine atom, and (b) the twelve PBDEs formed by loss of two bromine atoms, from a molecule of decabromodiphenyl ether. Assume that the two rings cannot rotate relative to each other around the intermediate oxygen atom.

The product called *Penta* is a mixture mainly of PBDEs having four or five bromine atoms. It was used as a fire retardant in polyurethane foams, such as those used in furniture upholstery and padding in vehicles. The Penta mixture constitutes up to 30% by mass of some polyurethane foams, products that easily deteriorate by outdoor weathering and break into small, easily transportable fragments that can eventually find their way into natural waters. From this source, PBDE molecules can enter the aquatic food chain. Indeed, it is tetra- and pentabromodiphenyl ethers that are most widely distributed in the environment and that are the most bioaccumulative and the most toxic. The commercial PBDE product called *Octa*, which consists mainly of congeners with six or seven bromine atoms, is used in thermoplastics.

PBDEs with a large amount of bromine (more than six Br atoms per molecule) probably do not bioaccumulate because they are not readily incorporated by organisms; instead they bind to particles and accumulate in sediments. However, congeners with four to six bromines are taken up by organisms and have the potential to biomagnify in the food chain. They may in the future represent a danger to human health due to our exposure to them in food, especially fish.

Although some human exposure to PBDEs occurs through the food we eat, inadvertent digestion of household dust appears to be an even more important source, especially in North America. The difference in exposure to this source may account for the higher blood levels of PBDEs found in Americans and Canadians compared to Europeans. One recent small-scale study found a definitive link between PBDE levels in human breast milk and their levels in the dust of the women's homes.

The concentrations of PBDEs in human blood, milk, and tissues had risen exponentially for three decades, until at least the early 2000s, with a doubling time of about five years. Though few data are available, the concentration of PBDEs in human breast milk is known to have risen sharply in the 1990s in both U.S. and Canadian women—by more than a factor of 10 from 1982 to 2002 for the latter—and is approaching that of PCBs, though the levels in milk samples from European women are much lower. The main human health concern is that PBDEs having relatively few bromine atoms may affect hormone and liver systems and interfere with neurodevelopment.

PROBLEM 12-7

The concentration of PBDEs in herring gull eggs from the Great Lakes was about 1100 ppb in 1990 and about 7000 ppm in 2000. What is the doubling time for the PBDEs in this source? If past trends continue, what will be the concentration in 2010? [Hint: For exponential growth Ae^{kt}, the doubling time is equal to $0.69/k$.]

Because of environmental concerns, the European Union banned the Penta and Octa products in 2006. The sole North American manufacturer of Penta and Octa voluntarily ceased production of these products at the end of 2004. However, the Deca product is still available and widely used in Europe and North America, though there is much controversy within the European Union concerning whether or not it should be banned. The argument against banning brominated fire retardants is that they are vital in reducing losses of human lives and of property in fires.

Other Brominated Fire Retardants

Two non-PBDE brominated organic compounds are also widely used as fire retardants. Indeed, the retardant of largest usage volume of all is TBBPA, **tetrabromobisphenol-A,** a compound composed of molecules in which all four carbon atoms that stand ortho to the two hydroxyl groups of bisphenol-A (see Figure 12-2) have been brominated:

TBBPA

Commonly, TBBPA is incorporated chemically into the structure of polymers by covalent coupling at the two hydroxyl groups; retardants that are covalently bonded to polymers are called *reactive* ones. Printed circuit boards are a major reactive use for TBBPA. When incorporated into materials in this manner, retardants are much less likely to leach or volatilize into the environment compared to those such as PBDEs, which are only physically dissolved in materials and are called *additive* substances. In some products, however, TBBPA is used as an additive rather than a reactive retardant. The compound itself is not very toxic (its LD_{50} is several grams per kilogram), and although it has been found in some biota, it decomposes in air, water, and sediment in weeks or a few months.

The other important brominated fire retardant is the half-brominated cyclic hydrocarbon **hexabromocyclododecane,** HBCD, which is used primarily as an additive in polystyrene foams used in building materials and upholstery, though it is also finding new applications in replacing PBDEs. It is not used as a reactive retardant, since it contains no reactive groups that can bond to the polymer chain. Like TBBPA, it is of low acute toxicity. However, it is now a ubiquitous contaminant in the environment, and it undergoes biomagnification in top predators such as birds of prey and marine mammals. To date, it has been detected only in much lower levels in humans. Its environmental levels in Europe are higher than those in North America, owing to its greater use there.

PROBLEM 12-8

Draw the structure of HBCD, given that its name is 1,2,5,6,9,10-hexabromo-cyclododecane. Then, knowing that Br$_2$ molecules will add across C=C bonds, deduce the structure of the cyclic triene that could be brominated to produce HBCD. [*Hint: Dodeca means "twelve."*]

Polybrominated biphenyls, PBBs, were also used as fire retardants but are now banned in some countries, including the United States. A 1973 industrial accident in Michigan resulted in the widespread contamination of the food supply there with PBBs.

Perfluorinated Sulfonates

All the organic compounds discussed previously in this chapter act either as hydrophobic (water-repelling) or as *oleophobic* (oil-repelling) substances, but not as both. There exists a small class of organic compounds that will dissolve in neither of these classes. **Fluorinated surfactants** are compounds that consist of molecules and ions having a long perfluorinated carbon tail; i.e., a hydrocarbon chain in which each hydrogen atom has been replaced by a fluorine atom. The best-known example of such a molecule is **perfluorooctane sulfonate** (PFOS):

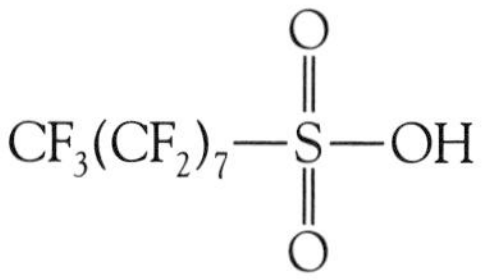

$$CF_3(CF_2)_7 - \overset{\displaystyle O}{\underset{\displaystyle O}{\overset{\|}{\underset{\|}{S}}}} - OH$$

Notice the similarity in its structure to that of sulfuric acid: One —OH group in the latter has been replaced by an unbranched, perfluorinated eight-carbon octane chain. This substance was used to make the 3M product *Scotchgard*, a fabric protector that, because of the characteristics of the perfluoro chain, repelled both water and oily spills and potential stains. Other compounds based upon PFOS were used in fire-fighting foams, pesticide formulations, cosmetics, lubricants, grease-resistant coatings for paper products, adhesives, and paints and polishes.

The 3M company has voluntarily phased out the production of PFOS because it persists long enough in the environment to eventually be detected in human blood samples. Although it is not very toxic, its concentration in some wildlife had reached levels of concern to some scientists. Since 2003, 3M has used the corresponding perfluorosulfonate having a chain of only four, rather than eight, carbon atoms, since such chains do not seem to either bioaccumulate or be toxic.

The fully fluorinated eight-carbon carboxylic acid compound **perfluorooc-tanoic acid,** PFOA ($CF_3(CF_2)_6COOH$), and its associated carboxylate salts have also become of environmental concern since they have no environmental or metabolic degradation pathways. Indeed, the very lack of reactivity that makes perfluorinated compounds so appealing in practical uses also results in their persistence in the environment. Although its lifetime in rats is only several hours, PFOA is only slowly eliminated in humans, resulting in an average lifetime of about four years. Due to its slow elimination, the acid is now found at the parts-per-billion level in the blood of most humans and wild animals worldwide. It is potentially acutely toxic, potentially carcinogenic, and may cause developmental problems. Human exposure to PFOA results from its use in producing polymers used to coat surfaces of nonstick cookware, including frying pans, as well as for the membranes of breathable outdoor garments. Fully fluorinated carbon chains are added by covalent bonds to polymer chains in order to make the materials stain-resistant. PFOA has been detected in samples of drinking water in several U.S. states; there is no federal standard for it. In 2006, the U.S. EPA announced a voluntary program, requesting companies using PFOA in consumer products to stop using the compound.

PFOA is the most prominent member of the family of **perfluoroalkyl acids,** PFAAs. In general, the longer the carbon chain in such molecules, the more persistent is the acid in the human body. 3M is formulating products that use PFAAs with relatively short chains in order to overcome the persistence problems of eight-carbon-chain substances.

PFOA and similar compounds are now found even in remote regions such as the Arctic, being transported there by LRTAP. Apparently such PFOA results from the atmospheric reaction of **fluorotelomer alcohols,** $CF_3(CF_2)_nCH_2CH_2OH$, industrial compounds that are used to make stain repellants. Unfortunately, a small fraction of the alcohols are released inadvertently into the atmosphere during the manufacturing process. In addition, the very small concentration of the reactant that was not converted into a repellant but was instead weakly incorporated into the material slowly degasses from it, adding to the atmospheric load of the alcohol.

In air, the fluorotelomer alcohols are converted to the carboxylic acids by a chain reaction that begins when a hydroxyl radical, OH, in air abstracts a hydrogen atom from the —CH_2— group bonded to OH. This process initiates a sequence of free-radical reactions, the net result of which is the oxidation of the terminal —CH_2CH_2OH group to COOH, producing the final perfluorocarboxylic acid, $CF_3(CF_2)_nCOOH$. It is also thought that microbial action and animal metabolism play a role in converting the alcohols to acids. Fluorotelomer alcohols have become the main remaining source of PFOA in the environment. In 2006, the Canadian government proposed to ban fluorotelomer polymers that can decompose into long-chain perfluorinated carboxylic acids.

Review Questions

1. What does *PAH* stand for? Draw the structures of two examples.

2. In what processes are PAHs commonly formed?

3. By means of a structural diagram, show what is meant by the *bay region* present in certain PAHs. How is the presence of this region related to the health effects of PAHs?

4. Define the term *environmental estrogen*. Give two chloroorganic and two nonchloroorganic examples.

5. Recount some of the evidence that environmental estrogens affect the health of wildlife and of humans.

6. What is a *phytoestrogen*?

7. What does *LRTAP* stand for?

8. Which three physical properties are used to predict the ultimate deposition zone of volatile chemicals?

9. What does *PBDE* stand for? Draw the structure of any PBDE. What are some of the uses of this class of compound?

10. Draw the structure of a *perfluorinated sulfonate*. What are such substances used for?

11. What do PFOA and PFFA stand for? What is the molecular structure of PFOA? What is meant by the term *fluorotelomer alcohol*? What are such alcohols used for?

Additional Problems

1. In an experiment, the level of benzo[*a*]pyrene in hamburgers was found to depend significantly on the cooking method and cooking time. For oven-broiled hamburgers, levels of 0.01 ng/g were found for both medium and very-well-done burgers. For barbecued burgers, levels of 0.09 and 1.52 ng/g were found for medium and very-well-done burgers, respectively.

(a) Explain the observed difference in benzo[*a*]pyrene formation by the two cooking methods, and explain the difference in medium versus very-well-done barbecued burgers.

(b) What does 1.52 ng/g translate to on a "parts-per" scale?

(c) How many micrograms of benzo[*a*]pyrene would be ingested in the consumption of a typical "quarter pounder" hamburger if it was barbecued to the very-well-done stage and there was no loss in mass during its cooking?

2. Which octabromo diphenylether isomers identified in Problem 12-6b would not correspond to isolatable compounds if free rotation were to exist about the C—O bonds?

Further Readings

1. C. Maczka et al., "Evaluating Impacts of Hormonally Active Agents in the Environment," *Environmental Science and Technology* (1 March 2000): 136A; G. M. Solomon and T. Schletter, "Environment and Health: 6. Endocrine Disruption and Potential Human Health Implications," *Canadian Medical Association Journal* 163 (2000): 1471; S. H. Safe, "Endocrine Disruptors and Human Health: Is There a Problem?" *Environmental Health Perspectives* 108 (2000): 487.

2. P. H. Jongbloet et al., "Where the Boys Aren't: Dioxin and the Sex Ratio," *Environmental Health Perspectives* 110 (2002): 1.

3. World Health Organization, "Global Assessment of the State-of-the-Science of Endocrine Disruptors" (2007): available at www.who.int/ipcs/publications/new_issues/endocrine_disruptors/en/

4. P. A. Darnreud et al., "Polybrominated Diphenyl Ethers: Occurrence, Dietary Exposure,

and Toxicology," *Environmental Health Perspectives* 109 (supplement 1) (2001): 49.

5. R. Renner, "The Long and Short of Perfluorinated Replacements," *Environmental Science and Technology* 40 (2006): 12.

6. K. S. Betts, "Perfluoroalkyl Acids: What Is the Evidence Telling Us?" *Environmental Health Perspectives* 115 (2007): A250.

Websites of Interest

Log on to www.whfreeman.com/envchem4/ and click on Chapter 12.

Environmental Instrumental Analysis III	Electron Capture Detection of Pesticides

Chlorine-containing organic compounds of the type that have been discussed in the preceding chapters usually occur in the environment at very small concentrations, but they can be detected and quantified by techniques such as the one discussed in this box.

The widespread occurrence of pesticides in the environment makes their detection an important task, but their often-low concentration makes this job difficult. One of the solutions to detecting very small amounts of environmentally important chemicals is to use very sensitive chromatographic detectors. In the case of methane, this is accomplished with the flame ionization detector (see Environmental Instrumental Analysis Box II).

The most common gas chromatographic (GC) detector used for halogen-containing pesticides is the **electron capture detector** (ECD). Since many important pesticides contain chlorine, a detector system that responds to molecules that contain this element is the key to sensitive biospheric analysis. Examples of target chlorinated pesticides are DDT (and its breakdown product DDE), lindane, and chlordane. The only chlorine-containing compounds unsuitable for this technique are those whose high boiling points make them unsuitable for gas chromatographic analysis.

The electron capture detector, like all GC detectors, is located at the end of the chromatographic column (see Environmental Instrumental Analysis Box II) located in a temperature-controlled (and programmable) oven. When analytes (compounds that have been separated by the chromatographic process)

exit the column, they enter the ECD and are detected.

The principle upon which the ECD works involves the disruption of a detector's electronic standing current by the arrival in the ECD of an analyte containing *electron-loving* (electrophilic) atoms such as halogens, which is the basis for the ECD signal. The standing current is generated by a piece of radioactive nickel-63 fixed on the wall of the detection chamber. This unstable element (half-life 92 years) continuously emits beta particles (β particles, high-energy electrons from nuclear decay, as described in Chapter 9) at a relatively constant rate. The GC carrier gas used in this analysis is usually a mixture of helium and a small amount (say, 5%) of another volatile compound, such as methane, at a constant concentration. Because the carrier gas mixture is homogeneous, a constant ratio of helium and CH_4 flows into the ECD. The flow of β particles from ^{63}Ni collides with some of the methane molecules in the carrier gas and create a "cloud" of slow (or thermal) electrons in the detection chamber. This cloud creates an electrical potential between the two electrodes placed in the detection chamber; the resultant current is amplified and sent to the computer (or integrator). Since this constant standing current is present whenever the detector is on and the carrier gas is flowing, the computer receives a constant detector signal. The figure on the top of next page shows the major components of the ECD.

The ECD's standing current changes when an analyte arrives in the ECD from the end of the GC column after chromatographic

(continued on p. 538)

| Environmental Instrumental Analysis III | **Electron Capture Detection of Pesticides** (*continued*) |

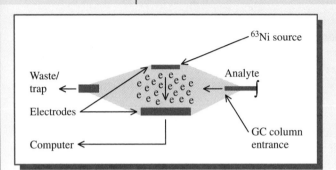

separation: Target compounds decrease the standing current because some of the electrons are captured by electrophilic atoms that are present in the analyte. The more of the compound that arrives, the larger the decrease in current. The computer measures the amount of this decrease and correlates detector signal with analyte concentration; however, unlike the positive FID signal (in which more analyte means more signal), the ECD signal is "upside down"—its information is a measure of *missing* signal. The result is, however, the same: The amount of each target analyte can be sensitively and reproducibly determined by the ECD. Furthermore, like other chromatographic systems, the *time* that each compound exits the column and generates the detector signal can be used as a means of identification if other analyses and chemical standards are used.

Among many other applications, the ECD has been used to measure the presence and amount of DDT and the related substance DDE (Chapter 10) in the tissue of Mexican free-tailed bats (*Tadarida*

brasiliensis). These animals absorb DDT and DDE from their diet of insects that have been exposed to DDT in the environment. Although the DDT content of the bats is very low, the breakdown product p,p'-DDE remains detectable. The figure below shows two (superimposed) chromatograms resulting from ECD analysis of carcasses of female bats collected from two southwestern U.S. caves, Carlsbad Caverns, New Mexico, and Vickery Cave, Oklahoma (Thies and McBee, 1994). Although no DDT itself was detected in either animal, the DDE content of the bat carcass (all tissue except brain and intestines) from Vickery Cave was approximately 41.9 μg DDE per gram of total fat.

Seabirds collected in the Barent Sea (in the Arctic, above 75°N) have also been analyzed using this instrument. A number of chlorinated pesticides were determined for two bird species whose diets were known to consist

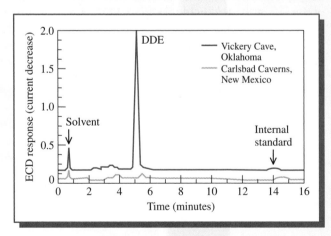

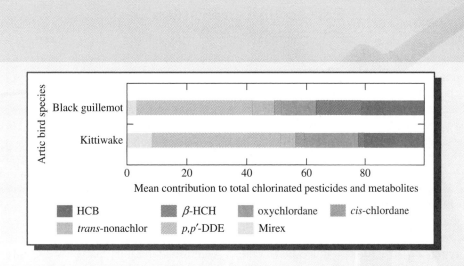

entirely of polar codfish. DDT's metabolite DDE was again found, along with six or seven chlorinated pesticides. The mean DDE concentrations were 608 ($\pm$43) ppb for Black guillemot and 1168 ($\pm$231) ppb for Kittiwake, expressed as mass of DDE per mass of bird lipid (Borga et al., 2007). The figure above shows the mean distribution of seven chlorinated compounds examined using this method.

By comparing the ratio of the concentration (in these birds' fat) of a specific chlorinated pesticide such as *cis*-chlordane to that of a highly bioaccumulated polychlorinated biphenyl (PCB 153), workers have been able to determine a relative measure of bioaccumulation for the chlorinated pesticides and metabolites examined. In comparing the bird species Black guillemot and Kittiwake, *cis*-chlordane was better eliminated by both species when compared

with the DDT metabolite DDE. But for two other bird species examined, DDE was very slowly eliminated and, in fact, is bioaccumulating in relation to the PCB-153 standard in those species. This result means that the biotransformation of chlorinated compounds is highly bird-species-specific.

References: K. Borga, H. Hop, J. U. Skaare, H. Wolkers, and G. W. Gabrielsen, "Selective Bioaccumulation of Chlorinated Pesticides and Metabolites in Artic Seabirds," *Environmental Pollution,* 145 (2007): 545–553.

M. L. Thies and K. McBee, "Cross-Placental Transfer of Organochlorine Pesticides in Mexican Free-Tailed Bats from Oklahoma and New Mexico," *Archives of Environmental Contamination and Toxicology* 27 (1994): 239–242.

Chemistry-Based Animations, 2006. http://www.shsu.edu/~chm_tgc/sounds/sound.html.

Environmental Instrumental Analysis IV	Gas Chromatography/Mass Spectrometry (GC/MS)

Analytical identification of volatile compounds in samples taken from the environment often relies on this extremely powerful technique. The heart of this method is mass spectrometry, a method of identifying molecules by their unique "fingerprints."

As we saw in our discussion concerning FID identification (page 537), gas chromatography (GC) is a very powerful tool for separating the components of a mixture. In identifying the structure of individual compounds in a separated mixture, one of the most powerful gas chromatography detectors is the *mass spectrometer* (MS); it is one the few GC detectors whose analytical signals actually probe the structure and elemental composition of the molecules it analyzes. Since the MS can be used as a stand-alone analytical tool, the combination of GC and MS is called a "hyphenated technique": **gas chromatography/ mass spectrometry, GC/MS.**

Like the FID and ECD, the mass spectrometric detector is positioned at the end of the gas chromatographic column and analyzes compounds one by one, as they exit the GC column in the gas phase. The MS can be divided into three parts: the ionization/fragmentation source, the mass analyzer, and the mass detector.

As the molecules enter the low-pressure ionization/fragmentation chamber of the mass spectrometer, they are bombarded with high-energy electrons, which causes many of the molecules to lose an electron, to form free-radical cations. Only charged molecular fragments are accelerated from there to the mass analyzer; all un-ionized particles are sucked out, by the vacuum system, to waste. A simple

example of ionization is the formation of CH_4^+ cations from methane molecules. This ion has a charge of +1 and a mass of 16, giving a mass/charge (m/z) ratio of 16.

After their formation, the ions are separated from one another according to their m/z ratio in the mass analyzer and then enter the mass detector. In the figure on the top of next page, depicting a 70-electron-volt ionization source, the GC column effluent enters on the right, ionization occurs where the stream of electrons cross its path, and ionized fragments are accelerated by the charged plates and exit from the ionization chamber and into the mass analyzer on the left. The intensity of the mass detector's signal versus m/z is recorded in each mass spectral scan, the *mass spectrum*.

The chromatographic plot for GC/MS is called the *total ion chromatogram*; it displays the total ion current on the y axis versus time on the x axis. The individual peaks correspond to different compounds in the mixture that were separated by the GC. This plot is comparable to the FID's chromatogram, which has the FID signal on the y axis and time on the x axis. (In contrast to GC/MS, the FID technique requires chemical standards—known compounds—whose retention-time data can be compared to the unknown compounds for identification purposes.) In the figure on the bottom of next page, the ionization source has been miniaturized and shown above and feeding molecular fragments to the quadrupole mass analyzer. The mass detector is shown below the mass analyzer. An example of a GC/MS total ion chromatogram is shown at the right in the figure.

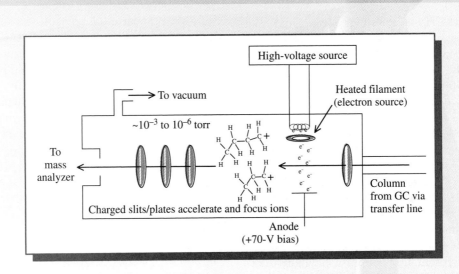

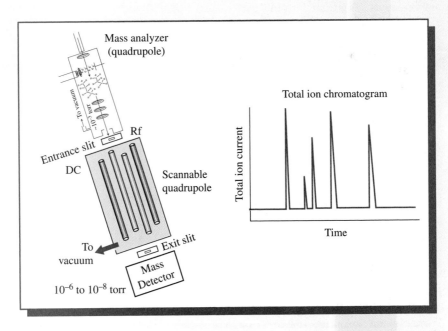

(continued on p. 542)

Environmental Instrumental Analysis **IV**	Gas Chromatography/Mass Spectrometry (GC/MS) *(continued)*

The CH_4^+ species that is generated from methane is know as the *molecular* (M+) or *parent ion*; from its m/z ratio of 16, one can determine the molecular weight of the molecule, a parameter that would aid in the identification of this compound if it were an unknown. Because the electrons that impact the molecules have more than enough energy to cause the formation of the molecular ions, some of them generally break down by cleavage of bonds to form *fragment ions* having lower m/z ratios. In the case of CH_4, since there are only C—H bonds, the fragmentation pattern is very simple, producing CH_3^+ ($m/z = 15$), CH_2^+ (14), CH^+ (13), and finally C^+ (12). Such fragmentation patterns yield more information about the molecular structure of the parent compound; its structural formula can often then be identified, particularly when the

fragmentation pattern is compared to a library of GC/MS patterns from known molecules.

For instance, the mass spectral fragmentation patterns of two chlorinated phenol isomers—which, of course, have the same molecular weight—can be used to differentiate between them. For example, although *2,3,4 trichlorophenol* and *2,4,5 trichlorophenol* have vanishingly small differences in boiling point and other physical properties, their electron impact mass spectra are significantly different since to some extent they fragment differently, and therefore they can be used to discriminate between these two isomers. Databases containing electron impact mass spectra are available free online (NIST, 2005) and are for sale from GC/MS instrument manufacturers.

The figure below shows the mass spectrum, reconstructed from data in the NIST

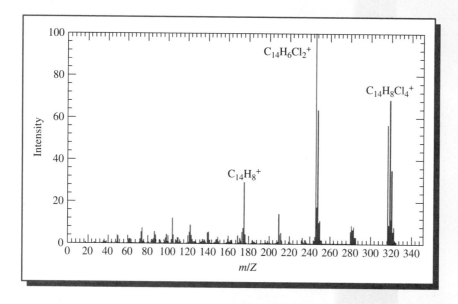

database, for *p,p'*-DDE, the environmental breakdown product of *p,p'*-DDT. The formulas corresponding to the three most intense mass spectral peaks in the DDE mass spectrum are shown above their appropriate *m/z* values. The $C_{14}H_8Cl_4^+$ peak corresponds to the parent ion, whereas the other two labeled peaks correspond to fragment ions formed by the loss of two or of all four chlorine atoms. The fragmentation pattern for other DDE isomers, such as *o,p'*-DDE, would have different relative intensities of the ion peaks.

References: Chemistry-Based Animations, 2006. http://www.shsu.edu/~chm_tgc/sounds/sound.html.

P. Janos and P. Aczel, "Ion Chromatographic Separation of Selenate and Selenite Using a Polyanionic Eluent," *Journal of Chromatography* A 749 (1996): 115–122.

NIST Chemistry WebBook, 2005. http://webbook.nist.gov/chemistry.

Tackling Mal

Interventions available today could lead to decisive gains in prevention and treatment—if only the world would apply them

By Claire Panosian Dunavan

Claire Panosian Dunavan, "Tackling Malaria," *Scientific American*, December 2005, 76–83.

Long ago in the Gambia, West Africa, a two-year-old boy named Ebrahim almost died of malaria. Decades later Dr. Ebrahim Samba is still reminded of the fact when he looks in a mirror. That is because his mother—who had already buried several children by the time he got sick—scored his face in a last-ditch effort to save his life. The boy not only survived but eventually became one of the most well-known leaders in Africa: Regional Director of the World Health Organization.

Needless to say, scarification is not what rescued Ebrahim Samba. The question is, What did? Was it the particular strain of parasite in his blood that day, his individual genetic or immunological makeup, his nutritional state? After centuries of fighting malaria—and conquering it in much of the world—it is amazing what we still do not know about the ancient scourge, including what determines life and death in severely ill children in its clutches. Despite such lingering questions, however, today we stand on the threshold of hope. Investigators are studying malaria survivors and tracking many other leads in efforts to develop vaccines. Most important, proven weapons—principally, insecticide-treated bed nets, other antimosquito strategies, and new combination drugs featuring an ancient Chinese herb—are moving to the front lines.

In the coming years the world will need all the malaria weapons it can muster. After all, malaria not only kills, it holds back human and economic development. Tackling it is now an international imperative.

A Villain in Africa

FOUR PRINCIPAL SPECIES of the genus *Plasmodium*, the parasite that causes malaria, can infect humans, and at least one of them still plagues

BITE OF INFECTED MOSQUITO begins the deadly cycle of malaria—a disease that kills one million to two million people annually, mainly young children in sub-Saharan Africa.

every continent save Antarctica to a lesser or greater degree. Today, however, sub-Saharan Africa is not only the largest remaining sanctuary of *P. falciparum*—the most lethal species infecting humans— but the home of *Anopheles gambiae,* the most aggressive of the more than 60 mosquito species that transmit malaria to people. Every year 500 million falciparum infections befall

disease's trademark fever and chills are followed by dizzying anemia, seizures and coma, heart and lung failure—and death. Those who survive can suffer mental or physical handicaps or chronic debilitation. Then there are people like Ebrahim Samba, who come through their acute illness with no residual effects. In 2002, at a major malaria conference in Tanzania where I met the

stream infection. Furthermore, experts believe that antibodies and immune cells that build up over time eventually protect many Africans from malaria's overt wrath. Ebrahim Samba is a real-life example of this transformed state following repeated infection; after his early brush with death, he had no further malaria crises and to this day uses no preventive measures to stave off new attacks.

> # Malaria not only kills, it holds back human and economic development.

(As a tropical medicine doctor, all I can say is: Don't try this on safari, folks, unless you, too, grew up immunized by hundreds of malarial mosquitoes every year.)

Africans, leaving one million to two million dead—mainly children. Moreover, within heavily hit areas, malaria and its complications may account for 30 to 50 percent of inpatient admissions and up to 50 percent of outpatient visits.

The clinical picture of falciparum malaria, whether in children or adults, is not pretty. In the worst-case scenario, the

surgeon-turned-public health leader, this paradox was still puzzling researchers more than half a century after Samba's personal clash with the disease.

That is not to say we have learned nothing in the interim regarding inborn and acquired defenses against malaria. We now know, for example, that inherited hemoglobin disorders such as sickle cell anemia can limit blood-

Samba's story also has another lesson in it. It affirms the hope that certain vaccines might one day mimic the protection that arises naturally in people like him, thereby lessening malaria-related deaths and complications in endemic regions. A different malaria vaccine might work by blocking infection altogether (for a short time, at least) in visitors such as travelers, aid workers or military peacekeepers, whose need for protection is less prolonged.

On the other hand, the promise of vaccines should not be overstated. Because malaria parasites are far more complex than disease-causing viruses and bacteria for which vaccines now exist, malaria vaccines may never carry the same clout as, say, measles or polio shots, which protect more than 90 percent of recipients who

Overview/*Where We Stand Today*

■ Researchers are laboring to create vaccines that would prevent malaria or lessen its severity.

■ But existing interventions could fight the disease now. They include insecticide-treated bed nets, indoor spraying and new combination drugs based on an ancient Chinese herb.

■ The question comes down to one of will and resources: In view of all the competing scourges—in particular, HIV/AIDS—is the world ready to take on malaria in its principal remaining stronghold, sub-Saharan Africa?

complete all recommended doses. And in the absence of a vaccine, Africa's malaria woes could continue to grow like a multi-headed Hydra. Leading the list of current problems are drug-resistant strains of *P. falciparum* (which first developed in South America and Asia and then spread to the African continent), followed by insecticide resistance among mosquitoes, crumbling public health infrastructures, and profound poverty that hobbles efforts to prevent infections in the first place. Finally, the exploding HIV/AIDS pandemic in Africa competes for precious health dollars and discourages the use of blood transfusions for severe malarial anemia.

Where does this leave us? With challenges, to be sure. But challenges should not lead to despair that Africa will always be shackled to malaria. Economic history, for one, teaches us it simply isn't so.

Lessons of History

WHEN I LECTURE about malaria to medical students and other doctors, I like to show a map of its former geography. Most audiences are amazed to learn that malaria was not always confined to the tropics—until the 20th century, it also plagued such unlikely locales as Scandinavia and the American Midwest. The events surrounding malaria's exit from temperate zones and, more recently, from large swaths of Asia and South America reveal as much about its perennial ties to poverty as about its biology.

Take, for example, malaria's flight from its last U.S. stronghold—the poor, rural South. The showdown began in the wake of the Great Depression when the U.S. Army, the Rockefeller Foundation and the Tennessee Valley Authority (TVA) started draining and oiling thousands of mosquito breeding sites and distributing quinine (a plant-based antimalarial first discovered in South America) to purge humans of parasites that might otherwise sustain transmission. But the efforts did not stop there. The TVA engineers who brought hydroelectric power to the South also regulated dam flow to maroon mosquito larvae and installed acres of screen in windows and doors. As malaria receded, the local economies grew.

Then came the golden days of DDT (dichlorodiphenyltrichloroethane). After military forces used the wettable powder to aerially bomb mosquitoes in the malaria-ridden Pacific theater during World War II, public health authorities took the lead. Five years later selective spraying within houses became the centerpiece of global malaria eradication. By 1970 DDT spraying, elimination of mosquito breeding sites and the expanded use of antimalarial drugs freed more than 500 million people, or roughly one third of those previously living under malaria's cloud.

Sub-Saharan Africa, however, was always a special case: with the exception of a handful of pilot programs, no sustained eradication efforts were ever mounted there. Instead the availability of chloroquine—a cheap, man-made relative of quinine introduced after World War II—enabled countries with scant resources to replace large, technical spraying operations with solitary health workers. Dispensing tablets to almost anyone with a fever, the village foot soldiers saved millions of lives in the 1960s and 1970s. Then chloroquine slowly began to fail against falciparum malaria. With little remaining infrastructure and expertise to counter Africa's daunting mosquito vectors, a rebound in deaths was virtually ordained.

Along the way, economists learned their lesson once again. Today in many African households, malaria not only limits income and robs funds for basic necessities such as food and youngsters' school fees, it fuels fertility because victims' families assume they will always lose children to the disease. On the regional level, it bleeds countries of foreign investment, tourism and trade. Continentwide, it costs up to $12 billion a year, or 4 percent of Africa's gross domestic product. In short, in many places malaria remains entrenched because of poverty and, at the same time, creates and perpetuates poverty.

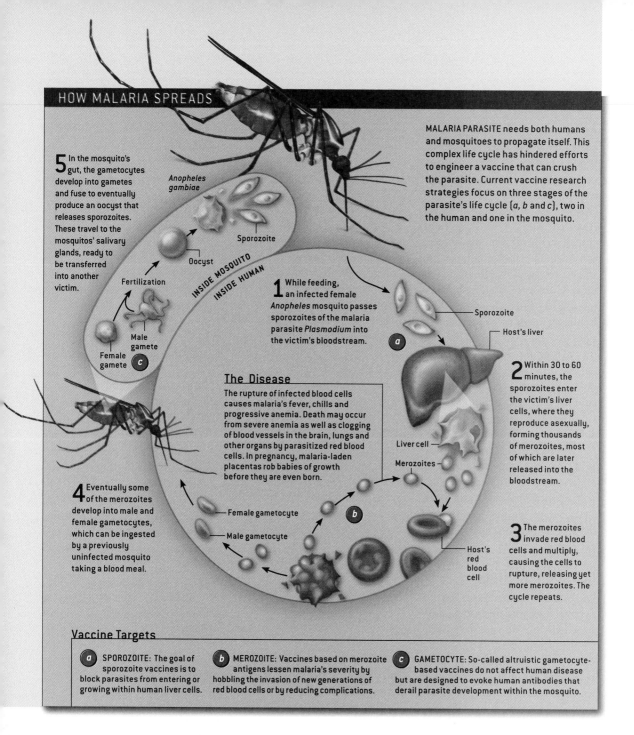

HOW MALARIA SPREADS

MALARIA PARASITE needs both humans and mosquitoes to propagate itself. This complex life cycle has hindered efforts to engineer a vaccine that can crush the parasite. Current vaccine research strategies focus on three stages of the parasite's life cycle (*a, b* and *c*), two in the human and one in the mosquito.

5 In the mosquito's gut, the gametocytes develop into gametes and fuse to eventually produce an oocyst that releases sporozoites. These travel to the mosquitos' salivary glands, ready to be transferred into another victim.

Anopheles gambiae

Sporozoite

Oocyst

Fertilization

INSIDE MOSQUITO

INSIDE HUMAN

Male gamete

Female gamete

1 While feeding, an infected female *Anopheles* mosquito passes sporozoites of the malaria parasite *Plasmodium* into the victim's bloodstream.

Sporozoite

Host's liver

2 Within 30 to 60 minutes, the sporozoites enter the victim's liver cells, where they reproduce asexually, forming thousands of merozoites, most of which are later released into the bloodstream.

The Disease

The rupture of infected blood cells causes malaria's fever, chills and progressive anemia. Death may occur from severe anemia as well as clogging of blood vessels in the brain, lungs and other organs by parasitized red blood cells. In pregnancy, malaria-laden placentas rob babies of growth before they are even born.

Liver cell

Merozoites

4 Eventually some of the merozoites develop into male and female gametocytes, which can be ingested by a previously uninfected mosquito taking a blood meal.

Female gametocyte

Male gametocyte

Host's red blood cell

3 The merozoites invade red blood cells and multiply, causing the cells to rupture, releasing yet more merozoites. The cycle repeats.

Vaccine Targets

a SPOROZOITE: The goal of sporozoite vaccines is to block parasites from entering or growing within human liver cells.

b MEROZOITE: Vaccines based on merozoite antigens lessen malaria's severity by hobbling the invasion of new generations of red blood cells or by reducing complications.

c GAMETOCYTE: So-called altruistic gametocyte-based vaccines do not affect human disease but are designed to evoke human antibodies that derail parasite development within the mosquito.

Battling the Mosquito

YEARS AGO I THOUGHT everyone knew how malaria infected humans: nighttime bites of parasite-laden *Anopheles* mosquitoes. Today I know better. Some highly intelligent residents of malaria-plagued communities still believe that an evil spirit or certain foods cause the illness, a fact that underscores yet another pressing need: better malaria education. Nevertheless, long before Ronald Ross and Giovanni Batista Grassi learned in the late 19th century that mosquitoes transmit malaria, savvy humans were devising ways to elude mosquito bites. Writing almost five centuries before the common era, Herodotus described in *The Histories* how Egyptians living in marshy lowlands protected themselves with fishing nets: "Every man has a net which he uses in the daytime for fishing, but at night he finds another use for it: he drapes it over the bed.... Mosquitoes can bite through any cover or linen blanket... but they do not even try to bite through the net at all." Based on this passage, some bed-net advocates view nets steeped in fish oil as the world's earliest repellent-impregnated cloth.

It was not until World War II, however, when American forces in the South Pacific dipped nets and hammocks in 5 percent DDT, that insecticides and textiles were formally partnered. After public opinion swung against DDT, treating bed nets with a biodegradable class of insecticides—the pyrethroids— was the logical next step. It proved a breakthrough. The first major use of pyrethroid-treated nets paired with antimalarial drugs, reported in 1991, halved mortality in children younger than five in the Gambia, and

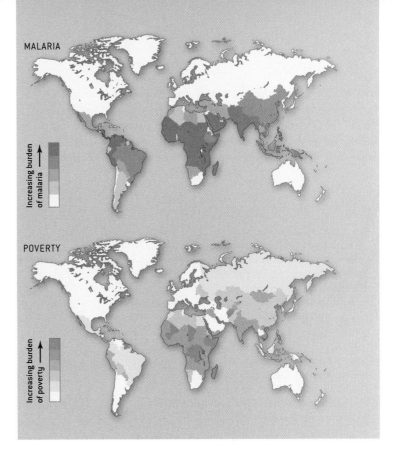

MALARIA

POVERTY

Increasing burden of malaria

Increasing burden of poverty

MALARIA AND POVERTY cover common ground. Costs levied by the disease go far beyond expenditures on prevention and treatment to include lost income, investment and tourism revenue. Annual economic growth in countries with endemic malaria averaged 0.4 percent of per capita GDP between 1965 and 1990, compared with 2.3 percent in the rest of the world.

later trials, without the drugs, in Ghana, Kenya and Burkina Faso confirmed a similar lifesaving trend, plus substantial health gains in pregnant women. Moreover, with wide enough use, whole families and communities benefited from the nets—even people who did not sleep under them.

But insecticide-treated bed nets also have drawbacks. They work only if malaria mosqui-

THE AUTHOR

CLAIRE PANOSIAN DUNAVAN, a tropical medicine specialist at the David Geffen School of Medicine at the University of California, Los Angeles, is co-editor of a recently published Institute of Medicine report, *Saving Lives, Buying Time: Economics of Malaria Drugs in an Age of Resistance.* A graduate of Stanford University, Northwestern University Medical School and the London School of Hygiene and Tropical Medicine, she is an avid teacher and clinician whose second career as a medical journalist spans nearly two decades.

toes bite indoors during sleeping hours—a behavior that is not universal. Nets make sleepers hot, discouraging use. Until recently, when PermaNet and Olyset—two long-lasting pyrethroid-impregnated nets—became available, nets had to be redipped every six to 12 months to remain effective. Finally, at $2 to $6 each, nets with or without insecticide are simply unaffordable for many people. A recent study in Kenya found that only 21 percent of households had even one bed net, of which 6

DDT: A Symbol Gone Awry

In the 1950s a worldwide campaign to eradicate malaria had as its centerpiece the spraying of houses with DDT (dichlorodiphenyltrichloroethane). In less than two decades, the pesticide enabled many countries to control the disease. In India, for example, deaths from malaria plummeted from 800,000 annually to almost zero for a time.

Then, in 1972, the U.S. government banned DDT for spraying crops—although public health and a few other minor uses were excepted. Rachel Carson's eloquent book *Silent Spring,* published a decade earlier, is often said to have sparked the ban. Carson meticulously charted the way DDT travels up the food chain in increasing concentrations, killing insects and some animals outright and causing genetic damage in others. DDT became a symbol of the dangers of playing God with nature, and the developed countries, having got rid of malaria within their borders, abandoned the chemical. Most of Europe followed the U.S. in banning the pesticide for agricultural applications in the 1970s.

For sub-Saharan Africa, where malaria still rages, these decisions have meant the loss of a valuable weapon. Most countries there go without DDT not because they have banned it themselves—in fact, it is allowed for public health uses in most areas of the world where malaria is endemic—but because wealthy donor nations and organizations are resistant to funding projects that would spray DDT even in responsible ways.

Many malaria researchers think DDT should be given another look. In addition to being toxic to mosquitoes, they note, it drives the insects off sprayed walls and out of doors before they bite, and it deters their entry in the first place. It is a toxin, irritant and repellent all rolled into one. Moreover, it lasts twice as long as alternatives, and it costs a quarter as much as the next cheapest insecticide.

The chemical's deadly trajectory through the food chain had its roots in massive agricultural spraying (mainly of cotton fields)—not in its much more moderate use inside dwellings to repel mosquitoes. Dusting a 100-hectare cotton field required some 1,100 kilograms of DDT over four weeks.

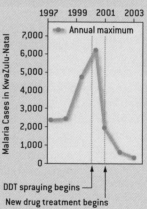

MALARIA CASES DECLINED dramatically in KwaZulu-Natal when the South African government sprayed dwellings with DDT and later also treated patients with an artemisinin-based combination treatment (graph). One of the few African countries wealthy enough to fund its own program, it did not have to rely on aid from donors reluctant to use the chemical. The eaves of a typical African house, such as those in the photograph, provide many points of entry for mosquitoes.

DDT alone will not save the world from malaria; for instance, spraying houses works only against mosquitoes that bite indoors. Effective drugs for patients already infected are essential, as are other measures to control mosquitoes. But most malaria health professionals support the targeted use of DDT as an important part of the tool kit.   —*The Editors* percent were insecticide-treated. A summary of 34 surveys conducted between 1999 and 2004 reached an even more depressing conclusion: a mere 3 percent of African youngsters were protected by insecticidal nets, although reports on the ground now suggest that use is quickly rising.

Insecticide resistance could also undermine nets as a long-term solution: mosquitoes genetically capable of inactivating pyrethroids have now surfaced in several locales, including Kenya and southern Africa, and some anophelines are taking longer to succumb to pyrethroids, a worrisome adaptive behavior known as knockdown resistance. Because precious few new insecticides intended for public health use are in sight (largely because of paltry economic incentives to develop them), one solution is rotating other agricultural insecticides on nets. Decoding the olfactory clues that attract mosquitoes to humans in the first place is another avenue of research that could yield dividends in new repellents. (Ironically, a change in body odor when *P. falciparum* parasites are present in the blood may also attract mosquito bites; according to a recent report, Kenyan schoolchildren harboring gametocytes—the malaria stage taken up by mosquitoes—drew twice as many bites as their uninfected counterparts.)

How about harnessing the winged creatures themselves to kill malaria parasites? In theory, genetic engineering could quell parasite multiplication before the protozoa ever left the insects' salivary glands. If such insects succeeded in displacing their natural kin in the wild, they could halt the spread of malaria parasites to people. Recently native genes hindering malaria multiplication within *Anopheles* mosquitoes have been identified, and genetically reengineered strains of several important species are now on the drawing board. Once they are reared in the laboratory, however, releasing these Trojan insects into the real world poses a whole new set of challenges, including ethical ones.

Bottom line: for the time being, old-fashioned, indoor residual spraying with DDT remains a

and killing others that perch on treated walls after feeding. A stunning example of its effectiveness surfaced in KwaZulu-Natal in 1999 and 2000. Pyrethroid-resistant *A. funestus* plus failing drugs had led to the largest number of falciparum cases there since the South African province launched its malaria-control program years ago. Reintroduction of residual spraying of DDT along with new, effective drugs yielded a 91 percent drop in cases within two years.

Treating the Sick

ANTIMOSQUITO MEASURES alone cannot win the war against malaria—better drugs and health services are also needed for the millions of youngsters and adults who, every year, still walk the malaria tightrope far from med-

An ancient disease that is both preventable and curable still claims at least one million lives every year.

valuable public health tool in many settings in Africa and elsewhere [*see box on facing page*]. Applied to surfaces, DDT is retained for six months or more. It reduces human-mosquito contact by two key mechanisms—repelling some mosquitoes before they ever enter a dwelling

ical care. Some are entrusted to village herbalists and itinerant quacks. Others take pills of unknown manufacture, quality or efficacy (including counterfeits) bought by family members or neighbors from unregulated sources. In Africa, 70 percent of antimalarials come from the informal private sector—in other words, small roadside vendors as opposed to licensed clinics or pharmacies.

Despite plummeting efficacy, chloroquine, at pennies per course, remains the top-selling antimalarial pharmaceutical downed by Africans. The next most affordable drug in Africa is sulfadoxine-pyrimethamine, an antibiotic that interferes with folic acid synthesis by the parasite. Unfortunately, *P. falciparum* strains in Africa and elsewhere are also sidestepping this compound as they acquire sequential mutations that will ultimately render the drug useless.

Given the looming specter of drug resistance, can lessons from other infectious diseases guide future strategies to beef up malaria drug therapy? In recent decades resistant strains of the agents responsible for tuberculosis, leprosy and HIV/AIDS triggered a switch to two- and three-drug regimens, which then helped to forestall further emergence of "superbugs." Now most experts believe that multidrug treatments can also combat drug resistance in falciparum malaria, especially if they include a form of *Artemisia annua,* a medicinal herb once used as a generic fever remedy in ancient China. *Artemisia*-derived drugs (collectively termed "artemisinins") beat back malaria parasites more quickly than any other treatment does and also block transmission from humans to mosquitoes. Because of these unequaled advan-

tages, combining them with other effective antimalarial drugs in an effort to prevent or delay artemisinin resistance makes sense, not just for Africa's but for the entire world's sake. After all, there is no guarantee malaria will not return someday to its former haunts. We know it can victimize global travelers. In recent years *P. falciparum*–infected mosquitoes have even stowed away on international flights, infecting innocent bystanders within a few miles of airports, far from malaria's natural milieu.

Yet there is a hitch to the new combination remedies: their costs—currently 10 to 20 times higher than Africa's more familiar but increasingly impotent malaria drugs—are hugely daunting to most malaria victims and to heavily affected countries. Even if the new cocktails were more modest in price, the global supply of artemisinins is well below needed levels and requires donor dollars to jump-start the 18-month production cycle to grow, harvest and process the plants. Novartis, the first producer formally sanctioned by the WHO to manufacture a co-formulated artemisinin combination treatment (artemether plus lumefantrine), may not have enough funding and raw material to ship even a portion of the 120 million treatments it once hoped to deliver in 2006.

The good news? Cheaper, synthetic drugs that retain the distinctive chemistry of plant-

based artemisinins (a peroxide bond embedded in a chemical ring) are on the horizon, possibly within five to 10 years. One prototype originating from research done in the late 1990s entered human trials in 2004. Another promising tactic that could bypass botanical extraction or chemical synthesis altogether is splicing *A. annua's* genes and yeast genes into *Escherichia coli,* then coaxing pharmaceuticals out of the bacterial brew. The approach was pioneered by researchers at the University of California, Berkeley.

Preventing, as opposed to treating, malaria in highly vulnerable hosts—primarily African children and pregnant women—is also gaining adherents. In the 1960s low-dose antimalarial prophylaxis given to pregnant Nigerians was found, for the first time, to increase their newborns' birthweight. Currently this approach has been superseded by a full course of sulfadoxine-pyrimethamine taken several times during pregnancy, infancy and, increasingly, childhood immunization visits. Right now the recipe works well in reducing infections and anemia, but once resistance truly blankets Africa, the question is, What preventive treatment will replace sulfadoxine-pyrimethamine? Although single-dose artemisinins might seem the logical answer at first blush, these agents are not suitable for prevention, because their levels in blood diminish so quickly. And repeated

EBRAHIM SAMBA, who recently retired as the WHO's Regional Director for Africa, still bears delicate hatch marks incised on his cheeks at the age of two, when he was close to death from severe malaria.

dosing of artemisinins in asymptomatic women and children—an untested practice so far—could also yield unsuspected side effects. In an ideal world, prevention equals vaccine.

Where We Stand on Vaccines

THERE IS NO DOUBT that creating malaria vaccines that deliver long-lasting protection has proved more difficult than scientists first imagined, although progress has occurred over several decades. At the root of the dilemma is malaria's intricate life cycle, which encompasses several stages in mosquitoes and humans; a vaccine effective in killing one stage may not inhibit the growth of another. A second challenge is malaria's complex genetic makeup: of the 5,300 proteins encoded by *P. falciparum's* genome, fewer than 10 percent trigger protective responses in naturally exposed individuals—the question is, Which ones? On top of that, several arms of the human immune system—antibodies, lymphocytes and the spleen, for starters—must work together to achieve an ideal response to malaria vaccination. Even in healthy people, much less populations already beset with malaria and other diseases, such responses do not always develop.

So far most experimental *P. falciparum* vaccines have targeted only one of malaria's three biological stages—sporozoite, merozoite or gametocyte [*see box on page 548*], although multistage vaccines, which could well prove more effective in the end, are also planned. Some of the earliest insights on attacking sporozoites (the parasite stage usually inoculated into humans through the mosquito's proboscis) came in the 1970s, when investigators at the University of Maryland found that x-ray-weakened falciparum sporozoites protected human volunteers, albeit only briefly. Presumably, the vaccine worked by inducing the immune system to neutralize naturally entering parasites before they escaped an hour later to their next way station, the liver.

The demonstration that antibodies artificially elicited against sporozoites could help fend off malaria prompted further work.

Three decades later, in 2004, efforts bore fruit when a sporozoite vaccine more than halved serious episodes of malaria in 2,000 rural Mozambican children between the ages of one and four, the years when African children are most susceptible to dying from the disease. The formula used in this clinical trial (the most promising to date) included multiple copies of a *P. falciparum* sporozoite protein fragment fused to a hepatitis B viral protein added for extra potency. Even so, subjects required three separate immunizations, and the period of protection was short (only six months). Realistically, the earliest that an improved version of the vaccine known as RTS,S (or any of its roughly three dozen vaccine brethren currently in clinical development) might come to market is in 10 years, at a final price tag that leaves even Big Pharma gasping for air. Because of the anticipated costs, public-private partnerships such as the Seattle-based Malaria Vaccine Initiative are now helping to fund ongoing trials.

There is just one more thing to keep in mind about malaria vaccines. Even when they do become available—with any luck, sooner rather than later—effective treatments and antimosquito strategies will still be needed. Why? First of all, because rates of protection will never reach anywhere near 100 percent in those who actually receive the vaccines. Other malaria-prone individuals, especially the rural African poor, may not have access to the shots at all. Therefore, at least for the foreseeable future, all preventive and salvage measures must remain in the arsenal.

Investing in Malaria

ONCE AGAIN THE WORLD is coming to terms with the truth about malaria: the ancient enemy still claims at least one million lives every year while, at the same time, imposing tremendous physical, mental and economic hardships. Given our current tools and even more promising weapons on the horizon, the time has come to fight back.

The past decade has already witnessed significant milestones. In 1998 the WHO and the World Bank established the Roll Back Malaria partnership. In 2000 the G8 named malaria as one of three pandemics they hoped to curb, if not vanquish. The United Nations subsequently created the Global Fund to Fight AIDS, Tuberculosis and Malaria and pledged to halt and reverse the rising tide of malaria within 15 years. In 2005 the World Bank declared a renewed assault on malaria, and President George W. Bush announced a $1.2-billion package to fight malaria in Africa over five years, using insecticide-treated nets, indoor spraying of insecticides and combination drug treatments. More recently, the World Bank has begun looking for ways to subsidize artemisinin combination treatments. As this issue of *Scientific American* went to press, the Bill and Melinda Gates Foundation announced three grants totaling $258.3 million to support advanced development of a malaria vaccine, new drugs and improved mosquito-control methods.

Despite these positive steps, the dollars at hand are simply not equal to the task. Simultaneously with the announcement from the Gates Foundation, a major new analysis of global malaria research and development funding noted that only $323 million was spent in 2004. This amount falls far short of the projected $3.2 billion a year needed to cut malaria deaths in half by 2010. Perhaps it is time to mobilize not only experts and field-workers but ordinary folk. At roughly $5, the price of a lunch in the U.S. could go a long way toward purchasing an insecticide-treated bed net or a three-day course of artemisinin combination treatment for an African child.

In considering their potential return on investment, readers might also recall a small boy with scars on his cheeks who made it through malaria's minefield, then devoted his adult life to battling disease. Decades from now, how many other children thus spared might accomplish equally wondrous feats?   SA

MORE TO EXPLORE

What the World Needs Now Is DDT. Tina Rosenberg in *New York Times Magazine,* pages 38–43; April 11, 2004.

Medicines for Malaria Venture; **www.mmv.org/**

World Health Organization, Roll Back Malaria Department: **www.who.int/malaria**

WATER CHEMISTRY AND WATER POLLUTION

Contents of Part IV

Environmental Instrumental Analysis V

- Ion Chromatography of Environmentally Significant Anions

CHAPTER 13

THE CHEMISTRY OF NATURAL WATERS

In this chapter, the following introductory chemistry topics are used:

- Concepts of oxidation and reduction as electron loss/gain; half-reactions; redox reactions; oxidizing and reducing agents; electrode potentials
- Solubility product and weak acid/weak base calculations; water constant K_w
- Oxidation numbers and the balancing of redox reactions (reviewed in Box 13-1)

Background from previous chapters used in this chapter:

- Equlibria involving gases dissolved in water: Henry's law (Chapter 3)

Introduction

All life forms on Earth depend on water. Each human being needs to consume several liters of fresh water daily to sustain life. Much more water is used for other domestic activities: Typical daily usages for showering/bathing, washing, and toilets each amount to about 50 L, in addition to about 20 L for dishwashing and 10 L for cooking. (A hose delivers 10 L or more of water per minute, so watering gardens and lawns can easily double average domestic consumption.) Vastly larger amounts per capita are used by industry and especially for irrigation in agriculture. For example, thousands of liters of fresh water are required to produce one kilogram of beef or cotton or even of rice.

However, fresh water is at a premium. Over 97% of the world's water is seawater, unsuitable for drinking and for most agricultural purposes. Three-quarters of the fresh water is trapped in glaciers and ice caps. Lakes and rivers are one of the main sources of drinking water, even though taken together they constitute less than 0.01% of the total water on the planet. About half

FIGURE 13-1 Global pools and fluxes of water on Earth, showing the size of groundwater storage relative to other major water sources and fluxes. All pool volumes (green) are in cubic kilometers, and all fluxes (black) are in cubic kilometers per year.
[Source: W. H. Schlesinger, *Biogeochemistry—An Analysis of Global Change,* 2nd ed. (San Diego: Academic Press, 1997), Chapter 10.]

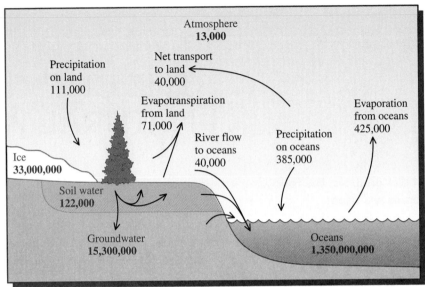

Atmosphere
13,000

Precipitation on land
111,000

Net transport to land
40,000

Evapotranspiration from land
71,000

River flow to oceans
40,000

Precipitation on oceans
385,000

Evaporation from oceans
425,000

Ice
33,000,000

Soil water
122,000

Groundwater
15,300,000

Oceans
1,350,000,000

of drinking water is obtained from *groundwater*, the fresh water that lies underground and that is discussed in detail in Chapter 14. The annual fluxes of water between its global pools in the oceans, the air, and beneath the ground are shown in Figure 13-1, along with the sizes of each pool.

Humanity currently consumes, mostly for agriculture, about one-fifth of the accessible runoff water that travels through rivers to the seas; this fraction is predicted to rise to about three-quarters by 2025. Runoff water is highly variable in both location and time in terms of its availability unless storage and transport are available. Although only 10% of the world's population in 2000 lived under conditions of water stress or scarcity, this figure is expected to rise to 38% by 2025 (Figure 13-2).

It is important to understand the types of chemical activities that prevail in natural waters and how the science and application of chemistry can be employed to purify water intended for drinking purposes. Although some discussion of pollution problems is contained in this chapter, the remediation of contaminated water is considered in detail in Chapter 14.

It will be convenient to divide our considerations of water chemistry in this chapter into the two common reaction categories: acid–base reactions and oxidation–reduction (redox) reactions. Acid–base and solubility phenomena control the concentrations of dissolved inorganic ions such as carbonate in waters, whereas the organic content of water is dominated by redox reactions. The pH and concentrations of the principal ions in most natural water systems are controlled by the dissolution of atmospheric carbon dioxide

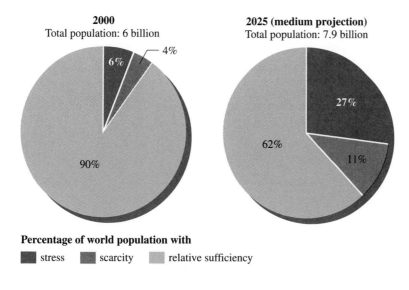

2000
Total population: 6 billion

2025 (medium projection)
Total population: 7.9 billion

Percentage of world population with

■ stress   ■ scarcity   ■ relative sufficiency

FIGURE 13-2 Global water supply in 2000 and projection for 2025. [Source: R. Engelman et al., *People in the Balance* (Washington, DC: Popular Action International, 2000) as reproduced in *Nature* 422 (2003): 252.]

and soil-bound carbonate ions; such reactions are considered in detail later in the chapter. Before considering these acid–base processes, we consider some important redox processes, especially those involving dissolved oxygen. For clarity, the phase (aq) for ions and molecules dissolved in aqueous solution will not be shown in equations but simply assumed.

Oxidation–Reduction Chemistry in Natural Waters

Dissolved Oxygen

By far the most important oxidizing agent (i.e., substance that extracts electrons from other species) in natural waters is dissolved **molecular oxygen,** O_2. Upon reaction, each of the oxygen atoms in O_2 is reduced from the zero oxidation number to -2, in H_2O or OH^-. (The concept and calculation of oxidation numbers is reviewed in Box 13-1, as is the balancing of redox equations.) The half-reaction that occurs in acidic solution is

$$O_2 + 4\,H^+ + 4e^- \longrightarrow 2\,H_2O$$

whereas that which occurs in basic aqueous solution is

$$O_2 + 2\,H_2O + 4e^- \longrightarrow 4\,OH^-$$

The concentration of dissolved oxygen in water is small and therefore precarious from the ecological point of view. As discussed in Chapter 3, for the dissolution of a gas in water such as the process

$$O_2(g) \rightleftharpoons O_2(aq)$$

BOX 13-1	Redox Equation Balancing Reviewed

Assigning Oxidation Numbers

A simple way to determine the extent (if any) to which an element is oxidized or reduced in a reaction is to deduce the change in its *oxidation number*, O.N., in the product as compared to that in the reactant. The oxidation number of the elements in most compounds and ions can be determined by applying, in sequence, the following set of rules, keeping in mind that *the sum of all the oxidation numbers in a substance must equal its net charge*. The rules are listed in terms of priority so that, e.g., if for a compound rule (iv) is inconsistent with rule (iii), then rule (iii) takes precedence since it is higher in the order.

(i) Elements appearing in the free, unbonded form have an O.N. equal to their ionic charge, which is zero if the element is uncharged.

(ii) Fluorine has an O.N. of −1 in compounds. Group I and II metals have O.N. values corresponding to their ionic charges +1 and +2, respectively, and Al is +3.

(iii) Hydrogen has an O.N. of +1, except when bonded to a metal, where it is −1.

(iv) Oxygen has an O.N. of −2 (except when overridden by a rule higher in the sequence, as an example below illustrates).

(v) Chlorine, bromine, and iodine have O.N.'s of −1 (except when overridden by a rule higher in the sequence, as an example below illustrates).

Some examples:

HF: F is −1 (rule ii) and H is +1 (rule iii); the sum is zero, as required.

H_2O_2: H is +1 (rule iii), but O here cannot be −2 (rule iv) since the sum of charges would be $2(+1) + 2(-2) = -2$; the charges must add up to zero for the molecule as a whole. Since rule (iii) takes precedence, each H must be +1, so each O here must be −1 in order that the sum be zero.

ClO_2^-: Each O is −2 (rule iv), for a total of −4, so Cl here cannot be −1 (rule v) since the sum of charges would then be $-1 + 2(-2) = -5$, as compared to the actual net charge of −1. Since $O.N._{Cl} + 2 \times (-2) = -1$, it follows that $O.N._{Cl} = +3$ here.

As an example of the use of oxidation numbers in reactions, consider the half-reaction in which nitrate ion is converted into nitrous oxide:

$$NO_3^- \longrightarrow N_2O$$

the appropriate equilibrium constant is the Henry's law constant K_H, which for oxygen at 25°C has the value 1.3×10^{-3} mol L^{-1} atm^{-1}:

$$K_H = [O_2(aq)]/P_{O_2} = 1.3 \times 10^{-3} \text{ mol L}^{-1} \text{ atm}^{-1} \text{ at } 25°C$$

Since in dry air at sea level the partial pressure, P_{O_2}, of oxygen is 0.21 atmospheres (atm), it follows that the solubility of O_2 is 8.7 milligrams per liter of water (see Problem 13-1). This value can also be stated as 8.7 ppm since, as discussed in Chapter 10, ppm concentrations for condensed phases are based

Since in nitrate ion, each O is -2, the sum of which is -6, and the charge on the ion is only -1, it follows that N here is $+5$.

In nitrous oxide, the O is -2, and the sum of oxidation numbers is zero, so each N must be $+1$.

Keeping in mind that the reaction requires 2 nitrate ions to supply enough nitrogen for one nitrous oxide molecule, we see that 2 $+5$ N's, total $+10$, here become 2 $+1$ N's, total $+2$. Thus the half-reaction must be a $(10 - 2 =)$ 8-electron reduction:

$$2\,NO_3^- + 8\,e^- \longrightarrow N_2O$$

If it is required to know the amounts of water molecules and H^+ or OH^- ions involved, the detailed balancing scheme discussed below must be employed.

Balancing Redox Equations

There are many equivalent schemes to completely balance redox half-reactions and overall reactions, of which the following is one example.

• To balance a half-reaction, first deduce the number of electrons involved in the process, as in the scheme above, by balancing atoms other than H and O.

• Next, to balance charge, add sufficient H^+ ions to the side having excess negative charge; note that only real charges on ions and electrons are considered here, *not* oxidation numbers.

• Finally, balance the number of oxygen atoms by adding H_2O molecules to the side deficient in oxygen.

Consider, for example, the nitrate to nitrous oxide example given above:

$$2\,NO_3^- + 8\,e^- \longrightarrow N_2O$$

The actual charge on the left-hand side is $2 \times (-1) + (-8) = -10$, but that on the right side is zero. Thus we should add 10 positive charges, each in the form of H^+, to the left side, so that its charge also becomes zero:

$$2\,NO_3^- + 8\,e^- + 10\,H^+ \longrightarrow N_2O$$

Finally, since we now have $2 \times 3 = 6$ O atoms on the left side and only one on the right side, we need to add 5 O each in the form of H_2O molecules to the right side:

$$2\,NO_3^- + 8\,e^- + 10\,H^+ \longrightarrow N_2O + 5\,H_2O$$

The half-reaction is now balanced.

---

on mass rather than moles. (Note that for simplicity, molar concentrations rather than activities are used in all equilibrium calculations in this book, since in general we are considering only very dilute solutions.)

Because the solubilities of gases increase with decreasing temperature, the amount of O_2 that dissolves at $0°C$ (14.7 ppm) is greater than the amount that dissolves at $35°C$ (7.0 ppm). The median concentration of oxygen found in natural, unpolluted surface waters in the United States is about 10 ppm.

PROBLEM 13-1

Confirm by calculation the value of 8.7 mg/L for the solubility of oxygen in water at 25°C.

PROBLEM 13-2

Given the solubility quoted above for O_2 at 0°C, calculate the value of K_H for it at this temperature.

River or lake water that has been artificially warmed can be considered to have undergone **thermal pollution** in the sense that, at equilibrium, it will contain less oxygen than colder water because of the decrease in gas solubility with increasing temperature. To sustain their lives, most fish species require water containing at least 5 ppm of dissolved oxygen; consequently, their survival in warmed water can be problematic. Thermal pollution often occurs in the region of electric power plants (whether fossil fuel, nuclear, or solar), since they draw cold water from a river or lake, use it for cooling purposes, and then return the warmed water to its source.

Oxygen Demand

The most common substance oxidized by dissolved oxygen in water is organic matter having a biological origin, such as dead plant matter and animal wastes. If, for the sake of simplicity, the organic matter is assumed to be entirely polymerized carbohydrate (e.g., plant fiber) with an approximate empirical formula of CH_2O, the oxidation reaction would be

$$CH_2O(aq) + O_2(aq) \longrightarrow CO_2(g) + H_2O(aq)$$
$$\text{carbohydrate}$$

Dissolved oxygen in water is also consumed by the oxidation of dissolved **ammonia,** NH_3, and **ammonium ion,** NH_4^+—substances that, like organic matter, are present in water as a result of biological activity—eventually to **nitrate ion,** NO_3^- (see Problem 13-4).

PROBLEM 13-3

Show that 1 L of water saturated with oxygen at 25°C is capable of oxidizing 8.2 mg of polymeric CH_2O.

PROBLEM 13-4

Determine the balanced redox reaction for the oxidation of ammonia to nitrate ion by O_2 in alkaline (basic) solution. Does this reaction make the water more alkaline or less? [Hint: Recall the redox balancing procedure in Box 13-1.]

Water that is aerated by flowing in shallow streams and rivers is constantly replenished with oxygen. However, stagnant water or that near the bottom of a deep lake is usually almost completely depleted of oxygen because of its reaction with organic matter and the lack of any mechanism to replenish it quickly, diffusion being a slow process and turbulent mixing being absent.

The capacity of the organic and biological matter in a sample of natural water to consume oxygen, a process catalyzed by bacteria present, is called its **biochemical oxygen demand,** BOD. It is evaluated experimentally by determining the concentration of dissolved O_2 at the beginning and at the end of a period in which a sealed water sample seeded with bacteria is maintained in the dark at a constant temperature, usually either 20°C or 25°C. A neutral pH is maintained by use of a buffer consisting of two ions of phosphoric acid, namely $H_2PO_4^-$ and HPO_4^{2-}:

$$H_2PO_4^- \rightleftharpoons HPO_4^{2-} + H^+$$

The BOD equals the amount of oxygen consumed as a result of the oxidation of dissolved organic matter in the sample. The oxidation reactions are catalyzed in the sample by the action of microorganisms present in the natural water. If it is suspected that the sample will have a high BOD, it is first diluted with pure, oxygen-saturated water so that sufficient O_2 will be available overall to oxidize all the organic matter; the results are corrected for this dilution. Usually the reaction is allowed to proceed for five days before the residual oxygen is determined. The oxygen demand determined from such a test, often designated BOD_5, corresponds to about 80% of that which would be determined if the experiment were allowed to proceed for a very long time—which of course is not very practical. The median BOD for unpolluted surface water in the United States is about 0.7 mg O_2 per liter, which is considerably less than the maximum solubility of O_2 in water (of 8.7 mg/L at 25°C). In contrast, the BOD values for sewage are typically several hundreds of milligrams of oxygen per liter.

A faster determination of oxygen demand of a water sample can be made by evaluating its **chemical oxygen demand,** COD. **Dichromate ion,** $Cr_2O_7^{2-}$, can be dissolved as one of its salts, such as $K_2Cr_2O_7$, in sulfuric acid: The result is a powerful oxidizing agent. It is this solution, rather than O_2, that is used to ascertain COD values. The reduction half-reaction for dichromate when it oxidizes the organic matter is

$$Cr_2O_7^{2-} + 14\,H^+ + 6\,e^- \longrightarrow 2\,Cr^{3+} + 7\,H_2O$$

(In practice, excess dichromate is added to the sample and the resulting solution is back-titrated with Fe^{2+} to the end point.) The number of moles of O_2

that the sample would have consumed in accomplishing the oxidation of the same material equals 6/4 (= 1.5) times the number of moles of dichromate, since the latter accepts six electrons per ion whereas O_2 accepts only four:

$$O_2 + 4H^+ + 4e^- \longrightarrow 2H_2O$$

Thus the moles of O_2 required for the oxidation is 1.5 times the number of moles of dichromate actually used. (See Problems 13-5 and 13-6.)

PROBLEM 13-5

A 25-mL sample of river water was titrated with 0.0010 M $K_2Cr_2O_7$ and required 8.3 mL to reach the end point. What is the chemical oxygen demand, in milligrams of O_2 per liter, of the sample?

PROBLEM 13-6

The COD of a water sample is found to be 30 mg of O_2 per liter. What volume of 0.0020 M $K_2Cr_2O_7$ will be required to titrate a 50-mL sample of the water?

The difficulty with the COD index as a measure of oxygen demand is that acidified dichromate is such a strong oxidant that it oxidizes substances that are very slow to consume oxygen in natural waters and that therefore pose no real threat to their dissolved oxygen content. In other words, dichromate oxidizes substances that would not be oxidized by O_2 in the determination of the BOD. Because of this excess oxidation, namely of stable organic matter such as cellulose to CO_2, and of Cl^- to Cl_2, the COD value for a water sample as a rule is slightly greater than its BOD value. Neither method of analysis oxidizes aromatic hydrocarbons or many alkanes, which, in any event, resist degradation, and therefore oxygen consumption, in natural waters.

It is not uncommon for water polluted by organic substances associated with animal or food waste or sewage to have an oxygen demand that exceeds the maximum equilibrium solubility of dissolved oxygen. Under such circumstances, unless the water is continuously aerated, it will soon be depleted of its oxygen, and fish living in the water will die. The treatment of wastewater to reduce its BOD is discussed in Chapter 14.

Finally, we note that there are two other measures used for the amount of organic substances present in natural waters. The **total organic carbon,** TOC, is used to characterize the dissolved *and* suspended organic matter in raw water. For example, the TOC usually has a value of approximately 1 milligram per liter, i.e., 1 ppm carbon, for groundwater. The parameter **dissolved organic carbon,** DOC, is used to characterize only organic material that is actually dissolved, not suspended. For surface waters, the DOC averages about 5 ppm, although bogs and swamps can have DOC values that are ten times this amount, and untreated sewage typically has a DOC value of hundreds of ppm. The largest component of organic carbon in natural waters is usually carbohydrates, but many other types including proteins and low-molecular-weight aldehydes, ketones, and carboxylic acids are also present. The *humic* materials in water are discussed in Chapter 16.

 ## Green Chemistry: Enzymatic Preparation of Cotton Textiles

Globally over 40 billion pounds (20 million kilograms) of cotton are produced each year. Even with the invasion of synthetic fibers such as nylon, polyester, and acrylics, cotton still holds over 50% of the market share for apparel and home furnishings that are sold in the United States. In preparing raw cotton for use as a fiber, several steps—including desizing, scouring, and bleaching—are required, leaving a fiber that is 99% cellulose. These steps use copious amounts of chemicals, water, and energy, and they produce millions of pounds of aqueous waste that is high in BOD and COD.

Raw cotton is composed of several concentric layers. The outermost layer is composed of fats, waxes, and pectin, while the inner layers consist primarily of cellulose. The fats and waxes make the raw cotton fiber waterproof, and the pectin acts as a powerful glue to hold the layers together. In order to prepare cotton for use as a fiber for bleaching and dyeing, the outer layer must be removed. This process, which is known as scouring, has traditionally been done by immersing the cotton in 18–25% aqueous **sodium hydroxide** solution at elevated temperatures. This results in hydrolysis of the fats (saponification) and waxes, which solubilizes the components of the outer layer so that they can simply be rinsed away. Scouring results in fibers of even and high wettability. In addition to sodium hydroxide, chelating agents (Chapter 14) and emulsifiers are added during the scouring process. To end the process, the mixture is neutralized with acetic acid and rinsed several times.

The scouring process is estimated by the U.S. EPA to account for about half of the total BOD produced in the preparation of cotton fibers. The BOD in wastewater from cotton production generally exceeds 1100 mg/L, which is several times that of raw sewage. In addition to the large amounts of chemicals, energy, and water that are used, and the concomitant pollution that is produced, another disadvantage of this process is the unintended weakening and loss of some of the cellulose fibers.

An alternative to the traditional scouring process, known as *Bio-Preparation*, was developed by Novozymes-North America Inc. and won a Presidential Green Chemistry Challenge Award in 2001. BioPreparation employs an enzyme (a pectin *lyase*) that selectively degrades pectin at ambient temperatures. Since pectin acts as a glue to hold the outer layer of the cotton fiber together, destruction of the pectin results in disintegration of this layer. Because the *lyase* is selective for only pectin, its use is much less aggressive than the typical scouring process and removes much less organic material (including cellulose) from the cotton. Since the dissolved organic materials are what contribute to the high BOD and COD of the wastewater, this enzymatic treatment lowers the BOD by 20% and the COD by 50%.

In addition to these environmental advantages, BioPreparation eliminates the use of sodium hydroxide solutions at elevated temperatures, which in turn

- lowers the pH of the wastewater,

- eliminates the need for neutralization with acetic acid and the concomitant wastes,

- lowers energy requirements, and

- lowers rinsing requirements, reducing water consumption by 30–50%.

Elimination of the use of sodium hydroxide also reduces the risk to workers, and the reduced degradation of cellulose provides for more robust fibers and higher yield.

Decomposition of Organic Matter in Water

Dissolved organic matter will decompose in water under anaerobic (oxygen-free) conditions if appropriate bacteria are present. Anaerobic conditions occur naturally in stagnant water such as swamps and at the bottom of deep lakes. The bacteria operate on carbon to disproportionate it; that is, some carbon is oxidized to **carbon dioxide,** CO_2, and the rest is reduced to **methane,** CH_4:

$$2\ CH_2O \xrightarrow{\text{bacteria}} CH_4 + CO_2$$

organic matter

$$\text{C oxidation number}\quad 0 \qquad\qquad -4 \quad +4$$

This is an example of a **fermentation** reaction, which in chemistry is defined as one in which both the oxidizing and the reducing agents are organic materials. Since methane is almost insoluble in water, it forms bubbles that can be seen rising to the surface in swamps and sometimes catches fire; indeed, methane was originally called *marsh* or *swamp gas.* The same chemical reaction shown above occurs in *digestor* units used by rural inhabitants in semi-tropical developing countries (India, for instance) to convert animal wastes into methane gas that can be used as a fuel. The reaction also occurs in landfills, as discussed in Chapter 16.

Since anaerobic conditions are reducing conditions in the chemical sense, insoluble Fe^{3+} compounds that are present in sediments at the bottom of lakes are converted into soluble Fe^{2+} compounds, which then dissolve into the lake water:

$$Fe^{3+} + e^- \longrightarrow Fe^{2+}$$

insoluble Fe(III)   soluble Fe(II)

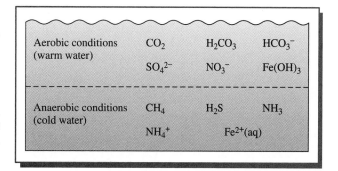

FIGURE 13-3 The stratification of a lake in the summer, showing the typical forms of the major elements it contains at different levels.

It is not uncommon to find aerobic and anaerobic conditions in different parts of the same lake at the same time, particularly in the summertime when a stable stratification of distinct layers often occurs (see Figure 13-3). Water at the top of the lake is warmed by the absorption of sunshine by biological materials, while that below the level of penetration of sunlight remains cold. Since warm water is less dense than cold (at temperatures above 4°C), the warm upper layer "floats" on the cold layer below, and little transfer between them occurs. The top layer, called the *epilimnium*, usually contains near-saturation levels of dissolved oxygen, due both to its contact with air and to the O_2 produced in photosynthesis by algae. Since conditions in the top layer are aerobic, elements exist there in their *most oxidized* forms:

- carbon, with an oxidation number (O.N.) of +4, as CO_2 or H_2CO_3 or HCO_3^-;

- sulfur, O.N. of +6, as SO_4^{2-};

- nitrogen, O.N. of +5, as NO_3^-; and

- iron, as Fe(III), in the form of insoluble $Fe(OH)_3$.

Near the bottom, in the *hypolimnium*, the water is oxygen-depleted since it has no contact with air and since O_2 is consumed when biological material, such as the dead algae that have sunk to these depths, decomposes. Under such anaerobic conditions, elements exist in their *most reduced* forms:

- carbon, with an O.N. of −4, as CH_4;

- sulfur, O.N. of −2, as H_2S;

- nitrogen, O.N. of −3, as NH_3 and NH_4^+; and

- iron, as Fe(II), in the form of soluble Fe^{2+}.

Anaerobic conditions usually do not last indefinitely. In the fall and winter, the top layer of water is cooled by cold air passing over it, so that eventually the oxygen-rich water at the top becomes more dense than that below it and gravity induces mixing between the layers. Thus in the winter and early spring the environment near the bottom of a lake usually is aerobic.

TABLE 13-1	Common Oxidation Numbers for Sulfur				
	Increasing Levels of Sulfur Oxidation →				
Oxidation Number of S	−2	−1	0	+4	+6
Aqueous solution and salts	H_2S			H_2SO_3	H_2SO_4
	HS^-			HSO_3^-	HSO_4^-
	S^{2-}	S_2^{2-}		SO_3^{2-}	SO_4^{2-}
Gas phase	H_2S			SO_2	SO_3
Molecular solids			S_8		

Sulfur Compounds in Natural Waters

The common inorganic oxidation numbers in which sulfur is encountered in the environment are illustrated in Table 13-1; they range from the highly reduced −2 form that is found in **hydrogen sulfide gas,** H_2S, and insoluble minerals containing the **sulfide ion,** S^{2-}, to the highly oxidized +6 form that is encountered in **sulfuric acid,** H_2SO_4, and in salts containing the **sulfate ion,** SO_4^{2-}. In organic and bioorganic molecules such as amino acids, intermediate levels of sulfur oxidation are present. When such molecules decompose anaerobically, hydrogen sulfide and other gases such as CH_3SH and CH_3SSCH_3 containing sulfur in highly reduced forms are released, thereby giving swamps their unpleasant odor. The occurrence of such gases as air pollutants was mentioned in Chapter 3.

As discussed in Chapter 3, hydrogen sulfide is oxidized in air first to **sulfur dioxide,** SO_2, and then fully to sulfuric acid or a salt containing the sulfate ion. Similarly, hydrogen sulfide dissolved in water can be oxidized by certain bacteria to **elemental sulfur** or more completely to sulfate. Overall the complete oxidation reactions correspond to

$$H_2S + 2\,O_2 \longrightarrow H_2SO_4$$

Some anaerobic bacteria are able to use sulfate ion as the oxidizing agent to convert organic matter, such as polymeric CH_2O, to carbon dioxide when the concentration of oxygen in the water is very low; the SO_4^{2-} ions are reduced in the process to elemental sulfur or even to hydrogen sulfide:

$$2\,SO_4^{2-} + 3\,CH_2O + 4\,H^+ \longrightarrow 2\,S + 3\,CO_2 + 5\,H_2O$$

Such reactions are especially important in seawater, for which the sulfate ion concentration is much higher than the average for fresh-water systems.

Acid Mine Drainage

One characteristic reaction of groundwater, which by definition is not well-aerated since it has spent much time not exposed to air, is that when it reaches the surface and O_2 has an opportunity to dissolve in it, its rather high level of soluble Fe^{2+} is converted to insoluble Fe^{3+}, and an orange-brown deposit of $Fe(OH)_3$ is formed.

$$4\ Fe^{2+} + O_2 + 2\ H_2O \longrightarrow 4\ Fe^{3+} + 4\ OH^-$$

$$4\ [Fe^{3+} + 3\ OH^- \longrightarrow Fe(OH)_3(s)]$$

The overall reaction is

$$4\ Fe^{2+} + O_2 + 2\ H_2O + 8\ OH^- \longrightarrow 4\ Fe(OH)_3(s)$$

An analogous reaction occurs in some underground coal and metal (especially copper) mines, especially abandoned ones, and in piles of mined coal left open to the environment. Normally FeS_2, called **iron pyrites,** or *fool's gold*, is a stable, insoluble component of underground rocks as long as it does not come into contact with air. However, as a result of the mining of coal and certain ores—and especially after underground mines have been abandoned and spontaneously fill with groundwater—some of it is exposed to water, oxygen, and certain bacteria and becomes partially solubilized as a result of its oxidation. The **disulfide ion,** S_2^{2-}, in which sulfur has an average oxidation number of -1, is oxidized to sulfate ion, SO_4^{2-}, which contains sulfur in the $+6$ form:

$$S_2^{2-} + 8\ H_2O \longrightarrow 2\ SO_4^{2-} + 16\ H^+ + 14\ e^-$$

The main oxidizing agent acting on the sulfur is atmospheric O_2:

$$7\ [O_2 + 4\ H^+ + 4\ e^- \longrightarrow 2\ H_2O]$$

When this ($28\ e^-$) half-reaction is added to twice the ($14\ e^-$) oxidation half-reaction, the net redox reaction is obtained:

$$2\ S_2^{2-} + 7\ O_2 + 2\ H_2O \longrightarrow 4\ SO_4^{2-} + 4\ H^+$$

Since the sulfate salt of the ferrous ion, Fe^{2+}, is soluble in water, the iron pyrites are effectively solubilized by the reaction. More importantly, the reaction produces a large amount of concentrated acid (note the H^+ product), only a portion of which is consumed by the air oxidation of Fe^{2+} to Fe^{3+} that accompanies the process:

$$4\ Fe^{2+} + O_2 + 4\ H^+ \longrightarrow 4\ Fe^{3+} + 2\ H_2O$$

In acidic environments, this reaction is catalyzed by bacteria (*Thiobacillus ferrooxidans*); the resulting Fe^{3+} can oxidize various metal sulfides, liberating the metal ions.

Combining the last two reactions in the correct ratio, i.e., 2:1, we obtain the overall reaction for the oxidation of both the iron and the sulfur:

$$4\,FeS_2 + 15\,O_2 + 2\,H_2O \longrightarrow 4\,Fe^{3+} + 8\,SO_4^{2-} + 4\,H^+$$

i.e., $2\,Fe_2(SO_4)_3 + 2\,H_2SO_4$

In other words, the oxidation of the fool's gold produces soluble **iron(III) sulfate** (also called ferrous sulfate), $Fe_2(SO_4)_3$, and sulfuric acid. The Fe^{3+} ion is soluble in the highly acidic water that is first produced, the pH of which can be less than zero, with the usual range being 0 to 2. However, once the drainage from the highly acidic mine water becomes diluted and its pH consequently rises, a yellowish-brown precipitate of $Fe(OH)_3$ forms from Fe^{3+}, discoloring the water and waterway and smothering plant and animal life (including fish) in it (see Problem 13-7). Thus the pollution associated with acid mine drainage is characterized in the first instance by the seeping from the mine of copious amounts of both acidified water and a rust-colored solid. Unfortunately, the concentrated acid can liberate toxic heavy metals—especially zinc, copper, and nickel, but also lead, arsenic, manganese, and aluminum—from their ores in the rock of the mine, further adding to the pollution of the waterway.

Interestingly, the oxidation of disulfide ion to sulfate ion in the above process is accomplished to some extent by the action of Fe^{3+} as the oxidizing agent, rather than by O_2:

$$S_2^{2-} + 14\,Fe^{3+} + 8\,H_2O \longrightarrow 2\,SO_4^{2-} + 14\,Fe^{2+} + 16\,H^+$$

The phenomenon of acid drainage is currently of particular importance in the many abandoned mines in the mountains of Colorado. However, the most acidic water in the world comes from the Richmond Mine at Iron Mountain, California. There the pH reaches as low as -3.6 because the high temperatures (up to 47°C) of the mine water cause much of it to evaporate, thus concentrating the acid. By comparison, the most acidic natural waters occur near the Ebeko volcano in Russia, with a pH as low as -1.7; the acidity is due to hydrochloric and sulfuric acids in the hot springs water.

The acid produced by acid mine drainage is spontaneously neutralized if the soil contains limestone, in the same way we encountered for acid rain in Chapter 4. For example, some of the coal mines in Pennsylvania discharge water that is acidic (pH < 5), whereas others discharge alkaline water resulting from dissolution of limestone. Powdered or chipped limestone can also be added to water exiting the mine, although the insoluble calcium sulfate that forms on the surface of the limestone particles prevents full reaction. Calcium oxide or hydroxide can also be used to neutralize the acid. Raising the pH precipitates most of the liberated heavy metals as their insoluble hydroxides in a sludge that can be removed from the water. An alternative method of remediation is the introduction into the waters of anaerobic bacteria that reverse the oxidation of sulfate ion back to sulfide, thereby precipitating the heavy metals as insoluble sulfides as well as raising the pH. In some instances, the sulfides are rich enough in metals to be used as ores. Some abandoned mines have been sealed to prevent further intrusions of water and oxygen, but this means of preventing further production of acid is sometimes unsuccessful.

PROBLEM 13-7

The K_{sp} values for $Fe(OH)_2$ and $Fe(OH)_3$ are 7.9×10^{-15} and 6.3×10^{-38}, respectively. Calculate the solubilities of Fe^{2+} and Fe^{3+} at a pH of 8, assuming they are controlled by their hydroxides. Also calculate the pH value at which the ion solubilities reach 100 ppm.

The pE Scale

Environmental scientists sometimes use the concept of pE to characterize the extent to which natural waters are chemically reducing in nature, by analogy to the way in which pH is used to characterize their acidity. In particular, **pE** is defined as the negative base-10 logarithm of the *effective* concentration—i.e., of the *activity*—of electrons in water, notwithstanding the fact that free electrons do not exist in solution (any more than do bare protons, H^+ ions). pE values are dimensionless numbers; like pH, they have no units.

- *Low* pE values signify that electrons are readily available from substances dissolved in the water, so the medium is very reducing in nature.

- *High* pE values signify that the dominant dissolved substances are oxidizing agents, so that few electrons are available for reduction purposes.

When several acids or bases are present in a water sample, one of them usually makes a dominant contribution to the hydrogen or hydroxide ion concentration. In such situations, the position of equilibrium for the other, less dominant weak acids or bases is determined by the H^+ or OH^- level set by the dominant process. In a similar way, in natural waters one or another redox equilibrium reaction is dominant, and it determines the electron availability for the other redox reactions that occur simultaneously. If we know the position of equilibrium for the dominant process, we can calculate the pE and from it the position of equilibrium—and hence the dominant species—in the other reactions.

When a significant amount of O_2 is dissolved in water, the reduction of the oxygen to water is the dominant reaction determining overall electron availability:

$$\tfrac{1}{4} O_2 + H^+ + e^- \rightleftharpoons \tfrac{1}{2} H_2O$$

In such circumstances, the pE of the water is related to the acidity and to the partial pressure of oxygen and its acidity by the following equation, the origin of which is discussed subsequently:

$$pE = 20.75 + \log ([H^+] P_{O_2}^{1/4})$$

$$= 20.75 - pH + \tfrac{1}{4} \log (P_{O_2})$$

For a neutral sample of water that is saturated by oxygen from air, i.e., when $P_{O_2} = 0.21$ atm and is free of carbon dioxide so that its pH = 7, the pE value is calculated from this equation to be 13.9. If the concentration of dissolved oxygen is less than the equilibrium amount, then the equivalent partial pressure of atmospheric oxygen is less than 0.21 atm, so the pE value is smaller than 13.9 and in some cases even negative.

The pE expression given above looks very much like the Nernst equations encountered in the study of electrochemistry. Indeed the pE value for a water sample is simply the electrode potential, E, for whatever process determines electron availability, but divided by RT/F (where R is the gas constant, T the absolute temperature, and F the Faraday constant), which at 25°C has the value 0.0591:

$$pE = E/0.0591$$

Thus the pE expression for any half-reaction in water can be obtained from its standard electrode potential E^0, corrected by the usual concentration and/or pressure terms and evaluated for a one-electron reduction. For example, for the half-reaction linking the reduction of nitrate ion to ammonium ion, we first write the process as a (balanced) one-electron reduction:

$$\tfrac{1}{8} NO_3^- + \tfrac{5}{4} H^+ + e^- \rightleftharpoons \tfrac{1}{8} NH_4^+ + \tfrac{3}{8} H_2O$$

For this reaction, $E^0 = + 0.836$ volts (from standard tables), so $pE^0 = E^0/0.0591 = + 14.15$. The equation for pE involves the subtraction from the standard pE^0 of the logarithm of the ratio of concentrations of products to reactants, each raised to its coefficient in the one-electron half-reaction:

$$pE = pE^0 - \log ([NH_4^+]^{1/8}/[NO_3^-]^{1/8} [H^+]^{5/4})$$

$$= 14.15 - \tfrac{5}{4} pH - \tfrac{1}{8} \log ([NH_4^+]/[NO_3^-])$$

(Here we have used the properties of logarithms that $\log a^x = x \log a$, and that $\log (1/b) = -\log b$.) As usual, the concentration of water does not appear in the expression as its effect is already included in pE^0.

PROBLEM 13-8

Deduce the equilibrium ratio of NH_4^+ to NO_3^- at a pH of 6.0 (a) for aerobic water having a pE value of +11, and (b) for anaerobic water for which the pE value is 3.

Returning to the subject of the dominant reaction that determines pE in natural waters, we note that low values of dissolved oxygen usually are caused by the operation of microorganism-catalyzed organic decomposition reactions, and their dissolved products, rather than O_2, can determine electron availability. For example, in cases of low oxygen availability, the pE of water may be determined by ions such as nitrate or sulfate. In the extreme case of the anaerobic conditions found at the bottoms of lakes in the summer and in swamps and rice paddies, the electron availability is determined by the ratio of dissolved methane, a reducing agent, to dissolved carbon dioxide, an oxidizing agent, both of which are produced by the fermentation of organic matter, discussed above. They are connected in the redox sense by the half-reaction

$$\tfrac{1}{8} CO_2 + H^+ + e^- \rightleftharpoons \tfrac{1}{8} CH_4 + H_2O$$

The pE value for water controlled by this half-reaction is

$$pE = 2.87 - pH + \tfrac{1}{8} \log (P_{CO_2}/P_{CH_4})$$

For example, if the partial pressures of the two gases are equal and the water is neutral, the pE value is -4.1. Thus the lower levels of a stratified lake are characterized by negative pE values, whereas the oxygenated top layer has a substantially positive pE.

The pE concept is useful in predicting the ratio of oxidized to reduced forms of an element in a water body when we know how the electron availability is controlled by another species. Consider, for example, the equilibrium between the two common ions of iron:

$$Fe^{3+} + e^- \rightleftharpoons Fe^{2+}$$

For this reaction,

$$pE = 13.2 + \log ([Fe^{3+}]/[Fe^{2+}])$$

If the pE is determined by another redox process and its value is known, the ratio of Fe^{3+} to Fe^{2+} can be deduced. Thus for the oxygen-depleted water discussed above that has a pE value of -4.1,

$$-4.1 = 13.2 + \log ([Fe^{3+}]/[Fe^{2+}])$$

so

$$\log ([Fe^{3+}]/[Fe^{2+}]) = -17.3$$

and hence

$$[Fe^{3+}]/[Fe^{2+}] = 5 \times 10^{-18}$$

In contrast, for a sample of aerobic water that has a pE of 13.9, the calculated ratio is 5:1 in favor of the Fe^{3+} ion. The transition between dominance of the two forms occurs when their concentrations are equal:

$$pE = 13.2 + \log (1) = 13.2 + 0 = 13.2$$

PROBLEM 13-9

Find the pE value for acidic water at which the ratio of concentrations of Fe^{3+} to Fe^{2+} is 100:1.

PROBLEM 13-10

Assuming that dissolved oxygen determines the electron availability in an aqueous solution and that the partial pressure equivalent to the amount dissolved is 0.10 atm, deduce the ratio of dissolved CO_2 to dissolved CH_4 at a pH of 4.

pE–pH Diagrams

A visual representation of the zones of dominance for the various oxidation states in water of an element can be displayed in a **pE–pH diagram,** as shown in Figure 13-4 for iron. It is clear from the diagram that the situation is more complicated than that included in our example above, since, in moderately acidic or alkaline environments, the solid hydroxides $Fe(OH)_2$ and $Fe(OH)_3$ also come into play in the equilibria. The solid lines in the diagrams indicate combinations of pE and pH values where the concentrations of the two species listed on either side of the line are equal. Thus we see from the top left side of Figure 13-4 that equilibrium between dissolved Fe^{2+} and Fe^{3+} is important only for pH < 3. Equality in the concentrations of these two dissolved forms corresponds to the short horizontal line in the figure. As expected from our calculations above, the transition occurs at pE $= 13.2$ regardless of pH; hence the line is horizontal.

If iron is in the 3+ state at higher pH, it exists predominantly as the solid $Fe(OH)_3$, whereas solutions containing iron in the 2+ state are not mainly precipitated until the solution becomes basic (Figure 13-4).

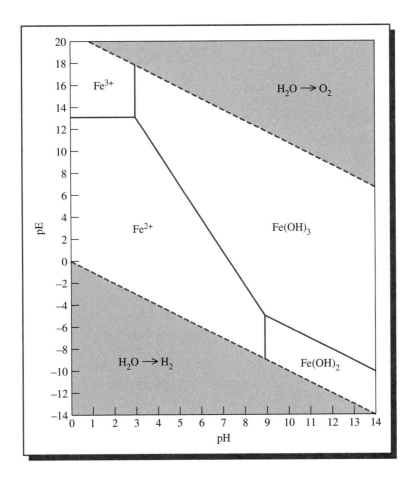

FIGURE 13-4 The pE–pH diagram for the iron system at 10^{-5} M concentration. [Source: Adapted from S. E. Manahan, *Environmental Chemistry,* 4th ed. (Boston, MA: Willard Grant Press/PWS Publishers, 1984).]

The shaded regions at the top right and bottom left side of the pE–pH diagram represent extreme conditions, under which water itself is unstable to decomposition, being oxidized or reduced to yield O_2 or H_2, respectively:

$$2\,H_2O \longrightarrow O_2 + 4\,H^+ + 4\,e^-$$

and

$$2\,H_2O + 2\,e^- \longrightarrow H_2 + 2\,OH^-$$

Nitrogen Compounds in Natural Waters

In some natural waters, nitrogen occurs in inorganic and organic forms that are of concern with respect to human health. As discussed in Chapter 6, there are several environmentally important forms of nitrogen that differ in the extent of oxidation of the nitrogen atom.

TABLE 13-2	Common Oxidation Numbers for Nitrogen						
	Increasing Levels of Nitrogen Oxidation →						
Oxidation Number of N	−3	0	+1	+2	+3	+4	+5
Aqueous solution and salts	NH_4^+ NH_3				NO_2^-		NO_3^-
Gas phase	NH_3	N_2	N_2O	NO		NO_2	

The common oxidation numbers of nitrogen, along with the most important examples for each, are illustrated in Table 13-2. The most reduced forms all have the −3 oxidation number, as occurs in ammonia, NH_3, and its conjugate acid, the ammonium ion, NH_4^+. The most oxidized form, having an oxidation number of +5, occurs as the nitrate ion, NO_3^-, which exists in salts, aqueous solutions, and **nitric acid,** HNO_3. In solution, the most important intermediates between these extremes are the **nitrite ion, NO_2^-,** and **molecular nitrogen, N_2.**

The pE–pH diagram for the existence of these forms in aqueous solution is shown in Figure 13-5. Notice the relatively small field of dominance (the small triangle on the right side of the diagram) for the nitrite ion, NO_2^-, in which nitrogen's oxidation number has the intermediate value of +3. In particular, it is the predominant species only under alkaline conditions that are intermediate in oxygen content (small positive pE values).

The equilibrium between the most highly reduced and the most highly oxidized forms of nitrogen is given by the half-reaction

$$NH_4^+ + 3\,H_2O \rightleftharpoons NO_3^- + 10\,H^+ + 8\,e^-$$

Previously we derived the equation relating pE to pH for this reaction written as a reduction; this equation defines the diagonal line in Figure 13-4 that separates these two ions when their concentrations are equal, so that

$$pE = 14.15 - \tfrac{5}{4}\,pH - \tfrac{1}{8}\log(1) = 14.15 - \tfrac{5}{4}\,pH$$

We might conclude from this equation, and from the slope of the diagonal line in Figure 13-5 separating NH_4^+ and NO_3^-, that since the oxidation of ammonium ion is highly pH dependent, it would not occur under highly acidic conditions. However, electron availability from the reduction of dissolved oxygen also decreases as the pH is lowered, so the pE value in such

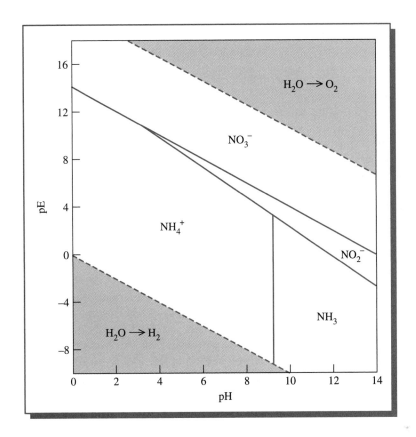

FIGURE 13-5 pE–pH diagram for inorganic nitrogen in an aqueous system. [Source: Adapted from C. N. Sawyer, P. L. McCarty, and G. F. Parkin, *Chemistry for Environmental Engineering,* 4th ed. (New York: McGraw-Hill, 1994).]

water is quite high, and nitrate still predominates. For example, the pE value calculated for oxygen in water at pH of 1 is about 20, so nitrate still predominates.

Recall from Chapter 6 that in the microorganism-catalyzed process of nitrification, ammonia and ammonium ion are oxidized to nitrate, whereas in the corresponding denitrification process, nitrate and nitrite are reduced to molecular nitrogen. Both processes are important in soils and in natural waters. In aerobic environments such as the surface of lakes, nitrogen exists as the fully oxidized nitrate, whereas in anaerobic environments such as the bottom of stratified lakes, nitrogen exists as the fully reduced forms ammonia and ammonium ion (Figure 13-3). Nitrite ion occurs in anaerobic environments such as waterlogged soils that are not sufficiently reducing to convert the nitrogen all the way to ammonia. Most plants can absorb nitrogen only in the form of nitrate ion, so any ammonia or ammonium ion used as fertilizer must first be oxidized via microorganisms before it is useful to such plant life.

Consider the reduction of nitrate ion to nitrite ion in a natural water system.

(a)   Write the balanced one-electron half-reaction for the process if it occurs in acidic media.
(b)   Given that for this reaction, $E^0 = +0.881$ volts, calculate pE^0.
(c)   From your answer to (a), deduce the expression relating pE to pE^0 and ion concentrations.
(d)   From your result in part (c), obtain an equation relating the pE and pH conditions under which the ratio of nitrate to nitrite is 100:1.
(e)   From your result in part (c), deduce the ratio of nitrite to nitrate under conditions of $pE = 12$, $pH = 5$.

Acid–Base Chemistry in Natural Waters: The Carbonate System

Natural waters, even when "pure," contain significant quantities of dissolved carbon dioxide and of the anions it produces, as well as cations of calcium and magnesium. In addition, the pH of such natural water is rarely equal to exactly 7.0, the value expected for pure water. In this section, the natural processes that involve these substances in natural waters are analyzed.

The CO_2–Carbonate System

The acid–base chemistry of many natural water systems, including both rivers and lakes, is dominated by the interaction of the **carbonate ion,** CO_3^{2-}, a moderately strong base, with the weak acid H_2CO_3, **carbonic acid.** Loss of one hydrogen ion from the acid produces the **bicarbonate ion,** HCO_3^- (also called *hydrogen carbonate ion*):

$$H_2CO_3 \rightleftharpoons H^+ + HCO_3^- \tag{1}$$

The acid dissociation constant for this process, K_1, is numerically much greater than K_2, the constant for the second stage of ionization, which produces carbonate ion:

$$HCO_3^- \rightleftharpoons H^+ + CO_3^{2-} \tag{2}$$

In order to discover the dominant form at any given pH, it is instructive to consider a *species diagram* for the CO_2–bicarbonate–carbonate system in water, such as that shown in Figure 13-6. In it, the fraction of the total inorganic carbon that is present in each of the three forms is shown as a function of the master variable, the pH of the solution. Clearly, carbonic acid is the dominant species at low pH (< 5), carbonate is dominant at high pH (> 12), and bicarbonate is the predominant—but not the only—species present in the pH range of most natural waters, i.e., from 7 to 10. At the pH of natural rainwater, 5.6, most of the dissolved carbon dioxide exists as carbonic acid, but a measurable fraction is bicarbonate ion (Figure 13-6).

The curves in Figure 13-6 were constructed by solving for the three unknowns $[H_2CO_3]$, $[HCO_3^-]$, and $[CO_3^{2-}]$, relative to their total concentration, C, from individual equilibrium constant expressions, as detailed in Box 13-2.

$$[H_2CO_3]/C = [H^+]^2/D$$

$$[HCO_3^-]/C = K_1 [H^+]/D$$

and

$$[CO_3^{2-}]/C = K_1 K_2/D$$

where the common denominator $D = [H^+]^2 + K_1[H^+] + K_1 K_2$.

These expressions clearly show that H_2CO_3 is the dominant species under conditions of high acidity and determine the other conditions as discussed above.

The carbonic acid in natural waters results from the dissolution of carbon dioxide gas in water, the gas originating either in the air or from the decomposition of organic matter in the water. The gas in the air and the acid in water in contact with the surface usually are at equilibrium:

$$CO_2(g) + H_2O(aq) \rightleftharpoons H_2CO_3(aq) \qquad (3)$$

The relevant equilibrium constant for this reaction is the Henry's law constant, K_H, for CO_2. [In fact, much of the dissolved carbon dioxide exists as CO_2 (aq) rather than as H_2CO_3 (aq), but following conventional practice we collect the two forms together and represent it all as the latter.] The pH of pure water in equilibrium with the current level of atmospheric CO_2 is 5.6, according to the methods discussed in Chapter 3 (see Problem 3-10).

The sequestration of carbon dioxide, as such, into oceans would result in an increase in the acidity of the surrounding waters since the resulting increase in H_2CO_3 concentration [reaction (3)] would give rise to further ionization [reaction (1)], which is called *ocean acidic* for that reason (Chapter 7). The resulting decrease in pH could affect biological life in the vicinity. Indeed, the increase in the *atmospheric* concentration of CO_2 that has already occurred has resulted in a drop of about 0.1 pH unit in surface ocean water worldwide.

The predominant source of the carbonate ion in natural waters is limestone rocks, which are largely made up of **calcium carbonate, $CaCO_3$**.

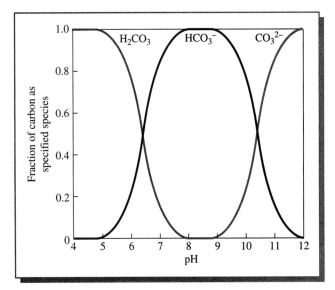

FIGURE 13-6 Species diagram for the aqueous carbon dioxide–bicarbonate ion–carbonate system. [Source: S. E. Manahan, *Environmental Chemistry*, 6th ed. (Boca Raton, FL: Lewis Publishers, 2000), Figure 3.3, p. 54.]

| BOX 13-2 | Derivation of the Equations for Species Diagram Curves |

The final equations relating the concentrations of carbonic acid, bicarbonate ion, and carbonate ion to the pH, the equilibrium constants, and their total concentration C were obtained by algebraic manipulation of the three simultaneous equations.

Mass balance:

$$[H_2CO_3] + [HCO_3^-] + [CO_3^{2-}] = C$$

Acid dissociation constants:

$$K_1 = [HCO_3^-][H^+]/[H_2CO_3]$$

and

$$K_2 = [CO_3^{2-}][H^+]/[HCO_3^-]$$

From the second and third equations, respectively, we can express both $[H_2CO_3]$ and

$[CO_3^{2-}]$ in terms of $[HCO_3^-]$ and $[H^+]$, and substitute these relationships solutions into the mass balance equation:

$$[H_2CO_3] = [HCO_3^-][H^+]/K_1$$
$$[CO_3^{2-}] = K_2[HCO_3^-]/[H^+]$$

Thus

$$([HCO_3^-][H^+]/K_1) + ([HCO_3^-])$$
$$+ (K_2[HCO_3^-]/[H^+]) = C$$

Solving this equation for bicarbonate and substituting the solution into the preceding pair of equations yields the expressions given in the main text for the fraction of each species present at any pH.

Although this salt is almost insoluble, a small amount of it dissolves when water passes over it:

$$CaCO_3(s) \rightleftharpoons Ca^{2+} + CO_3^{2-} \qquad (4)$$

Natural waters that are exposed to limestone are called **calcareous waters.** The dissolved carbonate ion acts as a base, producing its conjugate weak acid, the bicarbonate ion, as well as hydroxide ion in the water:

$$CO_3^{2-} + H_2O \rightleftharpoons HCO_3^- + OH^- \qquad (5)$$

These reactions that occur in the natural three-phase (air, water, rock) system are summarized pictorially in Figure 13-7; the reactions of the carbon dioxide–carbonate system are summarized for convenience in Table 13-3.

In the discussions that follow, we analyze the effects on the composition of a body of water of the simultaneous presence of both carbonic acid and calcium carbonate. We shall see that the presence of each of these substances increases the solubility of the other, and that the hydrogen ion and hydroxide ion produced indirectly from their dissolution largely neutralize each other, yielding water with almost neutral pH.

To obtain a qualitative understanding of this rather complicated system, the effect of the carbonate ion alone is first considered.

Water in Equilibrium with Solid Calcium Carbonate

For simplicity, we first consider a (hypothetical) body of water that is in equilibrium with excess solid calcium carbonate and in which all other reactions are of negligible importance. The only process of interest in this case is reaction (4) (see Table 13-3). Recall from introductory chemistry that the appropriate equilibrium constant for processes that involve the dissolution of slightly soluble salts in water is the **solubility product**, K_{sp}, which equals the concentrations of the product ions, each raised to its coefficient of the balanced equation. Thus, for reaction (4), K_{sp} is related to the equilibrium concentration of the ions by the equation

$$K_{sp} = [Ca^{2+}][CO_3^{2-}]$$

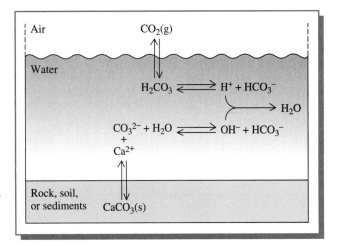

FIGURE 13-7 Reactions among the three phases (air, water, rocks) of the carbon dioxide–carbonate system.

TABLE 13-3	Reactions in the CO_2–Bicarbonate–Carbonate System		
Reaction Number	Reaction	Equilibrium Constant	K Value at 25°C
1	$H_2CO_3 \rightleftharpoons H^+ + HCO_3^-$	K_{a1} (H_2CO_3)	4.5×10^{-7}
2	$HCO_3^- \rightleftharpoons H^+ + CO_3^{2-}$	K_{a2} (H_2CO_3)	4.7×10^{-11}
3	$CO_2(g) + H_2O(aq) \rightleftharpoons H_2CO_3(aq)$	K_H	3.4×10^{-2}
4	$CaCO_3(s) \rightleftharpoons Ca^{2+} + CO_3^{2-}$	K_{sp}	4.6×10^{-9}
5	$CO_3^{2-} + H_2O \rightleftharpoons HCO_3^- + OH^-$	K_b (CO_3^{2-})	2.1×10^{-4}
6	$CaCO_3(s) + H_2O(aq)$ $\rightleftharpoons Ca^{2+} + HCO_3^- + OH^-$		
7	$H^+ + OH^- \rightleftharpoons H_2O(aq)$	$1/K_w$	1.0×10^{14}
8	$CaCO_3(s) + CO_2(g) + H_2O(aq)$ $\rightleftharpoons Ca^{2+} + 2\ HCO_3^-$		

It follows from the stoichiometry of reaction (4) that as many calcium ions are produced as carbonate ions, and that in this simplified system both ion concentrations are equal to S, the solubility of the salt:

$$S = \text{solubility of } CaCO_3 = [Ca^{2+}] = [CO_3^{2-}]$$

After substituting S for the ion concentrations in the K_{sp} equation and inserting the K_{sp} value from Table 13-3, we obtain

$$S^2 = 4.6 \times 10^{-9}$$

Taking the square root of each side of this equation, a value for S can be extracted:

$$S = 6.8 \times 10^{-5} \text{ M}$$

Thus the solubility in water of calcium carbonate is estimated to be 6.8×10^{-5} mol/L, assuming that all other reactions are negligible.

PROBLEM 13-12

Consider a body of water in equilibrium with solid calcium sulfate, $CaSO_4$, for which $K_{sp} = 3.0 \times 10^{-5}$ at 25°C. Calculate the solubility, in g/L, of calcium sulfate in water, assuming that other reactions are negligible.

According to reaction (5), dissolved carbonate ion acts as a base in water. The relevant equilibrium constant for this process is the base ionization constant, K_b, where

$$K_b(CO_3^{2-}) = [HCO_3^-] [OH^-]/[CO_3^{2-}]$$

Since the equilibrium in this reaction lies to the right in solutions that are not very alkaline, an approximation of the overall effect resulting from the simultaneous occurrence of reactions (4) and (5) can be obtained by adding together the equations for the two individual reactions. The overall reaction is

$$CaCO_3(s) + H_2O(aq) \rightleftharpoons Ca^{2+} + HCO_3^- + OH^- \tag{6}$$

Thus the dissolution of calcium carbonate in neutral water results essentially in the production of calcium ion, bicarbonate ion, and hydroxide ion.

It is a principle of equilibrium that if several reactions are added together, the equilibrium constant K for the *combined* reaction is the *product* of the equilibrium constants for the individual processes. Thus, since reaction (6) is the sum of reactions (4) and (5), its equilibrium constant K_6 must equal $K_{sp}K_b$, the product of the equilibrium constants for reactions (4) and (5).

Since the acid and base ionization constants for any acid–base conjugate pair such as HCO_3^- and CO_3^{2-} are simply related by the equation

$$K_a K_b = K_w = 1.0 \times 10^{-14} \text{ at } 25°C$$

it follows that for the conjugate base CO_3^{2-}

$$K_5 = K_b(CO_3^{2-}) = K_w/K_a(HCO_3^-)$$

Since K_a for HCO_3^- is the K_{a2} value for the carbonic acid system, then from Table 13-3,

$$K_b = 1.0 \times 10^{-14}/4.7 \times 10^{-11}$$
$$= 2.1 \times 10^{-4}$$

Thus, since K_6 for the overall reaction (6) is $K_{sp}K_b$, its value is $(4.6 \times 10^{-9}) \times (2.1 \times 10^{-4}) = 9.7 \times 10^{-13}$.

The equilibrium constant for reaction (6) is related to the ion concentrations by the equation

$$K_6 = [Ca^{2+}][HCO_3^-][OH^-]$$

If we make the approximation that reaction (6) is the *only* process of relevance in the system, then from its stoichiometry we have a new expression for the solubility of $CaCO_3$, namely

$$S = [Ca^{2+}] = [HCO_3^-] = [OH^-]$$

Upon substitution of S for the concentrations, we obtain

$$S^3 = 9.7 \times 10^{-13}$$

Taking the cube root of both sides of this equation, we find

$$S = 9.9 \times 10^{-5} \text{ M}$$

Thus the estimated solubility for $CaCO_3$ is 9.9×10^{-5} M, in contrast to the lesser value of 6.8×10^{-5} M that we obtained when the reaction of carbonate ion was ignored. The $CaCO_3$ solubility here is greater than estimated from reaction (4) alone, since much of the carbonate ion it produces subsequently disappears by reacting with water molecules. In other words, the equilibrium in reaction (4) is shifted to the right since a large fraction of its product reacts further [reaction (5)].

From these results, it is clear that the saturated aqueous solution of calcium carbonate is moderately alkaline; its pH can be obtained from the hydroxide ion concentration of 9.9×10^{-5} M:

$$pH = 14 - pOH = 14 - \log_{10}[OH^-] = 14 - \log_{10}(9.9 \times 10^{-5}) = 10.0$$

That the solution is alkaline is not surprising, given that the carbonate ion, as weak bases go, is a moderately strong one.

PROBLEM 13-13

Repeat the calculation of the solubility of calcium carbonate by the approximate single equilibrium method using a realistic wintertime water temperature of 5°C; at that temperature, $K_{sp} = 8.1 \times 10^{-9}$ for $CaCO_3$, $K_a = 2.8 \times 10^{-11}$ for HCO_3^-, and $K_w = 0.2 \times 10^{-14}$. By comparing the result with that in the foregoing text for 25°C, decide whether the solubility of calcium carbonate increases or decreases with increasing temperature.

PROBLEM 13-14

What is the net reaction when reactions (1) and (3) are added together? How is the equilibrium constant for this combined process related to K_1 and K_3? Show that the pH of the aqueous solution resulting from a CO_2 partial pressure of 0.00037 atm in the combined process has the same value of 5.6 as is determined by considering the individual reactions consecutively, as in Problem 3-10.

In the analysis above, it has been assumed that calcium carbonate is dissolving in pure water and that the reaction of carbonate with water determines the pH. In some real-world situations, the pH of the aqueous solution is predetermined by the presence of some dominant source of H^+ or OH^-, so the contribution from calcium carbonate is negligible. In such cases, the ratio of bicarbonate to carbonate ion can be determined from K_b and the known, fixed $[OH^-]$:

$$[HCO_3^-]/[CO_3^{2-}] = K_b/[OH^-]$$

and thus

$$[HCO_3^-] = K_b [CO_3^{2-}]/[OH^-]$$

The solubility S of calcium carbonate, and the resulting dissolved calcium ion concentration, here equal the sum of the carbonate and bicarbonate ion concentrations, so

$$S = [Ca^{2+}] = [CO_3^{2-}] + [HCO_3^-]$$
$$= [CO_3^{2-}] + K_b [CO_3^{2-}]/[OH^-]$$
$$= [CO_3^{2-}] (1 + K_b/[OH^-])$$

If, for convenience, we temporarily define $f = (1 + K_b/[OH^-])$, then

$$S = f [CO_3^{2-}]$$

Substituting this equation for $[Ca^{2+}]$ into the K_{sp} expression gives

$$K_{sp} = f[CO_3^{2-}]^2$$

Solving for the carbonate concentration, we obtain

$$[CO_3^{2-}] = (K_{sp}/f)^{1/2}$$

and hence

$$S = f[CO_3^{2-}] = (f K_{sp})^{1/2}$$
$$= \{K_{sp}(1 + K_b/[OH^-])\}^{1/2}$$

Thus, as expected by application of Le Châtelier's principle to the reactions, the solubility of calcium carbonate decreases as the (fixed) hydroxide concentration increases, the limit at high levels of OH^- being $(K_{sp})^{1/2}$, the value we obtained assuming no reaction of carbonate ion with water. In contrast, for water that is neutral or acidic and therefore low in hydroxide ion, the $CaCO_3$ solubility is much larger than this value, as illustrated in Figure 13-8, where the logarithm of S is plotted against the pH of the water body.

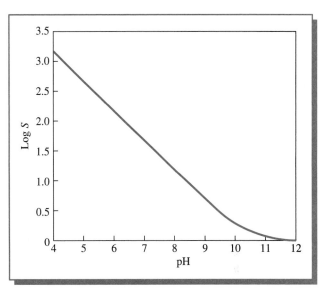

FIGURE 13-8 Molar solubility, S, in units of $K_{sp}^{0.5}$, of $CaCO_3$ in CO_2-free water versus pH.

Water in Equilibrium with Both CaCO₃ and Atmospheric CO₂

The systems discussed above are somewhat unrealistic since they fail to consider the other important carbon species in water—namely, carbon dioxide and carbonic acid—and the reactions that involve them. These reactions will now be considered in the context of a body of water that is also in equilibrium with solid calcium carbonate, i.e., the three-phase system illustrated in Figure 13-7.

At first sight, it might seem that since reaction (1) provides another source of bicarbonate ion, then by Le Châtelier's principle the production of bicarbonate from the reaction (5) of carbonate with water should be suppressed. However, a more important consideration is that reaction (1) produces hydrogen ion, which combines with the hydroxide ion that is produced in reaction (4) by the interaction of carbonate ion with water:

$$H^+ + OH^- \rightleftharpoons H_2O(aq) \qquad (7)$$

Consequently, the equilibrium positions of both reactions that produce bicarbonate ion are shifted to the right due to the disappearance of one of their products by the above reaction.

If reactions (1), (3), (4), (5), and (7) are all added together to deduce the net process, then after canceling common terms the net result is

$$CaCO_3(s) + CO_2(g) + H_2O(aq) \rightleftharpoons Ca^{2+} + 2\ HCO_3^- \qquad (8)$$

In other words, combining equimolar amounts of solid calcium carbonate and atmospheric carbon dioxide yields aqueous **calcium bicarbonate,** $Ca(HCO_3)_2$, without any apparent production or consumption of acidity or alkalinity:

calcium carbonate (rock) + carbon dioxide (air) = calcium bicarbonate (in solution)

Natural waters in which this overall process occurs can be viewed as the site of a giant titration of an acid that originates with CO_2 from air with a base that originates with carbonate ion from rocks. [Note that we need not consider reaction (2) in this analysis, since reaction (5) of the conjugate base of bicarbonate with water was included.] In the *ocean neutral* scheme of sequestration (Chapter 7), bulk carbon dioxide is reacted with solid calcium carbonate or with some other calcium-containing salt; the resulting slurry of calcium bicarbonate (or other salt) is then transported to and deposited in the ocean. This technique avoids the pH-lowering side effect of the *ocean acidic* scheme in which CO_2 is directly dissolved in the ocean water.

It should be noted that each of the individual reactions added together is itself an equilibrium that does not lie entirely to the right. Since the reactions differ in their extent of completion, it is an approximation to state that the overall reaction shown above is the only resulting reaction. Nevertheless, it is the dominant process, and it is mathematically convenient to first consider this process alone in estimating the extent to which $CaCO_3$ and CO_2 dissolve in water when both are present.

Since reaction (8) equals the sum of reactions (1), (3), (4), (5), and (7), its equilibrium constant K_8 is the product of their equilibrium constants:

$$K_8 = K_{sp}K_bK_HK_{a1}/K_w$$

Here K_{a1} is the first acid dissociation constant for carbonic acid. K_H is the Henry's law constant (see Chapter 3) for reaction (3). Since K_w is the ion product for water, the equilibrium constant for reaction (7) is $1/K_w$. The other constants in the equation for K_8 have been defined previously (see Table 13-3). Thus at 25°C for the overall reaction (8), it follows that

$$K_8 = 1.5 \times 10^{-6}$$

From the balanced equation for the reaction, it follows that the expression for K_8 is

$$K_8 = [Ca^{2+}]\,[HCO_3^-]^2/P_{CO_2}$$

If the calcium concentration again is called S, then from the stoichiometry of reaction (8) the bicarbonate concentration must be twice as large, equal to $2S$; after substitution for the concentrations in the equation for K_8 and rearrangement, we obtain

$$S\,(2S)^2 = K_8\,P_{CO_2}$$

or

$$4\,S^3 = K_8\,P_{CO_2}$$

Thus the solubility of calcium carbonate increases as the cube root of the partial pressure of carbon dioxide to which the water is exposed:

$$S = (K_8\,P_{CO_2}/4)^{1/3}$$

Substituting the current partial pressure of CO_2 in the atmosphere, 0.00036 atm, corresponding to an atmospheric concentration of CO_2 of 365 ppm (Chapter 6), and the numerical value of K_8 into this equation yields

$$S = 5.1 \times 10^{-4}\,mol/L^{-1} = [Ca^{2+}]$$

and thus

$$[HCO_3^-] = 2S = 1.0 \times 10^{-3}\,M$$

The amount of CO_2 dissolved is also equal to S and is 35 times that which dissolves without the presence of calcium carbonate (see results of Problem 3-10). Furthermore, the calculated calcium concentration is four times that calculated without the involvement of carbon dioxide. Thus the acid reaction of dissolved CO_2 and the base reaction of dissolved carbonate have a synergistic effect on each other that increases the solubilities of both the gas and the solid (see Table 13-4). In other words, water that contains carbon dioxide more readily dissolves calcium carbonate. In fact, groundwater may

TABLE 13-4	Calculated Ion Concentrations for Aqueous Equilibrium Systems	
Ion	$CaCO_3$	CO_2 and $CaCO_3$
$[HCO_3^-]$	$9.9 \times 10^{-5}\,M$	$1.0 \times 10^{-3}\,M$
$[CO_3^{2-}]$	—	$8.8 \times 10^{-6}\,M$
$[Ca^{2+}]$	$9.9 \times 10^{-5}\,M$	$5.2 \times 10^{-4}\,M$
$[OH^-]$	$9.9 \times 10^{-5}\,M$	$1.8 \times 10^{-6}\,M$
$[H^+]$	$1.0 \times 10^{-10}\,M$	$5.6 \times 10^{-9}\,M$
pH	10.0	8.3

become supersaturated with carbon dioxide as a result of biological decom-position processes; and in that case, the calcium carbonate solubility increases even more—at least until the water reaches the surface, when degassing of the excess CO_2 would occur.

PROBLEM 13-15

Repeat the above calculation for the solubility of $CaCO_3$ in water that is also in equilibrium with atmospheric CO_2 for a water temperature of 5°C. At this temperature, $K_H = 0.065$ for CO_2 and K_1 for H_2CO_3 is 3.0×10^{-7}; see Problem 13-13 for other necessary data.

Finally, the residual concentrations of CO_3^{2-}, of H^+, and of OH^- in the system can be deduced from equilibrium constants for reactions (4), (5), and (7), since equilibria in these processes are in effect, notwithstanding the over-all reaction (8). Thus from reaction (4),

$$[CO_3^{2-}] = K_{sp}/[Ca^{2+}] = 4.6 \times 10^{-9}/5.1 \times 10^{-4} = 9.0 \times 10^{-6}\,M$$

From reaction (5),

$$[OH^-] = K_b\,[CO_3^{2-}]/[HCO_3^-]$$
$$= (2.1 \times 10^{-4}) \times (9.0 \times 10^{-6})/1.0 \times 10^{-3} = 1.9 \times 10^{-6}$$

and finally from reaction (7)

$$[H^+] = K_w/[OH^-] = 1.0 \times 10^{-14}/1.9 \times 10^{-6} = 5.3 \times 10^{-9}$$

From this value for the hydrogen ion concentration, we conclude that according to this calculation, river and lake water at 25°C whose pH is deter-mined by saturation with CO_2 and $CaCO_3$ should be slightly alkaline, with a pH of about 8.3.

Typically, the pH values of calcareous waters lie in the range from 7 to 9, in reasonable agreement with our calculations. Due to the smaller amount of bicarbonate in noncalcareous waters, their pH values are usually close to 7. Of course, if natural waters are subject to acid rain, the pHs can become sub-stantially lower since there is little HCO_3^- or CO_3^{2-} readily available with which to neutralize the acid.

About 80% of natural surface waters in the United States have pH val-ues between 6.0 and 8.4. Lakes and rivers into which acid rain falls will have elevated levels of sulfate ion and perhaps of nitrate ion since the principal acids in the precipitation are H_2SO_4 and HNO_3 (see Chapter 4).

PROBLEM 13-16

In waters subject to acid rain, the pH is determined not by the CO_2–carbonate system but rather by the strong acid from the precipitation. Assuming that equilibrium with atmospheric carbon dioxide is in effect, calculate the concentration of HCO_3^- in natural waters with pH = 6, 5, and 4 at 25°C. (See Table 13-3 for data.)

PROBLEM 13-17

Using algebraic expressions and numerical values for K_{a1} and K_{a2} of H_2CO_3 (Table 13-3), calculate the pH values for which $[H_2CO_3] = [HCO_3^-]$ and for which $[HCO_3^-] = [CO_3^{2-}]$.

Ion Concentrations in Natural Waters and Drinking Water

The Abundant Ions in Fresh Water

As is evident from Table 13-5, the most abundant ions found in samples of unpolluted fresh calcareous water usually are calcium and bicarbonate, as expected from our previous analysis. Commonly, such water also contains **magnesium ion,** Mg^{2+}, principally from the dissolution of $MgCO_3$; plus some sulfate ion, SO_4^{2-}; smaller amounts of **chloride ion,** Cl^-, and **sodium ion,** Na^+; and even smaller levels of fluoride ion, F^-, and potassium ion, K^+. The overall reaction (8) of carbon dioxide and calcium carbonate implies that the ratio of bicarbonate ion to calcium ion should be 2:1, and this is indeed a rule that is closely obeyed on average in river water in North America and Europe. The calculated calcium ion concentration, 5.1×10^{-4} M, agrees well with the North American river-water average value of 5.3×10^{-4} M, and similarly for the bicarbonate ion data. The close agreement between the calculated and the experimental results is somewhat fortuitous because river-water temperatures on average lie below 25°C—which results in a higher CO_2 solubility than has been assumed—and because several minor factors have been oversimplified in the calculation. In fact, even calcareous river water is usually unsaturated with respect to $CaCO_3$.

Water in rivers and lakes that is not in contact with carbonate salts contains substantially fewer dissolved ions than are present in calcareous waters. The concentration of sodium and potassium ions may be as high as those of calcium, magnesium, and bicarbonate ions in these fresh waters. Even in areas with no limestone in the soil, the waters contain some bicarbonate ion due to the weathering of *aluminosilicates* in submerged soil and rock in the

TABLE 13-5	River Water Concentrations and Drinking Water Standards for Ions				
	River Water Molar Concentration		Drinking Water Concentration in ppm		
				Maximum Recommended Concentration	
Ion	Average for World	Average for U.S.	Average U.S.	U.S.	Canada
*HCO_3^-	9.2×10^{-4}	9.6×10^{-4}	60		
Ca^{2+}	3.8×10^{-4}	3.8×10^{-4}	15		
Mg^{2+}	1.6×10^{-4}	3.4×10^{-4}	8		
Na^+	3.0×10^{-4}	2.7×10^{-4}	6		200
Cl^-	2.3×10^{-4}	2.2×10^{-4}	8	250	250
SO_4^{2-}	1.1×10^{-4}	1.2×10^{-4}	12	250	500
K^+	5.4×10^{-5}	5.9×10^{-5}	2		
F^-	—	5.3×10^{-6}	0.1	0.8–2.4	1.5
NO_3^-	1.4×10^{-5}	—			
Fe^{3+}	7.3×10^{-6}	—			

*Note: The value for bicarbonate is actually the total alkalinity.
Sources: World data from R. A. Larson and E. J. Weber, *Reaction Mechanisms in Environmental Organic Chemistry* (Boca Raton, FL: Lewis Publishers).

presence of atmospheric carbon dioxide. The weathering reaction can be written in general terms as

$$M^+(\text{Al-silicate}^-)(s) + CO_2(g) + H_2O \longrightarrow M^+ + HCO_3^- + H_4SiO_4$$

Here M is a metal such as potassium, and the anion is one of the many aluminosilicate ions found in rocks (see Chapter 16). The weathering of *potassium feldspar* is an example of one of the most important sources of potassium ion in natural waters:

$$3\,KAlSi_3O_8(s) + 2\,CO_2(g) + 14\,H_2O \longrightarrow$$
$$2\,K^+ + 2\,HCO_3^- + 6\,H_4SiO_4 + KAl_3Si_3O_{10}(OH)_2(s)$$

Thus bicarbonate normally is the predominant anion in both calcareous and noncalcareous waters since it is produced by the dissolution of limestone and aluminosilicates, respectively.

The average compositions of river water in the United States and in the world as a whole are given in Table 13-5. As discussed, the values for the calcium and magnesium ion concentrations vary significantly from place to place, depending upon whether or not the underlying soil is calcareous.

Fresh water in which the concentration of ions is abnormally high is called **saline water** and is usually unsuitable for drinking. Most saline water is the result of irrigation, in which water is transported into land where little rainfall occurs. The water largely evaporates if the climate is hot and dry, leaving behind salts of the ions that were present in the irrigation water. The runoff water from rainfall and irrigation into water supplies consequently is saline. If the irrigation water is recycled, it becomes more and more saline as time goes on. The wintertime de-icing of roads in northern climates also contributes salinity to water bodies.

Fluoride Ion in Water

The level of **fluoride ion, F^-**, in water also displays substantial variations, from less than 0.01 ppm to more than 20 ppm, in different regions of the world. The source of most F^- is weathering of the mineral **fluorapatite,** $Ca_5(PO_4)_3F$.

In Mexico and some European countries, **sodium fluoride,** NaF, is added to table salt. In many communities of English-speaking countries—including the United States (about half the population), Canada, Australia, and New Zealand in which the F^- concentration in the drinking-water source is low— a soluble fluoride compound such as **fluorosilicic acid,** H_2SiF_6, or its sodium salt, both of which react with water to release fluoride ion, is often added in order to bring the fluoride level up to about 1 ppm, i.e., 5×10^{-5} M. This value was considered at least in the past to be optimum in strengthening children's teeth against decay while providing a margin of safety. If the fluoride level is in excess of this value, as it is in some natural waters, deleterious effects on teeth such as mottling can occur. The maximum contaminant level (MCL) of fluoride in U.S. drinking water is 4 ppm. Almost all brands of toothpaste available in developed countries contain added fluoride in the form of sodium fluoride, **stannous fluoride,** SnF_2, or **sodium monofluorophosphate.** Most children in North America also receive topical fluoride not only from their toothpaste, but in some cases by application from their dentists.

The addition of fluoride ion to public supplies of drinking water continues to be a controversial subject because at high concentrations fluoride is known to be poisonous and perhaps carcinogenic, and because some people feel that it is immoral to force everyone to drink water to which a substance has been added. In fact, for many people, the total amount of fluoride ion ingested from food and beverages (especially tea) exceeds that from water.

Bottled Drinking Waters

The maximum concentration of ions recommended for drinking water in the United States and in Canada is also listed in Table 13-5. The concentration of sodium ion, Na^+, in water is of interest since high consumption of it from water and salted food is believed to increase blood pressure, which may lead to cardio-vascular disease. Excessive sulfate, beyond 500 mg/L, causes a laxative effect in some people. It is interesting to note that some varieties of bottled drinking water, which people presumably drink in preference to tap water due to health concerns about the latter, exceed the recommended values for some ions. Several well-known bottled waters exceeded the drinking water standards for arsenic and/or fluoride in a 1999 survey by the U.S. National Resources Defense Council. However, they were remarkably free of chloroform, a substance that plagues municipal water supplies, as discussed in Chapter 14. A more recent survey found that most brands would meet the new 10-ppm standard for arsenic (Chapter 15) but that bisphenol-A (Chapter 12) is leached from the plastic into the water contained in most large polycarbonate jugs.

Suppliers of bottled water sometimes advertise their products as having "zero" concentrations of fluoride and/or sodium ions or as being *sodium-free* or *fluoride-free*. These are misleading statements since in reality the actual concentrations are not zero but below the level of detection in the analytical method used by the bottler or below a threshold specified by government. "Zero" is not a meaningful chemical concept to answer the question of "how much" of a substance is present in a sample.

Seawater

The total concentration of ions in seawater is much higher than that in fresh water since it contains large quantities of dissolved salts. The predominant species in seawater are sodium and chloride ions, which occur at about 1000 times their average concentration in fresh water. Seawater also contains some Mg^{2+} and SO_4^{2-} and lesser amounts of many other ions. If seawater is gradually evaporated, the first salt to precipitate is $CaCO_3$ (present to the extent of 0.12 g/L), followed by $CaSO_4 \cdot H_2O$ (1.75 g/L), then NaCl (29.7 g/L), $MgSO_4$ (2.48 g/L), $MgCl_2$ (3.32 g/L), NaBr (0.55 g/L), and finally KCl (0.53 g/L). Thus "sea salt" is a mixture of all these salts, which together constitute about 3.5% of the mass of seawater. Due primarily to the operation of the CO_2–bicarbonate–carbonate equilibrium system discussed previously for fresh water, the average pH of surface ocean water is about 8.1. Seawater has a low organic content, its DOC value being about 1 mg/L.

Alkalinity Indices for Natural Waters

The actual concentrations of the cations and anions in a real water sample cannot simply be assumed to be the theoretical values calculated above for calcium, carbonate, and bicarbonate for two reasons:

- the water may not be in equilibrium with either solid calcium carbonate or with atmospheric CO_2; and

- other acids or bases may also be present.

The index devised by analytical chemists to represent the actual concentration in water of the anions that are basic is provided by the **alkalinity** value for the sample. *Alkalinity is a measure of the ability of a water sample to act as a base by reacting with hydrogen ions.* In practical use, the alkalinity of a body of water is a handy measure of the capacity of the water body to neutralize acids and hence to resist acidification when acid rain falls into it. (Alkalinity differs from [OH$^-$], or pOH, in that the latter equals only the hydroxide concentration a water sample has at a particular moment and does not include its ability to generate additional OH$^-$ if acid is added to it.) From an operational viewpoint, alkalinity, more properly termed **total alkalinity,** is the number of moles of H$^+$ required to titrate 1 liter of a water sample to the end point. For a solution containing carbonate and bicarbonate ions, as well as OH$^-$ and H$^+$, by definition

$$\text{(total) alkalinity} = 2\,[CO_3^{2-}] + [HCO_3^{-}] + [OH^-] - [H^+]$$

The factor of 2 appears in front of carbonate ion concentration since in the presence of H$^+$ it is first converted by the ion to bicarbonate ion, which is then converted by a second hydrogen ion to carbonic acid:

$$CO_3^{2-} + H^+ \rightleftharpoons HCO_3^{-}$$
$$HCO_3^{-} + H^+ \rightleftharpoons H_2CO_3$$

Minor contributors to the alkalinity of fresh-water systems can include dissolved ammonia and the anions of *phosphoric, boric,* and *silicic* acids, and H$_2$S, as well as natural organic matter.

The alkalinities of natural waters range from less than 5×10^{-5} M (50 μM) to more than 2×10^{-3} M (2000 μM), compared to the value of about 3×10^{-4} M that corresponds to our estimate for water in contact with atmospheric CO$_2$ (see Problem 13-19). Lakes having alkalinities less than about 200 μM are considered "high" in sensitivity to acid rain, those from 200 to 400 μM are classified as "moderate" in sensitivity, and those with alkalinity greater than 400 μM are considered to have "low" sensitivity. Alkalinity values are sometimes reported as milligrams of CaCO$_3$ equivalent, rather than moles of H$^+$, per liter in a manner similar to that explained below for the hardness concept.

By convention in analytical chemistry, **methyl orange** is used as the indicator in titrations by which total alkalinity is determined. Methyl orange is chosen because it does not change color until the solution is slightly acidic (pH = 4); under such conditions, not only has all the carbonate ion in the sample been transformed to bicarbonate, but virtually all the bicarbonate ion has been transformed to carbonic acid (see Problem 13-18 and Figure 13-6).

Another index encountered in the analysis of natural waters is the **phenolphthalein alkalinity** (also called *carbonate alkalinity*), which is a measure of the concentration of the carbonate ion and of other similarly basic anions.

In order to titrate only CO_3^{2-} and not HCO_3^- as well, the indicator **phe-nolphthalein** or one with similar characteristics is used. Phenolphthalein changes color in the pH range 8 to 9, so it provides a fairly alkaline end point. At such pH values, only a negligible amount of the bicarbonate ion has been converted to carbonic acid, but the majority of CO_3^{2-} has been converted to HCO_3^- (Figure 13-6). Thus,

$$\text{phenolphthalein alkalinity} = [CO_3^{2-}]$$

PROBLEM 13-18

Calculate the value of the ratios $[HCO_3^-]/[CO_3^{2-}]$ and $[H_2CO_3]/[HCO_3^-]$ at pH values of 4 and 8.5 to confirm the statements made above concerning the nature of the species present at the methyl orange and phenolphthalein end points of the titrations. [*Hint: Use the equilibrium constant expressions and K values for reactions (1) and (2).*]

PROBLEM 13-19

Calculate the value expected for the total alkalinity and for the phenolph-thalein alkalinity of a 25°C saturated solution of calcium carbonate in water that is also in equilibrium with atmospheric carbon dioxide. Use the concentrations quoted in the last column of Table 13-4.

PROBLEM 13-20

Calculate the total alkalinity for a sample of river water whose phenolph-thalein alkalinity is known to be 3.0×10^{-5} M, whose pH is 10.0, and whose bicarbonate ion concentration is 1.0×10^{-4} M.

The alkalinity value for a lake is sometimes used by biologists as a measure of its ability to support aquatic plant life, a high value indicating a high potential fertility. The reasons for such a situation are often the following ones. Algae extract the carbon dioxide they need for photosynthesis from bicarbonate ion, which is plentiful in calcareous waters, by a reversal of the CO_2–$CaCO_3$ reaction discussed previously:

$$Ca^{2+} + 2\,HCO_3^-(aq) \longrightarrow CO_2 + CaCO_3(s) + H_2O$$

Indeed, small crystals of calcium carbonate are sometimes observed in lakes where there is active photosynthesis.

$$CO_2 + H_2O + \text{sunlight} \longrightarrow CH_2O \text{ polymer} + \tfrac{1}{2}O_2$$
$$\text{(as algae)}$$

In noncalcareous waters, which have low alkalinity and low calcium content, dissociation of the bicarbonate ion in the water forms not only carbon dioxide but also hydroxide ion:

$$HCO_3^- \rightleftharpoons CO_2 + OH^-$$

The algae readily exploit this CO_2 for their photosynthetic needs, at the cost of allowing a buildup of hydroxide ion to such an extent that the lake water becomes quite basic—with a pH as high as 12.3 in some cases.

The Hardness Index for Natural Waters

As a measure of certain important cations present in samples of natural waters, analytical chemists often use the **hardness index,** which measures the total concentration of the ions Ca^{2+} and Mg^{2+}, the two species that are principally responsible for hardness in water supplies. Chemically, the hardness index is defined in this way:

$$hardness = [Ca^{2+}] + [Mg^{2+}]$$

Experimentally, hardness can be determined by titrating a water sample with **ethylenediaminetetraacetic acid** (EDTA), a substance that forms very strong complexes with metal ions other than those of the alkali metals (see Chapter 15 for details). Traditionally, hardness is expressed not as a molar concentration of ions but as *the mass in milligrams (per liter) of calcium carbonate that contains the same total number of dipositive (2+) ions*. For example, a water sample that contains a total of 0.0010 mole of $Ca^{2+} + Mg^{2+}$ per liter would possess a hardness value of 100 mg of $CaCO_3$, since the molar mass of $CaCO_3$ is 100 g and 0.0010 mole of it weighs 0.1 g, or 100 mg.

Most calcium enters water from either $CaCO_3$ in the form of limestone or from mineral deposits of $CaSO_4$. The source of much of the magnesium is *dolmitic limestone*, $CaMg(CO_3)_2$. Hardness is an important characteristic of natural waters, since calcium and magnesium ions form insoluble salts with the anions present in soaps, thereby forming a scum in wash water. Water is termed "hard" if it contains substantial concentrations of calcium and/or magnesium ions; thus calcareous water is "hard." Some scientists define water as being hard if its hardness index exceeds 150 mg/L.

Many areas possess soils that contain little or no carbonate ion, and thus its dissolution and reaction with CO_2 to produce bicarbonate do not occur. Such "soft" water typically has a pH much closer to 7 than does hard water, since it contains few basic anions. However, there are lakes with little dissolved calcium or magnesium but relatively high concentrations of dissolved **sodium carbonate,** Na_2CO_3; such lakes have a very low degree of hardness but are high in alkalinity.

Interestingly, people who live in hard-water areas are found to have a lower average death rate from ischemic heart disease than do people living in areas with very soft water. Recent research in rural Finland found the risk of heart attacks decreased continuously as the magnesium concentration in the local water supply increased.

PROBLEM 13-21

What is the value of the hardness index in milligrams $CaCO_3/L$ for a 500-mL sample of water that contains 0.0040 g of calcium ion and 0.0012 g of magnesium ion?

PROBLEM 13-22

Calculate the hardness, in milligrams $CaCO_3/L$, of water that is in equilibrium at 25°C with carbon dioxide and calcium carbonate, using results in the last column of Table 13-4. Assume the water is free of magnesium. Is the calculated value greater or less than the median hardness value found for surface waters of 37 mg/L?

Aluminum in Natural Waters

The concentration of **aluminum ion,** Al^{3+}, in natural waters normally is quite small, typically about 10^{-6} M. This low value is the consequence of the fact that in the typical pH range for natural waters (6 to 9), the solubility of the aluminum contained in rocks and soils to which the water is exposed is very small. The solubility of aluminum in water is controlled by the insolubility of **aluminum hydroxide,** $Al(OH)_3$. Given that the K_{sp} of the hydroxide is about 10^{-33} at usual water temperatures, then for the reaction

$$Al(OH)_3 \rightleftharpoons Al^{3+} + 3\,OH^-$$

it follows that

$$[Al^{3+}][OH^-]^3 = 10^{-33}$$

Take, for instance, a sample of water whose pH is 6. Since the hydroxide concentration in such water is 10^{-8} M, it follows that

$$[Al^{3+}] = 10^{-33}/(10^{-8})^3 = 10^{-9}\,M$$

Although this value is very small, for every one-unit decrease of the pH, the concentration of aluminum ion increases by a factor of 1000, so it reaches 10^{-6} M at pH = 5 and 10^{-3} M at a pH of 4. Thus aluminum is much more soluble in highly acidified rivers and lakes than in those where pH values do not fall below 6 or 7. Indeed, Al^{3+} is usually the principal cation in waters whose pH is less than 4.5, exceeding even the concentrations of Ca^{2+} and Mg^{2+}, which are the dominant cations at pH values greater than 4.5.

In the recent past, fears arose that human ingestion of aluminum from drinking water and from the use of aluminum cooking pots was a major cause of Alzheimer's disease; however, the research upon which this conclusion was reached could not be reproduced. Today many neuroscientists do not believe that there is a strong connection between the disease and intake of the metal, since past epidemiological studies on this matter have not been definitive or consistent. However, Canadian and Australian research reported in the mid-1990s indicates that consumption of drinking water with greater than 100 ppb aluminum—not an uncommon level in drinking water purified by aluminum sulfate (see Chapter 14)—can lead to neurological damage such as memory loss and perhaps to a small increase in the incidence of Alzheimer's disease.

It is thought that the principal deleterious effect of acid waters upon fish arises from the solubilization of aluminum from soil and its subsequent existence as a free ion in the acidic water, as discussed in Chapter 4. Unfortunately, the $Al(OH)_3$ then precipitates as a gel on contact with the less acidic gills of the fish, and the gel prevents the normal intake of oxygen from water, thus suffocating the fish.

It is also believed that aluminum mobilization in soils is one of the stresses that acid rain places on trees, which in turn results in the dieback of forests. Soils that contain limestone are usually considered to be buffered against much change in pH, due to the ability of carbonate and bicarbonate ion to neutralize H^+, but over a period of decades, surface soil may gradually lose its carbonate content due to a continual bombardment by acid rain. Thus soils receiving acid rain eventually become acidified. When the pH of the soil drops below about 4.2, aluminum leaching from soil and rocks becomes particularly appreciable. Such acidification has occurred in some regions of central Europe, including Poland, the former Czechoslovakia, and eastern Germany, and the resulting solubilization of aluminum may have contributed to the forest diebacks observed there in the 1980s.

PROBLEM 13-23

What is the concentration, in grams per liter, of dissolved aluminum in water having a pH of 5.5?

PROBLEM 13-24

Calculate the pH value at which the aluminum ion concentration dissolved in water is 0.020 M, assuming that it is controlled by the equilibrium with solid aluminum hydroxide.

Review Questions

1. Write the balanced half-reaction involving O_2 when it oxidizes organic matter in acidic waters.

2. How does temperature affect the solubility of O_2 in water? Explain what is meant by *thermal pollution*.

3. Define *BOD* and *COD*, and explain why their values for the same water sample can differ slightly. Explain why natural waters can have a high BOD.

4. What do the acronyms *TOC* and *DOC* stand for, and how do they differ in terms of what they measure?

5. Write the half-reaction, used in the COD titration, which converts dichromate ion to Cr^{3+} ion, and balance it.

6. Write the balanced chemical reaction by which organic carbon, represented as CH_2O, is disproportionated by bacteria under anaerobic conditions.

7. Draw a labeled diagram classifying the top and bottom layers of a lake in summer as either oxidizing or reducing in character, and show the stable forms of carbon, sulfur, nitrogen, and iron in the two layers.

8. What are some examples of highly reduced and highly oxidized sulfur in environmentally important compounds? Write the balanced reaction by which sulfate can oxidize organic matter.

9. Explain the phenomenon of *acid mine drainage*, writing balanced chemical equations as appropriate. How does Fe^{3+} also act as an oxidizing agent here?

10. What is meant by the *pE* of an aqueous solution? What does a low (negative) pE value imply about the solution? What species determines the pE value in aerated water?

11. What are the acid and the base that dominate the chemistry of most natural water systems and whose interaction produces bicarbonate ion?

12. What is the source of most of the carbonate ion in natural waters? What name is given to waters that are exposed to this source?"

13. Write the approximate net reaction between carbonate ion and water in a system that is *not* also exposed to atmospheric carbon dioxide. Is the resulting water acidic, alkaline, or neutral?

14. Write the approximate net reaction between carbonate ion and water in a system that *is* exposed to atmospheric carbon dioxide. Is the resulting water mildly acidic or mildly alkaline? Explain why the production of bicarbonate ion from carbonate ion does not inhibit its production from carbon dioxide, and vice versa.

15. If two equilibrium reactions are added together, what is the relationship between the equilibrium constants for the individual reactions and that for the overall reaction?

16. What is the natural source of fluoride ion in water? How and why is the fluoride level in drinking water artificially increased to about 1 ppm in many municipalities?

17. Define the *total alkalinity index* and the *phenolphthalein alkalinity index* for water.

18. Define the *hardness index* for water.

19. Which are the most abundant ions in clean, fresh water?

20. Explain why aluminum ion concentrations in acidified waters are much greater than those in neutral water. How does the increased aluminum ion level affect fish and trees?

Green Chemistry Questions

See the discussion of focus areas and the principles of green chemistry in the Introduction before attempting these questions.

1. What takes place during the scouring of cotton, and why is this process necessary for the production of finished cotton fibers?

2. BioPreparation (an enzymatic process) replaced the use of large amounts of sodium hydroxide in the scouring of cotton.

(a) Describe any environmental problems or worker hazards that would be associated with the use of sodium hydroxide solutions in the scouring of cotton.

(b) Would these same environmental problems or worker hazards be eliminated by the use of BioPreparation?

3. The development of BioPreparation by Novozymes won a Presidential Green Chemistry Challenge Award.

(a) Into which of the three focus areas for these awards does this award best fit?

(b) List at least three of the twelve principles of green chemistry that are addressed by the green chemistry developed by Novozymes.

Additional Problems

See Table 13-3 for data.

1. Over a period of several days, estimate your approximate daily water usage in the categories of showering/bathing, clothes washing, toilets, dishwashing, and cooking. (Many flush toilets display volume-per-flush data. Washing machine water volume capacity can be estimated from the dimensions of the washer cavity: $1 \text{ L} = 10 \text{ cm} \times 10 \text{ cm} \times 10 \text{ cm}$. Using a measuring cup, discover how long it takes your shower to deliver 1 liter of water, and adjust the data accordingly for the length of your average shower.)

2. The TOC parameter for water samples is measured by oxidizing the organic material to carbon dioxide and then measuring the amount of this gas evolved from the solution. If a 5.0-L sample of wastewater produced 0.25 mL of carbon dioxide gas, measured at a pressure of 0.96 atm and a temperature of 22°C, calculate the TOC for the sample. Assuming the average composition of the organic matter to be CH_2O, calculate what the chemical oxygen demand for the water sample would be due to its organic content. [The gas constant $R = 0.082 \text{ L atm mol}^{-1} \text{ K}^{-1}$.]

3. (a) Balance the reduction half-reaction that converts SO_4^{2-} to H_2S under acidic conditions.

(b) Deduce the expressions relating pE to pH, the concentration of sulfate ion, and the partial pressure of hydrogen sulfide gas, given that for the half-reaction, $pE^0 = -3.50$ V when the pH is 7.0.

(c) Deduce the partial pressure of hydrogen sulfide when the sulfate ion concentration is 10^{-5} M and the pH is 6.0 for water that is in equilibrium with atmospheric oxygen.

4. Calculate the solubility of lead(II) carbonate, $PbCO_3$ ($K_{sp} = 1.5 \times 10^{-13}$) in water, given that most of the carbonate ion it produces subsequently reacts with water to form bicarbonate ion. Recalculate the solubility assuming that none of the carbonate ion reacts to form bicarbonate. Is your result significantly different from that calculated assuming complete reaction of carbonate with water?

5. The bicarbonate ion, HCO_3^-, can potentially act as an acid or as a base in water. Write the chemical equations for these two processes, and from the information given in this chapter, determine the corresponding acid and base dissociation constants. Given the relative magnitudes of the dissociation constants, decide whether the dominant reaction of bicarbonate in water will be as an acid or as a base. Calculate the pH of an aqueous 0.010 M solution of sodium bicarbonate in water using the dominant reaction alone and assuming that the amounts of carbonate ion and carbonic acid from other sources are negligible in this case.

6. A sample of lake water at 25°C is analyzed and the following parameters are found:

total alkalinity = 6.2×10^{-4} M

phenolphthalein alkalinity = 1.0×10^{-5} M

pH = 7.6

hardness = 30.0 mg/L

$[Mg^{2+}] = 1.0 \times 10^{-4}$ M

Extract all possible single-ion concentrations that you can by combining one or more of these data. Also determine whether or not the water is at equilibrium with respect to the carbonate–bicarbonate system and whether or not it is saturated with calcium carbonate.

7. The O_2 concentration of a water sample can be determined using the so-called Winkler titration method. In it, the oxygen in a small sample of the water is reacted with $MnSO_4$ in a basic solution. The reaction precipitates the manganese as MnO_2, which converts added I^- to I_2. Molecular iodine is then quantitatively determined by titrating it in acidic solution with a standardized solution of sodium thiosulfate, $Na_2S_2O_3$. The set of equations for the reactions is:

$$2\,Mn^{2+} + 4\,OH^- + O_2(aq) \longrightarrow$$
$$2\,MnO_2(s) + 2\,H_2O$$

$$MnO_2(s) + 4\,H^+ + 2\,I^- \longrightarrow Mn^{2+} + I_2 + 2\,H_2O$$

$$I_2(aq) + 2\,S_2O_3{}^{2-} \longrightarrow S_4O_6{}^{2-} + 2\,I^-$$

In determining the BOD of a sample of water, a chemist tested two 10.00-mL samples of the water, one before and one after the five-day incubation period. They required 10.15 and 2.40 mL of a 0.00100 M standard solution of $K_2S_2O_3$. Calculate the BOD, in units of milligrams per liter, of this water sample. On the basis of these results, would you consider this water to be polluted?

Further Readings

1. T. Oki and S. Kanae, "Global Hydrological Cycles and World Water Resources," *Science* 313(2006): 1968.

2. W. Stumm and J. J. Morgan, *Aquatic Chemistry: Chemical Equilibria and Rates in Natural Waters*, 3rd ed. (New York: Wiley-Interscience, 1996).

3. A. Kousa et al., "Calcium: Magnesium Ratio in Local Groundwater and Incidence of Acute Myocardial Infarction Among Males in Rural Finland," *Environmental Health Perspectives* 114 (2006): 730.

4. "What's in That Bottle?" *Consumer Reports* (January 2003): 38.

5. B. Hileman, "Fluoridation of Water," *Chemical and Engineering News* (August 1, 1998): 26.

6. F. M. M. Morel and J. G. Hering, *Principles and Applications of Aquatic Chemistry* (New York: Wiley, 1993).

7. G. Sposito, ed., *The Environmental Chemistry of Aluminum*, 2nd ed. (Boca Raton, FL: Lewis Publishers, 1996).

Websites of Interest

Log on to www.whfreeman.com/envchem4/ and click on Chapter 13.

CHAPTER 14

THE POLLUTION AND PURIFICATION OF WATER

In this chapter, the following introductory chemistry topics are used:

- Acid–base and equilibrium concepts and calculations; pH
- Basic structural organic chemistry (as in the Appendix)
- Oxidation numbers; redox half-reactions
- Catalysis
- Distillation

Background from previous chapters used in this chapter:

- Maximum contaminant levels (Chapter 10)
- BOD (Chapter 13)
- VOCs (Chapter 3)
- Adsorption (Chapter 4)
- Photochemical reactions; UV light (Chapters 1–5)
- Free radicals (Chapter 1)
- BTX hydrocarbons (Chapter 7)
- ppm concentration scale in water (Chapter 10)
- No effects level, NOEL (Chapter 11)

Introduction

The pollution of natural waters by both biological and chemical contaminants is a worldwide problem. There are few populated areas, whether in developed or undeveloped countries, that do not suffer from one form of water pollution or another. In this chapter, we shall survey the various methods—both traditional and innovative—by which water can be purified. We begin by discussing techniques that are used to purify drinking water from relatively uncontaminated sources, and then consider the pollution and remediation of groundwater and of sewage and wastewater. Lastly, we investigate modern advanced techniques whereby polluted air and water can be cleansed.

Water Disinfection

The quality of "raw" (untreated) water, whether drawn from surface water or groundwater, that is intended eventually for drinking varies widely, from almost pristine to highly polluted. Because both the type and quantity of pollutants in raw water vary, the processes used in purification also vary from place to place. The most commonly used procedures are shown in schematic form in Figure 14-1. Before discussing the major topic of disinfection, we shall discuss the various nondisinfection steps that are often taken in the overall purification process.

Aeration of Water

Aeration is commonly used in the improvement of water quality. Municipalities aerate drinking water that is drawn from underground aquifers in order to remove dissolved gases such as the foul-smelling **hydrogen sulfide, H_2S,** and *organosulfur* compounds, as well as volatile organic compounds, some of which may have a detectable odor. Aeration of drinking water also results in reactions that produce CO_2 from the most easily oxidized organic material. If necessary for reasons of odor, taste, or health, most of the remaining organics can be removed by subsequently passing the water over activated carbon, although this process is relatively expensive, so rather few communities use it (see Box 14-1). Another advantage to aeration is that the increased oxygen content of water oxidizes water-soluble Fe^{2+} to Fe^{3+}, which then forms insoluble hydroxides (and related species) that can be removed as solids.

$$Fe^{3+} + 3\,OH^- \longrightarrow Fe(OH)_3(s)$$

(Recall that ions listed in equations without a state specified are assumed to be in aqueous solution.)

After aeration, colloidal particles in the water are removed. If the water is *excessively* hard, calcium and magnesium are removed from it before the final stages of disinfection and the addition of fluoride. All these procedures are described below (see Figure 14-1).

Removal of Calcium and Magnesium

If the water comes from wells in areas having limestone bedrock, it will contain significant levels of Ca^{2+} and Mg^{2+} ions, which are usually removed during processing since these ions can interfere with soaps and detergents used by consumers for washing. Calcium can be removed from water by addition of phosphate ion in a process analogous to that discussed later for phosphate removal; here, however, phosphate is *added* in order to precipitate the

BOX 14-1 | Activated Carbon

Activated carbon (activated charcoal) is a very useful solid for purifying water of small organic molecules present in low concentrations. The ability of this material to remove contaminants from water and to improve its taste, color, and odor has been known for a long time; indeed, the ancient Egyptians used charcoal-lined vessels to store water for drinking purposes.

Activated carbon is produced by anaerobically charring a high-carbon-content material such as peat, wood, or lignite (a soft brown coal) at temperatures below 600°C, followed by a partial oxidation process using carbon dioxide or steam at a slightly higher temperature.

The removal of contaminants by activated carbon is a physical adsorption process and therefore is reversible if sufficient energy is applied. The characteristic that makes activated carbon such an excellent adsorber is its huge surface area, about 1400 m^2/g. This surface is internal to the individual carbon particles, so that crushing the material neither

increases nor decreases the area. The internal structure of the solid involves series of channels (pores) of progressively decreasing size that are produced by the charring and partial oxidation processes. The internal sites where adsorption occurs are large enough only for small molecules, including chlorinated solvents. At the typical ppm concentrations found for organic contaminants in water, each gram of activated carbon can adsorb a few percent of its mass in contaminants such as chloroform and the dichloroethenes as well as much higher masses of TCE, PCE, and pesticides such as dieldrin, heptachlor, and DDT.

Once a sample of activated carbon has reached near-saturation in terms of adsorbed organics, three alternatives are available. It can be simply disposed of in a landfill, it can be incinerated to destroy it and the adsorbed contaminants, or it can be heated to rejuvenate the surface by driving off the organic pollutants, which can then be incinerated or catalytically oxidized.

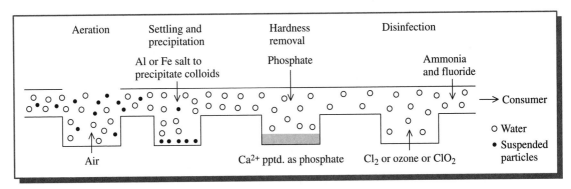

FIGURE 14-1 The common stages of purification of drinking water.

calcium ion. More commonly, calcium ion is removed by precipitation and filtering of the insoluble salt **calcium carbonate,** $CaCO_3$. The carbonate ion is either added as **sodium carbonate,** Na_2CO_3, or if sufficient HCO_3^- is naturally present in the water, hydroxide ion, OH^-, is added in order to convert dissolved bicarbonate ion to carbonate:

$$OH^- + HCO_3^- \longrightarrow CO_3^{2-} + H_2O$$

$$Ca^{2+} + CO_3^{2-} \rightleftharpoons CaCO_3(s)$$

Magnesium ion precipitates as **magnesium hydroxide,** $Mg(OH)_2$, when the water is made sufficiently alkaline, i.e., when the OH^- ion content is increased. After removal by filtration of the solid $CaCO_3$ and $Mg(OH)_2$, the pH of the water is readjusted to near-neutrality by bubbling carbon dioxide into it.

PROBLEM 14-1

Ironically, calcium ion is often removed from water by adding hydroxide ion in the form of $Ca(OH)_2$. Deduce a balanced chemical equation for the reaction of calcium hydroxide with dissolved calcium bicarbonate, $Ca(HCO_3)_2$, to produce insoluble calcium carbonate. What molar ratio of $Ca(OH)_2$ to dissolved calcium should be added to ensure that almost all the calcium is precipitated?

Disinfection to Prevent Illness

In terms of causing immediate sickness and even death, biological contaminants of water are almost always much more important than chemical ones. For that reason, we begin our discussion of the purification of water by extensively discussing its disinfection, i.e., the elimination of microorganisms that can cause illness.

Many of the microorganisms in raw water are present as a result of contamination by human and animal feces. The microorganisms are principally

- **bacteria,** including those of the *Salmonella* genus, one species of which causes typhoid. In this category is also *Escherichia coli O157:H7*, whose transmission in water has caused a number of deaths in recent years, including those from an outbreak in Walkerton, Ontario, in 2000;

- **viruses,** including polio viruses, the hepatitus-A virus, and the Norwalk virus; and

- **protozoans** (single-celled animals), including *Cryptosporidium* and *Giardia lamblia.*

Because many microorganisms of these three types are pathogenic, causing mild to serious and sometimes fatal illnesses, they must be largely removed from water before it is suitable for drinking.

Notwithstanding well-known techniques for water disinfection, many of which have been used extensively for more than a century in developed countries, there are still about 1 billion people in the world who do not yet have access to safe drinking water. According to the World Health Organization, about 4500 children die *daily* from the consequences of polluted water and inadequate sanitation.

Filtering of Water

In addition to dissolved chemicals, the raw water that is obtained from rivers, lakes, or streams contains a multitude of tiny particles, some of which consist of or contain microorganisms. Many of the small, suspended particles consist of clay, resulting from the erosion of soil and rock, whether by natural forces or due to plowing of land for agriculture, mining, or commercial or housing development. The suspended particles increase water's turbidity and thereby reduce the ability of light to penetrate deeply enough to support photosynthesis.

The larger of the particles suspended in water are often removed by simply filtering it. Indeed, the filtration of water by passing it through a bed of sand is the oldest form of water purification known, dating back to ancient times. The sand retains suspended solids of all types, including microorganisms, down to about 10 μm in size.

Recently it was realized that forcing raw water through filters having especially small openings can be used instead of chemicals or light to disinfect water of some viruses and bacteria, and even some dissolved chemicals, by just removing them.

Removal of Colloidal Particles by Precipitation

Most municipalities allow raw water to settle, since this permits large particles to settle out or to be readily separated. However, much of the insoluble matter originating from rocks and soil, and from the disintegration and decomposition of water-based plants and animals, will not precipitate spontaneously since it is suspended in water in the form of **colloidal particles.** These are particles that have diameters ranging from 0.001 to 1 μm and consist of *groups* of molecules or ions that are weakly bound together. These groups dissolve as a unit, rather than breaking up and dissolving as individual ions or molecules. In many cases, the individual units within a colloidal particle are spatially organized such that the surface of the particles contains ionic groups. The ionic charges on the surface of one particle repel those of like charge on neighboring particles, preventing their aggregation and subsequent precipitation.

Colloidal particles must be removed from drinking water for both aesthetic and health and safety reasons. To capture the colloidal particles, a small amount of either **iron(III) sulfate,** $Fe_2(SO_4)_3$, or **aluminum sulfate,**

$Al_2(SO_4)_3$ ("alum"), is deliberately dissolved in the water. By then making the water neutral or alkaline in pH (7 and up), both the Fe^{3+} and Al^{3+} ions produced from the salts form gelatinous hydroxides that physically incorporate the colloidal particles and form a removable precipitate. The water is greatly clarified once this precipitate has been removed. Commonly, after the removal of the colloidal particles, the water is filtered through sand and/or some other granular material.

Although the idealized formulas of the precipitates are $Fe(OH)_3$ and $Al(OH)_3$, the actual situation is much more complex. For example, aluminum actually forms a polymeric cation, $Al_{13}O_4(OH)_{24}^{7+}$, which produces a loose network structure that is held together by hydrogen bonds. This network entraps the colloidal particles and forms the precipitate. Only if the pH rises to quite a high value does the aluminum in solution form the expected hydroxide $Al(OH)_3$. Since the concentration of aluminum sulfate added to the water is only about 10 μm/L, very little residual aluminum ion is left in the treated water.

PROBLEM 14-2

Calculate the approximate number of atoms contained in colloidal particles of (a) 1-μm and (b) 0.01-μm diameters, assuming that their densities are similar to that of water and that the atomic mass of the atoms averages 10 g/mol.

Disinfection of Water by Membrane Technology

Water can be purified of most contaminant ions, molecules, and small particles, including viruses and bacteria, by passing it through a membrane in which the individual holes, called *pores*, are of uniform and microscopic size. The range of sizes of the various contaminants in raw water are summarized in Figure 14-2. Clearly, for a technique to be effective in providing a barrier, the pore size of the membrane must be smaller than the contaminant size.

In the processes of **microfiltration** and **ultrafiltration,** a membrane or some other analogous barrier containing pores of 0.002- to 10-μm diameter (2–10,000 nm) is employed to remove larger constituents from water. The water can be forced through the barrier by pressure or can be drawn through it by suction, leaving behind the larger impurities. In one modern version of this technology, the barrier is composed of thousands of strands of plastic tubing having walls that are pierced with thousands of tiny pores of similar size.

Some bacteria and colloid particles are as small as 0.1 μm and so can pass through conventional filters and even some microfilters (Figure 14-2). Viruses can be as small as 0.01 μm and therefore require at least the ultrafiltration level to eliminate them. However, filtration using membranes can be

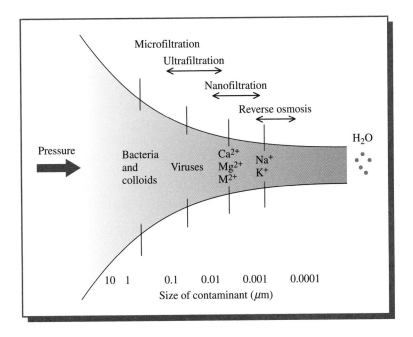

FIGURE 14-2 Filtration of contaminants by various methods.

used to disinfect water if a sufficiently small pore size is used and if the water is later irradiated with ultraviolet light to eliminate any microbes that have passed through the filtration stage.

Neither microfiltering nor ultrafiltering removes dissolved ions or small organic molecules. Generally speaking, before water is treated using membranes with even smaller pores (see below), it must be pretreated to remove the larger particles—especially colloids—which would otherwise foul the finer membrane by leaving deposits.

Membrane systems have been developed recently that purify water of virtually all contaminants by **nanofiltration.** Water is pumped under pressure through fine membranes that have pores only about 1 nm wide, which therefore remove not only most bacteria and viruses, but also any larger organic molecules that would nourish the regrowth of bacteria. These **nanofilters** still allow water molecules to pass through the filter, since the molecules are only a few tenths of a nanometer in size. Unlike ultrafiltration, nanofiltration can be used to soften water, since hydrated divalent ions such as Ca^{2+} and Mg^{2+} are larger than the pores and so do not pass through. Hydrated monovalent ions such as sodium and chloride also pass through some nanofilters, but not through ones with subnanometer pore sizes. As a consequence, some nanofilter membrane systems can be used to desalinate seawater and to help purify wastewater, as discussed later in this chapter.

Reverse Osmosis

The ultimate in membrane filtration occurs in the widely used technique called **reverse osmosis,** sometimes called *hyperfiltration*. Here, water is forced under high pressure to pass through the pores in a *semipermeable membrane*, composed of an organic polymeric material such as cellulose acetate or triacetate or a polyamide. Since only water (and other molecules of its small size) can pass efficiently through the pores, the liquid on the other side of the membrane is purified water. The solution on the impact side of the membrane becomes more and more concentrated in contaminants as time goes on and is discarded. The procedure is called *reverse* osmosis because, by use of pressure, the natural phenomenon of osmosis—by which pure water would spontaneously migrate through the membrane *into* solution, thereby diluting it—is reversed.

Particles, molecules (including small organic molecules), and ions down to less than 1 nm (0.001 μm) in size, or about 150 g/mol in mass, are removed by reverse osmosis. It is particularly useful for removing alkali and alkaline earth metal ions, as well as salts of heavy metals. Thus it is employed in hospitals and renal units to produce water that is particularly free of ions. Reverse osmosis is used on large scale for the *desalination*, i.e., the removal of salts, from seawater and brackish water, a topic considered in Box 14-2.

BOX 14-2 | The Desalination of Salty Water

Desalination is the production of fresh water from salty water, often seawater, by the removal of its ions. There are more than 15,000 large-scale desalination plants in operation, located in more than 125 countries. Reverse osmosis is widely used in some areas of the world, such as the Middle East, to generate drinking water from salt water.

The other main commercial desalination process is the *thermal distillation*—evaporation—of seawater or brackish water. Desalination of seawater by evaporation is a technique that goes back to ancient times; it is especially suited even today for seawater that contains particularly high levels of dissolved salts and suspended solids in areas such as the Persian Gulf. The evaporation method is even more energy-intensive than is reverse osmosis. Modern, large-scale thermal distillation plants use energy to raise salty water to the boiling point, then reduce the air pressure above the liquid to create a partial vacuum into which the liquid readily "flash" evaporates, leaving the salt behind in the remaining liquid. The vapor is removed and condensed as desalted water. Thermal distillation plants are often incorporated within electricity-generating plants to use the low-grade waste steam from the latter as their energy source.

Desalination of water is also sometimes accomplished using the technique of electrodialysis, which is described later in this chapter.

Water destined for drinking purposes is commonly pretreated, e.g., by filtering it through sand and gravel, and passing it over activated carbon to remove the larger particles such as bacteria, etc., and treating it with chlorine, before subjecting it to reverse osmosis in order to minimize fouling and degradation of the membrane.

Because of the high pressures needed to force water through the small pores in the membrane, reverse osmosis is an energy-intensive process. A pressure of about 2 atm is sufficient for portable and domestic units, but a greater force must be applied to brackish or salty water. However, advances in the engineering of large-scale desalination plants have markedly reduced energy consumption by redirecting pressure from waste brine to low-pressure incoming water.

Reverse osmosis tends to be wasteful of water, since so much of it—a third to a half—is discarded. Also, the accumulated discharges of brine— sometimes called *concentrate*—from desalination processes of any kind can cause cumulative environmental problems, such as harming fish populations, in the immediate area of the seacoast into which it is deposited if it is not first treated. In some locations, the brine is injected into an underground saltwater aquifer. In others, the brine is left to evaporate in large outdoor pools, and the salts disposed of later.

Some domestic consumers of drinking water have installed small under-the-sink reverse osmosis units to further purify their water by removing unwanted contaminants such as heavy metal cations (e.g., lead), hard water cations (calcium and magnesium), anions (e.g., nitrate and fluoride), and organic molecules from water obtained from domestic supplies. Small reverse osmosis units are also used in medical facilities for producing water that is particularly ion-free.

Some bottled water has been purified and deionized by reverse osmosis, but small amounts of salts are reintroduced into it before it is sold to consumers. Drinking large amounts of deionized water is not healthy, since the ion balance in the body can be upset as a consequence.

Disinfection by Ultraviolet Irradiation

Ultraviolet light can also be used to disinfect and purify water. Powerful lamps containing mercury vapor whose excited atoms emit UV-C light (see Chapter 1) centered at a wavelength of 254 nm are immersed in the water flow. About 10 seconds of irradiation are usually sufficient to eliminate the toxic microorganisms, including *Cryptosporidium*, which is resistant to treatment by some other methods. The germicidal action of the light disrupts the DNA in microorganisms, preventing their subsequent replication and thereby inactivating the cells. At the molecular level, absorption of UV-C light results in the formation of new covalent bonds between nearby thymine units on the same strand of DNA. If sufficient such thymine dimers are formed, the DNA molecule becomes so distorted that subsequent replication of the organism is prevented.

The use of ultraviolet light to purify water is complicated by the presence of dissolved iron and humic substances, both of which absorb the UV light and thus reduce the amount available for disinfection. Small solid particles suspended in the water also inhibit the action of the UV light since they can shade or absorb bacteria and also scatter or absorb the light. An advantage of UV disinfection technology is that small units can be employed to serve small population bases, whether in the developed or developing world, so the continuous monitoring activity of chemical systems is avoided. As discussed later, UV light can also be used to purify water of dissolved organic compounds, but by a different mechanism.

Disinfection by Chemical Methods: Ozone and Chlorine Dioxide

To rid drinking water of harmful bacteria and viruses, especially those arising from human and animal fecal matter, by use of a chemical agent requires an oxidizing agent more powerful than O_2. In some localities, particularly in France and other parts of western Europe but also in some North American cities—Montreal and Los Angeles are examples—**ozone** is used for this purpose. Since O_3 cannot be stored or shipped because of its very short lifetime, it must be generated on-site by a relatively expensive process involving electrical discharge (20,000 V) in dry air. The resulting ozone-laden air is bubbled through the raw water; about 10 minutes of contact is usually sufficient for disinfection. Since the lifetime of ozone molecules is short, there is no residual protection in the purified water to ensure that it will not be subject to future contamination. Some pollutants in water react with the ozone itself and others with free radicals such as hydroxyl and hydroperoxy (Chapters 1–5) that are produced when ozone reacts with water.

Unfortunately, the reaction of ozone with bromine in water leads to the formation of oxygen-containing organic compounds, particularly those containing the *carbonyl group*, $\diagup C{=}O$, such as formaldehyde and other low-molecular-weight aldehydes and various other compounds, some of which are toxic. In addition, ozone reacts with bromide ion, Br^-, present in the water to produce the **bromate ion, BrO_3^-**, a carcinogen in test animals and probably also in humans. The reaction of ozone with bromide, a natural constituent of water that is often present at ppm concentrations, occurs in several steps; the overall reaction is

$$Br^- + 3\,O_3 \longrightarrow BrO_3^- + 3\,O_2$$

The bromate ion produced by ozonation may subsequently react with organic matter in the water to produce toxic organobromine compounds, though experiments have shown that the only brominated product under water treatment conditions is **dibromoacetonitrile,** $CHBr_2CN$, produced by the reaction of bromate ion with acetonitrile. The MCL (maximum contaminant level) of bromate ion in drinking water is set at 10 ppb (0.010 ppm) by the U.S. EPA. Substances such as bromate ion that are produced during water purification are called **disinfection by-products,** or DBPs. *All* known chemical methods of disinfecting water produce DBPs of one type or another.

Similarly, **chlorine dioxide** gas, ClO_2, is used in more than 300 North American and several thousand European communities to disinfect water. The ClO_2 molecules, themselves free radicals, operate to oxidize organic molecules by extracting electrons from them:

$$ClO_2 + 4\,H^+ + 5\,e^- \longrightarrow Cl^- + 2\,H_2O$$

The organic cations created in the accompanying oxidation half-reaction subsequently react with oxygen and eventually become more fully oxidized.

Since chlorine dioxide is *not* a chlorinating agent—it does not generally introduce chlorine atoms into the substances with which it reacts—and since it oxidizes the dissolved organic matter, much smaller amounts of toxic organic chemical by-products are formed than if molecular chlorine were used (see below).

As is the case with ozone, ClO_2 cannot be stored since it is explosive in the high concentrations that its practical use calls for, so it must be generated on-site. This is accomplished by oxidizing its reduced form, the **chlorite ion,** ClO_2^-, from the salt **sodium chlorite,** $NaClO_2$:

$$ClO_2^- \longrightarrow ClO_2 + e^-$$

Some of the chlorine dioxide in these processes is converted to **chlorate ions,** ClO_3^-. The presence of chlorite and chlorate ions as residuals in the final water has raised health concerns due to their potential toxicity. The U.S. EPA has set an MCL of 1.0 ppm for chlorite ion, and a MRDL (*maximum residual disinfectant level*) of 0.8 ppm for chlorine dioxide, in drinking water.

Disinfection by Chlorination: History

The most common water purification agent used in North America is **hypochlorous acid,** HOCl. About half the U.S. population uses surface water, and one-quarter of the population uses groundwater, that is disinfected by HOCl. This neutral, covalent compound kills microorganisms, as it readily passes through their cell membranes. In addition to being effective, disinfection by **chlorination** is relatively inexpensive. Incorporating a small excess of the chemical in the treated water provides it with residual disinfection power during its subsequent storage and transmission to the consumer. Chlorination is more common than ozonation in North America because generally the raw water is less polluted. Chlorination of public water supplies in the United States, Canada, and Great Britain began in the early years of the twentieth century. For the previous 50 years, chlorination had been practiced on an emergency basis during epidemics caused by water-borne pathogens.

Disinfection by Chlorination: Production of Hypochlorous Acid

Like ozone, HOCl is not stable in concentrated form and so cannot be stored. For large-scale installations, e.g., municipal water treatment plants, it is generated by dissolving **molecular chlorine** gas, Cl_2, in water. At moderate pH values, the equilibrium in the reaction of chlorine with water lies far to the right and is achieved in a few seconds:

$$Cl_2(g) + H_2O(aq) \rightleftharpoons HOCl(aq) + H^+ + Cl^-$$

Thus a dilute aqueous solution of chlorine in water contains very little aqueous Cl_2 itself. If the pH of the reaction water were allowed to become too high, the result would be the ionization of the weak acid HOCl to the **hypochlorite ion,** OCl^-, which is less able to penetrate bacteria on account of its electrical charge. Once chlorination is complete, the pH of the water is adjusted upward, if necessary, by the addition of *lime*, CaO.

In small-scale applications of chlorination, as in swimming pools, the handling of cylinders of Cl_2 is inconvenient and dangerous. The chlorine gas can be produced as needed on the spot by the electrolysis of salty water. More commonly, hypochlorous acid instead is generated from the salt **calcium hypochlorite,** $Ca(OCl)_2$, or is supplied as an aqueous solution of **sodium hypochlorite,** NaOCl. In water, an acid–base reaction occurs to convert most of the OCl^- in these substances to HOCl:

$$OCl^- + H_2O \rightleftharpoons HOCl + OH^-$$

Close control of the pH in an environment like a swimming pool is necessary to avoid the shift to the left side in the position of equilibrium for this reaction that occurs if a very alkaline condition is permitted to prevail. On the other hand, corrosion of pool construction materials can occur in acidic water, so the pH is usually maintained above 7 to prevent such deterioration. Maintenance of an alkaline pH also prevents the conversion of dissolved **ammonia,** NH_3, to the **chloramines** NH_2Cl, $NHCl_2$, and especially NCl_3, which is a powerful eye irritant:

$$NH_3 + 3\ HOCl \longrightarrow NCl_3 + 3\ H_2O$$

Significant respiratory and eye irritation problems from exposure to chloramines in the air around indoor swimming pools has been reported when appropriate ventilation is unavailable.

It is desirable to adjust the equilibrium point in the $OCl^- \longrightarrow HOCl$ reaction to favor the predominance of the disinfectant molecular species, HOCl. Since the equilibrium between HOCl and OCl^- shifts rapidly in favor of the ion between pH values of 7 and 9, however, the acidity level must be meticulously controlled. Swimming pool acidity can be adjusted by the addition of acid (in the form of *sodium bisulfate*, $NaHSO_4$, which contains the acid HSO_4^-) or a base (sodium carbonate, Na_2CO_3) or a buffer (sodium bicarbonate, $NaHCO_3$, which contains the amphoteric anion HCO_3^-). Chlorine must be constantly replenished in outdoor pools since UV-B and the short-wavelength components of UV-A light in sunshine are absorbed by and decompose the hypochlorite ion, thereby affecting the equilibrium in the $OCl^- \longrightarrow HOCl$ process toward the ion:

$$2\ ClO^- \xrightarrow{\ UV\ } 2\ Cl^- + O_2$$

Hypochlorous acid can also be generated by the reaction with water of the chlorine-containing compound **isocyanuric acid,** $C_3N_3O_3H_3$:

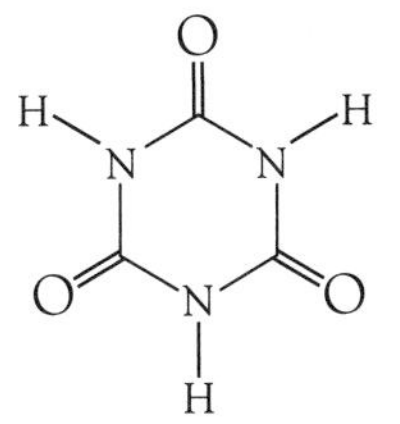

Either the trichloro derivative, in which each hydrogen is replaced by Cl to give $C_3N_3O_3Cl_3$, or the sodium dichloro derivative, $C_3N_3O_3Cl_2Na$, is used. In either case, the OH group from water combines with the chlorine to produce HOCl and the hydrogen of H_2O becomes bonded to the nitrogen, giving **isocyanuric acid,** $C_3N_3O_3H_3$:

$$C_3N_3O_3Cl_3 + 3\ H_2O \rightleftharpoons C_3N_3O_3H_3 + 3\ HOCl$$

Since this process is an equilibrium, not all the compound is immediately converted to hypochlorous acid. As HOCl is used up, both by its use as a disinfectant and through dissociation in sunlight of its ionic form, the equilibrium shifts to the right and more HOCl is produced. None of the various forms of isocyanuric acid absorb UV light, so its chlorine is "protected" against decomposition by sunlight. Since the bulk chlorinated forms of isocyanuric acid are expensive, it is common to supply hypochlorite from a cheaper source and to add isocyanuric acid as a stabilizer, temporarily reversing the above reaction to "store" the chlorine until it is needed.

Disinfection by Chlorine: By-Products and Their Health Effects

An important drawback to the use of chlorination in disinfecting water is the concomitant production of chlorinated organic substances, some of which are toxic, since HOCl is not only an oxidizing agent but also a chlorinating agent. Examples of these important by-products are the group of *halogenated acetic acids* (haloacetic acids), such as $CH_2Cl-COOH$, which the U.S. EPA restricts to 60 ppb as an MCL annual average for drinking water, and *haloacetonitriles*, such as CH_2Cl-CN. **Dichloroacetic acid,** $CHCl_2-COOH$, is a more potent carcinogen than is chloroform.

If the water to be disinfected contains **phenol,** C_6H_5OH, or a derivative thereof, chlorine readily substitutes for the hydrogen atoms on the ring to give rise to chlorinated phenols: These compounds have an offensive odor and taste and are toxic. Some communities switch from chlorine to chlorine dioxide when their supply of raw water is temporarily contaminated with phenols to avoid the formation of chlorinated phenols.

A more general problem with chlorination of water lies in the production of **trihalomethanes,** THMs. Their general formula is CHX_3, where the three X atoms can be chlorine or bromine or a combination of the two. The THM of principal concern is **chloroform,** $CHCl_3$, which is produced when hypochlorous acid reacts with organic matter dissolved in the water (see Box 14-3). Chloroform is a suspected liver carcinogen in humans, and it may also give rise to negative reproductive and developmental effects. Its

---

BOX 14-3	The Mechanism of Chloroform Production in Drinking Water

Humic acids, with which HOCl reacts to form chloroform, are water-soluble, nonbiodegradable components of decayed plant matter. Of particular importance are humic acids that contain 1,3-dihydroxybenzene rings. The carbon atom (#2) located between those carrying the —OH groups is readily chlorinated by HOCl, as in this elementary case:

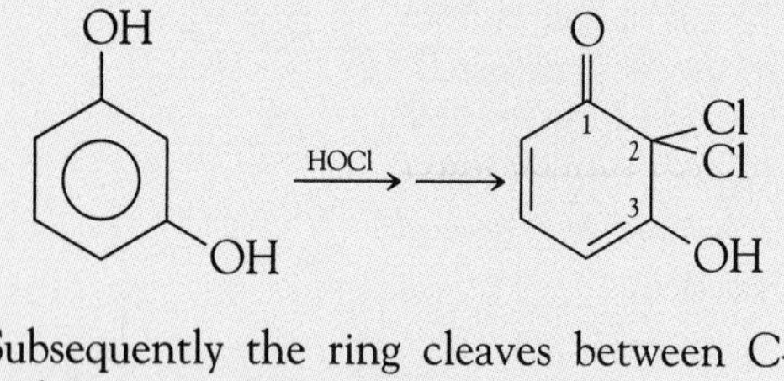

Subsequently the ring cleaves between C-2 and C-3 to yield a chain:

$$R-\underset{\underset{O}{\|}}{C}-CHCl_2$$

In the presence of the HOCl, the terminal carbon becomes trichlorinated, and the —CCl$_3$ group is readily displaced by the OH$^-$ in water to yield chloroform:

$$R-\underset{\underset{O}{\|}}{C}-CHCl_2 \xrightarrow{\text{HOCl}} R-\underset{\underset{O}{\|}}{C}-CCl_3$$

$$\xrightarrow[H^+]{OH^-} R-\underset{\underset{O}{\|}}{C}-OH + CHCl_3$$

Analogous sequences of reactions produce bromoform, $CHBr_3$, and mixed chlorine–bromine trihalomethanes from the action on humic materials of hypobromous acid, HOBr, which is formed when bromide ion in water displaces chlorine from HOCl:

$$HOCl + Br^- \rightleftharpoons HOBr + Cl^-$$

presence, even at very low levels of approximately 30 ppb, raises the specter that chlorinated drinking water may pose a health hazard, though one that pales by comparison with the benefits that it confers in the elimination of fatal waterborne diseases. The annual average limit of total THMs in drinking water in the United States and the European Union (EU) has been reduced to 80 ppb. The previous limit of 100 ppb is still used in Canada. In fact, these 80–100-ppb limits are set not only to regulate the THM chemicals themselves but also as an indicator that the production of *other* chlorinated organic DBPs (see below) is not excessive.

The U.S. EPA has set a **maximum contaminant level goal,** MCLG, of 70 ppb for THMs in drinking water. An MCLG is the *maximum level at which the contaminant is believed to be safe*, allowing for adequate margins of safety, but unlike the MCL (Chapter 10), it is not an enforceable standard. A nonzero goal is considered by some scientists and policymakers to be appropriate to substances, like chloroform, that are believed to operate indirectly as carcinogens. They do not damage DNA directly but cause tissue damage that leads to rapid cell proliferation, which in turn increases the likelihood that cancer will form in the damaged tissue. A threshold below which no effects are likely to be observed is expected for carcinogens that operate in this manner.

The level of trihalomethanes formed in water depends sharply on the organic content of the raw water since the THMs are formed from the reaction of organics with HOCl (see Box 14-3). THM levels can reach 250 ppb in areas of Scotland and Northern Ireland that have peat moorlands. Water exposed to bogs in Newfoundland, Canada, has generated THM levels in excess of 400 ppb. As of the early 1990s, about 1% of the larger U.S. drinking water utilities that used surface waters, but none that used groundwater, had average THM levels exceeding 100 ppb. The THM content of chlorinated water could be decreased by using activated carbon either to remove dissolved organic compounds before the water is chlorinated or to remove THMs and other chlorinated organics after the process, although THMs are not very efficiently adsorbed by the carbon and it is an expensive process.

An analysis of epidemiological studies relating the chlorination of water to cancer rates in various communities in the United States led to the conclusion that the risk of bladder cancer in humans increased by 21%, and that of rectal cancer by 38%, for Americans who drank chlorinated surface water in the past. A similar study in Ontario found even higher bladder cancer risk factors for people who drank water for 35 years or more that had THM levels greater than 50 ppb, and for colon cancer when the concentration exceeded 75 ppb, but found no correlations of THMs with rectal cancer rates. A recent study found no increased risk for pancreatic cancer from lifetime exposure to the by-products of chlorinated water.

Given that slightly more than half the population of the United States drinks surface water, one effect of chlorination is to have increased bladder cancer incidence by about 4200 cases per year and rectal cancer incidence by about 6500 cases annually. Because of these risks, some communities are considering a switch, or have already switched, to water disinfection by ozone or chlorine dioxide, since these agents produce little or no chloroform. The extent of chlorination has already been reduced in most American communities relative to the levels that led to these statistics.

Several other mutagenic chlorinated organic DBPs formed during chlorination have been detected in water, in addition to chloroform. It is not clear whether the main carcinogens in the chlorinated drinking water are the THMs themselves or some nonvolatile, higher-molecular-weight mutagenic by-product present at still lower concentrations but whose concentration would presumably be proportional to THM. The same risks do not usually apply to chlorinated well water, since its organochlorine content is much less (only 0.8 ppb on average, versus 51 ppb for surface water) because it contains much smaller amounts of organic matter that could become chlorinated. Exposure to chloroform by dermal contact and inhalation of the gases deabsorbed from the hot water during showers, baths, and hand-washing of dishes contribute more to one's intake of THMs, as measured by the blood levels of these compounds, than does drinking the water itself. Swimming in pools in which the water is chlorinated for disinfection also contributes significantly to dermal exposure. Recently it has been found that use of the disinfectant *triclosan*, a phenol-based compound present in some hand soaps, in chlorinated water can produce additional chloroform and chlorinated phenols, to which the user is then exposed.

Recently, public health officials have expressed concern about the possible link between THMs and adverse human reproductive outcomes, including first-term miscarriages, stillbirths, impaired fetal growth, and certain birth defects. Even though the existing research in this area is not yet definitive, some officials suggest that women drink bottled water rather than chlorinated tap water during their first three months of pregnancy.

Disinfection by Chlorine: Advantages over Other Methods

Notwithstanding the preceding discussion of chlorination by-products, it is important to point out that the disinfection of water by chlorine is extremely important in protecting public health and saves many more lives—by a very wide margin—than are affected negatively. For example, both typhoid and cholera were widespread in both Europe and North America a century ago but have been almost completely eradicated in the developed world, thanks to chlorination and the other disinfection methods for drinking water and to improved sanitation in general. The same is not true in many developing countries; e.g., there were more than half a million cases of cholera in Peru in the early 1990s. Overall, about 20 million people, most of them infants, die from waterborne diseases annually worldwide in underdeveloped countries, where water purification is often erratic or even nonexistent. Under no circumstances should effective disinfection of water be abandoned because of concern for the by-products of chlorination!

An advantage chlorination has over disinfection by chlorine dioxide or ozone or by UV is that some chlorine remains dissolved in water once it has left the purification plant, so that the water is protected from subsequent bacterial contamination before it is consumed. Indeed, some chlorine is usually added to water purified by the other methods to provide this protection. There is very little danger of significant chloroform production in the purified water since its organic content has been virtually eliminated before the chlorine is introduced. If the chlorine level in water purified by chlorination is too high, it can be lowered by the addition of sulfur dioxide.

The residual chlorine in water often exists in the form of the chloramines NH_2Cl, $NHCl_2$, and NCl_3, which are produced from reaction with dissolved ammonia gas, NH_3. Although not as fast as HOCl in disinfecting water, the mono- and dichloroamines especially are good disinfectants. The mixture of chloramines, called **combined chlorine,** is longer-lived in water than is hypochlorous acid and thus provides longer residual protection. Indeed, ammonia is often added to purified drinking water in order to convert the residual chlorine to the combined form (Figure 14-1). Chloramines are sometimes used, rather than chlorine or ozone or chlorine dioxide, as the main disinfectant in the purification of drinking water. They have the advantage over chlorine of producing little (though not zero) amounts of THMs and haloacetic acids. The U.S. EPA has set MRDLs of 4.0 ppm for both chlorine and chloramine in drinking water.

Bromine rather than chlorine is sometimes used as the disinfectant in swimming pools. The main disinfecting agent in bromination is **hypobromous acid,** HOBr, in analogy with the role of hypochlorous acid in chlorination. HOBr reacts more rapidly with dissolved ammonia than does HOCl, producing mainly NH_2Br, which is also a good disinfectant.

To disinfect water for drinking purposes, hikers either boil raw water or treat it chemically with either chlorine, in the form of bleach, which provides HOCl, or iodine, as **elemental I_2** or **hypoiodous acid,** HIO. Concerns have been expressed about chronic health problems such as thyroid dysfunction associated with long-term usage of iodine, however. Treating the water with elemental iodine tends to make it unpalatable as well.

PROBLEM 14-3

Assuming that the nitrogen atom in monochloramine, NH_2Cl, has an oxidation number of -3, calculate that of the chlorine atom. Using the principle that unlike charges attract, predict whether it will be the hydrogen ion or the hydroxide ion from dissociated water molecules that will extract the Cl from NH_2Cl; from your result, predict the products of the decomposition reaction of chloramine in water.

A drinking-water quality issue of current concern involves the pathogenic protozoa called *Cryptosporidium*, which was responsible for the death of 100 people and for about 400,000 cases of watery diarrhea in Milwaukee in 1993. Less serious *Cryptosporidium* outbreaks occurred in Oxford, England, in 1989, and in Saskatchewan, Canada, in 2001. This deadly parasite is resistant to standard methods of disinfection, such as chlorination at normal levels, and is so small (about 5-μm diameter) that it easily passes through the standard filters used to separate sediments. Several possible solutions have been advanced, including ozonation or the use of ultrafiltration or UV irradiation or the application of monochloramine following chlorination. A longer-than-usual exposure of water containing *Cryptosporidium* is necessary with ozonation, since the activation energy for inactivation of protozoa by ozone is about twice as large as that for bacteria (80 vs. about 40 kJ/mol).

Another protozoa, *Giardia lamblia*, also causes many instances of waterborne disease. Like *Cryptosporidium*, it is also somewhat resistant to chlorination, but since it is larger (about 10-μm diameter), it is more easily removed by filtration through sand.

Groundwater: Its Supply, Chemical Contamination, and Remediation

The Nature and Supply of Groundwater

The great majority of the available fresh water on Earth lies underground, half of it at depths exceeding a kilometer. As one digs into the ground below the initial belt of soil moisture, the **aeration** or **unsaturated zone,** where the particles of soil are covered with a film of water but in which air is present between the particles, is next encountered. At lower depths is the **saturated zone,** in which water has displaced all the air from these **pore spaces. Groundwater** is the name given to the fresh water in the saturated zone (see Figure 14-3); it makes up 0.6% of the world's total water supply. The ultimate source of groundwater is precipitation that falls onto the surface; a small fraction of it eventually filters down to the saturated zone. Underground water ranges in "age" from a few years to millions of years. For example, in zones that are now arid, much of the groundwater currently being accessed has been present since the wetter conditions of the last ice age and will not be quickly replaced.

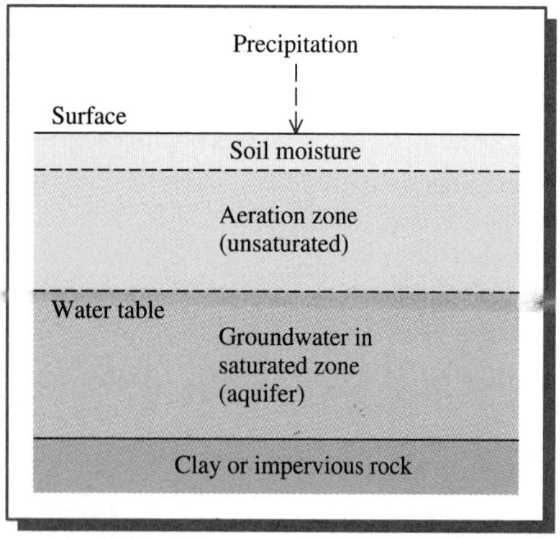

The top of the groundwater (saturated) region is called the **water table.** In some places it occurs right at the surface of the soil, a phenomenon that gives rise to swamps. Where the water table lies above the soil, we encounter lakes and streams.

FIGURE 14-3 Groundwater location in relation to regions in the soil.

Precipitation

Surface

Soil moisture

Aeration zone (unsaturated)

Water table

Groundwater in saturated zone (aquifer)

Clay or impervious rock

If groundwater is contained in soil that is composed of porous rocks such as sandstone, or in highly fractured rock such as gravel or sand, and if the water is bounded at its lower depths by a layer of clay or impervious rocks, then it constitutes a permanent reservoir—a sort of underground lake—called an **aquifer.** Some aquifers lie below several layers of impermeable rock or soil; these are called *confined* or *artesian* aquifers.

Groundwater in aquifers can be extracted by wells, and it is the main supply of drinking water for almost half the population of North America and over 1.5 billion people worldwide. In the United States in 1990, groundwater supplied 39% of the water used for public supplies and 96% of that withdrawn for individual domestic systems, the latter being very common in rural homes. In Europe the proportion of public drinking water extracted from aquifers ranges from nearly 100% for Denmark, Austria, and Italy, to about two-thirds in Germany, Switzerland, and the Netherlands, to less than one-third in Great Britain and Spain.

In the United States, the majority of groundwater usage is for irrigation purposes, almost all of it in the western states. The massive extraction of water from American aquifers has given rise to fears about future supplies of fresh water (and about the sinking of land above the aquifers), since such aquifers are replenished only very slowly. In the High Plains of the central United States, more than half the groundwater in storage has been depleted in some areas. In northern China, the depletion of shallow aquifers is forcing the sinking of wells more than 1 km deep in order to reach a new supply of groundwater. Indeed, groundwater depletion—along with the buildup of salts in the soil—is now the dominant threat to irrigated agriculture. In addition, the contamination of groundwater by chemicals is becoming a serious concern in many areas. A side effect of the rise in sea levels that will accompany global warming is the intrusion of salt water into aquifers near coasts.

The Contamination of Groundwater

Groundwater has been traditionally considered to be a pure form of water. Because of its filtration through soil and its long residence time underground, it contains much less natural organic matter and many fewer disease-causing microorganisms than water from lakes or rivers, although the latter point may be a misconception, according to recent evidence. Some groundwater is naturally too salty or too acidic for either drinking or irrigation purposes and may contain too much sodium, sulfide, or iron for many uses.

Humans have been concerned about the pollution of surface water in rivers and lakes for a long time. Indeed, a recent survey indicated that stream water in both agricultural and urban areas of the United States contains pesticide concentrations that exceed human-health benchmark standards. In contrast, the contamination by chemicals of groundwater was not recognized as a serious environmental problem until the 1980s, notwithstanding the fact that it had been occurring for half a century. To a large extent, groundwater contamination was neglected because it was not immediately visible—it was "out of sight, out of mind"—even though groundwater is a major source of drinking water. We were ignorant of the long-range consequences of our waste disposal practices. Ironically, surface water can be cleaned up relatively easily and quickly, whereas groundwater pollution is a much harder, much more expensive, long-range problem to solve.

Because we are now aware of the consequences—including high remediation costs—of the uncontrolled disposal of organic chemical wastes, most large corporations in developed countries have become much more responsible in their disposal of chemicals. Unfortunately, the collective discharges from smaller sources, including many municipalities, small industries, and farms, have not yet been controlled in like manner. Similarly, the huge number of septic tanks that exist are collectively a major source of nitrate, bacteria, viruses, detergents, and household cleaners to groundwater.

Nitrate Contamination of Groundwater

The inorganic contaminant of greatest concern in groundwater is the **nitrate ion,** NO_3^-, which commonly occurs in both rural and suburban aquifers. Although uncontaminated groundwater generally has nitrate nitrogen levels of 4–9 ppm, about 9% of shallow aquifers—from which water is often extracted via privately owned wells—in the United States now have nitrate levels that exceed the 10-ppm nitrogen MCL value. Indeed, elevated levels of about 100 ppm can result from agricultural activity. The location of areas in the United States that have a high risk of nitrate contamination of groundwater is shown in Figure 14-4. Exceeding the 10-ppm MCL limit is much rarer (1%) for public U.S. groundwater supplies, partially because they are drawn from deeper aquifers; these are generally less contaminated because of their depth, because their location is remote from large sources of contamination, and because natural remediation via denitrification of nitrate in the low-oxygen conditions can occur.

The expenditure of public money on nitrate-level reductions in drinking water has become a controversial subject. In Great Britain, in particular, hundreds of millions of dollars have been spent on achieving the 50-ppm maximum level of nitrate ion set by the European Union. Because nitrate removal from well water is very expensive, water contaminated with high levels of the ion are not normally used for human consumption, at least in public supplies.

PROBLEM 14-4

Convert the EU nitrate standard of 50 ppm to its nitrogen content alone. Is the EU standard more or less stringent than the U.S. regulatory limit of 10 ppm nitrogen as nitrate per liter?

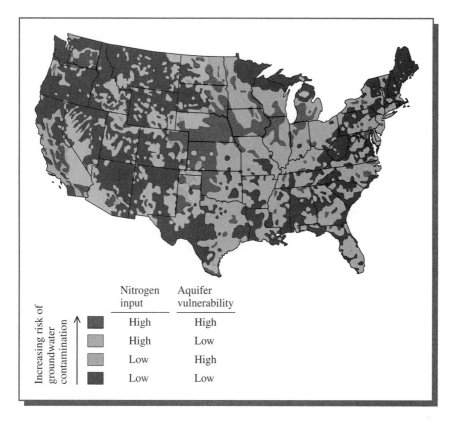

FIGURE 14-4 The risk of nitrate contamination of the groundwaters in the United States. [Source: B. T. Nolan et al., "Risk of Nitrate in Groundwaters of the United States—A National Perspective," *Environmental Science and Technology* 31 (1997): 2229.]

Nitrate in groundwater originates mainly from four sources:

- application of nitrogen fertilizers, both inorganic and animal manure, to cropland,

- cultivation of the soil,

- human sewage deposited in septic systems, and

- atmospheric deposition.

Concern has been expressed about the increasing levels of nitrate ion in drinking water, particularly in well water in rural locations; the main source of this NO_3^- is runoff from agricultural lands into rivers and streams. Almost 12 million tons of nitrogen is applied annually as fertilizer for agriculture in the United States, and manure production contributes almost 7 million tons more. Initially, oxidized animal wastes (manure), unabsorbed **ammonium nitrate,** NH_4NO_3, and other nitrogen fertilizers were thought to be the culprits in nitrate contamination of groundwater, since reduced nitrogen unused by plants is converted naturally to nitrate, which is highly soluble in water and can easily leach down into groundwater. It now appears that intensive cultivation of land, even without the application of fertilizer or manure, also facilitates the oxidation of reduced nitrogen to nitrate in decomposed organic matter in the soil by providing aeration and moisture. The original, reduced forms of nitrogen become oxidized in the soil to nitrate, which, being mobile, then migrates down to the groundwater, where it dissolves in water and is diluted. Denitrification of nitrate to nitrogen gas (see Chapter 6) and uptake of nitrate by plants can occur in forested areas that separate agricultural farms from streams, thereby lowering the risk of contamination in areas with significant woodland. However, rural areas with high nitrogen input, well-drained soil, and little woodland are at particular risk for nitrate contamination of groundwater.

The atmospheric deposition of nitrate results from its production in the atmosphere when NO_X emissions from vehicles and power plants, and its natural sources in thunderstorms, are oxidized in air to nitric acid and then neutralized to ammonium nitrate (see Chapters 3 and 5).

In urban areas, the use of nitrogen fertilizers on domestic lawns and golf courses, parks, etc. contributes nitrate to groundwater. Septic tanks and cesspools also are significant contributors where they exist.

Excess nitrate ion in wastewater flowing into seawater, e.g., the Baltic Sea, has resulted in algal blooms that pollute the water after they die. Nitrate ion normally does not cause this effect in bodies of fresh water, where phosphorus rather than nitrogen is usually the limiting nutrient; increasing the nitrate concentration there without an increase in phosphate levels does not lead to an increased amount of plant growth. There are, however, instances where nitrogen rather than phosphorus temporarily becomes the limiting nutrient even in fresh waters.

PROBLEM 14-5

The nitrate concentration in an aquifer is 20 ppm, and its volume is 10 million liters. What mass of ammonia upon oxidation would have produced this mass of nitrate?

Health Hazards of Nitrates in Drinking Water

Excess nitrate ion in drinking water is a potential health hazard since it can result in **methemoglobinemia** in newborn infants as well as in adults with a specific enzyme deficiency. The pathological process, in brief, runs as follows.

Bacteria, e.g., in unsterilized milk-feeding bottles or in the baby's stomach, reduce some of the nitrate to **nitrite ion,** NO_2^-:

$$NO_3^- + 2\,H^+ + 2\,e^- \longrightarrow NO_2^- + H_2O$$

The nitrite combines with and oxidizes the iron ions in the hemoglobin in blood from Fe^{2+} to Fe^{3+} and thereby prevents the proper absorption and transfer of oxygen to cells. The baby turns blue and suffers respiratory failure. (In almost all adults, the oxidized hemoglobin is readily reduced back to its oxygen-carrying form, and the nitrite is readily oxidized back to nitrate; also, nitrate is mainly absorbed in the digestive tract of adults before reduction to nitrite can occur.) Methemoglobinemia, or *blue-baby syndrome*, is now relatively rare in industrialized countries. It was a serious problem in Hungary up until the late 1980s and in Romania.

The U.S. EPA MCL of 10 ppm of nitrate nitrogen was set in order to avoid blue-baby syndrome. Since the syndrome is now almost nonexistent in the United States (only two cases since the mid-1960s), some policy analysts think this value is too stringent.

Recently, an increase in the risk of acquiring non-Hodgkin's lymphoma has been found for persons in some communities in Nebraska who consume drinking water having the highest levels (long-term average of 4 ppm or more of nitrogen as nitrate) of nitrate. As discussed in the next section, excess nitrate ion in drinking water is also of concern because of its potential link with stomach cancer. Recent epidemiological investigations have, however, failed to establish any positive, statistically significant relationship between nitrate levels in drinking water and the incidence of stomach cancer. A study reported in 2001 found that older women in Iowa who drank water from municipal supplies having elevated nitrate levels (> 2.46 ppm) were almost three times as likely to be diagnosed with bladder cancer than those least exposed (< 0.36 ppm in their drinking water). However, a recent large-scale study from the Netherlands failed to find an association between nitrate exposure and the risk of bladder cancer. A review of the current literature concluded that there is also no association between nitrate exposure from drinking water and adverse reproductive effects.

Nitrosamines in Food and Water

Some scientists have warned that excess nitrate ion in drinking water and foods could lead to an increase in the incidence of stomach cancer in humans, since some of it is converted in the stomach to nitrite ion. The nitrites could subsequently react with amines to produce **N-nitrosamines,** compounds that are known to be carcinogenic in animals. N-nitrosamines are amines in which two organic groups and an $-N=O$ unit are bonded to the central nitrogen:

N-nitrosamines   NDMA

Of concern not only with respect to its production in the stomach and its occurrence in foods and beverages (e.g., cheeses, fried bacon, smoked and/or cured meat and fish, and beer), but also as an environmental pollutant in drinking water, is the compound in which R in the above structure is the methyl group CH_3; it is called **N-nitrosodimethylamine,** or NDMA for short. This organic liquid is somewhat soluble in water (about 4 g/L) and somewhat soluble in organic liquids. It is a probable human carcinogen, and a potent one if extrapolation from animal studies is a reliable guide. It can transfer a methyl group to a nitrogen or oxygen of a DNA base, thereby altering the instructional code for protein synthesis in the cell.

In the early 1980s, it was found that NDMA was present in beer to the extent of about 3000 ppt. Since that time, commercial brewers have modified the drying of malt so that the current levels of NDMA in American and Canadian beers are now only about 70 ppt.

Large quantities of nitrate are used to "cure" pork products such as bacon and hot dogs. In these foods, some of the nitrate ion is biochemically reduced to nitrite ion, which prevents the growth of the organism responsible for botulism. Nitrite ion also gives these meats their characteristic taste and color by combining with hemoproteins in blood. Nitrosamines are produced from excess nitrite during frying (e.g., of bacon) and in the stomach, as discussed. Government agencies have instituted programs to decrease the residual nitrite levels in cured meats. Some manufacturers of these foods now add vitamin C or E to the meat in order to block the formation of nitrosamines. Based upon average levels of NDMA in various foods and the average daily intake for each of them, most of us now ingest more NDMA from consumption of cheese (which is often treated with nitrates) than from any other source.

Even though the commercial production of NDMA has been phased out, it can be formed as a by-product due to the use of amines in industrial processes such as rubber tire manufacturing, leather tanning, and pesticide production.

The levels of NDMA in drinking water drawn from groundwater is of concern in some localities that have industrial sources of the compound. For example, following the discovery that the water supply of one town had been contaminated by up to 100 ppt NDMA from a tire factory, Ontario, Canada, adopted a guideline maximum of 9 ppt of NDMA in drinking water, which corresponds to a lifetime cancer risk of 1 in 100,000. By contrast, the guideline for water in the United States is set at 0.68 ppt, which corresponds to a cancer risk of 1 in a million, but which actually lies considerably below the detection limit (about 5 ppt) for the compound.

PROBLEM 14-6

Write balanced redox half-reactions (assuming acidic conditions) for the conversion of NH_4^+ to NO_3^-, and of NO_2^- to N_2.

Perchlorates

Perchlorate ion, ClO_4^-, is analogous to nitrate ion in that both oxyanions involve nonmetals in their highest common oxidation numbers (+7 in the case of perchlorate; see Additional Problem 13 to generate a summary of chlorine with its many oxidation numbers). For that reason, both ions are oxidizing agents, and both have been used in explosives and propellants. Both have both natural and anthropogenic sources. Perchlorate is a newly discovered (late 1990s) pollutant in the drinking-water supply of about 15 million Americans. Large quantities of **ammonium perchlorate,** NH_4ClO_4, are manufactured for use as oxidizing agents in solid rocket propellants, fireworks, batteries, and automobile air bags. Because rocket fuel has a limited shelf life, it must be replaced regularly. Large amounts of perchlorates were washed out of missiles and rocket boosters onto the ground or into holding lagoons in the second half of the twentieth century.

Perchlorate contamination has been established in 23 U.S. states, including much of the Colorado River and aquifers in the deserts of the Southwest. Concentrations of perchlorate in drinking water in the U.S. Southwest range from 5 to 20 ppb. The map (Figure 14-5) of perchlorate releases in the United States indicates that most occur in the south-central and western states, especially California. Perchlorate has also been found in garden fertilizers at concentrations approaching 1%. It occurs naturally in some Chilean deposits of nitrate which are exported to the United States and elsewhere as fertilizers. Perchlorate also exists naturally in some minerals

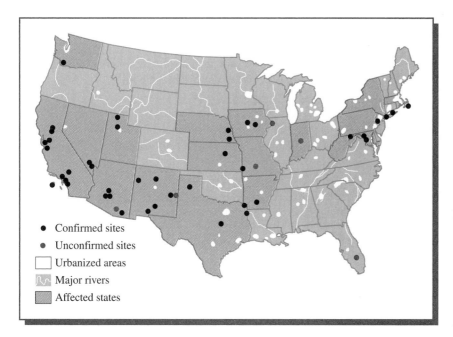

FIGURE 14-5 Regions of perchlorate use and contamination in the United States. [Source: B. E. Logan, "Assessing the Outlook for Perchlorate Remediation," *Environmental Science and Technology* (1 December 2001): 484A.]

- ● Confirmed sites
- ● Unconfirmed sites
- ▢ Urbanized areas
- Major rivers
- Affected states

found in the U.S. Southwest. A recent analysis indicates that both its use as an oxidizer, in fireworks, rockets, etc., and its presence in fertilizer make important contributions to its contamination of foodstuffs in the United States.

At high doses, perchlorate affects human health by reducing hormone production in the thyroid, where it competes with iodide ion. Its hazard at low concentrations, if any, is not known, making the development of a drinking-water standard for the ion a difficult problem. No federal MCL for the ion has yet been set, but several states have set their own limits, ranging from 1 to 18 ppb; e.g., California's limit is 6 ppb. In 2002, a draft report by the U.S. EPA proposed a drinking-water standard of 1 ppb as safe for human health, but this value has been criticized as too low by the Department of Defense and companies that make or use perchlorates. Research reported in 2002 on volunteers indicated that the no-effect level (NOEL; Chapter 10) for the inhibition of iodine uptake corresponds to a drinking-water concentration of at least 180 ppb. The EPA based its 1-ppb recommendation on studies indicating that mothers who drink perchlorate-contaminated water could give birth to children whose IQs would be affected negatively because correct maternal thyroid hormone levels are vital to fetal brain development.

Like nitrate, perchlorate is a difficult ion to remove from water supplies since it is a highly water-soluble anion that is very inert and does not adsorb readily either to mineral surfaces or to activated carbon. Barrier methods, using elemental iron, etc. are not successful because the anion is so unreactive. The primary technologies currently in use to remediate perchlorate in water are ion exchange and biological treatment. Some ion exchange resins successfully remove perchlorate, though it tends to remain in solution until all other anions have been absorbed. Ion exchange is used especially when perchlorate concentrations are low to begin with.

Certain bacteria found naturally in many soils, sediments, and natural waters biodegrade perchlorate by reduction to chloride ion:

$$ClO_4^- \longrightarrow \longrightarrow Cl^- + 2\,O_2$$

Perchlorate can be biodegraded in bioreactors, large vats that are engineered to maintain a high concentration of the appropriate bacteria in contact with the water.

Groundwater Contamination by Organic Chemicals

The contamination of groundwater by organic chemicals is a major concern. Many organic substances decay rapidly or are immobilized in the soil, so the number of compounds that are sufficiently persistent and mobile to travel to the water table and to contaminate groundwater there is relatively small.

The compounds that are most often detected in groundwater-based U.S. community public water supplies, including those near hazardous waste sites, are summarized in Table 14-1. Municipal landfills as well as industrial waste

TABLE 14-1	Organic Compounds Commonly Found in U.S. Groundwater-Based Community Water Supplies and Their Properties		
Chemical		Density (g/mL)	Water Solubility (g/L)
Present at 25–50% of sites:			
Chloroform (trichloromethane)		1.48	8.2
Bromodichloromethane		1.98	4.4
Dibromochloromethane		2.45	2.7
Bromoform (tribromomethane)		2.89	3.0
Present at a smaller fraction of sites:			
Trichloroethene		1.46	1.1
Tetrachloroethene (perchloroethene)		1.62	0.15
1,1,1-Trichloroethane		1.34	1.5
1,2-Dichloroethenes		1.26, 1.28	3.5, 6.3
1,1-Dichloroethane		1.18	5.5
Carbon tetrachloride		1.46	0.76
Dichloroiodomethane		1.58	
Xylenes		0.86–0.88	0.18 (o)
1,2-Dichloropropane		1.16	2.8
Benzene		0.88	1.8
Toluene		0.87	0.54
Also commonly present at wells close to hazardous waste sites:			
Methylene chloride		1.33	20
Ethylbenzene		0.87	0.15
Acetone		0.79	sol
1,1-Dichloroethene		1.22	2.3
1,2-Dichloroethane		1.24	8.5
Vinyl chloride (chloroethene)		gas	8.8
Methyl ethyl ketone		0.80	268
Chlorobenzene		1.11	0.47
1,1,2-Trichloroethane		1.44	4.5
Chloroethane		0.90	5.7
Fluorotrichloromethane		gas	1.1
1,1,2,2-Tetrachloroethane		1.60	2.7
Methyl isobutyl ketone		0.80	19

Source: Based on U.S. EPA surveys of about 2% of U.S. water supplies.

disposal sites are often the source of the contaminants. Liquid that contains dissolved matter that drains from a terrestrial source, such as a landfill, is called a **leachate.** In rural areas, the contamination of shallow aquifers by organic pesticides, such as *atrazine* (Chapter 10) leached from the surface, has become a concern. The insecticide *dieldrin* (Chapter 10), which has been banned since 1992, is the pesticide found most often to exceed human-health guideline levels in U.S. groundwater. Ironically, shallow groundwater aquifers used to supply drinking water are often more polluted by pesticides at greater than acceptable levels than are those in agricultural areas in the United States.

The typical organic contaminants in most major groundwater supplies are:

- Chlorinated solvents, especially **trichloroethene** (TCE, "tric," also called *trichloroethylene*), C_2HCl_3, and **perchloroethene** (PCE, "perc," also called *perchloroethylene* or *tetrachloroethene*), C_2Cl_4. These molecules contain a $C=C$ bond, with three or four of the four hydrogen atoms of ethene (ethylene) replaced by chlorine:

$$
\begin{array}{cc}
\underset{Cl}{\overset{H}{\diagdown}}C=C\underset{Cl}{\overset{Cl}{\diagup}} & \underset{Cl}{\overset{Cl}{\diagdown}}C=C\underset{Cl}{\overset{Cl}{\diagup}} \\
\text{TCE} & \text{PCE}
\end{array}
$$

By a large margin, chlorinated solvents are the most prevalent organic pollutants in groundwater.

- Hydrocarbons from the BTX component of gasoline and other petroleum products: **benzene**, C_6H_6, and its methylated derivatives **toluene**, $C_6H_5(CH_3)$, and the three isomers of **xylene**, $C_6H_4(CH_3)_2$. (See Chapter 7 for structures.)

- MTBE (methyl *tertiary-butyl* ether) from gasoline (see Chapter 8)

The chemicals in the groups mentioned above occur commonly in groundwater at sites where manufacturing and/or waste disposal occurred, especially from 1940 to 1980. In that period, little attention was paid to the ultimate fate and residence following the in-ground injection of these chemicals. The sources of these organic substances also include leaking chemical waste dumps, leaking underground gasoline storage tanks, leaking municipal landfills, and accidental spills of chemicals on land.

Trichloroethene is an industrial solvent, used to dissolve grease on metal, as is perchloroethene. The U.S. MCL for TCE in drinking water is 5 ppb, and the same limit is now used in Canada as well. A 2006 report by the U.S. National Academy of Sciences concluded that TCE is a possible cause of kidney cancer, can impair neurological function, and can cause reproductive and developmental damage. A link between TCE exposure and an abnormally low sperm count in males has been established. The International Agency for Research on Cancer has classified TCE as "probably carcinogenic to humans."

PCE is used not only in metal degreasing but also finds wide application as the solvent in dry-cleaning operations, so it is released from a large number of small sources. A group of women in Cape Cod, Massachusetts, who were inadvertently exposed over several decades to high levels of PCE in their drinking water were found to have small to moderate increases in their risk of contracting breast cancer.

Gasoline enters the soil via surface spills, leakage from underground storage tanks, and pipeline ruptures. Before 1980, underground gasoline storage tanks were made from steel; almost half of them were sufficiently corroded to leak by the time they were 15 years old. Once they descend to groundwater, the water-soluble components of the gasoline are preferentially leached into the water and can migrate rapidly in the dissolved state. The BTX component, which is the most soluble of the hydrocarbons, often occurs at concentrations of 1–50 ppb in groundwater. However, the alkylated benzenes are rapidly degraded by aerobic bacteria and consequently are not long-lasting.

The MTBE component of gasoline (Chapter 7) is more water-soluble than the hydrocarbons, but unlike them, it is not readily biodegraded. It is not highly toxic. The main problem is the odor and taste that it gives to water; as little as 15 ppb in water can be tasted or smelled. MTBE contamination of well water, albeit at low levels, has become of concern in the United States since it has occurred at about a quarter of a million sites.

The Ultimate Sink for Organic Contaminants in Groundwater

The subsequent behavior of the organic compounds that do migrate to the water table depends significantly upon their density relative to that of 1.0 g/mL of water. Liquids that are *less* dense ("lighter") than water and have low solubility in it form a mass that floats on the top of the water table. All hydrocarbons having a small or medium molecular mass belong to this group, including the BTX fraction of gasolines and other petroleum products (see Table 14-1). In contrast, polychlorinated solvents are *more* dense ("heavier") than water and insoluble in it, so they tend to sink deeply into aquifers; important examples are *methylene chloride, chloroform, carbon tetrachloride, 1,1,1-trichloroethane,* TCE, and PCE (see Table 14-1). Nonchlorinated but insoluble high-molecular-weight organic materials, such as *creosote* and *coal tar*, also belong to the heavier-than-water group. These substances are sometimes referred to as **dense nonaqueous-phase liquids,** DNAPLs.

Although the oily liquid blobs that these organic compounds form generally are found in an aquifer at a position either directly below their original point of entry into the soil or close to it, the implication that they are horizontally immobile is misleading. Very slowly—in a process that often takes decades or centuries to complete—these low-solubility compounds gradually dissolve in the water that passes over the blob and so provide a continuous supply of contaminants to the groundwater. The complete removal of such deposits usually is not feasible since they may exist as several blobs whose exact location is difficult to pinpoint. In addition, disturbance of the deposit during removal or treatment may increase its net exposure to the water phase. Even removing 90% of the substance does not necessarily result in reduction

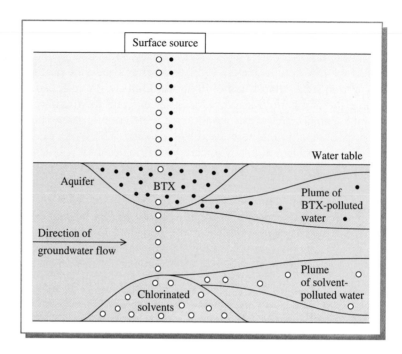

Surface source

Water table

Aquifer

BTX

Plume of
BTX-polluted
water

Direction of
groundwater flow

Plume
of solvent-
polluted water

Chlorinated
solvents

FIGURE 14-6 The contamination of groundwater by organic chemicals.

of its groundwater concentration. Thus, **plumes** of polluted water grow, in the direction of the water's flow, and thereby contaminate the bulk of the aquifer (see Figure 14-6). Because of such contamination, many wells used for drinking water have had to be closed.

Decontamination of Groundwater: Physical and Chemical Processes

In the last two decades, considerable energy and money have been spent in the United States on attempts to control aquifer pollution by the oily liquids discussed above. Dense organic leachates, especially PCE and TCE, have contaminated the groundwater that lies below the waste sites associated with the U.S. Superfund remediation initiative (see Chapter 16). Unfortunately, no easy cure to the problem of contamination has been found. Control usually consists of **pump-and-treat** systems that pump contaminated water from the aquifer, treat it to remove its organic contaminants (using methods of the type described later in this chapter), and return the cleaned water to the aquifer or to some other water body. Alternatively, a fine mist of the contaminated groundwater is sprayed into the air above agricultural land using a long, moveable sprinkler system; the contaminant volatile organic compounds (VOCs) evaporate into the air and the cleansed water is used for irrigation.

The volume of water that must be pumped and treated in a given aquifer is huge. For organic contaminants with low water solubility, recontamination of water returned to the aquifer by additional dissolution from the blob will occur. Consequently, the treatment systems must operate in perpetuity, and there are already thousands of them spread across the United States.

Both *in situ* heating, to vaporize the organic liquids so their vapors rise to the soil surface, and the addition of oxidants to convert the substances to products such as carbon dioxide have been tried in some locations. Typically, temperatures close to 100°C are used in heating, though it is not known if this is optimal in most cases. Heating and/or the production of gases or precipitates by oxidation may inadvertently change the geologic and biological conditions in the immediate vicinity of the treatment, with unforeseen effects on the distribution and mobility of the remaining pollutant.

Decontamination of Groundwater: Bioremediation and Natural Attenuation

Bioremediation is the term applied to the decontamination of water or soil using biochemical rather than chemical or physical processes. Recently there has been interesting progress reported in using bioremediation to cleanse water of chlorinated ethene solvent contamination.

The biodegradation of chloroethenes by *aerobic* bacteria becomes less and less efficient as the extent of chlorination increases, so it is ineffective for perchloroethene. However, under *anaerobic* conditions, the reductive biodegradation of PCE and TCE proceeds more quickly, particularly if an easily oxidized substance such as methanol is added to supply electrons for the reduction processes. Unfortunately, the stepwise dechlorination of these compounds proceeds through **vinyl chloride,** CH_2=CHCl, a known carcinogen. Recently, a bacterium has been discovered that removes all the chlorine from organic solvents such as TCE and PCE.

Owing to the high cost and limited effectiveness of many groundwater cleanup technologies, the inexpensive process of natural attenuation—allowing natural biological, chemical, and physical processes to treat groundwater contaminants—has become popular. Indeed, it is now used at more than 25% of Superfund program sites in the United States and is the leading method to remedy the contamination of groundwater from leaking underground storage sites.

However, there is great controversy about whether or not natural attenuation is an appropriate strategy for managing groundwater contamination: Many environmentalists feel that it is a cheap way for industry to avoid expensive cleanup costs. The U.S. National Research Council in 1997 appointed a committee to determine which pollutants could be treated successfully by this technique. Table 14-2 summarizes their results.

TABLE 14-2	Likelihood of Success of Groundwater Remediation by Natural Attenuation for Various Substances	
Chemical Class	**Dominant Attenuation Processes**	**Likelihood of Success Given Current Level of Understanding**
Organic Compounds		
Hydrocarbons		
BTEX	Biotransformation	High
Gasoline, fuel oil	Biotransformation	Moderate
Nonvolatile aliphatic compounds	Biotransformation, immobilization	Low
PAHs	Biotransformation, immobilization	Low
Creosote	Biotransformation, immobilization	Low
Oxygenated hydrocarbons		
Low-molecular-weight alcohols, ketones, esters	Biotransformation	High
MTBE	Biotransformation	Low
Halogenated aliphatics		
PCE, TCE, carbon tetrachloride	Biotransformation	Low
TCA	Biotransformation, abiotic transformation	Low
Methylene chloride	Biotransformation	High
Vinyl chloride	Biotransformation	Low
Dichloroethylene	Biotransformation	Low
Halogenated aromatics		
Highly chlorinated PCBs, tetrachlorodibenzofuran, pentachlorophenol, multichlorinated benzenes	Biotransformation, immobilization	Low
Less chlorinated PCBs, dioxins	Biotransformation	Low
Monochlorobenzene	Biotransformation	Moderate
Inorganic Substances		
Metals		
Ni	Immobilization	Moderate
Cu, Zn	Immobilization	Moderate
Cd	Immobilization	Low
Pb	Immobilization	Moderate

TABLE 14-2	Likelihood of Success of Groundwater Remediation by Natural Attenuation for Various Substances	
Chemical Class	Dominant Attenuation Processes	Likelihood of Success Given Current Level of Understanding
Cr	Biotransformation, immobilization	Low to moderate
Hg	Biotransformation, immobilization	Low
Nonmetals		
As	Biotransformation, immobilization	Low
Se	Biotransformation, immobilization	Low
Oxyanions		
Nitrate	Biotransformation	Moderate
Perchlorate	Biotransformation	Low

Source: Adapted from J. A. Macdonald, "Evaluating Natural Attenuation for Groundwater Cleanup," *Environmental Science and Technology* (1 August 2000): 346A.

Only three pollutants are highly likely to be successfully treated by natural attenuation:

- BTEX hydrocarbons (i.e., BTX hydrocarbons plus ethylbenzene),

- low-molecular-weight oxygen-containing organics, and

- methylene chloride.

In all three cases, biotransformation is the dominant process by which attenuation occurs. Notice that neither MTBE nor highly chlorinated organics, including TCE and PCE, are usually successfully treated in this way, nor is mercury or perchlorate ion.

Decontamination of Groundwater: *In Situ* Remediation

Scientists have developed a promising *in situ* technique for treating groundwater contaminated by volatile (mainly C_1 and C_2) chlorinated organics. They construct an underground permeable "wall" of material (mostly coarse sand) along the path of the water. The water is cleansed as a result of its passage through the wall and never has to be pumped out of the ground (see Figure 14-7).

The ingredient that is placed within the sand bed and that chemically cleans the water is **metallic iron,** Fe^0, in the form of small granules, a common

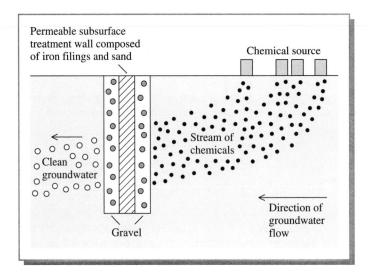

FIGURE 14-7 *In situ* purification of groundwater using an "iron wall."

waste product of manufacturing processes. When placed in contact with certain chlorinated organics dissolved in water, the iron acts as a reducing agent, giving up electrons to form the **ferrous,** or Fe(II), ion Fe^{2+}, which dissolves in the water:

$$Fe(s) \longrightarrow Fe^{2+}(aq) + 2\ e^-$$

Usually, these electrons are donated to chloroorganic molecules that are temporarily adsorbed onto the metal's surface; the chlorine atoms contained in these molecules are consequently reduced to chloride ions, Cl^-, which are released into aqueous solution. This technique is an example of **reductive degradation.** For example, the reduction of trichloroethene to its completely dechlorinated form, ethene, can be written in unbalanced form as

$$C_2HCl_3 \longrightarrow C_2H_4 + 3\ Cl^-$$

Upon application of a standard redox balancing technique for alkaline solution, we obtain the balanced half-reaction

$$C_2HCl_3 + 3\ H_2O + 6\ e^- \longrightarrow C_2H_4 + 3\ Cl^- + 3\ OH^-$$

Combination of the half-reactions (after tripling the oxidation step to ensure the number of electrons lost and gained is the same) yields the overall reaction

$$3\ Fe(s) + C_2HCl_3 + 3\ H_2O \longrightarrow 3\ Fe^{2+}(aq) + C_2H_4 + 3\ Cl^- + 3\ OH^-$$

One of the by-products of the reaction is hydroxide ion, OH^-. Recall from Chapter 13 that in limestone-rich areas, the groundwater contains significant concentrations of dissolved **calcium bicarbonate,** $Ca(HCO_3)_2$. The hydroxide ions produced in the groundwater remediation reaction react with bicarbonate to produce carbonate ion, $CO_3{}^{2-}$, which combines with dissolved calcium ions to produce insoluble calcium carbonate, $CaCO_3$, which then precipitates in the sand–iron mixture.

Field trials indicate that this new technology can work successfully for several years at least, and it may replace the pump-and-treat methods for many applications involving chlorinated methanes and ethanes dissolved in underground water.

Recently, it has been found that coating the iron filings with nickel speeds up the rate of degradation of the organic compounds by a factor of 10; with this modification, the technique may be even more useful than first imagined. In addition, it has been discovered that the elemental iron in the barriers will reduce soluble Cr^{6+} ions to insoluble Cr^{3+} oxides and can therefore remediate groundwater contaminated by Cr^{6+}. (The environmental chemistry of chromium is discussed in more detail in Chapter 15). A technique for the *in situ* creation of elemental iron from its ions (Fe^{2+} and Fe^{3+}) by the injection of aqueous reducing agents has also been tested.

An *in situ* technique of treating TCE and PCE by hydrogenation has been developed. The process uses dissolved H_2 gas to rapidly dechlorinate these two organics, eventually forming ethane and HCl. The reaction, which uses a palladium catalyst, can be done within a well bore so that the water need not be brought up to the surface.

PROBLEM 14-7

The dissolution of iron in the process described above produces some molecular hydrogen gas, H_2. Show by balanced equation(s) how the hydrogen could arise from the reduction of water rather than of TCE.

PROBLEM 14-8

Suppose that this "iron wall" technology reduced an appreciable fraction of TCE to vinyl chloride rather than completely to ethene. Why would this be an unacceptable result environmentally? (Note that in practice a sufficiently thick wall of iron is used to convert any vinyl chloride by-product to ethene.)

PROBLEM 14-9

Deduce the overall reaction by which perchloroethene is converted to ethene by metallic iron.

At one test site for this remediation process, the water contained 270 ppm TCE and 53 ppm perchloroethene. Calculate the mass of iron required to remediate 1 L of this groundwater.

The Chemical Contamination and Treatment of Wastewater and Sewage

Most municipalities treat the raw sewage collected from homes, buildings, and industries (including food processing plants) through a **sanitary sewer** system before the liquid residue is deposited into a nearby source of natural waters, whether a river, lake, or ocean. In contrast, since the rainwater and melted snow that drain from streets and other paved surfaces are usually not highly contaminated, they are often collected separately by *storm sewers* and deposited directly into a body of natural water. Unfortunately, in some municipalities, storm-driven overflow occurs in the sanitary sewer system and the overflow is combined with storm water and deposited, untreated, into waterways.

The main component of sewage—other than water—is organic matter of biological origin. It occurs mainly as particles—ranging from those of macroscopic size large enough to be trapped (together with such *objets d'art* as facial tissues, stones, socks, tree branches, condoms, and tampon applicators) by mesh screens to those which are microscopic in size and are suspended in the water as large colloids.

Sewage Treatment

In the **primary** (or mechanical) treatment stage of wastewater (see the schematic diagram in Figure 14-8), the larger particles—including sand and silt—are removed by allowing the water to flow across screens and slowly along a lagoon or settling basin. A **sludge** of insoluble particles forms at the bottom of the lagoon, while "liquid grease" (a term which here includes not only fat, oils, and waxes but also the products formed by the reaction of soap with calcium and magnesium ions) forms a lighter-than-water layer at the top and is skimmed off. About 30% of the biochemical oxygen demand (BOD, Chapter 13) of the wastewater is removed by the primary treatment process, even though this stage of the process is entirely mechanical in nature. The treatment and disposal of the sludge is discussed later.

After passing through conventional primary treatment, the sewage water has been much clarified but still has a very high BOD—typically several hundred milligrams per liter (ppm)—and is detrimental to fish life if released at this stage (as occurs in some jurisdictions that discharge into the ocean). The high BOD is due mainly to organic colloidal particles. In the **secondary** (biological) treatment stage, most of this suspended organic matter as well as

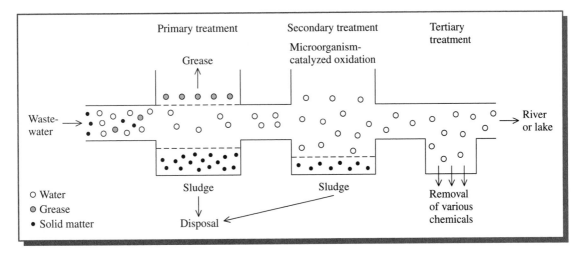

FIGURE 14-8 The common stages in the treatment of wastewater.

that actually dissolved in the water is biologically oxidized by microorganisms to carbon dioxide and water or converted to additional sludge, which can readily be removed from the water. Either the water is sprinkled onto a bed of sand and gravel or of plastic covered with the aerobic bacteria (*trickling filters*), or it is agitated in an aeration reactor (*activated sludge process*) in order to effect the microorganism-driven reaction. The system is kept well-aerated to speed the oxidation. In essence, by deliberately maintaining in the system a high concentration of aerobic microorganisms, especially bacteria, the same biological degradation processes that would require weeks to occur in open waters are accomplished in several hours.

The biological oxidation processes of secondary treatment reduce the BOD of the polluted water to less than 100 ppm, which is about 10% of the original concentration in the untreated sewage. For comparison, Canadian wastewater quality standards require that the BOD in treated water be 20 ppm or less. The process of nitrification also occurs to some extent, converting organic nitrogen compounds to nitrate ion and carbon dioxide (see below). In summary, the secondary treatment of wastewater involves biochemical reactions that oxidize much of the oxidizable organic material that was not removed in the first stage. Treated water diluted with a greater amount of natural water can support aquatic life. Normally, municipalities take the water produced by secondary treatment and disinfect it by chlorination or irradiation with UV light before pumping it into a local waterway. Recent research in Japan has shown that chlorination of the effluent before its release produces mutagenic compounds, presumably by interaction of chlorine-containing substances with the organic matter that remains in the water.

A few municipalities employ **tertiary** (or *advanced* or *chemical*) wastewater treatment as well as primary and secondary ones. In the tertiary phase, specific substances are removed from the partially purified water before its final disinfection. In some cases, the water produced by tertiary treatment is of sufficiently high quality to use as drinking water. Alternatively, river water into which the effluent from sewage treatment plants has been deposited is used by municipalities downstream as drinking water. The reuse of water after it has been cleansed is particularly prevalent in Europe, where less fresh water is available than in North America and the population density is high.

Depending upon locale, tertiary treatment can include some or all of the following chemical processes:

- further reduction of BOD by removal of most remaining colloidal material, using an aluminum salt in a process that operates in the same manner as described previously for the purification of drinking water;

- removal of dissolved organic compounds (including chloroform) and some heavy metals by their adsorption onto activated carbon, over which the water is allowed to flow (see Box 14-1);

- phosphate removal (as discussed in the next section; some phosphorus is removed in the secondary treatment stage since microbes incorporate it as a nutrient for their growth);

- heavy metal removal by the addition of hydroxide or **sulfide ions,** S^{2-}, to form insoluble metal hydroxides or sulfides, respectively (see Chapter 15);

- iron removal by aeration at a high pH to oxidize it to its insoluble Fe^{3+} state, possibly in combination with use of a strong oxidizing agent to destroy organic ligands bound strongly to the Fe^{2+} ion, which would otherwise prevent its oxidation; and

- removal of excess inorganic ions, as discussed below.

In some wastewaters, the further removal of **nitrogen** compounds—usually either ammonia or organic nitrogen compounds—is deemed necessary. Ammonia removal can be achieved by raising the pH to about 11 (with lime) to convert most ammonium ion to its molecular form, ammonia, NH_3, followed by bubbling air through the water to air-strip it of its dissolved ammonia gas. This process is relatively expensive, however, because it is energy-intensive. Ammonium ion can also be removed by ion exchange, using certain resins that have their exchange sites initially populated by sodium or calcium ions.

Alternatively, both organic nitrogen and ammonia can be removed by first using nitrifying bacteria to oxidize all the nitrogen to nitrate ion. Then the nitrate is subjected to denitrification by bacteria to produced **molecular nitrogen,** N_2, which bubbles out of the water. Since this reduction step requires a substance to be oxidized, **methanol,** CH_3OH, is added if necessary to the water and is converted to carbon dioxide in the process:

$$5\ CH_3OH + 6\ NO_3^- + 6\ H^+ \xrightarrow{\text{bacteria}} 5\ CO_2 + 3\ N_2 + 13\ H_2O$$

BOX 14-4	Time Dependence of Concentrations in the Two-Step Oxidation of Ammonia

The bacteria-catalyzed oxidation of ammonia (or of other reduced organic nitrogen compounds) to nitrate is a reaction with two main steps, with nitrite ion, NO_2^-, an intermediate:

Step 1 $NH_3 + \frac{3}{2}O_2 \longrightarrow$
$$NO_2^- + H^+ + H_2O$$

Step 2 $NO_2^- + \frac{1}{2}O_2 \longrightarrow NO_3^-$

If sufficient oxygen is available, the rate of each reaction is first-order only in the concentration of the nitrogen reactant, so the sequence can be represented as

$$A \xrightarrow{k_1} B \xrightarrow{k_2} C$$

where A stands for ammonia, B for nitrite ion, and C for nitrate ion, and k_1 and k_2 are the pseudo-first-order rate constants. Since the rate of step 1 depends on the first power of the ammonia concentration, then the rate of disappearance of this species is

$$\frac{d[A]}{dt} = -k_1[A]$$

Since B (nitrite) is produced at this rate by step 1, but is consumed in step 2 by a process whose rate is proportional to the first power of its concentration, we can write

$$\frac{d[B]}{dt} = +k_1[A] - k_2[B]$$

These differential equations can be coupled and integrated to yield the following expressions for the evolution of [A] and [B] with time, relative to $[A]_0$, the original concentration of A:

$$[A]/[A]_0 = e^{-k_1 t}$$
$$[B]/[A]_0 = k_1(e^{-k_1 t} - e^{-k_2 t})/(k_2 - k_1)$$

As can be seen from the solution to Problem 1, the concentration of B (nitrite) rises exponentially at first, reaches a peak value, then declines slowly. Thus significant concentrations of nitrite ion occur in water undergoing the two-step conversion of ammonia to nitrate.

PROBLEM 1

(a)   Derive a general expression relating the time at which [B] reaches a peak to k_1 and k_2.
(b)   Draw a graph showing the evolution of [A] and of [B] with time for the values $k_1 = 1$ and $k_2 = 2$.

Of course, water contaminated by nitrate ion can also be treated by this latter step. A mathematical analysis of the kinetics of the transformations is given in Box 14-4.

PROBLEM 14-11

Given that the K_b value for ammonia is 1.8×10^{-5}, deduce a formula giving the ratio of ammonia to ammonium ion as a function of the pH of water. What is the value of this ratio at pH values of 5, 7, 9, and 11?

The Origin and Removal of Excess Phosphate

One of the world's most famous cases of water pollution involves Lake Erie, which in the 1960s was said to be dying. Indeed, one of the authors of this book can recall visiting a once-popular beach on Lake Erie's north shore in the early 1970s and being repulsed by the sight and smell of dead, rotting fish on the shoreline. Lake Erie's problems stemmed primarily from an excess input of **phosphate ion,** PO_4^{3-}, in the waters of its tributaries. The phosphate sources were the polyphosphates in detergents (as explained in detail later), raw sewage, and the runoff from farms that used phosphate products. Since there is commonly an excess of other dissolved nutrients in lakes, phosphate ion usually functions as the *limiting* (or controlling) *nutrient* for algal growth: The larger the supply of the ion, the more abundant the growth of algae—and its growth can be quite abundant indeed. When the vast mass of excess algae eventually dies and starts to decompose by oxidation, the water becomes depleted of dissolved oxygen, with the result that fish life is adversely affected. The lake water also becomes foul-tasting, green, and slimy, and masses of dead fish and aquatic weeds rot on the beaches. The series of changes, including rapid degradation and aging, that occur when lakes receive excess plant nutrients from their surroundings is called **eutrophication.** When the enrichment arises from human activities, it is called **cultural eutrophication.**

To correct the regional problem, the United States and Canada in 1972 signed the *Great Lakes Water Quality Agreement.* Since that time, over $8 billion has been spent in building sewage treatment plants to remove phosphates from wastewater before it reaches the tributaries and the lake itself. In addition, the levels of polyphosphates in laundry detergents were restricted in Ontario and in many of the states that border the Great Lakes. The total amount of phosphorus entering Lake Erie has now decreased by more than two-thirds. As a result, Lake Erie has sprung back to life: Its once-fouled beaches are regaining popularity with tourists, and its commercial fisheries have been revived.

As we have pointed out, the presence of excess phosphate ion in natural waters can have a devastating effect on an aquatic ecology because it overfertilizes plant life. Formerly, one of the largest sources of phosphate as a pollutant was detergents, and in the material that follows, the role of such phosphates is discussed.

The reaction of synthetic detergents with calcium and magnesium ions, forming complex ions, diminishes the cleansing potential of the detergent. **Polyphosphate ions,** which are anions containing several phosphate units linked by shared oxygens, are added to detergents as *builders* that preferentially form soluble complexes with these metal ions and thereby allow the molecules of the detergent to operate as cleansing agents rather than being complexed with the Ca^{2+} and Mg^{2+} naturally present in the water. Another role of the builder is to make the wash water somewhat alkaline, which helps remove the dirt from certain fabrics. With soap itself, the ions form insoluble complexes that foul the cleaning water.

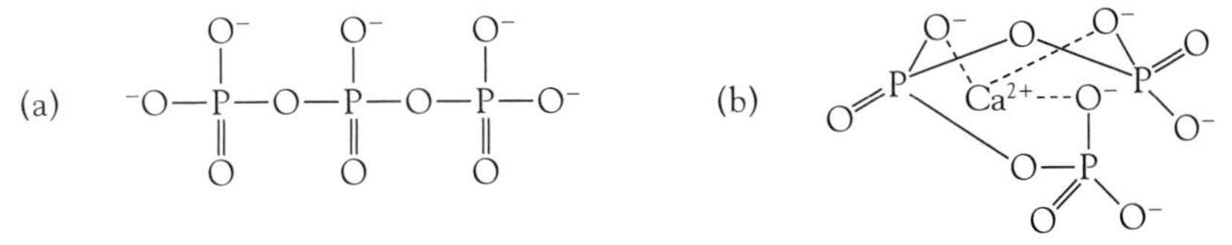

FIGURE 14-9 Structure of the polyphosphate ion: (a) uncomplexed and (b) complexed with calcium ion.

Great quantities of **sodium tripolyphosphate** (STP), $Na_5P_3O_{10}$, were formerly added as the builder in most synthetic detergent formulations. As shown in Figure 14-9a, STP contains a chain of alternating phosphorus and oxygen atoms, with one or two additional oxygens attached to each phosphorus. In solution, one tripolyphosphate ion can form a complex with one calcium ion by forming interactions between three of its oxygen atoms and the metal ion (Figure 14-9b).

Substances like STP, which have more than one site of attachment to the metal ion and thereby produce ring structures that each incorporate the metal, are called **chelating agents** (from the Greek word for "claw"). Because several bonds are formed, the resulting chelates are very stable and do not normally release their metal ions back into the free form. The use of chelating agents to remove metals from the human body is discussed in Chapter 15.

Tripolyphosphate ion, $P_3O_{10}^{5-}$, like phosphate ion itself, is a weak base in aqueous solution and thus provides the alkaline environment that is required for effective cleaning:

$$P_3O_{10}^{5-} + H_2O \longrightarrow P_3O_{10}H^{4-} + OH^-$$

Unfortunately, when wash water containing STP is discarded, the excess tripolyphosphate enters waterways, where it slowly reacts with water and is transformed into phosphate ion (sometimes called *orthophosphate*):

$$P_3O_{10}^{5-} + 2 H_2O \longrightarrow 3 PO_4^{3-} + 4 H^+$$

Note that when tripolyphosphate decomposes, STP behaves as an acid rather than a base (since H^+ is formed in the reaction).

Because of environmental concerns, polyphosphates are now used only sparingly as builders in detergents in many areas of the world. In Canada and parts of Europe, STP was replaced largely by **sodium nitrilotriacetate** (NTA) (see Figure 14-10a). The anion of NTA acts in a similar fashion to that of STP, chelating calcium and magnesium ions using three of its oxygen atoms and the nitrogen atom (Figure 14-10b). NTA is not used as a builder in the United States because of concerns that its slow rate of degradation might lead to health hazards in drinking water. However, the early experiments with test animals that led to this concern are open to question, as are fears about its persistence and its tendency to solubilize heavy metals into water supplies.

Other builders now used include *sodium citrate*, sodium carbonate (*washing soda*), and *sodium silicate*. Currently, substances called **zeolites** are also employed as detergent builders. Zeolites are abundant aluminosilicate minerals

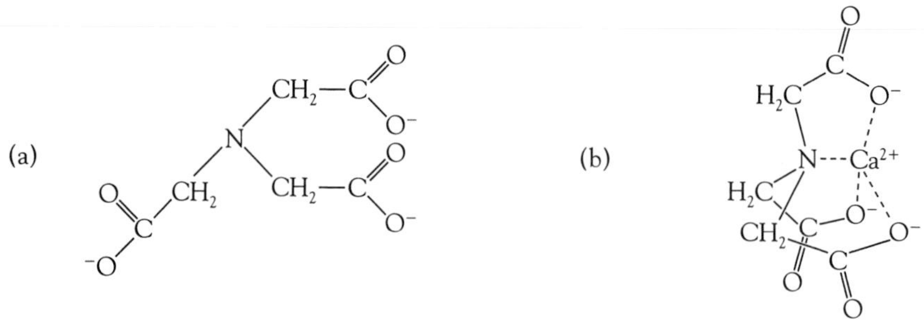

FIGURE 14-10 Structure of the nitrilotriacetate ion: (a) uncomplexed and (b) complexed with calcium ion.

(a)

(b)

(see Chapter 16) consisting of sodium, aluminum, silicon, and oxygen. The latter three elements are bonded together to form cages, which the sodium ions can enter. In the presence of calcium ion, zeolites exchange their sodium ions for Ca^{2+} (though not for Mg^{2+}), thereby sequestering it in a manner similar to polyphosphates. Like polyphosphates, they also control pH. One disadvantage to the use of zeolites is that they are insoluble, so their use increases the amount of sludge that must be removed at wastewater treatment plants.

Phosphate ion can be removed from municipal and industrial wastewater by the addition of sufficient calcium as the hydroxide $Ca(OH)_2$, so that insoluble calcium phosphates such as $Ca_3(PO_4)_2$ and $Ca_5(PO_4)_3OH$ are formed as precipitates that can then be readily removed. Phosphate removal could be a standard practice in the treatment of wastewater, but it is not yet practiced in all cities. Some policymakers believe that the optimum environmental solution is to use polyphosphates, rather than some other builder, in detergents and then to efficiently remove phosphates at wastewater treatment plants.

Geographically, phosphate ion enters waterways from both point and nonpoint sources. **Point sources** are specific locations such as factories, landfills, and sewage treatment plants that discharge pollutants. **Nonpoint sources** are numerous large land areas such as farms, logged forests, septic tanks, golf courses and individual domestic lawns, stormwater runoff, and atmospheric deposition. Although each nonpoint source may provide a small amount of pollution, on account of the large number of them involved they can generate larger *total* quantities than do point sources. For example, now that sewage treatment plants and detergent controls have been instituted, much of the remaining phosphate arises from nonpoint agricultural sources in many areas.

 ## Green Chemistry: Sodium Iminodisuccinate— A Biodegradable Chelating Agent

Because most chelating agents are not biodegradable or are only slowly biodegradable, not only do they place a load on the environment (e.g., phosphates acts as nutrients), but it may be necessary to remove them during treatment of wastewater in a wastewater treatment plant. Unlike many chelating agents, **sodium iminodisuccinate** (IDS, see Figure 14-11) [also known as D,L-aspartic-N-(1,2-dicarboxylethyl) tetrasodium salt] readily degrades in the environment. Not only is IDS biodegradable, it is also nontoxic.

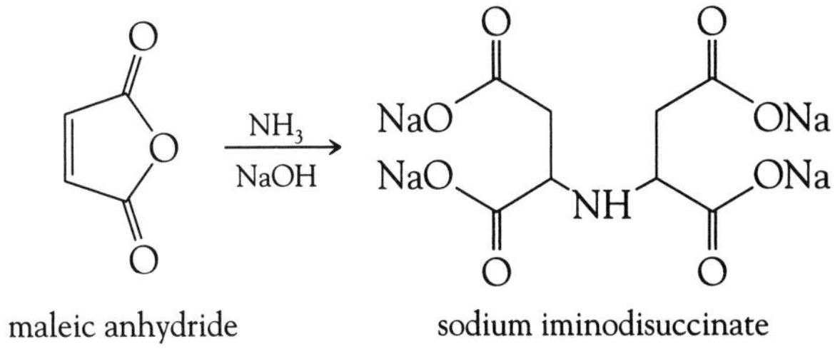

maleic anhydride   sodium iminodisuccinate

FIGURE 14-11 Synthesis and structure of sodium iminodisuccinate, a biodegradable chelating agent.

IDS can be used as an effective chelating agent for absorption of agricultural nutrients, metal ion scavenging in photographic processing, groundwater remediation, and as a builder in detergents and household and industrial cleaners. The *Bayer Corporation* won a Presidential Green Chemistry Challenge Award in 2001 for the development of IDS as a chelating agent and for its synthesis from maleic anhydride (Figure 14-11). This synthesis is accomplished under mild conditions, in water as the only solvent. The excess ammonia is recycled back into the production of more IDS. This synthesis stands in stark contrast to typical syntheses of aminocarboxylate chelating agents, which employ hydrogen cyanide as a reagent. Bayer markets IDS as a chelating agent under the name Baypure.

Reducing the Salt Concentration in Water

The decomposition of organic and biological substances during the secondary phase of wastewater treatment usually results in the production of inorganic salts, many of which remain in the water even after the techniques listed above have been applied. Water can also become salty due to its use in irrigation or because water softener units have been recharged and their discharge disposed of as sewage. Inorganic ions can be removed from water by desalination by using one of the techniques listed below or using the precipitation methods mentioned above.

- **Reverse osmosis** As previously mentioned, this technique is also used to produce drinking water from salty water, such as seawater.

- **Electrodialysis** Here a series of membranes permeable either only to small inorganic cations or only to small inorganic anions are set up vertically in an alternating fashion (see Figure 14-12) within an electrochemical cell. Direct current is applied across the water, so cations migrate toward the cathode and anions toward the anode. The liquid in alternating zones becomes more concentrated (enriched) or less concentrated (purified) in ions; eventually the ion-concentrated water can be disposed of as brine and the purified water released into the environment. This technology is also used to desalinate seawater for drinking purposes.

- **Ion exchange** Some polymeric solids contain sites that hold ions relatively weakly, so one type of ion can be exchanged for another of the same charge that

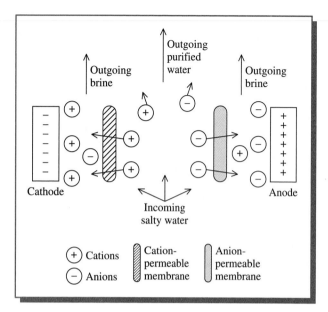

FIGURE 14-12 Electro-dialysis unit (schematic) for the desalination of water. [Adapted from S. E. Manahan. 1994. *Environmental Chemistry,* 6th ed. Boca Raton, FL: Lewis Publishers.]

happens to pass by it. Ion exchange resins can be formulated to possess either cationic or anionic sites that function in this manner. The exchange sites of a cationic resin of this type are initially occupied by H^+ ions, and the exchange sites of an anionic exchange resin are occupied by OH^- ions. When water polluted by M^+ and X^- ions is passed sequentially through the two resins, the H^+ ions on the first are replaced by M^+, and then the OH^- ions on the second resin are replaced by X^-. Thus the water that has passed through contains H^+ and OH^- ions, rather than those of the salt, which remain behind in the resins. Of course, these two ions immediately combine to form more water molecules. Thus ion exchange can be used to remove salts, including those of heavy metals, from wastewater.

PROBLEM 14-12

Water polluted by inorganic ions could be purified by distilling it or by freezing it. Why do you think such techniques are not generally used on a mass scale to purify water?

Transition metal cations can be removed from water using either precipitation or reduction techniques, in either case to form insoluble solids. Precipitation of sulfides or hydroxides has already been mentioned; a disadvantage of the latter is the production of a voluminous sludge that must be disposed of in an acceptable manner. Electrolytic reduction of metals leads to their deposition on the cathode. If, instead of the elemental metal, a concentrated aqueous solution of it is desired, the deposited metal can be reoxidized chemically by adding hydrogen peroxide or electrolytically by reversing the polarity of the cell.

The Biological Treatment of Wastewater and Sewage

An alternative to the processing of wastewater through a conventional treatment plant in small communities is biological treatment in an **artificial marsh** (also called a *constructed wetland*) that contains plants such as bullrushes and reeds. The decontamination of the water is accomplished by the bacteria and other microbes that live among the plants' roots and rhizomes. The plants themselves take up metals through their root systems and concentrate contaminants within their cells. In facilities that have been constructed to deal with sewage, primary treatment to filter out solids, etc. in a lagoon is usually implemented before the wastewater is pumped to the marsh, where the equivalent of secondary and tertiary treatment occurs. The plant growth uses up the pollutants and increases the pH—which serves to destroy some harmful microorganisms.

One advantage of biological treatment is that great amounts of sludge are not generated, in contrast to conventional treatments. Furthermore, it requires neither the addition of synthetic chemicals nor the input of commercial energy. Among the problems in such facilities are decaying vegetation, which must be limited so that the BOD of the processed water does not rise too much, and the fact that the marshes usually require a great deal of land unless they are constructed so that part of the routing is vertical.

In many rural and small communities, **septic tanks** are used to decontaminate sewage since central sewage facilities are not available. These underground concrete tanks receive the wastewater, often from only one home. Although solids settle in the tank, grease and oil rise to the top, from which they are periodically removed. The bacteria in the wastewater feed on the bottom sludge, thereby liquefying the waste. Partially purified water flows out of the tank into an underground drain, where further decontamination takes place. The system is relatively passive, compared to central facilities, and as in the case of artificial marshes, time is required for the processes to occur. In addition, nitrogen compounds are converted to nitrate, but the latter is not reduced to molecular nitrogen, so groundwater under the septic system can become contaminated by nitrate, as discussed earlier in this chapter.

Drugs in Wastewater from Sewage Treatment Plants

In recent years, trace concentrations of various drugs—prescription, over-the-counter, illegal, and veterinary—have been detected in the waters leading from sewage treatment plants, and in rivers and streams into which this water then flows, at concentrations up to the ppb level. About 100 substances have been detected in various rivers, lakes, and coastal waters. The substances—commonly including *estradiol, ibuprofen*, the antidepressant drug Prozac (*fluoxetine*), the anti-epileptic drug *carbamazepine*, and degradation products associated with cholesterol-reducing pharmaceuticals—are present in raw sewage after their excretion in urine or feces from humans and animals since most drugs are poorly absorbed and metabolized by the body. They also result from the disposal of unused or expired medication in toilets.

Most commonly, concentrations of drugs in drinking water are at the parts-per-trillion level, so their risk to human health is probably small. Research is under way to determine whether there could be effects on human health from sustained exposure to a combination of these substances. The synthetic hormones are thought to pose the greatest risk to aquatic species. Certain fish have been found to undergo some skewed sexual development due to exposure to sewage effluent containing the synthetic estrogen in birth control pills (see Chapter 12).

The Treatment of Cyanides in Wastewater

The **cyanide ion,** CN^-, binds strongly to many metals, especially those of the transition series, and is often used to extract them from mixtures. Consequently, cyanide is widely used in mining, refining, and electroplating metals such as gold, cadmium, and nickel. Unfortunately, cyanide ion is very poisonous to animal life since it binds strongly to metal ions in living matter, e.g., to the iron in proteins that are necessary for molecular oxygen to be utilized by cells.

Cyanide is a very stable species and does not quickly decompose on its own or in the environment. Thus it is an important water pollutant and should be destroyed chemically rather than simply disposed of in a waterway.

We can deduce the type of treatment that will be effective for cyanide by considering its acid–base and redox characteristics. Cyanide ion is the conjugate base of the weak acid **hydrocyanic acid,** HCN, which has limited solubility in water. Thus, acidification of cyanide solutions will result in the release of poisonous HCN gas from it and therefore is not a good solution to the problem of cyanide contamination.

The redox chemistry of cyanide ion can be predicted by considering the oxidation numbers of the two atoms involved. If nitrogen, the more electronegative atom, is considered to be in its fully reduced -3 oxidation form, then the carbon must be $+2$. Thus one way to destroy cyanide ion is to oxidize the carbon more fully, to $+4$ as it is in CO_2 and HCO_3^-. This oxidation can be accomplished by dissolved molecular oxygen if high temperatures and elevated air pressures are used:

$$\overset{+2}{C}N^- + \overset{0}{O_2} + 4\,H_2O \longrightarrow 2\,\overset{+4}{H}\overset{-2}{CO_3^-} + 2\,NH_3$$

The use of stronger oxidizing agents, such as Cl_2 or ClO^-, not only oxidizes the carbon from $+2$ to $+4$, but can also oxidize the nitrogen from the -3 state to the zero oxidation number of molecular nitrogen:

$$2\,\overset{+2\ -3}{CN^-} + 5\,\overset{0}{Cl_2} + 8\,OH^- \longrightarrow 2\,\overset{+4}{CO_2} + \overset{0}{N_2} + 10\,\overset{-1}{Cl^-} + 4\,H_2O$$

(Four of the ten electrons gained collectively by the chlorines are used by the carbons and six by the nitrogens in this overall process.) Other oxidizing agents that are used in cyanide treatment include **hydrogen peroxide,** H_2O_2, and/or molecular oxygen, in both cases with a copper salt added as a catalyst.

The process can also be carried out electrochemically for high cyanide concentrations; the remaining low concentration can be subsequently oxidized by ClO^-.

Sodium cyanide is now used in some shallow tropical waters such as those in Indonesia to stun reef fish so that they can be captured and sold live as seafood or pets. Unfortunately, the cyanide kills smaller fish and destroys the coral.

PROBLEM 14-13

If an oxidizing agent even more powerful than chlorine or hypochlorite were to be used in the treatment of cyanide, what other possibilities for the nitrogen-containing product would there be?

PROBLEM 14-14

For HCN in water, $K_a = 6.0 \times 10^{-10}$. Calculate the fraction of cyanide that exists as the anion rather than in the molecular form at pH values of 4, 7, and 10.

The Disposal of Sewage Sludge

The **sludge** from both the primary and secondary treatment stages of sewage is principally water and organic matter. It can be digested anaerobically, in a process that takes several weeks to complete. Bacteria levels in the sludge are not thereby completely eliminated, but the levels are reduced about a thousandfold. The sludge that remains after this further organic decomposition has occurred and after the supernatant water is removed is sometimes then incinerated or simply dumped into a landfill or into a water body such as the ocean. However, sludge is high in plant nutrients, so about half the sewage sludge in North America and Europe is spread on farm fields, golf courses, and even residential lawns as low-grade fertilizer sometimes called *biosolid*.

Unfortunately, sewage sludge may contain toxic substances, which potentially could be incorporated into food grown on the land or could contaminate groundwater under the fields. In particular, heavy metal concentrations often are higher in sewage sludge than in soil, principally because industrial wastes are sometimes released directly into sewage lines shared by households. For example, the lead level in municipal sludge can range from several hundred to several thousand parts per million, compared to an average of about 10 ppm in the Earth's crust. In a few communities, an attempt is made to eliminate these toxic materials before final disposal occurs. Some scientists have worried that food crops grown in soil fertilized by sewage sludge may incorporate some of the increased amounts of heavy metals. Control experiments indicate that vegetables vary greatly in the extent to which they will absorb increased amounts of the metals; e.g., the uptake of lead by lettuce is particularly large, but that by cucumbers is negligible. The concentration of arsenic in agricultural soils is greatly increased if arsenic pesticides are applied to them; crops planted on these soils subsequently absorb some of the adsorbed arsenic. Other substances of concern in using sewage sludge as fertilizer for food are alkylphenols from detergents, brominated fire retardants, and pharmaceuticals—especially antibiotics given to farm animals.

Modern Wastewater and Air Purification Techniques

The most important chemical (as opposed to biological) pollutants dissolved in wastewater are usually chloroorganics, phenols, cyanides, and heavy metals. Below we describe some of the high-tech methods that have recently been developed and put into practice to purify wastewater, particularly for removal of chloroorganics. Some of these same techniques are also used to cleanse the compounds from contaminated air.

The Destruction of Volatile Organic Compounds

The major stationary sources in North America of VOCs (Chapter 3) are the evaporation of organic solvents, the manufacture of chemicals, and the petroleum industry and its storage activities. Wastewater effluent that is contaminated with VOCs, e.g., the water emanating from chemical or petrochemical plants, is commonly treated by a two-step process:

1. The VOCs are removed from the wastewater by **air stripping.** In this process, air is passed upward into a downward stream of the water, and the volatile materials are transferred from the liquid to the gas phase. This technique does not work well for compounds that are highly water soluble.

2. The resulting VOCs, now present in low concentration in a contained mass of humid air, are destroyed by a process of **catalytic oxidation.** For example, air heated to 300–500°C is passed for a short time over platinum or, depending upon the VOC, some other precious metal that is supported on alumina. The energy costs of this step are very high since it involves heating a large volume of humid air. Note that the outlet air from such processes contains **hydrogen chloride,** HCl, if the VOCs originally contained chlorine; this compound must be removed by scrubbing with a basic substance before the air is released into the atmosphere.

The removal of VOCs from *gaseous* emissions from industries usually operates by the same catalytic oxidation process; typically the concentration of VOCs in the air stream is thereby reduced by 95%. A primary heat exchanger recovers and reuses the VOCs' heat of combustion to warm incoming gases to the operating temperature.

The **adsorption** of compounds onto activated carbon (see Box 14-1) or onto synthetic carbonaceous adsorbents is a cost-effective technology used for the removal of low-level VOC concentrations from both liquid and vapor streams; it is also useful for nonvolatile organic compounds. These adsorbents can be easily regenerated by treatment with steam or by other thermal techniques as well as by solvents; the concentrated pollutants can be subsequently destroyed by catalytic oxidation.

Advanced Oxidation Methods for Water Purification

Conventional water purification methods often do not successfully deal with synthetic organic compounds such as chloroorganics that are dissolved at low concentrations; examples include the common groundwater pollutants trichloroethene and perchloroethene. The conventional method for the treatment of water containing such pollutants is adsorption of the chloro-organics onto activated carbon; this removes the compounds but doesn't destroy them. The wastewater from pulp-and-paper mills also contains organochlorines that are resistant to conventional treatments.

In order to cleanse water of these extra-stable organics, so-called **advanced oxidation methods** (AOMs) have been developed and deployed. The aim of these methods is to **mineralize** the pollutants, i.e., to convert them entirely to CO_2, H_2O, and mineral acids such as HCl. Most AOMs are ambient-temperature processes that use energy to produce highly reactive intermediates of high oxidizing or reducing potential, which then attack and destroy the target compounds. The majority of the AOMs involve the generation of significant amounts of the **hydroxyl free radical,** OH, which in aqueous solution is a very effective oxidizing agent, as it is in air (see Chapters 1–5). The hydroxyl radical can initiate the oxidation of a molecule by extraction of a hydrogen atom or addition to one atom of a multiple bond, as it does in air (Chapter 5); in water, as an additional alternative, it can also extract an electron from an anion.

Since the generation of OH in solution is a relatively expensive process, it is economical to use AOMs to treat only the components of the wastes that are resistant to the cheaper, conventional treatment processes. Thus, integrating an AOM with pretreatment of the wastewater by biological or other processes to first dispose of the easily oxidized materials is often appropriate.

Ultraviolet (UV) light is often used to initiate the production of hydroxyl radicals and thus to begin the oxidations. Commonly, hydrogen peroxide, H_2O_2, is added to the polluted water and UV light from a strong source in the 200–300-nm range is shone on the solution. The hydrogen peroxide absorbs the ultraviolet light (especially that closer to 200 nm than to 300 nm) and uses the energy obtained to split the O—O bond, resulting in the formation of two OH radicals:

$$H_2O_2 \xrightarrow{\text{UV}} 2\,OH$$

Alternatively, and less commonly, ozone is produced and then photochemically decomposed by UV light. The resulting oxygen atom reacts with water to efficiently produce OH via the intermediate production of hydrogen peroxide, which is photolyzed:

$$O_3 \xrightarrow{\text{UV}} O_2^* + O^*$$

$$O^* + H_2O \xrightarrow{\text{UV}} H_2O_2 \longrightarrow 2\,OH$$

A fraction of the oxygen atoms produced by ozone photolysis are electronically excited, and these react with water to directly produce hydroxyl radicals, as discussed in Chapter 1.

PROBLEM 14-15

Given that the enthalpies of formation for H_2O_2 and OH are, respectively, -136.3 and $+39.0$ kJ mol^{-1}, calculate the heat energy required to dissociate one mole of hydrogen peroxide into hydroxyl free radicals. What is the maximum wavelength of light that could bring about this transformation? *[Hint: See Chapter 1].* Given that light of 254-nm wavelength is usually used, and that all the energy of each photon that is in excess of that required to dissociate one molecule is lost as waste heat, calculate the maximum percentage of the input light energy that can be used for dissociation itself.

Hydroxyl radicals for wastewater treatment can also be efficiently produced *without* the use of UV light by combining hydrogen peroxide with ozone. The chemistry of the intermediate processes is complex, but the overall reaction between these two species is

$$H_2O_2 + 2\,O_3 \longrightarrow 2\,OH + 3\,O_2$$

This *ozone/H_2O_2 method* is more cost-effective and easier to adapt to existing water treatment systems than is any other AOM system.

It is also possible to generate the hydroxyl radical electrolytically. In most such applications, a metal ion (such as Ag^+ or Ce^{3+}) is first oxidized to a more positively charged ion (Ag^{2+} or Ce^{4+} in our examples) that will subsequently oxidize water to H^+ and OH.

The biggest liability associated with advanced oxidation processes is that their action produces toxic chemical by-products. For example, in the ozone/peroxide and peroxide/UV treatments of groundwater contaminated with trichloroethene and perchloroethene, the toxic intermediates *trichloroacetic acid*, CCl_3COOH, and *dichloroacetic acid*, $CHCl_2COOH$, are formed in about 1% yield.

Photocatalytic Processes

Another innovative technology for wastewater treatment involves the irradiation by UV light of solid semiconductor photocatalysts such as **titanium dioxide,** TiO_2, small particles of which are suspended in solution. Titanium dioxide is chosen as the semiconductor for such applications since it is non-toxic, is very resistant to photocorrosion, is cheap and plentiful, absorbs light efficiently in the UV-A region, and can be used at room temperature. Irradiation at wavelengths less than 385 nm produces electrons, e^-, in the conduction band and *holes*, h^+, in the valence band of the metal oxide. The holes in the semiconductor can react with surface-bound hydroxide ions or with water molecules, thereby producing hydroxyl radicals in both cases:

$$h^+ + OH^- \longrightarrow OH$$

$$h^+ + H_2O \longrightarrow OH + H^+$$

The holes can also react directly with adsorbed pollutants, producing radical cations that readily engage in subsequent degradation reactions.

Normally, O_2 molecules dissolved in the water react with the electron produced at the semiconductor surface, a process that eventually produces more reactive free radicals but is relatively slow. If hydrogen peroxide is added to the water instead, it will react with the electron to form the anion radical and generate reactive radicals more quickly.

The cost of the electrical energy required to generate the needed UV light is usually the major expense in operation of AOM systems. On this basis, the titanium dioxide methods are even less cost-effective than those described previously, since considerably more electricity is required per pollutant molecule destroyed. Sunlight could be used to supply the UV light, but only about 3% of its light lies in the appropriate UV-A range and is absorbed by the solid. Another problem with the TiO_2 processes is the difficulty in separating the various reactants and products from the TiO_2 particle if the metal oxide has been used in the form of a fine powder. However, there are now closed systems in which the titanium dioxide slurry is efficiently separated from the purified water and recycled back to the inlet stream.

Some scientists have experimented with immobilizing TiO_2 as a thin film (1 μm thick) on a solid surface such as glass, tile, or alumina. Indeed, TiO_2-coated tiles are now used on walls and floors in some buildings. The low-level UV light from fluorescent lighting in such rooms is sufficient to allow the destruction of gaseous and liquid-phase pollutants that touch the oxide on the tiles! For example, odors that upset people are usually present in air at concentrations of only about 10 ppm; at such levels, the UV from normal fluorescent lighting should be sufficient to destroy them with TiO_2 photocatalysts. Bacterial infections such as those that cause many secondary infections in hospitals can also be eliminated by spraying walls and floors (in rooms lit by fluorescent bulbs) to give them a titanium dioxide film. Photocatalysts are quiet, unobtrusive cleansing materials.

Other Advanced Oxidation Methods

A process called **direct chemical oxidation** has been proposed for the destruction of solid and liquid organic wastes in the aqueous phase, particularly in environments such as those under buildings, where the light required for UV processes cannot conveniently be supplied. It uses one or another of the strongest known chemical oxidants—e.g., acidified **peroxydisulfate anion,** $S_2O_8^{2-}$, under ambient pressure and moderate temperatures to oxidize the wastes. Such a process needs no catalysts and produces no secondary wastes of concern. The sulfate that results from the peroxydisulfate can be recycled back to the oxidant. Other very strong oxidizing agents that have been tested are the **peroxymonosulfate anion,** HSO_5^-, and the **ferrate ion,** FeO_4^{2-}; in the latter, iron has an oxidation number of $+6$, so it is not surprising that it is a strong oxidizing agent. Unfortunately, ferrate ion suffers from the problem of instability.

PROBLEM 14-16

Deduce the balanced half-reaction (acidic media) in which the peroxydisulfate ion is converted into sulfate ion. Repeat the exercise for the conversion of oxalic acid, $C_2H_2O_4$, into carbon dioxide. Combine these half-reactions into a balanced equation, and calculate the volume of 0.010 M peroxydisufate that is required to oxidize one kilogram of oxalic acid.

Review Questions

1. Describe the function of **(a)** aeration and **(b)** addition of aluminum or iron sulfate in the purification of drinking water.

2. Describe the chemistry underlying the removal of excess calcium and magnesium ions from drinking water.

3. Describe how water can be disinfected by **(a)** membrane filtration and **(b)** ultraviolet irradiation.

4. What two other chemical methods, other than chlorination, are used to disinfect water? What are some advantages and disadvantages to these alternatives?

5. Explain the chemistry underlying the disinfection of water by chlorination. What is the active agent in the destruction of the pathogens? What are the practical sources of the active ingredient?

6. Explain why pH control of water in swimming pools is important. What compounds are formed when the chlorinated water reacts with ammonia?

7. Discuss the advantages and disadvantages of using chlorination to disinfect water, including the nature of the THM compounds.

8. What is meant by the terms *groundwater* and *aquifer*? How does the *saturated zone* of soil differ from the *unsaturated*?

9. Why did concern about groundwater pollution lag far behind that about surface water?

10. Name three important sources of nitrate ion to groundwater.

11. Construct a table that shows the common oxidation numbers for nitrogen. Deduce in which column the following environmentally important compounds belong: HNO_2, NO, NH_3, N_2O, N_2, HNO_3, $NO_3{}^-$. Which of the species become prevalent in aerobic conditions in a lake? Under anaerobic conditions? What is the oxidation number of nitrogen in NH_2OH?

12. Explain why excess nitrate in drinking water or food products can be a health hazard; include the relevant balanced chemical reaction showing how nitrate becomes reduced.

13. What is an *N-nitrosamine*? Write the structure and the full name for NDMA.

14. What is the formula for the perchlorate ion? What is the origin of perchlorate ion in U.S. drinking water?

15. Define *leachate*.

16. Name two types of organic contaminants found in groundwater, and give two examples of each type.

17. Explain the difference in vertical location in an aquifer between compounds such as chloroform and those such as toluene.

18. Define the term *plume* and describe how it forms in an aquifer.

19. Why are the *BTX* and *MTBE* components of gasoline the ones that are most often found in groundwater? Are both components easily biodegraded?

20. What is meant by *reductive degradation*? Describe the *in situ* technique by which chloroorganics in aquifers can be destroyed by reductive dechlorination.

21. What procedures are involved in *primary wastewater treatment*? In *secondary treatment*?

22. List five possible water purification processes that are associated with the *tertiary treatment* of wastewater, including one that removes phosphate ion.

23. What polyphosphate was commonly used in detergents, and why did its use lead to environmental problems? What are the other main sources of phosphate to natural waters? What other builders are used in detergents?

24. Describe two important methods that are used to *desalinate* wastewater.

25. Describe the chemical processes by which cyanide ion can be removed from wastewater.

26. Describe how VOCs dissolved in wastewater are usually removed and destroyed.

27. Describe what *AOMs* stands for, and state the most common reactive agent in such processes. Describe three methods by which this reactive species can be generated.

28. Describe two *photocatalytic* methods that can destroy organic wastes.

29. What is *direct chemical oxidation*? What are two of the strong oxidizing agents that can be used for such procedures?

Green Chemistry Questions

See the discussion of focus areas and the principles of green chemistry in the Introduction before attempting these questions.

1. What are the environmental advantages of using iminodisuccinate compared to most chelating agents?

2. The development of iminodisuccinate by Bayer won a Presidential Green Chemistry Challenge Award.

(a) Into which of the three focus areas for these awards does this award best fit?

(b) List at least three of the twelve principles of green chemistry that are addressed by the green chemistry developed by Bayer.

Additional Problems

1. Given that for HOCl, $K_a = 2.7 \times 10^{-8}$, deduce the fraction of a sample of the acid in water that exists in the molecular form at pH values (predetermined by the presence of other species) of 7.0, 7.5, 8.0, and 8.5. [*Hint: Derive an expression that relates the fraction of HOCl that is ionized to the concentration of hydrogen ions.*] Would it be a good idea to allow the pool water's pH to rise to 8.5?

2. The equilibrium constant for the reaction of dissolved molecular chlorine with water to give hydrogen ions, chloride ions, and HOCl is 4.5×10^{-4}, where as usual the concentration of water is included in the K value. If the pH of the solution is determined by other processes so that the amount of hydrogen ion contributed by the chlorine reaction is negligible, calculate the fraction of the original 50 ppm chlorine which remains as Cl_2 at pH values of 0, 1, and 2. [Notes: (1) The dissociation of HOCl into ions is negligible at these low pH values. (2) Approximate solutions to the quadratic equation involved in these calculations will not be accurate due to the high percentage of reaction.]

3. Calculate the volume of $Ca_5(PO_4)_3OH$, the density of which is 3.1 g/mL, which is produced for each gram of sodium tripolyphosphate present in a detergent when it is removed in tertiary wastewater treatment. Estimate the annual mass of detergent used for laundry purposes for a typical household of four persons. Assuming that the phosphate levels in

laundry detergents used were about 50%, calculate the volume that was required annually to dispose of its waste laundry phosphate.

4. Calculate the oxidation number of the chlorine in molecular chlorine, HOCl, chlorine dioxide, monochloramine (NH_2Cl), and sodium chloride. Given that the last item is the most stable form of chlorine, predict whether the other substances mentioned are likely to be oxidizing agents or reducing agents, and rank them in likely order of this redox behavior. Using this analysis and the section on chlorine compounds in your introductory chemistry textbook, suggest other compounds that might be useful to disinfect water.

5. Given their names, can you deduce the nature of the similarity in molecular structure between hydrogen peroxide and the sulfur compounds that are used in direct chemical oxidation methods? By calculating the oxidation number of the atoms in hydrogen peroxide, deduce why it can act as an oxidizing agent.

6. What could be done to dispose of the solvents that are used to extract VOCs from adsorbents?

7. Write the initial reaction step that occurs if methyl chloroform, CH_3CCl_3, were to be destroyed by (a) reductive degradation and (b) hydroxyl radical attack.

8. Water samples from three wastewater streams were analyzed, and the important pollutants determined to be those listed below. In each case, devise economical, practical processes (other than activated carbon treatment) for purifying the water of the three pollutants:

(a) Phosphate ion, ammonium ion, and salt (in water containing bicarbonate ion)

(b) Nitrite ion, PCE, and Fe(II)

(c) Cadmium ion, carbon tetrachloride, and glucose

9. Write the reaction that occurs between hydroxyl radical and carbonate ion dissolved in water. What are two alternative substances that you could add to the water to decrease the carbonate ion concentration, one that operates by elimination of carbon dioxide and the other by precipitation of the ion, so as to decrease the amount of hydroxyl radical destroyed by this reaction? Given the solubility product constants in this chapter, estimate the lowest practical carbonate concentration you could reach by this process, and comment on the desirability of the ions that you have introduced into the water.

10. Treated drinking water should contain 0.5 mg/L of Cl_2 after most of the chlorine has been converted to HOCl. What pressure of $Cl_2(g)$ is required to maintain this concentration? $K_H = 8.0 \times 10^{-3}$ M for Cl_2.

11. At a particular temperature, K_a for HOCl is 3.5×10^{-8}. What would be the pH values for 1.00 M and 0.100 M concentrations of HOCl at this temperature? What percentages of the HOCl is undissociated at these two concentrations?

12. To desalinate seawater by reverse osmosis, pressure in excess of a solution's osmotic pressure, π, must be applied across the membrane. The total osmotic pressure exerted in a solution is determined by the total molar concentration, M, of its solutes, and is given by the equation $\pi = MRT$. Using the composition of seawater listed in Chapter 13, determine the minimum pressure that must be exerted on seawater to desalinate it using reverse osmosis at 20°C. Recall that $R = 0.082$ L atm mol^{-1} K^{-1}.

13. Chlorine-containing substances covering a wide variety of oxidation numbers have been encountered, some within this chapter and others previously in the book. For each substance in the list below, write out its formula, deduce the oxidation number of its chlorine, and fill in the appropriate row of the table below, as in the example shown:

chlorite ion   perchlorate ion

molecular chlorine   hypochlorous acid

chlorine dioxide   chlorine monoxide

chloride ion   chlorate ion

Oxidation Number	Formula of Example	Name of Example
−1		
0		
+1		
+2		
+3		
+4		
+5		
+6	ClO_3	Chlorine trioxide
+7		

Further Readings

1. A. Kolch, "Disinfecting Drinking Water with UV Light," *Pollution Engineering* (October 1999): 34.

2. F. Bove et al., "Drinking Water Contaminants and Adverse Pregnancy Outcomes: A Review," *Environmental Health Perspectives* 110 (supplement 1) (2002): 61.

3. J. R. Nuckols et al., "Influence of Tap Water Quality and Household Use Activities on Indoor Air and Internal Dose Levels of Trihalomethanes," *Environmental Health Perspectives* 113 (2005): 863. See also ibid, 114 (2006): 514.

4. U. van Gunten, "Ozonation of Drinking Water," Parts I and II, *Water Research* 37 (2003): 1443, 1469.

5. S. D. Richardson et al., "Identification of New Ozone Disinfection Byproducts in Drinking Water," *Environmental Science and Technology* 33 (1999): 3368.

6. R. F. Service, "Desalination Freshens Up," *Science* 313 (2006): 1088.

7. "Chlorinated Solvent Source Zones," *Environmental Science and Technology* (June 1, 2003): 225A.

8. P. B. Hatzinger, "Perchlorate Biodegradation for Water Treatment," *Environmental Science and Technology* (June 1, 2005): 239A.

9. L. Fewtrell, "Drinking-Water Nitrate, Methemoglobinemia, and Global Burden of Disease: A Discussion," *Environmental Health Perspectives* 112 (2004): 1371.

10. M. P. Zeegers et al., "Nitrate Intake Does Not Influence Bladder Cancer Risk: The Netherlands Cohort Study," *Environmental Health Perspectives* 114 (2006): 1527.

11. M. A. Montgomery and M. Elimelech, "Water and Sanitation in Developing Countries," *Environmental Science and Technology* 41 (2007): 17.

12. B. T. Nolan et al., "Risk of Nitrate in Groundwaters of the United States—A National Perspective, *Environmental Science and Technology* 31 (1997): 2229.

13. J. A. MacDonald, "Evaluating Natural Attenuation for Groundwater Cleanup," *Environmental Science and Technology* (August 1, 2000): 346A.

14. L. W. Canter, R. C. Knox, and D. M. Fairchild, *Groundwater Quality Protection* (Boca Raton, FL: Lewis Publishers, 1987).

15. E. K. Nyer, *Groundwater Treatment Technology,* 2nd ed. (New York: Van Nostrand Reinhold, 1992).

16. D. M. Mackay and J. A. Cherry, "Groundwater Contamination: Pump-and-Treat Remediation," *Environmental Science and Technology* 23 (1989): 630.

17. R. J. Gilliom, "Pesticides in U.S. Streams and Groundwater," *Environmental Science and Technology* 41 (2007): 3409.

18. "Environmental Processes '96: A Special Report," *Hydrocarbon Processing Magazine (International Edition)* 75 (1996): 85 [reviews many emerging technologies that can handle water and air pollution problems].

19. D. Simonsson, "Electrochemistry for a cleaner environment," *Chemical Society Reviews* 26 (1997): 181.

20. N. C. Baird, "Free Radical Reactions in Aqueous Solutions: Examples from Advanced Oxidation Processes for Wastewater and from the Chemistry in Airborne Water Droplets," *Journal of Chemical Education* 74 (1997): 817.

Websites of Interest

Log on to www.whfreeman.com/envchem4/ and click on Chapter 14.

| Environmental Instrumental Analysis **V** | Ion Chromatography of Environmentally Significant Anions |

The quantitative determination of levels of environmentally important ions, such as those discussed in the preceding chapters, can be accomplished using chromatographic methods described in this box.

The need to determine the prevalence of common anions like phosphate (PO_4^{3-}), nitrate (NO_3^-), or fluoride (F^-) isn't immediately clear. The biospheric significance of these ubiquitous ions is not as obvious as is, for example, the presence of PCBs, pesticides, or toxic metals like lead, mercury, or cadmium. These ionic components are important because they give an indication of the relative reduction–oxidation potential in an aqueous sample taken from an environment such as a stagnant lake (PO_4^{3-}), or of the contamination of groundwater from fertilizer runoff (NO_3^-), or of whether municipal water supplies need to be supplemented with fluoride (F^-) for the health of children's teeth. Although these charged ions can be detected by widely available ultraviolet detectors common in most high-performance liquid chromatographic systems (Janos and Aczel, 1996), a more sensitive means of detection involves ionic conductivity. This chromatographic method is called *ion chromatography with ionic conductivity detection*. Although cations can also be separated by ion chromatography (IC), only anionic separations will be discussed here.

The heart of the separation process in an ion chromatograph is a short column (10–15 cm) packed with small-diameter particles called *ion exchange* resins. These are often made of a styrene/divinylbenzene polymer or microparticles of silica coated with compounds containing an anionic functional group such as a quaternary amine, $-N(CH_3)_3^+OH^-$, or a primary amine, $-NH_3^+OH^-$, when they are to be used for anion separation.

The actual process of chromatographic separation occurs after a sample containing analyte anions (and their associated cations) is injected onto the chromatographic column. With gas chromatography (see Environmental Instrumental Analysis Box II), the mobile phase is an inert gas that does not chemically interact with the chromatographic surface. The mobile phase in ion chromatography, on the other hand, is a solution of cations and anions with a carefully controlled pH; often buffers are used. This complex mixture of mobile-phase ions—carefully chosen for each group of analyte ions to be separated—interacts with the analyte ions and the functional groups of the column's chromatographic surface. That interaction involves competition of the mobile-phase anions and the analyte anions for chromatographic sites on the packing material (the charged functional groups such as $-N(CH_3)_3^+$ or $-SO_3^-$). This competition yields different overall travel times for each of the analytes as they pass down the column; some are retained longer than others. (The overall down-column movement is provided by a pumping of the mobile phase by an external pump.) Different analyte travel times—as in gas chromatography—translate into different exit (or retention) times for each anion in the original mixture. The result is chromatographic separation of anions.

(continued on p. 658)

Ion Chromatography of Environmentally Significant Anions (continued)

The process for anionic ion chromatographic retention by ion exchange resins can be represented by the equation

$$RN(CH_3)_3{}^+HCO_3{}^-(s) + \textbf{anion}^-(aq) \longrightarrow$$
$$RN(CH_3)_3{}^+ \textbf{anion}^-(s) + HCO_3{}^-(aq)$$

In this equation, the term **anion**$^-$ represents any of the analyte anions mentioned above. When a sample is injected onto the column, this anion is quickly retained by complexation with the stationary phase near the head of the column. The next step in the chromatographic process takes place as a mobile phase, with a carefully controlled amount of anionic ion such as bicarbonate, $HCO_3{}^-$, is pumped through the column. The presence of the bicarbonate anion in the mobile phase forces the equilibrium in the above equation to the left; the retained analyte anion is freed and moves down the column in the flowing mobile phase. As the analyte moves along, it repeatedly undergoes this same process of retention and movement (or *exchange* between the stationary and mobile phases). Most importantly, different analyte anions (e.g., fluoride or phosphate or chloride) undergo this exchange process to differing degrees and therefore travel at different overall rates during their time in the IC column. The result is that different analytes exit the chromatographic column at different times, i.e., separation has taken place.

The task of detecting analyte anions in the presence of the anions always present in the mobile phase is by no means a trivial one. Since both kinds of anions—the analytes' and the mobile phase's—conduct electricity, using an ordinary conductivity cell as a detector at the end of the IC column is normally not practical. The problem is especially difficult because, in order to obtain adequate separation of some important anions, the mobile phase often has to have high ionic content to displace the analyte anions from the chromatographic surface—something that is obviously required for separation. Therefore, most of the ionic conductivity passing through the detector is attributable to the mobile-phase ions and not the analyte—an unworkable situation when trying to detect the analyte anions by their conductivity.

An ingenious solution to this problem is called *conductivity suppression* or *eluent (mobile-phase) suppression*. This technology converts the mobile-phase anions from an easily dissociated ionic form to a (soluble) molecular form that does not strongly influence the signal produced by the conductivity detector. The suppression module is placed after the chromatographic column but before the conductivity detector. In an anion exchange system, the suppression module might carry out the following reaction:

$$Na^+(aq) + HCO_3{}^-(aq) + \textbf{resin}^-\textbf{H}^+(s) \longrightarrow$$
$$resin^-Na^+(s) + H_2CO_3(aq)$$

Here **resin**$^-$**H**$^+$ represents a *cation* exchange resin that will exchange cations—instead of anions as in the chromatographic column described. This process basically prevents (or suppresses) the mobile phase's anions from contributing to the conductivity by converting current-conducting bicarbonate anion into relatively undissociated H_2CO_3. Therefore, the conductivity detector's signal is based almost

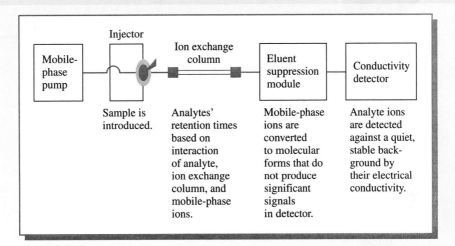

completely on the passage of analyte anions through the detector cell. (Those anions are not affected by the cation ion exchange resin.) This results in lower (i.e., better) detection limits for the analytes of interest and a more stable baseline (less noise and drift) than a similar system without eluent suppression.

The figure above is a schematic of an ion chromatographic system. Detailed are an injector, chromatographic column, eluent suppression module, and conductivity detector, as well as the processes that occur at each step.

The figure at right is an example of the kind of chromatogram that a system of this type would generate. The anions detected are fluoride, chloride, phosphate, and nitrate. As with all chromatograms, detector signal intensity is plotted versus time.

Researchers have used this method recently to determine nitrate and nitrite anions in dew, rain, and snow collected in Massachusetts (Zuo et al., 2006). Instead of conductivity detection, these authors used a UV absorption detector and chose an analytical wavelength at which neither the mobile phase nor other anions would absorb (205 nm). Surprisingly, dew had the highest concentrations of these

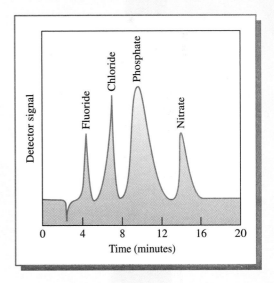

(continued on p. 660)

Ion Chromatography of Environmentally Significant Anions (continued)

Sample	Date	Nitrite (ppb)	Nitrate (ppm)
Dew 1	27.09.2005	640	4.87
2	27.09.2005	620	4.79
3	26.09.2005	830	5.99
Rain 1	02.10.2005	< DL*	2.63
2	02.09.2005	< DL*	2.62
3	28.05.2005	140	1.20
Snow 1	29.01.2005	21	0.320
2	18.01.2005	32	0.376
3	12.01.2005	32	0.60
4	12.01.2005	26	0.56

* Detection limit for nitrite = 10 ppb.

anions compared to rain and snow collected at the same site (see table above). The authors proposed that these dew nitrate concentrations, ranging from 4.79 to 5.99 $\mu g/mL$, suggest that dew is acting as a nighttime sink for these anionic species; they also note that this may be important for vegetation because these anions are held in contact with the leaf surface for long time periods as dew forms, and the concentration may spike as dew evaporates in the morning.

Since photolysis of both these anions in shallow aqueous solution can lead to the formation of hydroxyl radical and hydrogen peroxide, this may be a source of oxidative stress for plants on which the dew forms (Kobayashi et al., 2002):

$$NO_2^- + H_2O + light \longrightarrow OH + NO + OH^-$$

$$NO_3^- + H_2O + light \longrightarrow OH + NO_2 + OH^-$$

Studies of this process on red pine needles on trees on Mount Gokurakuji in western Japan have concluded that 40% of hydroxyl radial production in dew on those trees originates from nitrite and nitrate (Nakatani et al., 2001).

References: P. Janos and P. Aczel, "Ion Chromatographic Separation of Selenate and Selenite Using a Polyanionic Eluent," *Journal of Chromatography* A 749 (1996): 115–122.

T. Kobayashi, N. Natanani, T. Hirakawa, M. Suzuki, T. Miyake, M. Chiwa, T. Yuhara, N. Hashimoto, K. Inoue, K. Yamamura, N. Agus, J. R. Sinogaya, K. Nakane, A. Kume, T. Arakaki, and H. Sakugawa, *Environmental Pollution* 118 (2002): 383–391.

N. Nakatani, T. Miyake, M. Chiwa, M. Hashimoto, T. Arakaki, and H. Sakugawa, "Photochemical Formation of OH Radicals in Dew Formed on the Pine Needles at Mt. Gokurakuji," *Water, Air and Soil Pollution* 130 (2001) 397–402.

Y. Zuo, C. Wang, and T. Van, "Simultaneous Determination of Nitrite and Nitrate in Dew, Rain, Snow and Lake Water Sample by Ion-Pair High-Performance Liquid Chromatography," *Talanta* (2006): 281–285.

PART V

METALS, SOILS, SEDIMENTS, AND WASTE DISPOSAL

Contents of Part V

Environmental Instrumental Analysis VI

- Inductively Coupled Plasma Determination of Lead

Scientific American Feature Article

- Mapping Mercury

CHAPTER **15**

TOXIC HEAVY METALS

In this chapter, the following introductory chemistry topics are used:

- Redox half-reactions
- Electrolysis
- Half-life calculations
- Solubility product and acid–base equilibrium constant calculations, including manipulations for multiple equilibria

Background from previous chapters used in this chapter:

- Steady state; UV and visible light wavelengths (Chapter 1)
- Synergism (Chapter 4)
- Bioaccumulation; LD_{50}; dose–response curves; maximum contaminant level (Chapter 10)
- Aerobic, anaerobic, and calcareous waters (Chapter 13)

Introduction

In chemistry, **heavy metal** refers not to a type of rock music but rather to a type of chemical element, many examples of which are poisonous to humans. The five main ones discussed here—**mercury** (Hg), **lead** (Pb), **cadmium** (Cd), **chromium** (Cr), and **arsenic** (As)—present the greatest environmental hazard due to their extensive use, their toxicity, and their widespread distribution. None have yet pervaded the environment to such an extent as to constitute a widespread danger. However, each one has been found at toxic levels in certain locales in recent times. Metals differ from the toxic organic compounds

TABLE 15-1	Densities of Some Important Heavy Metals and Other Substances	
Substance	**Density (g/cm^3)**
Hg	13.5
Pb	11.3
Cu	9.0
Cd	8.7
Cr	7.2
Sn	5.8–7.3
As	5.8
Al	2.7
Mg	1.7
H_2O	1.0

we discussed in Chapters 10–12 in that they are totally nondegradable to nontoxic forms, although they ultimately may be transformed to insoluble forms, which therefore are biologically unavailable unless they are again converted into more soluble substances. The ultimate sinks for heavy metals are soils and sediments.

The heavy metals occur near the middle and bottom of the periodic table. Their densities are high compared to those of other common materials. The densities of the metals of interest here are collected in Table 15-1, as are values for water and two common "light" metals for contrast.

Although we commonly think of heavy metals as water pollutants, they are for the most part transported from place to place via the air, either as gases or as species adsorbed on, or absorbed in, suspended particulate matter. As we shall see later in the chapter, the deposition of airborne lead into European lake sediments dates back to the time of the ancient Greeks. These days, over half the heavy-metal input into the waters of the Great Lakes, for example, is due to deposition from the air.

Speciation and the Toxicity of Heavy Metals

Although mercury *vapor* is highly toxic, the heavy metals Hg, Pb, Cd, Cr, and As are not particularly toxic as the *condensed* free elements. However, all are dangerous in the form of their cations and most are also highly toxic when bonded to short chains of carbon atoms. Biochemically, the mechanism of the toxic action usually arises from the strong affinity of the cations for sulfur. Thus, *sulfhydryl groups*, —SH, which occur commonly in the enzymes that control the speed of critical metabolic reactions in the human body, readily attach themselves to ingested heavy-metal cations or to molecules that contain the metals. Because the resultant metal–sulfur bonding affects the entire enzyme, it cannot act normally and, as a result, human health is adversely affected, sometimes fatally. The reaction of heavy-metal cations M^{2+} (where M is Hg, Pb, or Cd) with the sulfhydryl units of enzymes R—S—H to produce stable systems such as R—S—M—S—R is analogous to their reaction with the simple inorganic chemical **hydrogen sulfide,** H_2S, with which they yield the insoluble solid MS.

PROBLEM 15-1

Write the balanced chemical reactions that correspond to the reaction of an M^{2+} ion (a) with H_2S and (b) with R—S—H to produce hydrogen ions and the products mentioned above.

A common medicinal treatment for acute heavy-metal poisoning is the administration of a compound that binds to the metal even more strongly than does the enzyme; subsequently the metal–compound combination is solubilized and excreted from the body. One compound used to treat mercury and lead poisoning is **British Anti-Lewisite** (BAL); its molecules contain two —SH groups that together capture the metal. Also useful for this purpose is the calcium salt of **ethylenediaminetetraacetic acid** (EDTA), a well-known compound that extracts and solubilizes most metal ions. The metal ions are complexed by the two nitrogens and the charged oxygens to form a chelate (Chapter 14) which is subsequently excreted from the body.

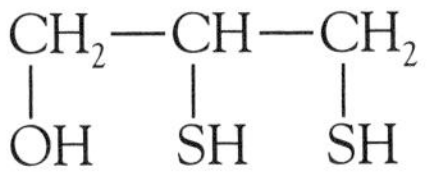

$$\begin{array}{ccc} CH_2 & -CH- & CH_2 \\ | & | & | \\ OH & SH & SH \end{array}$$

British Anti-Lewisite

Treatment of heavy-metal poisoning by chelation therapy is best begun early, before neurological damage has occurred. The calcium rather than the sodium salt is used in order that calcium ion is not inadvertently leached from the body by the EDTA.

The toxicity for all four heavy metals depends very much on the chemical form of the element, i.e., upon its **speciation.** Substances that are almost totally insoluble pass through the human body without doing much harm. The most devastating forms of the metals

- cause immediate sickness or death (e.g., a sufficiently large dose of arsenic oxide) so that therapy cannot exert its effects in time, and

- can pass through the membrane protecting the brain—the blood–brain barrier—or the placental barrier that protects the developing fetus.

For mercury and lead, the forms that have alkyl groups attached to the metal are highly toxic. Because such compounds are covalent molecules, they are soluble in animal tissue and can pass through biological membranes, whereas charged ions are less able to do so; e.g., the toxicities of lead as the ion Pb^{2+} and in covalent molecules differ substantially.

The toxicity of a given concentration of a heavy metal present in a natural waterway depends not only on its speciation but also on the water's pH and on the amounts of dissolved and suspended carbon in it, since interactions such as complexation and adsorption may well remove some of the metal ions from potential biological activity.

Bioaccumulation of Heavy Metals

Recall from Chapter 10 that some substances display the phenomenon of biomagnification: Their concentrations increase progressively along an eco-logical food chain. The only one of the five heavy metals under consideration that is indisputably capable of doing this is mercury. Many aquatic organisms do, however, bioconcentrate (but do not biomagnify) heavy metals. For example, oysters and mussels can contain levels of mercury and cadmium that are 100,000 times greater than those in the water in which they live.

The concentrations of most heavy metals in drinking water are usually small and cause no acute health problems; however, exceptions do occur and will be discussed later. As is the case with toxic organic chemicals, the amounts of metals that are ingested through our food supply are usually of much greater concern than is the intake attributable to drinking water. Paradoxically, the heavy metals in the fish that we ingest usually originate in fresh water.

Mercury

Mercury Vapor

Elemental mercury was employed in hundreds of applications, many of which (e.g., electrical switches) took advantage of the unusual property that it is a liquid that conducts electricity well. In automobiles built before 2000, electri-cal switches that operate convenience and trunk lighting contained mercury, as did instrument panels and antilock brakes; all of this mercury is lost to the environment when the cars are recycled for their steel unless the element is specifically collected, as is required in some locales.

Elemental mercury is still used in fluorescent light bulbs, including the small ones now used domestically to replace incandescent bulbs, and in the mercury lamps employed for street lighting. Although energized mercury atoms emit light in the ultraviolet rather than the visible wavelength region of the spectrum (Chapter 1), the bulbs are coated with a material that absorbs the UV and re-emits it from the bulb as visible light. The metal is released into the environment if such light bulbs are broken, though the mercury content of fluorescent lamps has been reduced by about 80% since the mid-1980s, down to 5–10 mg each today. This amount of mercury is less than the additional quantity of the metal that would have been emitted into the air by a coal-fired power plant if an incandescent light bulb, with its much lower efficiency in converting electricity to light, had been used instead. Fluorescent bulbs are virtually the only use of mercury for which a suitable alternative has not been found. For street lighting, there has been a shift toward the use of sodium vapor lamps, since such bulbs present a lower toxicity hazard and are even more efficient light sources than mercury bulbs.

Mercury is the most volatile of metals, and its vapor is highly toxic. Adequate ventilation is required whenever mercury is used in closed quarters, since the equilibrium vapor pressure of mercury is hundreds of times the maximum recommended exposure. Mercury vapor consists of free, neutral atoms. If inhaled, the atoms diffuse from the lungs into the bloodstream; then, because they are electrically neutral, they readily cross the blood–brain barrier to enter the brain. The result is serious damage to the central nervous system, manifested by difficulties with coordination, eyesight, and tactile senses. Liquid mercury itself is not highly toxic, and most of what is ingested is excreted. Nevertheless, children should not be allowed to play with droplets of the metal because of the danger from breathing the vapor.

The latter part of the twentieth century saw a substantial decline in anthropogenic emissions of mercury into the water and land environments from many sources in developed countries, resulting from governmental attempts to reduce its uses and emissions. Emissions of mercury from large industrial operations in developed countries have been successfully curtailed. The overall use of mercury has decreased in the United States by more than 95% in the last three decades. For example, mercury has been eliminated from batteries. In the United States, the reduction of mercury emissions arising from disposal of mercury-containing products has resulted mainly from

- emission controls on municipal and medical waste incinerators, and

- removal of batteries and paint from the waste stream.

In Canada, emission reductions have come from controls on metal smelting, the near-complete closure of the chlor-alkali industry (which had used mercury electrodes), and controls on waste incineration.

Over the 1990s, the global concentration of airborne mercury increased by about 1.5% per year, notwithstanding the decline in industrial emissions. Large amounts of mercury vapor are released into the air as a result of the unregulated burning of coal and fuel oil, both of which always contain trace amounts of the element, and of incinerating municipal waste that contains mercury in products such as batteries. Currently, coal-fired power plants and municipal and medical waste incinerators are the biggest sources of mercury emissions to the atmosphere in North America. The vaporized mercury is eventually oxidized and returns in rain, often falling far from the site of the original emissions. The issue of *Mercury Emissions from Power Plants* is discussed in detail in the online Case Study associated with this chapter.

In air, the great majority of elemental mercury is in the vapor (gaseous) state, with only a fraction of it bound to airborne particles. Airborne gaseous mercury can travel long distances before being oxidized and then dissolving in rain and subsequently being deposited on land or in water bodies. This global cycling of mercury results in its being distributed even to remote parts of the planet.

U.S. Emission Controls on Power Plants

The 1100 coal-burning power plants in the United States are its last unregulated major emitters of mercury into the environment. As a result of legal pressure from environmental groups, by court order the U.S. EPA was forced in 2005 to issue regulations that will reduce these emissions. The reductions are scheduled to occur in two phases:

- By 2010, total mercury emissions are to be reduced from the current 48 tons a year to 38 tons. This will be accomplished as a "co-benefit" of the new reductions in sulfur dioxide and nitrogen oxide emissions from the plants that must be implemented by that date, as discussed in Chapters 3 and 4.

- By 2018, annual mercury emissions must be reduced to 15 tons.

Each state has been assigned an emissions budget and is required to develop a plan for meeting its assigned emission reduction. States and power plants can trade reductions among themselves, as long as overall goals are achieved. Critics of the new program point out that averaging of reductions between years is allowed, so that higher emissions could occur in later years if greater-than-regulated reductions occur early. They also point out that "hot spots" of mercury deposition may be perpetuated because of the trading provisions and that the overall reduction in emissions will be quite small for many years to come.

Mercury Amalgams

Mercury readily forms **amalgams,** which are solutions or alloys with almost any other metal or combination of metals. The dental amalgam that has been used to fill cavities in teeth for more than 150 years initially has a putty-like consistency. It is prepared by combining approximately equal proportions of liquid mercury and a solid mixture that is mainly silver with variable amounts of copper, tin, and zinc. The slight expansion of volume that accompanies its solidification ensures that the final amalgam fills the cavity. When a filling is first placed in a tooth, and whenever the filling is involved in the chewing of food, a tiny amount of the mercury is vaporized. Some scientists believe that mercury exposure from this source causes long-term health problems in some individuals, but an expert panel of the U.S. National Institutes of Health concluded that dental amalgams do not pose a health risk. A recent study of adults found that no measure of exposure to dental mercury—neither the concentration of the element in the urine nor the number of dental fillings—correlated with any measure of mental functioning or fine-motor control; another study found that no IQ or other neurological problems correlated with the use of amalgams for filling teeth in children.

Some countries in Europe such as Germany have banned the use of mercury in fillings, at least for pregnant women and small children. Mercury-free "amalgams" for use in dentistry are under development; porcelain fillings are already common, though expensive. Some fears have been expressed about the release of elemental mercury vapor into the atmosphere when cremating deceased persons who had amalgam-filled teeth, since the amalgam decomposes at high temperatures. In countries such as Sweden, crematoria are fitted with selenium filters that remove most mercury from emissions by forming mercury selenide crystals.

In some areas, dentists are now required to install a separator to capture mercury from their wastewater rather than have it flow down drains and become part of municipal sewage. On average, each dentist produces about one kilogram of mercury waste per year. Dentists collectively release about the same amount of the metal as is emitted by coal-fired power plants. Because of the control of emissions by dentists, the sewage sludge used by farmers as fertilizer (Chapter 16) will have a much lower concentration of mercury in the future.

In working some ore deposits, tiny amounts of elemental gold or silver are extracted from much larger amounts of the denser particles of soil or sediment by adding elemental liquid mercury to the mixture. The mercury extracts the gold or silver by forming an amalgam, which is then roasted to distill off the mercury. From 1570 until about 1900, this process was used to extract silver from ores in Central and South America. About one gram of mercury was lost to the environment for every gram of silver so produced, resulting in the release of almost 200,000 tonnes of mercury. The mercury was shipped to these regions from Almaden, Spain, and from Peru. Until recently, the corresponding process of extracting gold by amalgamation with mercury was carried out in China in both large-scale and small-scale operations. The ratio of mercury to gold in such workshops, some of which continue illegally today in remote regions, averaged 15 to 1.

Today, the gold extraction procedure using mercury is carried out on a large scale in Brazil to obtain gold from muddy sediments; it results in substantial mercury pollution both in the air and, because of careless handling practices, in the Amazon River itself. The health hazards to those whose work involves the vaporization of mercury are significant, since the element is so toxic in its gaseous form. Indeed, mercury vaporized from such operations currently makes up more than 10% of the anthrogogenic emissions of mercury in air. People who live in mining regions often inhale air in which the concentration of elemental mercury exceeds 50 $\mu g/m^3$, which is 50 times the public exposure guideline of the World Health Organization (WHO). As a consequence, many "amalgam burner" workers exhibit tremors and other signs of mercury poisoning. In addition, mercury in surface sediments disturbed by slash-and-burn deforestation and agriculture in the region enters the aquatic environment, where some of it enters the food chain. The European Union has undertaken initiatives to incorporate inexpensive technology into the process to prevent the massive release of mercury into the air and to the Amazon river during the extraction of gold.

Mercury and the Industrial Chlor-Alkali Process

An amalgam of sodium and mercury is used in some industrial *chlor-alkali* plants in the process that converts aqueous sodium chloride into the commercial products **chlorine,** Cl_2, and **sodium hydroxide,** NaOH, (and gaseous hydrogen) by electrolysis. In order to form a concentrated, pure solution of NaOH, flowing mercury is used as the negative electrode (cathode) of the electrochemical cell. The metallic sodium that is produced by reduction in the electrolysis immediately combines with the mercury and is removed from the NaCl solution without having reacted with the aqueous medium:

$$Na^+(aq) + e^- \xrightarrow{\text{Hg}} Na \text{ (in a Na–Hg amalgam)}$$

When metals such as sodium are dissolved in amalgams, their reactivity is greatly lessened compared to that for the free state, so that the otherwise highly reactive elemental sodium in the Na–Hg amalgam does not react with the water in the original solution. Instead, the amalgam is removed and later induced by the application of a small electrical current to react with water in a separate chamber, thereby producing sodium hydroxide that is free of salt.

The mercury is recovered after NaOH production and is recycled back to the original cell. The recycling of mercury is not complete, however, and some finds its way into the air and into the water body from which the plant's cooling water is obtained and to which it is returned. Although liquid mercury is not soluble in water or in dilute acid, it can be oxidized to soluble form by the intervention of bacteria that are present in natural waters. By this means, the mercury becomes accessible to fish.

The mass of mercury lost to the environment from the average chlor-alkali plant has decreased enormously since the problem was identified in the 1960s. Nevertheless, installations in North America that use mercury electrodes have largely been phased out. They were replaced by ones that use a fluorocarbon membrane that separates the NaCl solution from the chloride-free solution at the negative electrode. The membrane is designed such that Na^+, but not anions, can pass through it. In both types of cells, the overall reaction is

$$2\,NaCl(aq) + H_2O(l) \longrightarrow 2\,NaOH(aq) + Cl_2(g) + H_2(g)$$

The 2+ Ion of Mercury

Like its partners zinc and cadmium in the same subgroup of the periodic table, the common ion of mercury is the 2+ species, Hg^{2+}, the **mercuric or mercury(II)** ion. An example of a compound containing the mercuric ion is the red ore *cinnabar*, HgS, i.e., $Hg^{2+}S^{2-}$. Like most sulfides, this salt is very insoluble in water; indeed, the wastewater at chlor-alkali plants is sometimes treated by adding a soluble salt such as Na_2S that contains the sulfide ion, since this action precipitates ionic mercury as HgS.

$$Hg^{2+} + S^{2-} \longrightarrow HgS(s)$$

PROBLEM 15-2

The solubility product, K_{sp}, for HgS is 3.0×10^{-53}. Calculate the solubility of HgS in water in moles per liter and transform your answer into the number of mercuric ions per liter. According to this calculation, what volume of water in equilibrium with solid HgS contains a single Hg^{2+} ion?

Most of the mercury in the environment is inorganic, in the form of the Hg^{2+} ion. The levels of ionic mercury even in remote areas are two to five times as great as preindustrial values, with local polluted sites having levels 10 or more times greater. In natural waters, much of the Hg^{2+} is attached to suspended particulates, so it is eventually deposited in sediments—a topic considered in further detail when soil and sediment chemistry is discussed (Chapter 16).

The nitrate salt of Hg^{2+} is water soluble and was at one time used to treat the fur used to make felt for hats. The fur was immersed in a hot solution of mercuric nitrate, which made the fibers rough and twisted so they would then mat together easily. As a consequence of this constant exposure to mercury, workers in the felt trade often displayed nervous disorders: muscle tremors, depression, memory loss, paralysis, and insanity (giving rise to the expression "mad as a hatter," a concept familiar to fans of Lewis Carroll's *Alice in Wonderland*). Mercury vapor and, to a lesser extent, mercury salts attack the central nervous system, but the main target organs for Hg^{2+} are the kidney and the liver, where it can cause extensive damage.

Mercuric oxide, HgO, is present in a paste in *mercury cell* batteries such as those used in hearing aids. If the discarded spent batteries are subsequently incinerated as garbage, the volatile mercury can be released into the air. The amount of mercury used in ordinary flashlight batteries, added as a minor constituent in the zinc electrode to prevent its corrosion and thereby extend the shelf life of the product, was first drastically curtailed—typically from about 10,000 ppm to about 300 ppm in alkaline batteries—and in many cases has been completely eliminated, thereby halving the mercury in domestic garbage. In North America, only some "button batteries" used in watches, calculators, hearing aids, etc. still have significant mercury content.

The other inorganic ion of mercury, Hg_2^{2+}, is not very toxic since it combines in the stomach with chloride ion to produce insoluble Hg_2Cl_2.

Methylmercury Toxicity

When in combination with anions that are capable of forming covalent bonds, the mercuric ion Hg^{2+} forms covalent molecules rather than an ionic solid. For example, $HgCl_2$ is a molecular compound, not a salt of Hg^{2+} and Cl^-. As chloride ion forms a covalent compound with Hg^{2+}, so does the *methyl anion*, CH_3^-, yielding the volatile molecular liquid **dimethylmercury, $Hg(CH_3)_2$.** The process of dimethylmercury formation occurs in the muddy sediments of rivers and lakes, especially under anaerobic conditions, when anaerobic bacteria and microorganisms convert Hg^{2+} into $Hg(CH_3)_2$. The active agent in the biomethylation process is a common constituent of microorganisms; it is a derivative of vitamin B_{12} with a CH_3^- anion bound to cobalt and is called *methylcobalamin.*

The less volatile mixed compounds CH_3HgCl and CH_3HgOH, collectively called **methylmercury** (or *monomethylmercury*), are often written as CH_3HgX, or somewhat misleadingly as $CH_3Hg^+X^-$, since these substances, like most of those written as Hg^{2+}, consist of covalent molecules, not ionic lattices. In fact, the methylmercury ion CH_3Hg^+ exists as such only in compounds with anions such as nitrate or sulfate. Methylmercury compounds are even more readily formed in the same way as dimethylmercury at the surface of sediments in anaerobic water. Methylmercury production predominates over dimethylmercury formation in acidic or neutral aqueous systems. **Sulfate ion, SO_4^{2-},** stimulates the sulfate-reducing bacteria that methylate mercury; in contrast, the presence of sulfide ion results in formation of mercury sulfide complexes that do not undergo methylation.

Due to its volatility, dimethylmercury evaporates from water relatively quickly unless it is transformed by acidic conditions into the monomethyl form. The pathways for the production and fate of dimethylmercury and other mercury species in a body of water are illustrated in Figure 15-1. Methylation of inorganic mercury does occur in anaerobic regions of lakes, especially near the interface of the epilimnion and the hypolimnion, and at the interface of the latter with sediments, but not in aerobic water. Organic-rich sediments at the bottom of warm, shallow lakes are important sites of methylmercury production. Wetlands are also active sites of methylmercury production. Methylmercury in surface water is photodegraded (to as yet unknown products) and is the most important sink for this substance in some lakes.

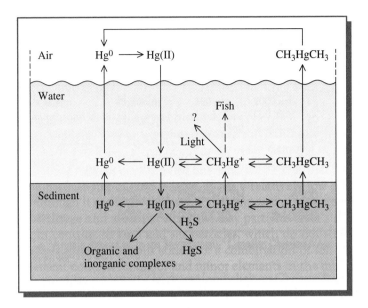

FIGURE 15-1 The cycling of mercury in fresh-water lakes. [Source: Adapted from M. R. Winfrey and J. W. M. Rudd, "Environmental Factors Affecting the Formation of Methylmercury in Low pH Lakes," *Environmental Toxicology and Chemistry* 9 (1990): 853–869.]

Mercuric ion itself is not readily directly transported across biological membranes. Methylmercury is a more potent toxin than are salts of Hg^{2+} because it is soluble in fatty tissue in animals, bioaccumulates and biomagnifies there, and is more mobile. Once ingested, the original covalent CH_3HgX compound is converted to substances in which X is a sulfur-containing amino acid; in some of these forms, it is soluble in biological tissue and can cross both the blood–brain barrier and the human placental barrier, presenting a twofold hazard. Methylmercury is, in fact, the most hazardous form of mercury, followed by the vapor of the element. The main toxicity of methylmercury occurs in the central nervous system. In the brain, methylmercury is converted to Hg^{2+}, which is probably responsible for the brain damage. Mercury vapor is also oxidized to this ion once it has entered the cell. Thus the usual barriers in the cell to Hg^{2+} are circumvented by Hg^0 and CH_3HgX, which by their electric neutrality can penetrate through the defenses and which later can be converted to the highly toxic +2 ionic form.

Most of the mercury present in humans is in the form of methylmercury. Almost all methylmercury originates from the fish in our food supply: Mercury in fish is usually at least 80% methylmercury. Mercury contamination is the reason behind about 97% of the advisories against eating fish caught in various regions of North America. In contrast to organochlorines, which predominate in the fatty portions of fish, methylmercury can bind to the sulfhydryl group in proteins and so is distributed throughout the fish. Consequently, the mercury-containing part cannot be cut away before the fish is eaten.

Fish absorb methylmercury that is dissolved in water as it passes across their gills (bioconcentration); they also absorb it from their food supply (biomagnification). The ratio between methylmercury in fish muscle and that dissolved in the water in which the fish swims is often about 1 million to 1, and can exceed 10 million to 1. The highest methylmercury concentrations (over 1 ppm) are usually found in large, long-lived predatory marine species such as shark, king mackerel, tilefish, swordfish, and large tuna (sold as steaks and sushi), as well as in fresh-water species such as bass, trout, and pike. Indeed, the U.S. Food and Drug Administration warns women of childbearing age not to eat the first four types of marine fish in the list above. On average, the older the fish, the more methylmercury it will have bioaccumulated. Noncarnivorous species such as whitefish do not accumulate very much mercury since biomagnification in their food chain operates to a much lesser extent than in carnivorous fish. On average, most Americans take in almost half their methylmercury from tuna (mostly of the canned variety), followed by swordfish, pollock, shrimp, and cod. Canada has recently limited the maximum concentration of mercury in six species of ocean fish to one part per million; this concentration is not uncommon in swordfish, although most fish have levels of 0.10–0.15 ppm. The U.S. EPA has set a criterion of 0.3 ppm maximum for methylmercury in fish tissue.

In lakes, the mercury content in fish is generally greater in acidic water, probably because both the solubility of mercury is greater and the methylation of mercury is faster at lower pH. In this way, the acidification of natural waters indirectly increases the exposure of fish-eaters to methylmercury. The relationship between mercury levels in small fish and the pH of the water is illustrated in Figure 15-2. The data in this figure are from a collection of lakes in Wisconsin, eastern Ontario, and Nova Scotia; most of the acidic lakes are in Nova Scotia.

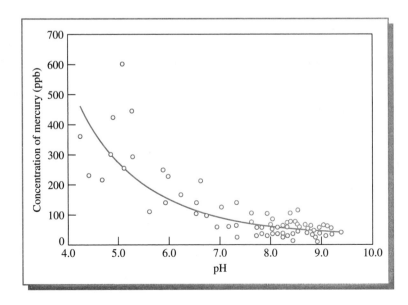

FIGURE 15-2 The relationship between pH and mercury concentration in a fish for a standard fish length for 48 lakes from Ontario, Nova Scotia, and Wisconsin.
[Source: D. Lean, "Mercury Pollution," *Canadian Chemical News* (January 2003): 23.]

Methylmercury Accumulation in the Environment and in the Human Body

The half-life of methylmercury compounds in humans, about 70 days, is much longer than that for Hg^{2+} salts, due in part to the compounds' greater solubility in a lipid environment. Consequently, methylmercury can accumulate in the body to a much higher steady-state concentration, even if on a daily basis a person consumes amounts that individually would not be harmful.

PROBLEM 15-3

If the half-life of methylmercury in the human body is 70 days, what is its steady-state accumulation in a person who consumes daily 1.0 kg of fish containing 0.5 ppm methylmercury? [Hint: Recall the discussion of steady-state concentrations in Chapter 6.]

Most of the well-publicized environmental problems involving mercury have arisen from the fact that in the methylated form it is a cumulative poison. However, in high enough concentration it can be acutely fatal. In 1997, cancer researcher Karen Wetterhahn of Dartmouth College died from mercury poisoning several months after a drop or two of pure dimethylmercury apparently seeped through latex gloves she was wearing while using the compound in experiments. Dialkylmercury compounds, including dimethylmercury, are sometimes called *supertoxic* because they are lethal even in small amounts.

At the fishing village of Minamata, Japan, a chemical plant employing Hg^{2+} as a catalyst in a process that produced polyvinyl chloride discharged mercury-containing residues into Minamata Bay. The methylmercury compounds, mainly $CH_3Hg-SCH_3$, that subsequently formed from the inorganic mercury by biomethylation by microorganisms in the bay's sediments then bioaccumulated. Concentrations were as high as 100 ppm in the fish, which were the main component of the diet for many local residents. (By way of contrast, the current U.S. recommended limit for methylmercury in fish to be consumed by humans is 0.3 ppm.) Thousands of people in Minamata were affected in the 1950s by mercury poisoning from this source, and hundreds of them died from it. Because the onset of symptoms in humans is delayed, the first signs of *Minamata disease* were observed in cats who ate discarded fish: They began jumping around and twitching, ran in circles, and finally threw themselves into the water and drowned. Symptoms in humans arise from dysfunctions of the central nervous system, since the target organ for methylmercury is the brain; they include numbness in arms and legs, blurring and even loss of vision, loss of hearing and muscle coordination, and lethargy and irritability.

Since methylmercury can be passed to the fetus, children born to Minamata mothers poisoned even slightly by mercury showed severe brain damage, some to a fatal extent. The infants showed symptoms similar to those of cerebral palsy: mental retardation, seizures, motor disturbance, and even paralysis. Just as in the case of high PCB levels, discussed in Chapter 11, the developing fetuses were much more affected by methylmercury than were the mothers themselves. The poisonings at Minamata must surely rank as one of the major environmental disasters of modern times.

Other Sources of Methylmercury

Organic compounds of mercury have been used as fungicides in agriculture and in industry and enter the environment as a side effect of these applications. However, as a result of contact with soil, the compounds are eventually broken down and the mercury becomes trapped as insoluble compounds by attachment to sulfur ions present in clays and organic matter.

Hundreds of deaths in Iraq in 1956, 1960, and 1972, and a few in China and the United States, resulted from the consumption of bread made from seed grain (intended for planting) that had been treated with mercury-based fungicides to reduce seedling losses from fungus attack. The fungicides contained compounds of **ethylmercury,** $CH_3CH_2Hg^+$, the toxicity of which is presumed to be similar to that of methylmercury. In Sweden and Canada, the use of mercury compounds to treat seeds led to a significant reduction in the number of birds of prey that consumed the smaller birds and mammals that fed on the scattered seed. The use of ethylmercury products in agriculture has now been curtailed in North America and western Europe.

Mercury is leached from rocks and soil into water systems by natural processes, some of which are accelerated by human activities. Flooding of vegetated areas can release mercury into water. For example, after the flooding of huge areas of northern Quebec and Manitoba in constructing hydroelectric power dams, the newly submerged surface soils (and to a lesser extent the vegetation) released a considerable quantity of soluble methylmercury, formed from the "natural" mercury content of these media. The additional methylmercury resulted from contact of soil-bound Hg^{2+} with anaerobic bacteria produced by the decomposition of the immersed organic matter. In this way, previously insoluble inorganic mercury was converted to methylmercury, which readily dissolved in the water. The methylmercury subsequently entered the food chain through its absorption by fish, and native persons who ate fish from these flooded areas now have substantially elevated levels of mercury in their bodies. Indeed, the methylmercury concentration in fish from these areas, 5 ppm or more, approaches that previously associated only with regions of industrial mercury pollution.

In 1999, the safety of using a mercury-based preservative to prevent microbial growth in many vaccine preparations administered to infants was questioned by the American Academy of Pediatrics and the U.S. Public Health Service. The preservative, *Thimerosal*, is $CH_3CH_2\text{—Hg—S—}C_6H_4COOH$; it is sometimes said to contain the ethylmercury ion, but presumably, like most methylmercury systems, it is actually a covalent compound. This substance has now been removed by its manufacturer in all vaccines destined for use in young children in the United States, with the exception of inactivated influenza vaccine. The preservative was also widely used as a topical disinfectant.

The Use of Mercury in Preservatives and as Medications

Compounds of the **phenylmercury** ion, $C_6H_5Hg^+$, with acetate or nitrate as the anion, have been used to preserve paint while in the can and to prevent mildew after application of latex paint, particularly in humid areas. The phenylmercury salts are not as toxic to humans as are methylmercury compounds, since they break down quickly into compounds of the less toxic Hg^{2+}. However, mercury compounds have been banned from indoor latex paints in North America for more than a decade because some ingestion of the element from this source is inevitable. Phenylmercury compounds were also formerly used as *slimicides* in the pulp-and-paper industry in order to prevent the growth of slime on wet pulp; because this practice has now been curtailed and because mercury-containing wastes are now usually treated, mercury releases from such sources have greatly decreased.

Because of their antiseptic and preservative qualities, however, some mercury compounds are still used in pharmaceuticals (especially topical antiseptics) and cosmetics. Elemental mercury was also used in some pharmaceuticals in the old days. Indeed, the antidepressant pills that Abraham Lincoln took, mainly in the years before becoming president, contained the element; indeed, some medical historians think the leaching of mercury into his bloodstream can account for his often bizarre behavior in that period. The mercury ore cinnabar is still used today in China, as a drug, pigment, and preservative.

PROBLEM 15-4

A quantity of a mercury–chlorine compound is included in a shipment of waste to a toxic waste disposal dump. Before it can be disposed of properly, the owners of the dump need to know whether it is $HgCl_2$, or Hg_2Cl_2, or some other compound. They send a sample of it for analysis and find that it contains 26.1% chlorine by mass. What is the empirical formula of the compound?

Safe Level of Mercury in the Body

It is somewhat reassuring that both the direct effects of methylmercury on humans and the prenatal effects probably have thresholds below which no effects are observed. Currently the daily methylmercury intake of 99.9% of Americans lies below the WHO's "safe limit." Nonetheless, some effects of methylmercury consumption on human vision are observed even when the concentration of (total) mercury in hair lies below the generally recognized threshold of 50 ppm.

However, if prenatal health is the main consideration and if a safety factor of 10 is applied, a substantial fraction of the population of the United States would exceed the safe limit. The WHO has concluded that levels of 10–20 ppm of methylmercury in hair indicate that a pregnant woman has sufficient methylmercury in her blood to represent a threat to a developing fetus. This places at risk the developing fetus of more than 30% of the women in some native communities in northern Canada, for example, in which fish play a large part in the diet. Although it is clear that high levels of methylmercury can result in developmental disabilities, there is continuing controversy over whether methylmercury acquired through a diet high in fish and marine mammals can cause significant neurological damage to an adult or a developing fetus. The epidemiological studies of this question have, to this point, produced inconsistent results.

PROBLEM 15-5

What is the mass, in milligrams, of mercury in a 1.00-kg lake trout which just meets the Northern American standard of 0.50 ppm Hg? What mass of fish, each at the 0.50-ppm Hg level, would you have to eat in order to ingest a total of 100 mg of mercury?

PROBLEM 15-6

The new U.S. EPA oral reference dose for methylmercury is 0.1 micrograms per kilogram body weight per day. What mass of fish can a 60-kg woman safely eat each week if the average methylmercury level in the fish is 0.30 ppm? Approximately how many average servings of fish does this correspond to?

International Controls on Mercury

Although atmospheric emissions now dominate mercury concerns in most countries, especially developed ones, other sources also contribute significantly elsewhere. Mercury is still used extensively in the extraction of gold and silver, as well as in the production of chlorine in chlor-alkali plants, in developing countries.

In 2005 the United Nations Environment Programme considered devising a global treaty to curb the production of mercury and to ban completely the export of mercury between countries. However, the United States led a movement, which ultimately was successful, that instead proposed voluntary partnerships between countries to improve their management of mercury.

Lead

Although the environmental concentration of lead, Pb, is still increasing in some parts of the world, the uses that result in its uncontrolled dispersion have been greatly reduced in the last few decades in many developed countries. Consequently, its concentration in soil, water, and air has decreased substantially.

Lead's relatively low melting point of 327°C allows it to be readily worked—it was the first metal to be extracted from its ores—and shaped. Lead was used as a structural metal in ancient times as well as for weatherproofing buildings, in water pipes and ducts, and for cooking vessels. Lead is still used for roofing and flashing and for soundproofing in buildings. When combined with tin, it forms *solder*, the low-melting alloy used in electronics and in other applications (e.g., tin cans) to connect solid metals.

Analysis of ice-core samples from Greenland indicates that atmospheric lead concentration reached a peak in Roman times that was not equaled again until the Renaissance. The history of lead's presence in the environment can be seen in Figure 15-3, in which the ratio of two stable lead isotopes

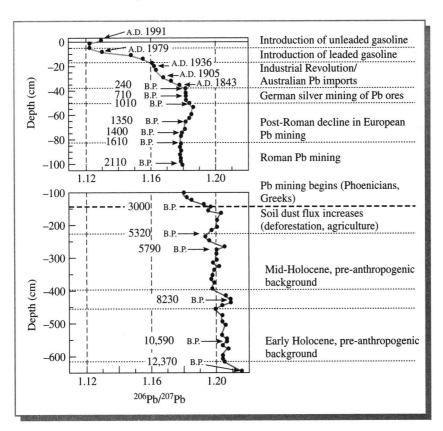

FIGURE 15-3 Isotopic composition of lead in a Swiss peat bog and the chronology of atmospheric lead deposition. Notice the change in depth scale at 100 cm. [Source: W. Shotyk et al., "History of Atmospheric Lead Deposition Since ^{14}C BP from a Peat Bog, Jura Mountain, Switzerland," *Science* 281 (1998): 1635, Figure 3B.]

in samples taken from a peat bog in Switzerland is plotted against the depth at which the sample was taken. The layers of the bog were laid down gradually over millennia, and each layer incorporated lead-containing dust particles deposited from the air at the time. Lead originating in different geographic locales has different isotope ratios, so we can tell the origin of atmospheric lead at different times in the past from the graph. The $^{206}Pb/^{207}Pb$ ratio is close to 1.20 for depths exceeding 145 cm, or 3000 years ago (sediment ages being determined by ^{14}C dating of the peat); the variations in the ratio below that depth reflect changes in the dominance of weathering of soils and rocks in different areas over time. Beginning about 3000 years ago, the ratio fell to 1.18, reflecting the isotopic composition of European lead ores mined during Roman times and thereafter. Previous to that, in Greek times, silver was first mass-produced for use in coins; apparently the substantial amount of lead contaminant in the crude silver escaped into the air during the refining of the metal. In about 1860, the isotope ratio of the lead deposited in Europe began a continual decrease, with the rate of change increasing with the introduction of leaded gasoline in about 1940, probably as a consequence of the extensive use of lead in it, first from Australia (ratio of 1.04) and then also from Canada. Recently the ratio has begun to increase due to the decreased use of lead in European gasoline.

Elemental Lead as an Environmental Risk

Elemental lead is also found in ammunition ("lead shot") used in huge amounts by hunters, especially of waterfowl. Many ducks and geese are injured or die from chronic lead poisoning after ingestion of lead shot, which dissolves in the acidic environment inside them. In addition, ducks consume the pellets left lying on the ground or at the bottom of ponds, since they look like food or grit. Birds (such as bald eagles) sometimes prey on ducks and other waterfowl that have been shot by hunters but not harvested by them, or they sometimes eat lead shot to help grind food in their gizzard; these predators become victims of lead poisoning. For these reasons, lead shot has been banned in the United States, Canada, the Netherlands, Norway, and Denmark. However, in North America, many loons die because they swallow and are subsequently poisoned by lead sinkers and jigs still used in sport fishing.

Lead ammunition in the form of bullets and shotgun shells (used for shooting wild game) also poses an environmental problem. Condors in California suffer from lead poisoning, sometimes fatally, when they eat deer that have been shot and then abandoned by hunters; the lead bullets explode into many fragments on impact and contaminate the meat.

Ionic 2+ Lead in Water and Food as an Environmental Hazard to Humans

Although elemental lead is not an environmental problem to most life forms, it does become a real concern when it dissolves to yield an ionic species.

The stable ion of lead is the 2+ species, Pb(II) as Pb^{2+}. For example, lead forms the ionic **lead sulfide,** PbS, $Pb^{2+}S^{2-}$, which is the metal-bearing component of the highly insoluble ore *galena,* from which almost all lead is extracted.

Lead does not react on its own with dilute acids. Indeed, elemental lead is stable as an electrode in the *lead storage battery,* even though it is in contact with fairly concentrated **sulfuric acid,** H_2SO_4. However, some lead in the solder that was commonly used in the past to seal tin cans will dissolve in the dilute acid of fruit juices and other acidic foods if air is present—that is, once the can has been opened—since lead is oxidized by oxygen in acidic environments:

$$2\,Pb(s) + O_2 + 4\,H^+ \longrightarrow 2\,Pb^{2+}(aq) + 2\,H_2O$$

The Pb^{2+} produced by this half-reaction contaminates the contents of the can; for this reason, lead solder is not usually used any more for food containers in North America. Partially as a result of this change, the average daily intake of lead for two-year-old children dropped from about 30 μg in 1982 to about 2 μg in 1991.

The 1845 Franklin Expedition to find a Northwest Passage across the Arctic is thought to have failed because the members all died from lead poisoning from the solder in the tin cans that held their food. Canadian writer Margaret Atwood has written eloquently about the incident in her short story "The Age of Lead":

> It was the tin cans that did it, a new invention back then, a new technology, the ultimate defence against starvation and scurvy. The Franklin Expedition was excellently provisioned with tin cans, stuffed full of meat and soup and soldered together with lead. The whole expedition got lead poisoning. Nobody knew it. Nobody could taste it. It invaded their bones, their lungs, their brains, weakening them and confusing their thinking, so that at the end those that had not yet died in the ships set out in an idiotic trek across the stony, icy ground, pulling a lifeboat laden down with toothbrushes, soap, handkerchiefs and slippers, useless pieces of junk. When they were found (ten years later, skeletons in tattered coats, lying where they'd collapsed) they were headed back toward the ships. It was what they'd been eating that had killed them.
>
> [Margaret Atwood, "The Age of Lead," in *Wilderness Tips,* copyright 1991 by O. W. Toad Limited]

The recommended maximum levels for important heavy metal ions in drinking water are summarized in Table 15-2. The limit for lead, 10–15 ppb, is sometimes exceeded in water delivered to the consumer even though it was sufficiently pure when it left the water treatment plant. Lead used in the solder in the joints of domestic copper water pipes, and lead used in previous decades and centuries to construct the pipes themselves, can dissolve in drinking water during its transport to the point of consumption, particularly if the water is quite acidic or particularly soft. This problem of contamination of

TABLE 15-2	Drinking-Water Guidelines for Heavy Metals		
Metal	U.S. EPA Maximum Contaminant Level (ppb)	Canadian Maximum Acceptable Concentration (ppb)	World Health Organization Guideline (ppb)
As	10	10	10
Cd	5	5	3
Cr	100	50	50
Hg (inorganic)	2	1	6
Pb	15	10	10

water by lead during transit became a controversial issue in 2007 in the home-town (London, Ontario) of one of the authors of this book, with the concern quickly spreading to older homes in other cities in Ontario as well. In general, it is a good idea not to drink water that has been standing overnight in older drinking fountains or in the pipes of older dwellings; water in such plumbing systems should be allowed to run for a minute or so.

The contamination of water by lead is less of a problem in areas of cal-careous water, since an insoluble layer containing compounds such as $PbCO_3$ forms on the surface of the lead by reaction of the metal with dissolved oxygen and the **carbonate ion,** CO_3^{2-}, in the water (Chapter 13). This layer prevents the metal underneath from dissolving in the water that passes over it. In some regions of England and in some cities in the northeast United States that have soft water and networks of old lead pipes, phosphates are added to drinking water in order to form a similar insoluble protective coat-ing of lead phosphate on the inside of lead pipes and so reduce the concentra-tion of dissolved lead.

Lead in water is more fully absorbed by the body than is lead in food. Now that many other sources of lead have been phased out, drinking water accounts for about one-fifth of the collective lead intake of Americans, whose major source is from food. Many domestic water treatment systems successfully remove the great majority of lead from drinking water. Bottled water sold in plastic containers usually has very low levels of lead, averaging 16 ppt in one recent survey, which is not much higher than those in groundwater taken from pristine deep aquifers. Bottled water in glass containers has more lead, up to about 1 ppb, since tiny amounts of the metal are leached from the glass.

PROBLEM 15-7

According to an informal 1992 survey, the drinking water in about one-third of the homes in Chicago had lead levels of about 10 ppb. Assuming that an adult drinks about 2 L of water a day, calculate the total lead that residents of these Chicago homes obtain daily from their drinking water.

Lead Salts as Glazes and Pigments

One form of the oxide PbO is a yellow solid that has been used at least as far back in history as ancient Egypt to glaze pottery. In glazing, the material is fused as a thin film to the surface of the pottery in order to make it waterproof and to give it a brilliant high gloss. The oxide becomes a hazard if applied incorrectly: Some of it will dissolve over a period of hours and days if acidic foods and acidic liquids, such as cider, are stored in pottery containers, giving dissolved Pb^{2+}, up to hundreds or even thousands of parts per million, in the food:

$$PbO(s) + 2\,H^+(aq) \longrightarrow Pb^{2+}(aq) + H_2O$$

Indeed, lead-glazed dishware is still a major source of dietary lead, especially, but not exclusively, in developing countries. The leaching of lead from glazed ceramics used to prepare food is one of the major sources of the element for children in Mexico, where lead contamination is a leading public health problem. Nowadays, lead silicate rather than oxide or sulfate is used for glazing in most countries since it is almost insoluble and thus much safer.

Various salts of lead have been used as pigments for millennia, since they give stable, brilliant colors. **Lead chromate,** $PbCrO_4$, is the yellow pigment used in paints applied to school buses and for the yellow stripes on roads. **Red lead,** Pb_3O_4, is used in corrosion-resistant paints and has a bright red color. It was used in great quantity in the past to produce a rust-resistant surface coating for iron and steel. **Lead acetate** is often used in preparations to cover gray hair, since the Pb^{2+} ion of this soluble salt will combine with the SH group of hair proteins to give a dark color.

Lead pigments have been used to produce the colors used in glossy magazines and food wrappers. In past centuries, lead salts were used as coloring agents in various foods. **White lead,** $Pb_3(CO_3)_2(OH)_2$, was extensively used until the middle of the twentieth century as a major component of white indoor paint. Since it was more durable than unleaded paint, it was often used on surfaces subject to punishment such as kitchen cabinets and window trim. However, when the paint peels off, small children may eat the paint flecks since Pb^{2+} has a sweet taste. Persons who renovate old homes are urged to ensure that dust from layers of old paint is properly contained. Children in inner-city slums, in which old coats of paint continue to peel, are often found to have elevated blood levels of lead. In indoor paint, white lead has now been replaced by the pigment **titanium dioxide,** TiO_2. Although now banned from use in indoor paints (since 1978 in the United States), lead pigments continue to be used in exterior paints, with the result that soil around houses may eventually become contaminated. Some of this lead-contaminated soil may be ingested by small children because of its sweet taste. Lead is still widely used in indoor paint sold in China, India, and some other Asian countries, sometimes at levels exceeding 180,000 ppm (as compared to the U.S. standard of 600 ppm maximum for new paints).

An additional source of sweet lead-containing dust was the surface of some types of PVC miniblinds that had lead incorporated as a stabilizer in the plastic and that underwent partial decomposition from exposure to UV in sunlight. Lead is used as a stabilizer in a variety of other PVC products as well, including children's toys.

Lead dust, which originates as soil containing tiny particles of lead compounds, is now the biggest source of lead for children in U.S. inner cities. The lead collectively originates from individually small but numerous contributions from many sources already mentioned—paint flakes, ceramics, plastics, gasoline, recycling plants, and even lead salts used in hair coloring preparations for people with graying hair. The use of **lead arsenate,** $Pb(AsO_4)_2$, as a pesticide was another former source of Pb^{2+} in soil.

Green Chemistry: Replacement of Lead in Electrodeposition Coatings

Sheet metal surfaces made of steel undergo corrosion very rapidly unless they are covered with a protective coating. Since the 1960s a technique called *electrodeposition* has competed with spray painting for coating steel. In 1976 the first automobiles were treated by electrodeposition. In this technique, the surface to be treated is dipped in a bath, with the surface acting as a cathode or anode, and the coating is deposited electrophoretically. Electrodeposition has many advantages over spray painting, including:

- lower air pollution, due to decreased solvent emissions;

- better corrosion protection, due to better coverage of poorly accessible areas;

- reduced waste, due to high transfer efficiency; and

- more uniform coating thickness.

Virtually all primer coats for automobiles are done using this method. Red lead, mentioned previously, offers significant corrosion resistance, and primer coats use large amounts of this material. Although lead has been banned from house paints in the United States since 1972, the demand for corrosion resistance for motor vehicles has resulted in exemptions from environmental regulations regarding lead in automobile and truck paints.

PPG Industries has discovered that *yttrium oxide* serves as an excellent replacement for lead as a corrosion inhibitor and won a Presidential Green Chemistry Challenge Award in 2001. On a weight basis, yttrium is twice as effective in inhibiting corrosion as red lead but is only 1/120th as toxic. An additional consideration is the pretreatment process that is used to assist in adhesion and corrosion resistance prior to the application of the electrocoat. The use of yttrium eliminates chromium from metal pretreatments and reduces the amount of nickel compared to the lead process. It is estimated that employing yttrium in automobile electrodeposition will eliminate not only the use of 1 million pounds of lead but also 25,000 pounds of chromium and 50,000 pounds of nickel on an annual basis. As of September 2006, more than 38 million motor vehicles have been coated with the yittrium-containing product since its introduction in 2001. According to PPG Industries, no customers in either the United States or Europe purchase any lead-containing coating product for any application, including automotive.

Dissolution of Otherwise Insoluble Lead Salts

The presence of significant concentrations of lead in natural waters is seemingly paradoxical, given that both its sulfide, PbS, and its carbonate, $PbCO_3$, are highly insoluble in water:

$$PbS(s) \rightleftharpoons Pb^{2+} + S^{2-} \qquad K_{sp} = 8.4 \times 10^{-28}$$

$$PbCO_3(s) \rightleftharpoons Pb^{2+} + CO_3^{2-} \qquad K_{sp} = 1.5 \times 10^{-13}$$

However, the anions in both salts are fairly strong bases. Thus both of the above dissolution reactions are followed by the reaction of the anions with water:

$$S^{2-} + H_2O \rightleftharpoons HS^- + OH^-$$

$$CO_3^{2-} + H_2O \rightleftharpoons CO_3^- + OH^-$$

Because these reactions reduce the concentrations of the original anions produced by dissolution of the salt PbS or $PbCO_3$, the position of equilibrium in the original reactions shifts to the right side, thereby dissolving more of the salt, in analogy with the process involving $CaCO_3$ that we analyzed in Chapter 13. Thus the solubilities of PbS and $PbCO_3$ in water are substantially increased by the reaction of the anion with water.

If highly acidic water comes into contact with minerals such as PbS, the "insoluble" solid dissolves to a much greater extent than in neutral waters. This occurs because the sulfide ion initially produced is subsequently converted almost entirely to **bisulfide ion,** HS^-, which in turn is converted by the acid to dissolved hydrogen sulfide gas, H_2S, since both S^{2-} and HS^- act as bases in the presence of acid:

$$S^{2-} + H^+ \rightleftharpoons HS^- \qquad K = 1/K_a\,(HS^-) = 7.7 \times 10^{12}$$

$$HS^- + H^+ \rightleftharpoons H_2S \qquad K' = 1/K_a\,(H_2S) = 1.0 \times 10^7$$

When these two reactions are added to that for the dissolution of PbS into Pb^{2+} and S^{2-}, the overall reaction is seen to be

$$PbS(s) + 2\,H^+ \rightleftharpoons Pb^{2+} + H_2S(aq)$$

Since the equilibrium constant $K_{overall}$ for an overall process which is the sum of several others is the product of their equilibrium constants, in this case $K_{overall} = K_{sp}KK' = 6.5 \times 10^{-8}$. By application of the *law of mass action* to this reaction, we find the expression for the equilibrium constant in terms of concentrations:

$$K_{overall} = [Pb^{2+}]\,[H_2S]/[H^+]^2$$

Under conditions in which no significant amount of hydrogen sulfide gas is vaporized, but which are sufficiently acidic that almost all the sulfur exists as H_2S rather than as S^{2-} or HS^-, the stoichiometry of the reaction allows us to write that $[Pb^{2+}] = [H_2S]$. By substitution of this relationship into the above equation, we obtain

$$[Pb^{2+}]^2 = 6.5 \times 10^{-8}\,[H^+]^2$$

or

$$[Pb^{2+}] = 2.5 \times 10^{-4}\,[H^+]$$

Thus the solubility of PbS increases linearly with the H^+ concentration in acidic water. At pH = 4, the solubility of PbS and the concentration of Pb^{2+} ion in water is calculated to be 2.5×10^{-8} M, whereas at pH = 2, the solubility is 2.5×10^{-6} M. We conclude that dangerous concentrations of lead ion can occur in highly acidic bodies of water that are in contact with "insoluble" lead minerals.

PROBLEM 15-8

By calculations similar to those for PbS above, deduce the relationship between the solubility of mercuric sulfide, HgS ($K_{sp} = 3.0 \times 10^{-53}$), and the hydrogen ion concentration in acidic water. Is the solubility of HgS increased by exposure to acid?

Ionic 4+ Lead in Automobile Batteries

In highly oxidizing environments, lead can occur as the 4+ ion, one form of Pb(IV). Thus the oxide PbO_2, written in ionic form as $Pb^{4+}(O^{2-})_2$, exists, as do the mixed oxides Pb_2O_3 and Pb_3O_4, which are combinations of Pb(II) as PbO and Pb(IV) as PbO_2.

The elemental lead and the lead oxide PbO_2 employed as the two electrodes in storage batteries in vehicles together now constitute the major use of the element. Storage batteries that are not recycled constitute the main source of lead in municipal waste; some states and countries have banned the discarding of such batteries. The majority of used lead storage batteries are recycled for their lead content. During the recycling operation, lead can be expelled into the environment if careful controls are not maintained. Indeed, such recycling operations often constitute urban "hot spots" of lead emission into the surrounding communities. Although lead recycling operations in the United States are carried out under strict control, this is not necessarily the case in developing countries, where batteries are often shipped for recycling.

Tetravalent Organic Lead Compounds as Gasoline Additives

Whereas the compounds of Pb(II) are ionic, most Pb(IV) compounds are covalent molecules rather than ionic compounds of Pb^{4+}. In this respect, tetravalent lead is similar to the corresponding form of the other elements (C, Si, Ge, Sn) in its group of the periodic table.

Commercially and environmentally, the most important covalent compounds of lead(IV) are *tetraalkyl* compounds, PbR_4, especially those formed with the methyl group, CH_3, and the ethyl group, CH_2CH_3—namely **tetramethyllead,** $Pb(CH_3)_4$, and **tetraethyllead,** $Pb(C_2H_5)_4$. In the past, both compounds found widespread use as additives to gasoline—about a gram per liter—to produce leaded gasoline. As discussed in Chapter 7, this practice has now been phased out in North America and in many other developed countries, except in some types of aviation fuel, for which no acceptable substitute for lead has yet been found.

Since tetraalkyl lead compounds are volatile, they evaporate to some extent from gasoline and enter the environment in gaseous form. They are not water-soluble, but they are readily absorbed through the skin. In the human liver, PbR_4 molecules are converted into the more toxic compounds of PbR_3^+, which are neurotoxins because they can cross the blood–brain barrier. In substantial doses, these organic compounds of lead cause symptoms that mimic psychosis. It is not clear what the effects may be, if any, of chronic low-level exposure to them. At very high exposures, tetraalkyllead compounds are fatal, as was discovered many years ago when several employees of the companies that originally produced these compounds died. In contrast to mercury, little or no methylation of inorganic lead occurs in nature. Thus almost all the tetraalkylated lead in the environment probably originated from leaded gasoline.

Environmental Lead from Leaded Gasoline

When these additives are used in gasoline, the atoms of lead that are liberated by the combustion of the tetraalkyl compounds must be removed before they form metallic deposits and damage the vehicle's engine. In order to convert the combustion products into volatile forms that can leave the engine in the exhaust gases, small quantities of **ethylene dibromide** and **ethylene dichloride** are also added to the leaded gasoline. As a result, the lead is removed from the engine and enters the atmosphere from the tailpipe in the form of a mixture of the mixed dihalide PbBrCl and the dihalides $PbBr_2$ and $PbCl_2$. Subsequently, under the influence of sunlight, these compounds form PbO, which then exists in particulate form as an aerosol in the atmosphere for hours or days. Consequently, not all of it is deposited in the immediate surroundings of the roadway. As a result, PbO can enter the food chain at more distant sites if deposited on vegetables or on fields used by grazing animals. Furthermore, a small fraction of the ethylene dihalides are converted into dioxins and furans and enter the environment in these forms.

A high proportion of environmental lead in many parts of the world is that emitted from vehicles; it occurs in the environment mainly in inorganic form. The conversion to nonleaded gasoline in North America and Europe, the initial impetus for which was the interference of lead in exhaust gases with the proper functioning of catalytic converters, has had the welcome side effect of greatly decreasing the average amount of lead ingested by urban inhabitants. Indeed, the noted environmentalist Barry Commoner has called the elimination of lead from gasoline "one of the [few] environmental success stories." European scientists have traced the rise and fall of atmospheric alkylated lead by analyzing different vintages of a French red wine (*Chateauneuf-du-Pape*) that used grapes grown near two busy auto routes. They found that the concentration of trimethyllead, PbR_3^+—the degradation product of the tetramethyl compound—rose steadily to a maximum in the mid-1970s, which was followed by a steady decline to about one-tenth of the peak concentration by the early 1990s as the compound was phased out of gasoline. This pattern of usage is consistent with the variation in the U.S. consumption of lead for gasoline use; it is plotted in Figure 15-4 and shows a sharp rise from 1930 to 1970, followed by an even sharper decline thereafter.

Many countries still use leaded gasoline. In these areas, the air is the major source of lead ingested by humans, as it was in the past in North America and Europe. For example, in Mexico airborne lead from vehicular emissions is a major source of the elevated lead levels found in the

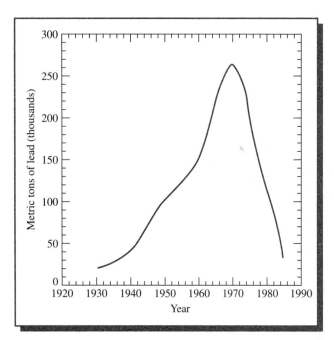

FIGURE 15-4 The historical consumption of lead in gasoline in the United States. [Source: C. E. Dunlop et al., "Past Leaded Gasoline Emissions as a Nonpoint Source Tracer in Riparian Systems," *Environmental Science and Technology* 34 (2000): 1211.]

bloodstream of many Mexican children. Some of the gasoline-based lead enters the body directly from inhaled air, and some enters indirectly from food into which lead has been incorporated. Microorganisms do bioconcentrate lead, but in contrast to mercury, lead does not undergo biomagnification in the food chain.

Lead's Effects on Human Reproduction and Intelligence

Most ingested lead in humans is initially present in the blood, but that amount eventually reaches a plateau. Any excess enters the soft tissues, including the organs, particularly the brain. Eventually lead becomes deposited in bone, where it replaces calcium, because Pb^{2+} and Ca^{2+} ions are similar in size. Indeed, lead absorption by the body increases in persons having a calcium (or iron) deficiency and is much higher in children than in adults. A study in Mexico indicated that pregnant women can decrease the lead levels in their blood—and presumably in the blood of their developing fetus—by taking calcium supplements.

At high levels, inorganic lead (Pb^{2+}) is a general metabolic poison. The toxicity of lead is proportional to the amount present in the soft tissues, not to that in blood or bone. Lead remains in human bones for decades; thus it can accumulate in the body. The dissolving of bone, as occurs with old age or illness such as osteoarthritis and advanced periodontal disease or in times of stress such as pregnancy and menopause, results in the remobilization of bone-stored lead back into the bloodstream where it can produce toxic effects. Excess lead may lead to the deterioration of bones in adults. Recently a correlation has been found between periodontal bone loss and blood lead levels in U.S. adults, particularly in those who smoke. Children exposed to environmental lead also have more dental cavities.

Although there is some evidence that too much lead can slightly increase the blood pressure of adults, the humans most at risk from Pb^{2+} even at relatively low levels are fetuses and children under the age of about seven years. Both these groups are more sensitive to lead than are adults, partly because they absorb a greater percentage of dietary lead and partly because their brains are growing rapidly. The metal readily crosses the placenta and thus is passed from mother to unborn child. Because of the immaturity of the fetus's blood–brain barrier, there is little to prevent the entry of lead into its brain. In addition, lead is transferred postnatally from the mother in her breast milk and/or from the tap water used to prepare formula for bottle-fed babies.

The principal risk to children from lead is interference with the normal development of their brains. A number of studies have found small but consistent and significant neuropsychological impairment in young children due to environmental lead absorbed either before or after birth. Lead appears to have deleterious effects on children's behavior and attentiveness, and possibly also on their IQs. This is illustrated in Figure 15-5, where a mental

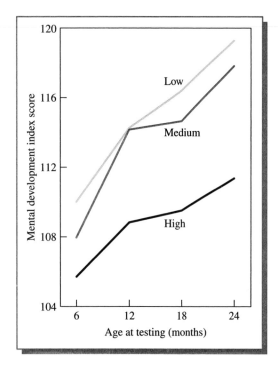

FIGURE 15-5 The effect of prenatal exposure to lead on the mental development of infants. Lead exposure is measured by its concentration in the blood of the child's umbilical cord. [Source: D. Bellinger et al., "Longitudinal Analyses of Prenatal and Postnatal Lead Exposure and Early Cognitive Development," *New England Journal of Medicine* 316 (1987): 1037–1043. Reprinted by permission of the *New England Journal of Medicine*.]

development score is plotted as a function of age for groups of young children differentiated by the amount of lead in their umbilical cord at birth. A study of children in a lead-smelting community (Port Pirie) in Australia indicates that children with a blood lead level of 300 ppb had an average IQ 4 to 5 points lower than those whose level was 100 ppb. This result is consistent with other studies that indicate an IQ deficit of about 2–3 points for each increase by 100 ppb of blood lead. Some studies indicate that prenatal exposure to lead—especially during the first trimester of pregnancy—has the greatest detrimental effect on the IQs measured in children in primary grades, some indicate it is the lead level at the age of two (when blood concentrations usually peak) that is predominant, and others that it is the concurrent lead level—even if lower than that in early childhood—that is the dominant factor. No threshold for the effects of lead upon IQ are apparent in the studies.

A survey in 1976–1980 of American children aged six months to five years found that about 4% of them had blood lead levels in excess of 300 ppb and that an additional 20% had levels over 200 ppb (see Figure 15-6a). These concentrations represent two of the cutoffs that had been proposed in the past as "safe" levels, but it appears that there may be no level at which lead does not produce a deleterious effect (i.e., there is no threshold) in young and unborn children. A second survey, in 1988–1991, indicated that blood lead levels in U.S. children had fallen substantially; less than 9% of those aged one to five years had blood lead levels greater than 100 ppb (see Figure 15-6b). The average American adult blood lead level fell from about 150 ppb in the 1970s to about 10–20 ppb today.

The effects of lead poisoning were known to the ancient Greeks, who realized that drinking acidic beverages from containers coated with lead-containing substances could result in illness. This information was not available to the Romans. Indeed, they sometimes deliberately adulterated overly acidic wine with sweet lead salts to improve the flavor. The concentration of lead in the bones of Romans is almost 100 times that found in modern North Americans. Some historians have hypothesized that chronic lead poisoning of upper-class Romans, from wine and other sources, contributed to the eventual downfall of the Roman Empire because of the metal's detrimental effects on the neurological and reproductive systems. The latter effects include dysfunctional sperm in males and an inability to bring the fetus to term in females. Indeed, in the past women who worked in the lead industry suffered higher-than-average rates of miscarriages and stillbirths. Due primarily to the contamination of beverages by lead from the distillation of alcohol in lead vessels, episodes

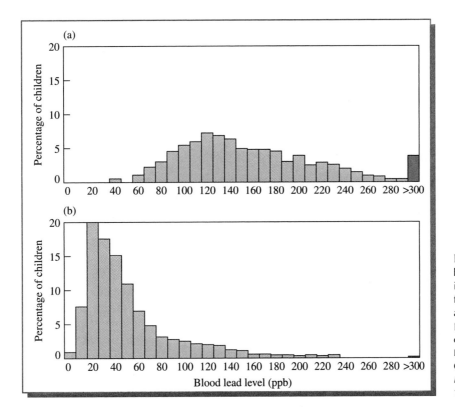

FIGURE 15-6 The distribution of blood lead levels in U.S. children aged one to five years (a) in 1976–1980 and (b) in 1988–1991. [Source: R. A. Goyer, "Results of Lead Research: Prenatal Exposure and Neurological Consequences," *Environmental Health Perspectives* 104 (1996): 1050–1054.]

of colic and gout due to lead poisoning were recorded through the Middle Ages and even until recent times.

In summary, on an atom-for-atom basis, lead is not as dangerous as mercury. However, the general population is exposed to lead from a greater variety of sources and generally at higher levels than those associated with mercury. Overall, more people are adversely affected by lead, though on average to a lesser extent, than those fewer individuals exposed to mercury. Both metals are more toxic in their organic compound form than in the simple inorganic cation form. In terms of its environmental concentration, lead is much closer—within a factor of 10—to the level at which overt signs of poisoning become manifest than is any other substance, including mercury. Thus it is appropriate that society continues to take steps to further reduce human exposure to lead.

PROBLEM 15-9

The concentrations of lead in blood samples are often reported in units of micrograms of Pb per deciliter of blood or in micromoles of lead per liter of blood. Calculate the value of the concentration in these units of a blood sample containing 60 ppb lead, assuming that the density of blood is one gram per milliliter.

Cadmium

Cadmium, Cd, lies in the same subgroup of the periodic table as zinc and mercury, but it is more similar to the former. Like zinc, the only common ion of cadmium is the 2+ species. In contrast to mercury, cadmium's compounds with simple anions such as chloride are ionic salts rather than covalent molecules.

Environmental Sources of Cadmium

Most cadmium is produced as a by-product of zinc smelting, since the two metals usually occur together. Some environmental contamination by cadmium often occurs in the areas surrounding zinc, lead, and copper smelters. As is the case for the other heavy metals, burning coal introduces cadmium into the environment. The disposal by incineration of waste materials that contain cadmium is also an important source of the metal to the environment.

A major use of cadmium is as an electrode in rechargeable *nicad* (nickel–cadmium) batteries used in calculators and similar devices. When current is drawn from the battery, the solid elemental metal cadmium electrode partially disintegrates to form insoluble **cadmium hydroxide,** $Cd(OH)_2$, by incorporating hydroxide ions from the medium into which it dips. When the battery is being recharged, the solid hydroxide, which was deposited on the metal electrode, is converted back to cadmium metal:

$$Cd(s) + 2\ OH^- \rightleftharpoons Cd(OH)_2(s) + 2\ e^-$$

Each nicad battery contains about 5 g of cadmium, much of which is volatilized and released into the environment if the spent batteries are incinerated in garbage. The metallic cadmium preferentially condenses on the smallest particles in the incinerator smoke stream, which are precisely the ones that are difficult to capture by pollution-control devices inserted in the gas stack. In order to avoid releasing airborne cadmium into the environment upon combustion, some municipalities require nicad batteries to be separated from other garbage. The recycling of metals from such batteries has also begun in some areas. However, the European Union has banned the use of nicad batteries, except in cordless power tools and systems used for safety and medical purposes. Some U.S. states have banned the disposal of nicad batteries. Battery manufacturers hope to replace nicad batteries soon with ones that do not contain cadmium.

In ionic form, the main use of cadmium is as a pigment. Because the color of **cadmium sulfide,** CdS, depends on the size of the particles, cadmium pigments of many hues can be prepared. Both CdS and CdSe have been used extensively to color plastics. For several centuries, painters have used cadmium sulfide pigments in paints to produce brilliant yellow colors and thus oppose any ban on them, since at present there are no suitable replacements.

Van Gogh could not have painted his famous *Sunflowers* canvas without cadmium yellows, although it is speculated that cadmium poisoning may have contributed to the painter's anguished mental state.

Cadmium is released into the environment during the incineration of plastics and other materials that contain it as a pigment or as a stabilizer. It is also released into the atmosphere when cadmium-plated steel is recycled, since the element is fairly volatile when heated (its boiling point is 765°C).

Human Intake of Cadmium

Cd^{2+} is rather soluble in water, unless sulfide ions are also present to precipitate the metal as CdS. Thus humans usually ingest only a small proportion of their cadmium directly from drinking water or from air, except for individuals who live near mines and smelters, particularly those that process zinc. The maximum containment level (MCL) for cadmium in drinking water is 5 ppb in the United States and Canada (Table 15-2).

Smokers are also exposed to cadmium that is absorbed from soil and irrigation water by tobacco leaves and then released into the smoke stream when a cigarette is burned. Heavy smokers have approximately double the net cadmium intake of nonsmokers.

Owing to its similarity to zinc, plants absorb cadmium from irrigation water. The use on agricultural fields of phosphate fertilizers, which contain ionic cadmium as a natural contaminant, and of sewage sludge contaminated with cadmium from industrial releases increases the cadmium level in soil and subsequently in plants grown in it. In the future, cadmium may be removed from phosphate fertilizer before it is sold to the consumer (see also Chapter 16). Soil also receives cadmium from atmospheric deposition. Since cadmium uptake in plants increases with decreasing soil pH, one effect of acid rain is to increase cadmium levels in food.

For most of us, the greatest proportion of our exposure to cadmium comes from our food supply. Seafood and organ meats, particularly kidneys, have higher cadmium levels than do most other foods. However, the majority of cadmium in the diet usually comes from potatoes, wheat, rice, and other grains, since most people consume so much more of them than of seafood and kidneys. An exception is the Inuit people of Canada's Northwest Territories; a prized component of their diet is caribou kidneys, organs which are highly contaminated by cadmium that has reached the Arctic regions on the wind from industrial regions in Europe and North America.

Historically, all episodes of serious cadmium contamination resulted from pollution from nonferrous mining and smelting. The most acute environmental problem involving cadmium occurred in the Jintsu River Valley region of Japan, where rice for local consumption was grown with the aid of irrigation water drawn from a river that was chronically contaminated with dissolved cadmium from a zinc mining and smelting operation upstream.

Hundreds of people in this area, particularly older women who had borne many children and who had poor diets, contracted a degenerative bone disease called *itai-itai* or "ouch-ouch," so named because it causes severe pain in the joints. In this disease, some of the Ca^{2+} ions in the bones are replaced by Cd^{2+} ions since they have the same charge and are virtually the same size. The bones slowly become porous and can subsequently fracture and collapse. The intake of cadmium by itai-itai sufferers was estimated at about 600 μg per day, about 10 times the average ingestion of North Americans.

Protection Against Low Levels of Cadmium

Cadmium is acutely toxic: The lethal dose is about 1 g. Humans are protected against chronic exposure to low levels of cadmium by the presence of the sulfur-rich protein **metallothionein,** the usual function of which is the regulation of zinc metabolism. Because it has many sulfhydryl groups, metallothionein can complex almost all ingested Cd^{2+}; the complex is subsequently eliminated in the urine. If the amount of cadmium absorbed by the body exceeds the capacity of metallothionein to complex it, the metal is stored mainly in the liver and kidneys. Indeed, there is evidence that chronic exposure to cadmium eventually leads to an increased chance of acquiring kidney disease.

The average cadmium burden in humans is increasing. Although cadmium is not biomagnified, it is a cumulative poison since, if not eliminated quickly (by metallothionein, as discussed above), its lifetime in the body is several decades. The geographic areas at greatest risk from cadmium exposure are Japan and central Europe; in both regions, the pollution of the soil by cadmium is particularly high due to contamination from industrial operations. Rice grown in many areas of Japan is often contaminated with rather high cadmium levels. As a consequence, the dietary intake of cadmium by residents of Japan is substantially greater than for peoples of other developed countries. Indeed, in Japan the average daily amount of ingested cadmium is beginning to approach the maximum level recommended by health authorities, although this limit has a large built-in safety factor relative to levels at which health effects would occur.

Arsenic

Arsenic is not actually a metal; it is a metalloid—its properties are intermediate between those of metals and nonmetals. However, for convenience we discuss it in this chapter.

Arsenic compounds such as the oxide As_2O_3, **white arsenic,** were common poisons used for murder and suicide from Roman times through the Middle Ages. In the seventeenth century, arsenic was believed in some European societies to be not only a poison but also a magical substance that was a cure for certain ailments, including impotence, and to be a prophylactic against the plague. Indeed, arsenic compounds have been used therapeutically for 2000 years, and even today about 50 Chinese drugs contain the element. There are small background levels of arsenic in many foods, and a trace amount of this element apparently is essential to good human health.

Arsenic(III) Versus Arsenic(V) Toxicity

Arsenic occurs in the same group of the periodic table as phosphorus and so also has an s^2p^3 electron configuration in its valence shell. Loss of all three p electrons gives the 3+ ion, whereas sharing of the three electrons gives trivalent arsenic; collectively, these two forms are designated As(III). Arsenic(III) commonly exists in aqueous solution and in solids as the **arsenite ion,** AsO_3^{3-} (which can be considered to be As^{3+} bonded to three surrounding O^{2-} ions), or one of its successively protonated forms: $HAsO_3^{2-}$, $H_2AsO_3^-$, or H_3AsO_3.

Alternatively, loss of all five valence shell electrons gives the 5+ ion, and sharing them all gives pentavalent arsenic; collectively, these two forms are designated As(V). Arsenic(V) also commonly exists as an oxyanion, the **arsenate ion,** AsO_4^{3-} (equivalent to As^{5+} bound to four O^{2-} ions), or one of its successively protonated forms: $HAsO_4^{2-}$, $H_2AsO_4^-$, or H_3AsO_4.

arsenite
AsO_3^{3-}

arsenate
AsO_4^{3-}

Overall, arsenic acts much like phosphorus, which commonly exists in the analogous oxyanion forms PO_3^{3-} and PO_4^{3-}, called *phosphite* and *phosphate,* respectively. However, arsenic has more of a tendency than phosphorus to form ionic rather than covalent bonds, since it is more metal-like. Due to the similarity in properties, arsenic compounds coexist with those of phosphorus in nature. Consequently, arsenic often contaminates phosphate deposits and commercial phosphates.

Arsenic's lethal effect when consumed in an acute dose is due to gastrointestinal damage, resulting in severe vomiting and diarrhea. Inorganic As(III) is more toxic than As(V), although some of the latter is converted by reduction to the former in the human body. It is thought that the greater toxicity of As(III) is due to its ability to be retained in the body longer since it becomes bound to sulfhydryl groups in one of several enzymes. Due to the subsequent inactivity of the enzymes, energy production in the cell declines and the cell is damaged. Once the arsenic becomes methylated in the liver, it does not bind tightly to enzymes and hence is largely detoxified.

Anthropogenic Sources of Arsenic to the Environment

Anthropogenic environmental sources of arsenic stem from

- the continuing use of its compounds as pesticides;

- its unintended release during the mining and smelting of gold, lead, copper, and nickel, in whose ores it commonly occurs (the leachate from abandoned gold mines of previous decades and centuries can still be a significant source of arsenic pollution in water systems);

- the production of iron and steel;

- the combustion of coal, of which it is a contaminant; and

- arsenic-contaminated water brought to ground level by wells.

The arsenic present in raw coal can become a serious pollutant, especially around areas where the fossil fuel is burned. The total pollution from arsenic can be substantial where the coal is burned in small, unventilated stoves rather than in large power plants. In these cases, which occur in some developing countries, the arsenic not only becomes an indoor air pollutant but also contaminates the food and water stored indoors. A particularly acute example occurs in the Chinese province of Guizhou, where arsenic levels in the coal are extraordinarily high, exceeding 1% (i.e., 10,000 ppm) in some cases. Many of the residents of Guizhou suffer arsenic-related health problems, since they use this coal for domestic cooking and heating. By contrast, the level of arsenic in U.S. coal averages about 22 ppm, and most coal worldwide has arsenic levels of less than 5 ppm.

Arsenic compounds found widespread use as pesticides before the modern era of organic chemicals. Although its use in these applications has decreased, arsenic contamination from pesticides remains an environmental problem in some areas of the world. The common arsenic-based pesticides include the insecticide lead arsenate, $Pb_3(AsO_4)_2$, and the herbicide **calcium arsenate,** $Ca_3(AsO_4)_2$, both of which contain As(V) as AsO_4^{3-}. The herbicides **sodium arsenite,** Na_3AsO_3, and **Paris Green,** $Cu_3(AsO_3)_2$, both contain As(III) as AsO_3^{3-}. An organic compound containing As(V) is routinely used in chicken feed to stimulate growth and prevent disease; some scientists have worried about the contamination of soil and water by arsenic leached from chicken litter. Some methylated derivatives of arsenic acids are still used as herbicides, even in developed countries. The sodium salt of the arsenate ion in which one —OH has been replaced by a methyl group, producing the **methanearsonate ion,** $O{=}As(OH)\,(CH_3)O^-$, is a herbicide widely used on golf courses and cotton fields in the United States. Such As(V) compounds act as weedkillers because they enter into plant metabolism in place of phosphate ion. The environmental consequences of using another heavy metal, tin, in a pesticide are discussed in Box 15-1.

BOX 15-1	Organotin Compounds

Although inorganic compounds of tin (Sn) are relatively nontoxic, the bonding of one or more carbon chains to the metal results in substances that are toxic. Such organotin compounds have some common uses, such as additives to stabilize PVC plastics and fungicides to preserve wood, and therefore are of environmental concern.

Tin forms a series of compounds of general formula R_3SnX, which are molecular substances though often shown in formulas as if they were ionic, e.g., $(R_3Sn^+)(X^-)$, where R is a hydrocarbon group and X is a monatomic anion; corresponding compounds such as $(R_3Sn)_2O$ also occur. All these compounds are toxic to mammals when R is a very short alkyl chain; maximum toxicity occurs when R is the ethyl group, C_2H_5, and decreases progressively with increasing chain length.

For fungi the greatest toxic activity is attained when each hydrocarbon chain has four carbons in an unbranched chain, i.e., when R is the n-butyl group, $—CH_2CH_2CH_2CH_3$ (or simply n-C_4H_9). *Tributyltin oxide*, $(R_3Sn)_2O$ where R = n-C_4H_9, and the corresponding fluoride have both been used as fungicides; commonly they are incorporated as antifouling agents in the paint applied to docks, to the hulls of boats, to lobster pots, and to fishing nets, etc. to prevent the accumulation of slimy marine organisms such as the larvae of barnacles. In recent years tributyltin has been incorporated into polymeric coatings for boat hulls; a thin layer of the compound subsequently forms around the hull. The tin compounds replaced copper(I) oxide, Cu_2O, in such applications since their effectiveness lasts longer than a single season.

Unfortunately, some of the tributyltin compound leaches into the surface waters in contact with the coatings or paint, particularly in harbors where the boats are moored, and subsequently enters the food chain via the microorganisms that live near the surface. This can lead to sterility or death for fish and some types of oysters and clams that feed on these microorganisms. Some countries have restricted the use of tributyltin compounds to large ships. Thus, although the concentration of tributyltin has decreased in the waters of small harbors and marinas, the pollutant still tends to concentrate in marine coastal regions due to its use on large vessels. Scientists are worried that the presence of tributyltin compounds in these waters could affect fish reproduction.

For this reason, the International Maritime Organization banned new applications of tributyltin to ships of any size effective 2003 and required that this material be removed from all old applications by 2008. Ironically, the triazine herbicide added to copper-based antifoulant paints that were introduced to replace those based upon tributyltin degrades only slowly in water and has now begun to accumulate there.

Higher organisms have enzymes that break down tributyltin fairly rapidly, so it is not very toxic to humans. However, most humans now have detectable levels of tributyltin in their blood.

Since the 1970s, arsenic has been used in the form of the compound **chromated copper arsenate**, CCA, to pressure-treat lumber in order to prevent rot and termite damage. Unfortunately, some of the arsenic leaches out of the wood over time. U.S. and Canadian producers of CCA-treated wood voluntarily phased out use of the arsenic compound at the end of 2003 for wood destined for residential structures such as decks, picnic tables, fences, and playground equipment. CCA is discussed in further detail in the section on chromium. The U.S. EPA has already banned arsenic in all other pesticides.

Arsenic in Drinking Water

Arsenic—much of it from natural sources—is one of the most serious environmental health hazards. The presence of significant levels of arsenic in drinking-water supplies is a significant and controversial environmental issue. Natural levels of arsenic in water can be quite high, and it is more common for health problems to arise from this source than from anthropogenic arsenic. Although arsenic has been used for millennia as a poison, the major health problem stemming from its presence at low levels in drinking water is cancer. Drinking arsenic-contaminated water has also been linked to diabetes and cardiovascular disease, perhaps by disrupting a hormonal process associated with both conditions.

Arsenic is carcinogenic in humans. Lung cancer results from the inhalation of arsenic and probably also from its ingestion. Cancers of the lung, bladder, and skin, and perhaps also of the kidney, arise from ingested arsenic, including that in water. The mechanism by which arsenic causes cancer is not clear. Evidence suggests that it acts as a *cocarcinogen,* inhibiting the DNA repair mechanism and thereby enhancing the cancer-causing abilities of other carcinogens. There is evidence from Chile that smoking and simultaneous exposure to high levels of arsenic in drinking water act synergistically in causing lung cancer; i.e., their effect taken together is greater than the sum of their individual effects if each acted independently, as discussed in Chapter 4. Other data from Chile show that exposure to arsenic during early childhood or even *in utero* increases subsequent mortality in young adulthood from both malignant and nonmalignant lung diseases. Indeed, arsenic seems to act synergistically with several *cofactors*—i.e., factors whose presence negatively affects the health of an individual to an extent greater than if it or the arsenic operated independently. Exposure to excessive levels of UV from sunlight and a lack of selenium in the diet (stemming from malnutrition and/or low selenium levels in local foods) are other cofactors with arsenic. The protective effect of selenium in reducing the amount of active arsenic in the body may arise from the formation of a biomolecule containing an $As=Se$ bond. Research is under way to determine whether selenium supplementation of the diet would be effective in countering the negative health effects of excess arsenic in the drinking water of Bangladesh and the Bengal region of India.

Drinking water, especially that derived from groundwater, is a major source of arsenic for many people. Although anthropogenic uses of arsenic can result in its contamination of water, by far the greatest problems occur with that produced by natural processes. Groundwater in several parts of the world is highly contaminated with inorganic arsenic. Unfortunately, the arsenic is tasteless, odorless, and invisible, so its presence is not easily detected.

Major problems from high arsenic levels occur in the Bengal Delta, with the result that tens of millions of people in Bangladesh and in the West Bengal region of India drink arsenic-laced water. The World Health Organization has called this the "largest mass poisoning of a population in history." The problem arose from the creation of tens of millions of tube wells, which mine groundwater that was previously inaccessible. The concrete tube wells extend 20 m (60 ft) or more into the ground. Ironically, the wells were constructed by UNICEF in the 1970s and early 1980s in an otherwise highly successful project to eliminate epidemics of diarrhea, cholera, and other waterborne diseases and to reduce the high child-mortality rate caused by use of microbially unsafe water drawn from streams, ponds, and shallow wells used in the past. About half the tube wells—affecting about 50 million people in Bangladesh—produce water with arsenic levels as high as 500–1000 ppb, greatly exceeding the 10 ppb WHO guideline for drinking water (Table 15-2). The sediments through which the groundwater travels contain the arsenic. Generally, the deeper the well beyond about 20 m, the lower the concentration of arsenic.

Several million people living in the Bengal Delta region will probably contract skin disorders from drinking arsenic-laced groundwater if remedial action is not taken; a fraction of them will also suffer from the more serious ailment of *arsenicosis*, which can cause cancer of the skin, bladder, kidneys, and lungs. Skin lesions appear after 5–15 years of exposure to high levels of arsenic in drinking water. A large number of residents of West Bengal, India, have already developed such lesions—the usual outward sign of chronic exposure to arsenic—that may develop into skin cancer because they consumed arsenic-laced groundwater from underground wells. The main cause of arsenic-related deaths among these people is lung cancer. It has also been established that rice and vegetables grown in Bangladesh using irrigation water from tube wells are also contaminated by arsenic, and this may be the dominant source of the element for some people. Grains and beans absorb additional arsenic from the water they absorb when they are cooked.

Recent research from Bangladesh indicates that increasing levels of arsenic and/or of manganese in drinking water confer progressively more and more negative effects on the intellectual levels of six- and ten-year-old children. The Mn levels in one such study averaged 1.4 ppb, compared to the WHO standard of 0.5 ppb. Elevated manganese levels are also present even in the United States: Approximately 6% of domestic wells exceed the U.S. EPA lifetime health advisory concentration of 0.3 ppb Mn in drinking water.

The origin of the dissolved arsenic in the water in Bangladesh and India is controversial. Normally the element, as arsenate ion, is coprecipitated with and adsorbed on the surface of iron oxides in the soil, as would have occurred in ancient times when sediments were being laid down. However, the arsenic, along with the iron, dissolves when insoluble Fe(III) is reduced by natural organic carbon to the more soluble Fe(II) state. Indeed, the higher the concentration of dissolved iron, the higher the arsenic concentration found in the water. The controversy centers around whether the dominant process is the natural one, by which buried peat acts as the reducing agent and has been doing so for millennia, or whether the release has been greatly accelerated in recent years as an indirect effect of annually lowering the water table by extracting massive amounts of water for crop irrigation. In the latter mechanism, the subsequent recharge of the depleted aquifer transports carbon in the water drawn down from the surface, resulting in further reduction of iron oxides and solubilization of the arsenic. Reduction of arsenic from As(V), as it exists when adsorbed to the iron mineral, to the more soluble As(III) form is also believed to be a factor in solubilizing the element. The water obtained from adjacent wells separated even by only tens of meters from each other can differ enormously in arsenic content, apparently as a result of being drawn from sediments initially laid down in ancient times by different streams that had different sources of organic carbon being deposited simultaneously. Widespread testing in 1999 of tube wells in Bangladesh identified those delivering high arsenic levels, and the handles on such wells were painted red to warn people of the danger. Thousands of larger, deeper wells that draw water from less contaminated aquifers have subsequently been installed as centralized facilities in many villages.

Arsenic-contaminated drinking water is also a major problem in Chile, Argentina, Mexico, Nepal, Taiwan, Cambodia, Vietnam, and large areas of China. Indeed, 8% of the deaths of Chilean adults over 30 are attributable to arsenic poisoning. In a study of residents of Taiwan who were exposed to high levels of the element in their well water, a relationship between arsenic exposure and skin cancer incidence has been established. As in Bangladesh, arsenic only became a problem when people began to drink groundwater, which was touted as being purer than surface water, since the latter is often contaminated by sewage.

Drinking-Water Standards for Arsenic

Drinking water, especially groundwater, is a major source of arsenic for most people. The global average inorganic arsenic content of drinking water is about 2.5 ppb. The World Health Organization has set 10 ppb as the acceptable limit for arsenic in drinking water, and the European Union adopted this standard in 2003 (Table 15-2). The standard in many developing countries is still 50 ppb, which is no longer considered to be protective.

In the last days (2000) of the Clinton administration, the maximum contaminant level for arsenic in U.S. drinking water was lowered from 50 ppb to 10 ppb. Although the Bush administration at first withdrew this regulation, it later concluded that the reduction was warranted. As a result, the 10-ppb limit became law in February 2002, and the compliance date was set for 2006.

The shape of the dose–response curve (see Chapter 10) at such low concentrations of arsenic is unknown. Assuming that no threshold exists, linear extrapolations of human cancer incidence from populations that were exposed to high levels of arsenic leads to the conclusion that there is a 1-in-1000 lifetime risk of dying from cancer induced by normal background levels of arsenic. This estimate makes arsenic almost equivalent to environmental tobacco smoke and radon exposure as an environmental carcinogen. Drinking water over a lifetime at the 50-ppb level, the old U.S. standard, would cause bladder or lung cancer in about 1% of the population, a much greater risk than continuously consuming any other water-based contaminant at its MCL. Some environmentalists argue that the arsenic standard should be lowered still further, to 3 ppb, at which the risk is 1 in 1000, whereas at 10 ppb it is about 3 per thousand. About 57 million Americans currently drink water containing more than 1 ppb of arsenic; areas of the contiguous United States whose groundwater sometimes contains more than 10 ppb As are shown in dark green in Figure 15-7. Most affected systems lie in the West, Midwest, Southwest, and New England and use groundwater having naturally occurring arsenic.

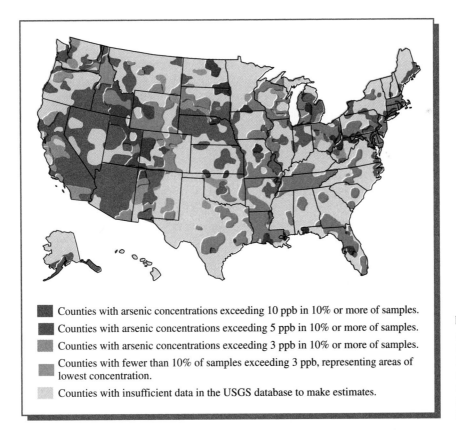

■ Counties with arsenic concentrations exceeding 10 ppb in 10% or more of samples.

■ Counties with arsenic concentrations exceeding 5 ppb in 10% or more of samples.

■ Counties with arsenic concentrations exceeding 3 ppb in 10% or more of samples.

■ Counties with fewer than 10% of samples exceeding 3 ppb, representing areas of lowest concentration.

Counties with insufficient data in the USGS database to make estimates.

FIGURE 15-7 Average arsenic concentrations in U.S. drinking water. [Source: "Pressure to Set Controversial Arsenic Standard Increases," *Environmental Science and Technology* 34 (2000): 208A.]

One of the difficulties in setting a standard for arsenic levels in drinking water is deciding the manner in which the element operates as a carcinogen. For carcinogens that induce cancer directly—by damaging DNA—the assumption is made that no amount of exposure to the substance is safe, since the risk from it rises from zero in proportion to exposure. However, as mentioned previously, there is evidence that arsenic does not act directly but indirectly, by inducing cell damage and regrowth or by inhibiting repair of DNA damage caused by other carcinogens such as UV light or tobacco smoke. For carcinogens that act indirectly, there can be a threshold, a level below which the substance can be considered safe and not cause damage.

Some scientists are not convinced that the estimates of cancer risk quoted above are at all realistic, since the extrapolation of the cancer incidence from high arsenic levels to the low environmental concentrations may not be valid if arsenic acts indirectly as a carcinogen. It will be difficult to resolve this issue by analyzing cancer trends in different parts of the United States, however, since the predicted fraction of bladder and lung cancers caused by arsenic is still a small percentage of the total for these diseases.

One argument that was advanced against making the arsenic standard even as low as 10 ppb in the United States is that it forces some small-scale suppliers of drinking water to shut down since they cannot afford the cleanup costs associated with introducing equipment to remove the element. Such shutdowns may lead consumers to turn to water supplies that are even more unsafe in other respects. Indeed, lowering the standard to 10 ppb is estimated to cost users of small water utilities, i.e., many of those in rural areas, several hundred dollars per year, whereas it will cost users of large facilities only a few dollars annually.

Removal of Arsenic from Water

The most widely used process for removing arsenic is to flow the drinking water over activated *alumina* (aluminum oxide), onto the surface of which the arsenic is adsorbed. The surface requires periodic cleaning of adsorbed species to remain effective. Reverse osmosis can also be used to remove arsenic, although, as previously discussed (Chapter 14), the process is expensive.

Because arsenic readily adsorbs onto iron oxide, water can be passed through a bed of ferric oxide to remove most of the arsenic. Alternatively, arsenic can be captured when iron hydroxide is precipitated from water, in a technique similar to the removal of colloids described in Chapter 14. Some of the other techniques used in villages in India and Bangladesh use the alumina method described above or filtration of the water through sand. All removal techniques require regular maintenance of the equipment and the proper periodic disposal of the arsenic-laden wastes. Some analysts believe that none of the arsenic-removal techniques work reliably in many areas, partly due to poor maintenance; instead, people should be directed to deeper wells with low arsenic contamination rather than trying to clean arsenic out of water from shallower wells that contains high levels of the element. Centralized water treatment plants using surface water are being constructed in some areas to overcome dependence on groundwater.

Like calcium and magnesium, arsenic can be removed from drinking water at large treatment facilities by precipitating it in the form of one of its insoluble salts. The arsenic in surface water normally exists as As(V). Since the salt formed between the ferric ion, Fe^{3+}, and arsenate ion is insoluble, the soluble salt **ferric chloride,** $FeCl_3$, can be dissolved in the water and the precipitated **ferric arsenate,** $FeAsO_4$, can be filtered from the resulting mixture:

$$Fe^{3+} + AsO_4{}^{3-} \longrightarrow FeAsO_4(s)$$

Arsenic in groundwater often exists as As(III) since reducing conditions occur underground. Such arsenic must be oxidized to As(V) before this removal process can occur.

Arsenic cannot be removed from water by cation exchange, since it occurs as an anion, not a cation. However, anion exchange can be used to remove arsenic from drinking water. Anion exchange also works better for As(V) than As(III), since the latter exists partially as the neutral H_3PO_3, rather than an anionic form at normal water pH values (6.5–8.5), whereas As(V) is completely ionic in that range (see Additional Problem 3). Anion exchange is problematic if appreciable amounts of sulfate ion, $SO_4{}^{2-}$, are also present in the water, as these are exchanged preferentially to arsenate, thereby tying up many sites and leaving fewer at which arsenic can exchange.

Steady State of Arsenic Levels in Natural Waters

A model for the mass balance of arsenic in a typical large water body, in this case Lake Ontario, is shown in Figure 15-8. The lake receives 161 tonnes of As per year, almost all of it from river and lake flows originating with land-based sources; the rest comes from the atmosphere, mainly in the form of arsenic dissolved in rain and snow. About three-quarters of the annual input quantity leaves the lake by outflow (to the St. Lawrence River). The other quarter corresponds to the net deposited into the surface sediment, after correction for the arsenic redissolved in the water column from this source. Over time, the sediment arsenic, with a concentration of about 10 ppm, becomes buried. The net input and output of arsenic into Lake Ontario are equal, so it is in a steady state, and the concentration of the element in the water, about 0.5 ppb, remains constant with time.

Arsenic in Organic and Other Molecular Forms

The common environmental organic forms of arsenic are not simple methyl derivatives, as with mercury and lead. Rather, they are water-soluble oxyacid derivatives that can be excreted by the body and thus are less toxic than some

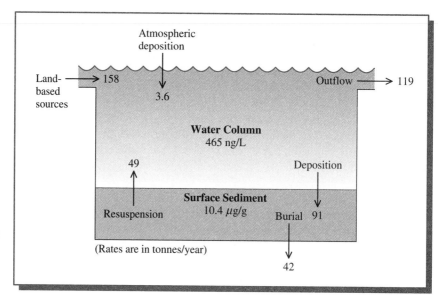

FIGURE 15-8 Steady-state model mass-balance diagram for arsenic in Lake Ontario. [Source: Adapted from S. Thompson et al., "A Modeling Strategy for Planning the Virtual Elimination of Persistent Toxic Chemicals from the Great Lakes," *Journal of Great Lakes Research* 25 (1999): 814.]

inorganic forms. As previously mentioned, in water arsenic occurs most commonly as the As(V) acid H_3AsO_4, i.e., $(OH)_3As\!=\!O$, or one of its deprotonated forms. Biological methylation in the environment by methylcobalamin initially involves the replacement of one or more —OH groups of the acid by —CH_3 groups. Monomethylation by the human liver and kidneys converts most but not all ingested inorganic arsenic to $(CH_3)(OH)_2As\!=\!O$ and then to the corresponding dimethyl acid, which is then readily excreted.

Although most daily exposure to arsenic by North American adults is due to food intake, especially meat and seafood, much of the arsenic present in food sources occurs in the organic form and is readily excreted. In seafood, the common forms of arsenic are either the $(CH_3)_4As^+$ ion, a form of As(III) itself, or this ion with one methyl group replaced by —CH_2CH_2OH or —CH_2COOH. The organic forms of arsenic found in seafood are probably noncarcinogenic and are much less toxic than inorganic forms, as illustrated in dramatic fashion by their high LD_{50} values, which lie in the thousands of milligrams per kilogram, compared to those for inorganic arsenic, whose LD_{50}'s are about 1% of these values (see Table 15-3).

In contrast to the compounds discussed above, neutral As(III) compounds such as **arsine,** AsH_3, and **trimethylarsine,** $As(CH_3)_3$, are the most toxic forms of arsenic. Curiously, the trimethyl compound is produced by the reaction, under humid conditions, of molds in wallpaper paste with the arsenic-containing green pigment $CuHAsO_3$ in wallpaper. Instances of mysterious illnesses and even of human "death by wallpaper" due to chronic exposure to the $As(CH_3)_3$ gas released into rooms by this mechanism have been reported. Some historians believe Napoleon was fatally poisoned by the

TABLE 15-3	LD$_{50}$ Values for Some Common Forms of Arsenic	
Name	Formula	LD$_{50}$ (mg/kg)
Arsenous acid	H_3AsO_3	14
Arsenic acid	H_3AsO_4	20
Methylarsonic acid	$CH_3AsO(OH)_2$	700–1800
Dimethylarsonic acid	$(CH_3)_2AsO(OH)$	700–1800
Arsenocholine	$(CH_3)_3As^+CH_2CH_2OH$	6500
Arsenobetaine	$(CH_3)_3As^+CH_2COO^-$	>10,000

Source: X. C. Le, "Arsenic Speciation in the Environment," *Canadian Chemical News* (September 1999): 18.

trimethylarsine emitted from the wallpaper in his chronically damp house on the island of St. Helena, where he had been exiled. There have also been episodes of human poisoning from gaseous arsine that was accidentally generated and released when aqueous solutions of As(III) in the form of $HAsO_2$ came into contact with an easily oxidized metal such as aluminum or zinc and the arsenic was further reduced to As($-$III):

$$2\, Al(s) + HAsO_2 + 6\, H^+ \longrightarrow 2\, Al^{3+} + AsH_3 + 2\, H_2O$$

Chromium

Chromium normally occurs in the form of inorganic ions. Its common oxidation states are +3 and +6, i.e., Cr(III) and Cr(VI), known as *trivalent* and *hexavalent* chromium, respectively.

Under oxidizing (i.e., aerobic) conditions, chromium exists in the (VI) state, usually as the **chromate ion,** CrO_4^{2-}, though under even slightly acidic conditions this oxyanion protonates to $HCrO_4^-$.

$$H^+ + CrO_4^{2-} \rightleftharpoons HCrO_4^-$$

The oxyanions of Cr(VI) are highly soluble in water. Both the Cr(VI) ions mentioned above are yellow, and they impart a yellowish tinge to water even at chromium levels as low as 1 ppm. (At high concentrations not encountered in the environment, chromate dimerizes to give the orange *dichromate ion*, $Cr_2O_7^{2-}$, familiar as a strong oxidizing agent in the laboratory and used in the determination of the chemical oxygen demand (COD) of water samples, as discussed in Chapter 13.)

Under reducing (i.e., anaerobic) conditions, chromium exists in the (III) state. In aqueous solution, this state occurs as the +3 ion, i.e., Cr^{3+}.

However, the aqueous solubility of this ion is not high, and Cr(III) is often precipitated as its hydroxide, $Cr(OH)_3$, under alkaline, neutral, or even slightly acidic conditions:

$$Cr^{3+} + 3\,OH^- \rightleftharpoons Cr(OH)_3(s)$$

Thus, whether chromium occurs as an ion dissolved in water or as a precipitate depends on whether the aqueous environment is oxidizing or reducing. The difference is important, since hexavalent Cr(VI) is toxic and a suspected carcinogen, whereas trivalent Cr(III) is much less toxic and even acts as a trace nutrient. Chromate ion readily enters biological cells, apparently because of its structural similarity to sulfate ion, SO_4^{2-}. Inside the cell, chromate can oxidize DNA and RNA bases. Because hexavalent chromium is more toxic, more soluble, and more mobile than trivalent chromium, it is considered to pose a greater health risk. (The term *hexavalent chromium* was made famous a few years ago in the movie *Erin Brockovich*, the story of how a legal assistant battled successfully against pollution of local groundwater by this substance.)

Chromium Contamination of Water

Chromium is widely used for electroplating, corrosion protection, and leather tanning. In tanning, Cr(III) binds to protein in animal skin to form leather that is resistant to water, heat, and bacteria. As a consequence of industrial emissions, chromium is a common water pollutant, especially of groundwater beneath areas with metal-plating industries. It is also the second most abundant inorganic contaminant of groundwater under hazardous waste sites. The MCL for total chromium in U.S. drinking water is 100 ppb (Table 15-2).

Most dissolved heavy metals can be removed from wastewater by simply increasing the pH, since their hydroxides are insoluble. However, Cr(VI) does not precipitate out at any pH since it does not exist as a cation but rather as an oxyanion in water. Owing to the low solubility and hence the low mobility of Cr(III), however, the usual way to extract Cr(VI) from water is to first use a reducing agent to convert Cr(VI) to Cr(III):

$$\underset{\text{(soluble)}}{CrO_4^{2-}} + 3\,e^- + 8\,H^+ \rightleftharpoons \underset{\text{(insoluble)}}{Cr^{3+}} + 4\,H_2O$$

Reducing agents commonly employed for this conversion are gaseous SO_2 or a solution of *sodium sulfite*, Na_2SO_3. In addition, reducing the Cr(VI) to Cr(III) by adding iron in the form of Fe(II) and then adding base to precipitate Cr(III) is a common practice in purifying Cr-contaminated wastewater. Fine-grained elemental iron placed in permeable underground walls positioned in the path of flowing polluted groundwater is another application of this technique. The iron reduces the chromium, and then as Fe^{3+} it forms an insoluble Fe(III)–Cr(III) compound. This reduction process can occur spontaneously in soils with, e.g., Fe^{2+} or organic carbon as the reducing agent.

Hexavalent chromium is quite mobile in soils, since it is not strongly absorbed by many types of soil. However, it can be reduced to the less mobile trivalent form by the humic substances in soils that are rich in organic matter.

The Wood Preservative CCA

Another potentially significant source of chromium to the environment stems from its presence in *chromated copper arsenate* (CCA), the widely used wood preservative previously mentioned. CCA is a waterborne mixture of metal oxides with which wood is treated using a vacuum-pressure impregnation process. The amount of CCA forced into the wood is almost 10% of the mass of the lumber. The chromium used here originally is hexavalent. However, during a period of *fixation*, which lasts for several weeks after treatment, almost all the Cr(VI) is reduced to Cr(III) by reaction with carbon in the wood. This process produces insoluble complexes that are slow to leach from the treated wood over its lifetime, since the copper and chromium at least are bound to the wood. Leaching of heavy metals from the wood becomes very slow a few months after treatment, with more copper and arsenic than chromium being lost.

One use of CCA is to protect wooden structures, such as residential docks, that are destined to be used in aquatic environments. For environmental and human-health reasons, CCA had largely replaced organic preservatives such as creosote and pentachlorophenol (mentioned in Chapter 11) in such applications. However, not only chromium but also arsenic and copper leach from the structures into the water over time.

 ## Green Chemistry: Removing the Arsenic and Chromium from Pressure-Treated Wood

Wood that is used for exterior construction decays in about three to twelve years unless it is treated with pesticides that prevent destruction from termites, fungi, and other wood-destroying agents. Most of the preserved exterior wood that is presently used is commonly called *pressure-treated wood*. Found in over 50% of homes in the United States, pressure-treated wood is also used in decks, fences, retaining walls, piers, docks, wooden bridges, picnic tables, and playground equipment, and it lasts 20–50 years. Treatment of wood thus results in the conservation of millions of trees each year and limits the use of scarce woods that contain natural preservatives, such as redwoods.

Pressure-treated wood is produced by placing the wood in a horizontal cylinder and evacuating the cylinder, which draws out much of the moisture from the wood cells. An aqueous preservative solution is then pumped into the cylinder and the pressure is raised, forcing the preservative solution into the wood cells. Since the 1930s in the United States, the preservative solution used in 95% of pressure-treated wood was copper chromated arsenate (CCA), discussed in the previous section.

Although the exact composition varies, the most common formulation for the preservative solution is 35.3% CrO_3, 19.6% CuO, and 45.1% As_2O_3. Treatment with CCA results in wood with copper, chromium, and arsenic concentrations of 0.1–2.0%, 0.25–4.0%, and 0.15–4.0%, respectively. In 2001, 7 billion board feet of pressure-treated wood (enough to build 450,000 homes) was produced utilizing 150 million pounds of CCA. The CCA contained 64 million pounds of hexavalent chromium and 40 million pounds of arsenic. A 12-foot-long 2 × 6 board of CCA-treated wood contains from 16 to 300 g of arsenic. If all this arsenic were ingested, it would be enough to kill many people.

Although the preservatives are "locked" into the wood, health officials and environmentalists have long been concerned with the potential for leaching of arsenic and chromium from pressure-treated wood and the ingestion of these elements by infants and children from direct contact with the wood. Studies of the soils beneath decks made of pressure-treated wood gave copper, chromium, and arsenic concentrations averaging 75, 43, and 76 mg/kg, while control soils averaged 17, 20, and 4 mg/kg. Studies also indicate that measurable amounts of arsenic can be dislodged from the surfaces of pressure-treated wood by direct contact.

Because of the environmental and human-health concerns associated with CCA, the U.S. EPA announced that wood producers had voluntarily ceased production of CCA-treated wood on December 31, 2003, for products intended for residential use. Chemical Specialties, Inc. (CSI) in 1996 introduced a new wood preservative called *Preserve* to replace CCA, for which it earned a Presidential Green Chemistry Challenge Award in 2002. Preserve is formulated with an alkaline quaternary (ACQ) wood preservative. The active ingredients in the preparation are copper and a *quaternary ammonium salt*, $R_4N^+Cl^-$ (either didecyl dimethyl ammonium chloride or alkyl dimethyl benzyl ammonium chloride). According to the World Health Organization, none of these ingredients are mammalian or human carcinogens.

Because there are no environmental and health concerns for ACQ, under the EPA system, ACQ is registered as a nonrestricted pesticide for treatment of wood products. Analogous formulations of copper and ACQ are used as algaecides and fungicides in lakes, rivers, and streams, as well as in fish hatcheries and potable water supplies. Quaternary ammonium salts are also used as surfactants in typical household and industrial detergents and disinfectants, and unlike arsenic, they have low toxicity to mammals. It is also noteworthy that the copper that is used in the ACQ formulations is obtained from scrap copper. ACQ-treated wood not only eliminates the cancer and toxicity concerns associated with CCA, but it offers the advantages of simplified disposal of treated wood and elimination of hazardous waste generation at the approximately 450 treatment sites across the United States.

Review Questions

1. What is a *sulfhydryl group*, and how does it interact biochemically with heavy metals? How does the interaction affect processes in the body?

2. What is a *chelate*? What principle underlies the usual cure for heavy-metal poisoning?

3. Do heavy metals bioconcentrate? Do any biomagnify?

4. What are some important sources of airborne mercury?

5. Is the liquid or the vapor of mercury more toxic? Describe the mechanism by which mercury vapor affects the human body.

6. What is an *amalgam*? Give two examples and explain how they are used.

7. Explain how the *chlor-alkali process* led to the release of mercury in the environment.

8. Name two uses for mercury in batteries.

9. Write the formulas for the methylmercury ion, for two of its common molecular forms, and for dimethylmercury. What is the principal source of human exposure to methylmercury?

10. Explain why mercury vapor and methylmercury compounds are much more toxic than other forms of the element.

11. What is meant by *Minamata disease*? Explain its symptoms and how it first arose.

12. List several uses for organic compounds of mercury. Which ones have been phased out?

13. What are the two common ionic forms of lead?

14. Explain how lead can dissolve—e.g., in canned fruit juice—even though it is insoluble in mineral acids.

15. Explain why lead contamination of drinking water by lead pipes is less common in hard-water areas than in soft-water areas.

16. Why were lead compounds used in paints? Why were mercury compounds used in paints?

17. Explain why heavy-metal compounds such as PbS and $PbCO_3$ become much more soluble in acidic water.

18. In what forms does lead exist in the lead storage battery?

19. What are the formulas and names of the two organic compounds of lead that were used as gasoline additives? What was their function?

20. Discuss the toxicity of lead, especially with respect to its neurological effects. Which subgroups of the population are at particular risk from lead?

21. What are the main sources of cadmium in the environment?

22. Explain how *nicad* batteries operate. What other uses are made of cadmium?

23. What is the main source of cadmium to humans?

24. Describe what is meant by *itai-itai* disease, and relate where it arose and why.

25. What is *metallothionein*? What is its significance with respect to cadmium in the body?

26. What are some uses of arsenic that result in contamination of the environment?

27. What organic compounds of arsenic are of environmental significance? Why is arsenic in organic acid forms not very toxic to humans?

28. What are the main health concerns about arsenic in drinking water? Why is the drinking water in many regions of Bangladesh heavily polluted with arsenic?

29. Describe how arsenic can be removed from water.

30. What are the two important oxidation states of chromium? Which one is more toxic?

31. Explain how Cr(VI) can be removed from wastewater.

32. What is CCA? Name two toxic heavy metals it contains.

33. Complete the chart shown in outline below:

Element	Common ionic forms	Common organo-metallic forms	Most toxic forms
Mercury			
Lead			
Cadmium			
Arsenic			
Chromium			

 Green Chemistry Questions

See the discussion of focus areas and the principles of green chemistry in the Introduction before attempting these questions.

1. The replacement of lead with yttrium in electrodeposition coatings won PPG a Presidential Green Chemistry Challenge Award.

(a) Into which of the three focus areas does this award best fit?

(b) List one of the twelve principles that is addressed by the green chemistry developed by PPG.

2. What environmental advantages does the use of yttrium oxide have over the use of lead oxide in electrodeposition coatings?

3. What environmental advantages does electrodeposition offer over spray painting?

4. The removal of arsenic and chromium from pressure-treated wood won Chemical Specialties, Inc. a Presidential Green Chemistry Challenge Award.

(a) Into which of the three focus areas for these awards does this award best fit?

(b) List one of the twelve principles that is addressed by the green chemistry developed by Chemical Specialties, Inc.

Additional Problems

1. A man whose flesh weighs 50 kg eats 1 kg of fish a day. If the fish contains the legal limit of 0.5 ppm of methylmercury, and assuming that this substance becomes equally distributed within his flesh, calculate the steady-state concentration of methylmercury that his flesh will achieve. By comparison of this concentration to that of the fish, decide whether biomagnification is occurring in the transfer of methylmercury from the fish to the man. Would your answer to the latter question differ if he ate only 0.2 kg of fish a day?

2. (a) Approximately fit an exponential decay curve to the distribution of blood lead among children based on the portion of the curve in Figure 15-6b from 20 ppb (the zero point of the function) to higher levels. By integration, determine the total percentage of children having levels in excess of 100 ppb that your function predicts.

(b) What does the fact that the curve in Figure 15-6b does not continue to rise as the blood level comes close to zero tell us about the background level of lead in the environment?

3. (a) Since the ion AsO_4^{3-} is basic, the forms $HAsO_4^{2-}$, $H_2AsO_4^{-}$, and H_3AsO_4 will all be present in aqueous solutions of its salts. Given that for H_3AsO_4 the successive acid dissociation constants are 6.3×10^{-3}, 1.3×10^{-7}, and 3.2×10^{-12}, deduce the predominant form of arsenic in waters of pH = 4, 6, 8, and 10.

(b) Arsenic in its As(III) form exists in solution as H_3PO_3 or one of its ionized forms. Given that the acid dissociation constant for H_3PO_3 is 6×10^{-10}, calculate the ratio of its un-ionized molecular form to the ionized form $H_2PO_3^{-}$ at pH values of 8 and 10.

4. The object of this problem is to estimate the mass of lead that would have been deposited annually on each square meter of land near a typical, busy, six-lane freeway from the lead compounds emitted by cars using the roadway. Use reasonable estimates for the number of cars passing a point each day and their average mileage per liter or gallon of gasoline. Assume that the gasoline contained about 1 g of lead per gallon, or 0.2 g/L, and make the approximation that half the lead was evenly deposited within 1000 m on each side of the freeway.

5. How does the phenomenon of acid rain indirectly affect the risk to human health from mercury, lead, and cadmium?

6. The 2+ ions of mercury, lead, and cadmium each form a series of complexes by attaching to themselves in successive equilibrium reactions up to four chloride ions. Deduce the formulas for the species for one of these metals, including the net charges on the complexes. Would the complexes having three or four chlorines be more likely to be found in fresh water or in seawater?

7. By reference to reliable websites or textbooks in your library on heavy metals and/or water pollution, determine why copper is considered to be toxic, and find out what types of organisms are at risk from elevated levels in the environment. Does speciation affect the toxicity of copper?

8. Based on the material in this chapter, write a paragraph supporting your choice of which of the five metals you think still requires the most regulatory control for environmental reasons.

9. The equilibrium vapor pressure of mercury at room temperature is about 1.6×10^{-6} atm. Imagine an old chemistry lab that, through liquid mercury spills over the years, has accumulated enough mercury in cracks in the floor, etc. that mercury liquid–vapor equilibrium has been established. What is the concentration, in units of milligrams per cubic meter, of Hg^0 in the air of the room at 20°C? Does this value exceed the limit of 0.05 mg/m^3 established by the American Conference of Governmental Industrial Hygienists for safe exposure based on a 40-hour workweek?

Further Readings

1. J. E. Ferguson, *The Heavy Elements: Environmental Impact and Health Effects*, (Oxford: Pergamon Press, 1990).

2. T. W. Clarkson, "The Three Modern Faces of Mercury," *Environmental Health Perspectives* 110, supplement 1 (2002): 11.

3. G. J. Myers and P. W. Davidson, "Does Methylmercury Have a Role in Causing Developmental Disabilities in Children?" *Environmental Health Perspectives* 108, supplement 3 (2002): 413.

4. G.-B. Jiang, J.-B. Shi, and X.-B. Feng, "Mercury Pollution in China," *Environmental Science and Technology* 40 (2006): 3672.

5. T. W. Clarkson, "Mercury: Major Issues in Environmental Health," *Environmental Health Perspectives* 100 (1992): 31–38.

6. R. Hoffmann, "Winning Gold," *American Scientist* 82 (1994): 15–17.

7. R. A. Goyer, "Results of Lead Research: Prenatal Exposure and Neurological Consequences," *Environmental Health Perspectives* 104 (1996): 1050–1054.

8. R. L. Canfield et al., "Intellectual Impairment in Children with Blood Lead Concentrations Below 10 μg per Deciliter," *New England Journal of Medicine* 348 (2002): 1517.

9. A. Spivey, "The Weight of Lead: Effects Add Up in Adults," *Environmental Health Perspectives* 115 (2007): A31.

10. P. A. Baghurst et al. (a) "Environmental Exposure to Lead and Children's Intelligence at the Age of Seven Years," *New England Journal of Medicine* 327(18) (1992): 1279–1284; (b) "Exposure to Environmental Lead and Visual-Motor Integration at Age 7 Years: The Port Pirie Cohort Study," *Epidemiology* 6 (1995): 104.

11. W. Shotyk and M. Krachler, "Lead in Bottled Water: Contamination from Glass and Comparison with Pristine Groundwater," *Environmental Science and Technology* 41 (2007): 3508.

12. M. N. Mead, "Arsenic: In Search of an Antidote to a Global Poison," *Environmental Health Perspectives* 113 (2005): A378.

13. A. Lykknes and L. Kvittingen, "Arsenic: Not So Evil After All?" *Journal of Chemical Education* 80 (2003): 497.

14. J. A. Hingston et al., "Leaching of Chromated Copper Arsenate Wood Preservatives: A Review," *Environmental Pollution* 111 (2001): 53.

Websites of Interest

Log on to www.whfreeman.com/envchem4/ and click on Chapter 15.

CHAPTER 16

WASTES, SOILS, AND SEDIMENTS

In this chapter, the following introductory chemistry topics are used:

- Thermochemistry
- Concept of oxidation and reduction as electron loss or gain; oxidation number; basic electrochemistry
- Background organic chemistry (see Appendix)
- Concepts of acids and bases; pH
- Phase diagrams

Background from previous chapters used in this chapter:

- Adsorption; NO_X; particulates (Chapters 3 and 4)
- Aerobic and anaerobic decompositions; methane (Chapter 6)
- Ethanol; MTBE (Chapter 8)
- DDT; K_{ow} (Chapter 10)
- PCBs, dioxins, and furans (Chapter 11)
- PAHs; phthalates; BTEX, chlorinated solvents (Chapter 12)
- BOD, COD, and water carbonate chemistry (Chapter 13)
- Heavy-metal chemistry (Chapter 15)

Introduction

In this chapter, we turn our attention to the environmental aspects of the solid state—particularly of soil and of the sediments of natural water systems—and of ways that polluted soils and sediments can be remediated.

A tractor moving shredded paper in the warehouse of a paper-recycling plant. The paper can be reused or recycled in several different ways. (Digital Vision)

A closely related issue is the nature and disposal of concentrated wastes of all kinds, including both domestic garbage and hazardous waste, and their possible recycling.

The material in this chapter has been arranged in the order of generally increasing toxicity and hazard. Thus we begin with the least toxic substances—domestic and commercial garbage—and consider its disposal by landfilling, incineration, or recycling. We then consider soils and sediments and their contamination by chemicals. Finally, we look at hazardous wastes and some of the high-technology methods that are being developed to dispose of them.

Domestic and Commercial Garbage: Its Disposal and Minimization

The great majority of the material that we discard and that must be disposed of is not hazardous but is simply *garbage* or *refuse*. The greatest single constituent of this **solid waste** (defined as waste that is collected and transported by a means other than water) is construction and demolition debris, almost all of which is either reused or eventually buried in the ground. The second largest volume of waste is that generated by the commercial and industrial sectors, followed by the domestic waste generated by residences. Typically, a North American generates about 2 kg of domestic and commercial waste a day, twice as much as the average European. In these discussions, we will not consider the much larger amounts of waste generated by the petroleum industries, by agriculture, as ashes from power plants, or as sewage, which was discussed in Chapter 14.

A breakdown by the type of solid waste typically generated in countries at various levels of economic development is shown in Figure 16-1. Notice that the fraction of the waste that is vegetable matter declines as the level of economic development rises. The opposite is true of paper, which in industrialized countries is the largest single component of waste and dominates commercial-sector waste. Historically, the largest component of paper waste was newspapers; now the volume of paper packaging is similar. The amount of packaging has grown, in part, because so many goods are now produced far from their ultimate destination and must be transported safely over long distances.

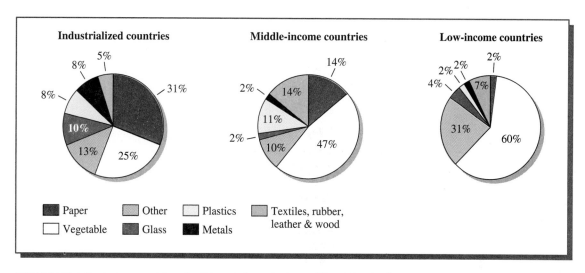

FIGURE 16-1 Typical composition of solid waste for countries at different levels of economic development. [Source: "Waste and the Environment," *The Economist* (29 May 1993): 5 (Environment Survey section).]

Plastics, glass, and metals each account for about one-tenth of the volume of solid waste in developed countries, whereas organic matter (food waste) accounts for about twice this value. These proportions would differ significantly in areas that collect materials for recycling or composting: The glass and metals components would be much smaller.

Burying Garbage in Landills

The main method used for disposal of **municipal solid waste,** MSW, is to place it in a **landfill** (also variously called a *garbage dump* or a *rubbish tip*), which is a large hole in the ground that is usually covered with soil and/or clay after it is filled. For example, 85–90% of domestic and commercial waste is currently landfilled in the United Kingdom, about 6% is incinerated, and the same fraction is recycled or reused; similar figures apply to many municipalities in North America. Landfilling dominates the disposal methods because its direct costs are substantially lower than disposal by any other means.

In the past, landfills were often simply large holes in the ground that had been created by mineral extraction—especially old sand or gravel pits. In many instances, they leaked and contaminated the aquifers that lay beneath them; this was especially true for landfills that used former sand pits, since water easily percolates through sand. These landfills were not designed, controlled, or supervised, and they accepted many types of wastes, including hazardous materials.

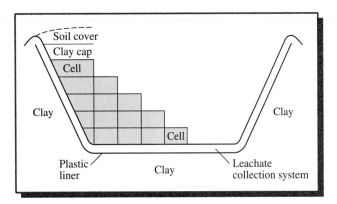

FIGURE 16-2 Components of a modern landfill (in the process of being filled).

Modern municipal landfills are much more elaborately designed and engineered, often accept no hazardous waste, and have their sites selected to minimize impact on the environment. The components of a typical modern landfill are illustrated in Figure 16-2.

In a **sanitary landfill,** the MSW is compacted in layers (to reduce its volume) and is covered with about 20 cm (8 in.) of soil at the conclusion of each day's operations. Thus the landfill consists of many adjacent *cells*, each corresponding to a day's waste (Figure 16-2). After one layer of cells is completed, another is begun, and the process is continued until the hole is filled. Usually, the landfill is eventually capped by a meter or so of soil, or preferably clay, a material that is fairly impervious to rain. A *geomembrane* made of plastic may be added on top as a liner instead of the clay, or over it. The system recommended by the U.S. EPA is illustrated in Figure 16-3.

During the time that municipal wastes in a landfill are decomposing—aerobically at first, then anaerobically after a few months or a year—water

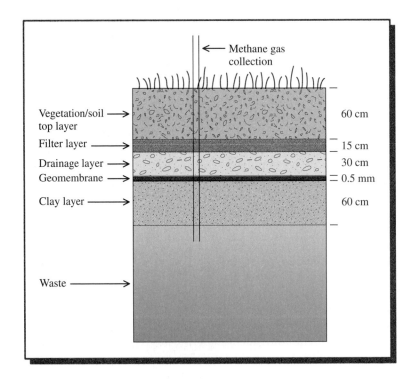

FIGURE 16-3 Landfill design with cover system recommended by the U.S. EPA.

from precipitation, liquid from the waste itself, and groundwater that seeps into the landfill all percolate through the garbage, producing a liquid called **leachate.** This liquid contains dissolved, suspended, and microbial contaminants extracted from the solid waste. Leachate volume is relatively high for the first few years after a site is covered. Typically, leachate contains

- volatile organic acids such as *acetic acid* and various longer-chain fatty acids;

- bacteria;

- heavy metals, usually in low concentration (those of most concern in leachate are lead and cadmium); and

- salts of common inorganic ions such as Ca^{2+}.

The **micropollutants** present even in MSW leachate include common volatile organic compounds such as *toluene* and *dichloromethane*.

Stages in the Decomposition of Garbage in a Landfill

There are three stages of decomposition in a municipal landfill. Operating landfills still receiving garbage undergo all three stages simultaneously in different regions or depths. In practice, only food and yard waste biodegrade. Rubber, plastics, and much of the paper content of garbage are very slow to degrade.

- In the first, short, **aerobic stage,** oxygen is available to the waste; it oxidizes organic materials to CO_2 and water with the release of heat. The internal temperature can rise to 70–80°C, since the reactions are exothermic. The carbon dioxide released from the organic matter as it decomposes makes the leachate acidic, thereby further facilitating its ability to leach metals encountered in the waste. Since the majority of biodegradable material is *cellulose*, whose empirical formula is approximately CH_2O, we can approximate this phase of the reaction as

$$CH_2O + O_2 \longrightarrow CO_2 + H_2O$$

Some organic matter is partially oxidized to aldehydes, ketones, and alcohols, which give fresh waste its characteristic sweet smell.

- In the second, **anaerobic acid phase,** the process of *acidic fermentation* occurs, generating *ammonia, hydrogen,* and **carbon dioxide** gases and large quantities of partially degraded organic compounds, especially organic acids. The pH of the leachate in this phase, 5.5–6.5, is chemically aggressive. Other organic and inorganic substances dissolve in this leachate due to its acidity. Again, carbon dioxide is released. This phase of the reaction can be approximated by the reaction

$$2 \, CH_2O \longrightarrow CH_3COOH$$

although longer-chain fatty acids, which subsequently decompose into acetic acid, are formed initially, as is hydrogen gas.

In this phase, the leachate has a high oxygen demand (see BOD and COD, Chapter 13), as well as relatively high heavy-metal concentrations. Anaerobic decomposition produces volatile carboxylic acids and esters, which dissolve in the water present. The sickly sweet smell that emanates from landfills during this phase is due to these esters and to thioesters.

• The third, **anaerobic**—or **methanogenic**—**stage** starts about six months to a year after coverage and can continue for very long periods of time. Anaerobic bacteria work slowly to decompose the organic acids and hydrogen that were produced in the second stage. Since the organic acids are consumed in the process, the pH rises to about 7 or 8, and the leachate becomes less reactive. The main products of this stage are carbon dioxide and **methane,** CH_4. To a first approximation, the overall reaction is

$$CH_3COOH \longrightarrow CH_4 + CO_2$$

Methane generation usually continues for a decade or two and then drops off relatively quickly. Some methane is also formed when hydrogen gas combines with carbon dioxide. Much lower BOD values and a smaller volume are associated with landfills in this phase. Because the leachate is not acidic in this phase, the heavy-metal concentrations drop since these substances are not as soluble at higher pH.

Often the methane gas produced by a landfill is vented to the atmosphere by being directed into wells or gravel-packed seams in the landfill. In some municipalities, the methane gas is burned as it is released through vents (see Figure 16-3) rather than being released into the air. This treatment of the methane is especially desirable, given that the greenhouse gas potential per molecule of CH_4 is much greater than that of the CO_2 produced by its combustion (Chapter 6). The heat produced from the combustion of this gas can be used for practical purposes. Indeed, the second and third stages of decomposition in landfills are identical to those used in the deliberate production of **biogas** (biomethane) for energy in reactors using municipal solid waste sludge, food processing waste, livestock waste, and other biodegradable materials.

PROBLEM 16-1

Calculate the volume of methane gas, at 15°C and 1.0 atm pressure, that is released annually by the anaerobic decomposition of 1 kg of garbage, assuming the latter is 20% biodegradable organic in nature and that decomposition occurs evenly over a 20-year period. *[Hint: Add together the equations for the two anaerobic stages of decomposition.]*

Leachate from a Landfill

Engineering is needed to control the leachate from a landfill. Otherwise, the liquid can flow out at the bottom of the landfill and percolate through porous soil to contaminate the groundwater below it. Alternatively, if the soil under the landfill is nonporous, the leachate can build up and gradually overflow the site (the overflowing bathtub effect), possibly contaminating nearby surface waters.

The typical components used to control the leachate consist of:

- A **leachate collection and removal system,** followed by treatment of the liquid. Often the potential effect of leachate on groundwater is monitored by digging and testing several wells in the vicinity.

- A **liner** placed around the walls and bottom of the landfill. The liner material is either synthetic (e.g., a plastic such as 2-mm-thick high-density polyethylene) or natural (e.g., compacted clay). The material chosen is impervious to water and will largely prevent the leakage of the contaminated leachate into the groundwater, especially if and when the collection system fails due to clogging, etc. Since 1991 new landfills in the United States must have at least six layers of protection between the garbage and the underlying groundwater! Liners have been developed that consist of *bentonite clay*—an excellent sealant that efficiently binds heavy metals, preventing their migration out of the landfill—sandwiched between two layers of a plastic such as polypropylene.

Leachate treatment systems must address all the liquid's major components. The treatment of leachate, usually at a sewage treatment plant, is accomplished by aerobic degradation to rapidly decrease the BOD, sometimes using advanced oxidation methods that employ ozone (Chapter 14). In the past, collected leachate was often simply returned to the top of the landfill, since, during its second percolation through the waste, much of its organic content would be biologically degraded; however, this practice is now discouraged in the United States.

Incineration of Garbage

Besides landfilling, the most common way to dispose of wastes, particularly organic and biological ones, is by **incineration**—the oxidation by controlled burning of materials to simple, mineralized products such as carbon dioxide and water. The primary incentive in the incineration of municipal solid waste is to substantially reduce the *volume* of material that must be landfilled. In the case of toxic or hazardous substances, an even more important goal is to eliminate the toxic threat from the material. Incineration of hospital wastes is done to sterilize them as well as to reduce their volume.

Many municipalities throughout the world burn domestic garbage in incinerators. For example, Japan and Denmark burn more than half their domestic waste, but the practice is banned in some countries. The combustible components of the garbage, such as paper, plastics, and wood, provide the fuel for the fire. The most common domestic MSW incinerators are one-stage **mass burn** units; the two-stage **modular** type is more modern. In the latter, wastes are placed in the primary chamber and burn at a temperature of about 760°C. The gases and airborne particles that result from the first stage are then burned more completely, at temperatures in excess of 870°C, in the secondary combustion chamber. The quantity of waste gases that must later be controlled is greatly reduced in the two-stage units compared to the one-stage unit, although the gases are further heated as they exit the one-stage unit to produce more complete combustion. In some incinerators, an attempt is made to recover some of the heat of the combustion processes and to convert it to steam, hot water, or even electricity.

The output from municipal incinerators includes not only the final gases but also solid residues that amount to about one-third of the initial weight of the garbage. **Bottom ash** is the noncombustible material that collects at the bottom of the incinerator. **Fly ash** is the finely divided solid matter that is usually trapped by environmental pollution controls in the stack to prevent it from being released into the outside air. Much of the ash consists of the inorganic constituents of the waste, which form solids rather than gases even when fully oxidized. Although fly ash accounts for only 10–25% of the total ash mass, it is generally the more toxic component, since heavy metals, dioxins, and furans readily condense onto its small particles. The low density and small-particle character of the ash make inadvertent dispersal into the environment a significant risk. Of particular concern are heavy metals in the ash, which could potentially be leached from it and pollute nearby surface water and groundwater. For many years, it was common for incinerator ash to be taken to a hazardous waste landfill. Techniques such as the addition of an adhesive or melting and vitrification have now been developed to solidify ash into a leach-resistant material that need not be classified as hazardous waste. In some countries such as Denmark and the Netherlands, the ash is mainly recycled into asphalt.

The main environmental concern about incineration is the air pollution that it generates, consisting of both gases and particulates. The emission controls on MSW incinerators can control a large fraction, but not all, of the toxic substances emitted into the air from the combustion process. About half the capital costs of new incinerators is spent on air pollution control equipment. Typically, the controls include a **baghouse filter,** which is made of woven fabric and is used to filter particulates, especially those with diameters over 0.5 μm, from the flow of output gas. Periodically, the bags are shaken or the air flow is reversed to collect the fly ash. Also typical is a **gas scrubber,** which is a stream of liquid or solid that passes through the gas stream, removing some particles and gases. If the liquid stream consists of **lime,** CaO, and water, which collectively form $Ca(OH)_2$, or if the solid stream consists of lime, acid gases such as HCl and SO_2 are efficiently removed since they are neutralized to salts by the lime. Heavy metals are also captured by the alkaline environment, since they form insoluble hydroxides. In some modern installations, *nitrogen oxides* are removed by spraying ammonia or *urea* into the hot exhaust gases (recall the chemistry explained in Chapter 3). In another new technology used in garbage incinerators, activated charcoal or lignite coke powder is blown into the exhaust gases, which are subsequently filtered by baghouse; much of the dioxin, furan, and mercury content of the exhaust gases is removed, since these components adsorb onto the charcoal or coke surface.

Although public concern has centered on emissions from *hazardous waste* incinerators (to be discussed later in this chapter), several U.S. surveys in the 1990s indicated that many more dioxin and furan emissions emanate from medical waste and municipal waste incinerators than from hazardous waste ones, although cement kiln units used for hazardous wastes (discussed in a later section) also make a significant contribution. There are more than 1000 medical waste incinerators in the United States. Emissions to the air from incinerators are most likely to happen during start-up and when equipment fails. Because medical waste and backyard barrel garbage incinerators operate in a start-and-stop mode, they tend to produce more airborne pollutants per unit mass of incinerated waste than do larger incinerators. Overall, municipal, backyard, and medical incinerators are believed to be a major anthropogenic source of both mercury and dioxins/furans in the U.S. environment and a moderately important source of cadmium and lead.

 ## Green Chemistry: Polyaspartate—A Biodegradable Antiscalant and Dispersing Agent

In pipes, boilers, water cooling systems, and other devices that handle water, scale buildup (Figure 16-4) tends to reduce water flow and heat transfer, thereby lowering efficiency. In addition, scale may lead to corrosion and damage of these devices. Scale is generally the result of the precipitation of insoluble compounds such as *calcium carbonate, calcium sulfate,* and *barium sulfate*. Compounds called *antiscalants* or *dispersants* are employed to prevent the buildup of scale. Whereas antiscalants prevent the formation of scale, dispersants allow its formation but maintain the scale in a state of suspension so that it can simply be washed away.

One of the most commonly used antiscalants and dispersants is the polyanion **polyacrylate** or PAC:

$$\left[-CH_2 - CH - \atop \underset{\displaystyle \underset{\displaystyle O^-}{\overset{\displaystyle \|}{C=O}}}{\;} \right]_n$$

Short chains of this polymer act as antiscalants while longer chains act as dispersants. The anionic *carboxylate groups*, —COO⁻, of PAC are able to form complexes with the cations (such as calcium and barium) normally found in scale, thus preventing the formation of scale or dispersing it. Globally, several hundred million kilograms of PAC are produced each year, a significant portion of which is used as a dispersant or antiscalant. Although PAC is nontoxic, it is nonvolatile and does not degrade in the environment. When used for water treatment, it builds up in lakes and streams, or at best it must be removed in wastewater treatment plants as a sludge and then landfilled.

To prevent this environmental burden, biodegradable antiscalants and dispersants such as **polyasparate** have been developed. Polyaspartate can be used to replace PAC, but because it undergoes biodegradation to innocuous products (such as carbon dioxide and water), it eliminates the need for removal in wastewater treatment plants and disposal in landfills.

Although the performance of polyaspartate is comparable to that of PAC, its price was formerly prohibitive. The Donlar Corporation developed a new synthesis of polyaspartate that lowered the cost of the polymer so that it was competitive with PAC. For this accomplishment, Donlar won a Presidential Green Chemistry Challenge Award in 1996. Donlar's synthesis (Figure 16-5) begins by heating *aspartic acid* (a naturally occurring amino acid) to produce *polysuccinimide*, followed by basic hydrolysis to produce polyaspartate. This straightforward synthesis is not only economically desirable but also environmentally sound. The first step simply requires heat and yields only water as a by-product, while the second step uses water under basic conditions to produce the desired product. The product of this synthesis is generally called **thermal polyaspartate** (TPA) because of the heat used

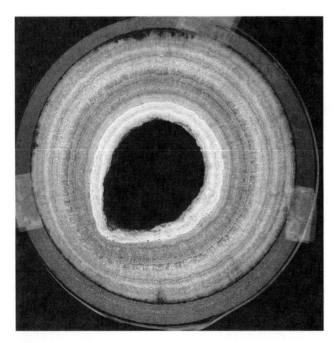

FIGURE 16-4 Scale buildup in a water pipe. (Ward Lopes)

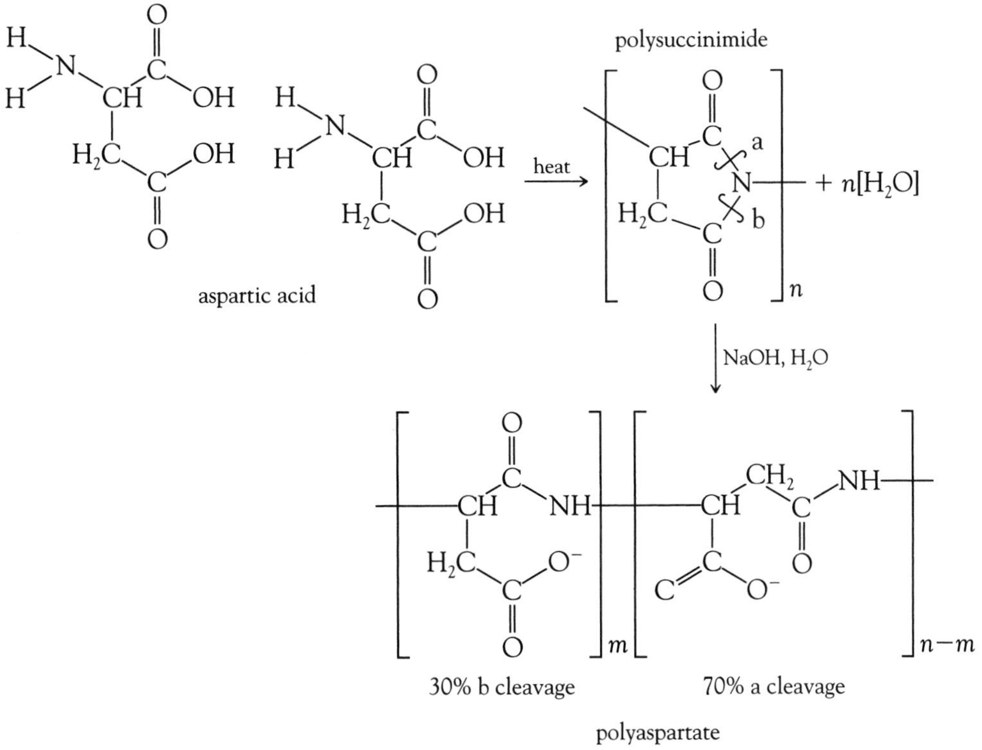

FIGURE 16-5 Donlar Corporation synthesis of polyaspartate.

in the synthesis. Polyaspartate can also be used in fertilizers (to enhance uptake of nutrients) and detergents (as builders).

The Recycling of Household and Commercial Waste

In the past few decades, there has been mounting pressure in developed countries to reduce the amount of material discarded as waste after a single use. The incentives here are to conserve the natural resources, including energy, from which the materials are produced and to reduce the volume of material that must be buried as garbage, incinerated, etc. The **four Rs** of such waste management philosophies are:

- **Reduce** the amount of materials used (sometimes called *source reduction*).

- **Reuse** materials once they are formulated.

- **Recycle** materials to recover components that can be refabricated.

- **Recover** the energy content of the materials if they cannot be used in any other way.

These principles can be and are applied to all types of wastes, including hazardous ones, but in the following discussions we concentrate on their application to domestic materials, particularly in regard to recycling.

A distinction is often made between **preconsumer recycling,** which involves the use of waste generated during a manufacturing process, and **postconsumer recycling,** which involves the reuse of materials that have been recovered from domestic and commercial consumers. The postconsumer items most often collected for recycling are

- paper (especially newspapers and cardboard),

- aluminum (especially beverage cans),

- steel (especially food cans), and

- plastic and glass containers.

The labor, energy, and pollution costs associated with collecting the materials, sorting them, and transporting them to facilities where they can be reused must be considered in any analysis of recycling. In addition, historically the demand for recycled materials in these categories has been unstable, with prices swinging wildly in response to changes in supply and demand. For these and similar reasons, the recycling of paper, glass, and plastics usually needs to be justified on noneconomic and nonenergy grounds—such as savings in landfill space. The most economically viable forms of recycling materials usually involve a minimum of chemical reprocessing and correspond more closely to **reuse**—e.g., using newspapers to make paperboard or insulation and reusing glass and plastic containers. Debates still rage both in the popular press and in the scientific literature as to whether or not recycling is a worthwhile activity.

The Recycling of Metals and Glass

From the viewpoints of both economics and energy conservation, the recycling of metals makes sense. Virgin metal must be obtained by reduction of the oxidized form of the element found in nature. The reduction process requires energy that does not need to be expended again when the metallic form of the element is recycled.

Consider the reduction of aluminum and iron from the oxide ores. By definition, the enthalpies of these processes equal the negative of their enthalpies of formation:

$$Al_2O_3 \longrightarrow 2\,Al + \tfrac{3}{2}O_2 \qquad \Delta H^\circ = -\Delta H_f^\circ\,(Al_2O_3)$$

$$= +1676 \text{ kJ/mol oxide}$$

$$= +31 \text{ kJ/g metal}$$

Recycling aluminum cans saves 95% of the energy that is needed to produce Al metal from bauxite ore. Since the energy required for the aluminum reduction must come in the form of electricity, and since this energy accounts for about 25% of the cost of its production, it makes good economic sense to recycle this metal. However, the recycling rate for aluminum cans dropped from 65% in 1992 to only 45% in 2005 in the United States. In contrast, Sweden, which requires deposits on beverage containers that are refunded upon their return, collects 85% of aluminum cans. Considerable savings in aluminum have resulted from the reduction by about one-third of the weight of individual cans over the last few decades. Aluminum can be recycled endlessly without loss in quality. Globally, recycled aluminum provides about one-third of the production of the metal. Recycling steel cans saves about two-thirds of the energy required to produce them from iron ore.

PROBLEM 16-2

The enthalpy of formation of the principal ore of iron, Fe_2O_3, is -824 kJ/mol. Calculate the enthalpy of the reaction in which 1.00 g of metallic iron is formed from the ore. Given your result, would you expect the price that recycling operators are willing to pay for scrap iron per kilogram to be greater or less than that for scrap aluminum?

In the case of paper, glass, and plastics, there is no significant change in the average oxidation number of the principal materials during their transformation from inexpensive raw material components—wood, sand and lime, and oil, respectively—to finished products; thus there are no great energy savings when they are recycled.

Modern, low-polluting, and energy-efficient electric furnaces cannot handle as high a proportion of used glass as can their more polluting, more energy-consuming counterparts that use fossil fuels. Consequently, the recycling of too much glass can produce more pollution and use more energy than would otherwise be the case! In any event, the use of glass containers for beverages has fallen sharply, and their recycling rate dropped from 31% in 1992 to 20% in 2003 in the United States.

The Recycling of Paper

People in developed countries throw away more paper than any of the other components of municipal solid waste (see Figure 16-1), and it seems an obvious material to recycle. However, the production of virgin paper, for example, uses only about one-quarter more energy than the recycling of old paper. The transportation of waste paper to recycling mills and the de-inking process itself are heavy consumers of energy. Notwithstanding these considerations, tremendous quantities of paper, especially newsprint, are currently recycled. Indeed, corrugated paper products are the most intensely recycled material in North America.

The first step in recycling paper is mechanical dispersal into its component fibers in water. Then it is cleaned to remove nonfibrous contaminants, followed by treatment with *sodium hydroxide* or *sodium carbonate* to de-ink it. A detergent is added to help disperse the pigment, and the ink particles are removed by washing or flotation on air bubbles, which rise to the top. The resulting de-inked stock is usually less white than virgin fiber, so the two types are often blended. If necessary, the whiteness of the recycled stock can be improved by bleaching, usually with *peroxides* and *hydrosulfites*. The used ink, which is recovered in a sludge with some pulp fibers, is later pressed to remove water and then can be burned to produce steam for use in mill operations, or it can be treated to detoxify it. In general, the use and release into the environment of materials such as *chlorine* or other bleaching agents, acids, and organic solvents are significantly less with the production of recycled paper than with the creation of the virgin material.

Paper of different types is composed of fibers of very different lengths (long ones in office paper, short ones in newsprint) that cannot be mixed in recycling to produce high-quality paper. In addition, there is a limit to the number of times that paper can be recycled, since with each cycle the pulp fibers become shorter and so lose some of their integrity. Newsprint can be recycled back into newsprint about six to eight times. Food boxes and egg cartons are usually made from recycled pulp fiber.

From 1985 to 2000, the paper industry in the United States spent almost $20 billion on the technology and capital investment required to recycle paper; the ultimate goal was to recycle about half of all paper used in the country. Indeed, by the mid-2000s, close to half the paper and paperboard products in North America was being recycled, compared to two-thirds in many European countries. Recycling rates in the United States in 2003 were 82% for newspapers, 71% for corrugated boxes, 56% for office paper, 33% for magazines, and 16% for phone directories.

The issue of whether it is preferable to recycle paper, rather than bury it in a landfill, is controversial. Since paper is made from wood, which has grown by extracting carbon dioxide from the atmosphere, landfilling paper that will never rot is a form of sequestering CO_2. Research indicates that about 70% of the carbon in paper—and more than 97% of that in other wood products—remains undecomposed in landfills. However, this argument is negated if any significant fraction of the paper decomposes to emit methane; nor does it take into account the larger amounts of energy and water required to produce virgin paper compared to recycled paper.

Perhaps a more clever use of waste paper in the future will be its conversion to fuel *ethanol*, as discussed in Chapter 8. Paper can be incinerated directly to recover its energy content, reducing the amount of fossil fuels burned in power stations. According to an analysis by Britain's *Centre for Environmental Technology*, recycling paper is environmentally superior to landfilling it but is actually inferior to burning it for its fuel value when all factors are taken into consideration.

The Recycling of Tires

Another consumer commodity that presents a waste-management headache is vehicle tires. In North America, about one 10-kg rubber tire per person per year on average is discarded; thus about one-third of a *billion* tires are added to the supply of approximately 3 billion tires presently stored in mountainous piles, awaiting ultimate disposal! Because the tires are made primarily from oil and consequently are flammable, tire fires in these huge piles are not uncommon and produce tremendous amounts of smoke, *carbon monoxide*, and toxins such as **polynuclear aromatic hydrocarbons (PAHs)** and **dioxins** (Chapters 11 and 12). The fires are difficult to extinguish because of air pockets in and between the tires.

There have been efforts to use tires either as fuel or as a filler for asphalt, but currently such applications consume only about 10% of the tires that are discarded annually. Some used tires are also utilized for their rubber content, to produce landscaping products.

A number of attempts have been made to commercially reprocess shredded scrap tires by **pyrolysis**—the thermal degradation of a material in the absence of oxygen. The resulting products are low-grade gaseous and liquid fuels and a **char** containing minerals and a low-grade version of the material called **carbon black,** which can be further treated and converted into *activated carbon* (Chapter 14). It may eventually be possible to convert the liquid component into high-grade char, thereby making the process economically profitable. The "rubber" in tires consists of about 62% of a hydrocarbon polymer and 31% of carbon black—added to strengthen the tires and reduce wear—so there is a ready market for the latter. Using the liquid component as a fuel is problematic because of its high content of aromatic hydrocarbons.

The Recycling of Plastics

One of the triumphs of industrial chemistry in the twentieth century was the development of a wide variety of useful **plastics.** All plastics are composed at the molecular level of polymeric organic molecules, very long units of matter in which a short structural unit is repeated over and over again. All the raw materials (except chlorine) from which the plastics are currently made are obtained from crude oil.

Conceptually, the simplest organic polymer is **polyethylene** (or **polyethene**), the molecules of which are composed of many thousands of $-CH_2-$ units bonded together:

$$\cdots -\overset{\overset{\displaystyle H}{|}}{\underset{\underset{\displaystyle H}{|}}{C}}-\overset{\overset{\displaystyle H}{|}}{\underset{\underset{\displaystyle H}{|}}{C}}-\overset{\overset{\displaystyle H}{|}}{\underset{\underset{\displaystyle H}{|}}{C}}-\overset{\overset{\displaystyle H}{|}}{\underset{\underset{\displaystyle H}{|}}{C}}-\overset{\overset{\displaystyle H}{|}}{\underset{\underset{\displaystyle H}{|}}{C}}-\overset{\overset{\displaystyle H}{|}}{\underset{\underset{\displaystyle H}{|}}{C}}- \cdots$$

This polymer is prepared by combining many molecules of *ethylene* (ethene) (hence the name) and is an example of an **addition polymer.** Depending on exactly how the polymerization takes place, either **low-density polyethylene (LDPE)** (the plastic given the recycling number 4) or **high-density polyethylene (HDPE)** (the cloudy white or opaque plastic given the recycling number 2) is formed.

There are several other addition polymers similar to polyethylene in which one (or more) of the four hydrogen atoms in each ethylenic $-CH_2-CH_2-$ unit is replaced by a group or atom X, giving the polymer

$$\cdots -\overset{\overset{\displaystyle H}{|}}{\underset{\underset{\displaystyle H}{|}}{C}}-\overset{\overset{\displaystyle H}{|}}{\underset{\underset{\displaystyle X}{|}}{C}}-\overset{\overset{\displaystyle H}{|}}{\underset{\underset{\displaystyle H}{|}}{C}}-\overset{\overset{\displaystyle H}{|}}{\underset{\underset{\displaystyle X}{|}}{C}}- \cdots$$

If the X attached to every second carbon atom in each chain is chlorine, then the clear (or blue-tinted) polymer **polyvinyl chloride (PVC)** (recycling number 3) is obtained. If the substituent X is a methyl group, we have **polypropylene** (recycling number 5), and if it is a benzene ring we have **polystyrene** (recycling number 6). The plastics formed from all these polymers are used extensively in packaging, as indicated by the original uses listed in Table 16-1.

The other plastic that is commonly recycled (number 1) is the clear plastic **poly(ethylene terephthalate) (PET).** Its structure is a chain of two CH_2 units alternating with a unit of the organic molecule *terephthalic acid*. PET is used in the form of film (for magnetic tape as well as photographic film), fiber, and molded resin (e.g., plastic bottles).

In the last quarter of the twentieth century, plastics became the symbol of a throwaway society, since much of the product—especially that used in packaging—was designed to be used once and then discarded. Many environmentalists believed that waste plastic was a major culprit in the garbage crisis. Indeed, molded plastics take up a greater percentage of volume in landfills than their percentage by mass because their densities are low, although they are eventually compressed by the weight of the materials piled on them, as well as by compacting machinery before they are placed in the landfill. Plastics are the second most common constituent of municipal garbage, following

TABLE 16-1	Commonly Recycled Plastics		
Plastic Recycling Number	Acronym and Name of Plastic	Original Use Examples	Recycle Use Examples
1	PET Poly(ethylene terephthalate)	Beverage bottles; food and cleanser bottles; drugstore product containers	Carpet fibers; fiber-fill insulation; non-food containers.
2	HDPE High-density polyethylene	Milk, juice, and water bottles; margarine tubs; crinkly grocery bags	Oil and soap bottles; trash cans; grocery bags; drain pipes
3	PVC (or V) Polyvinyl chloride	Food, water, and chemical bottles; food wraps; blister packs; construction material	Drainage pipes; flooring tiles; traffic cones
4	LDPE Low-density polyethylene	Flexible bags for trash, milk, and groceries; flexible wraps and containers	Bags for trash and groceries; irrigation pipes; oil bottles
5	PP Polypropylene	Handles, bottle caps, lids, wraps, and bottles; food tubs	Auto parts; fibers; pails; trash containers
6	PS Polystyrene	Foam cups and packaging; disposable cutlery; furniture; appliances	Insulation; toys; trays; packaging "peanuts"
7	Other	Various	Plastic "timber": posts, fencing, and pallets

paper and cardboard by quite a margin. The per capita annual use of plastics in North America is approximately 30 kg.

For a number of reasons, including the fact that landfills, especially throughout Europe, are reaching their capacity and that many citizens in developed countries are opposed to their incineration, many plastics are now collected from consumers and recycled. As of the mid-1990s, over 80% of the mass of plastics recycled in the United States consisted of PET and HDPE, in approximately equal amounts, with LDPE the only other one of significance. Some countries, such as Sweden and Germany, have made manufacturers legally responsible for the collection and recycling of the packaging used in their products.

There is little doubt that the public in many developed countries has embraced recyling of plastics. As of the late 1990s, about half the urban communities in the United States had curbside recycling programs that included plastics. However, the U.S. recycling rate for PET soda bottles dropped from a high of 40% in 1995 to only 22% (of the annual total of about 25 billion, corresponding to about 100 per capita) in 2004. As in the case of aluminum, Sweden achieves a much higher recycling rate (80%) by requiring refundable deposits on plastic bottles. The majority of recycled PET bottles are used to make fiber for carpet, strapping, and film, though 14% of it is used in food and beverage containers. Only a fraction of the demand for PET bottles is met by the current supply.

HDPE containers have achieved a somewhat higher recycling rate (26%) than PET ones in the United States, presumably because they are used for milk and for nonfood products that are primarily found at home. Recycled HDPE products include nonfood bottles, plastic garden products such as pots, and plastic lumber.

About 150 polyethylene carrier bags are produced annually for every man, woman, and child on the planet. About 80% of them are reused at least once, in kitchens as trash bags or by dog owners. Although they constitute less than 1% of garbage sent to landfills, and their incineration can recover the energy of the oil used to make them, they are highly visible forms of pollution when allowed to blow around and collect on streets and beaches. In addition, plastic bags can have a devastating effect on marine animals such as turtles and whales, whose stomachs can be blocked if they inadvertently consume the bags. Although plastic bags are collected for recycling in some locales, countries such as Denmark, Ireland, and Taiwan have imposed a tax on them in order to discourage their use.

There has been much resistance to plastics recyling in some quarters, including many in the plastics industry. Their argument is that virgin plastics are a low-cost product that is made from a relatively low-cost raw material (oil). The input energy for making plastics is very small compared to that used for making aluminum or steel from its raw materials. The cost of cleaning used plastic and converting it back into its monomers so that it can be polymerized again is substantial, compared to the current cost of oil. Some executives in the plastics industry believe that the best disposal method for plastics is simply to burn them and use the heat energy provided, especially given the fact that there is little objection by the public to simply burning most (over three-quarters) of the oil produced—in vehicles, domestic furnaces, and power plants. In addition, experiments indicate that the presence of plastics makes the other materials in domestic garbage burn more cleanly and reduces the need for supplemental fossil fuel to be added. Although plastics account for less than 10% of the mass of garbage, they make up more than one-third of its energy content.

Environmentalists counter these arguments by pointing out that if environmental impacts were to be included in determining the cost of virgin materials, recycled plastic would be the cheaper choice. Also, the combustion of some plastics, notably PVC, produces dioxins and furans and releases *hydrogen chloride* gas.

Ways of Recycling Plastics

There are four ways to recycle plastics, one physical and three chemical:

1. Reprocess the plastic (a physical process) by remelting or reshaping. Usually the plastics are washed, shredded, and ground up, so that clean, new products can then be made.

2. Depolymerize the plastic to its component monomers by a chemical or thermal process so that it can be polymerized again.

3. Transform the plastic chemically into a low-quality substance from which other materials can be made.

4. Burn the plastic to obtain energy (energy recycling).

Examples of the *reprocessing* option include the production of carpet fibers from recycled PET; of plastic trash cans, grocery bags, etc. from recycled HDPE; and of CD cases and office accessories such as trays and rulers from recycled polystyrene. Further examples of reprocessing are listed for each category of packaging plastic in the last column of Table 16-1.

The *depolymerization* option can be employed with PET and other polymers of the $-A-B-A-B-A-B-$ type, in which units of types A and B alternate in the structure. These **condensation polymers** are produced by combining small molecules that contain A and B units. During the polymerization process, A and B form the polymer and the remaining parts of the molecules combine. For example, in the production of PET, the molecule *methanol*, CH_3OH, is formed from the OH unit of one component and the CH_3 of the other. In the chemical *depolymerization* process, a catalyst and heat are applied to a mixture of methanol and the plastic to *reverse* the polymerization process and recover the original components:

$$CH_3-A-CH_3 + HO-B-OH + CH_3-A-CH_3 + HO-B-OH + \cdots$$

$$\underset{\text{depolymerization}}{\overset{\text{polymerization}}{\rightleftharpoons}} CH_3-A-B-A-B-A \cdots -A-B-A-B-OH + \text{many } CH_3OH$$

For PET, B is $-CH_2-CH_2-$ and A is

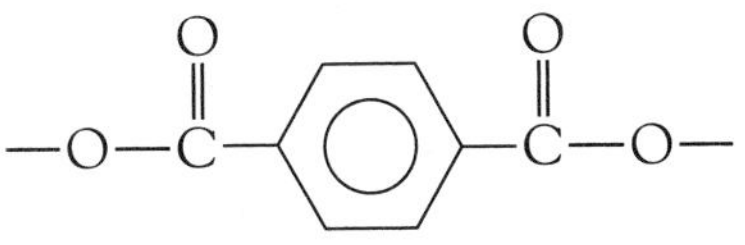

Physical recycling of PET is currently more economically viable than chemical reprocessing.

One of the difficulties in depolymerization of plastics is that the organic and inorganic compounds that are often added to the original polymer to modify the physical properties of the plastic, such as its flexibility, must be removed before the monomers can be reused.

For many addition polymers, it is difficult to devise a process by which the original monomers can be re-formed. For example, the monomer yield for the thermal depolymerization of polystyrene is about 40%, but it is close to zero for polyethylene because the chain will be broken at random positions and will not exclusively produce two-carbon units.

Examples of the *transformation* option are:

- *Reductive* processes such as the production of synthetic crude oil by hydrogenation of plastics or by heating them to a high temperature to "crack" the polymer molecules, a process that can be used even with mixed plastics. The pyrolysis of polyethylene to monomers that can be converted to lubricants has also been proposed.

- *Oxidative* processes such as the gasification of plastics by adding oxygen and steam to produce synthesis gas (a mixture of hydrogen and carbon monoxide, discussed in Chapter 8).

 ## Green Chemistry: Development of Recyclable Carpeting

In the United States, 2.3 billion square yards of carpeting were shipped in 2004. This is enough to cover the entire surface area of Washington, D.C., 11 times, or about 3/4 of the state of Rhode Island. The Carpet America Recovery Effort (CARE) reports that in 2005, 89% of used carpeting was disposed of in landfills, 3% was incinerated in cement kilns, and only 7% was recycled. Carpeting is not biodegradable, takes up valuable and rapidly declining landfill space, and is ultimately made from petroleum, a nonrenewable resource. As landfill fees escalate—estimates indicate they will double every five years—and the costs of transporting, installing, and replacing carpeting rise, the demand for an economical and environmentally responsible alternative to landfilling of this product has increased.

The two major components of carpeting are the backing and the face fiber. Since the 1970s, PVC has been the material of choice for carpet backing. Environmental and health concerns about PVC include *vinyl chloride* (the monomer used to produce PVC) and *phthalate* plasticizers. Vinyl chloride, a known carcinogen, is volatile. Some believe that this compound outgases from the polymer, whereas others argue that the high temperatures used to process PVC should eliminate virtually all of the volatile monomer. Phthalates, which are added to PVC to make it more flexible, migrate out of the polymer and have become widely dispersed in the environment. The growing concern over the effects of phthalates on the human reproductive system was discussed in Chapter 12. In addition, as also discussed in Chapter 12, when PVC burns, it produces toxic by-products including dioxins, furans, and hydrochloric acid.

Recycling is now commonplace when it comes to paper, glass, and plastic. However, most of us do not think of recycling carpeting. In the United States, only 7% of carpeting is recycled, in part because the PVC backing interferes with the recycling process. Shaw Industries won a Presidential Green Chemistry Challenge Award in 2003 for its development of a new type of carpeting that allows for *closed-loop recycling*, i.e., recycling of used carpeting back into carpeting with little loss of material. This new type of carpeting, known as *EcoWorx*, employs polyolefin backing and nylon-6 fiber. In addition to lending itself to recycling, the polyolefin backing has low toxicity and eliminates the significant environmental and health concerns related to PVC.

The process of recycling EcoWorx carpeting begins with grinding the used carpeting and separating the heavier particles (the polyolefin backing) from the lighter ones (the nylon-6 fibers) with a stream of air (a process known as *elutriation*). The polyolefin particles can then be reused in the extrusion process to produce new backing. The nylon-6 is depolymerized to its monomer (caprolactam) and repolymerized to virgin nylon-6 fibers. The fibers are then used to form new carpeting, completing the cycle. Both the backing and the fibers of EcoWorx carpeting can be used again and again to form new carpeting.

There are several additional environmental advantages of EcoWorx, as well as economic advantages:

• Recycling requires no additional petroleum feedstocks; this benefits not only the environment but also the economic bottom line.

• Recycling reduces the amount of landfill space needed.

• The polyolefin-backed carpeting is 40% lighter in weight than carpeting backed with PVC. More carpeting can be shipped by truck within weight limits, thus lowering fuel consumption, cost, and pollution.

• The use of polyolefins eliminates the energy-intensive heating process required for PVC. Again, this results in lowering fuel consumption, cost, and pollution.

• In carpet backing (including PVC backing), significant amounts of inorganic fillers are used to provide loft and bulk. Traditionally, virgin *calcium carbonate* was employed for this purpose. EcoWorx contains 60% class C fly ash (a waste by-product from the burning of lignite or sub-bituminous coal) as a filler. Using fly ash as a filler utilizes an unwanted by-product and precludes the use of a virgin chemical.

Returning goods for recycling is often a significant barrier to recycling. To overcome this difficulty, Shaw developed a system for returning the carpeting at the end of its useful life, at no cost to the consumer.

Life Cycle Assessments

One technique used in minimizing the production of wastes and in pollution prevention is the **life cycle assessment** (or **analysis**), LCA—an accounting of all the inputs and outputs in a product's life, from raw material extraction to final disposal. This cradle-to-grave analysis for a product (or process) can be used to identify the types and magnitudes of its environmental impacts, including both the natural resources used and the pollution produced.

The results of a life cycle assessment can be used in two ways:

- to identify opportunities within the life cycle to minimize the overall environmental burden of a product, and

- to compare two or more alternative products to determine which is more environmentally friendly.

An example of the first use is in the production of motor vehicles; the life cycle and the most important inputs and outputs are illustrated in Figure 16-6.

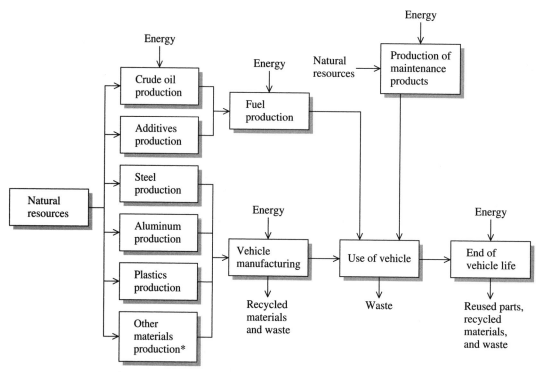

*For example, rubber, lead, glass, paints, and coolants.

FIGURE 16-6 Important input and output components in life cycle assessments of motor vehicles. [Source: M. Freemantle, "Total Life-Cycle Analysis Harnessed to Generate 'Greener' Automobiles," *Chemical and Engineering News* (27 November 1995): 25.]

In designing new cars, life cycle assessments are used to help minimize pollution while maintaining economic viability. The analysis is particularly useful for identifying new environmental burdens that would arise if others are decreased. For example, vehicles could be made much lighter by increased use of plastics; however, the types of plastics used are difficult to recycle and would increase the eventual burden of solid waste in landfills.

Soils and Sediments

The contamination of soil by wastes is not solely a phenomenon of modern times. In Roman times, metal ores were mined and the ores smelted, polluting the surrounding countryside. The production of materials and chemicals in Europe even at the start of the Industrial Revolution produced substantial pollution. However, the extent of contamination and the hazard from discarded materials expanded greatly in the last century, particularly in the period after World War II.

We begin this section by discussing the nature of soil and of sediments.

Basic Soil Chemistry

Soils are composed roughly equally of solid particles, about 90% of which are inorganic in nature and the rest organic matter, and of pore space, about half of which is air and half water. The inorganic particles are residues of weathered rock; chemically they are mainly **silicate minerals.** At the atomic level, these minerals consist of polymeric inorganic structures in which the fundamental unit is a silicon atom surrounded tetrahedrally by four oxygen atoms. Since these oxygen atoms are in turn each bonded to another silicon, etc., the resulting structure is an extended network. There are many variations on the silicate structural theme. Some networks have exactly twice as many oxygens (formally, O^{2-}) as silicons (formally, Si^{4+}) and correspond to electrically neutral SiO_2 polymers. In others, some of the tetrahedral sites are occupied by aluminum ions, Al^{3+}, instead of Si^{4+}; the extra negative charge in these networks is neutralized by the presence of other cations such as H^+, Na^+, K^+, Mg^{2+}, Ca^{2+}, and Fe^{2+}. Some common silicon–oxygen structural units are illustrated in Figure 16-7.

Over time, the weathering of the silicate minerals from rocks can involve chemical reactions with water and acids in which ion substitution occurs. Eventually, these reactions yield substances that are important examples of the class of soil materials known as **clay minerals.** A mineral having a particle size less than about 2 μm is by definition a component of the clay fraction of soil.

In addition to clay, there are several other soil types; the definition of each type depends on particle size, as indicated in Figure 16-8. Notice the factor-of-10 increase in size with each transition in type: The upper boundary

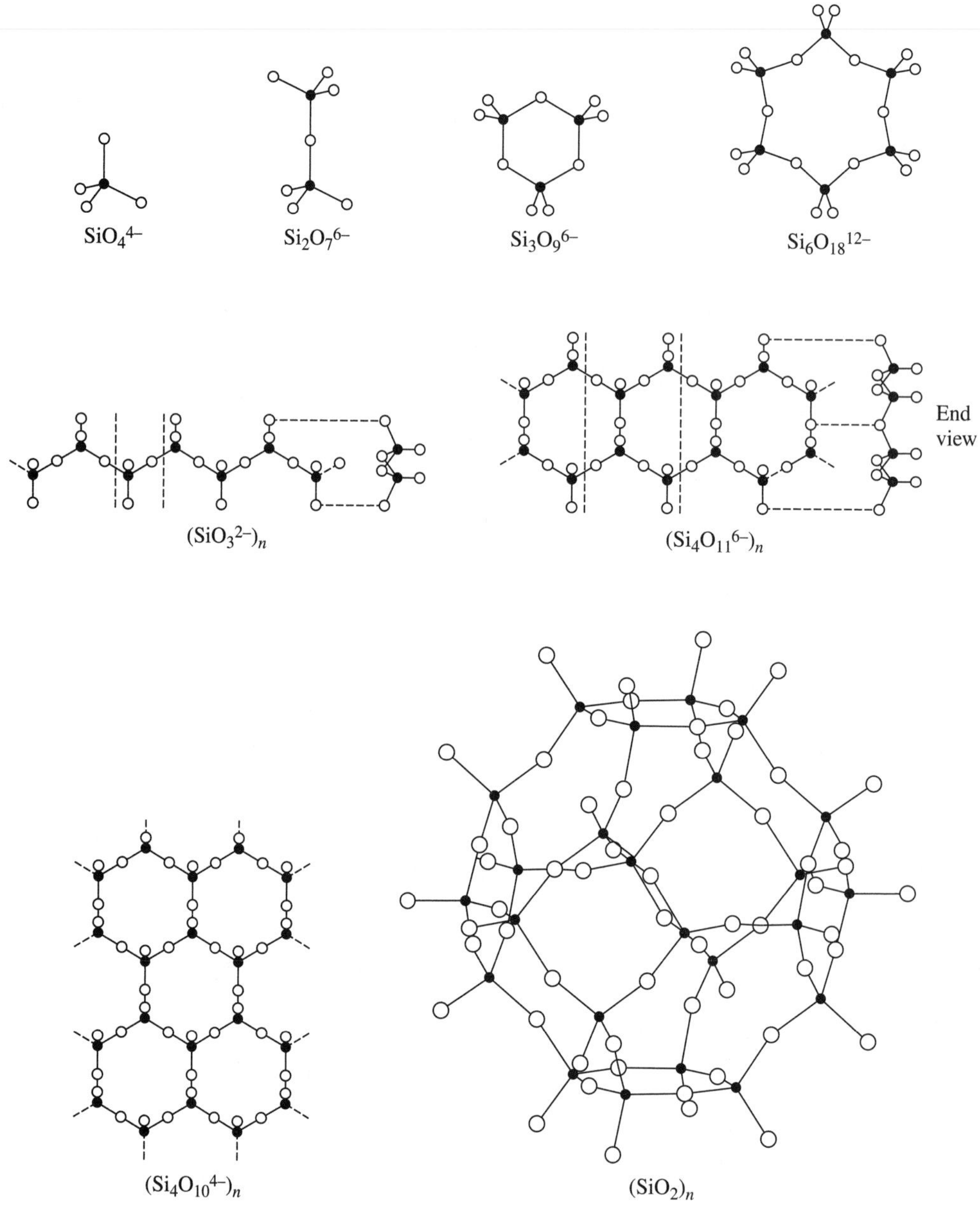

FIGURE 16-7 The common structural units in silicate minerals. Dark circles represent silicon atoms; open circles represent oxygens. [Source: R. W. Raiswell, P. Brimblecombe, D. L. Dent, and P. S. Liss, *Environmental Chemistry* (London: Edward Arnold Publishers, 1980).]

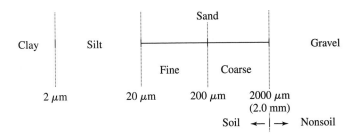

FIGURE 16-8 The soil particle size classification system of the International Society of Soil Science. [Source: G. W. vanLoon and S. J. Duffy, *Environmental Chemistry* (Oxford: Oxford University Press, 2000).]

for silt is 10 times that for clay, that for fine sand is 10 times that for silt, etc. Because the particle size of sand is large, it has a relatively low density and water runs through it easily. In contrast, soils composed of clay are dense and have poor drainage and aeration, since the clay particles form a sticky mass when wet, in contrast to sand and silt particles, which do not stick to each other. The best agricultural soils consist of a combination of soil types. The range of element content for major and minor elements in the mineral component of soil is given in Table 16-2.

Clay particles act as colloids in water. Because the clay particles are much smaller than those of sand or silt, their total surface area per gram is thousands of times larger. Consequently, most important processes in soil occur on the surface of colloidal clay particles.

Particles of clay possess an outer layer of cations that are bound electrostatically to an electrically charged inner layer, as illustrated in Figure 16-9. The most common cations in soil are H^+, Na^+, K^+, Mg^{2+}, and Ca^{2+}. Depending on the concentration of cations in the water surrounding the clay

TABLE 16-2	Element Content of the Mineral Components of Soils		
Major Elements (%)		**Minor Elements (ppm)**	
Si	30–45	Zn	10–250
Al	2.4–7.4	Cu	5–15
Fe	1.2–4.3	Ni	20–30
Ti	0.3–0.7	Mn	~400
Ca	0.01–3.9	Co	1–20
Mg	0.01–1.6	Cr	10–50
K	0.2–2.5	Pb	1–50
Na	Trace–1.5	As	1–20

Source: G. W. vanLoon and S. J. Duffy, *Environmental Chemistry* (Oxford: Oxford University Press, 2000).

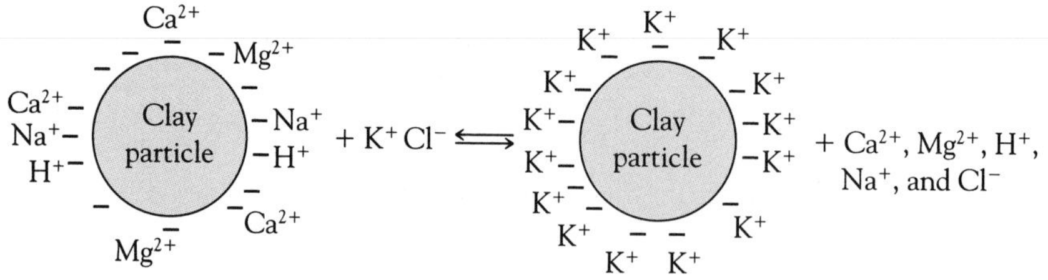

FIGURE 16-9 Ion-exchange equilibria on the surface of a clay particle. The addition of K^+ ions to the soil water displaces the exchange equilibria to the right, whereas removal of K^+ ions from solution displaces it to the left. [Source: R.W. Raiswell, P. Brimblecombe, D. L. Dent, and P. S. Liss, *Environmental Chemistry* (London: Edward Arnold Publishers, 1980).]

particle, the cations on the particle are capable of being exchanged for them. For example, in water rich in potassium ions but poor in other ions, K^+ ions will displace the ions bound to the surface of the clay particle (see Figure 16-9). If, on the other hand, the soil is acidic—i.e., rich in H^+ ions—the metal ions on the surface will be displaced by H^+ ions and the previously bound metal ions will enter the aqueous phase. Generally, the greater the positive charge on a cation, the more strongly it binds to the particle. Heavy metals dissolved in soil water are often bound to the surface of clay particles.

In addition to minerals, the other important components of soil are organic matter, water, and air. The proportion of each component varies greatly from one soil type to another. The organic matter (1–6%), which gives soil its dark color, is primarily a material called **humus.** Humus is derived principally from photosynthetic plants, some components of which (such as cellulose and hemicellulose) have previously been decomposed by organisms that live in the soil. The undecomposed plant material in humus is mainly protein and lignin, both polymeric substances that are largely insoluble in water. A significant amount of the carbon in lignin exists in the form of six-membered aromatic benzene rings connected by chains of carbon and oxygen atoms (see Figure 11-4). Much of the organic matter in soil also consists of colloidal particles.

As a result of the partial oxidation of some of the lignin, many of the resulting polymeric strands contain *carboxylic acid* groups, —COOH. This dark-colored portion of humus consists of **humic** and **fulvic acids** and is soluble in alkaline solutions due to the presence of the acid groups. By definition, humic acid is *insoluble* in acid solution, whereas fulvic acid is *soluble*. The humic acid is less soluble in acid than the fulvic not only because its molecular weight is much greater (100 to 1000 times greater), but also because its oxygen content is lower, so there are fewer —OH groups per carbon to form hydrogen bonds with the water. The acid groups are often adsorbed onto the surfaces of the clay minerals, to an extent dependent on the distribution of surface charge on the particles. Humic and fulvic acids form colloids that are hydrophilic, whereas those of clay are hydrophobic. Generally speaking, nonpolar organic molecules are more strongly attracted to the organic matter in soil than to the surfaces of particles derived from minerals.

Since some of the carbon in the original plant material has been transformed to carbon dioxide and thus lost as a gas to the surroundings, humus has relatively more nitrogen than the original plants; its other main components are carbon, oxygen, and hydrogen. Due to decomposition processes occurring in the organic component of soil, the O_2 content of soil air is often only 5–10% rather than 20%, and its CO_2 concentration is often several hundred times that in the atmosphere.

PROBLEM 16-3

The percentage composition of a typical fulvic acid is 50.7% carbon by mass, 45.1% oxygen, and 4.22% hydrogen. Derive the (simplest) empirical formula for the substance.

The Acidity and Cation-Exchange Capacity of Soil

If the soil at the surface contains minerals with elements in a reduced state, their oxidation by atmospheric oxygen can produce acid. An example is the oxidation of sulfur in *pyrite*, discussed as acid mine drainage in Chapter 13. Acid rain, of course, provides another source of acidity in certain areas (Chapter 4).

Quantitatively, the ability to exchange cations is expressed as the soil's **cation-exchange capacity,** CEC, which is defined as the quantity of cations that are reversibly adsorbed per unit mass of the (dry) material. The quantity of cations is given as the number of moles of positive charge (usually expressed as centimoles or millimoles), and the soil mass is usually taken to be 100 g or 1.00 kg. Typical values of the CEC for common clay minerals range from 1 to 150 centimoles per kilogram (cmol/kg). The CEC values are determined in large part by the surface area per gram of the mineral. CEC values for the organic component of soil are high, due to the large number of —COOH groups that can bind to and exchange cations; e.g., the CEC of peat can be as high as 400 cmol/kg.

PROBLEM 16-4

The CEC for a soil sample is found to be 20 cmol/kg. What is the CEC value for this sample in units of millimoles per 100 grams?

Biologically, the exchange of cations by soils is the mechanism by which the roots of plants take up metal ions such as potassium, calcium, and magnesium. Although the roots release hydrogen ions to the soil in exchange for the metal ions, this is not the main reason why soil in which plants grow is often somewhat acidic. Most of the acidity is due to metabolic processes involving the roots and microorganisms in the soil, which result in the production of *carbonic acid*, H_2CO_3, and of weak organic acids.

Rainwater that is acidic releases base cations from soil particles by exchanging them for H^+ ions. The acidity of water flowing through the soil stays low for this reason. However, once the base cations have been exhausted, aluminum ions are released, as discussed in Chapter 4. It is now known that in the past, base cations from dust particles, especially those containing calcium and magnesium carbonates, neutralized some of the acidity in precipitation; but along with *sulfur dioxide*, their emission from industrial sources has been curtailed in recent decades.

The pH of soil can vary over a significant range for a variety of reasons. For example, soils in areas of low rainfall but high concentrations of the soluble salt *sodium carbonate*, Na_2CO_3, become alkaline due to the (hydrolysis) reaction of the **carbonate ion,** CO_3^{2-}, with water, as discussed in Chapter 13.

Soils that are too alkaline for agricultural purposes can be remediated either by the addition of elemental sulfur, which releases hydrogen ions as it is oxidized by bacteria to *sulfate ion*, or by the addition of the soluble sulfate salt of a metal, e.g., iron(III) or aluminum, which reacts with the soil's water to extract hydroxide ions and thereby release hydrogen ions:

$$2\,S(s) + 3\,O_2 + 2\,H_2O \longrightarrow 4\,H^+ + 2\,SO_4^{2-}$$
$$Fe^{3+} + 3\,H_2O \longrightarrow Fe(OH)_3(s) + 3\,H^+$$

The pH of water present in the soil is determined by the concentration of hydrogen and hydroxide ions. However, soil has **reserve acidity** due to the large number of hydrogen atoms in the $-$COOH and $-$OH groups in the organic fraction and on the cation-exchange sites on minerals that are occupied by H^+ ions. In other words, soils act as weak acids, retaining their H^+ ions in a bound condition until acted upon by bases. Thus the pH of soil tends to be buffered against large increases in pH, since this bound hydrogen ion can be slowly released into the aqueous phase. In the process of **liming,** which is the addition of salts such as calcium carbonate to soil, carbonate ions neutralize acids present in the uppermost soil zones, producing carbon dioxide and water. Once this process has occurred, calcium ions can replace hydrogen ions in the organic matter or clays. The additional carbonate ions that enter the aqueous phase combine with the newly released H^+ ions, again to produce weakly acidic carbonic acid, which dissociates into carbon dioxide gas and water. Thus liming is a procedure by which the pH of a soil can be raised somewhat, and it is the practical method by which acidic soils can be remediated.

Soil Salinity

In hot, dry climates, salts and alkalinity tend to accumulate in soil since there is little rainfall to leach ions from it. In contrast to other climates, the net movement of water in arid climates is upward rather than downward in the soil: Water evaporation and loss by transpiration of plants exceeds rainfall. The salts that accompany the upward migration of water remain at or near the surface when the water has escaped. Salt accumulation at the surface also occurs in semiarid regions due to the use of poor-quality irrigation water, whose salt content remains after the water has evaporated.

Ions are also liberated at the surface of soil in the weathering of otherwise insoluble minerals. A simple example is the reaction of *olivine*:

$$Mg_2SiO_4 + 4\,H_2O \longrightarrow Si(OH)_4 + 2\,Mg^{2+} + 4\,OH^-$$

Additional hydroxide ion is produced when the *silicate ion*, SiO_4^{4-}, produced by mineral weathering reacts as a strong base with water.

As a general rule, hydrolysis (reaction with water) of silicate minerals at the surface produces cations and hydroxide ions. In nonarid climates, the hydroxide is neutralized by acids that are naturally produced in the soil (see later section), but this does not occur in arid regions. There is very little organic matter in the soils of arid areas. The hydroxide reacts with atmospheric carbon dioxide that dissolves in water to produce *bicarbonate* and carbonate ions:

$$OH^- + CO_2 \longrightarrow HCO_3^-$$
$$HCO_3^- \longrightarrow H^+ + CO_3^{2-}$$

Consequently, bicarbonate and carbonate salts accumulate in arid soils. If the predominant cations in the soil are calcium and magnesium, most of the carbonate ion will be locked away as their insoluble carbonate salts. However, if the predominant cations are sodium and potassium, the soil when moist will have a high pH, since the carbonate salts of these ions are soluble and the free CO_3^{2-} will act as a base:

$$CO_3^{2-} + H_2O \rightleftharpoons HCO_3^- + OH^-$$

Increasing soil salinity is a major problem in Australia, especially in regions where wheat and other shallow-rooted crops have replaced natural, long-rooted vegetation. This replacement, plus irrigation of crops including rice and cotton, has resulted in a rise of the water table and, with it, the salt that was formerly deep in the soil.

Sediments

Sediments are the layers of mineral and organic particles, often fine-grained, that are found at the bottom of natural water bodies such as lakes, rivers, and

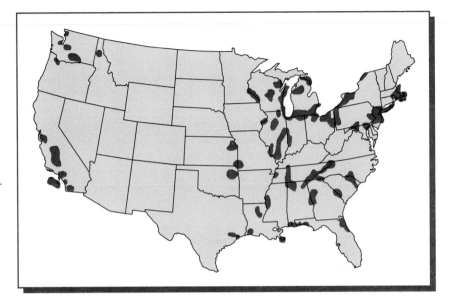

FIGURE 16-10 U.S. watersheds where contaminated sediments may pose environmental risks. [Source: U.S. EPA, in B. Hileman, "EPA Finds 7% of Watersheds Have Polluted Sediments," *Chemical and Engineering News* (26 January 1998): 27.]

oceans. The ratio of minerals to organic matter in sediments varies considerably, depending on location. Sediments are of great environmental importance because they are sinks for many chemicals, especially heavy metals and organic compounds such as PAHs and pesticides, from which they can be transferred to organisms that inhabit this region. Thus the protection of sediment quality is a component of overall water management.

The map in Figure 16-10 shows watersheds in the continental United States where sediments are sufficiently contaminated to pose environmental risks. According to a U.S. EPA report, 7% of all watersheds pose a risk to people who eat fish from them and to the fish and wildlife themselves. The two pollutants found at high levels most frequently at contaminated sites are **PCBs** and mercury, although *DDT* (and its metabolites) and PAHs were also found at high concentrations at many sampling stations.

The transfer of hydrophobic organic pollutants to organisms may proceed by intermediate transfer to **pore water,** which is the water present in the microscopic pores that exist within the sediment material. Organic chemicals equilibrate between being adsorbed on the solid particles and being dissolved in the pore water. For this reason, pore water is often tested for toxicity in determining sediment contamination levels.

The Binding of Heavy Metals to Soils and Sediments

The ultimate sink for heavy metals, and for many toxic organic compounds as well, is deposition and burial in soils and sediments. Heavy metals often accumulate in the top layer of the soil and are therefore accessible for uptake by the roots of crops. For these reasons, it is important to know the nature of these systems and how they function.

Humic materials have a great affinity for heavy-metal cations and extract them from the water that passes through by the process of ion exchange. The binding of metal cations occurs largely by the formation of complexes with the metal ions by —COOH groups in the humic and fulvic acids. For example, for fulvic acids the most important interactions probably involve a —COOH group and an —OH group on adjacent carbons of a benzene ring in the polymer structure, where the heavy-metal M^{2+} ion replaces two H^+ ions:

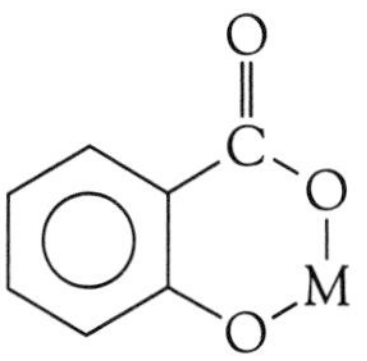

M = heavy metal

Humic acids normally yield water-insoluble complexes, whereas those of smaller fulvic acids are water-soluble.

PROBLEM 16-5

Draw the structure that would be expected if a dipositive metal ion M^{2+} were to be bound to two —COO⁻ groups on adjacent carbons in a benzene ring.

Heavy metals (Chapter 15) are retained by soil in three ways:

- by adsorption onto the surfaces of mineral particles,

- by complexation by humic substances in organic particles, and

- by precipitation reactions.

The precipitation processes for mercury and cadmium ions involve the formation of the insoluble sulfides HgS and CdS when the free ions in solution encounter *sulfide ion*, S^{2-}. Significant concentrations of aqueous sulfide ion occur near lake bottoms in summer months when the water is usually oxygen-depleted, as discussed in Chapter 13. However, the total concentration of mercury in soil water can exceed the limits set by the solubility product of HgS because some of the mercury will take the form of the moderately soluble molecular compound $Hg(OH)_2$ and does not participate in the equilibrium with the sulfide.

In acidic soils the concentration of Cd^{2+} can be substantial, since this ion adsorbs only weakly onto clays and other particulate materials. Above a pH of 7, however, Cd^{2+} precipitates as the sulfide, carbonate, or phosphate, since the concentration of these ions increases with increasing hydroxide ion levels. Thus the liming of soil to increase its pH is an effective way of tying up cadmium ion and thereby preventing its uptake by plants.

Like many other chemicals, heavy-metal ions are often adsorbed onto the surfaces of particulates, especially organic ones that are suspended in water, rather than simply being dissolved in water as free ions or as complexes with soluble biomolecules such as fulvic acids. The particles eventually settle to the bottom of lakes and are buried when other sediments accumulate on top of them. This burial represents an important sink for many water pollutants and is a mechanism by which the water is cleansed. Before they are covered by subsequent layers of sediments, however, freshly deposited matter at the bottom of a body of water can recontaminate the water above it by desorption of the chemicals, since adsorption and desorption establish an equilibrium. Furthermore, the adsorbed pollutants can enter the food web if the particles are consumed by bottom-growing and -feeding organisms.

Just as the total concentration of organic material in sediments may not be a good measure of the amounts that are biologically available, the same is true for the levels of heavy-metal ions present. Different sediments with the same total concentration of the ions of a heavy metal can vary by a factor of at least 10 in terms of the toxicities to organisms arising from the metal. This variation occurs principally because sulfides in the sediments control the availability of the metals. If the concentration of sulfide ions *exceeds* the total of that of the metals, virtually all the metal ions will be tied up as insoluble sulfide salts such as HgS, CdS, etc. and will be unavailable biologically at normal pH values. However, if the sulfide concentration is *less* than that of the metals, the difference is biologically available. The sulfide ion that is available to complex with metals is the amount that will dissolve in cold aqueous acid and is termed **acid volatile sulfide,** AVS. Industrially polluted sediments may have AVS concentrations of hundreds of micromoles of sulfur per gram, whereas uncontaminated sediments from oxidizing environments can have values as low as 0.01 μmol/g.

Although mercury in the form of Hg^{2+} is firmly bound to sediments and does not readily redissolve into water, environmental problems have arisen in several bodies of water due to the conversion of the metal into *methylmercury* and its subsequent release into the aquatic food web. The overall cycling of mercury species among air, water, and sediments was illustrated in Figure 15-1.

As previously discussed, anaerobic bacteria methylate the *mercuric ion* to form $Hg(CH_3)_2$ and CH_3HgX, which then rapidly desorb from sediment particles and dissolve in water, thereby entering the food web. Although the levels of methylmercury dissolved in water can be extremely low (of the order of hundredths of a part per trillion), a biomagnification factor of 10^8 results in ppm-range concentrations in the flesh of some fish. The devastating

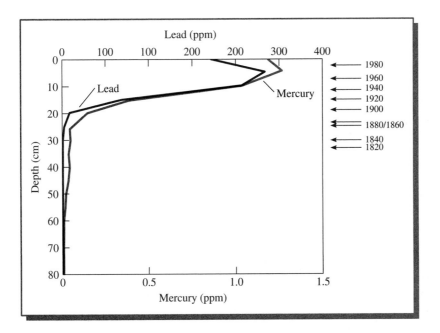

FIGURE 16-11 Lead and mercury concentrations in the sediments of Halifax Harbor versus depth (and therefore year of deposit). [Source: D. E. Buckley, "Environmental Geochemistry in Halifax Harbour," *WAT on Earth* (1992): 5.]

consequences of methylmercury poisoning have already been described (Chapter 15).

Excavation and analysis of the sediments at the bottom of a body of water can yield a historical record of contamination by various substances. For example, the curves in Figure 16-11 show the levels of mercury and of lead in the sediments of the harbor in Halifax, Nova Scotia, as a function of depth and therefore of year. For decades raw sewage has been dumped into this harbor; consequently, its sediments are a historical record of the levels of pollutants in sewage. The metal pollution peaked about 1970, having begun to increase dramatically about 1900. These trends are also typical of heavy-metal levels in other water bodies, such as the Great Lakes; the characteristic decreases of the past few decades are due to the imposition of pollution controls.

In summary, then, both soils and sediments act as vast sinks and reservoirs in the containment of heavy metals.

Mine Tailings

In modern times, many minerals (and in some cases fossil fuels) are extracted from much, much larger quantities of rock (or sand, etc.) than in the past, since the remaining supplies of the minerals occur in dilute form. This practice produces huge quantities of unwanted crushed rock in the form of dry, coarse-grained waste that must be disposed of, usually in slag heaps or landfills close to the mine. Eventually these waste piles are covered with soil and vegetation.

A more important environmental problem arises from disposal of the **tailings** of mining processing—fine-grained slurries that are more mineral-rich and often contain chemicals such as *cyanide* (Chapter 14) that were used to extract or process the ore. The toxic components of tailings are a potential source of pollution to local surface water, groundwater, and soil.

To prevent their dispersal into the environment, tailings are usually deposited as slurries in dams constructed on-site for the purpose. To prevent leakage into the soil, a clay or geomembranic liner is usually incorporated at the dam's base. Over time, the solid settles to the bottom of the dam and the water evaporates or is drained off, i.e., the tailings become *dewatered*, although this can take a long time to achieve. In some cases, toxic materials such as heavy metals, nitrates, or excess acidity is removed by treatment of the tailings. Eventually, to establish a vegetated cover, organic matter and fertilizer must be added, since the dried tailings themselves contain little or no organic matter and consequently are sterile as well as hostile to plants.

If the crushed rock or tailings contain *iron pyrites*, exposure to oxygen will produce *sulfuric acid* by the series of reactions discussed as acid mine drainage in Chapter 13. The acidity can be neutralized by the continuous addition of limestone.

The most important environmental problem associated with a tailings dam is its potential failure, resulting in a catastrophic discharge of the tailings into a waterway and/or onto land. The failure might be due to flood, earthquake, or simply the loss of stability of the dam to pressure over time. A number of such incidents have occurred within the last decade—e.g., in Spain—with devastating results to wildlife, fish, and, in some cases, agricultural land.

An alternative to the storage of tailings on land is disposal in the deep ocean by using pipelines reaching down 100 m or more. Lack of oxygen there will slow the process of oxidation, and within a few years the tailings will be covered by other debris. Some biologists are dubious about this plan, because the environment for bottom-dwelling organisms will suffer. Heavy-metal contamination of fish has occurred in areas where this means of disposal is practiced. Due to the fine-grained nature of the tailings, dispersal over wide areas of the ocean floor occurs.

The Remediation of Contaminated Soil

Even areas thought to be rather pristine can have localized areas of contaminated soil. For example, the large-scale forestry industry in New Zealand has resulted in the contamination of several hundred sites where lumber is treated with the preservative *pentachlorophenol*, PCP (Chapter 11). Furan and dioxin contaminants of the PCP are also found in the soils. All three contaminant types have now leached into the groundwater at some of the sites and have begun to bioacccumulate in the food chain.

Contaminated soil is found most often not only near waste disposal sites and chemical plants, but also near pipelines and gasoline stations. The three main types of technologies currently available for the remediation of contaminated sites are

- containment or immobilization,

- mobilization, and

- destruction.

In general, the technologies can be applied *in situ,* i.e., in the place of contamination, or *ex situ,* i.e., after removing the contaminated matter to another location. Owing to the costs and risks, e.g., air pollution arising from excavation, *in situ* processes are usually preferred.

Among the techniques associated with *in situ* **containment** (i.e., the isolation of wastes from the environment) are capping of the contaminated site, especially with clay, and/or the imposition of cut-off walls of low permeability that prevent the lateral spread of contaminants. *Ex situ* containment would consist of the placing of the excavated soil in a special landfill. **Immobilization** techniques, including solidification and stabilization, are especially useful for inorganic wastes, which tend to be difficult to treat by other methods. Stabilization can often be achieved by adding a substance to convert a heavy-metal ion into one of its insoluble salts, such as the sulfide in the case of mercury and lead, or the oxide in the case of chromium. A concentrated waste can be solidified by reaction with *portland cement,* for example, or by entombing the wastes in molten glass in the process of **vitrification.** By these techniques, the solubility and mobility of the contaminants are reduced.

Mobilization techniques are mainly accomplished *in situ* and include soil washing and the extraction of contaminant vapor from soil for highly volatile, water-insoluble contaminants such as gasoline. Heating of the soil to increase the rate of evaporation and air injection wells are sometimes used in conjunction with **soil vapor extraction,** in which the contaminants are removed by drilling wells in the soil and applying vacuum extraction. As indicated in Table 16-3, this technique is the most frequently used innovative technology at *Superfund* sites (Box 16-1) in the United States. A related technology is **thermal desorption,** in which wastes are heated to cause volatile organic compounds to vaporize. Both soil vapor extraction and thermal desorption are useful to remediate both volatile and semivolatile organic compounds, the latter including many PAHs.

TABLE 16-3	Common Innovative Remediation Technologies in Projects at U.S. Superfund Sites (as of 1996)		
Technology	Sites in Design or Installation	Sites Operational or Completed	Total Number
Soil vapor extraction	69	70	139
Thermal desorption	22	28	50
Bioremediation (ex situ)	24	19	43
Bioremediation (in situ)	14	12	26
In situ flushing	9	7	16
Soil washing	8	1	9
Solvent extraction	4	1	5

Source of data: U.S. EPA, *Innovative Treatment Technologies: Annual Status Report*, 8th ed., 1996.

BOX 16-1 | The Superfund Program

In 1980 the federal government of the United States established a program now known as *Superfund* to clean up abandoned and illegal toxic waste dumps, since dangerous chemicals from many such sites were polluting groundwater. The cleanup costs are shared by chemical companies, the current and past owners of the sites, and the government. Many billions of dollars have already been spent on remediation, and many billions more will eventually be required. Progress in the cleanups has been rather slow on account of the litigation involved and the huge amounts of money at stake. Many decades will pass before even the highest-priority sites are all cleaned up.

The Superfund program is administered by the Environmental Protection Agency, which has identified nearly 1,300 waste sites having such serious potential to cause harm to humans and the environment that they have been placed on a National Priorities List. New Jersey, Pennsylvania, and California have the greatest number of priority sites.

By the late 1990s, the EPA had finished cleanup work at 300 sites, begun work at more than 700 others, and conducted emergency removal of materials at more than 3000 additional locations. The rate at which cleanup work was completed at additional sites slowed appreciably in the mid-2000s. In all, over 30,000 sites have been identified as potentially in need of cleanup.

The most common contaminants at the Superfund sites are the heavy metals lead, cadmium, and mercury, and the organic compounds benzene, toluene, ethylbenzene, and trichloroethylene.

In situ **soil washing** is accomplished by injecting fluids through wells into subsurface soil and collecting them in other wells. The fluid can simply be water, which will remove water-soluble constituents, or an aqueous solution that is acidic or basic in order to remove basic and acidic contaminants, respectively. Other options in soil washing include the use of solutions containing chelating agents such as *EDTA* (see Chapter 15) to remove metals and of oxidizing agents to oxidize and thereby solubilize previously insoluble species. The solvents used to extract the metal–organic complex from the aqueous environment of the soil include hydrocarbons and supercritical carbon dioxide (a substance described later in this chapter). In order for the resulting complex to be electrically neutral and therefore preferentially soluble in the organic phase, a chelating agent with acidic hydrogens that are substituted by the metal is usually employed. If the metal exists as an oxyanion, e.g., chromium as Cr(VI), it may first have to be reduced in oxidation number before it will bind to a chelating agent.

Sometimes the washing solution uses **surfactants,** *surface-active* agents. These are substances such as detergents that possess both hydrophobic and hydrophilic components within the same molecule and therefore can increase the mobilization of hydrophobic contaminants into the aqueous phase. **Biosurfactants** produced by microbes have recently been discovered that can selectively remove certain heavy metals such as cadmium from soil. Currently, soil washing and flushing are the most common innovative technologies used to remove metals at Superfund sites.

The containment, mobilization, and immobilization techniques by themselves do not result in the *elimination* of the hazardous contaminants. **Destruction** techniques, principally incineration and bioremediation, do result in permanent elimination because they chemically or biochemically transform the contaminants. Organic contaminants in soil can be oxidized (mineralized) by feeding the excavated soil into the combustion chamber of an incinerator or by using incineration or one of the specialty oxidation techniques, to be discussed later, to treat the substances that have been extracted from the soil. Bioremediation uses the metabolic activities of microorganisms to destroy toxic contaminants and is also discussed in detail later.

Electrochemical techniques are sometimes used to remediate soil. Placing electrodes in the contaminated ground and applying a dc voltage between them results in ion transport within the soil: The ions travel within the groundwater electrolyte. If heavy-metal ions are dissolved in the electrolyte water, they will eventually move to the (negatively charged) cathode and be deposited on it. Indeed, other metal ions tend to be desorbed from their positions on negatively charged clay surfaces in the process, since hydrogen ion is released at the anode (as water is electrolyzed) and subsequently migrates in the groundwater toward the cathode (see Figure 16-12). Recall that heavy-metal ions are much more soluble in an acidic than in a neutral or alkaline environment.

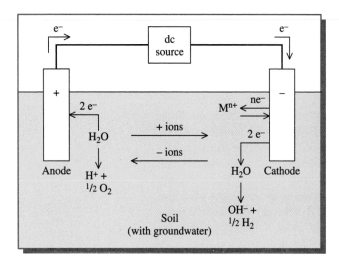

FIGURE 16-12 Electro-chemical remediation of metal-contaminated soil.

In some applications of such **electrokinetic** methods, the process is stopped before the metals are deposited on the cathode but after they have migrated most of the distance toward it. The flow of hydroxide ion from the cathode (see Figure 16-12) precipitates many metals in any event. The metal-rich soil surrounding the electrode is then excavated and cleansed by washing it or by other techniques. By repeating this procedure and inserting the electrodes at different locations in the soil, more efficient extraction of the heavy metals is possible. The electrokinetic technique has been used successfully for copper, lead, cadmium, mercury, chromium, and some radioactive metals. The method is high in energy demand, and hence in cost, because so much of the applied electrical potential is lost by the electrolysis of soil water.

In situ **chemical oxidation** can often be used to remediate soils (and groundwater) contaminated with chlorinated solvents and/or with *BTEX* (Chapter 14). The oxidizing agent is injected directly by means of a well into the underground waste and may or may not be extracted at the other side of the contaminated zone. Typically, a salt of *permanganate ion*, MnO_4^-, is used for *TCE*, *PCE*, and *MTBE* deposits, whereas *ozone* or *hydrogen peroxide* is used for BTEX and PAHs or, in some cases, for the C_2 chlorinated solvents as well. The MnO_2 product of oxidation by permanganate is a natural constituent of soil. Hydrogen peroxide is often supplemented with ferrous salts (together called *Fenton's Reagent*) to create hydroxyl radical by a reaction described in Chapter 14.

The Analysis and Remediation of Contaminated Sediments

We now realize that many river and lake sediments are highly contaminated by heavy metals and/or toxic organic compounds and that such sediments act as sources to recontaminate the water that flows above them.

One way to determine the extent of contamination of a sediment is to analyze a sample of it for the total amounts of lead, mercury, and other heavy metals that are present. However, this technique fails to distinguish between toxic materials already present in either their toxic form or in a form that can be resolubilized into the water and those that are firmly bound to sediment particles and unlikely to become resolubilized. Thus a more meaningful test involves extracting from a sediment sample the substances that are soluble in water or in a weakly acidic solution and analyzing the resulting liquid. In this way, the permanently bound and therefore inactive toxic agents can be left out of the account. Finally, the effect of sediments on organisms that usually dwell in or on them can be determined by adding the organisms to a sediment sample and observing whether they survive and reproduce normally.

Several types of remediation have been used for highly contaminated sediments. The simplest solution is often to simply cover the contaminated sediments with clean soil or sediment, thereby placing a barrier between the contaminants and the water system. In other instances, the contaminated sediments are dredged from the bottom of the water body to a depth below which the contaminant concentration is acceptable. If the sediments are high in organic content and inorganic nutrients, they are often used to enhance soil used for nonagricultural purposes. In some cases, the sediment can be used for cropland provided that its heavy metals and other contaminants will not enter the growing food. Cadmium is usually the heavy metal of greatest concern in such sediments; if the pH of the resulting soil is 6.5 or greater, most of the cadmium will not be soluble, so a higher total concentration is often tolerated.

Several chemical and biological methods of decontaminating sediments are in use. For example, treatment with calcium carbonate or lime increases the pH of the sediments and thereby immobilizes the heavy metals. In some situations, contaminated sediments are simply covered with a chemically active solid, such as limestone (calcium carbonate), *gypsum* (calcium sulfate), *iron(III) sulfate*, or *activated carbon*, that gradually detoxifies the sediments. In other cases, the sediments are first dredged from the bottom of the water body and then treated. Heavy metals are often removed by acidifying the sediments or treating them with a chelating agent; in both cases, the heavy metals become water-soluble and leach from the solid. For organic contaminants, extraction of toxic substances using solvents and destruction by either heat treatment of the solid or the introduction of microorganisms that consume them are the main options. The cleaned sediments can then be returned to the water body or spread on land. These techniques for removing metals and organics from sediments are also often useful on contaminated soils.

Bioremediation of Wastes and Soil

Recall from Chapter 14 that *bioremediation* involves the use of living organisms, especially microorganisms, to degrade environmental wastes. It is a rapidly growing technology, especially in collaboration with genetic engineering, which is used to develop strains of microbes with the ability to deal with specific pollutants. Bioremediation is used particularly for the remediation of waste sites and soils contaminated with semivolatile organic compounds such as PAHs. It is a popular method for use at Superfund sites (see Table 16-3).

Bioremediation exploits the ability of microorganisms, especially bacteria and fungi, to degrade many types of wastes, usually to simpler and less toxic substances. Indeed, for many years it was thought that microorganisms could and would eventually biodegrade *all* organic substances, including all pollutants, that entered natural waters or the soil. The discovery that some compounds, chloroorganics especially, were resistant to rapid biodegradation was responsible for correcting that misconception. Substances resistant to biodegradation are termed **recalcitrant** or **biorefractory.** In addition, other substances, including many organic compounds, biodegrade only partially; they are transformed instead into other organic compounds, some of which may be biorefractory and/or even more toxic than the original substances. An example of the latter phenomenon is the potential conversion of the once widely used solvent *1,1,1-trichloroethane* (methyl chloroform, now banned as an ozone-depleting substance, as discussed in Chapter 2) into carcinogenic *vinyl chloride*, $CHCl=CH_2$, by a combination of abiotic and microbial steps.

If a bioremediation technique is to operate effectively, several conditions must be fulfilled:

- The waste must be susceptible to biological degradation and in a physical form that is susceptible to microbes.

- The appropriate microbes must be available.

- The environmental conditions, such as pH, temperature, and oxygen level, must be appropriate.

An example of biodegradation is the degradation of aromatic hydrocarbons by soil microorganisms when land is contaminated by gasoline or oil. The largest bioremediation project in history was the treatment of some of the oil spilled by the *Exxon Valdez* tanker in Alaska in 1989. The bioremediation consisted of adding nitrogen-containing fertilizer to more than 100 km of the shoreline that had been contaminated, thereby stimulating the growth of indigenous microorganisms, including those that could degrade hydrocarbons. Both surface and subsurface oil was biodegraded in this operation. Some of the aromatic components in crude oil in marine spills become more susceptible to biodegradation once they are photooxidized by sunlight into more polar molecules.

In contexts such as contaminated soils, the biodegradation of PAHs is slow since they are strongly adsorbed onto soil particles and are not readily released into the aqueous phase, where biodegradation could occur. PAH-contaminated soils are especially prevalent at gasworks sites used in the 1850–1950 period for the production of "town gas" from coal or oil. The pollution is mainly in the form of deposits of **tars,** which are waste products that are high-molecular-weight organic liquids denser than water that are mixed with the soil and contain high levels of both BTEX and PAHs. Unfortunately, groundwater that comes into contact with the tar can become contaminated if some of its more soluble constituents such as benzene and naphthalene dissolve, although most of the tar is insoluble in water. The other common soil contaminants at gasworks sites are *phenols* and cyanide.

Bioremediation processes can take place under either aerobic or anaerobic conditions. In the **aerobic treatment** of wastes, aerobic bacteria and fungi that utilize oxygen are employed; chemically, the processes are oxidations, as the microorganisms use the wastes as food sources. In some of the bioremediation procedures used for aerobic soil, oxygen-saturated water is pumped through the solid to ensure that O_2 availability remains high. For example, about 85,000 tonnes of soil contaminated by gasoline, oil, and grease from a fuel plant in Toronto were decontaminated by first enclosing it in plastic and then pumping air, water, and fertilizer into it to encourage the population of aerobic bacteria to multiply and devour the hydrocarbon pollutants. The bioremediation process took only three months.

There are many examples of wastes that can be degraded by anaerobic microorganisms, although usually the most rapid and complete biodegradations are obtained with aerobic microorganisms. The anaerobic process usually works best when there are some oxygen atoms in the organic wastes themselves. In general, the process corresponds to the *fermentation* discussed in Chapter 13, in which biomass with an approximate empirical formula of CH_2O decomposes ultimately to methane and carbon dioxide:

$$2\ CH_2O \longrightarrow CH_4 + CO_2$$

An advantage of anaerobic biodegradation is its production of *hydrogen sulfide*, which *in situ* precipitates heavy-metal ions as the corresponding sulfides.

Several strategies used in bioremediation are based on the fact that microorganisms evolve quickly—due to their short reproductive cycle—and they develop the ability to use the food source at hand, even if it is chemical wastes. One remediation strategy is to isolate the most efficient degrading biomicroorgansims flourishing at a contaminated site, grow a large population of them in the laboratory, and finally return the enhanced population to the site. Another strategy is to introduce microorganisms that were found to be useful at other sites in degrading a particular type of waste, rather than waiting for them to evolve. Unfortunately, microorganisms adapted to one environment may not be capable of surviving in another if additional hostile contaminants are present. A third strategy is to encourage an increase in the population of indigenous microorganisms at the site by adding nutrients to the wastes and ensuring that the acidity and moisture levels are optimum.

Rather than waiting for microorganisms to evolve spontaneously, an alternative strategy is to use genetic engineering to develop microbes specifically designed to attack common organic pollutants. However, regulatory authorities have thus far been reluctant to allow genetically modified organisms to be released into the environment, since public opposition to such a move would be substantial.

In addition to bacteria, **white-rot fungi** can be used in biodegradation. This species protects itself from pollutants by degrading them outside its cell wall by secreting enzymes there that catalyze the production of *hydroxyl*

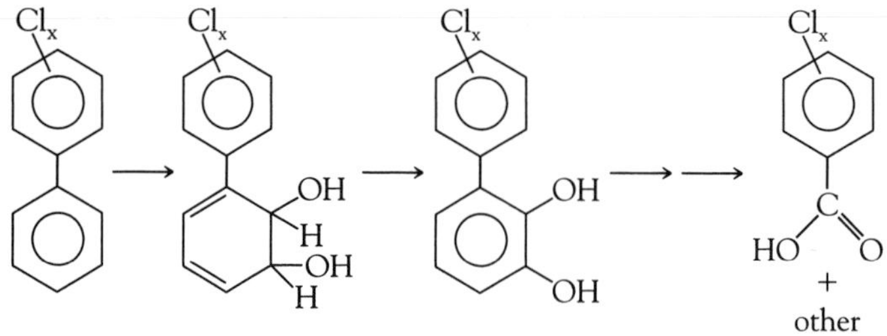

FIGURE 16-13 Example
of the aerobic degradation
of PCB molecules.

radicals and other reactive chemicals. Since hydroxyl radical in particular is quite nonspecific about which substances it oxidizes, the fungi are useful in degrading mixtures of waste, including various chlorinated substances such as DDT and *2,4,5-T*, as well as PAHs.

Bioremediation of Organochlorine Contamination

It has been discovered that PCBs (Chapter 11) in sediments undergo some biodegradation. PCB molecules with relatively few chlorine atoms undergo oxidative aerobic biodegradation by a variety of microorganisms. For the reaction to begin, a pair of nonsubstituted carbons—one ortho to the point of connection between the rings and one meta site next to it—must be available on one of the benzene rings. After 2,3 hydroxylation at these two sites, the 1,2 carbon–carbon bonds to the ortho carbons split in sequence, thereby destroying the second ring and producing compounds that readily degrade (see Figure 16-13).

PROBLEM 16-6

Deduce the chlorine substitution positions on the benzene ring in the benzoic acid that results from the aerobic degradation of 2,3′,5-trichlorobiphenyl.

Although PCB molecules that are heavily substituted by chlorines will not undergo this process, since they are unlikely to have adjacent unsubstituted carbons, they will instead undergo **anaerobic degradation,** as will perchlorinated organic compounds such as TCE and *HCB* (Chapter 10). In the absence of oxygen, anaerobic microorganisms facilitate the removal of chlorine atoms and their replacement by hydrogen atoms, apparently by a reductive dechlorination mechanism that initially involves the addition of an electron to the molecule. In the case of PCBs, this reductive dechlorination occurs most readily with meta and para chlorines. Apparently, steric effects block the ortho position from being attacked in most anaerobic mechanisms.

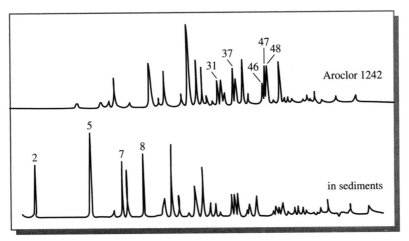

FIGURE 16-14 Concentration profiles (chromatograms) for various PCB congeners of commercial Aroclor 1242 before and after biodegradation (different scales) in a sediment. The major components of the peaks correspond to the following positions of chlorine substitution: 2 = 2-chlorobiphenyl; 5 = 2,2' and some 2,6; 7 = 2,3'; 8 = 2,4' and some 2,3; 31 = 2,2', 5,5'; 46 = 2,4,4', 5; 47 = 2,3', 4',5; 48 = 2,3', 4,4'. Notice that biodegradation produces ortho-substituted congeners that are present in low concentrations or absent in the original sample. [Source: D. A. Abramowicz and D. R. Olson, *CHEMTECH* (July 1995): 36–40.]

Thus the products of anaerobic treatment here are ortho-substituted congeners, ultimately 2-chlorobiphenyl and 2,2'-dichlorobiphenyl especially.

Since dioxin-like toxicity of PCBs requires several meta and para chlorines (see Chapter 11), the anaerobic degradation process significantly reduces the health risk from PCB contamination. Of course, once adjacent ortho and meta sites without chlorine are available, aerobic microorganisms—if available—could degrade the biphenyl structure, as already discussed. Figure 16-14 illustrates the change in composition of a commercial PCB sample after it has resided for some time in sediment that contains anaerobic bacteria.

Phytoremediation of Soils and Sediments

The technique of **phytoremediation,** the use of vegetation for the *in situ* decontamination of soils and sediments of heavy metals and organic

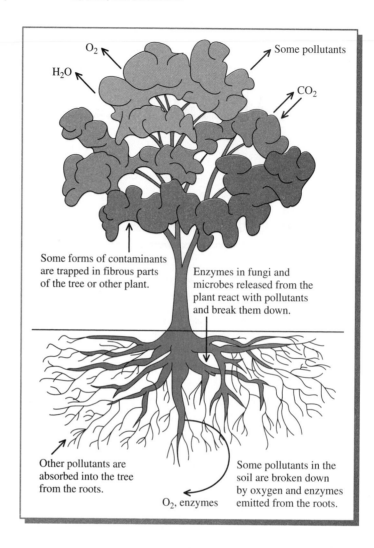

O$_2$

Some pollutants

H$_2$O

CO$_2$

Some forms of contaminants are trapped in fibrous parts of the tree or other plant.

Enzymes in fungi and microbes released from the plant react with pollutants and break them down.

Other pollutants are absorbed into the tree from the roots.

O$_2$, enzymes

Some pollutants in the soil are broken down by oxygen and enzymes emitted from the roots.

FIGURE 16-15
Mechanisms of phytoremediation by a plant.

pollutants, is an emerging technology. As illustrated in Figure 16-15, plants can remediate pollutants by three mechanisms:

- the direct uptake of contaminants and their accumulation in the plant tissue (phytoextraction),

- the release into the soil of oxygen and biochemical substances such as enzymes that stimulate the biodegradation of pollutants, and

- the enhancement of biodegradation by fungi and microbes located at the root–soil interface.

Advantages of phytoremediation include its relatively low cost, aesthetic benefits, and nonobtrusive nature.

Certain plants are **hyperaccumulators** of metals, i.e., they are able to absorb through their roots much higher than average levels of these contaminants (by a factor of at least 10–100, to yield a contaminant concentration of 0.1% or more) and to concentrate them much more than do normal plants. This ability probably evolved over long periods of time as the plants grew on natural soils that contained high concentrations of pollutants, especially heavy metals. In bioremediation, these plants are deliberately planted on contaminated sites and then harvested and burned. In some instances, the resulting ash is so concentrated in metal that it can be mined!

Phytoremediation is an attractive technique because metals are often difficult to extract with other technologies since their concentration is usually so small. For example, the shrub called the *Alpine pennycress* has the ability to hyperaccumulate cadmium, zinc, and nickel. Phytoremediation has been successfully used for extraction, e.g., of cadmium by both water hyacinths and various grasses; of lead and copper by alfalfa; and of chromium by Indian mustard, sunflowers, and buckwheat. In some cases, a chelating agent is added to the soil to enhance the accumulation of metals by the plant. Scientists are experimenting with various types of plants that can extract lead from soils. One difficulty with phytoremediation is that hyperaccumulators are usually plants that are slow to grow and therefore slow to accumulate metals. However, fast-growing poplar trees show promise of being efficient phytoremediators. One recently developed type of hybrid poplar efficiently absorbs TCE from hazardous waste sites and from groundwater. In general, there is a need to harvest the plants before they lose their leaves or begin to decay, so that contaminants do not become dispersed or return to the soil.

Plants can efficiently take up organic substances that are moderately hydrophobic, with log K_{ow} values (Chapter 10) from about 0.5 to 3, a range that includes BTEX components and some chlorinated solvents. Substances that are more hydrophobic than the upper limit of this range bind so strongly to roots that they are not easily taken up within the plant. Once taken up, the plant may store the substance by transformation within its lignin component or it may metabolize it and release the products into the air.

Substances that plants release into the soil include chelating ligands and enzymes; by complexing a metal, the former can decrease its toxicity, and the latter in some cases can biodegrade pollutants. For example, it has been found that the plant-derived enzyme *dehalogenase* can degrade TCE. Plants also release oxygen at the roots, thereby facilitating aerobic transformations. As previously discussed, the fungi that exist in symbiotic association with a plant also have enzymes that can assist in the degradation of organic contaminants in soil.

Bioremediation in general and phytoremediation in particular are rapidly emerging technologies. The long-term potential for these techniques to be used at many sites that require decontamination is apparent. Experiments at several test sites have shown that phytoremediation can be used successfully to degrade petroleum products in soils. The number of Superfund sites applying bioremediation technology is shown in Table 16-3.

Hazardous Wastes

In this section, we consider the nature of various types of hazardous wastes and discuss how individual samples of such wastes can be destroyed as an alternative to simply dumping them and thereby deferring the problem to a later date.

Currently there are more than 50,000 hazardous waste sites and perhaps 300,000 leaking underground storage tanks in the United States alone. The Superfund program of the U.S. EPA was created to remediate waste sites; its eventual cost is estimated to be $31 billion (see Box 16-1).

The Nature of Hazardous Wastes

A substance can be said to be a hazard if it poses a danger to the environment, especially to living things. Thus **hazardous wastes** are substances that have been discarded or designated as waste and that pose a danger. Most of the hazardous wastes with which we shall deal are commercial chemical substances or by-products from their manufacture; biological materials are not considered here.

In Chapter 10, we paid a great deal of attention to substances that were **toxic,** i.e., they threaten the health of an organism when they enter its body. Other common types of hazardous materials include those that are

- **ignitable** and burn readily and easily;

- **corrosive** because their acid or base character allows them to easily corrode other materials;

- **reactive** in senses not covered by ignition or corrosion, i.e., by explosion; and

- **radioactive.**

Some waste materials are hazardous in more than a single category.

The Management of Hazardous Wastes

There are four strategies in the management of hazardous waste. In order of decreasing preference, they are:

- **Source reduction:** The deliberate minimization, through process planning, of hazardous waste generation in the first place. The green chemistry cases presented throughout this text provide many examples of this strategy.

- **Recycling and reuse:** The use in a different process as raw materials, whether by the same company or a different one, of hazardous wastes generated in a process.

- **Treatment:** The use of any physical, chemical, biological, or thermal process—including incineration—that reduces or eliminates the hazard from the waste. Examples of such technology are discussed in subsequent sections.

- **Disposal:** Burial of the nonliquid waste in a properly designed landfill. In the past, liquid hazardous wastes were often injected into deep underground wells.

Landfills that are specially designed to accommodate hazardous wastes have several characteristics in addition to those discussed for sanitary landfills. The locations of such landfills should be

- in an area with clay or silt soil, to provide an additional barrier to leachate dispersal, and

- away from groundwater sources.

Often the hazardous wastes in such landfills are grouped according to their physical and chemical characteristics so that incompatible materials are not placed near each other.

Toxic Substances

As mentioned, toxic wastes are those that can cause a deterioration in the health of humans or other organisms when they enter a living body. Their characteristics, as well as many examples, were discussed in Chapters 10–12 and 15 particularly and so will not be reviewed in detail here. Those of main concern are heavy metals, organochlorine pesticides, organic solvents, and PCBs.

As an example of the magnitude of the problem of toxic substance waste management, consider the PCBs that are still used in the capacitors in the ballast tubes of fluorescent light fixtures. Within the sealed capacitor container is a thick, gel-like liquid of concentrated PCB oil that is absorbed in several layers of paper. A typical capacitor contains about 20 g—about a tablespoon—of liquid PCB. Although each ballast does not contain much PCB, the number of these lighting fixtures in use in the developed world is huge. Thus the ultimate collection and disposal of PCBs from these sources will be a task requiring many years and many dollars. Disposal methods for toxic organic compounds are discussed later in this chapter.

Incineration of Toxic Waste

Incinerators that deal with hazardous waste are often more elaborate than those that burn municipal waste because it is important that the material be more completely destroyed and that emissions be more tightly controlled.

In some cases, the waste (e.g., PCBs) will not ignite on its own and must be added to an existing fire fueled by other wastes or by supplemental fuel such as natural gas or petroleum liquids. Modern facilities employ very hot flames, ensure that there is sufficient oxygen in the combustion zone, and keep the waste compounds in the combustion region long enough to ensure that their **destruction and removal efficiency,** DRE, is essentially complete, i.e., >99.9999%, called "six nines." The presence of carbon monoxide at a concentration greater than 100 ppm in the gaseous emissions is often used as an indicator of incomplete combustion.

About 3 million tonnes of hazardous wastes are burned annually in the United States, although this is only 2% of the amount generated. Three-quarters of the hazardous waste is dealt with by aqueous treatment, and 12% is disposed of on land or injected into deep wells.

The two most common forms of toxic waste incinerators are the rotary kiln and the liquid injection types. The **rotary kiln incinerator** can accept wastes of all types, including inert solids such as soil and sludges. The wastes are fed into a long (>20 m) cylinder that is inclined at a slight angle (about 5° from the horizontal) away from the entrance end and slowly rotates so that unburned material is continually exposed to the oxidizing conditions of 650–1100°C; see Figure 16-16. Over a period of about an hour, the waste makes its way down the cylinder and is largely combusted. The hot exit gases from the kiln are sent to a secondary (nonrotating) combustion chamber equipped with a burner in which the temperature is 950–1200°C. The gas stays in the chamber for at least two seconds so that destruction of organic molecules is essentially complete. In some installations, liquid wastes can be fed directly into this chamber as the fuel. The gases exiting the secondary chamber are rapidly cooled to about 230°C by an evaporating water spray (in some cases with heat recovery), since they would otherwise destroy the air pollution equipment that they next enter.

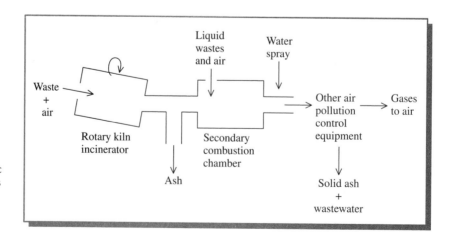

FIGURE 16-16 Schematic diagram of the components of a rotary kiln incinerator, including air pollution equipment.

Rotary kiln and other types of incinerators usually employ the same series of steps to purify the exhaust of particulates and of acid gases before it is released into the air, as do garbage incinerators, e.g., a gas scrubber and a baghouse filter.

Cement kilns are a special type of large rotary kiln used to prepare cement from limestone, sand, clay, and shale. Very high temperatures of 1700°C or more are generated in cement kilns in order to drive off the carbon dioxide from limestone, $CaCO_3$, in the formation of lime, CaO. In addition to hotter combustion temperatures (compared to incinerators), wastes in cement kilns are burned with the fuel right in the flame; the residence time of the material in the kiln is also longer. Liquid hazardous wastes are sometimes used as part of the fuel (up to 40%) for these units, the remainder being a fossil fuel, usually coal. Recently, techniques have been developed that allow kilns to handle sludges and solids. Cement kilns burn more hazardous waste (about a million tonnes a year) than do commercial incinerators in the United States, although even more waste is incinerated on-site by chemical industries.

In the **liquid injection incinerator**—a vertical or horizontal cylinder—pumpable liquid wastes are first dispersed into a fine mist of small droplets. The finer these droplets, the more complete their subsequent combustion, which occurs at about 1600°C with a waste residence time of a second or two. A fuel or some "rich" (easily combustible) waste is used to produce the high temperatures of the combustion. In contrast to the rotary kiln, only a single combustion chamber is used, although in some modern versions a secondary input of air is introduced to improve oxygen distribution and generate more complete combustion. The exit gases can be passed through a spray dryer to neutralize and remove acid gases, followed by a baghouse filter to remove particulates, before being released into the outside air.

The incineration of hazardous waste has garnered much attention from environmentalists and some of the general public because of the potential release of toxic substances—particularly from the stack into the air—resulting from the operation of these units. Of special concern are the organic **products of incomplete combustion,** PICs, that have been found both in gases and adsorbed on particles emitted from incinerators. The PICs must be formed in the post-flame region because they could not survive the temperatures of the flame. Some of the most prevalent PICs are methane and benzene.

In their research to understand the production of these pollutants, scientists and engineers have discovered that reactions can occur downstream of the flame in "quench" zones and in pollution control devices, where temperatures fall below 600°C. Both gas-phase and surface-catalyzed processes apparently occur. For example, trace amounts of various dioxins and furans form at 200–400°C on fly ash and soot surfaces where the processes may be catalyzed by transition metal ions. A temperature of about 400°C is optimal for dioxin formation; below this the formation reaction is slow and above it they are quickly decomposed. It has not been established whether the dioxins and furans result from the coupling on surfaces of precursor compounds such as chlorobenzenes and chlorophenols, or from the so-called de novo synthesis involving chlorine-free furan- or dioxin-like structures reacting with inorganic chlorides.

Concern has also been expressed about increased emissions that could occur when the incinerator is being closed down and during accidents or power failures, when lower temperatures would result for some time, since much greater quantities of dioxins and furans could presumably form under such conditions. **Fugitive emissions,** which include emissions from valves, minor ruptures, incidental spills, etc. are also a concern. The dust emitted from cement kiln incinerators has been found to contain toxic metals and some PICs. In fact, the health risk from toxic metal ion emissions (usually as oxides or chlorides) from the hazardous waste incinerators is found to exceed that from the toxic organics. As in the case of municipal incinerators, the solid residue from hazardous waste units can amount to one-third of the original waste volume and contains traces of toxic materials, as does the wastewater from the scrubber units.

Concerns about incineration have spurred the development of other technologies for disposing of hazardous waste. In **molten salt combustion,** wastes are heated to about 900°C and destroyed by being mixed with molten sodium carbonate. The spent carbonate salt contains NaCl, NaOH, and various metals from the combusted waste, which can be recovered so that the Na_2CO_3 can be reused. No acidic gas is evolved, since it reacts within the salt. In **fluidized-bed incinerators,** a solid material such as limestone, sand, or alumina is suspended in air (fluidized) by means of a jet of air, and the wastes are combusted in the fluid at about 900°C. A secondary combustion chamber completes the oxidation of the exhaust gases. **Plasma incinerators** can achieve temperatures of 10,000°C by passing a strong electrical current through an inert gas such as argon. The plasma consists of a mixture of electrons and positive ions, including nuclei, and can successfully decompose compounds, producing much lower emissions than traditional incinerators. In such a **thermal** or **hot** plasma, all the particles travel at high speeds and are thermally hot. In a variant that is used to treat municipal solid waste, plasma is first created in air, which is then used to heat a mixture of waste, coke, and limestone to 1500°C or more in a second, oxygen-starved chamber. The inorganic compounds are converted into a slag that is innocuous enough to be used as a construction material. The organic compounds are broken down into syngas, the combination of carbon monoxide and hydrogen discussed in Chapter 8, which is then used as a fuel.

Supercritical Fluids

The use of **supercritical fluids** is another modern alternative to incineration. The supercritical state of matter is produced when gases or liquids are subjected to very high pressures and, in some cases, to elevated temperatures. At pressures and temperatures at or beyond the **critical point,** separate gaseous and liquid phases of a substance no longer exist. Under these conditions, only the supercritical state, with properties that lie between those of a gas and those of a liquid, exists. For example, for water, the critical pressure is 218 atm (22.1 megapascals) and the critical temperature is 374°C, as illustrated in the phase diagram in Figure 16-17. Depending on exactly how much pressure is applied, the physical properties of the supercritical fluid vary between those of a gas (at relatively lower pressures) and those of a liquid (at higher pressures); the variation of properties with

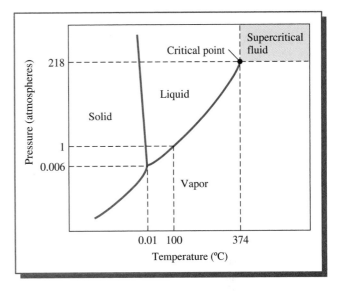

FIGURE 16-17 Phase diagram for water (not to scale). Notice the region for the supercritical state (shaded light green), which exists at temperatures and pressures beyond the critical point.

changes in pressure or temperature is particularly acute near the critical point. Thus the density of supercritical water can vary over a considerable range, depending upon how much pressure (beyond 218 atm) is applied. Other substances that readily form useful supercritical fluids are carbon dioxide (see Chapters 6 and 7; it is used for many extractions in the food industry, including the decaffeination of coffee beans), xenon, and argon. See Table 16-4 for their critical temperatures and pressures.

One rapidly developing innovative technology for destruction of organic wastes and hazardous materials such as phenols is **supercritical water oxidation** (SCWO). Initially, the organic wastes to be destroyed are either dissolved in aqueous solution or suspended in water. The liquid is then subjected to very high pressure and a temperature in the 400–600°C range so that the water lies beyond its critical conditions and so is a supercritical fluid. The solubility characteristics of supercritical water differ markedly from those of normal liquid water: Most organic substances become much *more* soluble, and

TABLE 16-4	Supercritical Fluid Characteristics	
Substance	Critical Temperature (°C)	Critical Pressure (atm)
Water	374.1	217.7
Carbon dioxide	31.3	72.9
Argon	150.9	48.0
Xenon	16.6	58.4

many ionic substances become much *less* soluble. Similarly, and also because very high pressures are applied, O_2 is much more soluble in supercritical than in liquid water.

At the elevated temperatures associated with supercritical water, the dissolved organic materials are readily oxidized by the ample amounts of O_2 that are pumped into and dissolve in the fluid. Hydrogen peroxide may be added to generate hydroxyl radicals, which initiate even faster oxidation. Because materials diffuse much more rapidly in the supercritical state than in liquids, the reaction is generally complete within seconds or minutes. One practical problem with the SCWO method is that insoluble inorganic salts that are formed in the reactions can corrode the high-pressure equipment used in the process and so shorten its lifetime; this problem can be solved by designing the reactor so that there are no zones where salts can build up.

The advantages of the SCWO technology include the rapidity of the destruction reactions and the lack of the gaseous NO_X by-products that are characteristic of gas-phase combustion. The required pressure and temperature conditions are readily accessible with available high-pressure equipment. However, some intermediate products of oxidation—mainly organic acids and alcohols, and perhaps also some dioxins and furans—are formed with the SCWO method, which raises concerns about the toxicity of the effluent from the process. In the variant of this technology in which a catalyst is used, the percentage conversion to fully oxidized products is increased and the amount of intermediates that persist is decreased.

In the **wet air oxidation** process, temperatures (typically 120–320°C) and pressures lower than those required to achieve supercritical conditions for water are used to efficiently oxidize aqueous wastes (often catalytically). The oxidation is efficient because the amount of oxygen that dissolves at the enhanced pressures favors the reaction. The process is generally much slower than in supercritical water, requiring about an hour. The method is cheaper to operate, however, than the relatively expensive SCWO technology.

Supercritical carbon dioxide has been used to extract organic contaminants such as the gasoline additive MTBE from polluted water. After extraction, the pressure can be lowered, at which point the carbon dioxide becomes a gas, leaving behind the liquid contaminants to be incinerated or otherwise oxidized. Similarly, supercritical fluids could be used to extract contaminants such as PCBs and DDT from soils and sediments.

Nonoxidative Processes

All the processes just described employ oxidation as the means of destroying the hazardous wastes. However, a closed-loop **chemical reduction process** has been devised that has no uncontrolled emissions, using a *reducing* rather than an oxidizing atmosphere to destroy hazardous wastes. One advantage to the absence of oxygen is that there is no opportunity for the incidental formation of dioxins and furans. The reducing atmosphere is achieved by using hydrogen gas at about 850°C as the substance with which the preheated mist of wastes reacts. The carbon in the wastes is converted to methane (and some transiently to other hydrocarbons such as benzene, which are subsequently hydrogenated to produce additional methane). The oxygen, nitrogen, sulfur, and chlorine are converted into their hydrides. The process is actually enhanced by the presence of water, which under these reaction conditions can act as a reducing agent and form additional hydrogen by the water-gas shift reaction with methane (see Chapter 8). PAH formation, which is characteristic of other processes at high temperatures in the absence of air, is suppressed by maintaining the hydrogen level at more than 50%. The gas output is cooled and scrubbed to remove particulates. The hydrocarbon output from the process is subsequently burned to provide heat for the system; thus there are no direct emissions to the atmosphere.

PROBLEM 16-7

Construct and balance chemical equations for the destruction of the PCB molecule with the formula $C_{12}H_6Cl_4$ (a) by combustion with oxygen to yield CO_2, H_2O, and HCl and (b) by hydrogenation to yield methane and HCl.

Chemical dechlorination methods for the treatment of chlorine-containing organic wastes, especially PCBs from transformers, have been developed and used in various parts of the world, though they are rather expensive to operate. The basic idea is to substitute a hydrogen atom or some other nonhalogen group for the covalently bound chlorine atoms on the molecules, thereby detoxifying them. The mostly dechlorinated wastes can then be incinerated or disposed of in landfills. The commonly used reagent for this purpose is MOR, the alkali metal (M = sodium or potassium) salt of a polymeric alcohol. In the reaction, an —OR group replaces each of the chlorines, which depart as the salt MCl:

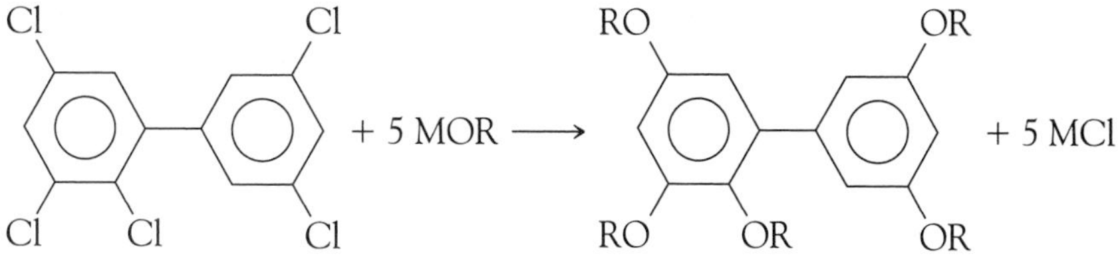

The process is carried out at temperatures above 120°C in the presence of *potassium hydroxide*, KOH, and occurs most efficiently for highly chlorinated PCBs. An alternative dechlorination method is the reaction of the PCBs with dispersions of metallic sodium to give sodium chloride and a polymer containing many biphenyl units joined together.

Review Questions

1. Define the term *solid waste* and name its five largest categories for developed countries.

2. Describe the components and steps in the creation of a *sanitary landfill*.

3. Describe the three stages of waste decomposition that occur in a sanitary landfill, including the products of each stage. Are all stages equal in production or consumption of acidity?

4. Define the term *leachate*, explain how this substance arises, and list several of its common components. How can leachate be controlled and how is it treated?

5. Explain the difference between the two common types of MSW incinerators.

6. What is the difference between *bottom ash* and *fly ash* in an incinerator? Describe some of the air pollution control devices found on incinerators.

7. What is meant by the *four Rs* in waste management?

8. Why can the recycling of metals often be justified by economics alone?

9. Describe the processes by which paper and rubber tires can be recycled.

10. What are the common types of consumer packaging plastics that can be recycled? What four ways are used to recycle plastics?

11. What are some of the arguments for and against the recycling of plastics?

12. What is meant by a *life cycle assessment* and what are the two main uses for LCAs?

13. Describe the main inorganic constituents of soil. How do clay, sand, and silt particles differ in size?

14. What are the names and origins of the principal organic constituents of soil? Are both types of acids soluble in base? In acid?

15. What is meant by a soil's *cation-exchange capacity*? What are its common units?

16. What is meant by a soil's *reserve acidity*? How does it arise?

17. Describe several methods, including chemical equations, by which soils that are too acid or too alkaline can be treated.

18. Describe the processes by which soil in arid areas becomes salty and alkaline.

19. What do the terms *sediments* and *pore water* mean?

20. By what three ways are heavy metals bound to sediments?

21. How can mercury stored in sediments be solubilized and enter the food chain?

22. Describe how mine tailings are usually stored and how this represents a potential environmental problem.

23. How can sediments contaminated by heavy metals be remediated so they can be used on agricultural fields?

24. Describe two ways by which contaminated sediments can be treated without removing the sediments themselves.

25. List the three categories of technologies commonly used to remediate contaminated soils. Give examples of each.

26. List three conditions that must be fulfilled if bioremediation of soil is to be successful.

27. Describe the two ways in which PCBs in sediments are bioremediated.

28. Define *phytoremediation* and list the three mechanisms by which it can operate. What are *hyperaccumulators*?

29. Define the term *hazardous waste*. What are the five common types?

30. List, in order of decreasing desirability, the four strategies used in the management of hazardous waste.

31. Name and describe the three common types of incinerators used to destroy hazardous waste. What does *DRE* stand for and how is it defined?

32. Define the term *PICs* and describe how they are formed.

33. Explain the advantages and disadvantages of supercritical water oxidation for the destruction of hazardous wastes.

34. How is the chemical reduction process carried out? What advantages does it have over oxidation methods?

Green Chemistry Questions

See the discussion of focus areas and the principles of green chemistry in the Introduction before attempting these questions.

1. What are the environmental advantages of using polyaspartate as an antiscalant/dispersant versus polyacrylate?

2. The synthesis of polyaspartate as developed by Donlar won a Presidential Green Chemistry Challenge Award.

(a) Into which of the three focus areas for these awards does this award best fit?

(b) List at least three of the twelve principles of green chemistry that are addressed by the chemistry developed by Donlar Corporation.

3. The development of recyclable carpeting by Shaw Industries won a Presidential Green Chemistry Challenge Award.

(a) Into which of the three focus areas for these awards does this award best fit?

(b) List three of the twelve principles of green chemistry that are addressed by the chemistry developed by Shaw Industries.

4. What are the environmental advantages of using polyolefin-backed nylon fiber carpeting in place of PVC-backed carpeting?

Additional Problems

1. Consider a small city of 300,000 people in the northern United States, southern Canada, or central Europe, and suppose that its residents on average produce 2 kg/day of MSW, about one-quarter of which will decompose anaerobically to release methane and carbon dioxide evenly over a 10-year period. Calculate how many homes could be heated by burning the methane from the landfill, given that residential requirements are about 10^8 kJ/year in that climate zone. Consult Chapter 7 for data on methane combustion energetics.

2. By reference to the general solubility rules for sulfides found in introductory chemistry textbooks, deduce which metals would *not* have their sediment availabilities determined by AVS. Which of these metals would occur in the form of insoluble carbonates instead and thus be biologically unavailable in marine environments?

3. The acid volatile sulfide concentration in a sediment is found to be 10 μmol/g. The majority of the sulfide is present in the form of insoluble FeS, since the Fe^{2+} concentration in the sediment is 450 μg/g. What mass of mercury in the form of Hg^{2+} can be tied up by the remaining sulfide in 1 tonne of such sediment? Assume that a negligible amount of iron is not tied up as FeS.

4. For the PCB congener 2,4,4′,5-tetrachloro-biphenyl, deduce **(a)** the chlorine substitution pattern on the benzoic acid that results from its aerobic degradation and **(b)** the various PCB congeners with one or two chlorines that could result from its anaerobic degradation.

5. Generate lists of the aspects of production, distribution, and disposal that you would employ in a life cycle assessment to decide which is the more environmentally friendly container for beer: glass bottles or aluminum cans. Do you think the conclusions of such an analysis would depend significantly on the extent of recycling of the containers?

6. Waste having a high cellulose content, such as paper, wood scraps, and corn husks, can be converted to ethanol for use as a fuel. One way this can be done is to first perform an acid hydrolysis of the cellulose to convert it into glucose. This is then followed by fermentation of the glucose to produce ethanol. Write a balanced equation for the fermentation of glucose, $C_6H_{12}O_6$, into ethanol, C_2H_5OH. Given that the pyranose monomer unit in cellulose has the formula $C_6H_{10}O_5$ (i.e., glucose − H_2O), determine the volume of ethanol that could be produced in this way from 1.00 tonne of hardwood scraps (46% cellulose), assuming 100% conversion in both steps. Ethanol's density is 0.789 g/mL.

7. One way to reduce the lifetime of plastic wastes is to make them photodegradable, so that some of the bonds in the polymers will break on absorption of light. When designing a photodegradable plastic, what range of sunlight wavelengths would it be best for the plastic to absorb? What is the main limitation of the breakdown of these plastics? If you were to design a photodegradable plastic that would break down on absorption of 300-nm light, what would be the maximum bond energy of the bonds to be cleaved?

Further Readings

1. M. B. McBride, *Environmental Chemistry of Soils* (New York: Oxford University Press, 1994).

2. P. S. Phillips and N. P. Freestone, "Managing Waste—The Role of Landfill," *Education in Chemistry* (January 1997): 11.

3. F. Pearce, "Burn Me," *New Scientist* (22 November 1997): 31.

4. B. Piasecki et al., "Is Combustion of Plastics Desirable?" *American Scientist* 86 (1998): 364.

5. K. Tuppurainen et al., "Formation of PCDDs and PCDFs in Municipal Waste Incineration and Its Inhibition Mechanisms: A Review," *Chemosphere* 36 (1998): 1493.

6. T. E. McKone and S. K. Hammond, "Managing the Health Impacts of Waste Incineration," *Environmental Science and Technology* (1 September 2000): 380A.

7. R. Brown et al., "Bioremediation," *Pollution Engineering* (October 1999): 26.

8. M. E. Watanabe, "Phytoremediation on the Brink of Commercialization," *Environmental Science and Technology* 31 (1997): 182A.

9. P. B. A. Kumar et al., "Phytoextraction: The Use of Plants to Remove Heavy Metals from Soils," *Environmental Science and Technology* 29 (1995): 1232.

10. D. A. Wolfe et al., "The Fate of the Oil Spilled from the *Exxon Valdez*," *Environmental Science and Technology* 28 (1994): 561A–568A.

11. F. Pearce, "Tails of Woe," *New Scientist* (11 November 2000): 46.

12. M. L. Hitchman et al., "Disposal Methods for Chlorinated Aromatic Waste," *Chemical Society Review* (1995): 423.

13. D. Simonsson, "Electrochemistry for a Cleaner Environment," *Chemical Society Reviews* 26 (1997): 181.

14. D. Amarante, "Applying in Situ Chemical Oxidation," *Pollution Engineering* (February 2000): 40.

15. H. Black, "The Hottest Thing in Remediation," *Environmental Health Perspectives* 110 (2002): A146.

16. D. M. Roundhill, "Novel Strategies for the Removal of Toxic Metals from Soils and Waters," *Journal of Chemical Education* 81 (2004): 275.

Websites of Interest

Log on to www.whfreeman.com/envchem4/ and click on Chapter 16.

| Environmental Instrumental Analysis **VI** | **Inductively Coupled Plasma Determination of Lead** |

The analysis of heavy metals in environmental samples is now routinely accomplished by the spectroscopic method discussed in this box.

As discussed in the text, anthropogenic lead contamination in the environment in North America and Europe has been decreasing since tetraethyllead and tetramethyllead were phased out of gasolines in the 1970s and 1980s. Presently, major sources of environmental Pb are residual lead in aerosol particles and dust near roadways (sometimes originating in decaying structures originally painted with lead-based paints), smelting ash, and lead pipes in plumbing (*plumbum* is Latin for "lead"). Other sources include automobile battery manufacture and disposal as well as cigarette smoke. In the developed countries, exposure to lead is dropping, but in developing countries this exposure is increasing (Ahmed and Ishiga, 2006).

Like some of the other toxic metals discussed in this book, lead concentrations can be determined quickly and accurately using an atomic emission technique called **inductively coupled plasma spectroscopy** (ICP).

Each chemical element has a unique atomic structure with electrons in well-defined (quantized) energy levels. The movement of electrons between these levels, which requires the absorption or emission of energy, is also well defined, and therein lies the key to atomic emission spectroscopy. If the atoms in a sample are excited using a very-high-energy source—such as a flame, spark, or plasma— many of the atoms' electrons will be excited to higher energy levels. Almost immediately,

these excited-state electrons will relax by returning to the ground state; but that return is accompanied by the *emission* of a photon whose energy corresponds to the difference between the excited-state and ground-state energy levels. And just as the energy of the promotion is well defined, meaning that only specific energies can be absorbed by a particular atom, the energy released by this relaxation—and the emitted photon containing that energy—is very specific to the atom. Since photon energy is related to wavelength (Chapter 1), a means of elemental detection can be based on detecting the light emitted from a sample after the atoms in it are excited by some means: That light is characteristic of the atoms excited in the sample.

In the case of ICP, the excitation source is a very-high-temperature plasma (see Chapter 16 for a definition of plasmas). Light emitted by sample atoms injected into the plasma is collected via lenses and mirrors and focused onto a diffraction grating. This grating separates individual wavelengths in space and focuses the light on a photomultiplier tube (PMT) or other detector that converts light into electronic signals. The wavelength of the light identifies the elements in the sample that emitted the photons, and the intensity of the light as measured by the PMT specifies the concentration of that element in the sample. Atomic identity and amount are the two parameters determined by emission spectrometry that enable it to be used as an analytical tool in environmental analysis.

The specific components of an ICP spectrometer depend on which of two basic designs is used. The first type is called a *sequential spectrometer* and the second a *simultaneous spectrometer*. The sequential spectrometer uses only one PMT (detector) and requires a process of scanning through the emission wavelengths to determine multiple elements in one sample. This scanning process is usually accomplished by very exacting rotation of the diffraction grating, which separates the light emissions. This rotation is carefully controlled by a computer so that the signal generated by the PMT can be correlated with the wavelength of light that falls on it; i.e., the computer knows by the grating position—which it is controlling—which wavelength is striking the PMT. Sequential ICPs require time to scan and sample the light for each element's emission, but the cost of a sequential instrument is modest compared to the second, more powerful spectrometer design, the simultaneous ICP spectrometer.

Simultaneous spectrometers have a similar configuration to the simpler sequential instruments—they both have a plasma source (also called a torch) and a monochromator to separate the light into individual wavelength components. However, in the simultaneous instrument the spatially separated wavelength beams are simultaneously dispersed onto a plane, by a grating and a prism, and focused on a detector similar to that of a digital camera. This detector, called a *charge-coupled device* (CCD), can simultaneously record the intensities of each of the different wavelengths for all of the elements emitting in the hot environment of the plasma. In this way, many

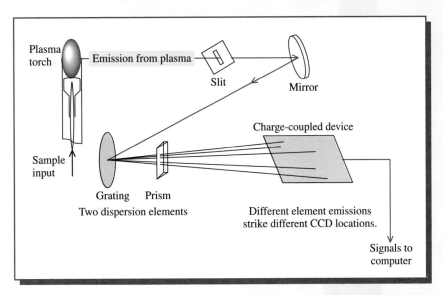

(continued on p. 772)

Environmental Instrumental Analysis **VI**	**Inductively Coupled Plasma Determination of Lead** *(continued)*

elemental emission wavelengths can be detected individually—but all at one time—without having to scan between them; therefore, many chemical elements can be determined in the same sample simultaneously. Once again, the wavelength identifies the element emitting it, and the light's intensity conveys the concentration of that element in the sample injected into the plasma.

Street dust is a measure, to some degree, of the amount of lead to which people in cities become exposed via inhalation. It is thought that most of the lead in urban dust originated in the past from the particulates released by the combustion of leaded gasoline. With that in mind, a comparison has been made of the lead determined in dust samples taken from the streets of Manchester, England, over time (Nageotte and Day, 1998). Samples of dust, dirt, and soil were scraped off streets with a spatula and analyzed by atomic absorption spectrometry (AAS) and by a modified ICP technique. In AAS, a high-temperature flame or hot carbon tube is used to atomize the element of interest (lead, in our case) in a dust sample that has been dissolved in an acid solution. Once the lead ions have been atomized (converted to elemental Pb) and vaporized by the high temperature, a source lamp with a specific wavelength is shone through the vaporized atoms. Since only specific energies can be absorbed by Pb atoms (since the gaps between the atomic energy levels are quantized and specific to that element), the amount of absorption of the source lamp's light can be used as a measure of the amount of lead in the beam. A PMT on the opposite side of the sample atoms from the source lamp records a decrease in source-lamp light intensity when the sample is introduced. The larger the absorption, the more lead atoms are present in the sample. AAS therefore significantly differs from ICP: One is an absorbance method and the other depends on emission.

The following table shows abridged results from both the English studies, in 1975 and 1997. Lead amounts are reported for three different sampling categories: high-traffic areas, low-traffic areas, and areas where children played (Nageotte and Day, 1998). The significant improvement in lowered lead levels from 1975 to 1997 certainly seems to stem from England's phase-out of leaded gasoline.

Category	ppm Pb (number of samples)	
	1975	**1997**
> 100 cars/hour	1001 ± 40 (180)	577 ± 53 (17)
< 10 cars/hour	993 ± 186 (53)	536 ± 93 (13)
Playgrounds, parks, gardens	1014 ± 206 (49)	572 ± 77 (47)

Childrens' blood lead levels are also a measure of exposure to this toxic metal. The U.S. Centers for Disease Control (2006) notes that lead-associated intellectual deficits are found in children with less than 100 μg/L of lead in their blood. In many developing countries, leaded gasoline and paint are still common. A meta-study incorporating 315 published papers and involving 11,272 children in provinces all over China (in the period 1994–2004) reported a weighted mean of 92.9 μg Pb/L of blood in children of ages 1 to 12 years. Blood concentration levels of lead in children living in industrial areas were significantly higher than those in suburban or rural areas (Wang and Zhang, 2006).

References: F. Ahmed and H. Ishiga, "Trace Metal Concentrations in Street Dusts of Dhaka City, Bangladesh," *Atmospheric Environment* 40 (2006): 3835–3844.

Chemistry-Based Animations, 2006: http://www.shsu.edu/~chm_tgc/sounds/sound.html.

S. M. Nageotte and J. P. Day, "Lead Concentration and Isotope Ratios in Street Dust Determined by Electrochemical Atomic Absorption Spectrometry and Inductively Coupled Plasma Mass Spectrometry," *Analyst* 123 (1998): 59–62.

U.S. Centers for Disease Control (2006), www.cdc.gov/nceh/lead.

S. Wang and J. Zhang, "Blood Lead Levels in Children, China," *Environmental Research* 101 (2006): 412–418.

ENVIRONMENT

Mapping Mercury

HOT-SPOT UNKNOWNS COMPLICATE MERCURY REGULATIONS BY REBECCA RENNER

In issuing the Clean Air Mercury Rule this past March, the Bush administration hoped to ease health concerns about mercury from coal-fired power plants. The White House enacted a "cap and trade" approach to reduce emissions of the element nationwide by about 20 percent in five years and 70 percent by 2018. In formulating its rule, the administration noted that power plants emit only 48 tons of the metal every year—just a small fraction of the total amount of mercury in the atmosphere. Mandating further emission cuts, it argued, would not solve the problem of human exposure to the neurotoxin.

Eleven states and four public health groups are challenging this approach, arguing that cap-and-trade does not address areas particularly vulnerable to mercury pollution. Not so, says the Environmental Protection Agency. When the cap-and-trade proposals were announced, the EPA's head of air regulations, Jeffrey Holmstead, said, "We don't think there will be any hot spots." The hot-spot standoff arises from big gaps in mercury science, according to environmental researchers, and the lack of

METALLIC SPREAD: Distinct environmental conditions help to amplify local concentrations of methyl mercury, leading to health warnings such as this one in the Florida Everglades.

Rebecca Renner, "Mapping Mercury," *Scientific American*, September 2005, 20–22.

news SCAN

MERCURY'S TIME TO FALL

The Bush administration argues that mercury from power plants is a small fraction of that already in the atmosphere. The comparison is misleading, argues Praveen Amar, science director for NESCAUM, an association of air-quality regulators in the northeastern states. That's because most of the mercury in the atmosphere is in a gaseous elemental state that remains there for about a year and should not be a major contributor to rapid changes in mercury deposition. On the other hand, about 50 percent of the mercury emitted by power plants is oxidized mercury that rains down within a few days, he says.

Many mercury scientists agree, but it is hard to prove. The best evidence comes from recent EPA monitoring of Ohio Valley power plants. It showed that the depositions of oxidized mercury and sulfur dioxide, a tracer for combustion, increase together, and back trajectories based on meteorological data implicate the plants.

comprehensive data on mercury deposition means that a consensus about emissions control will not likely emerge soon.

Theoretically, mercury should preferentially rain down in areas near to power plants. But attempts to determine the fallout have proved incomplete. For instance, the Mercury Deposition

Percent of Total Mercury as Methyl Mercury

■ Greater than 8   ■ Less than 8 but greater than 4
■ Less than 4 but greater than 2   □ Less than 2

HOT SPOTTING: Measurements reveal areas that readily convert mercury to methyl mercury.

Network, which measures the metal in rainwater in many parts of the country, does not account for mercury particulates that settle dry onto vegetation, a form of deposition that could be equal to the wet variety, according to Oak Ridge National Laboratory scientist Steve Lindberg.

And just because a region receives above-average deposition doesn't mean that it will have high levels of methyl mercury, the form that builds up in long-lived predatory fish such as trout, pike, tuna and swordfish. "The areas with the most problems may not have the highest levels of deposition," explains mercury expert David Krabbenhoft of the U.S. Geological Survey branch office in Middleton, Wis. Indeed, the Southeast has greater measured deposi-

tion than the Northeast, but both regions have serious methyl mercury problems.

A partial explanation for this dichotomy is the process by which elemental mercury becomes methyl mercury. For mercury to get methylated and enter the food web, it must be processed by bacteria that thrive on sulfate, a sulfur compound. This means that dissolved organic matter and sulfur enhance methylation, as do acidic waters such as those in the Northeast. The methylation process changes the conclusions drawn simply from examining mercury fallout. For instance, Krabbenhoft recently completed a study of New England lakes from locations near Boston and up to Maine. Mercury emissions and deposition are highest in the urban area, but methyl mercury in fish is low. The fish problems occur in Vermont and New Hampshire, where the conditions are right for methylation. "What do we want to protect?" he asks. "If it's beautiful fishing spots like these lakes, then we need to look at more than deposition."

Scientists finally understand methylation well enough to map out vulnerable areas at a national level, according to Krabbenhoft, who is currently working on such a map for the EPA. The work should, for the first time, combine deposition and vulnerability to identify unambiguous hot spots—regions where it makes the most sense to take action to limit the effects of mercury. Scientists hope that this information and other advances may resolve the hot-spot debate and demonstrate the wisdom or folly of the administration's approach to mercury regulation.

Rebecca Renner is a writer based in Williamsport, Pa.

APPENDIX
Background Organic Chemistry

In Chapters 10–12, the most important environmental problems caused by toxic organic chemicals are discussed in detail. In this appendix, we provide some necessary background in organic chemistry for those students whose previous education has not included this subject.

The organic compounds of interest environmentally are mostly electrically neutral molecules containing covalent bonds. Stable compounds of this type inevitably involve the formation of *four* bonds by carbon; in carbon-centered free radicals, they form three bonds. Conceptually at least, chemists view all organic chemicals as "derived" from those simple organic compounds that contain only carbon and hydrogen, i.e., **hydrocarbons.** We shall follow this convention and divide our discussion into several sections, most of which deal with specific types of hydrocarbons.

Alkanes

The simplest hydrocarbons are those that contain strings of carbon atoms, each one singly bonded to its closest neighboring carbon atom(s) and to several hydrogen atoms. Such hydrocarbons are called **alkanes,** of which the simplest are *methane*, CH_4; *ethane*, C_2H_6; and *propane*, C_3H_8. Commercial supplies of all three are readily available from natural gas wells. Structural formulas for these three alkanes are:

$$
\begin{array}{ccc}
\quad\quad H & \quad\quad H \quad\; H & \quad\quad H \quad\; H \quad\; H \\
\quad\quad | & \quad\quad | \quad\;\; | & \quad\quad | \quad\;\; | \quad\;\; | \\
H-\!\!\overset{\displaystyle}{\underset{\displaystyle}{C}}\!\!-H & H-\!\!\overset{\displaystyle}{\underset{\displaystyle}{C}}\!\!-\!\!\overset{\displaystyle}{\underset{\displaystyle}{C}}\!\!-H & H-\!\!\overset{\displaystyle}{\underset{\displaystyle}{C}}\!\!-\!\!\overset{\displaystyle}{\underset{\displaystyle}{C}}\!\!-\!\!\overset{\displaystyle}{\underset{\displaystyle}{C}}\!\!-H \\
\quad\quad | & \quad\quad | \quad\;\; | & \quad\quad | \quad\;\; | \quad\;\; | \\
\quad\quad H & \quad\quad H \quad\; H & \quad\quad H \quad\; H \quad\; H \\
\text{methane} & \text{ethane} & \text{propane}
\end{array}
$$

For convenience, chemists often write the formulas for such species by gathering together in one unit all the hydrogens bonded to a given carbon and displaying only the carbon–carbon bonds; thus ethane is represented as CH_3-CH_3 and propane as $CH_3-CH_2-CH_3$. In another common representation, called a **condensed formula,** the $C-C$ single bonds are not shown; rather the formula lists each carbon and the atoms attached to it. For example, the condensed formula for ethane is CH_3CH_3. Each carbon atom in an alkane molecule forms four equiangular single bonds, so the geometry about each carbon is tetrahedral. Thus all alkanes are three-dimensional, nonplanar molecules, even though, for the sake of clarity, their structural

TABLE 1	Some of the Simple Unbranched Alkanes		
Molecular Formula	Name	Condensed Formula	Boiling Point (°C)
CH_4	Methane	CH_4	−164
C_2H_6	Ethane	CH_3CH_3	−89
C_3H_8	Propane	$CH_3CH_2CH_3$	−42
C_4H_{10}	Butane	$CH_3CH_2CH_2CH_3$	−0.5
C_5H_{12}	Pentane	$CH_3(CH_2)_3CH_3$	36
C_6H_{14}	Hexane	$CH_3(CH_2)_4CH_3$	69
C_7H_{16}	Heptane	$CH_3(CH_2)_5CH_3$	98
C_8H_{18}	Octane	$CH_3(CH_2)_6CH_3$	126
C_9H_{20}	Nonane	$CH_3(CH_2)_7CH_3$	151
$C_{10}H_{22}$	Decane	$CH_3(CH_2)_8CH_3$	174
$C_{11}H_{24}$	Undecane	$CH_3(CH_2)_9CH_3$	196
$C_{12}H_{26}$	Dodecane	$CH_3(CH_2)_{10}CH_3$	216

formulas seem to represent them as planar molecules involving bond angles of 90° and 180°.

Table 1 lists the alkanes having one through twelve carbon atoms in a continuous string; more complex alkanes contain branches, as we shall see shortly. The alkane with four carbon atoms, called *butane*, is a gas; longer alkanes are liquids or solids under ordinary conditions. When five or more carbons are present in the alkane, the Latin-based abbreviation for that number (*pent* for 5, *hex* for 6, *hept* for 7, *oct* for 8, etc.) is employed as the prefix to the ending *-ane* in its name. For example, the molecule $CH_3—CH_2—CH_2—CH_2—CH_3$, which can be written more simply as $CH_3(CH_2)_3CH_3$, is called *pentane* since it has 5 carbon atoms; its formula is C_5H_{12}. When all the carbons lie in one continuous chain (without branches), the molecule is said to be *straight-chained* or *unbranched*, and often the prefix *n-* is added to the name; thus the pentane molecule described is called *n-pentane*.

The constituent atoms of many organic compounds may be "reshuffled" in chemical reactions to yield new structures—the ingredient atoms remain exactly the same, but the way in which they are linked together (their "order of linkage") can be altered. So we get distinctly different sets of compounds with the same molecular formula but different structures, which are called **structural isomers;** there are other types of isomerism, too, which we need not consider here. Sometimes the difference between isomers is slight—a matter of small variations in physical properties; sometimes it is enormous—the biological activity of two isomers may differ profoundly. Alkanes with four or more carbon atoms have isomers in which the chain of carbon atoms is branched: Not all the carbons are part of an unbroken path of bonded atoms. An example is the isomer of n-pentane illustrated below:

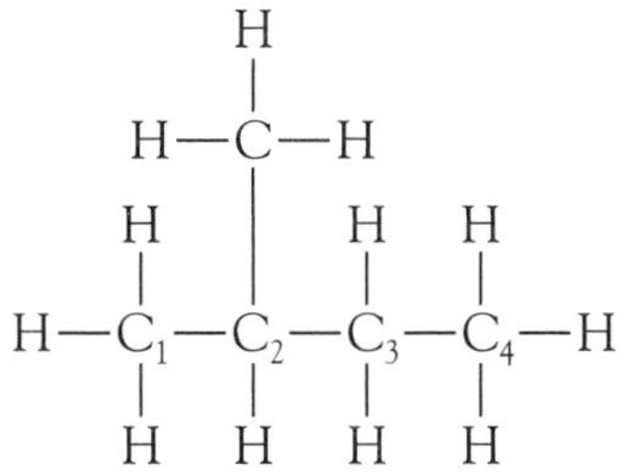

$$
\begin{array}{c}
\text{H} \\
| \\
\text{H}-\text{C}-\text{H} \\
\quad\quad\text{H} \quad | \quad\; \text{H} \quad \text{H} \\
\quad\quad | \quad\; | \quad\; | \quad\; | \\
\text{H}-\text{C}_1-\text{C}_2-\text{C}_3-\text{C}_4-\text{H} \\
\quad | \quad\; | \quad\; | \quad\; | \\
\quad \text{H} \quad \text{H} \quad \text{H} \quad \text{H}
\end{array}
$$

2-methylbutane

In naming such alkanes and other organic molecules, the short chains of carbon atoms that comprise the branches are assigned group names ending in *-yl*, which are derived by deleting the *-ane* ending from the name of the alkane hydrocarbon that has the same length (in terms of linked carbon atoms). Thus the CH_3— group is called the *methyl* group, CH_3CH_2— is called *ethyl*, etc. The names of these groups are listed as prefixes to the name for the longest continuous chain of carbon atoms, and each is preceded by a number that indicates the carbon atom of the chain to which the group is attached. For example, the molecule shown above is called *2-methylbutane*, since butane is the alkane consisting of four carbons in an unbranched chain; the —CH_3 group is bonded at the second carbon atom.

There are compounds in which one or more of the hydrogen atoms in hydrocarbons such as the alkanes have been replaced by another atom such as fluorine, chlorine, or bromine. Things that can be substituted for hydrogen atoms are called **substituents.** Some simple substituted hydrocarbons are encountered in Chapter 2. Examples include the substituted methanes CF_2Cl_2 (dichlorodifluoromethane), CHF_2Cl (chlorodifluoromethane), and CF_3Br (bromotrifluoromethane) as well as the substituted ethane CHF_2—CH_2F, which is called *1,1,2-trifluoroethane*, where the numbers refer to the carbon numbers to which the fluorines are bonded. Its structure is:

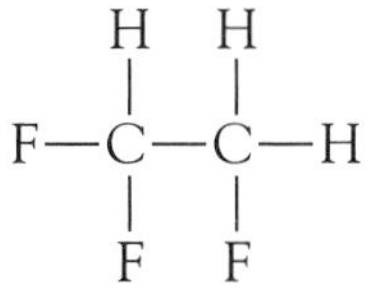

$$
\begin{array}{c}
\text{H} \quad \text{H} \\
| \quad\; | \\
\text{F}-\text{C}-\text{C}-\text{H} \\
| \quad\; | \\
\text{F} \quad \text{F}
\end{array}
$$

1,1,2-trifluoroethane

PROBLEM 1

Write out the structural formula and the condensed formula for each of the following alkanes: (a) *n*-pentane; (b) 3-ethylhexane; (c) 2,3-dimethylbutane.

Alkenes and Their Chlorinated Derivatives

In some organic molecules, one or more pairs of the carbon atoms are joined by double bonds; since each carbon atom forms a total of four bonds, there are only two additional bonds formed by such carbon atoms. The simplest hydrocarbon of this type is a colorless gas called *ethene*, usually known by its older name *ethylene*:

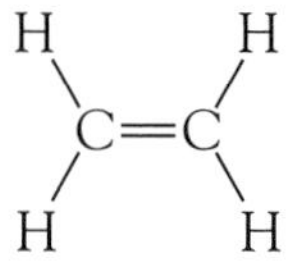

ethene (ethylene)

Notice that the actual planar geometry of this molecule, with bond angles of about 120° around each carbon, can be shown in the structural formula. Condensed formulas normally show the double bond: $CH_2{=}CH_2$ or $H_2C{=}CH_2$.

A $C{=}C$ bond can be a part of a longer sequence of carbon atoms that are joined together by other single, double, or triple C–C bonds. For example, *propene* is a three-carbon chain with one adjacent pair of carbons joined by a double bond:

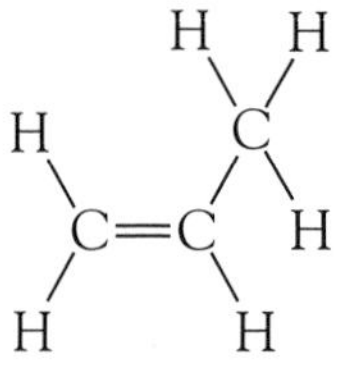

or $CH_2{=}CH{-}CH_3$

propene   propene

The name for a hydrocarbon chain containing a $C{=}C$ bond is the same as that used for the alkane of the same length, except that the *-ane* ending of the alkane is replaced by *-ene*. The molecule is numbered such that the $C{=}C$ unit is part of the continuous chain and such that the $C{=}C$ unit is at the lower-numbered end of the chain. Collectively, hydrocarbons containing $C{=}C$ bonds are called **alkenes.** If there are two $C{=}C$ bonds in a hydrocarbon, the prefix *di-* is placed before the *-ene* ending; thus the hydrocarbon below is called *1,3-pentadiene*:

1,3-pentadiene

The numbers preceding the name are those assigned to the first carbon atom that participates in each of the double bonds. The alternative numbering scheme, i.e., assigning the CH_3 carbon on the right to be #1, is not used since the first double bond would then start at carbon #2 and the name would be *2,4-pentadiene*; thus the first double bond would not have the lowest possible number.

In some derivatives of ethene, one or more of its hydrogen atoms have been replaced by chlorine atoms. The chloroethenes, like ethene itself, are planar molecules. The simplest example is $CH_2 = CHCl$, called *chloroethene* but known in the chemical industry as *vinyl chloride*; it is produced in huge quantities since the common plastic material polyvinyl chloride (PVC) is subsequently prepared from it.

A number is usually placed in front of the name of the substituent to indicate the specific carbon atom to which it is bonded; thus $Cl_2C = CH_2$ is called *1,1-dichloroethene* to distinguish it from *1,2-dichloroethene*, $CHCl = CHCl$. The molecule *1,1,2-trichloroethene*, $CCl_2 = CHCl$, is a liquid solvent that has extensive uses. Note that the prefix numbers in the compound name here are superfluous since it has no isomers and thus there is no need to distinguish one isomer from another. This compound is usually referred to by its traditional name *trichloroethylene*. The structural formulas of a few substitution products of ethene are:

1,1-dichloroethene   tetrachloroethene

The liquid compound *tetrachloroethene*, $CCl_2 = CCl_2$, is used on a large scale as the dry-cleaning solvent used commercially to remove grease spots and other stains on clothing. The prefix 1,1,2,2- is not used as part of its name since it is superfluous (no other arrangements of chlorine being possible).

Note that when all the hydrogens in a molecule have been replaced by a given atom or group, the prefix *per* can be used instead of the actual number; thus tetrachloroethylene is also called *perchloroethylene*, giving rise to its nickname "perc."

PROBLEM 2

Write structural formulas for each of the following: (a) 1,1-dichloropropene, (b) perchloropropene, (c) 2-butene.

PROBLEM 3

Determine the correct name for each of the following: (a) $CHCl_2CHCl_2$, (b) $CH_3-CH_2-CH=CH_2$, (c) $CH_2=CH-CH=CH_2$.

Symbolic Representations of Carbon Networks

Organic molecules often contain extensive networks of carbon atoms. Chemists find it convenient to construct shorthand visual representations of such molecules using a symbolic system of lines that indicate only the position of the *bonds* (not including bonds to hydrogen atoms), rather than writing out a structure in which the C and H atoms are shown explicitly. To indicate the presence of a carbon atom, a "kink" is shown in the chain's representation. For example, the molecule *n*-butane can be represented as (a) or (b) below:

$$CH_3-CH_2-CH_2-CH_3$$

(a)   (b)

The hydrogen atoms are not shown at all in the "stripped-down" version; the number of them at any carbon atom can be deduced by subtracting from 4 the number of bonds to that carbon that are displayed explicitly. Thus in the representation below for 2-chloropropane, carbons #1 and #3 must possess 3 hydrogens, since they are shown as forming one other bond, whereas carbon #2 has one hydrogen, since it is shown as forming three other bonds:

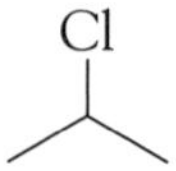

Cl

2-chloropropane

PROBLEM 4

Write out the full structural formulas for each of the following molecules:

(a) 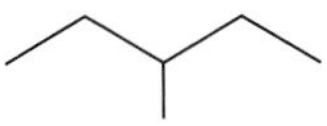   (b) 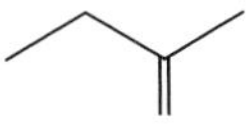   (c)

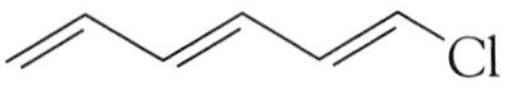

PROBLEM 5

Draw symbolic ("kinky") bond diagrams for each of the following molecules:

(a) 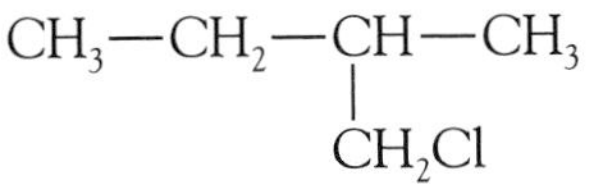 $CH_3-CH_2-CH-CH_3$
$\qquad\qquad\qquad |$
$\qquad\qquad\quad CH_2Cl$

(b) 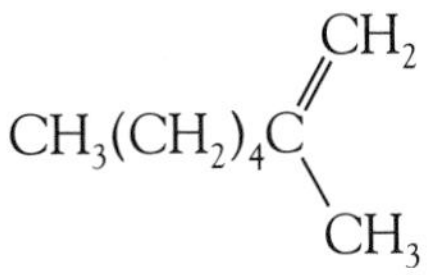 $CH_3(CH_2)_4C\begin{smallmatrix}\diagup CH_2\\ \diagdown CH_3\end{smallmatrix}$

(c) 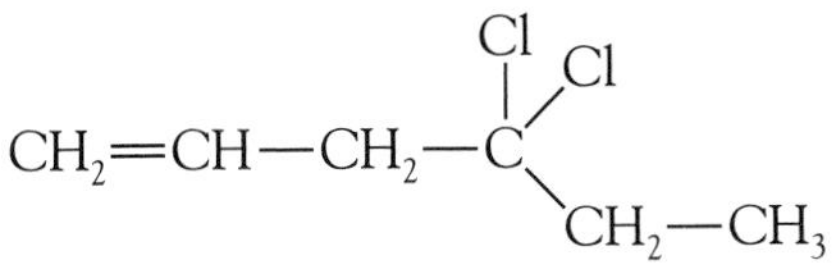 $CH_2=CH-CH_2-C\begin{smallmatrix}Cl\\ |\diagup Cl\\ \diagdown CH_2-CH_3\end{smallmatrix}$

Common Functional Groups

In addition to being replaced by simple single-atom substituents like Cl and F, the hydrogen atoms in alkanes and alkenes can be replaced by more complex "attachments" called **functional groups**—these are typically headed by oxygen or nitrogen atoms. The common functional groups are listed in Table 2. The simplest such polyatomic group is —O—H, usually simply shown as —OH; it is called the **hydroxyl group.** Compounds that correspond to alkanes or alkenes with the hydrogen of one C—H bond replaced by an —OH group are called **alcohols.** Familiar examples are methyl alcohol or *methanol* (also called *wood alcohol*) and ethyl alcohol or *ethanol* (grain alcohol):

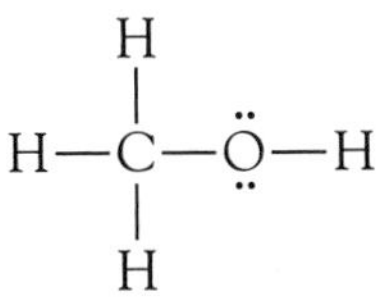

methanol, CH_3OH 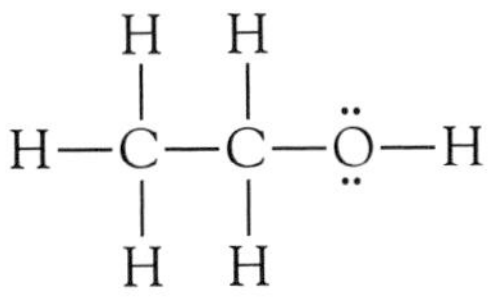   ethanol, CH_3CH_2OH

The use of alcohols as fuels is discussed in Chapter 8.

TABLE 2	Some Common Functional Groups
Name of Compound Type	**Functional Group**
Chloride	—Cl
Fluoride	—F
Alcohol	—OH
Ether	—O—
Aldehyde	$-C\diagup^{\displaystyle O}_{\displaystyle H}$
Carboxylic acid	$-C\diagup^{\displaystyle O}_{\displaystyle OH}$
Amine	$-N\diagup^{\diagdown}$

Compounds called **ethers** contain an oxygen atom connected on both sides to a carbon atom or chain:

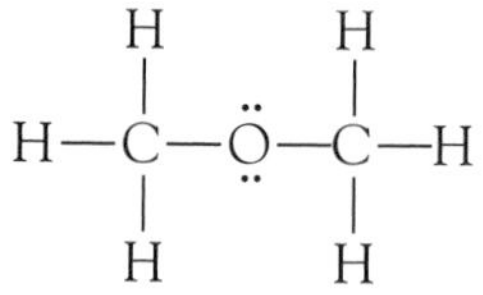

$$
\begin{array}{ccc}
\text{H} & & \text{H} \\
| & \ddot{} & | \\
\text{H}-\text{C}-\ddot{\text{O}}-\text{C}-\text{H} \\
| & & | \\
\text{H} & & \text{H}
\end{array}
$$

dimethyl ether, $(CH_3)_2O$

In more formal names for such compounds, the $-OCH_3$ group is known as *methoxy*, and the $-OCH_2-CH_3$ group is known as *ethoxy*, so that dimethyl ether would be named *methoxymethane*. The use of ethers as gasoline fuel additives is discussed in Chapter 8.

There are organic compounds analogous to alcohols and ethers in which sulfur occurs in the position otherwise occupied by oxygen. The prefix *thio-* is used to denote this substitution; thus we have thioalcohols, or just **thiols,** such as CH_3SH, and **thioethers,** such as CH_3-S-CH_3.

As discussed in Chapter 3, carbon–oxygen double bonds are found in some organic molecules. Molecules that contain the $H-C=O$ group bonded to hydrogen or to a carbon are known as **aldehydes;** the important examples encountered in polluted air are *formaldehyde,* $H_2C=O$, and *acetaldehyde,* $CH_3C(H)=O$. (Atoms or groups shown inside parentheses are bonded to the preceding carbon but do not themselves participate in the bond displayed next in the formula.)

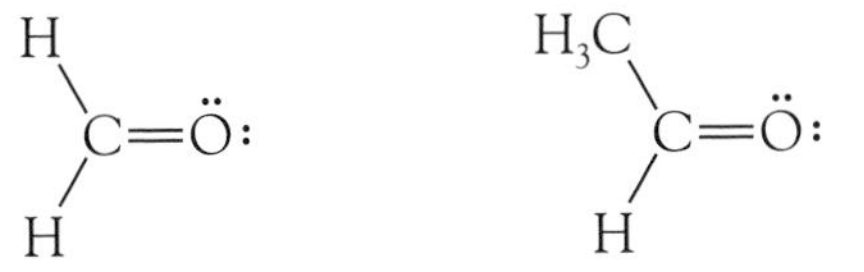

formaldehyde, H_2CO acetaldehyde, CH_3CHO

If the $C=O$ group is connected to an $-OH$ group, the system is called a **carboxylic acid;** examples are *formic acid* and *acetic acid:*

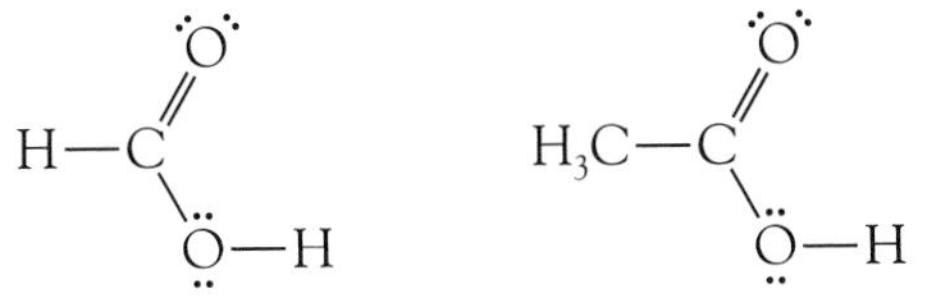

formic acid, HCOOH   acetic acid, CH_3COOH

If the hydrogen atom of the $-OH$ group is replaced by an organic group, the compound is called an **ester.**

Groups headed by nitrogen atoms are known as **amino groups;** they are found attached to carbon chains in some organic molecules. Compounds in which the amino group is bonded to a hydrocarbon chain are called **amines.** Note that nitrogen atoms form a total of three bonds, some (or all) of which can be directed to carbons. Two examples are:

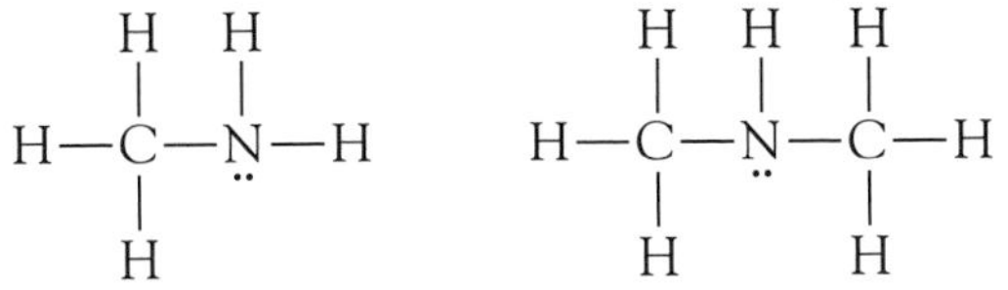

methylamine, CH_3NH_2 dimethylamine, $(CH_3)_2NH$

Molecules that contain *both* the carboxylic acid group and an amino group are called **amino acids.** They are the most important groups in proteins, which include the enzymes that accelerate specific biological reactions. The amino acid called *cysteine* is illustrated below; notice that it contains a thiol group as well as an amino and a carboxylic acid group.

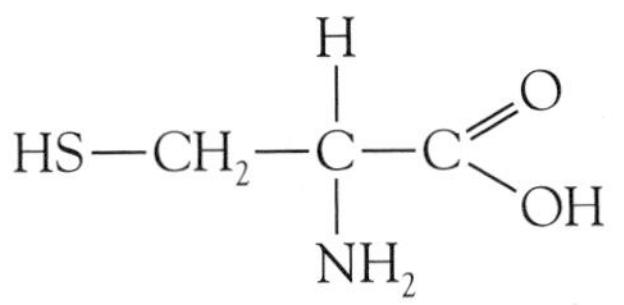

cysteine

Alcohols, acids, and amines that contain short carbon chains are quite soluble in water. The reason is that molecules of these three types contain O—H or N—H bonds, which possess a hydrogen atom that is partially depleted of electron density by the highly electronegative atom (O or N) to which it is bonded. The partial positive charge δ^+ of the hydrogen is attracted to regions of unbonded electron density—lone pairs—on atoms of adjacent molecules:

hydrogen
bond

Such interactions are called **hydrogen bonds;** the forces holding the two atoms together—and therefore also holding together the two molecules to which the atoms belong—are not nearly as strong as those of a regular bond within a molecule, but they are much stronger than the forces that operate between molecules in hydrocarbons. Water molecules illustrate this situation. They stick together because each hydrogen atom is hydrogen-bonded to the lone pair of the H_2O molecule closest to it. The attraction between H_2O molecules due to these interactions results in a relatively high boiling point for liquid water, much higher than anticipated for a molecule of its mass. For a (nonionic) substance to be freely soluble in water, these secondary bonds between adjacent water molecules must be replaced by similar interactions between the substance and the water molecules. Consequently, molecules that contain N–H or O–H bonds and a short chain of carbons are soluble in water because the hydrogen bonds they form with H_2O molecules replace those that are broken when the substance is incorporated into the liquid.

Hydrogen atoms bonded to carbon cannot form hydrogen bonds with water molecules, since the carbon is not sufficiently electronegative to produce much of a positive charge on a hydrogen atom bonded to it. In addition, there are no lone pairs of electrons on the carbon atoms. Consequently, there is no driving force that can disrupt the extensive network of hydrogen bonding within liquid water in order to incorporate a large number of molecules of hydrocarbons or chlorinated organic molecules. In both hydrocarbon and chlorinated organic molecules, all the hydrogens are bonded to carbon; for this reason, such molecules are not very soluble in water. Even molecules with one O—H or N—H group and many carbon atoms are insoluble in water, since their overall character is dominated by the large number of carbons. The forces of attraction that do exist between organic molecules that contain no hydrogen-bonding capacity are quite nonspecific and nondirectional; consequently, different hydrocarbons are quite soluble in each other, and organochlorine molecules are soluble in hydrocarbons. We can restate the familiar generalization that "like dissolves like": Compounds tend to dissolve in other substances having the same types of intermolecular interactions.

PROBLEM 6

Write the structural formulas and symbolic diagrams for each of the following: (a) ethyl alcohol, (b) ethylamine, (c) acetic acid (the carboxylic acid with a methyl group bonded to the carbon).

Rings of Carbon Atoms

Networks of carbon atoms exist as rings in many organic molecules. The most common rings are those that contain five, six, or seven carbon atoms. Molecules containing rings are named by placing the prefix *cyclo* in front of the usual name for the carbon chain of that length. Thus, a ring of six carbons, all joined by single C–C bonds, is called *cyclohexane*. The molecule shown at the right below is called *methylcyclopentane*.

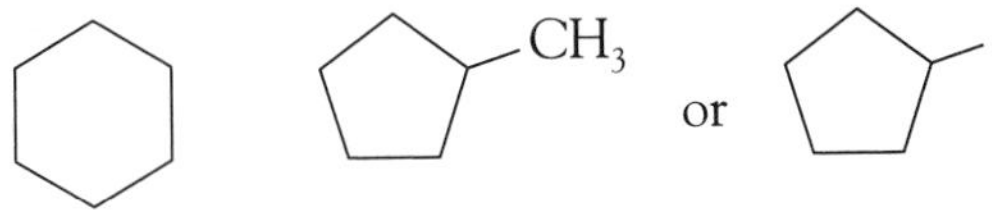

cyclohexane   methylcyclopentane

PROBLEM 7

Write out both simple and symbolic bond diagrams for the following molecules: (a) cyclopropane, (b) chlorocyclobutane, (c) any isomer of dimethylcyclohexane.

Benzene

One of the most common and most stable organic structural units is the **benzene ring,** which is a planar hexagon of six carbon atoms. In the parent hydrocarbon, it also contains six hydrogen atoms, one bonded to each carbon and lying in the C_6 plane:

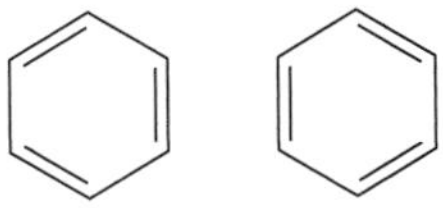

A   B

Each carbon in C_6H_6 is bonded to two carbons and to one hydrogen, so in order to form four bonds, it must be doubly bonded to one of its neighboring carbons. The two ways of achieving this result are shown in the so-called *Kekulé structures* (A and B at the bottom of the previous page). In fact, benzene molecules adopt neither of these two forms, each of which would have alternating short $C=C$ and long $C-C$ bonds; rather they exist in an averaged "resonance" structure in which all C–C bonds have the same intermediate length. (The term *resonance* here alludes to the mathematics of the bond description and is commonly interpreted as "blend" or "hybrid.") This result is represented by the structure shown below, with the hexagon containing an enclosed circle to represent the three double bonds; often, however, just one of the Kekulé structures is shown, it being understood, at least by chemists, that no actual alternation of bonds is meant.

Since the molecule is planar and has six equal sides, each $C-C-C$ and $C-C-H$ angle is 120°. When benzene occurs as a substituent group in another molecule, it is given the name *phenyl*.

Alkenes readily react by addition of molecules such as H_2, HCl, and Cl_2 across the double bonds—i.e., with one atom attaching itself to each of the two carbons on the double bond—and thereby convert the $C=C$ units to single bonds. For example, the addition of HCl to ethene produces chloroethane. The corresponding reactions do *not* readily occur to the bonds in benzene or its derivatives. Benzene can be hydrogenated, i.e., hydrogen can be added to its double bonds, but only under rather extreme conditions. This difference in the behavior of benzene compared to alkenes is an example of the special stability of a six-membered ring containing three sets of alternating double and single bonds. The electrons of the bonds interact with each other in a manner that makes the molecule energetically much more stable than would be expected from adding up single- and double-bond energies appropriate to alkanes and alkenes. The extra stability disappears if even one of the three double bonds is hydrogenated or otherwise added to by other molecules. Thus the six-membered benzene ring is a unit of great inherent stability and survives intact in media that would destroy other $C=C$ bonds. Benzene and other molecules that possess this extra stability are said to be **aromatic** systems.

An exception to the rule that benzene does not add to its double bonds occurs when the attacking atom or molecule is a free radical such as a chlorine atom; recall that such species possess an unpaired electron. Thus, whereas molecular chlorine, Cl_2, itself does not add to the double bonds in benzene, a single chlorine atom does. In fact, as early as 1825, Michael Faraday found that chlorine gas would react with benzene if the reaction mixture was exposed to strong light, which we now realize splits the Cl_2 molecules into free Cl atoms. Once such a reaction starts, it continues until all the carbon atoms have added one chlorine atom, and 1,2,3,4,5,6-hexachlorocyclohexane is produced:

$$C_6H_6 + 3\,Cl_2 \xrightarrow{\text{UV light}} C_6H_6Cl_6$$

Chlorinated Benzenes

Although benzene does not readily undergo addition reactions with molecules or ions that are not free radicals, it does participate in **substitution** reactions: One of its hydrogen atoms can be replaced by a group such as methyl, hydroxyl, etc., that forms a single bond. Of particular interest is substitution by chlorine, since many compounds of environmental concern contain chlorine-substituted benzene rings. When benzene is reacted with chlorine gas in the presence of a catalyst such as iron(III) chloride, $FeCl_3$, one of the hydrogens (explicitly drawn on the ring below for clarity) is replaced by chlorine, and HCl gas is released:

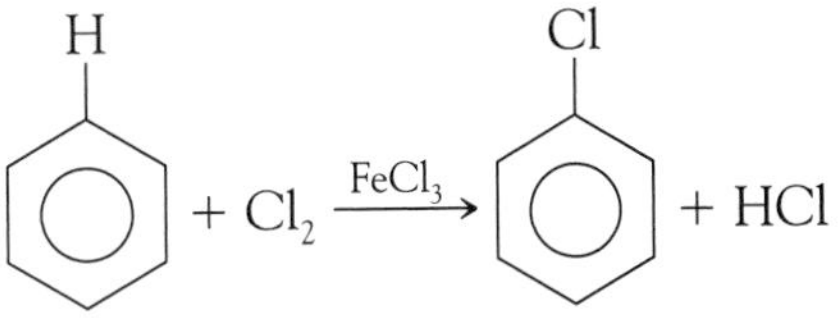

Notice that the aromatic C_6 ring survives intact. Notice also that in these structures the lone pairs on the chlorine atoms are not shown, as they were in the previous structures for alcohols and amines. Chemists use both types of structures—those that show the lone pairs and those that do not—for such compounds.

If the reaction is allowed to continue, i.e., if excess Cl_2 is available, one or more hydrogen atoms of the chlorobenzene molecules will in turn be replaced by chlorine. There are three isomeric dichlorobenzenes, all of which could in principle be produced in such a reaction:

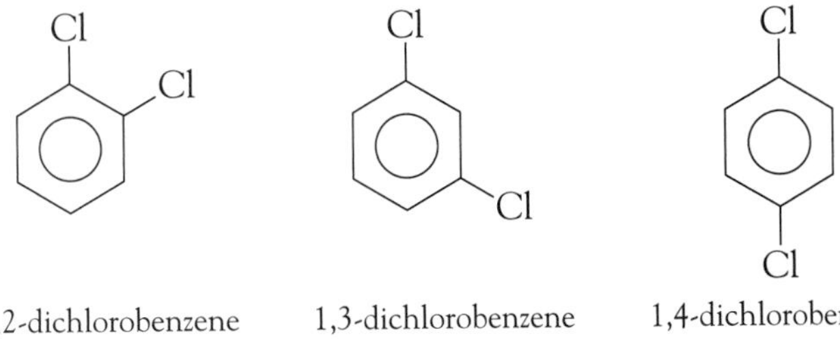

1,2-dichlorobenzene   1,3-dichlorobenzene   1,4-dichlorobenzene

The numbering scheme begins at one of the "substituted" carbons; the direction of numbering around the ring is chosen to yield the smallest possible number for the second substituent. In older nomenclature, 1,2-disubstituted benzene is called the *ortho* substituted isomer, 1,3-disubstitution calls for the prefix *meta*, and the 1,4 isomer is termed *para*. Thus the compound 1,4-dichlorobenzene, shown above at the right, is also called *para*-dichlorobenzene or *p*-dichlorobenzene.

When using the Kekulé structure for benzene in which double and single bonds are displayed, it is important to remember that the choice of a structure (A or B on page AP-11) for the positions of the double bonds is an arbitrary one. This has the consequence that there are only three, not five, isomeric dichlorobenzenes. For example, 1,6-dichlorobenzene is not different from the 1,2 isomer; they represent the same molecule viewed from different perspectives. (To avoid any such complications, many chemists use only the circle-containing hexagon symbol shown on page AP-12.)

In some derivatives of benzene, there are two or more types of substituents, even other benzene rings. In such cases, the carbon having the most important substituent is called *C-1*, and the numbering continues in the direction that gives the smallest number to the first substituent of the second kind. Many examples of multiply substituted benzenes are encountered in Chapters 10–12.

PROBLEM 8

Deduce the structures and names for the three chemically different trichlorobenzenes.

Review Questions

1. What is the name of the hydrocarbon $CH_3(CH_2)_4CH_3$? Is it an alkane? Draw its structural formula.

2. Draw the structural formula for 3-ethylheptane.

3. What would be the name for the substituent group $CH_3CH_2CH_2CH_2 —$?

4. Draw structural and condensed formulas for **(a)** trichloroethene and **(b)** 1,1-difluoroethene.

5. What is the main use for the compound $CCl_2 = CCl_2$? What two names are used for this compound?

6. Draw structural formulas for each of the following: methyl alcohol, methylamine, formaldehyde, and formic acid.

7. Draw the structural formulas and symbolic representations for cyclopentane and for cyclopentene.

8. Explain what the term *hydrogen bonding* means. Explain why a short-chain alcohol such as methanol is soluble in water whereas a long-chain one such as octanol is not.

9. Is the six-membered ring system in benzene particularly stable or unstable?

10. Do molecules such as H_2 readily add to benzene? Do free radicals such as atomic hydrogen readily add to benzene?

11. What is another name for perchlorobenzene? Do you expect it to be more soluble in aqueous or in hydrocarbon media?

Further Readings

For more extensive background in organic chemistry, consult a modern introductory textbook such as K. P. C. Vollhardt and N. E. Schore, *Organic Chemistry*, 4th ed. (New York: W. H. Freeman and Company, 2003).

Answers to Selected Odd-Numbered Problems

CHAPTER 1

Problems

1-1　(a) 427 kJ/mol; junction of UV-B with UV-C
　　(b) 299 kJ/mol; junction of UV with visible region
　　(c) 160 kJ/mol; junction of visible and infrared regions
　　(d) 29.9 kJ/mol; beginning of thermal IR region

1-3　390.7 nm; 127.5 nm

1-5　307 nm

1-7　$OH + O_3 \longrightarrow HOO + O_2$
　　$HOO + O \longrightarrow OH + O_2$

　　Overall: $O_3 + O \longrightarrow 2\,O_2$

Box 1-1, Problem 1　$[O^*] = k_1\,[O_2]/(k_2\,[M] + k_3\,[H_2O])$

Additional Problems

1. Net is $O_2 + UV$ photon $\longrightarrow 2\,O$
　Each O reacts as $O + O_2 \longrightarrow O_3$
　Overall: $3\,O_2 + UV$ photon $\longrightarrow 2\,O_3$

3. $ClONO_2 + $ photon $\longrightarrow Cl + NO_3$
　$NO_3 + $ photon $\longrightarrow NO + O_2$
　$Cl + O_3 \longrightarrow ClO + O_2$
　$NO + O_3 \longrightarrow NO_2 + O_2$
　$ClO + NO_2 \longrightarrow ClONO_2$
　Overall: $2\,O_3 + 2$ photons $\longrightarrow 3\,O_2$

5. 491 nm; visible

7. 3×10^3 molecules $cm^{-3}\,s^{-1}$; 2.5×10^{-12} g $cm^{-3}\,y^{-1}$

9. 1.5×10^{-6}

CHAPTER 2

Problems

2-1　$Cl + O_3 \longrightarrow ClO + O_2$
　　$Br + O_3 \longrightarrow BrO + O_2$
　　$ClO + BrO \longrightarrow Cl + Br + O_2$

　　Overall: $2\,O_3 \longrightarrow 3\,O_2$

2-5　$CF_3O + O_3 \longrightarrow CF_3OO + O_2$
　　$CF_3OO + O \longrightarrow CF_3O + O_2$

Green Chemistry Questions

1. (a) 2
　(b) 1, 4

3. No, not if the carbon dioxide is a waste by-product from another process.

5. Harpin is applied to the plant, which elicits the plant's own natural defenses.

Additional Problems

1. (b) 4×10^{15} g

3. (a) 140
　(b) 10
　(c) 141

7. Worse, since less Br is tied up in inactive forms.

CHAPTER 3

Problems

3-1　1.4×10^{-14} M; 0.35 ppt

3-3　Increases the ozone concentration.

3-5　$8\,NH_3 + 6\,NO_2 \longrightarrow 7\,N_2 + 12\,H_2O$
　　0.0092 g

3-7　1.2×10^{-7} M

3-9　0.59 ppm

3-11　156 kg

3-13　$I + O_3 \longrightarrow IO + O_2$
　　$2\,IO \longrightarrow I_2O_2$

3-15　3:1 ratio; area is bigger with smaller particles.

Box 3-1, Problem 1　(a) 0.032 ppm
　　　　　　　　　(b) 7.9×10^{11} molecules/cm^3
　　　　　　　　　(c) 1.3×10^{-9} M

Problem 3　(a) 9.3×10^{11} molecules/cm^3
　　　　　(b) 74.2 μg O_3/m^3 of air

Green Chemistry Questions

1. (a) Both are volatile and flammable hydrocarbons (VOCs); vapors would contribute to pollution of the troposphere; flammability would reduce worker safety.

   (b) It is a VOC and would contribute to tropospheric pollution, but it is not flammable.

   (c) It is a greenhouse gas, but waste carbon dioxide is used and recycled.

3. *First pair:* The cation has a delocalized charge and two nonpolar groups. *Second pair:* The charge on the anion is dispersed over the four chlorines, and the cation has a bulky nonpolar group. *Third pair:* Both the cation and anion have nonpolar bulky groups, and the charge on the anion is delocalized. *Fourth pair:* The charge on the anion is delocalized, and the cation has four nonpolar bulky groups.

Additional Problems

1. 4×10^{10} molecules cm^{-3} sec^{-1}; 4×10^{5} molecules cm^{-3} sec^{-1}; that with O_3

3. (a) $O + N_2 \longrightarrow NO + N$
   rate $= k [O] [N_2]$
   (b) Factor of 2.2×10^3

5. (a) $O_3 + 2 KI + H_2O \longrightarrow I_2 + O_2 + 2 KOH$
   (b) 120 ppb

7. 1.0%

9. 8 days

CHAPTER 4

Problems
4-1 $NH_4^+ + 3 H_2O \longrightarrow NO_3^- + 10 H^+ + 8 e^-$

Additional Problems
1. 4.3×10^{-5} M; 0.68; western
3. 3.7×10^9 g
5. 0.18 m^2/y
7. 0.024 g

CHAPTER 5

Problems

5-1 CFCs have no H or multiple bonds, so they don't react with OH or light and thus don't oxidize. No, since CH_2Cl_2 has H atoms and so it will react with OH.

5-3 Quite endothermic.

5-5 If a C-bonded H is abstracted by OH, the C-centered radical H_2COH will lose the hydroxyl H by O_2 abstraction, producing formaldehyde. If instead OH abstracts the OH hydrogen initially, the resulting C-centered H_3CO radical loses H to O_2 abstraction, producing formaldehyde.

5-7 Overall: $H_2CO + 2 NO + 2 O_2 +$ sunlight $\longrightarrow CO_2 + H_2O + 2 NO_2$
     No increase in the number of free radicals.

5-9 Overall: $CH_3(H)CO + 7 O_2 + 7 NO \longrightarrow 2 CO_2 + 7 NO_2 + 4 OH$

5-11 Same as for Problem 5-7.

5-13 CH_3OOH and O_2

5-15 (a) O_3, ClO, BrO, HOO
      (b) O_3, ClO, BrO, HOO, NO_2
      (c) O_3, ClO, BrO, HOO, NO_2
      (d) ClO, NO_2, and perhaps BrO
      (e) When 2 O_3 or 2 HOO react.

5-17 (a) $ClO + NO_2 \longrightarrow ClONO_2$
      (b) $2 ClO \longrightarrow ClOOCl$
      (c) Reaction of ClO with UV or O or NO; photolysis of ClOOCl.

5-19 Cycle destroys ozone if NO_2 reacts with O, but not if NO_2 decomposes in sunlight.

Additional Problems

1. $CO + OH \longrightarrow HOCO$
   $HOCO + O_2 \longrightarrow HOO + CO_2$
   $HOO + NO \longrightarrow OH + NO_2$
   $NO_2 + UV \longrightarrow NO + O$
   $O + O_2 \longrightarrow O_3$

   Overall: $CO + 2 O_2 + UV \longrightarrow CO_2 + O_3$

3. $\begin{aligned} CO + OH &\longrightarrow HOCO \\ HOCO + O_2 &\longrightarrow CO_2 + HOO \\ HOO + O_3 &\longrightarrow OH + 2\,O_2 \\ \hline \text{Overall: } CO + O_3 &\longrightarrow CO_2 + O_2 \end{aligned}$

5. NO_2: $\cdot\ddot{\text{O}}-\ddot{\text{N}}=\ddot{\text{O}}\cdot$

    HNO_2: $\text{H}-\ddot{\text{O}}-\ddot{\text{N}}=\ddot{\text{O}}\cdot$

    HNO_3: $\text{H}-\ddot{\text{O}}=\text{N}\begin{smallmatrix}\nearrow\ddot{\text{O}}\cdot\\[-2pt]\searrow\ddot{\text{O}}:\end{smallmatrix}$

CHAPTER 6

Problems

6-1 52°C

6-3 CO and NO; their stretching frequencies must lie outside the thermal IR range.

6-5 0.440 tonne; 0.27 g

6-7 1×10^8 kg

6-9 485 Tg

6-11 No; yes; no

Green Chemistry Questions

1. (a) 2
    (b) 1, 3, 4, 5, 6

Additional Problems

1. (a) Symmetric and antisymmetric stretch and bending vibrations for both.
    (b) Only the SO_2 symmetric stretch will contribute much.
    (c) Short atmospheric lifetimes.

3. 0.48

7. Sharply increased air temperature.

9. 1180 L; 13,500 L; 23,500 L

CHAPTER 7

Problems

7-1 $t = 0.69/k$; 17.3, 23.1, 46.2, and 69.3 years; 115 years; 4.6%

7-3 0.00112, 0.00152, and 0.00254 mol CO_2/kJ

7-5 1,2,3-; 1,2,4-; and 1,3,5-trimethylbenzene

7-7 2.274 tonnes

Green Chemistry Questions

1. (a) 2
    (b) 1, 5, 7, 9, 10

3. Growing crops requires fertilizers and pesticides. Energy is needed to plant, cultivate, and harvest; to produce, transport, and apply fertilizers and pesticides; to make and run tractors; and to transport seeds, biomass, monomers, and polymers. Use of land to produce crops for chemicals also removes land that could be used to produce food and feed.

Additional Problems

3. 1.4%

5. 1.83 m

7. (a) 16%
    (b) 22%
        For all kinds, 40%

CHAPTER 8

Problems

8-1 68%

8-3 387°C

8-5 Slightly exothermic, small loss.

8-7 -128.6 kJ mol^{-1}; increase

8-9 $\begin{aligned} O_2 + 2\,H_2O + 4\,e^- &\longrightarrow 4\,OH^- \\ 2\,OH^- + H_2 &\longrightarrow 2\,H_2O + 2\,e^- \\ O_2 + 2\,H_2 &\longrightarrow 2\,H_2O \end{aligned}$

8-11 23.7 kg; 12.0 kg; Mg

8-13 120 kJ g^{-1}; H_2 is superior by weight, but methane is superior by volume.

Green Chemistry Questions

1. 81%

3. Transesterfication (addition–elimination)

5. The high vapor pressure of methanol may result in vapor lock, whereas the low vapor pressure of both ethylene glycol and glycerin may cause difficulty in starting at cold temperatures.

Additional Problems

1. 5.6×10^{24} J; 0.007%
3. One-third of the CO
    $$3\,C + 4\,H_2O \rightleftharpoons 2\,CH_3OH + CO_2$$
7. 0.0011 EJ

CHAPTER 9

Problems

9-1 (a) $^{218}_{84}Po$

(b) $^{218}_{84}Po$

(c) $^{4}_{2}He$

(d) $^{238}_{92}U$

9-3 49%; 65%; yes, substantially above

9-5 $^{2}_{1}H + ^{3}_{2}He \longrightarrow ^{4}_{2}He + ^{1}_{1}p$ (or $^{1}_{1}H$)
     $2\,^{3}_{2}He \longrightarrow ^{4}_{2}He + 2\,^{1}_{1}p$ (or $2\,^{1}_{1}H$)

Additional Problems

1. $^{0}_{1}e$; $^{22}_{11}Na$ ¡ $^{22}_{10}Ne + ^{0}_{1}e$; $^{13}_{7}N$ ¡ $^{13}_{6}C + ^{0}_{1}e$
3. The effusion rate of $^{235}UF_6$ is 0.43% faster than that of $^{238}UF_6$. Since very little enrichment occurs in a single step, it is necessary to repeat the process hundreds of times to achieve suitable enrichment.

CHAPTER 10

Problems

10-1 (a) 4.0×10^{-5} ppm; 0.040 ppb
      (b) 3.0 μg/L
      (c) 300 ppb
10-3 5.0 ppm
10-5 (a) Yes
      (b) No
10-7 0.00010 mg/kg/day; 0.0055 mg/day
Box 10-3, Problem 1   In air: 9.9×10^{-11} mol/m^3
                      In water: 2.3×10^{-11} mol/m^3
                      In sediment: 5.7×10^{-7} mol/m^3

Green Chemistry Questions

1. (a) 3
    (b) 4
3. (a) A pesticide must meet one or more of the following requirements:
        (1) It reduces pesticide risks to human health.
        (2) It reduces pesticide risks to nontarget organisms.
        (3) It reduces the potential for contamination of valued environmental resources.
        (4) It broadens adoption of IPM or makes it more effective.
    (b) 1 and 2
5. (a) 3
    (b) 1, 4
7. 1, 2, and 3

Additional Problems

1. About 12 g
3. 7.4 μg
5. Yes, since concentration is 4.7 ppb.

CHAPTER 11

Problems

11-1 No, the same
      Yes
      The unique ones are 1,2; 1,3; 1,4; 1,6; 1,7; 1,8; 1,9; 2,3; 2,7; 2,8.
11-3 (a) 2,3,5,6- and 2,3,4,5-tetrachlorophenols
      (b) Two 2,3,5,6-tetrachlorophenols produce the 1,2,4,6,7,9-PCDD.
          Two 2,3,4,5-tetrachlorophenols produce the 1,2,3,6,7,8-PCDD.
11-5 1.87×10^5 molecules
11-7 2,3; 2,4; 2,5; 2,6; 3,4; 3,5; 2,2′; 2,3′; 2,4′; 2,5′; 2,6′; 3,3′; 3,4′; 3,5′; 4,4′
      With rotation, 2,2′ and 2,6′ interconvert, as do 2,3′ and 2,5′, as well as 3,3′ and 3,5′.
11-9 1,2; 1,3; 1,4; 2,3; 2,4; 3,4; 1,6; 1,7; 1,8; 1,9; 2,6; 2,7; 2,8; 3,6; 3,7; 4,6 dichlorodibenzofurans
11-11 28.6 pg
Box 11-1, Problem 1   2,6-dichlorophenol + 2,3,4-trichlorophenol, or 2,3-dichlorophenol + 2,3,6-trichlorophenol
Box 11-2, Problem 1   1-chlorodibenzofuran, and 1,4-, 1-6-, and 1-9-dichlorofurans

Green Chemistry Questions

1. (a) 2
   (b) 1, 4, and 6

Additional Problems

1. (a) One 2,3,4,5- and one 2,3,4,6-
      tetrachlorophenol
   (b) One 2,3,4,6- and one 2,3,5,6-
      tetrachlorophenol
   (c) One 2,3,4,6- and either 2,3,4,5- or
      2,3,5,6-tetrachlorophenol
3. Dechlorination; dioxins more toxic than the
   originals may be produced.
5. 2,3-, 2,2'-, and 2,3'-dichlorobiphenyls; 2,3,2'-
   trichlorobiphenyl
7. 2, 4, 5, 3' > 2, 4, 3', 4' > 2, 4, 2', 6'
9. 2.0×10^{-12} mol/m^3; 1.5×10^{-10} mol/m^3;
   4.9×10^{-5} mol/m^3

CHAPTER 12

Problems

12-5 Less; larger
12-7 3.7 years; 45,000

Additional Problems

1. (b) 1.52 ppb
   (c) 0.172 μg

CHAPTER 13

Problems

13-3 8.1 mg
13-5 16 mg/L
13-7 7.9×10^{-3} M; 6.3×10^{-20} M; 8.32; 2.51
13-9 15.2
13-11 (a) $NO_3^- + 2 H^+ + 2 e^- \longrightarrow$
                          $NO_2^- + H_2O$
   (b) +14.9
   (c) $14.9 - pH - 0.5 \log ([NO_2^-]/[NO_3^-])$
   (d) $pE + pH = 15.9$
   (e) 6×10^{-5}
13-13 8.3×10^{-5} M; solubility increases with
      increasing temperature.
13-15 8.0×10^{-4} M
13-17 10.3

13-19 1.02×10^{-3} M
13-21 30 mg CaCO$_3$/L
13-23 8.2×10^{-7} g

Green Chemistry Questions

1. It removes the outermost layer of the raw cotton fiber, know as the cuticle, which makes
   the fiber wettable for bleaching and dyeing.
3. (a) 2
   (b) 1, 3, 4, 6

Additional Problems

3. (a) $SO_4^{2-} + 10 H^+ + 8 e^- \longrightarrow H_2S + 4 H_2O$
   (b) $pE = 5.75 - (5/4) pH - (1/8) \log$
      $(P_{H_2S}/[SO_4^{2-}])$
   (c) 10^{-136} atm, absolutely negligible
5. 9.17
7. 6.1 mg/L; polluted

CHAPTER 14

Problems

14-1 $Ca(HCO_3)_2 + Ca(OH)_2 \longrightarrow$
                          $2 CaCO_3 \text{ (s)} + 2 H_2O$
      1:1 ratio
14-3 +1; OH$^-$; $NH_3 + HOCl$
14-5 5.5×10^4 g of NH$_3$
14-7 $Fe(s) \longrightarrow Fe^{2+}$
      $2 H_2O + 2 e^- \longrightarrow H_2 + 2 OH^-$
      $Fe(s) + 2 H_2O \longrightarrow Fe^{2+} + 2 OH^- + H_2(g)$
14-9 $4 Fe + C_2Cl_4 + 4 H_2O \longrightarrow$
                  $4 Fe^{2+} + C_2H_4 + 4 Cl^- + 4 OH^-$
14-11 5.5×10^{-5}; 0.0055; 0.55; 55
14-13 NO, NO$_2$, NO$_2^-$, NO$_3^-$
14-17 214.3 kJ/mol; 558 nm; 45%
Box 14-3, Problem 1 $t = [\ln (k_1/k_2)]/(k_1 - k_2)$

Green Chemistry Questions

1. It is biodegradable and its synthesis is performed
   under mild conditions; it uses only water as a
   solvent; and excess ammonia is recycled.

Additional Problems

1. 0.79; 0.54; 0.27; 0.10
3. 0.441 mL; 11 L
5. An O—O bond, as in HOSO$_2$OO$^-$; −1, so it
   needs to gain an electron to become −2.

7. (a) $CH_3CCl_3 + e^- \longrightarrow CH_3CCl_2 + Cl^-$
   (b) $CH_3CCl_3 + OH \longrightarrow CH_2CCl_3 + H_2O$
9. $CO_3^{2-} + OH \longrightarrow CO_3^- + OH^-$
   Strong acid or a soluble calcium or magnesium salt; about 10^{-4} M.
11. 3.73, 0.019%; 4.23, 0.059%

CHAPTER 15

Problems
15-1 $M^{2+} + H_2S \longrightarrow MS + 2 H^+$
   $M^{2+} + 2 R—S—H \longrightarrow$
   $\qquad\qquad R—S—M—S—R + 2 H^+$
15-3 0.051 g
15-5 0.5 mg; 2.0×10^5 g
15-7 2×10^{-5} g
15-9 6.0 $\mu g/dL$; 0.29 $\mu mol/L$

Green Chemistry Questions
1. (a) 3
   (b) 4
3. Lower air pollution, better corrosion protection, and reduced waste.

Additional Problems
1. Yes; yes, the answer changes.
3. H_3AsO_4 for pH < 2.20; $H_2AsO_4^-$ from 2.20 to 6.89; $HAsO_4^{2-}$ from 6.89 to 11.49; AsO_4^{3-} for pH > 11.49; 17; 0.17
5. H^+ liberates metals from sulfides; liberated metals enter the food chain and affect human health.
9. 13 mg m^{-3}, must be greater than the threshold limit.

CHAPTER 16

Problems
16-1 3.9 L
16-3 $C_3H_3O_2$
16-5

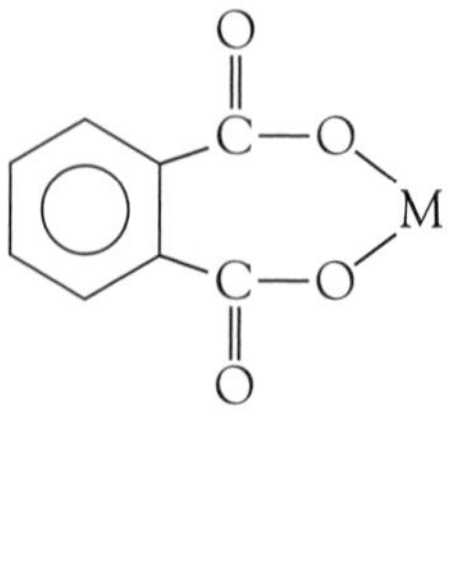

16-7 (a) $2 C_{12}H_6Cl_4 + 25 O_2 \longrightarrow$
   $\qquad\qquad 24 CO_2 + 2 H_2O + 8 HCl$
   (b) $C_{12}H_6Cl_4 + 23 H_2 \longrightarrow$
   $\qquad\qquad 12 CH_4 + 4 HCl$

Green Chemistry Questions
1. It is biodegradable; as a result, it places less of a load on the environment. The Donlar synthesis of polyaspartate has only two steps and requires only heat and hydrolysis with sodium hydroxide.
3. (a) 1
   (b) 1, 5, 6

Additional Problems
1. About 8000 homes
3. 1.14 kg
7. UV-B; need sunlight exposure to decompose; 399 kJ/mol

APPENDIX

1. (a)

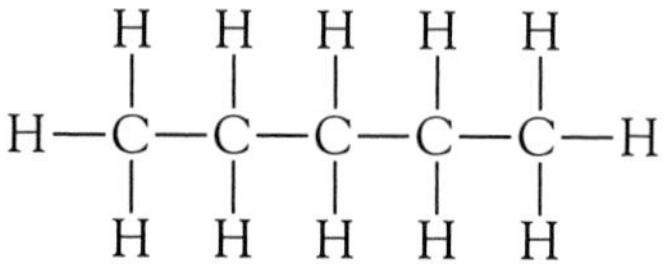

$CH_3CH_2CH_2CH_2CH_3$

(b)

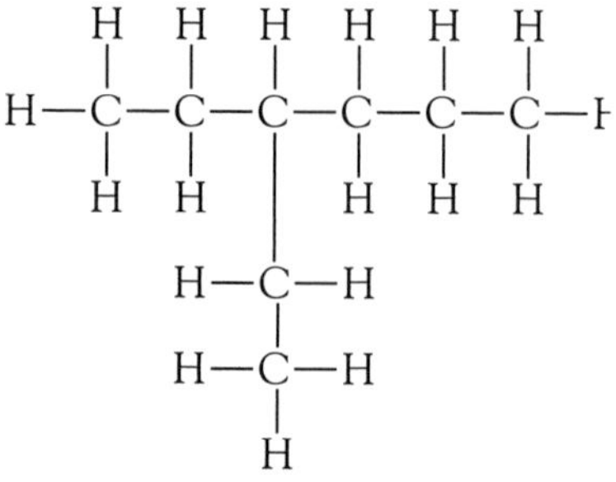

$(CH_3CH_2)_2CHCH_2CH_3$

(c)

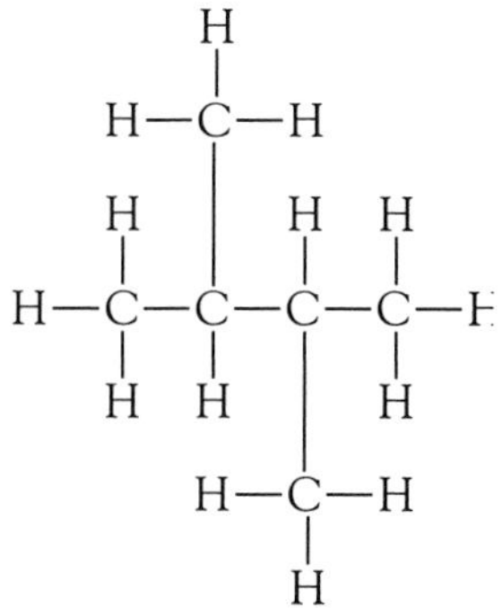

$(CH_3)_2CHCH(CH_3)_2$

3. (a) 1, 1, 1, 2-tetrachloroethane
   (b) 1-butene
   (c) 1, 4-butadiene

5. (a)

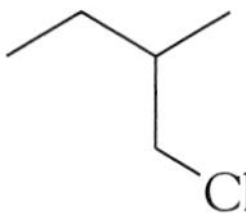

(b)

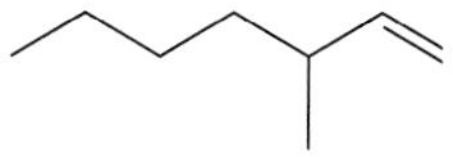

(c)

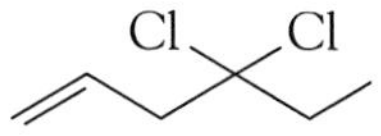

7. (a)

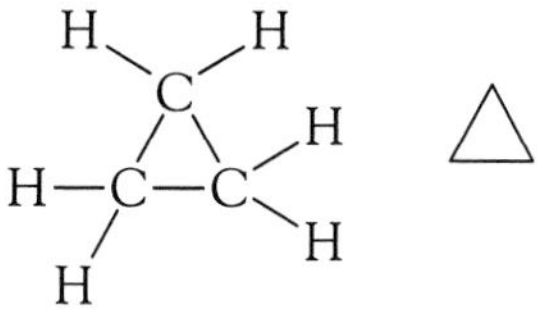

(b)

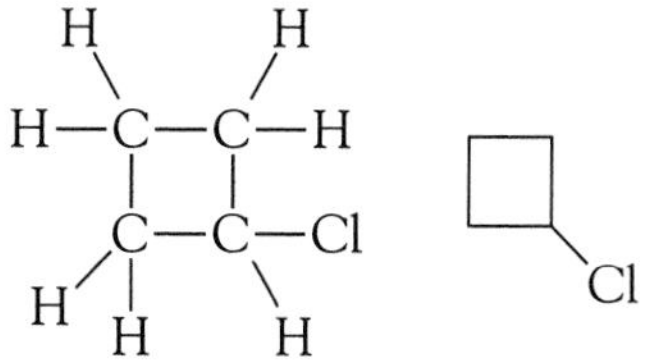

(c)

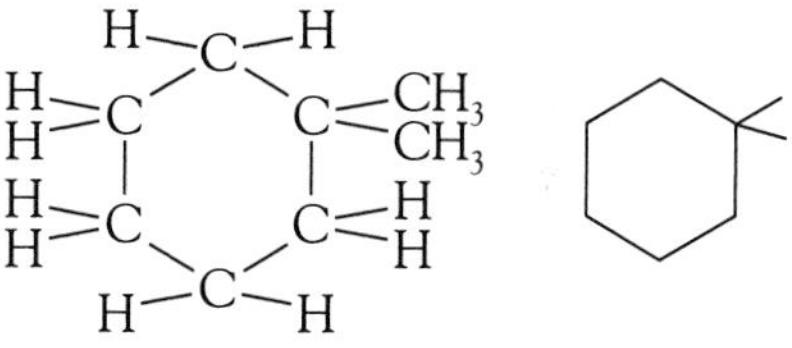

INDEX